Department of Mathematics
and Computer Science

Linear Algebra

AN INTRODUCTION WITH APPLICATIONS

Linear Algebra

AN INTRODUCTION WITH APPLICATIONS

Raymond A. Barnett
Merritt College

Michael R. Ziegler
Marquette University

Dellen Publishing Company
San Francisco, California

Collier Macmillan Publishers
London

divisions of Macmillan, Inc.

On the cover: "The Third Man" was executed by Los Angeles artist Peter Shire. The work measures 18 by 15 by 11 inches. One of the foremost ceramic sculptors in the United States, Shire is a key member of the Memphis Group. His work may be seen in the collections of the Los Angeles County Museum and the Los Angeles Museum of Contemporary Art. Shire is represented by Modernism in San Francisco, California.

Permissions: Dellen Publishing Company
400 Pacific Avenue
San Francisco, California 94133

Orders: Dellen Publishing Company
c/o Macmillan Publishing Company
Front and Brown Streets
Riverside, New Jersey 08075

Collier Macmillan Canada, Inc.

Library of Congress Cataloging-in-Publication Data

Barnett, Raymond A.
    Linear algebra.

    Includes index.
    1. Algebras, Linear.   I. Ziegler, Michael R.
II. Title.
QA184.B38   1987   512'.5   86-13594
ISBN 0-02-305960-5

PRINTING   456789   YEAR   23456789

ISBN 0-02-305960-5

Contents

Preface

This book presents a treatment of linear algebra suitable for students in a wide variety of disciplines, including mathematics, computer science, engineering, physics, economics, and the life sciences. The material is presented at a level appropriate for students in their freshman or sophomore year. **Calculus is not a prerequisite** (except for Section 9-3); however, optional examples and exercises are included for those students who have studied calculus. These examples and exercises are clearly marked and can be omitted by those students who have not had any previous calculus experience.

■ Emphasis

Teaching linear algebra requires a careful balance between theory, geometric intuition, computational skills, and applications. For most students, this is the first exposure to formal (abstract) mathematical structures (and these structures must be understood in order to perform the computations that arise naturally in the applications). To help students make the transition from the concrete to the abstract, most new concepts are introduced in terms of concrete examples and then generalized to a definition or a theorem. We have included those theorems that are vital to the subject and omitted those whose primary importance lies in their own abstraction. Proofs are included whenever they will lead to a better understanding of concepts under discussion or will contribute additional insight into the application of a concept. We have not hesitated to state theorems without proof if the proofs are too complicated or involve ideas that do not increase the student's understanding of the subject. In many instances, however, special cases of an omitted proof are discussed in the C-level exercises. We have avoided long and tedious arguments involving manipulation of summation notation and any proofs that require mathematical induction. Even with these omissions, there are a sufficient number of proved theorems and theoretical C-level exercises to provide a sound foundation in the theory of linear algebra.

■ Choice of Emphasis

The text provides an instructor or department with many choices in course design and emphasis—from a course strong in mechanics to a course strong in mathematical structure and theory. By deemphasizing proofs and restricting assigned problems to A and B levels in the exercise sets, a course emphasizing mechanics will evolve; by emphasizing proofs and problems in the C-level exercises, a fairly strong course emphasizing mathematical structure and theory will evolve; and, of course, a variety of courses combining these two emphases to varying degrees are possible. In short, an instructor can easily pitch a course to his or her own interests, class background, or department requirements.

■ Organization

The material is organized in three parts. The first four chapters present basic material concerning systems of linear equations, matrices, determinants, and vectors in the plane and in 3-space. The amount and depth of coverage here will depend on the background of the students. With the exception of Cramer's Rule (Section 3-4), students should be comfortable with all of this material before beginning Chapter 5.

Chapters 5 and 6 present a thorough coverage of the theory of finite-dimensional vector spaces and inner product spaces, culminating in a discussion of the Gram–Schmidt orthogonalization process. Eigenvalues and diagonalization techniques are covered in Chapter 7.

There are thirty-four sections in these first two parts of the book. Therefore, it should be possible to cover all of Chapters 1–7 comfortably in a one-quarter or one-semester course.

The third part of the text covers two topics: linear transformations (Chapter 8) and applications of diagonalization (Chapter 9). These topics are independent, so either chapter can directly follow Chapter 7.

Finally, a shorter course emphasizing applications can be formed by omitting Sections 3-4, 6-3 through 6-6, 7-3, 8-1 through 8-5, and 9-1.

■ Computers

One of the reasons for the growing importance of linear algebra is the rapid development of computers. Many problems that were impossible to solve by hand can now be solved routinely with the aid of a computer. Although we have not assumed that the students have had any previous experience with a computer or that they will be using a computer in this course, we have been careful to point out the relationship between linear algebra and computing. Whenever possible, procedures for solving problems have been presented in a manner that permits easy implementation on a computer. Procedures that are important for their theoretical implications and are not suited for use on a computer are clearly identified.

A computer supplement and an APPLE II or IBM PC microcomputer program disk are available for those who wish to incorporate the use of a computer into this course. The programs on the disk implement many of the procedures presented in this book and can be used to reinforce concepts and to permit the consideration of problems involving calculations too complicated to be done by hand. These programs are interactive, easy to use, and require no previous computer experience on the part of the student. The manual contains examples that illustrate the use of these programs, exercises for the student, discussions of some additional applications that are particularly suited to computer solutions, and discussions of some numerical techniques in linear algebra. Programs corresponding to these additional topics are also included on the program disk.

■ Examples and Matched Problems

We firmly believe that the best way to master this subject is to use it. To that end, the text contains over 350 completely worked examples. Each example is followed by a similar problem for the student to work while reading the material. The answers to these matched problems are included at the end of each section for easy reference.

■ Exercises

This book contains over 2,300 exercises. Each exercise set is designed so that an average or below-average student will experience success and a very capable student will be challenged. They are divided into three parts, A (routine, easy mechanics), B (more difficult mechanics), and C (difficult mechanics and theory).

■ Applications

Enough applications are included in this book to convince even the most skeptical student that linear algebra is really useful. The first three chapters contain applications of elementary linear algebra to a wide variety of areas, including engineering, physics, biology, economics, graph theory, and geometry. The last chapter contains detailed discussions of three applications of diagonalization, utilizing most of the theory developed in the book. With the exception of the section on differential equations (Section 9-3), no specialized background is required to solve any of the applications in this book.

■ Student Aids

Theorems, procedures for solving problems, and most important definitions are **displayed in boxes** for easy reference.

Examples and developments are often **annotated** to help students through critical stages.

A **second color** is used to indicate key steps, to delineate topics, and to increase the clarity of graphics.

Boldface type is used to introduce new terms and highlight important comments.

Answers to all odd-numbered, nontheoretical problems are included in the back of the book. If the answer to a theoretical problem is a proof, the proof is not included in the answer section (but some proofs are included in the student's solution manual).

Chapter review sections include a review of all important terms and symbols and a comprehensive review exercise. Answers to all (nontheoretical) review exercises are included in the back of the book.

A **solutions manual** by Robert Mullins is available through a book store. The manual includes solutions to all odd-numbered, nontheoretical problems, solutions to all nontheoretical review problems, and proofs (or outlines of proofs) for selected theoretical exercises. Each section of the solutions manual begins with a review of the important terms, theorems, and problem-solving procedures from the corresponding section in this book.

A **computer applications supplement** is available through a book store. This supplement is designed to be used with a **microcomputer program disk.** Copies of the program disk are available through instructors or departments using this book. (See the discussion earlier in this Preface.)

■ Instructor Aids

A unique **computer-generated random test system** is available to instructors without cost. The system, utilizing an IBM PC computer and a number of commonly used dot matrix printers, will generate an almost unlimited number of chapter tests and final examinations, each different from the other, quickly and easily. At the same time, the system produces an answer key and a student worksheet with an answer column that exactly matches the answer column on the answer key. Graphing grids are included on the answer key and on the student worksheet for problems requiring graphs.

A **printed and bound test battery** is also available to instructors without cost. The battery contains several chapter tests for each chapter, answer keys, and student worksheets with answer columns that exactly match the answer columns on the answer keys. Graphing grids are included on the answer key and on the student worksheet for problems requiring graphs.

An **instructor's answer manual** containing answers to the even-numbered problems not included in the text is available to instructors without charge.

A **student's solutions manual** by Robert Mullins (see Student Aids) is available to instructors without charge.

A **computer applications supplement** (see Student Aids) and an **APPLE II or**

IBM PC microcomputer program disk are available to instructors without charge. Copies of the program disk may be distributed to students at the discretion of the instructor.

■ Acknowledgments

In addition to the authors, many others are involved in the successful publication of a book. We wish to thank personally: C. Bandy, Southwest Texas State University; Gary Brown, College of St. Benedict; Bruce Edwards, University of Florida; Paul Eenigenburg, Western Michigan University; Garry Etgen, University of Houston; Frederic Gooding, Jr., Goucher College; Thomas Kearns, Northern Kentucky University; Stanley Lukawecki, Clemson University; Stephen Merrill, Marquette University; Robert Mullins, Marquette University; and John Spellmann, Southwest Texas State University.

We also wish to thank:

Janet Bollow for another outstanding book design

John Williams for a strong, effective cover design and for the original chapter-opening illustrations

Garry Etgen, Stephen Merrill, and Robert Mullins for carefully checking all examples and problems (a tedious but extremely important job)

Patricia Thomson for her careful and accurate typing of the manuscript

Phyllis Niklas for her ability to guide the book smoothly through all production details

Don Dellen, the publisher, who continues to provide all the support services and encouragement an author could hope for

Producing this new edition with the help of all these extremely competent people has been a most satisfying experience.

R. A. Barnett
M. R. Ziegler

Linear Algebra

AN INTRODUCTION WITH APPLICATIONS

|1| Systems of Linear Equations

|1| Contents

Linear algebra began as a study of systems of linear equations which occur naturally in a variety of areas, including mathematics, engineering, economics, and the physical, life, and social sciences. In some of these areas, it is quite common to find applications involving systems where the equations and the variables number in the thousands. Of course, these large-scale systems require computers for their solution. In this chapter we will introduce systematic procedures for solving systems of linear equations. These procedures have two important features. First, they can be applied to any linear system however large. Second, a digital computer can be programmed to carry out the calculations, making the solution of large systems practical as well as feasible. As a pleasant bonus, we will see that the techniques learned here have many other important applications. Thus, this chapter builds the foundation for much of what follows in this text and the material presented here deserves your most careful attention.

|1-1| Systems of Linear Equations—Introduction

- Linear Equations in Two Variables
- Systems of Linear Equations in Two Variables
- Nature of Solutions
- Systems of Linear Equations in Three Variables
- Application

■ Linear Equations in Two Variables

An equation of the form

$$ax + by = c \tag{1}$$

where a, b, and c are real constants (a and b not both zero), is called a **linear equation in x and y.** The **solution set** of a linear equation is the set of all ordered pairs of real numbers that satisfy the equation. Using set notation, the solution set can be expressed as

$$S = \{(x, y)|ax + by = c\} \qquad \text{Solution set}$$

The graph of the solution set (a and b not both zero) is a straight line. It may seem strange, but it is necessary to consider equations of the form (1) where a

and b are both zero. All possible types of solution sets for equation (1) are listed in Table 1.

Table 1
Solution Sets for $ax + by = c$

Equation	Solution Set	Graph
$ax + by = c$, $a \neq 0$ or $b \neq 0$	$\{(x, y) \mid ax + by = c\}$	A straight line
$0x + 0y = c$, $c \neq 0$	$\varnothing$	The empty set
$0x + 0y = 0$	$\{(x, y) \mid x$ and y are real numbers$\}$	The entire xy plane

Example 1   (A) Use set notation to describe the solution set of the linear equation

$$2x + 3y = 12$$

(B) List two explicit points in the solution set.
(C) Graph the solution set.

Solution   (A) $S = \{(x, y) \mid 2x + 3y = 12\}$ Solution set

(B) Let $x = 0$. Then

$$2 \cdot 0 + 3y = 12$$
$$3y = 12$$
$$y = 4$$

Let $y = 0$. Then

$$2x + 3 \cdot 0 = 12$$
$$2x = 12$$
$$x = 6$$

Thus, $(0, 4)$ and $(6, 0)$ are two of the infinite number of points in the solution set S. Other points in S can be determined by assigning any value to x (or to y), substituting this value in the equation $2x + 3y = 12$, and solving for y (or for x).

(C) Since two points determine a line, we complete the graph of $2x + 3y = 12$ by drawing a line through the two points found in part (B).

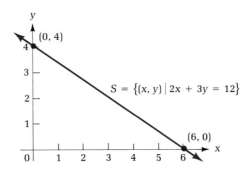

Problem 1   Repeat Example 1 for the linear equation

$$x - y = 1$$ ∎

▪ Systems of Linear Equations in Two Variables

To establish some basic concepts, consider the following simple example. A school has $13,000 to purchase ten microcomputers. Each available computer has either one or two built-in disk drives. If a computer with one disk drive costs $1,000 and one with two drives costs $1,500, how many of each type should the school purchase in order to use all of the $13,000?

Let

x = Number of computers with one drive

y = Number of computers with two drives

Then

$$x + y = 10 \qquad \text{Number of computers}$$
$$1{,}000x + 1{,}500y = 13{,}000 \qquad \text{Purchase cost}$$

We now have a **system of two linear equations in two variables.** The **solution set** of such a system is the set of all ordered pairs of real numbers that satisfy both equations at the same time. In elementary algebra, you may have learned to solve systems of this type by graphing or by substitution. Since these methods are not suitable for larger systems, we will present a new method here, called **elimination with back substitution.** This method involves replacement of systems of equations with simpler *equivalent systems* (by performing appropriate operations) until we obtain a system that is easy to solve. Equivalent systems are defined as follows:

Equivalent Systems of Linear Equations

Two systems of linear equations are **equivalent** if they have the same solution set.

Theorem 1 lists the operations that can be performed on a system of linear equations to produce an equivalent system. (The proof is omitted.)

Theorem 1   **Operations on Systems of Equations**

Equivalent systems of equations result if:

(A) Two equations are interchanged.
(B) An equation is multiplied by a nonzero constant.
(C) A constant multiple of one equation is added to another equation.

Returning to our example, we use Theorem 1 to eliminate one of the variables and obtain an equivalent system:

$$x + y = 10$$
$$1,000x + 1,500y = 13,000$$

If we multiply the top equation by $-1,000$ and add it to the bottom equation, we can eliminate x.

$$
\begin{array}{ll}
-1,000x - 1,000y = -10,000 & -1,000 \text{ times top equation} \\
\underline{1,000x + 1,500y = \quad 13,000} & \text{Bottom equation} \\
500y = \quad 3,000 & \text{Sum}
\end{array}
$$

If we replace the original bottom equation with this new equation, we obtain a system that is equivalent to the original system and that is easy to solve:

$$x + y = 10 \qquad \text{Equivalent system}$$
$$500y = 3,000$$

Now, solve this system for y and then x:

$$500y = 3,000 \qquad \text{Solve the bottom equation for } y.$$
$$y = 6$$

$$x + 6 = 10 \qquad \text{Substitute } y = 6 \text{ in the top equation and solve for } x.$$
$$x = 4$$

Thus, the school should purchase 4 computers with one disk drive and 6 with two disk drives.

Check
$$
\begin{array}{ll}
x + y = 10 & 1,000x + 1,500y = 13,000 \\
4 + 6 \overset{?}{=} 10 & 1,000(4) + 1,500(6) \overset{?}{=} 13,000 \\
10 \overset{\checkmark}{=} 10 & 13,000 \overset{\checkmark}{=} 13,000
\end{array}
$$

The first step in this process is called *elimination* and the second is called *back substitution*. The following example further illustrates the solution of a system by elimination with back substitution.

Example 2 Solve the system:
$$2x + 3y = 2$$
$$5x + 6y = 11$$

Solution First, eliminate x from the bottom equation:

$$
\begin{array}{ll}
-5x - \tfrac{15}{2}y = -5 & \text{Multiply the top equation by } -\tfrac{5}{2} \text{ and add to the bottom} \\
\underline{5x + 6y = \quad 11} & \text{equation.} \\
-\tfrac{3}{2}y = \quad 6 &
\end{array}
$$

This produces the following equivalent system:

$$2x + 3y = 2$$
$$-\tfrac{3}{2}y = 6$$

Now use back substitution to solve this system:

$$-\tfrac{3}{2}y = 6 \qquad \text{Solve the bottom equation for } y.$$
$$y = -4$$

$$2x + 3(-4) = 2 \qquad \text{Then substitute } y = -4 \text{ in the top equation and solve}$$
$$2x = 14 \qquad \text{for } x.$$
$$x = 7$$

$$\text{Check} \quad 2x + \quad 3y = 2 \qquad\qquad 5x + \quad 6y = 11$$
$$2(7) + 3(-4) \overset{?}{=} 2 \qquad\qquad 5(7) + 6(-4) \overset{?}{=} 11$$
$$2 \overset{\checkmark}{=} 2 \qquad\qquad\qquad 11 \overset{\checkmark}{=} 11$$

Problem 2   Solve the system using elimination with back substitution:

$$3x + 2y = 1$$
$$4x + 3y = 2$$

‖

▪ Nature of Solutions

So far, the systems of equations we have considered have had exactly one ordered pair in their solution set. Does every linear system have a solution set consisting of a single point? No, but it turns out that there are only two other possibilities: *no solution or an infinite number of solutions*. The next example illustrates all three possible types of solution sets.

Example 3   Find the solution set for each of the following systems:

(A) $x - 2y = -1$ (B) $x - 2y = -1$ (C) $2x + 3y = 12$
      $2x + 3y = 12$ $2x - 4y = -8$ $4x + 6y = 24$

Solution   (A) $x - 2y = -1$ To eliminate x from the bottom equation, multiply the
        $2x + 3y = 12$ top equation by -2 and add it to the bottom equation.

$$-2x + 4y = 2 \qquad (-2) \times \text{top equation}$$
$$\underline{2x + 3y = 12} \qquad \text{Bottom equation}$$
$$7y = 14 \qquad \text{Sum}$$

$$x - 2y = -1 \qquad \text{New equivalent system}$$
$$7y = 14$$

Now use back substitution to find the solution:

$$7y = 14$$
$$y = 2$$

$$x - 2(2) = -1$$
$$x - 4 = -1$$
$$x = 3$$

The solution set consists of the single point (3, 2), as shown in Figure 1(A). A linear system is said to be **consistent** whenever the solution set consists of one or more points.

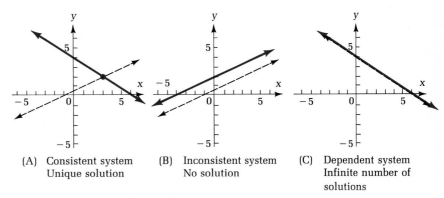

(A) Consistent system
     Unique solution

(B) Inconsistent system
     No solution

(C) Dependent system
     Infinite number of
     solutions

Figure 1 Types of solutions of linear systems

(B) $x - 2y = -1$ Original system
     $2x - 4y = -8$

$$
\begin{array}{ll}
-2x + 4y = 2 & (-2) \times \text{top equation} \\
\underline{2x - 4y = -8} & \text{Bottom equation} \\
0x + 0y = -6 & \text{Sum}
\end{array}
$$

   $x - 2y = -1$ New equivalent system
        $0 = -6$

There are no values of x and y that satisfy the second equation in the new system. Thus, the new system has no solution. Since Theorem 1 implies that the original system and the new system have the same solution sets, we can conclude that the original system also has no solution. Systems with no solutions are said to be **inconsistent.** As Figure 1(B) illustrates, this system is inconsistent because the graphs of the equations are two parallel lines that never intersect.

(C) $2x + 3y = 12$ Original system
     $4x + 6y = 24$

$$-4x - 6y = -24 \qquad (-2) \times \text{top equation}$$
$$\underline{4x + 6y = 24} \qquad \text{Bottom equation}$$
$$0x + 0y = 0 \qquad \text{Sum}$$

$$2x + 3y = 12 \qquad \text{New equivalent system}$$
$$0 = 0$$

This time the second equation in the new system is satisfied by all values of x and y. Thus, the solution set consists of all the ordered pairs that satisfy the first equation. That is, the solution set of both the new system and the original system is the infinite set given by

$$S = \{(x, y) | 2x + 3y = 12\}$$

Systems with an infinite number of solutions are said to be **dependent.** In this example, the system is dependent because both equations represent the same line in the xy plane, as shown in Figure 1(C). A convenient way to represent the solution set of a dependent system is to introduce a new variable t, called a **parameter.** If we let t be any real number and set $y = t$, then replacing y with t in $2x + 3y = 12$ and solving for x allows us to represent the general solution of the system (the set of all possible solutions) by the two equations

$$x = 6 - \tfrac{3}{2}t \qquad \text{General solution}$$
$$y = t \qquad t \text{ any real number}$$

If we substitute a specific value for t in the equations in the general solution and solve for x and y, we obtain a **particular solution** to the system. For example, if $t = 2$, then $(3, 2)$ is a (particular) solution; if $t = -4$, then $(12, -4)$ is a solution; and so on. Using set notation, the general solution is represented concisely by

$$S = \{(6 - \tfrac{3}{2}t, t) | t \text{ any real number}\}$$

Problem 3   Find the solution set for each of the following systems:

(A) $x - 3y = -4$ (B) $x - 3y = -4$ (C) $3x + 5y = 30$
      $3x + 5y = 30$ $2x - 6y = -8$ $6x + 10y = 30$ ■

Figure 1 illustrates the three possible types of solutions for a system of two linear equations in two variables. It is a surprising fact that the solution set of any linear system always falls into one of these three categories, no matter how many variables or equations are involved. This statement, which is restated as Theorem 2, will be proved in Chapter 8.

Theorem 2	**Nature of Solutions for a Linear System**
	Any system of linear equations must have no solution, one solution, or an infinite number of solutions. No other possibility exists.

■ Systems of Linear Equations in Three Variables

An equation of the form

$$ax + by + cz = d$$

where a, b, c, and d are real constants (a, b, and c not all zero), is called a **linear equation in three variables.** We want to use the method of elimination with back substitution to solve systems of three equations in three variables. We begin with an example that will lead to some general observations.

Example 4   Solve the system:
$$x + 2y - z = 9$$
$$2x + y + z = 6$$
$$3x - 2y - 2z = 2$$

Solution   In order to identify clearly the operations performed in the elimination process, we will label the equations in each system as E_1, E_2, and E_3:

$$x + 2y - z = 9 \quad E_1$$
$$2x + y + z = 6 \quad E_2$$
$$3x - 2y - 2z = 2 \quad E_3$$

Step 1.   Use equation E_1 to eliminate x in equations E_2 and E_3:

$$
\begin{array}{ll}
-2x - 4y + 2z = -18 & -2E_1 \\
\underline{2x + y + z = 6} & E_2 \\
-3y + 3z = -12 & -2E_1 + E_2
\end{array}
$$

$$
\begin{array}{ll}
-3x - 6y + 3z = -27 & -3E_1 \\
\underline{3x - 2y - 2z = 2} & E_3 \\
-8y + z = -25 & -3E_1 + E_3
\end{array}
$$

Step 2.   Consider the equivalent system formed by replacing the original equations E_2 and E_3 with the sums computed in Step 1. (We will still refer to these equations as E_1, E_2, and E_3.)

$$
\begin{array}{ll}
x + 2y - z = 9 & E_1 \\
-3y + 3z = -12 & E_2 \\
-8y + z = -25 & E_3
\end{array}
$$

If we mentally delete the first equation in this new system, we then have a system of two equations in two variables. Now we use E_2 to eliminate y in E_3:

$$
\begin{array}{ll}
8y - 8z = 32 & -\tfrac{8}{3}E_2 \\
\underline{-8y + z = -25} & E_3 \\
 -7z = 7 & -\tfrac{8}{3}E_2 + E_3
\end{array}
$$

Step 3.   Consider the equivalent system formed by replacing E_3 with the sum computed in Step 2:

$$
\begin{array}{ll}
x + 2y - z = 9 & E_1 \\
 -3y + 3z = -12 & E_2 \\
 -7z = 7 & E_3
\end{array}
$$

We now use back substitution to solve for x, y, and z:

$$-7z = 7 \qquad \text{Solve } E_3 \text{ for } z.$$
$$z = -1$$

$$-3y + 3z = -12 \qquad \text{Substitute for } z \text{ in } E_2 \text{ and solve for } y.$$
$$-3y + 3(-1) = -12$$
$$-3y = -9$$
$$y = 3$$

$$x + 2y - z = 9 \qquad \text{Substitute for } y \text{ and } z \text{ in } E_1 \text{ and solve for } x.$$
$$x + 2(3) - (-1) = 9$$
$$x = 2$$

Thus, the solution is $x = 2$, $y = 3$, $z = -1$ or $(2, 3, -1)$, and this is the only solution.

The steps we used in solving the system in Example 4 are summarized in the box.

Steps in Solving a System of Three Equations in Three Variables

Step 1.   If x is not present in the first equation (that is, the coefficient of x is zero), interchange the first equation with one in which x has a nonzero coefficient. Then use the first equation to eliminate x in all the subsequent equations.

Step 2.   Now consider the system formed by (mentally) deleting the first equation and repeat Step 1, but this time to eliminate y. If the coefficient of y is zero in all the equations considered in this step, you do not have to do anything at this step.

Step 3.   Use back substitution to find the solution.

Problem 4 Solve the system using elimination with back substitution:

$$x - 2y + z = -1$$
$$2x - 3y + z = -4$$
$$3x - 4y + 2z = -3$$

∎

In the process described above, if we encounter an equation of the form

$$0x + 0y + 0z = d \qquad d \neq 0 \qquad \text{Usually turns up as } 0 = d.$$

we can stop and conclude that the original system is inconsistent. On the other hand, if we encounter an equation of the form

$$0x + 0y + 0z = 0 \qquad \text{Usually turns up as } 0 = 0.$$

the original system may be inconsistent or dependent. We must proceed further to determine which. We will have more to say about these two cases in Example 5 and in later sections.

Example 5 Solve the following systems:

(A) $2x + y - 2z = 10$ (B) $y + 3z = 4$
$3x + 2y + z = 12$ $x - y + z = 1$
$5x + 4y + 7z = 30$ $3x - y + 9z = 11$

Solution (A) $2x + y - 2z = 10 \qquad E_1$
$3x + 2y + z = 12 \qquad E_2$
$5x + 4y + 7z = 30 \qquad E_3$

Step 1. Use E_1 to eliminate x in E_2 and E_3:

$$-3x - \tfrac{3}{2}y + 3z = -15 \qquad -\tfrac{3}{2}E_1$$
$$\underline{3x + 2y + z = 12} \qquad E_2$$
$$\tfrac{1}{2}y + 4z = -3 \qquad -\tfrac{3}{2}E_1 + E_2$$

$$-5x - \tfrac{5}{2}y + 5z = -25 \qquad -\tfrac{5}{2}E_1$$
$$\underline{5x + 4y + 7z = 30} \qquad E_3$$
$$\tfrac{3}{2}y + 12z = 5 \qquad -\tfrac{5}{2}E_1 + E_3$$

Step 2. Consider the following equivalent system and use E_2 to eliminate y in E_3:

$$2x + y - 2z = 10 \qquad E_1$$
$$\tfrac{1}{2}y + 4z = -3 \qquad E_2$$
$$\tfrac{3}{2}y + 12z = 5 \qquad E_3$$

$$-\tfrac{3}{2}y - 12z = 9 \qquad -3E_2$$
$$\underline{\tfrac{3}{2}y + 12z = 5} \qquad E_3$$
$$0 = 14 \qquad -3E_2 + E_3$$

Since this last equation is not satisfied by any values of x, y, and z, we can stop and conclude that the original system is inconsistent.

(B)
$$
\begin{array}{rlr}
y + 3z &= 4 & E_1 \\
x - y + z &= 1 & E_2 \\
3x - y + 9z &= 11 & E_3
\end{array}
$$

Step 1.   Interchange E_1 and E_2 to obtain an equation with a nonzero x coefficient:

$$
\begin{array}{rlr}
x - y + z &= 1 & E_1 \\
y + 3z &= 4 & E_2 \\
3x - y + 9z &= 11 & E_3
\end{array}
$$

Now use E_1 to eliminate x in E_3

$$
\begin{array}{rlr}
-3x + 3y - 3z &= -3 & -3E_1 \\
\underline{3x - y + 9z} &= \underline{11} & E_3 \\
2y + 6z &= 8 & -3E_1 + E_3
\end{array}
$$

Step 2.   Consider the following equivalent system and use E_2 to eliminate y in E_3:

$$
\begin{array}{rlr}
x - y + z = 1 & & E_1 \\
\boxed{\begin{array}{l} y + 3z = 4 \\ 2y + 6z = 8 \end{array}} & & \begin{array}{l} E_2 \\ E_3 \end{array}
\end{array}
$$

$$
\begin{array}{rlr}
-2y - 6z &= -8 & -2E_2 \\
\underline{2y + 6z} &= \underline{8} & E_3 \\
0 &= 0 & -2E_2 + E_3
\end{array}
$$

Step 3.   Solve the following equivalent system:

$$
\begin{array}{rlr}
x - y + z = 1 & & E_1 \\
\boxed{\begin{array}{l} y + 3z = 4 \\ 0 = 0 \end{array}} & & \begin{array}{l} E_2 \\ E_3 \end{array}
\end{array}
$$

The system in y and z consisting of equations E_2 and E_3 is a dependent system having infinitely many solutions. To represent the (general) solution, we introduce a parameter t. After substituting $z = t$ in the second equation and solving for y, we obtain

$$z = t$$
$$y = 4 - 3t$$

Using back substitution in E_1, we can express x in terms of the parameter t:

$$\begin{array}{cc} y & z \end{array}$$
$$x - (4 - 3t) + t = 1$$
$$x = 5 - 4t$$

Thus, the (general) solution to the original system is

$$x = 5 - 4t$$
$$y = 4 - 3t$$
$$z = t \qquad t \text{ any real number}$$

or

$$S = \{(5 - 4t,\ 4 - 3t,\ t)|t \text{ any real number}\}$$

For example, for $t = -1$, we obtain the particular solution $(9, 7, -1)$; for $t = 2$, we obtain the particular solution $(-3, -2, 2)$; and so on.

Problem 5   Solve the following systems:

(A) $\quad y + 3z = 1$ (B) $\quad 2x - y + z = 3$
$\quad x + y + z = 4$ $\quad -4x + 5y + 2z = 4$
$\quad x - y - 5z = 2$ $\quad 6x - 9y - 5z = 5$ ∎

▪ Application

Let us now consider a real-world problem that leads to a system of equations.

▎□▎ **Example 6**
Production Scheduling

A garment manufacturer produces three shirt styles. Each style shirt requires the services of three departments, as listed in the table. The cutting, sewing, and packaging departments have available a maximum of 1,160, 1,560, and 480 work-hours per week, respectively. How many of each style shirt must be produced each week for the plant to operate at full capacity?

	Style A	Style B	Style C
Cutting department	0.2 hr	0.4 hr	0.3 hr
Sewing department	0.3 hr	0.5 hr	0.4 hr
Packaging department	0.1 hr	0.2 hr	0.1 hr

Solution   Let

$$x = \text{Number of style } A \text{ produced per week}$$
$$y = \text{Number of style } B \text{ produced per week}$$
$$z = \text{Number of style } C \text{ produced per week}$$

Then

$$0.2x + 0.4y + 0.3z = 1,160 \qquad \text{Cutting department}$$
$$0.3x + 0.5y + 0.4z = 1,560 \qquad \text{Sewing department}$$
$$0.1x + 0.2y + 0.1z = 480 \qquad \text{Packaging department}$$

We clear the system of decimals by multiplying each side of each equation by 10:

$$2x + 4y + 3z = 11,600 \qquad E_1$$
$$3x + 5y + 4z = 15,600 \qquad E_2$$
$$x + 2y + z = 4,800 \qquad E_3$$

Step 1.   Eliminate x in E_2 and E_3:

$$-3x - 6y - \tfrac{9}{2}z = -17,400 \qquad -\tfrac{3}{2}E_1$$
$$\underline{3x + 5y + 4z = 15,600} \qquad E_2$$
$$- y - \tfrac{1}{2}z = -1,800 \qquad -\tfrac{3}{2}E_1 + E_2$$

$$-x - 2y - \tfrac{3}{2}z = -5,800 \qquad -\tfrac{1}{2}E_1$$
$$\underline{x + 2y + z = 4,800} \qquad E_3$$
$$- \tfrac{1}{2}z = -1,000 \qquad -\tfrac{1}{2}E_1 + E_3$$

Step 2.   Consider the equivalent system:

$$2x + 4y + 3z = 11,600 \qquad E_1$$
$$- y - \tfrac{1}{2}z = -1,800 \qquad E_2$$
$$- \tfrac{1}{2}z = -1,000 \qquad E_3$$

Since E_3 contains no y term, no further eliminations are necessary.

Step 3.   Solve by back substitution.

$$-\tfrac{1}{2}z = -1,000 \qquad \text{Solve } E_3 \text{ for z.}$$
$$z = 2,000$$

$$-y - \tfrac{1}{2}(2,000) = -1,800 \qquad \text{Substitute } z = 2,000 \text{ in } E_2 \text{ and}$$
$$y = 800 \qquad \text{solve for y.}$$

$$2x + 4(800) + 3(2,000) = 11,600 \qquad \text{Substitute } y = 800 \text{ and}$$
$$2x = 2,400 \qquad z = 2,000 \text{ in } E_1 \text{ and solve for x.}$$
$$x = 1,200$$

Thus, each week, the company should produce 1,200 style *A* shirts, 800 style *B* shirts, and 2,000 style *C* shirts to operate at full capacity. The check of the solution is left to the reader.

Problem 6   Repeat Example 6 with the cutting, sewing, and packaging departments having available a maximum of 1,180, 1,560, and 510 work-hours per week, respectively. ∎

Answers to Matched Problems

1. (A) $S = \{(x, y) | x - y = 1\}$

(B) $(0, -1)$ and $(1, 0)$ are two points in the solution set

(C)
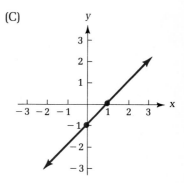

2. $x = -1$, $y = 2$ or $(-1, 2)$

3. (A) $x = 5$, $y = 3$ or $(5, 3)$
   (B) $x = 3t - 4$, $y = t$ (t any real number) or $\{(3t - 4, t) | t$ any real number$\}$
   (C) Inconsistent

4. $x = -1$, $y = 2$, $z = 4$ or $(-1, 2, 4)$

5. (A) $x = 3 + 2t$, $y = 1 - 3t$, $z = t$ (t any real number)
       or $\{(3 + 2t, 1 - 3t, t) | t$ any real number$\}$
   (B) Inconsistent

6. 900 style A, 1,300 style B, 1,600 style C

▌▌ Exercise 1-1

A *Graph each system of equations and discuss the nature of the solution set.*

1. $x + y = 5$
   $y = 1$

2. $x + y = 5$
   $2x - y = 1$

3. $x + y = 5$
   $3x + 3y = 9$

4. $x + y = 5$
   $2x - 3y = 5$

5. $2x - y = 1$
   $4x - 2y = 2$

6. $2x - y = 1$
   $-4x + 2y = 6$

7. $2x - 3y = 5$
   $3x - 2y = 0$

8. $2x - 3y = 5$
   $-4x + 6y = -10$

Use elimination with back substitution to find the solution of each system.

9. $x + 2y = 5$
   $3x - y = 8$

10. $x - 2y = 7$
    $3x + y = 7$

11. $2x - y = 4$
    $-2x + y = 6$

12. $4x + 2y = -1$
    $x - y = 2$

13. $-x + 3y = 2$
    $2x - 6y = -4$

14. $2x - 3y = 5$
    $-6x + 9y = -15$

15. $2x - 5y = -6$
    $5x + 7y = 11$

16. $4x + y = 5$
    $8x + 2y = 7$

B 17. $2x - 3y + z = 7$
        $2x - 4y + 3z = 11$
        $2x - 3y + 4z = 10$

18. $5x + 4y - 2z = 9$
    $5x + 5y + z = 16$
    $5x + 4y - z = 10$

19.
$$\begin{aligned} x - y + 5z &= 3 \\ -x + y - 3z &= 4 \\ 2x - 2y + 9z &= 18 \end{aligned}$$

20.
$$\begin{aligned} 4x + 2y - 6z &= 0 \\ 2x + y + z &= 5 \\ -6x - 3y + 2z &= 11 \end{aligned}$$

21.
$$\begin{aligned} 2y - z &= 5 \\ x - y + 2z &= 4 \\ x + y + z &= 9 \end{aligned}$$

22.
$$\begin{aligned} y - 7z &= 11 \\ x + 2y - 4z &= 7 \\ x + y + 3z &= -4 \end{aligned}$$

23.
$$\begin{aligned} 2x + 6y - 8z &= 15 \\ x + 3y - 5z &= 5 \\ 3x + 12y - 17z &= 26 \end{aligned}$$

24.
$$\begin{aligned} x + 2y - z &= 4 \\ -3y + 4z &= 12 \\ 2x + 4y - 2z &= 6 \end{aligned}$$

25.
$$\begin{aligned} x + 2y + z &= 15 \\ 2x + 5y - z &= 25 \\ x + 4y - 5z &= 5 \end{aligned}$$

26.
$$\begin{aligned} x + 4y - 5z &= 4 \\ x + y - z &= 3 \\ 2x - 4y + 6z &= 4 \end{aligned}$$

27.
$$\begin{aligned} 2x - y + 3z &= 7 \\ 3x + y - 5z &= -9 \\ -4x + 2y - 6z &= 12 \end{aligned}$$

28.
$$\begin{aligned} x + 2y - 2z &= 1 \\ -4x + 2y + 8z &= -1 \\ 3x + 4y - 2z &= 2 \end{aligned}$$

C   **29.** (A) Graph the system:

$$\begin{aligned} x + 2y &= 4 \qquad E_1 \\ 3x - y &= 5 \qquad E_2 \end{aligned}$$

(B) Replace E_2 with the equation formed by performing the operation $-3E_1 + E_2$, and graph this system.

(C) Refer to the system in part (A). Replace E_1 with the equation formed by performing the operation $2E_2 + E_1$, and graph this system.

30. (A) Graph the system:

$$\begin{aligned} x - y &= -3 \qquad E_1 \\ 2x + y &= 6 \qquad E_2 \end{aligned}$$

(B) Replace E_2 with the equation formed by performing the operation $-2E_1 + E_2$, and graph this system.

(C) Refer to the system in part (A). Replace E_1 with the equation formed by performing the operation $E_2 + E_1$, and graph this system.

31. Determine the values of a and b for which the system

$$\begin{aligned} x + ay &= 0 \\ x + y &= b \end{aligned}$$

will have:

(A) No solution

(B) An infinite number of solutions

(C) A unique solution

32. For which values of a and b will the system

$$x + ay + 3z = 2$$
$$x + y + 2z = 5$$
$$x + ay + bz = 3$$

have a unique solution?

33. Show that the system

$$ax + by = e$$
$$cx + dy = f$$

has a unique solution whenever $ad - bc \neq 0$. Find this solution. [*Hint:* Consider two cases, $a = 0$ and $a \neq 0$.]

34. Show that the system in Problem 33 does not have a unique solution when $ad - bc = 0$.

▌▢▌ Applications

35. *Chemistry.* A chemist has two saline solutions: one has a 10% concentration of saline and the other has a 35% concentration of saline. How many cubic centimeters of each solution should be mixed together in order to obtain 60 cubic centimeters of solution with a 25% concentration of saline?

36. *Chemistry.* Repeat Problem 35 if the chemist wants the final solution to have a 15% concentration of saline.

37. *Production scheduling.* A small manufacturing plant makes three types of inflatable boats: one-person, two-person, and four-person models. Each boat requires the services of three departments, as listed in the table. The cutting, assembly, and packaging departments have available a maximum of 380, 330, and 120 work-hours per week, respectively. How many boats of each type must be produced each week for the plant to operate at full capacity?

	One-Person Boat	Two-Person Boat	Four-Person Boat
Cutting department	0.6 hr	1.0 hr	1.5 hr
Assembly department	0.6 hr	0.9 hr	1.2 hr
Packaging department	0.2 hr	0.3 hr	0.5 hr

38. *Production scheduling.* Repeat Problem 37 assuming the cutting, assembly, and packaging departments have available a maximum of 260, 234, and 82 work-hours per week, respectively.

39. *Nutrition.* Animals in an experiment are to be kept under a strict diet. Each animal is to receive, among other things, 20 grams of protein and 6 grams of fat. The laboratory technician is able to purchase two food mixes of the following compositions:

Mix	Protein (%)	Fat (%)
A	10	6
B	20	2

How many grams of each mix should be used to obtain the right diet for a single animal?

40. *Diet.* In an experiment involving mice, a zoologist finds she needs a food mix that contains, among other things, 23 grams of protein, 6.2 grams of fat, and 16 grams of moisture. She has on hand mixes of the following compositions:

Mix	Protein (%)	Fat (%)	Moisture (%)
A	20	2	15
B	10	6	10
C	15	5	5

How many grams of each mix should she use to get the desired diet mix?

▌1-2▐ Augmented Matrices and Elementary Row Operations

- Augmented Coefficient Matrices
- Elementary Row Operations
- Solving Systems of Equations
- Application

Most linear systems of any consequence involve large numbers of equations and variables. These systems are solved with computers, since hand methods would be impractical. In this section we will introduce some new mathematical notation and operations which form the groundwork for solving larger systems of equations. In the next two sections we will use the ideas presented here to formulate two different procedures for solving systems of linear equations. The concepts discussed in this section are fundamental to the study of linear algebra and will be used frequently throughout the book.

▪ Augmented Coefficient Matrices

In solving systems of equations by elimination in the preceding section, the coefficients of the variables and constant terms played a central role. The process can be made more efficient for generalization and computer work by the introduction of a mathematical form called a *matrix*. A **matrix** is a rectangular array of numbers written within brackets. Some examples are

$$\begin{bmatrix} 3 & 5 \\ 0 & -2 \end{bmatrix} \qquad \begin{bmatrix} 2 \\ -3 \\ 0 \end{bmatrix} \qquad \begin{bmatrix} \frac{1}{2} & -\frac{5}{2} & 0 & \frac{2}{3} \end{bmatrix}$$

$$\begin{bmatrix} -1 & \sqrt{2} & -5 & 0 \\ 0 & \sqrt{3} & 2 & 7.9 \end{bmatrix} \qquad \begin{bmatrix} 1 & 0 & 0 \\ 0 & 1 & 0 \\ 0 & 0 & 1 \end{bmatrix}$$

Each number in a matrix is called an **element** of the matrix.

Associated with the linear system

$$\begin{aligned} 2x - 3y + 4z &= 7 \\ 5x - 6y &= 12 \\ -x + y - z &= 17 \end{aligned} \qquad (1)$$

is the **augmented coefficient matrix**

$$\begin{bmatrix} 2 & -3 & 4 & 7 \\ 5 & -6 & 0 & 12 \\ -1 & 1 & -1 & 17 \end{bmatrix} \qquad (2)$$

which contains the essential parts of the system—namely the coefficients of the variables and the constant terms. (The vertical bar is included only to separate the coefficients of the variables from the constants.)

For ease of generalization to larger systems we are now going to change the notation for the variables in (1) to a subscript form. That is, in place of x, y, and z, we will use x_1, x_2, and x_3 and (1) will be written as

$$\begin{aligned} 2x_1 - 3x_2 + 4x_3 &= 7 \\ 5x_1 - 6x_2 &= 12 \\ -x_1 + x_2 - x_3 &= 17 \end{aligned}$$

Notice that (2) is still the augmented coefficient matrix for this system. Changing notation has no effect on the coefficients and constants.

In order to write the general form of a linear system of three equations in three variables (and later for larger systems), we use *double subscript notation* for the coefficients. For example,

$$\begin{aligned} a_{11}x_1 + a_{12}x_2 + a_{13}x_3 &= c_1 & E_1 \\ a_{21}x_1 + a_{22}x_2 + a_{23}x_3 &= c_2 & E_2 \\ a_{31}x_1 + a_{32}x_2 + a_{33}x_3 &= c_3 & E_3 \end{aligned} \qquad (3)$$

In (3), a_{21} (read "a two one," *not* "a twenty-one") is the coefficient of variable x_1 in equation E_2. In general, a_{ij} is the coefficient of variable x_j in equation E_i.

The augmented coefficient matrix for system (3) is

Notice that the first subscript number for a coefficient element indicates the row in which it lies and the second subscript indicates the column. Thus, a_{32} is the element in the third row and second column. In general, a_{ij} is in the ith row and jth column.

The augmented coefficient matrix contains the essential parts of system (3). Our objective is to learn how to manipulate augmented coefficient matrices in order to solve system (3), if a solution exists. The manipulative process is a direct outgrowth of the elimination process discussed in the preceding section.

▪ Elementary Row Operations

Recall from Theorem 1 in Section 1-1 that the three basic operations that can be used to produce an equivalent system of equations are the following:

(A) Interchange two equations. $E_i \leftrightarrow E_j$
(B) Multiply an equation by a nonzero constant. $kE_i \rightarrow E_i$
(C) Add a constant multiple of one equation to another equa- $E_j + kE_i \rightarrow E_j$
     tion.

Table 2
Comparison of Operations on Systems and Matrices

System of Equations	Operation	Augmented Coefficient Matrix	Operation
$x_1 + 2x_2 - x_3 = 9$ $2x_1 + x_2 + x_3 = 6$ $3x_1 - 2x_2 - 2x_3 = 2$	$E_2 + (-2)E_1 \rightarrow E_2$ $E_3 + (-3)E_1 \rightarrow E_3$	$\begin{bmatrix} 1 & 2 & -1 & 9 \\ 2 & 1 & 1 & 6 \\ 3 & -2 & -2 & 2 \end{bmatrix}$	$R_2 + (-2)R_1 \rightarrow R_2$ $R_3 + (-3)R_1 \rightarrow R_3$
$x_1 + 2x_2 - x_3 = 9$ $-3x_2 + 3x_3 = -12$ $-8x_2 + x_3 = -25$	$E_3 + (-\frac{8}{3})E_2 \rightarrow E_3$	$\begin{bmatrix} 1 & 2 & -1 & 9 \\ 0 & -3 & 3 & -12 \\ 0 & -8 & 1 & -25 \end{bmatrix}$	$R_3 + (-\frac{8}{3})R_2 \rightarrow R_3$
$x_1 + 2x_2 - x_3 = 9$ $-3x_2 + 3x_3 = -12$ $-7x_3 = 7$		$\begin{bmatrix} 1 & 2 & -1 & 9 \\ 0 & -3 & 3 & -12 \\ 0 & 0 & -7 & 7 \end{bmatrix}$	

Since each row of the augmented coefficient matrix for a system represents one equation in the system, we can perform similar operations on the rows of the augmented coefficient matrix. Table 2 lists the operations and equivalent systems (rewritten in subscript notation) that arose in the solution of Example 4 in the preceding section, along with the augmented coefficient matrices for each system. Study this table carefully. [Notice that operation (C) is being used exclusively at this point. The importance of the other two operations will become evident as we proceed.]

When operations of this type are performed on matrices, they are referred to as *elementary row operations*. These are summarized in the box for easy reference.

Elementary Row Operations

(A) Interchange two rows. $\qquad\qquad\qquad\qquad\qquad\quad R_i \leftrightarrow R_j$
(B) Multiply a row by a nonzero constant. $\qquad\qquad\quad kR_i \rightarrow R_i$
(C) Add a constant multiple of one row to another row. $\quad R_j + kR_i \rightarrow R_j$

Performing a sequence of operations on a system of linear equations transforms the system into an equivalent system of linear equations. Thus, it is natural to say that performing a sequence of elementary row operations on a matrix transforms the matrix into an *equivalent* or, more formally, a *row equivalent* matrix.

Row Equivalent Matrices

Two matrices A and B are said to be **row equivalent,** denoted $A \sim B$, if A can be transformed into B by performing a sequence of elementary row operations.

Technically, if A is transformed into B, then A is said to be row equivalent to B. It can be shown (see Problem 32 in Exercise 1-2) that it is always possible to reverse these operations and transform B back into A. Thus, it is correct to say A and B are row equivalent if either can be transformed into the other. For example, all three of the matrices listed in Table 2 are row equivalent to each other.

Notice that the definitions of elementary row operations and row equivalent matrices are not restricted to augmented coefficient matrices. Elementary row operations can be performed on any matrix to produce a row equivalent matrix. Later in this book we will encounter many different problems that can be solved by performing elementary row operations on a matrix. For now, we are interested in using these operations to solve systems of linear equations.

▪ Solving Systems of Equations

If A and B are augmented coefficient matrices for two systems of equations and A and B are row equivalent, then the systems must also be equivalent. This is nothing more than Theorem 1 in Section 1-1 restated in terms of coefficient matrices. One of the fundamental principles of linear algebra is that the converse of this statement is also true. That is, if two systems of linear equations are equivalent (have the same solution), then their coefficient matrices are row equivalent (each can be transformed into the other by row operations). The proof of this statement is beyond the scope of this book. Theorem 3 uses the more formal mathematical phrase *if and only if* to combine these statements into a single statement.

Theorem 3	Two systems of linear equations are equivalent if and only if their augmented coefficient matrices are row equivalent.

The following examples will illustrate the procedure for solving linear systems by performing elementary row operations on the augmented coefficient matrix.

Example 7 Solve by using elementary row operations:

$$x_2 + x_3 = 2$$
$$2x_1 + 4x_2 + x_3 = -3$$
$$3x_1 + 2x_2 + x_3 = 5$$

Solution We start by writing the augmented coefficient matrix for the system:

$$\begin{bmatrix} 0 & 1 & 1 & 2 \\ 2 & 4 & 1 & -3 \\ 3 & 2 & 1 & 5 \end{bmatrix}$$

We now use elementary row operations to transform this matrix into one that corresponds to a system of linear equations that can be solved by back substitution. There are many different sequences of row operations that could be used to accomplish this task. The particular sequence of operations that we are going to use has been selected so that we will have a systematic procedure that will generalize to larger systems and that can be programmed on a computer.

Step 1. Interchange R_1 and R_2 to obtain a nonzero element in position 1,1 (the first row and first column).

$$\begin{bmatrix} 0 & 1 & 1 & 2 \\ 2 & 4 & 1 & -3 \\ 3 & 2 & 1 & 5 \end{bmatrix} \xrightarrow[\sim]{R_1 \leftrightarrow R_2} \begin{bmatrix} 2 & 4 & 1 & -3 \\ 0 & 1 & 1 & 2 \\ 3 & 2 & 1 & 5 \end{bmatrix}$$

Now use the nonzero element in position 1,1 to obtain 0's in the rest of column 1 (C_1). To obtain a 0 in the third row of C_1, we multiply R_1 by $-\frac{3}{2}$ and add to R_3; this changes R_3 but not R_1. Some people find it helpful to write $(-\frac{3}{2})R_1$ outside the matrix to help prevent errors in arithmetic, as shown:

$$
\begin{bmatrix} 2 & 4 & 1 & -3 \\ 0 & 1 & 1 & 2 \\ 3 & 2 & 1 & 5 \end{bmatrix}
\underset{R_3 + (-\frac{3}{2})R_1 \to R_3}{\sim}
\begin{bmatrix} 2 & 4 & 1 & -3 \\ 0 & 1 & 1 & 2 \\ 0 & -4 & -\frac{1}{2} & \frac{19}{2} \end{bmatrix}
$$

$$-3 \quad -6 \quad -\tfrac{3}{2} \qquad \tfrac{9}{2}$$

Now C_1 has 0's in all the positions below position 1,1. This completes the first step.

Step 2. Use the nonzero element in position 2,2 (second row and second column) to obtain 0's in all the positions in C_2 below position 2,2.

$$
\begin{bmatrix} 2 & 4 & 1 & -3 \\ 0 & 1 & 1 & 2 \\ 0 & -4 & -\frac{1}{2} & \frac{19}{2} \end{bmatrix}
\underset{R_3 + 4R_2 \to R_3}{\sim}
\begin{bmatrix} 2 & 4 & 1 & -3 \\ 0 & 1 & 1 & 2 \\ 0 & 0 & \frac{7}{2} & \frac{35}{2} \end{bmatrix}
$$

$$0 \quad 4 \quad 4 \qquad 8$$

Step 3. Now we write the linear system corresponding to the last matrix and complete the solution process using back substitution.

$$
\begin{array}{rcl}
2x_1 + 4x_2 + x_3 &=& -3 \\
x_2 + x_3 &=& 2 \\
\tfrac{7}{2}x_3 &=& \tfrac{35}{2}
\end{array}
\qquad
\begin{array}{l} E_1 \\ E_2 \\ E_3 \end{array}
\qquad
\begin{bmatrix} 2 & 4 & 1 & -3 \\ 0 & 1 & 1 & 2 \\ 0 & 0 & \frac{7}{2} & \frac{35}{2} \end{bmatrix}
$$

Solving E_3 for x_3, we have

$$\tfrac{7}{2}x_3 = \tfrac{35}{2}$$

$$x_3 = 5$$

Substituting for x_3 in E_2,

$$x_2 + 5 = 2$$

$$x_2 = -3$$

Finally, substituting for x_2 and x_3 in E_1, we have

$$2x_1 + 4(-3) + 5 = -3$$

$$2x_1 = 4$$

$$x_1 = 2$$

Thus, the solution to this system is $x_1 = 2$, $x_2 = -3$, and $x_3 = 5$ or $(2, -3, 5)$.

Since the augmented coefficient matrix for the system in Step 3 is row equivalent to the augmented coefficient for the original system, Theorem 3 implies that $(2, -3, 5)$ is also the solution of the original system. (Check this.)

Problem 7   Solve by using elementary row operations:

$$x_1 - 2x_2 + x_3 = -1$$
$$2x_1 - 3x_2 + x_3 = -4$$
$$3x_1 - 4x_2 + 2x_3 = -3$$

■■

Example 8   Solve each system using elementary row operations:

(A) $x_1 - x_2 + 4x_3 = -3$ (B) $x_1 - 2x_2 + x_3 = 1$
     $2x_1 - 2x_2 + 8x_3 = 1$ $2x_1 - 2x_2 + 4x_3 = -2$
     $4x_1 + 2x_2 + 10x_3 = 4$ $3x_1 - 7x_2 + 2x_3 = 5$

Solution   (A)

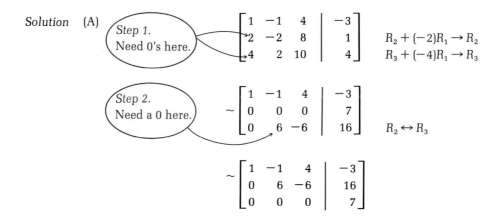

Step 1.
Need 0's here.

$$\begin{bmatrix} 1 & -1 & 4 & | & -3 \\ 2 & -2 & 8 & | & 1 \\ 4 & 2 & 10 & | & 4 \end{bmatrix} \quad \begin{array}{l} R_2 + (-2)R_1 \rightarrow R_2 \\ R_3 + (-4)R_1 \rightarrow R_3 \end{array}$$

Step 2.
Need a 0 here.

$$\sim \begin{bmatrix} 1 & -1 & 4 & | & -3 \\ 0 & 0 & 0 & | & 7 \\ 0 & 6 & -6 & | & 16 \end{bmatrix} \quad R_2 \leftrightarrow R_3$$

$$\sim \begin{bmatrix} 1 & -1 & 4 & | & -3 \\ 0 & 6 & -6 & | & 16 \\ 0 & 0 & 0 & | & 7 \end{bmatrix}$$

Step 3.   Solve the equivalent system:

$$x_1 - x_2 + 4x_3 = -3$$
$$6x_2 - 6x_3 = 16$$
$$0 = 7$$

The last equation indicates that this system is inconsistent. Thus, we can conclude that the original system is also inconsistent and has no solution (solution set is the empty set).

(B)

Step 1.
Need 0's here.

$$\begin{bmatrix} 1 & -2 & 1 & | & 1 \\ 2 & -2 & 4 & | & -2 \\ 3 & -7 & 2 & | & 5 \end{bmatrix} \quad \begin{array}{l} R_2 + (-2)R_1 \rightarrow R_2 \\ R_3 + (-3)R_1 \rightarrow R_3 \end{array}$$

Step 2.
Need a 0 here.

$$\sim \begin{bmatrix} 1 & -2 & 1 & | & 1 \\ 0 & 2 & 2 & | & -4 \\ 0 & -1 & -1 & | & 2 \end{bmatrix} \quad R_3 + \tfrac{1}{2}R_2 \rightarrow R_3$$

$$\sim \begin{bmatrix} 1 & -2 & 1 & | & 1 \\ 0 & 2 & 2 & | & -4 \\ 0 & 0 & 0 & | & 0 \end{bmatrix}$$

Step 3. Solve the equivalent system:

$$\begin{array}{rcll} x_1 - 2x_2 + x_3 =& 1 & E_1 \\ 2x_2 + 2x_3 =& -4 & E_2 \\ 0 =& 0 & E_3 \end{array}$$

This is a dependent system having infinitely many solutions. To represent the (general) solution, we introduce a parameter by letting $x_3 = t$. Substituting for x_3 in the second equation and solving for x_2, we have

$$\overset{x_3}{2x_2 + 2t = -4}$$
$$x_2 = -2 - t$$

Using back substitution in E_1, we can express x_1 in terms of t:

$$\overset{x_2 \quad\quad x_3}{x_1 - 2(-2 - t) + t} = 1$$
$$x_1 = -3 - 3t$$

Thus, the (general) solution to both this system and the original system is

$$x_1 = -3 - 3t$$
$$x_2 = -2 - t$$
$$x_3 = t \qquad t \text{ any real number}$$

or

$$S = \{(-3 - 3t, -2 - t, t) | t \text{ any real number}\}$$

Problem 8   Solve each system using elementary row operations:

(A) $\begin{aligned} x_1 + x_2 + x_3 &= 4 \\ x_2 - x_3 &= 4 \\ 2x_1 - x_2 + 5x_3 &= 6 \end{aligned}$ (B) $\begin{aligned} x_1 + x_2 - 3x_3 &= 5 \\ x_1 - x_2 + x_3 &= -1 \\ 3x_1 + x_2 - 5x_3 &= 9 \end{aligned}$

▌▌

▪ Application

An **interpolating polynomial** for a set of points is a polynomial whose graph passes through each of the given points. If we are given three noncollinear points with distinct x coordinates, then it can be shown that there is a unique second-degree polynomial

$$p(x) = ax^2 + bx + c$$

whose graph will pass through these three points. The values of a, b, and c can be determined by setting up and solving a system of linear equations.

▌□▌

Example 9   Find the second-degree polynomial $p(x)$ whose graph passes through the points
Interpolation $(-1, 5)$, $(1, -1)$, and $(3, 1)$.

Solution   Since the point $(-1, 5)$ is on the graph of $p(x) = ax^2 + bx + c$, the coordinates $(-1, 5)$ must satisfy this equation; that is,

$$5 = p(-1) = a(-1)^2 + b(-1) + c$$

or

$$a - b + c = 5 \qquad \text{A linear equation}$$

Using the other two points produces two more linear equations relating a, b, and c:

$$\begin{aligned} -1 = p(1) &= a + b + c \\ 1 = p(3) &= 9a + 3b + c \end{aligned}$$

Thus, a, b, and c must satisfy the system

$$\begin{aligned} a - b + c &= 5 \\ a + b + c &= -1 \\ 9a + 3b + c &= 1 \end{aligned}$$

Writing the augmented coefficient matrix for this system and proceeding as before, we have

$$\left[\begin{array}{ccc|c} 1 & -1 & 1 & 5 \\ 1 & 1 & 1 & -1 \\ 9 & 3 & 1 & 1 \end{array}\right] \qquad \begin{aligned} R_2 + (-1)R_1 &\rightarrow R_2 \\ R_3 + (-9)R_1 &\rightarrow R_3 \end{aligned}$$

$$\sim \begin{bmatrix} 1 & -1 & 1 & 5 \\ 0 & 2 & 0 & -6 \\ 0 & 12 & -8 & -44 \end{bmatrix} \quad R_3 + (-6)R_2 \to R_3$$

$$\sim \begin{bmatrix} 1 & -1 & 1 & 5 \\ 0 & 2 & 0 & -6 \\ 0 & 0 & -8 & -8 \end{bmatrix}$$

$a - b + c = 5$ Equivalent system

$2b = -6$

$-8c = -8$

$-8c = -8$ Solve the third equation for c.

$c = 1$

$2b = -6$ Solve the second equation for b.

$b = -3$

$a - (-3) + 1 = 5$ Solve the first equation for a.

$a = 1$

Thus, the graph of

$p(x) = x^2 - 3x + 1$

will pass through the given points $(-1, 5)$, $(1, -1)$, and $(3, 1)$, as shown in the figure.

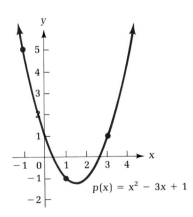

$p(x) = x^2 - 3x + 1$

Problem 9   Find the second-degree polynomial whose graph passes through the points $(-1, -6)$, $(1, 4)$, and $(2, 12)$.   ▮

Answers to   7. $x_1 = -1$, $x_2 = 2$, $x_3 = 4$ or $(-1, 2, 4)$
Matched Problems   8. (A) Inconsistent
    (B) $x_1 = 2 + t$, $x_2 = 3 + 2t$, $x_3 = t$ (t any real number)
    or $S = \{(2 + t, 3 + 2t, t)|t$ any real number$\}$
    9. $p(x) = x^2 + 5x - 2$

‖ Exercise 1-2

A   In Problems 1–4, find the augmented coefficient matrix for each system.

1. $\begin{aligned} 2x_1 - 3x_2 + x_3 &= 4 \\ -x_1 + 4x_2 - 3x_3 &= 7 \\ 8x_1 + 2x_2 + 5x_3 &= -11 \end{aligned}$
2. $\begin{aligned} -x_1 + x_2 - x_3 &= 1 \\ x_1 + 2x_2 + 3x_3 &= 4 \\ 5x_1 - 4x_2 + 3x_3 &= -2 \end{aligned}$

3. $\begin{aligned} x_1 \quad\quad - x_3 &= 0 \\ x_2 + x_3 &= 0 \\ -x_1 + 2x_2 \quad\quad &= 0 \end{aligned}$
4. $\begin{aligned} 2x_1 - 5x_2 + x_3 &= 9 \\ 3x_2 + 4x_3 &= -6 \\ 3x_3 &= 12 \end{aligned}$

In Problems 5–8, find the system of linear equations that corresponds to each augmented coefficient matrix.

5. $\left[\begin{array}{ccc|c} 2 & -3 & 1 & 4 \\ -1 & 2 & -1 & 5 \\ 4 & 6 & 7 & -9 \end{array}\right]$
6. $\left[\begin{array}{ccc|c} 3 & -1 & 4 & 2 \\ -2 & 5 & 9 & 11 \\ 6 & 5 & -3 & 10 \end{array}\right]$

7. $\left[\begin{array}{ccc|c} 2 & 0 & 1 & -1 \\ 0 & 1 & -3 & 0 \\ 1 & 0 & 4 & 2 \end{array}\right]$
8. $\left[\begin{array}{ccc|c} 1 & 0 & -1 & 0 \\ 0 & 2 & 3 & 0 \\ 0 & 0 & 4 & 5 \end{array}\right]$

In Problems 9–14, perform the indicated row operation on the following matrix:

$$\left[\begin{array}{ccc|c} 1 & -2 & 1 & 3 \\ 2 & 4 & -2 & 6 \\ -3 & 2 & 1 & 5 \end{array}\right]$$

9. $R_1 \leftrightarrow R_2$
10. $R_2 \leftrightarrow R_3$
11. $3R_1 \rightarrow R_1$

12. $\frac{1}{2}R_2 \rightarrow R_2$
13. $R_3 + 3R_1 \rightarrow R_3$
14. $R_2 + (-2)R_1 \rightarrow R_2$

B   In Problems 15–20, solve the system of linear equations that corresponds to the given augmented coefficient matrix.

15. $\left[\begin{array}{ccc|c} 2 & -1 & 3 & -8 \\ 0 & 1 & -4 & 12 \\ 0 & 0 & 3 & -6 \end{array}\right]$
16. $\left[\begin{array}{ccc|c} 3 & 4 & -2 & -6 \\ 0 & 2 & 1 & -1 \\ 0 & 0 & 2 & 2 \end{array}\right]$

17. $\left[\begin{array}{ccc|c} 1 & -1 & 1 & 0 \\ 0 & 1 & -2 & 1 \\ 0 & 0 & 0 & 0 \end{array}\right]$
18. $\left[\begin{array}{ccc|c} 2 & -3 & 5 & 1 \\ 0 & 2 & -1 & 1 \\ 0 & 0 & 0 & 2 \end{array}\right]$

$$19. \begin{bmatrix} 3 & 0 & 0 & | & 2 \\ 0 & 2 & 1 & | & 0 \\ 0 & 0 & 0 & | & 2 \end{bmatrix} \qquad 20. \begin{bmatrix} 1 & 2 & -3 & | & 4 \\ 0 & 3 & 0 & | & 9 \\ 0 & 0 & 0 & | & 0 \end{bmatrix}$$

In Problems 21–30, use elementary row operations to solve each system.

21. $x_1 - x_2 + x_3 = 2$
$2x_1 - x_2 + x_3 = 7$
$2x_1 + x_2 \qquad = 5$

22. $x_1 - 2x_2 + 3x_3 = 0$
$x_1 \qquad + 2x_3 = 4$
$x_1 - 2x_2 + 5x_3 = 4$

23. $2x_1 + 3x_2 + x_3 = 4$
$4x_1 + 9x_2 + x_3 = 12$
$2x_1 + 6x_2 + 2x_3 = 9$

24. $x_1 - 3x_2 + x_3 = 2$
$2x_1 - 5x_2 \qquad = 3$
$x_1 - 2x_2 - x_3 = -3$

25. $2x_1 + 3x_2 - 3x_3 = 0$
$2x_1 - 3x_2 + x_3 = 2$
$4x_1 + 3x_2 - 4x_3 = 1$

26. $x_1 + x_2 - 4x_3 = 1$
$x_1 - 2x_2 - x_3 = 4$
$2x_1 - x_2 - 5x_3 = 5$

27. $x_1 - 2x_2 + 3x_3 = 5$
$-2x_1 + 4x_2 - 6x_3 = -4$
$3x_1 - 4x_2 + 7x_3 = 9$

28. $x_1 \qquad - 2x_3 = -1$
$2x_1 + 3x_2 + 4x_3 = 2$
$4x_1 + 3x_2 + 4x_3 = 3$

29. $x_1 - x_2 + x_3 = 0$
$x_1 + x_2 - x_3 = 0$
$x_1 - x_2 - x_3 = 0$

30. $x_1 - x_2 + x_3 = 0$
$x_1 + x_2 - x_3 = 0$
$3x_1 - x_2 + x_3 = 0$

C   31. Suppose an augmented coefficient matrix A is transformed into a row equivalent matrix B by performing one of the elementary row operations. Find the operation that will transform B back into A if the original operation was:

(A) $R_i \leftrightarrow R_j$ (B) $kR_i \rightarrow R_i, k \neq 0$
(C) $R_i + kR_j \rightarrow R_i, k \neq 0$

32. Use Problem 31 to show that if an augmented coefficient matrix A is row equivalent to an augmented coefficient matrix B, then B is also row equivalent to A.

33. If A, B, and C are augmented coefficient matrices with the property that A is row equivalent to B and B is row equivalent to C, then show that A is row equivalent to C.

34. Show that any augmented coefficient matrix is row equivalent to itself.

35. Consider the system of linear equations

$$a_{11}x_1 + a_{12}x_2 + a_{13}x_3 = b_1$$
$$a_{22}x_2 + a_{23}x_3 = b_2$$
$$a_{33}x_3 = b_3$$

If a_{11}, a_{22}, and a_{33} are nonzero, show that this system has a unique solution.

36. Refer to the system of linear equations in Problem 35. If $a_{33} = 0$, show this system is either dependent or inconsistent.

▌▫▌ Applications

In Problems 37–40, find the second-degree polynomial whose graph passes through the indicated points.

37. $(-1, 2), (0, 1), (1, 2)$ **38.** $(-1, 0), (0, -1), (1, 0)$

39. $(1, 2), (2, 12), (3, 28)$ **40.** $(-2, -7), (1, 8), (4, 5)$

▌1-3▌ Triangularization with Back Substitution

- Introduction
- Triangular Form
- Triangularization with Back Substitution
- Application

▪ Introduction

Now that we have discussed the use of matrices and row operations in the solution of systems of three equations in three variables, we can generalize these procedures for larger systems of linear equations. A system of *n linear equations in n unknowns* has the form

$$
\begin{aligned}
a_{11}x_1 + a_{12}x_2 + \cdots + a_{1n}x_n &= c_1 \\
a_{21}x_1 + a_{22}x_2 + \cdots + a_{2n}x_n &= c_2 \\
&\;\;\vdots \\
a_{n1}x_1 + a_{n2}x_2 + \cdots + a_{nn}x_n &= c_n
\end{aligned}
\tag{1}
$$

Systems where the number of equations equals the number of variables are called **square systems.** In this section we will restrict our attention to square systems. More general systems will be considered in the next section.

Notice how simple it is to extend the double subscript notation for the coefficients of a linear system to larger systems. This notation applies equally well to augmented matrices. The augmented coefficient matrix for (1) is

$$
\left[
\begin{array}{cccc|c}
a_{11} & a_{12} & \cdots & a_{1n} & c_1 \\
a_{21} & a_{22} & \cdots & a_{2n} & c_2 \\
\vdots & \vdots & & \vdots & \vdots \\
a_{n1} & a_{n2} & \cdots & a_{nn} & c_n
\end{array}
\right]
\tag{2}
$$

▪ Triangular Form

We want to develop a procedure for transforming (2) into a matrix of a particular form that will always correspond to a system that can be solved by back substitution, if a solution exists. The following terminology will be helpful in describing this process: The numbers a_{11}, a_{22}, . . . , a_{nn} are called the **diagonal elements** of (2). The matrix (2) is said to be in **triangular form** if all the elements in each column *below the diagonal element* are 0's.

Example 10   The following augmented matrices are in triangular form:

(A)
$$\left[\begin{array}{cccc|c} -1 & 2 & -3 & -4 & 5 \\ 0 & 4 & 7 & -8 & 6 \\ 0 & 0 & 3 & 2 & 9 \\ 0 & 0 & 0 & 2 & 4 \end{array}\right]$$

Diagonal elements

(B)
$$\left[\begin{array}{ccccc|c} 2 & -1 & 3 & 0 & 4 & -7 \\ 0 & 0 & -4 & 0 & -6 & 3 \\ 0 & 0 & -1 & 1 & 9 & 0 \\ 0 & 0 & 0 & 0 & -3 & 2 \\ 0 & 0 & 0 & 0 & 2 & 0 \end{array}\right]$$

Diagonal elements

Problem 10   The augmented matrices below are not in triangular form. Indicate which non-zero elements violate the definition for each matrix.

(A)
$$\left[\begin{array}{cccc|c} 8 & 2 & -3 & 4 & 5 \\ 1 & 0 & 4 & -1 & 7 \\ 0 & 0 & 2 & 1 & 3 \\ 0 & 0 & 0 & 3 & 1 \end{array}\right]$$

(B)
$$\left[\begin{array}{ccccc|c} -2 & 1 & -4 & 6 & 9 & -2 \\ 0 & 0 & 0 & 3 & -2 & 4 \\ 0 & 0 & 0 & 4 & -6 & 7 \\ 0 & 0 & 0 & 3 & -1 & 2 \\ 0 & 0 & 0 & 7 & 9 & -5 \end{array}\right]$$

∎

▪ Triangularization with Back Substitution

We are now ready to outline the procedure for solving square systems by *triangularization with back substitution*. This procedure transforms an augmented coefficient matrix into triangular form, producing an equivalent system that can be solved by back substitution, if a solution exists.

Example 11   Solve by using elementary row operations:

$$x_1 + 2x_2 - x_3 + x_4 = 1$$
$$-x_1 - 2x_2 + 2x_3 + 2x_4 = 2$$
$$2x_1 + x_2 + 2x_3 - 3x_4 = 0$$
$$x_1 + 5x_2 + x_3 - 2x_4 = -5$$

Solution   The augmented coefficient matrix for this system is

$$\begin{bmatrix} 1 & 2 & -1 & 1 & | & 1 \\ -1 & -2 & 2 & 2 & | & 2 \\ 2 & 1 & 2 & -3 & | & 0 \\ 1 & 5 & 1 & -2 & | & -5 \end{bmatrix}$$

Our objective is to transform this matrix into one that is in triangular form.

Step 1.   Use the nonzero diagonal element in position 1,1 to obtain 0's in C_1 below the diagonal element.

$$C_1$$

Need 0's here.

$$\begin{bmatrix} 1 & 2 & -1 & 1 & | & 1 \\ -1 & -2 & 2 & 2 & | & 2 \\ 2 & 1 & 2 & -3 & | & 0 \\ 1 & 5 & 1 & -2 & | & -5 \end{bmatrix}$$

$$R_2 + R_1 \rightarrow R_2$$
$$R_3 + (-2)R_1 \rightarrow R_3$$
$$R_4 + (-1)R_1 \rightarrow R_4$$

$$\sim \begin{bmatrix} 1 & 2 & -1 & 1 & | & 1 \\ 0 & 0 & 1 & 3 & | & 3 \\ 0 & -3 & 4 & -5 & | & -2 \\ 0 & 3 & 2 & -3 & | & -6 \end{bmatrix}$$

Step 2.   Now we want to obtain 0's in column C_2 below the diagonal element in position 2,2. In order to do this, we first must have a nonzero diagonal element.

$$C_2$$

Need a nonzero element here.

$$\begin{bmatrix} 1 & 2 & -1 & 1 & | & 1 \\ 0 & 0 & 1 & 3 & | & 3 \\ 0 & -3 & 4 & -5 & | & -2 \\ 0 & 3 & 2 & -3 & | & -6 \end{bmatrix}$$

$$R_2 \leftrightarrow R_3$$
Note: We could have also used $R_2 \leftrightarrow R_4$ but not $R_2 \leftrightarrow R_1$. (Why?)

Need a 0 here.

$$\sim \begin{bmatrix} 1 & 2 & -1 & 1 & | & 1 \\ 0 & -3 & 4 & -5 & | & -2 \\ 0 & 0 & 1 & 3 & | & 3 \\ 0 & 3 & 2 & -3 & | & -6 \end{bmatrix}$$

$$R_4 + R_2 \rightarrow R_4$$

$$\sim \begin{bmatrix} 1 & 2 & -1 & 1 & | & 1 \\ 0 & -3 & 4 & -5 & | & -2 \\ 0 & 0 & 1 & 3 & | & 3 \\ 0 & 0 & 6 & -8 & | & -8 \end{bmatrix}$$

Step 3. Use the nonzero diagonal element in column C_3 to obtain 0's below the diagonal element.

$$C_3$$

Need a 0 here.
$$\begin{bmatrix} 1 & 2 & -1 & 1 & | & 1 \\ 0 & -3 & 4 & -5 & | & -2 \\ 0 & 0 & 1 & 3 & | & 3 \\ 0 & 0 & 6 & -8 & | & -8 \end{bmatrix} \qquad R_4 + (-6)R_3 \rightarrow R_4$$

$$\sim \begin{bmatrix} 1 & 2 & -1 & 1 & | & 1 \\ 0 & -3 & 4 & -5 & | & -2 \\ 0 & 0 & 1 & 3 & | & 3 \\ 0 & 0 & 0 & -26 & | & -26 \end{bmatrix}$$

Since this last matrix is in triangular form, we stop the transformation process and solve the corresponding system by back substitution.

$$
\begin{aligned}
x_1 + 2x_2 - x_3 + x_4 &= 1 \\
-3x_2 + 4x_3 - 5x_4 &= -2 \\
x_3 + 3x_4 &= 3 \\
-26x_4 &= -26
\end{aligned}
$$

$$-26x_4 = -26 \qquad \text{Solve the fourth equation for } x_4.$$
$$x_4 = 1$$

$$x_3 + 3(1) = 3 \qquad \text{Substitute for } x_4 \text{ in the third equation and}$$
$$x_3 = 0 \qquad \text{solve for } x_3.$$

$$-3x_2 + 4(0) - 5(1) = -2 \qquad \text{Substitute for } x_3 \text{ and } x_4 \text{ in the second}$$
$$x_2 = -1 \qquad \text{equation and solve for } x_2.$$

$$x_1 + 2(-1) - 0 + 1 = 1 \qquad \text{Substitute for } x_2, x_3, \text{ and } x_4 \text{ in the first}$$
$$x_1 = 2 \qquad \text{equation and solve for } x_1.$$

Solution. $x_1 = 2$, $x_2 = -1$, $x_3 = 0$, $x_4 = 1$ or $(2, -1, 0, 1)$ (Check this.)

The steps we used to transform the matrix in Example 11 to triangular form are summarized in the box at the top of the next page.

Transforming a Matrix to Triangular Form

Beginning with C_1 and proceeding from left to right, do the following for each column C_i:

1. If the diagonal element is the bottom element in the column, then stop—the transformation is complete.

2. If the diagonal element in C_i is 0, then find a row below R_i with a nonzero element in C_i and interchange this row with R_i. [*Note:* If all the elements below the diagonal are already 0, then go on to the next column.]

3. Use R_i and appropriate row operations to obtain 0's in all the positions below the diagonal element.

Solving systems by following the procedure outlined in the box is referred to as **triangularization with back substitution.** This is a very efficient method for solving square systems on a computer. (See the Preface for information concerning the computer supplement for this book.)

Problem 11 Solve by triangularization with back substitution:

$$
\begin{aligned}
x_1 - x_2 + x_3 + x_4 &= -2 \\
x_1 + x_2 + 2x_3 - x_4 &= 3 \\
-x_1 + x_2 - x_3 + 3x_4 &= 2 \\
3x_1 - 3x_2 + 4x_3 - 5x_4 &= -5
\end{aligned}
$$

∎

As we have seen, not all systems have unique solutions or even any solution at all. Systems may be dependent (infinitely many solutions) or inconsistent (no solution). We will postpone discussing dependent and inconsistent systems until the next section.

■ Application

Example 12
Heat Conduction

A metal grid consists of four thin metal bars. The temperature of the end of each bar is given in Figure 2. If the temperature at each point where two bars intersect is the average of the temperatures at the four adjacent points in the grid (adjacent points are either other intersection points or ends of bars), find the temperature at each intersection point.

Solution Let x_1, x_2, x_3, and x_4 be the temperatures at the intersection points, as indicated in Figure 3.

Each intersection point has four adjacent points, one to the left, one above, one to the right, and one below. The temperature at each intersection point is the average of the temperatures at these four points. Thus, if we begin at the

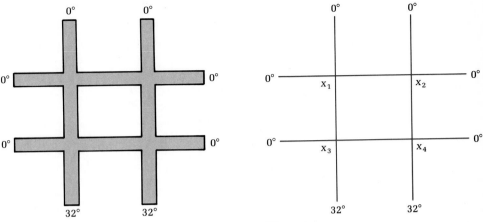

Figure 2   **Figure 3**

intersection point in the lower left-hand corner, we have

$$\begin{array}{cccccccc} & \text{Left} & & \text{Above} & & \text{Right} & & \text{Below} \\ x_3 = \tfrac{1}{4}(& 0 & + & x_1 & + & x_4 & + & 32 \quad) \end{array}$$

Applying the same procedure to the other three intersection points produces the following equations:

$$x_4 = \tfrac{1}{4}(x_3 + x_2 + 0 + 32)$$
$$x_1 = \tfrac{1}{4}(0 + 0 + x_2 + x_3)$$
$$x_2 = \tfrac{1}{4}(x_1 + 0 + 0 + x_4)$$

After simplifying, we have the following system:

$$\begin{array}{rrrrr} x_1 & -4x_3 & +x_4 & = & -32 \\ x_2 + & x_3 & -4x_4 & = & -32 \\ 4x_1 - x_2 - & x_3 & & = & 0 \\ x_1 - 4x_2 & & +x_4 & = & 0 \end{array}$$

We now form the augmented coefficient matrix for this system and solve by triangularization with back substitution:

$$\begin{bmatrix} 1 & 0 & -4 & 1 & \bigm| & -32 \\ 0 & 1 & 1 & -4 & \bigm| & -32 \\ 4 & -1 & -1 & 0 & \bigm| & 0 \\ 1 & -4 & 0 & 1 & \bigm| & 0 \end{bmatrix} \quad \begin{array}{l} \\ \\ R_3 + (-4)R_1 \to R_3 \\ R_4 + (-1)R_1 \to R_4 \end{array}$$

$$\sim \begin{bmatrix} 1 & 0 & -4 & 1 & \bigm| & -32 \\ 0 & 1 & 1 & -4 & \bigm| & -32 \\ 0 & -1 & 15 & -4 & \bigm| & 128 \\ 0 & -4 & 4 & 0 & \bigm| & 32 \end{bmatrix} \quad \begin{array}{l} \\ \\ R_3 + R_2 \to R_3 \\ R_4 + 4R_2 \to R_4 \end{array}$$

$$\sim \begin{bmatrix} 1 & 0 & -4 & 1 & \bigm| & -32 \\ 0 & 1 & 1 & -4 & \bigm| & -32 \\ 0 & 0 & 16 & -8 & \bigm| & 96 \\ 0 & 0 & 8 & -16 & \bigm| & -96 \end{bmatrix} \qquad R_4 + (-\tfrac{1}{2})R_3 \rightarrow R_4$$

$$\sim \begin{bmatrix} 1 & 0 & -4 & 1 & \bigm| & -32 \\ 0 & 1 & 1 & -4 & \bigm| & -32 \\ 0 & 0 & 16 & -8 & \bigm| & 96 \\ 0 & 0 & 0 & -12 & \bigm| & -144 \end{bmatrix} \qquad \begin{array}{l} \text{This matrix is in} \\ \text{triangular form.} \end{array}$$

The corresponding system of equations can be solved by back substitution, as follows:

$$\begin{aligned} x_1 \quad - \quad 4x_3 + \quad x_4 &= -32 \\ x_2 + \quad x_3 - \quad 4x_4 &= -32 \\ 16x_3 - \quad 8x_4 &= 96 \\ -12x_4 &= -144 \end{aligned}$$

$$\begin{aligned} -12x_4 &= -144 &&\text{Solve for } x_4. \\ x_4 &= 12 \\ 16x_3 - 8(12) &= 96 &&\text{Solve for } x_3. \\ x_3 &= 12 \\ x_2 + 12 - 4(12) &= -32 &&\text{Solve for } x_2. \\ x_2 &= 4 \\ x_1 - 4(12) + 12 &= -32 &&\text{Solve for } x_1. \\ x_1 &= 4 \end{aligned}$$

Thus, the temperature at the two upper intersection points is $4°$, and at the two lower points it is $12°$.

Problem 12   Repeat Example 12 if the temperature at the lower end of the vertical bar on the right in Figure 2 is changed from $32°$ to $8°$. ▮

Answers to Matched Problems

10. (A) $a_{21} = 1$ (B) $a_{54} = 7$

11. $x_1 = -1$, $x_2 = 2$, $x_3 = 1$, $x_4 = 0$ or $(-1, 2, 1, 0)$

12. Lower left-hand intersection point: $10°$; lower right-hand intersection point: $5°$; upper left-hand intersection point: $3°$; upper right-hand intersection point: $2°$

▌ **Exercise 1-3**

A   In Problems 1–8, determine whether the indicated augmented matrix is in triangular form. If the matrix is not in triangular form, list the nonzero elements that violate the definition of triangular form.

1. $\begin{bmatrix} 1 & -2 & \bigm| & 3 \\ 0 & 2 & \bigm| & 4 \end{bmatrix}$

2. $\begin{bmatrix} 0 & 0 & \bigm| & 0 \\ 2 & 0 & \bigm| & 0 \end{bmatrix}$

3. $\begin{bmatrix} 2 & 4 & -1 & \bigm| & 0 \\ 0 & 0 & 1 & \bigm| & 2 \\ 0 & 1 & 2 & \bigm| & 0 \end{bmatrix}$

4. $\begin{bmatrix} 2 & -1 & 3 & \bigm| & 4 \\ 0 & 2 & 1 & \bigm| & 0 \\ 0 & 0 & 3 & \bigm| & 4 \end{bmatrix}$

5. $\begin{bmatrix} 3 & -4 & 1 & 0 & \bigm| & 2 \\ 0 & 0 & 1 & 5 & \bigm| & 7 \\ 0 & 0 & 2 & 0 & \bigm| & 0 \\ 0 & 0 & 0 & 0 & \bigm| & 0 \end{bmatrix}$

6. $\begin{bmatrix} 3 & 0 & 0 & 1 & \bigm| & 2 \\ 1 & 0 & 2 & 1 & \bigm| & 2 \\ 0 & 0 & 0 & 2 & \bigm| & -3 \\ 0 & 0 & 0 & 4 & \bigm| & 6 \end{bmatrix}$

7. $\begin{bmatrix} 2 & 1 & -1 & 1 & 1 & \bigm| & 2 \\ 3 & -1 & 1 & 2 & -4 & \bigm| & 3 \\ 0 & 1 & -1 & 2 & 1 & \bigm| & 2 \\ 0 & 0 & 4 & 3 & 2 & \bigm| & 1 \\ 0 & 0 & 0 & 2 & 1 & \bigm| & 5 \end{bmatrix}$

8. $\begin{bmatrix} 1 & 0 & 0 & 0 & 0 & \bigm| & 1 \\ 0 & 1 & 0 & 0 & 0 & \bigm| & 1 \\ 0 & 0 & 1 & 0 & 0 & \bigm| & 1 \\ 0 & 0 & 0 & 1 & 0 & \bigm| & 1 \\ 0 & 0 & 0 & 0 & 0 & \bigm| & 0 \end{bmatrix}$

B In Problems 9–14, perform the row operations necessary to transform each augmented matrix to a triangular form.

9. $\begin{bmatrix} 1 & -1 & 1 & \bigm| & 2 \\ 0 & 0 & 2 & \bigm| & -1 \\ 0 & 3 & 2 & \bigm| & 1 \end{bmatrix}$

10. $\begin{bmatrix} 0 & 1 & -1 & \bigm| & 2 \\ 0 & 0 & 2 & \bigm| & -3 \\ 1 & -2 & 1 & \bigm| & 4 \end{bmatrix}$

11. $\begin{bmatrix} 1 & -1 & 1 & 2 & \bigm| & -1 \\ 0 & 1 & -1 & 1 & \bigm| & 2 \\ 0 & 2 & 1 & -2 & \bigm| & 1 \\ 0 & 0 & 0 & 0 & \bigm| & 0 \end{bmatrix}$

12. $\begin{bmatrix} 2 & -3 & 0 & 4 & \bigm| & 2 \\ 0 & 0 & -1 & 0 & \bigm| & 2 \\ 0 & 0 & 2 & 1 & \bigm| & -2 \\ 0 & 0 & 3 & 2 & \bigm| & 1 \end{bmatrix}$

13. $\begin{bmatrix} 3 & -1 & 4 & 2 & 5 & \bigm| & -7 \\ 0 & 1 & -1 & 1 & 2 & \bigm| & 2 \\ 0 & 3 & 2 & -1 & 4 & \bigm| & 6 \\ 0 & 0 & 0 & 0 & 3 & \bigm| & -5 \\ 0 & 0 & 0 & -2 & 5 & \bigm| & 6 \end{bmatrix}$

14. $\begin{bmatrix} 0 & 0 & -1 & 1 & 2 & \bigm| & 1 \\ 1 & -1 & 0 & 0 & 1 & \bigm| & 0 \\ 0 & 2 & 1 & 0 & 1 & \bigm| & 2 \\ 0 & 0 & 0 & 1 & -1 & \bigm| & 2 \\ 0 & 0 & 0 & 2 & -2 & \bigm| & 4 \end{bmatrix}$

In Problems 15–22, each matrix is an augmented coefficient matrix that is already in a triangular form. Solve the corresponding system of linear equations.

15. $\begin{bmatrix} 1 & 2 & -1 & \bigm| & 0 \\ 0 & 3 & 1 & \bigm| & 1 \\ 0 & 0 & 4 & \bigm| & 16 \end{bmatrix}$

16. $\begin{bmatrix} -1 & -1 & 0 & \bigm| & 3 \\ 0 & 2 & 5 & \bigm| & 31 \\ 0 & 0 & -3 & \bigm| & -27 \end{bmatrix}$

17. $\begin{bmatrix} 4 & -7 & 3 & 9 & \bigm| & 3 \\ 0 & 8 & -2 & 1 & \bigm| & 6 \\ 0 & 0 & 3 & -4 & \bigm| & 5 \\ 0 & 0 & 0 & 2 & \bigm| & 8 \end{bmatrix}$

18. $\begin{bmatrix} 2 & -1 & 0 & 1 & \bigm| & 5 \\ 0 & 1 & 3 & 0 & \bigm| & 2 \\ 0 & 0 & -1 & 2 & \bigm| & -1 \\ 0 & 0 & 0 & 1 & \bigm| & 0 \end{bmatrix}$

$$\textbf{19.} \begin{bmatrix} 2 & -1 & 1 & -7 & \bigm| & 5 \\ 0 & 1 & -2 & 3 & \bigm| & 2 \\ 0 & 0 & 2 & 4 & \bigm| & -6 \\ 0 & 0 & 0 & 2 & \bigm| & -1 \end{bmatrix} \qquad \textbf{20.} \begin{bmatrix} 3 & 1 & -1 & -2 & \bigm| & -3 \\ 0 & -2 & 1 & -1 & \bigm| & 4 \\ 0 & 0 & 1 & -3 & \bigm| & 4 \\ 0 & 0 & 0 & 3 & \bigm| & 0 \end{bmatrix}$$

$$\textbf{21.} \begin{bmatrix} 1 & 2 & 0 & 0 & -1 & \bigm| & -5 \\ 0 & 3 & 0 & 1 & 0 & \bigm| & 1 \\ 0 & 0 & -2 & 0 & -1 & \bigm| & -8 \\ 0 & 0 & 0 & 1 & 2 & \bigm| & 6 \\ 0 & 0 & 0 & 0 & 5 & \bigm| & 20 \end{bmatrix}$$

$$\textbf{22.} \begin{bmatrix} 2 & -1 & 1 & 2 & -1 & \bigm| & 6 \\ 0 & -3 & 0 & 2 & -1 & \bigm| & -11 \\ 0 & 0 & 1 & -4 & 6 & \bigm| & 5 \\ 0 & 0 & 0 & 2 & 0 & \bigm| & 4 \\ 0 & 0 & 0 & 0 & 1 & \bigm| & 1 \end{bmatrix}$$

In Problems 23–32, use elementary row operations to solve each system.

23.
$$\begin{aligned} x_1 + 2x_2 - 4x_3 &= -18 \\ -2x_1 + 3x_2 + 2x_3 &= -10 \\ 2x_1 + 4x_2 + 5x_3 &= 3 \end{aligned}$$

24.
$$\begin{aligned} x_1 + 4x_2 - 5x_3 &= 4 \\ x_1 + x_2 - x_3 &= 3 \\ 2x_1 - 4x_2 + 5x_3 &= 2 \end{aligned}$$

25.
$$\begin{aligned} x_1 + 4x_2 - 7x_3 &= 3 \\ 2x_1 - x_2 + x_3 &= 4 \\ -x_1 + 2x_2 + x_3 &= 1 \end{aligned}$$

26.
$$\begin{aligned} 2x_1 - 3x_2 + 5x_3 &= -2 \\ 3x_1 + 2x_2 - 5x_3 &= 19 \\ 5x_1 - 7x_2 + 4x_3 &= 8 \end{aligned}$$

27.
$$\begin{aligned} x_1 + 2x_2 - x_3 + x_4 &= 2 \\ -x_1 - x_2 + 2x_3 - 3x_4 &= -3 \\ 2x_1 - x_2 + x_3 - 2x_4 &= 1 \\ -x_1 - 2x_2 + x_3 + x_4 &= -2 \end{aligned}$$

28.
$$\begin{aligned} x_1 - 2x_2 - 3x_3 + x_4 &= 5 \\ 2x_2 + x_3 - x_4 &= 6 \\ x_1 + 2x_2 - 2x_3 &= 19 \\ -3x_1 + 7x_3 + 5x_4 &= 1 \end{aligned}$$

29.
$$\begin{aligned} 2x_1 + x_2 + 2x_3 - x_4 &= 6 \\ x_1 + 2x_2 + x_3 - x_4 &= 0 \\ -2x_1 + 4x_2 + 4x_3 + 3x_4 &= -2 \\ 3x_1 + x_3 &= 6 \end{aligned}$$

30.
$$\begin{aligned} x_1 + x_2 + 2x_3 - x_4 &= 0 \\ 2x_1 - x_2 - 3x_3 - 3x_4 &= 0 \\ -x_1 + x_4 &= 0 \\ x_2 + x_3 + x_4 &= 1 \end{aligned}$$

C   31.
$$\begin{aligned} x_1 + x_3 + x_5 &= 3 \\ x_2 + x_4 + x_5 &= 4 \\ x_1 + x_2 + x_5 &= 2 \\ x_1 + x_3 + x_4 &= -1 \\ x_1 + x_2 - x_4 + x_5 &= 2 \end{aligned}$$

32.
$$\begin{aligned} x_1 + x_2 &= 2 \\ x_1 + x_2 - x_3 &= 3 \\ x_2 + x_3 - x_4 &= 1 \\ x_3 + x_4 - x_5 &= -3 \\ x_4 + x_5 &= 0 \end{aligned}$$

▮▫▮ Applications

Heat Conduction   *In Problems 33–36, find the temperature at each intersection point in the indicated grid. [Hint: Use the averaging condition discussed in Example 12 to set up a system of linear equations and solve.]*

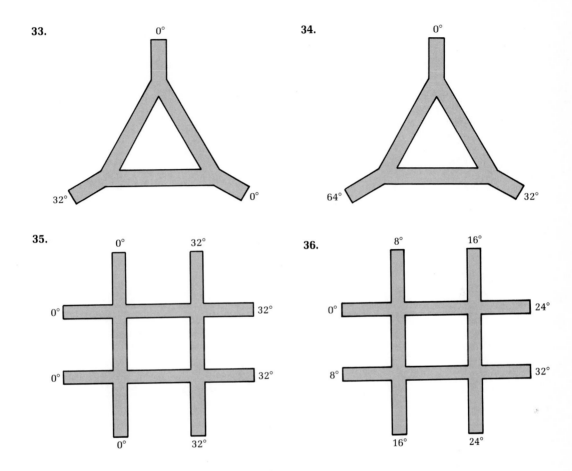

33. 0° 32° 0°

34. 0° 64° 32°

35. 0° 32° 0° 32° 0° 32° 0° 32°

36. 8° 16° 0° 24° 8° 32° 16° 24°

▌1-4▐ Gauss–Jordan Elimination

- Introduction
- Reduced Matrices
- Solving Systems Already in Reduced Form
- Solving Systems by Gauss–Jordan Elimination
- Application

▪ Introduction

In the preceding section we used triangularization with back substitution to solve square systems of linear equations. This is the simplest and most efficient method for solving square systems that have unique solutions. Triangularization with back substitution can also be used to solve nonsquare systems, but this is not often done. Instead, a more general method for solving systems is usually used. This method, called *Gauss–Jordan elimination*, works equally well for both square and nonsquare systems and for dependent and inconsistent systems. Furthermore, in later sections of this book we will see that the techniques used in Gauss–Jordan elimination are fundamental to the study of linear algebra. As the following example illustrates, this new method is a natural generalization of the triangularization procedure.

Example 13   Solve the system:

$$x_1 - 2x_2 + 3x_3 = 8$$
$$2x_1 - 3x_2 + 2x_3 = 7$$
$$-x_1 + x_2 + 2x_3 = 4$$

Solution   We begin by reducing the augmented coefficient matrix to triangular form.

$$\begin{bmatrix} 1 & -2 & 3 & | & 8 \\ 2 & -3 & 2 & | & 7 \\ -1 & 1 & 2 & | & 4 \end{bmatrix} \quad \begin{matrix} R_2 + (-2)R_1 \to R_2 \\ R_3 + R_1 \to R_3 \end{matrix}$$

$$\sim \begin{bmatrix} 1 & -2 & 3 & | & 8 \\ 0 & 1 & -4 & | & -9 \\ 0 & -1 & 5 & | & 12 \end{bmatrix} \quad R_3 + R_2 \to R_3$$

$$\sim \begin{bmatrix} 1 & -2 & 3 & | & 8 \\ 0 & 1 & -4 & | & -9 \\ 0 & 0 & 1 & | & 3 \end{bmatrix}$$

Instead of stopping at this point and using back substitution to find the solution, we will continue to use row operations to transform this last matrix into a very simple form where the solution can be read by inspection. First we will use the 1 in position 3,3 to obtain 0's in the rest of column 3:

$$\overset{C_3}{\begin{bmatrix} 1 & -2 & 3 & | & 8 \\ 0 & 1 & -4 & | & -9 \\ 0 & 0 & 1 & | & 3 \end{bmatrix}} \quad \begin{matrix} R_1 + (-3)R_3 \to R_1 \\ R_2 + 4R_3 \to R_2 \\ \sim \end{matrix} \quad \overset{C_3}{\begin{bmatrix} 1 & -2 & 0 & | & -1 \\ 0 & 1 & 0 & | & 3 \\ 0 & 0 & 1 & | & 3 \end{bmatrix}}$$

Next we use the 1 in position 2,2 to obtain 0's in the rest of column 2:

$$\overset{C_2}{\begin{bmatrix} 1 & -2 & 0 & | & -1 \\ 0 & 1 & 0 & | & 3 \\ 0 & 0 & 1 & | & 3 \end{bmatrix}} \quad \begin{matrix} R_1 + 2R_2 \to R_1 \\ \sim \end{matrix} \quad \overset{C_2}{\begin{bmatrix} 1 & 0 & 0 & | & 5 \\ 0 & 1 & 0 & | & 3 \\ 0 & 0 & 1 & | & 3 \end{bmatrix}}$$

Notice how simple this last matrix is. The linear system corresponding to this matrix is

$$x_1 = 5$$
$$x_2 = 3$$
$$x_3 = 3$$

The solution to this system is obvious: $x_1 = 5$, $x_2 = 3$, $x_3 = 3$ or $(5, 3, 3)$.

Problem 13 Use the procedure discussed in Example 13 to solve the system:

$$x_1 + 2x_2 - 4x_3 = 7$$
$$3x_1 + 7x_2 - 8x_3 = 14$$
$$-2x_1 - 5x_2 + 5x_3 = -9$$

∎

■ Reduced Matrices

Example 13 illustrates the procedure we now want to use to solve systems of equations. In general, we will use elementary row operations to transform the augmented coefficient matrix of a linear system into a simple form where the solution is very easy to find. The simple form we will obtain is called the *reduced form* (or, more formally, the *reduced row-echelon form*), and we define it as follows:

Reduced Matrix

A matrix is in **reduced form** if:

1. Each row consisting entirely of 0's is below any row having at least one nonzero element.
2. The leftmost nonzero element in each row is 1.
3. The column containing the leftmost 1 of a given row has 0's above and below the 1.
4. The leftmost 1 in any row is to the right of the leftmost 1 in the preceding row.

Example 14 Determine whether each matrix is in reduced form. If it is not, indicate which condition in the definition is violated and then perform the row operations necessary to transform the matrix to a reduced form.

(A) $\begin{bmatrix} 1 & 0 & 3 \\ 0 & 1 & 2 \end{bmatrix}$

(B) $\begin{bmatrix} 1 & 0 & 2 \\ 0 & 4 & 12 \end{bmatrix}$

(C) $\begin{bmatrix} 1 & 2 & 8 \\ 0 & 1 & 3 \\ 0 & 0 & 0 \end{bmatrix}$

(D) $\begin{bmatrix} 1 & 0 & -1 \\ 0 & 1 & 4 \\ 0 & 0 & 0 \end{bmatrix}$

(E) $\begin{bmatrix} 0 & 1 & 0 & 5 \\ 1 & 0 & 0 & 7 \\ 0 & 0 & 1 & 3 \end{bmatrix}$

(F) $\begin{bmatrix} 1 & 0 & 4 & 0 \\ 0 & 1 & 2 & 0 \\ 0 & 0 & 0 & 0 \end{bmatrix}$

(G) $\begin{bmatrix} 1 & 2 & 3 & 0 & 2 \\ 0 & 0 & 0 & 1 & 3 \end{bmatrix}$

(H) $\begin{bmatrix} 1 & 3 & 0 & 2 & 5 \\ 0 & 0 & 0 & 0 & 0 \\ 0 & 0 & 1 & 1 & 9 \end{bmatrix}$

Solution   (A) All conditions are satisfied. The matrix is in reduced form.
(B) Condition 2 is violated. The leftmost element in R_2 is 4, not 1.

$$\begin{bmatrix} 1 & 0 & 2 \\ 0 & 4 & 12 \end{bmatrix} \underset{\frac{1}{4}R_2 \to R_2}{\sim} \begin{bmatrix} 1 & 0 & 2 \\ 0 & 1 & 3 \end{bmatrix} \quad \text{This matrix is in reduced form.}$$

(C) Condition 3 is violated. Column 2 contains a nonzero element above the leftmost 1 in the second row.

$$\begin{bmatrix} 1 & 2 & 8 \\ 0 & 1 & 3 \\ 0 & 0 & 0 \end{bmatrix} \underset{\sim}{R_1 + (-2)R_2 \to R_1} \begin{bmatrix} 1 & 0 & 2 \\ 0 & 1 & 3 \\ 0 & 0 & 0 \end{bmatrix} \quad \text{This matrix is in reduced form.}$$

(D) All conditions are satisfied. The matrix is in reduced form.
(E) Condition 4 is violated. The leftmost 1 in R_2 is not to the right of the leftmost 1 in R_1.

$$\begin{bmatrix} 0 & 1 & 0 & 5 \\ 1 & 0 & 0 & 7 \\ 0 & 0 & 1 & 3 \end{bmatrix} \underset{\sim}{R_1 \leftrightarrow R_2} \begin{bmatrix} 1 & 0 & 0 & 7 \\ 0 & 1 & 0 & 5 \\ 0 & 0 & 1 & 3 \end{bmatrix} \quad \text{This matrix is in reduced form.}$$

(F) All conditions are satisfied. The matrix is in reduced form.
(G) All conditions are satisfied. The matrix is in reduced form.
(H) Condition 1 is violated. Row 2 consists entirely of 0's and R_3 contains nonzero elements.

$$\begin{bmatrix} 1 & 3 & 0 & 2 & 5 \\ 0 & 0 & 0 & 0 & 0 \\ 0 & 0 & 1 & 1 & 9 \end{bmatrix} \underset{R_3 \leftrightarrow R_2}{\sim} \begin{bmatrix} 1 & 3 & 0 & 2 & 5 \\ 0 & 0 & 1 & 1 & 9 \\ 0 & 0 & 0 & 0 & 0 \end{bmatrix} \quad \text{This matrix is in reduced form.}$$

Problem 14   Repeat Example 14 for the following matrices:

(A) $\begin{bmatrix} 1 & 0 & 0 & 2 \\ 0 & 0 & 1 & 3 \\ 0 & 1 & 0 & 5 \end{bmatrix}$
    (B) $\begin{bmatrix} 1 & 0 & 0 & 2 \\ 0 & 3 & -3 & 6 \end{bmatrix}$

(C) $\begin{bmatrix} 1 & -1 & 0 & 2 \\ 0 & 0 & 1 & 2 \\ 0 & 0 & 0 & 0 \end{bmatrix}$
    (D) $\begin{bmatrix} 1 & 0 & 2 \\ 0 & 0 & 0 \\ 0 & 1 & 4 \end{bmatrix}$

(E) $\begin{bmatrix} 1 & 2 & 0 & 0 \\ 0 & 0 & 1 & 0 \\ 0 & 0 & 0 & 1 \end{bmatrix}$
    (F) $\begin{bmatrix} 1 & -2 & 0 & 0 & 1 \\ 0 & 0 & 1 & -1 & 2 \\ 0 & 0 & 0 & 1 & 3 \end{bmatrix}$ ∎

▪ Solving Systems Already in Reduced Form

The solution of a system corresponding to a reduced matrix is easy to find, if it exists. The various possibilities are best illustrated by means of an example.

Example 15   Write the linear system corresponding to each reduced augmented matrix and solve.

$$\text{(A)} \begin{bmatrix} 1 & 0 & 0 & | & 2 \\ 0 & 1 & 0 & | & -1 \\ 0 & 0 & 1 & | & 3 \end{bmatrix} \quad \text{(B)} \begin{bmatrix} 1 & 0 & 4 & | & 0 \\ 0 & 1 & 3 & | & 0 \\ 0 & 0 & 0 & | & 1 \end{bmatrix}$$

$$\text{(C)} \begin{bmatrix} 1 & 0 & 2 & | & -3 \\ 0 & 1 & -1 & | & 8 \\ 0 & 0 & 0 & | & 0 \end{bmatrix} \quad \text{(D)} \begin{bmatrix} 1 & 4 & 0 & 0 & 3 & | & -2 \\ 0 & 0 & 1 & 0 & -2 & | & 0 \\ 0 & 0 & 0 & 1 & 2 & | & 4 \end{bmatrix}$$

Solution   (A) $x_1 \qquad = \quad 2$
$\qquad\qquad\quad x_2 \quad = -1$
$\qquad\qquad\qquad\quad x_3 = \quad 3$

The solution is obvious: $x_1 = 2$, $x_2 = -1$, $x_3 = 3$ or $(2, -1, 3)$.

(B) $\quad x_1 \qquad\quad + 4x_3 = 0$
$\qquad\qquad x_2 + 3x_3 = 0$
$\qquad 0x_1 + 0x_2 + 0x_3 = 1 \qquad$ Usually written as $0 = 1$.

The last equation implies $0 = 1$, which is a contradiction. Hence, the system is inconsistent and has no solution.

(C) $\quad x_1 \qquad\quad + 2x_3 = -3$
$\qquad\qquad x_2 - \quad x_3 = \quad 8 \qquad\qquad\qquad\qquad\qquad\qquad\qquad (1)$
$\qquad 0x_1 + 0x_2 + 0x_3 = \quad 0 \qquad$ We can disregard the equation corresponding to the third row in the matrix, since it is satisfied by all values of x_1, x_2, and x_3.

When a reduced system (a system corresponding to a reduced augmented matrix) has more variables than equations and contains no contradictions, the system is dependent and has infinitely many solutions. (In a later chapter we will prove this statement.) As we have already seen in Sections 1-1 and 1-2, the (general) solution of a dependent system is usually expressed in terms of parameters. The reduced form of a system provides an easy way to represent the (general) solution of a dependent system of any size. Since the first variable (with a nonzero coefficient) in each equation in a reduced system corresponds to a leftmost 1 in the reduced matrix, the first variable in each equation will be called a **leftmost variable.** The remaining variables will be referred to as **parametric variables.** Condition 3 in the definition of a reduced matrix implies that each leftmost variable will occur in exactly one equation in the reduced system. Thus, it is easy to solve for each leftmost variable in terms of the parametric variables. Examining system (1), we see that x_1 and x_2 are the leftmost variables and x_3 is a parametric variable. We solve for the leftmost variables x_1 and x_2 in terms of the parametric variable x_3:

$$x_1 = -2x_3 - 3$$
$$x_2 = \quad x_3 + 8$$

By introducing a parameter t and letting $x_3 = t$, the (general) solution of this system can be expressed by

$$x_1 = -2t - 3$$
$$x_2 = t + 8$$
$$x_3 = t \qquad\qquad t \text{ any real number}$$

or by

$$\{(-2t - 3, t + 8, t)|t \text{ any real number}\}$$

(D) $\quad x_1 + 4x_2 + 3x_5 = -2$
$x_3 - 2x_5 = 0$
$x_4 + 2x_5 = 4$

Solve for x_1, x_3, and x_4 (leftmost variables) in terms of x_2 and x_5 (parametric variables):

$$x_1 = -4x_2 - 3x_5 - 2$$
$$x_3 = 2x_5$$
$$x_4 = -2x_5 + 4$$

Since there are two parametric variables, the solutions to this system can be expressed in terms of two parameters. If we let $x_2 = s$ and $x_5 = t$, then the (general) solution is given by

$$x_1 = -4s - 3t - 2$$
$$x_2 = s$$
$$x_3 = 2t$$
$$x_4 = -2t + 4$$
$$x_5 = t \qquad\qquad s \text{ and } t \text{ any real numbers}$$

or

$$\{(-4s - 3t - 2, s, 2t, -2t + 4, t)|s \text{ and } t \text{ any real numbers}\}$$

Problem 15 Write the linear system corresponding to each reduced augmented matrix and solve.

(A) $\begin{bmatrix} 1 & 0 & 0 & | & -5 \\ 0 & 1 & 0 & | & 3 \\ 0 & 0 & 1 & | & 6 \end{bmatrix}$ (B) $\begin{bmatrix} 1 & 2 & -3 & | & 0 \\ 0 & 0 & 0 & | & 1 \\ 0 & 0 & 0 & | & 0 \end{bmatrix}$

(C) $\begin{bmatrix} 1 & 0 & -2 & | & 4 \\ 0 & 1 & 3 & | & -2 \\ 0 & 0 & 0 & | & 0 \end{bmatrix}$ (D) $\begin{bmatrix} 1 & 0 & 3 & 2 & | & 5 \\ 0 & 1 & -2 & -1 & | & 3 \\ 0 & 0 & 0 & 0 & | & 0 \end{bmatrix}$ ∎

■ Solving Systems by Gauss–Jordan Elimination

We now outline a step-by-step procedure, called *Gauss–Jordan elimination*, for solving systems of linear equations. The method provides us with a systematic way of transforming augmented matrices into a reduced form from which we can easily find the solution to the original system, if a solution exists.

Example 16   Solve by Gauss–Jordan elimination:
$$2x_1 - 2x_2 + x_3 = 3$$
$$3x_1 + x_2 - x_3 = 7$$
$$x_1 - 3x_2 + 2x_3 = 0$$

Solution   Write the augmented matrix and follow the steps indicated at the right.

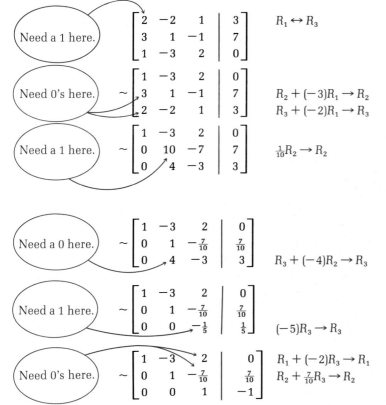

$R_1 \leftrightarrow R_3$

Step 1.   Choose leftmost nonzero column and get a 1 at the top.

$R_2 + (-3)R_1 \rightarrow R_2$
$R_3 + (-2)R_1 \rightarrow R_3$

Step 2.   Use multiples of the first row to get 0's below the 1 obtained in Step 1.

$\frac{1}{10}R_2 \rightarrow R_2$

Step 3.   Mentally delete R_1 and repeat Steps 1 and 2 with the submatrix (the matrix that remains after deleting the top row). Continue the above process (Steps 1–3) until it is not possible to go further; then proceed with Step 4.

$R_3 + (-4)R_2 \rightarrow R_3$

Mentally delete R_1 and R_2.

Since Steps 1–3 cannot be carried further, proceed to Step 4.

$(-5)R_3 \rightarrow R_3$

Step 4.   Return deleted rows. Begin with the bottom nonzero row and use appropriate multiples of it to get 0's above the leftmost 1. Continue the process, moving up row-by-row, until the matrix is in reduced form.

$R_1 + (-2)R_3 \rightarrow R_1$
$R_2 + \frac{7}{10}R_3 \rightarrow R_2$

$$\text{(Need a 0 here.)} \quad \sim \begin{bmatrix} 1 & -3 & 0 & \bigm| & 2 \\ 0 & 1 & 0 & \bigm| & 0 \\ 0 & 0 & 1 & \bigm| & -1 \end{bmatrix} \qquad R_1 + 3R_2 \to R_1$$

$$\sim \begin{bmatrix} 1 & 0 & 0 & \bigm| & 2 \\ 0 & 1 & 0 & \bigm| & 0 \\ 0 & 0 & 1 & \bigm| & -1 \end{bmatrix}$$

The matrix is in reduced form and we can write the solution to the original system by inspection.

Solution. $x_1 = 2$, $x_2 = 0$, $x_3 = -1$. It is left to the reader to check this solution.

Steps 1–4 outlined in the solution of Example 16 are referred to as *Gauss–Jordan elimination*. The steps are summarized in the box for easy reference:

Gauss–Jordan Elimination

Step 1. Choose the leftmost nonzero column and use appropriate row operations to get a 1 at the top.

Step 2. Use multiples of the first row to get 0's in all places below the 1 obtained in Step 1.

Step 3. Delete (mentally) the top row of the matrix. Repeat Steps 1 and 2 with the submatrix (the matrix that remains after deleting the top row). Continue this process (Steps 1–3) until it is not possible to go further.

Step 4. Consider the whole matrix obtained after mentally returning all the rows to the matrix. Begin with the bottom nonzero row and use appropriate multiples of it to get 0's above the leftmost 1. Continue this process, moving up row-by-row, until the matrix is finally in reduced form.

Note: If at any point in the above process we obtain a row with all 0's to the left of the vertical line and a nonzero number to the right, we can stop, since we will have a contradiction ($0 = n$, $n \neq 0$). We can then conclude that the system has no solution.

Following Steps 1–4 as outlined in the box will always transform a matrix into a reduced form. Furthermore, it can be shown that this reduced form is unique. That is, *every matrix has exactly one reduced form.* (A proof of this statement is beyond the scope of this book.) There are many different sequences of operations that will transform a matrix into its reduced form. The sequence we have outlined in the box is one that is frequently used in computer programs that find the reduced form of a matrix.

Problem 16 Solve by Gauss–Jordan elimination:
$$3x_1 + x_2 - 2x_3 = 2$$
$$x_1 - 2x_2 + x_3 = 3$$
$$2x_1 - x_2 - 3x_3 = 3$$

∎

Example 17 Solve by Gauss–Jordan elimination:
$$-x_1 - 2x_2 - 5x_3 \qquad = -3$$
$$2x_1 + x_2 + x_3 - 3x_4 = 9$$
$$3x_1 + 2x_2 + 3x_3 - 3x_4 = 10$$

Solution

Need a 1 here.
$$\begin{bmatrix} -1 & -2 & -5 & 0 & | & -3 \\ 2 & 1 & 1 & -3 & | & 9 \\ 3 & 2 & 3 & -3 & | & 10 \end{bmatrix} \qquad (-1)R_1 \to R_1$$

Need 0's here.
$$\sim \begin{bmatrix} 1 & 2 & 5 & 0 & | & 3 \\ 2 & 1 & 1 & -3 & | & 9 \\ 3 & 2 & 3 & -3 & | & 10 \end{bmatrix} \qquad \begin{matrix} R_2 + (-2)R_1 \to R_2 \\ R_3 + (-3)R_1 \to R_3 \end{matrix}$$

Need a 1 here.
$$\sim \begin{bmatrix} 1 & 2 & 5 & 0 & | & 3 \\ 0 & -3 & -9 & -3 & | & 3 \\ 0 & -4 & -12 & -3 & | & 1 \end{bmatrix} \qquad -\tfrac{1}{3}R_2 \to R_2$$

Need a 0 here.
$$\sim \begin{bmatrix} 1 & 2 & 5 & 0 & | & 3 \\ 0 & 1 & 3 & 1 & | & -1 \\ 0 & -4 & -12 & -3 & | & 1 \end{bmatrix} \qquad R_3 + 4R_2 \to R_3$$

Need a 0 here.
$$\sim \begin{bmatrix} 1 & 2 & 5 & 0 & | & 3 \\ 0 & 1 & 3 & 1 & | & -1 \\ 0 & 0 & 0 & 1 & | & -3 \end{bmatrix} \qquad R_2 + (-1)R_3 \to R_2$$

Need a 0 here.
$$\sim \begin{bmatrix} 1 & 2 & 5 & 0 & | & 3 \\ 0 & 1 & 3 & 0 & | & 2 \\ 0 & 0 & 0 & 1 & | & -3 \end{bmatrix} \qquad R_1 + (-2)R_2 \to R_1$$

$$\sim \begin{bmatrix} 1 & 0 & -1 & 0 & | & -1 \\ 0 & 1 & 3 & 0 & | & 2 \\ 0 & 0 & 0 & 1 & | & -3 \end{bmatrix}$$

The matrix is now in reduced form. Write the corresponding system and the solution.

$$x_1 \quad - \quad x_3 \quad = -1$$
$$x_2 + 3x_3 \quad = \quad 2$$
$$x_4 = -3$$

Solve for the leftmost variables x_1, x_2, and x_4 in terms of the parametric variable x_3:

$$x_1 = \quad x_3 - 1$$
$$x_2 = -3x_3 + 2$$
$$x_4 = \qquad -3$$

If $x_3 = t$, then the solutions are given by

$$x_1 = \quad t - 1$$
$$x_2 = -3t + 2$$
$$x_3 = \quad t$$
$$x_4 = \qquad -3 \qquad t \text{ any real number}$$

or

$$\{(t - 1, -3t + 2, t, -3)|t \text{ any real number}\}$$

Problem 17   Solve by Gauss–Jordan elimination: $-x_1 + x_2 - x_3 - 3x_4 = 3$
$$-2x_1 + x_2 - x_3 - 3x_4 = 1$$
$$2x_1 \quad + x_3 + \ x_4 = 4 \qquad ▌▌$$

Example 18   Solve by Gauss–Jordan elimination: $x_1 + x_2 + 3x_3 + \ x_4 + 4x_5 = 4$
$$-x_1 + x_2 + \ x_3 + 3x_4 + 2x_5 = 2$$
$$5x_1 - x_2 + 3x_3 - 7x_4 + 2x_5 = 3$$

Solution

$$\begin{bmatrix} 1 & 1 & 3 & 1 & 4 & | & 4 \\ -1 & 1 & 1 & 3 & 2 & | & 2 \\ 5 & -1 & 3 & -7 & 2 & | & 3 \end{bmatrix} \qquad \begin{array}{l} R_2 + R_1 \rightarrow R_2 \\ R_3 + (-5)R_1 \rightarrow R_3 \end{array}$$

$$\sim \begin{bmatrix} 1 & 1 & 3 & 1 & 4 & | & 4 \\ 0 & 2 & 4 & 4 & 6 & | & 6 \\ 0 & -6 & -12 & -12 & -18 & | & -17 \end{bmatrix} \qquad \tfrac{1}{2}R_2 \rightarrow R_2$$

$$\sim \begin{bmatrix} 1 & 1 & 3 & 1 & 4 & | & 4 \\ 0 & 1 & 2 & 2 & 3 & | & 3 \\ 0 & -6 & -12 & -12 & -18 & | & -17 \end{bmatrix} \qquad R_3 + 6R_2 \rightarrow R_3$$

$$\sim \begin{bmatrix} 1 & 1 & 3 & 1 & 4 & | & 4 \\ 0 & 1 & 2 & 2 & 3 & | & 3 \\ 0 & 0 & 0 & 0 & 0 & | & 1 \end{bmatrix}$$

We stop the Gauss–Jordan elimination, since the last row produces a contradiction.

The last row implies $0 = 1$, which is a contradiction; therefore, the system has no solution.

Problem 18   Solve by Gauss–Jordan elimination: $x_1 + x_2 - 3x_3 + 5x_4 + \ x_5 = \quad 2$
$$-2x_1 + x_2 + 3x_3 - \ x_4 + 4x_5 = -10$$
$$-x_1 + x_2 + \ x_3 + \ x_4 + 3x_5 = \ -3 \qquad ▌▌$$

We have now presented two methods for solving systems of linear equations, Gauss–Jordan elimination and triangularization with back substitution. In theory, either method can be used to solve any system of equations and both are frequently used in computer programs that solve systems of equations. In practice, triangularization with back substitution is usually used to solve square systems that are known in advance to have unique solutions. Such systems occur frequently in real-world applications. (Recall that we considered only square systems with unique solutions in Section 1-3.) If the system is not square or the nature of the solution set is unknown, then Gauss–Jordan elimination is the preferred method. Furthermore, in later chapters we will see that the reduced form of a matrix plays a fundamental role in the study of linear algebra and has applications to a wide variety of problems. Thus, it is important that you become proficient at finding the reduced form of a matrix.

■ Application

Example 19
Production Scheduling

Solution

A casting company produces three different bronze sculptures. The casting department has available a maximum of 350 work-hours per week and the finishing department has a maximum of 150 work-hours available per week. Sculpture A requires 30 hours for casting and 10 hours for finishing; sculpture B requires 10 hours for casting and 10 hours for finishing; and sculpture C requires 10 hours for casting and 30 hours for finishing. If the plant is to operate at maximum capacity, how many of each sculpture should be produced each week?

First, we summarize the relevant manufacturing data in a table:

	Work-Hours per Sculpture			Maximum Work-Hours Available per Week
	A	B	C	
Casting department	30	10	10	350
Finishing department	10	10	30	150

Let

$x_1 = $ Number of sculpture A produced per week

$x_2 = $ Number of sculpture B produced per week

$x_3 = $ Number of sculpture C produced per week

Then

$$30x_1 + 10x_2 + 10x_3 = 350 \quad \text{Casting department}$$
$$10x_1 + 10x_2 + 30x_3 = 150 \quad \text{Finishing department}$$

Now we can form the augmented matrix of the system and solve using Gauss–Jordan elimination:

$$\begin{bmatrix} 30 & 10 & 10 & | & 350 \\ 10 & 10 & 30 & | & 150 \end{bmatrix} \qquad \frac{1}{10}R_1 \rightarrow R_1 \qquad \text{Simplify each row.}$$
$$\frac{1}{10}R_2 \rightarrow R_2$$

$$\sim \begin{bmatrix} 3 & 1 & 1 & | & 35 \\ 1 & 1 & 3 & | & 15 \end{bmatrix} \qquad R_1 \leftrightarrow R_2$$

$$\sim \begin{bmatrix} 1 & 1 & 3 & | & 15 \\ 3 & 1 & 1 & | & 35 \end{bmatrix} \qquad R_2 + (-3)R_1 \rightarrow R_2$$

$$\sim \begin{bmatrix} 1 & 1 & 3 & | & 15 \\ 0 & -2 & -8 & | & -10 \end{bmatrix} \qquad -\frac{1}{2}R_2 \rightarrow R_2$$

$$\sim \begin{bmatrix} 1 & 1 & 3 & | & 15 \\ 0 & 1 & 4 & | & 5 \end{bmatrix} \qquad R_1 + (-1)R_2 \rightarrow R_1$$

$$\sim \begin{bmatrix} 1 & 0 & -1 & | & 10 \\ 0 & 1 & 4 & | & 5 \end{bmatrix} \qquad \text{The matrix is in reduced form.}$$

$$\begin{array}{lll} x_1 \quad - \quad x_3 = 10 & \text{or} & x_1 = \quad x_3 + 10 \\ x_2 + 4x_3 = \quad 5 & & x_2 = -4x_3 + \quad 5 \end{array}$$

Let $x_3 = t$. Then for any real number t,

$$x_1 = \quad t + 10$$
$$x_2 = -4t + \quad 5$$
$$x_3 = \quad t$$

is a solution—or is it? We cannot produce a negative or fractional number of sculptures. Thus, t must be a nonnegative whole number. And because of the middle equation $(x_2 = -4t + 5)$, t can only assume the values 0 and 1. Thus, for $t = 0$, we have $x_1 = 10$, $x_2 = 5$, $x_3 = 0$; and for $t = 1$, we have $x_1 = 11$, $x_2 = 1$, $x_3 = 1$. These are the only possible production schedules that utilize the full capacity of the plant.

Problem 19   Repeat Example 19 using a casting capacity of 400 work-hours per week and a finishing capacity of 200 work-hours per week.   ∎

Answers to
Matched Problems

13. $x_1 = -3$, $x_2 = 1$, $x_3 = -2$ or $(-3, 1, -2)$

14. (A) Condition 4 is violated. The reduced form is

$$\begin{bmatrix} 1 & 0 & 0 & 2 \\ 0 & 1 & 0 & 5 \\ 0 & 0 & 1 & 3 \end{bmatrix}$$

(B) Condition 2 is violated. The reduced form is

$$\begin{bmatrix} 1 & 0 & 0 & 2 \\ 0 & 1 & -1 & 2 \end{bmatrix}$$

(C) The matrix is in reduced form.

(D) Condition 1 is violated. The reduced form is

$$\begin{bmatrix} 1 & 0 & 2 \\ 0 & 1 & 4 \\ 0 & 0 & 0 \end{bmatrix}$$

(E) The matrix is in reduced form.

(F) Condition 3 is violated. The reduced form is

$$\begin{bmatrix} 1 & -2 & 0 & 0 & 1 \\ 0 & 0 & 1 & 0 & 5 \\ 0 & 0 & 0 & 1 & 3 \end{bmatrix}$$

15. (A)
$$\begin{aligned} x_1 &= -5 \\ x_2 &= 3 \\ x_3 &= 6 \end{aligned}$$

Solution. $x_1 = -5$, $x_2 = 3$, $x_3 = 6$ or $(-5, 3, 6)$.

(B)
$$\begin{aligned} x_1 + 2x_2 - 3x_3 &= 0 \\ 0x_1 + 0x_2 + 0x_3 &= 1 \\ 0x_1 + 0x_2 + 0x_3 &= 0 \end{aligned}$$

Inconsistent; no solution

(C)
$$\begin{aligned} x_1 \quad - 2x_3 &= 4 \\ x_2 + 3x_3 &= -2 \end{aligned}$$

Dependent: let $x_3 = t$. The (general) solution is $x_1 = 2t + 4$, $x_2 = -3t - 2$, $x_3 = t$ (t any real number) or $\{(2t + 4, -3t - 2, t)|t$ any real number$\}$.

(D)
$$\begin{aligned} x_1 \quad + 3x_3 + 2x_4 &= 5 \\ x_2 - 2x_3 - x_4 &= 3 \end{aligned}$$

Dependent: let $x_3 = s$ and $x_4 = t$. The (general) solution is $x_1 = -3s - 2t + 5$, $x_2 = 2s + t + 3$, $x_3 = s$, $x_4 = t$ (s and t any real numbers) or $\{(-3s - 2t + 5, 2s + t + 3, s, t)|s$ and t any real numbers$\}$

16. $x_1 = 1, x_2 = -1, x_3 = 0$ or $(1, -1, 0)$

17. $\{(2, 5 + 2t, -t, t)|t$ any real number$\}$

18. Inconsistent

19. $x_1 = t + 10$, $x_2 = -4t + 10$, $x_3 = t$ where $t = 0, 1, 2$; that is, $(x_1, x_2, x_3) = (10, 10, 0), (11, 6, 1),$ or $(12, 2, 2)$

‖ Exercise 1-4

A *In Problems 1–8, determine whether the matrix is in reduced form.*

1. $\begin{bmatrix} 1 & 0 & 2 \\ 0 & 1 & -1 \end{bmatrix}$

2. $\begin{bmatrix} 0 & 1 & 2 \\ 0 & 0 & -1 \end{bmatrix}$

3. $\begin{bmatrix} 1 & 0 & 2 & 3 \\ 0 & 0 & 0 & 0 \\ 0 & 1 & -1 & 4 \end{bmatrix}$

4. $\begin{bmatrix} 1 & 0 & 0 & -2 \\ 0 & 1 & 0 & 0 \\ 0 & 0 & 1 & 1 \end{bmatrix}$

5. $\begin{bmatrix} 0 & 1 & 0 & 2 \\ 0 & 0 & 3 & -1 \\ 0 & 0 & 0 & 0 \end{bmatrix}$

6. $\begin{bmatrix} 1 & 3 & 0 & 0 \\ 0 & 0 & 1 & 0 \\ 0 & 0 & 0 & 1 \end{bmatrix}$

7. $\begin{bmatrix} 1 & 2 & 0 & 3 & 2 \\ 0 & 0 & 1 & -1 & 0 \end{bmatrix}$

8. $\begin{bmatrix} 0 & 1 & 2 & 1 \\ 1 & 0 & -3 & 2 \end{bmatrix}$

In Problems 9–16, write the linear system corresponding to each reduced augmented matrix and solve.

9. $\begin{bmatrix} 1 & 0 & 0 & | & -2 \\ 0 & 1 & 0 & | & 3 \\ 0 & 0 & 1 & | & 0 \end{bmatrix}$

10. $\begin{bmatrix} 1 & 0 & 0 & 0 & | & -2 \\ 0 & 1 & 0 & 0 & | & 0 \\ 0 & 0 & 1 & 0 & | & 1 \\ 0 & 0 & 0 & 1 & | & 3 \end{bmatrix}$

11. $\begin{bmatrix} 1 & 0 & -2 & | & 3 \\ 0 & 1 & 1 & | & -5 \\ 0 & 0 & 0 & | & 0 \end{bmatrix}$

12. $\begin{bmatrix} 1 & -2 & 0 & | & -3 \\ 0 & 0 & 1 & | & 5 \\ 0 & 0 & 0 & | & 0 \end{bmatrix}$

13. $\begin{bmatrix} 1 & 0 & | & 0 \\ 0 & 1 & | & 0 \\ 0 & 0 & | & 1 \end{bmatrix}$

14. $\begin{bmatrix} 1 & 0 & | & 5 \\ 0 & 1 & | & -3 \\ 0 & 0 & | & 0 \end{bmatrix}$

15. $\begin{bmatrix} 1 & -2 & 0 & -3 & | & -5 \\ 0 & 0 & 1 & 3 & | & 2 \end{bmatrix}$

16. $\begin{bmatrix} 1 & 0 & -2 & 3 & | & 4 \\ 0 & 1 & -1 & 2 & | & -1 \end{bmatrix}$

B　In Problems 17–22, use row operations to change each matrix to reduced form.

17. $\begin{bmatrix} 1 & 2 & -1 \\ 0 & 1 & 3 \end{bmatrix}$

18. $\begin{bmatrix} 1 & 3 & 1 \\ 0 & 2 & -4 \end{bmatrix}$

19. $\begin{bmatrix} 1 & 0 & -3 & 1 \\ 0 & 1 & 2 & 0 \\ 0 & 0 & 3 & -6 \end{bmatrix}$

20. $\begin{bmatrix} 1 & 0 & 4 & 0 \\ 0 & 1 & -3 & -1 \\ 0 & 0 & -2 & 2 \end{bmatrix}$

21. $\begin{bmatrix} 1 & 2 & -2 & -1 \\ 0 & 3 & -6 & 1 \\ 0 & -1 & 2 & -\frac{1}{3} \end{bmatrix}$

22. $\begin{bmatrix} 0 & -2 & 8 & 1 \\ 2 & -2 & 6 & -4 \\ 0 & -1 & 4 & \frac{1}{2} \end{bmatrix}$

In Problems 23–40, solve the system by Gauss–Jordan elimination.

23. $2x_1 + 4x_2 - 10x_3 = -2$
$3x_1 + 9x_2 - 21x_3 = 0$
$x_1 + 5x_2 - 12x_3 = 1$

24. $3x_1 + 5x_2 - x_3 = -7$
$x_1 + x_2 + x_3 = -1$
$2x_1 + 11x_3 = 7$

25. $3x_1 + 8x_2 - x_3 = -18$
$2x_1 + x_2 + 5x_3 = 8$
$2x_1 + 4x_2 + 2x_3 = -4$

26. $2x_1 + 7x_2 + 15x_3 = -12$
$4x_1 + 7x_2 + 13x_3 = -10$
$3x_1 + 6x_2 + 12x_3 = -9$

27. $2x_1 - x_2 - 3x_3 = 8$
$x_1 - 2x_2 = 7$

28. $2x_1 + 4x_2 - 6x_3 = 10$
$3x_1 + 3x_2 - 3x_3 = 6$

29. $2x_1 + 3x_2 - x_3 = 1$
$x_1 - 2x_2 + 2x_3 = -2$

30. $x_1 - 3x_2 + 2x_3 = -1$
$3x_1 + 2x_2 - x_3 = 2$

31. $2x_1 + 2x_2 = 2$
$x_1 + 2x_2 = 3$
$-3x_2 = -6$

32. $2x_1 - x_2 = 0$
$3x_1 + 2x_2 = 7$
$x_1 - x_2 = -1$

33. $2x_1 - x_2 = 0$
  $3x_1 + 2x_2 = 7$
  $x_1 - x_2 = -2$

34. $x_1 - 3x_2 = 5$
  $2x_1 + x_2 = 3$
  $x_1 - 2x_2 = 5$

35. $3x_1 - 4x_2 - x_3 = 1$
  $2x_1 - 3x_2 + x_3 = 1$
  $x_1 - 2x_2 + 3x_3 = 2$

36. $3x_1 + 7x_2 - x_3 = 11$
  $x_1 + 2x_2 - x_3 = 3$
  $2x_1 + 4x_2 - 2x_3 = 10$

37. $x_1 + 2x_2 - x_3 + x_4 = 7$
  $2x_1 + 5x_2 + x_3 - x_4 = 9$

38. $2x_1 + 4x_2 + 5x_3 + 4x_4 = 8$
  $x_1 + 2x_2 + 2x_3 + x_4 = 3$

39. $x_1 + 2x_2 - 4x_3 - x_4 = 7$
  $2x_1 + 5x_2 - 9x_3 - 4x_4 = 16$
  $x_1 + 5x_2 - 7x_3 - 7x_4 = 13$

40. $x_1 + 4x_2 + 3x_3 - 2x_4 = 2$
  $2x_1 + 9x_2 + 8x_3 = 1$
  $x_1 - 4x_2 - 2x_3 + x_4 = -3$
  $x_1 + 5x_2 + 6x_3 + x_4 = 5$

C **41.** List all the possible reduced forms of

$$A = \begin{bmatrix} a_{11} & a_{12} \\ a_{21} & a_{22} \end{bmatrix}$$

42. List all the possible reduced forms of

$$A = \begin{bmatrix} a_{11} & a_{12} & a_{13} \\ a_{21} & a_{22} & a_{23} \end{bmatrix}$$

43. List all the possible reduced forms of

$$A = \begin{bmatrix} a_{11} & a_{12} \\ a_{21} & a_{22} \\ a_{31} & a_{32} \end{bmatrix}$$

44. List all the possible reduced forms of

$$A = \begin{bmatrix} a_{11} & a_{12} & a_{13} \\ a_{21} & a_{22} & a_{23} \\ a_{31} & a_{32} & a_{33} \end{bmatrix}$$

45. Find the reduced form of the augmented coefficient matrix for the system

$$ax + by = e$$
$$cx + dy = f$$

if $ad - bc \neq 0$ and $a \neq 0$.

46. Find the reduced form of the augmented coefficient matrix for the system in Problem 45 if $ad - bc = 0$ and $a \neq 0$. When will this system be dependent? Inconsistent?

47. The augmented coefficient matrix for a system of equations is

$$\begin{bmatrix} 1 & 0 & 1 & | & 1 \\ 0 & 1 & 1 & | & 1 \\ 0 & a & b & | & c \end{bmatrix}$$

Determine the values of a, b, and c so that this system will have:

(A) A unique solution   (B) An infinite number of solutions
(C) No solution

48. Repeat Problem 47 for the system with augmented coefficient matrix

$$\begin{bmatrix} 1 & 0 & | & 1 \\ 0 & 1 & | & 1 \\ a & b & | & c \end{bmatrix}$$

▮□▮ Applications

49. *Chemistry*. A chemist can purchase a 10% saline solution in 500 cubic centimeter containers, a 30% solution in 500 cubic centimeter containers, and a 50% solution in 1,000 cubic centimeter containers. He needs 8,000 cubic centimeters of 25% saline solution. How many containers of each type should he purchase to form this solution?

50. *Chemistry*. Repeat Problem 49 if the 50% solution is available only in 2,000 cubic centimeter containers.

51. *Production scheduling*. A small manufacturing plant makes three types of inflatable boats: one-person, two-person, and four-person models. Each boat requires the services of three departments, as listed in the table. The cutting, assembly, and packaging departments have available a maximum of 380, 330, and 120 work-hours per week, respectively. How many boats of each type must be produced each week for the plant to operate at full capacity?

	One-Person Boat	Two-Person Boat	Four-Person Boat
Cutting department	0.5 hr	1.0 hr	1.5 hr
Assembly department	0.6 hr	0.9 hr	1.2 hr
Packaging department	0.2 hr	0.3 hr	0.5 hr

52. *Production scheduling*. Repeat Problem 51 assuming the cutting, assembly, and packaging departments have available a maximum of 350, 330, and 115 work-hours per week, respectively.

53. *Production scheduling*. Work Problem 51 assuming the packaging department is no longer used.

54. *Production scheduling*. Work Problem 52 assuming the packaging department is no longer used.

55. *Production scheduling*. Work Problem 51 assuming the four-person boat is no longer produced.

56. *Production scheduling.* Work Problem 52 assuming the four-person boat is no longer produced.

57. *Nutrition.* A dietitian in a hospital is to arrange a special diet using three basic foods. The diet is to include exactly 340 units of calcium, 180 units of iron, and 220 units of vitamin A. The number of units per ounce of each special ingredient for each of the foods is indicated in the table. How many ounces of each food must be used to meet the diet requirements?

	Units per Ounce		
	Food A	Food B	Food C
Calcium	30	10	20
Iron	10	10	20
Vitamin A	10	30	20

58. *Nutrition.* Repeat Problem 57 if the diet is to include exactly 400 units of calcium, 160 units of iron, and 240 units of vitamin A.

59. *Nutrition.* Solve Problem 57 with the assumption that food C is no longer available.

60. *Nutrition.* Solve Problem 58 with the assumption that food C is no longer available.

61. *Nutrition.* Solve Problem 57 assuming the vitamin A requirement is deleted.

62. *Nutrition.* Solve Problem 58 assuming the vitamin A requirement is deleted.

│1-5│ Homogeneous Systems

- Homogeneous Systems
- Application

▪ Homogeneous Systems

A system of equations is called **homogeneous** if all the constant terms are 0. For example,

$$
\begin{aligned}
x_1 - x_2 + 2x_3 &= 0 \\
2x_1 + x_2 + x_3 &= 0
\end{aligned}
\tag{1}
$$

is a homogeneous system. Obviously, $x_1 = 0$, $x_2 = 0$, and $x_3 = 0$ is a solution to this system. In any homogeneous system, setting each variable equal to 0 will always produce a particular solution, called the **trivial solution.** Thus, homogeneous systems are never inconsistent and always have either one solution (which must be the trivial solution) or an infinite number of solutions. If there are solutions other than the trivial one, they are referred to as **nontrivial solutions.** You can verify that $x_1 = -1$, $x_2 = 1$, and $x_3 = 1$ is a nontrivial solution of (1). Gauss–Jordan elimination can always be used to solve a homogeneous system. However, in many cases, we will be interested in determining whether nontrivial solutions exist without actually finding them. Theorem 4 shows one way that this can be done.

Theorem 4

If the number of variables in a homogeneous system is greater than the number of equations, then the system always has nontrivial solutions.

Proof   The general form for a homogeneous system with m equations and n variables is

$$
m \text{ equations} \begin{cases} \overbrace{a_{11}x_1 + a_{12}x_2 + \cdots + a_{1n}x_n}^{n \text{ variables}} = 0 \\ a_{21}x_1 + a_{22}x_2 + \cdots + a_{2n}x_n = 0 \\ \quad\cdot \qquad\quad\cdot \qquad\qquad\quad\cdot \qquad\cdot \\ \quad\cdot \qquad\quad\cdot \qquad\qquad\quad\cdot \qquad\cdot \\ \quad\cdot \qquad\quad\cdot \qquad\qquad\quad\cdot \qquad\cdot \\ a_{m1}x_1 + a_{m2}x_2 + \cdots + a_{mn}x_n = 0 \end{cases} \tag{2}
$$

Since we know that any homogeneous system has at least one solution (the trivial solution), system (2) is equivalent to a reduced system that has no inconsistencies. The reduced system will have n variables and no more than m equations. Since each equation in the reduced system determines exactly one leftmost variable, the reduced system will have at most m leftmost variables. We are given that $n > m$. Thus, the remaining $n - m$ variables must be parametric variables and the system must have an infinite number of solutions. ∎

If the number of variables in a homogeneous system is less than or equal to the number of equations (that is, if $n \le m$), then the system may or may not have nontrivial solutions and Theorem 4 does not apply. At this point, the only way we can determine whether such a system has nontrivial solutions is to solve the system, as illustrated in Examples 20(B) and (C) below.

Example 20   Determine which of the following homogeneous systems have nontrivial solutions:

(A) $\begin{aligned} x_1 + \ x_2 + \ x_3 &= 0 \\ x_1 + 2x_2 - 2x_3 &= 0 \end{aligned}$ (B) $\begin{aligned} x_1 + \ x_2 &= 0 \\ x_1 + 2x_2 &= 0 \\ x_1 - 2x_2 &= 0 \end{aligned}$

(C) $\begin{aligned} x_1 + \ x_2 + \ x_3 &= 0 \\ x_1 + 2x_2 - 2x_3 &= 0 \\ 2x_1 + 3x_2 - \ x_3 &= 0 \end{aligned}$ (D) $\begin{aligned} x_1 + 2x_2 - \ x_3 + \ x_4 &= 0 \\ 2x_1 + \ x_2 - 3x_3 + \ x_4 &= 0 \\ -x_1 + \ x_2 - 5x_3 - 2x_4 &= 0 \end{aligned}$

Solution   (A) Since there are fewer equations than variables ($m = 2 < 3 = n$), Theorem 4 implies that this system has nontrivial solutions.

(B) Since there are more equations than variables ($m = 3 > 2 = n$), Theorem 4 does not apply. To see if there are nontrivial solutions, we must solve the system.

$$\begin{bmatrix} 1 & 1 & | & 0 \\ 1 & 2 & | & 0 \\ 1 & -2 & | & 0 \end{bmatrix} \begin{matrix} \\ R_2 + (-1)R_1 \rightarrow R_2 \\ R_3 + (-1)R_1 \rightarrow R_3 \end{matrix}$$

$$\sim \begin{bmatrix} 1 & 1 & | & 0 \\ 0 & 1 & | & 0 \\ 0 & -3 & | & 0 \end{bmatrix} \begin{matrix} R_1 + (-1)R_2 \rightarrow R_1 \\ \\ R_3 + 3R_2 \rightarrow R_3 \end{matrix}$$

$$\sim \begin{bmatrix} 1 & 0 & | & 0 \\ 0 & 1 & | & 0 \\ 0 & 0 & | & 0 \end{bmatrix}$$

The only solution to this system is the trivial one: $x_1 = 0$ and $x_2 = 0$.

(C) Since the numbers of equations and variables are the same ($m = 3 = n$), Theorem 4 does not apply. To see if there are nontrivial solutions, we must solve the system.

$$\begin{bmatrix} 1 & 1 & 1 & | & 0 \\ 1 & 2 & -2 & | & 0 \\ 2 & 3 & -1 & | & 0 \end{bmatrix} \begin{matrix} \\ R_2 + (-1)R_1 \rightarrow R_2 \\ R_3 + (-2)R_1 \rightarrow R_3 \end{matrix}$$

$$\sim \begin{bmatrix} 1 & 1 & 1 & | & 0 \\ 0 & 1 & -3 & | & 0 \\ 0 & 1 & -3 & | & 0 \end{bmatrix} \begin{matrix} R_1 + (-1)R_2 \rightarrow R_1 \\ \\ R_3 + (-1)R_2 \rightarrow R_3 \end{matrix}$$

$$\sim \begin{bmatrix} 1 & 0 & 4 & | & 0 \\ 0 & 1 & -3 & | & 0 \\ 0 & 0 & 0 & | & 0 \end{bmatrix}$$

The reduced system is

$$x_1 \quad + 4x_3 = 0$$
$$x_2 - 3x_3 = 0$$

If we let $x_3 = t$, then $x_1 = -4t$ and $x_2 = 3t$. Thus, $(-4t, 3t, t)$ is a nontrivial solution for any $t \neq 0$. (Check this in the original system.)

(D) Since there are fewer equations than variables ($m = 3 < 4 = n$), Theorem 4 applies and the system has nontrivial solutions.

Problem 20   Determine which of the following homogeneous systems have nontrivial solutions:

(A) $x_1 + 2x_2 = 0$
$\quad\;\, x_1 + 3x_2 = 0$
$\quad 2x_1 + 5x_2 = 0$

(B) $x_1 + 2x_2 + \;\; x_3 = 0$
$\quad\;\, x_1 + 3x_2 + 2x_3 = 0$

(C) $x_1 + 2x_2 + x_3 = 0$ (D) $x_1 + 2x_2 + x_3 + x_4 = 0$
      $x_1 + 3x_2 + 2x_3 = 0$ $x_1 + 3x_2 + 2x_3 + x_4 = 0$
      $2x_1 + 5x_2 + 3x_3 = 0$ $2x_1 + 5x_2 + 3x_3 + 2x_4 = 0$

▪ Application

A mobile is constructed by suspending objects from a wire. The mobile is said to be in *static equilibrium* if the wire between the two objects is horizontal (see Figure 4).

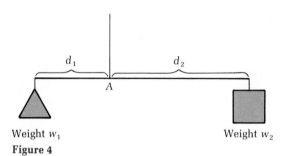

Weight w_1

Weight w_2

Figure 4

The *moment of force* of an object about a point is the product of the weight of the object and the horizontal distance between the object and the point. Thus, the moment of force of the triangular-shaped object in Figure 4 about the point A is $d_1 w_1$ and the moment of force of the rectangular object about the point A is $d_2 w_2$. The mobile will be in static equilibrium if these moments are equal, that is,

$d_1 w_1 = d_2 w_2$

Example 21
Static Equilibrium

A mobile with three suspended objects is illustrated in the figure. If this mobile is to be in static equilibrium, what should each object weigh?

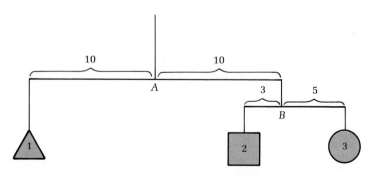

Solution   Let

$w_1 =$ Weight* of object 1   Triangle

$w_2 =$ Weight of object 2   Rectangle

$w_3 =$ Weight of object 3   Circle

First, we compute the moments of force about the point B. The horizontal distance from the rectangle to B is 3 units, so the moment of force of the rectangle about B is $3w_2$. The horizontal distance from the circle to B is 5 units, so the moment of force of the circle about B is $5w_3$. In order that the mobile be in static equilibrium at B, we must have

$$3w_2 = 5w_3 \tag{3}$$

Next, we compute the moments of force about the point A. The moment of force of the triangle about A is $10w_1$. Since the horizontal distance from A to the rectangle is $10 - 3 = 7$ units, the moment of force of the rectangle about A is $7w_2$. In the same manner, the moment of force of the circle about A is $15w_3$. In order for the mobile to be in static equilibrium at A, the combined moments of force of the rectangle and the circle must equal the moment of force of the triangle. Thus,

$$10w_1 = 7w_2 + 15w_3 \tag{4}$$

Combining (3) and (4), the mobile will be in static equilibrium if w_1, w_2, and w_3 satisfy the system

$$3w_2 = 5w_3$$
$$10w_1 = 7w_2 + 15w_3$$

or

$$3w_2 - 5w_3 = 0$$
$$10w_1 - 7w_2 - 15w_3 = 0$$

Since this is a homogeneous system with two equations and three variables, it will have nontrivial solutions.

$$\begin{bmatrix} 0 & 3 & -5 & | & 0 \\ 10 & -7 & -15 & | & 0 \end{bmatrix} \quad R_1 \leftrightarrow R_2$$

$$\sim \begin{bmatrix} 10 & -7 & -15 & | & 0 \\ 0 & 3 & -5 & | & 0 \end{bmatrix} \quad \begin{array}{l} \frac{1}{10}R_1 \to R_1 \\ \frac{1}{3}R_2 \to R_2 \end{array}$$

* Any units of weight and distance can be used in this type of problem, but all weights must be expressed in terms of the same unit of weight and all units of distance must be expressed in terms of the same unit of distance. For example, all weights could be in grams and all distances in centimeters, or all weights could be in pounds and all distances in inches.

$$\sim \begin{bmatrix} 1 & -\frac{7}{10} & -\frac{15}{10} & \Big| & 0 \\ 0 & 1 & -\frac{5}{3} & \Big| & 0 \end{bmatrix} \qquad R_1 + \frac{7}{10}R_2 \rightarrow R_1$$

$$\sim \begin{bmatrix} 1 & 0 & -\frac{8}{3} & \Big| & 0 \\ 0 & 1 & -\frac{5}{3} & \Big| & 0 \end{bmatrix}$$

The reduced system is

$$w_1 \quad - \tfrac{8}{3}w_3 = 0$$

$$w_2 - \tfrac{5}{3}w_3 = 0$$

Letting $w_3 = t$, the solution to this system is

$$w_1 = \tfrac{8}{3}t$$

$$w_2 = \tfrac{5}{3}t$$

$$w_3 = \quad t$$

Any positive value of the parameter t will produce a solution that would place the mobile in static equilibrium.

Problem 21   Repeat Example 21 for the mobile in the figure below.

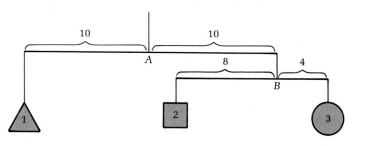

Answers to   **20.** (B), (C), and (D) have nontrivial solutions.
Matched Problems   **21.** $(\tfrac{3}{2}t, \tfrac{1}{2}t, t)$ where t is any positive number

∎ Exercise 1-5

A   *Determine which of the homogeneous systems in Problems 1–6 have nontrivial solutions.*

1. $x_1 - x_2 + x_3 = 0$
$ 2x_1 - x_2 + x_3 = 0$

2. $x_1 - x_2 + x_3 = 0$
$ 2x_1 - x_2 + x_3 = 0$
$ 4x_1 - 3x_2 + 3x_3 = 0$

3. $x_1 - x_2 + x_3 = 0$
    $2x_1 - x_2 + x_3 = 0$
    $-x_1 + x_2 + 2x_3 = 0$

4. $2x_1 - x_2 = 0$
    $x_1 + 3x_2 = 0$
    $3x_1 - 4x_2 = 0$

5. $x_1 + x_2 - x_3 + x_4 = 0$
    $2x_1 + x_2 + x_3 - x_4 = 0$
    $-x_1 - x_2 + 2x_3 + x_4 = 0$

6. $x_1 + x_2 - x_3 + x_4 = 0$
    $2x_1 + x_2 + x_3 - x_4 = 0$
    $-x_1 - x_2 + 2x_3 + x_4 = 0$
    $2x_1 + x_2 + 2x_3 - x_4 = 0$

B Solve the systems in Problems 7–20.

7. $x_1 - x_2 + x_3 = 0$
    $3x_1 - 2x_2 - 4x_3 = 0$

8. $x_1 - x_2 + x_3 = 0$
    $3x_1 - 3x_2 - 4x_3 = 0$

9. $x_1 - x_2 + x_3 = 0$
    $3x_1 - 2x_2 - 4x_3 = 0$
    $4x_1 - 3x_2 - 3x_3 = 0$

10. $x_1 - x_2 + x_3 = 0$
    $3x_1 - 2x_2 - 4x_3 = 0$
    $4x_1 - 3x_2 + 5x_3 = 0$

11. $x_1 - x_2 = 0$
    $3x_1 - 2x_2 = 0$
    $4x_1 - 3x_2 = 0$

12. $x_1 - x_2 = 0$
    $3x_1 - 3x_2 = 0$
    $-4x_1 + 4x_2 = 0$

13. $x_1 - 2x_2 + x_3 - x_4 = 0$
    $-x_1 + 3x_2 + x_3 - 2x_4 = 0$

14. $x_1 - 2x_2 + x_3 - x_4 = 0$
    $-x_1 + 2x_2 - x_3 + 2x_4 = 0$

15. $x_1 - 2x_2 + x_3 - x_4 = 0$
    $-x_1 + 2x_2 - x_3 + x_4 = 0$

16. $x_1 - 2x_2 + x_3 - x_4 = 0$
    $-x_1 + 3x_2 + x_3 - 2x_4 = 0$
    $2x_1 - 2x_2 + 2x_3 - 4x_4 = 0$

17. $x_1 - 2x_2 + x_3 - x_4 = 0$
    $-x_1 + 3x_2 + x_3 - 2x_4 = 0$
    $2x_1 - 2x_2 + 2x_3 - 4x_4 = 0$
    $x_1 - 2x_2 + x_3 = 0$

18. $x_1 - 2x_2 + x_3 - x_4 = 0$
    $-x_1 + 3x_2 + x_3 - 2x_4 = 0$
    $2x_1 - 2x_2 + 2x_3 - 4x_4 = 0$
    $2x_1 + 3x_2 + 4x_3 - 7x_4 = 0$

19. $x_1 + x_2 = 0$
    $x_2 + x_3 = 0$
    $x_3 + x_4 = 0$
    $x_4 + x_5 = 0$

20. $x_1 + x_2 = 0$
    $x_1 + x_2 + x_3 = 0$
    $x_2 + x_3 + x_4 = 0$
    $x_3 + x_4 + x_5 = 0$
    $x_4 + x_5 = 0$

C In Problems 21 and 22, find the (general) solution; then find the particular solution satisfying $x_1 + x_2 = 1$.

21. $\frac{1}{5}x_1 + \frac{2}{3}x_2 = x_1$
    $\frac{4}{5}x_1 + \frac{1}{3}x_2 = x_2$

22. $\frac{1}{2}x_1 + \frac{1}{4}x_2 = x_1$
    $\frac{1}{2}x_1 + \frac{3}{4}x_2 = x_2$

In Problems 23 and 24, find the (general) solution; then find the particular solution satisfying $x_1 + x_2 + x_3 = 1$.

23. $\frac{1}{2}x_1 + \frac{1}{3}x_2 = x_1$
    $\frac{1}{2}x_1 + \frac{1}{3}x_2 + \frac{1}{2}x_3 = x_2$
    $\frac{1}{3}x_2 + \frac{1}{2}x_3 = x_3$

24. $\frac{1}{2}x_1 + \frac{1}{2}x_2 + \frac{1}{2}x_3 = x_1$
    $\frac{1}{4}x_1 + \frac{1}{4}x_2 = x_2$
    $\frac{1}{4}x_1 + \frac{1}{4}x_2 + \frac{1}{2}x_3 = x_3$

25. If $a \neq 0$, when will the system below have nontrivial solutions?

$$ax_1 + bx_2 = 0$$
$$cx_1 + dx_2 = 0$$

26. When will the homogeneous system below have nontrivial solutions?

$$x_1 \quad + \; x_3 = 0$$
$$x_2 + \; x_3 = 0$$
$$ax_2 + bx_3 = 0$$

▮▫▮ Applications

Static Equilibrium　In Problems 27–30, determine the weights of the suspended objects that will place each mobile in static equilibrium.

27.

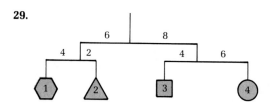

28.

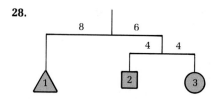

29.

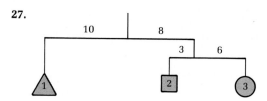

30.

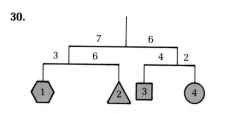

|1-6| Chapter Review

Important Terms and
Symbols

1-1. *Systems of linear equations—introduction.* Linear equation, solution set, system of linear equations, equivalent systems, elimination, back substitution, consistent system, unique solution, inconsistent system, dependent system, parameter, general solution, particular solution

1-2. *Augmented matrices and elementary row operations.* Matrix, element of a matrix, augmented coefficient matrix, double subscript notation, elementary row operations, row equivalent matrices, interpolating polynomial, a_{ij}, $R_i \leftrightarrow R_j$, $kR_i \rightarrow R_i$, $R_j + kR_i \rightarrow R_j$

1-3. *Triangularization with back substitution.* Square system, diagonal elements, triangular form, triangularization with back substitution

1-4. *Gauss–Jordan elimination.* Reduced (row-echelon) form, leftmost variable, parametric variable, Gauss–Jordan elimination

1-5. *Homogeneous systems.* Homogeneous system, trivial solution, nontrivial solution, static equilibrium

‖ Exercise 1-6 Chapter Review

Work through all the problems in this chapter review and check your answers in the back of the book. (Answers to most review problems are there.) Where weaknesses show up, review appropriate sections in the text.

A In Problems 1–3, graph each system and discuss the nature of the solution set.

1. $3x - y = 6$
$2x + y = -1$

2. $3x - y = 6$
$-3x + y = 3$

3. $6x + 3y = -3$
$2x + y = -1$

In Problems 4–7, solve each system by triangularization with back substitution.

4. $x_1 + 2x_2 = 9$
$3x_1 + x_2 = 2$

5. $2x_1 - 4x_2 = 11$
$-5x_1 + 3x_2 = 18$

6. $x_1 - 2x_2 + 3x_3 = 5$
$x_2 - 3x_3 = -6$
$x_1 + 4x_3 = 7$

7. $x_1 + 3x_2 - 5x_3 = -6$
$2x_1 + 8x_2 - 8x_3 = -14$
$-3x_1 - 9x_2 + x_3 = 4$

In Problems 8–11, write the linear system corresponding to each reduced augmented matrix and solve.

8. $\begin{bmatrix} 1 & 0 & | & 2 \\ 0 & 1 & | & 3 \end{bmatrix}$

9. $\begin{bmatrix} 1 & -1 & 2 & | & 0 \\ 0 & 0 & 0 & | & 1 \end{bmatrix}$

10. $\begin{bmatrix} 1 & 0 & 3 & | & 4 \\ 0 & 1 & -2 & | & 7 \end{bmatrix}$

11. $\begin{bmatrix} 1 & -2 & 0 & | & 2 \\ 0 & 0 & 1 & | & 4 \\ 0 & 0 & 0 & | & 0 \end{bmatrix}$

In Problems 12–14, use row operations to transform each matrix to reduced form.

12. $\begin{bmatrix} 1 & -1 & 2 \\ 0 & 1 & 4 \end{bmatrix}$

13. $\begin{bmatrix} 1 & 0 & 0 & 3 \\ 0 & 0 & 1 & 2 \\ 0 & 1 & 2 & 4 \end{bmatrix}$

14. $\begin{bmatrix} 1 & 0 & -1 & 2 & -1 \\ 0 & 1 & 1 & -1 & 0 \\ 1 & 0 & 0 & 1 & 1 \end{bmatrix}$

B   *Use Gauss–Jordan elimination to solve the systems in Problems 15–26.*

15. $\begin{aligned} 2x_1 + 3x_2 &= 5 \\ -3x_1 + 4x_2 &= 18 \end{aligned}$

16. $\begin{aligned} 2x_1 + 3x_2 &= 5 \\ -3x_1 + 4x_2 &= 18 \\ 7x_1 + 2x_2 &= 8 \end{aligned}$

17. $\begin{aligned} x_1 - 5x_2 + 2x_3 &= 4 \\ 3x_1 - 10x_2 + x_3 &= 2 \end{aligned}$

18. $\begin{aligned} 2x_1 + 4x_2 - 3x_3 &= 0 \\ 3x_1 + 8x_2 + 5x_3 &= 0 \end{aligned}$

19. $\begin{aligned} x_1 \quad - 4x_3 &= 4 \\ 2x_2 + x_3 &= -1 \\ x_1 + 2x_2 - 3x_3 &= 3 \end{aligned}$

20. $\begin{aligned} x_1 + 2x_2 + 4x_3 &= 0 \\ 2x_1 + 4x_2 + 5x_3 &= 0 \\ x_1 + 2x_2 + x_3 &= 0 \end{aligned}$

21. $\begin{aligned} x_1 - 2x_2 + x_3 &= 0 \\ -2x_1 + 3x_2 - x_3 &= 0 \\ -x_1 + 2x_2 + x_3 &= 0 \end{aligned}$

22. $\begin{aligned} 2x_1 + 4x_2 - 3x_3 &= 5 \\ -2x_1 - 8x_2 + 4x_3 &= -9 \\ 2x_1 + 6x_2 - 3x_3 &= 8 \end{aligned}$

C   23. $\begin{aligned} x_1 - x_2 + 2x_3 + 2x_4 &= -1 \\ -2x_1 + x_2 - x_3 \quad &= -4 \end{aligned}$

24. $\begin{aligned} 2x_1 - 8x_2 + x_3 - x_4 &= 0 \\ 2x_2 - x_3 + x_4 &= 0 \\ -2x_1 + 6x_2 + x_3 + x_4 &= 0 \end{aligned}$

25. $\begin{aligned} x_1 + 3x_2 - x_3 + 2x_4 - x_5 &= 0 \\ 2x_1 + 6x_2 - x_3 - x_4 + 2x_5 &= 0 \\ -x_1 - 3x_2 \quad - 2x_4 + x_5 &= 0 \end{aligned}$

26. $\begin{aligned} x_1 - 2x_2 + x_3 - 4x_4 + x_5 &= 2 \\ 4x_1 - 6x_2 + 5x_3 - 10x_4 + 3x_5 &= 11 \\ 3x_1 - 2x_2 + 5x_3 \quad + x_5 &= 6 \end{aligned}$

27. Solve the system below under the assumption that $ad - bc \neq 0$ and $a \neq 0$.

$$ax_1 + bx_2 + x_3 = 0$$
$$cx_1 + dx_2 + x_3 = 0$$

28. The augmented coefficient matrix for a system of linear equations is

$$\left[\begin{array}{ccc|c} 1 & 0 & -1 & 2 \\ 0 & 1 & 1 & -3 \\ 0 & 0 & a & b \end{array} \right]$$

Determine the values of a and b so that the system will have:

(A) A unique solution   (B) No solution
(C) An infinite number of solutions

29. If (u_1, u_2, u_3) is a particular solution of the homogeneous system

$$a_{11}x_1 + a_{12}x_2 + a_{13}x_3 = 0$$
$$a_{21}x_1 + a_{22}x_2 + a_{23}x_3 = 0$$

show that (tu_1, tu_2, tu_3) is also a particular solution for any real number t.

30. If (u_1, u_2, u_3) and (v_1, v_2, v_3) are both particular solutions of the homogeneous system

$$a_{11}x_1 + a_{12}x_2 + a_{13}x_3 = 0$$
$$a_{21}x_1 + a_{22}x_2 + a_{23}x_3 = 0$$

show that $(u_1 + v_1, u_2 + v_2, u_3 + v_3)$ is also a particular solution.

▐▫▌ Applications

31. *Chemistry.* A chemist has two saline solutions; one has a 20% concentration of saline and the other has a 45% concentration of saline. How many cubic centimeters of each solution should be mixed together in order to obtain 100 cubic centimeters of solution with a 30% concentration of saline?

32. *Scheduling.* A college mathematics department wants to hire 6 teachers to teach a total of 20 sections each semester. The teachers can be hired at three different ranks: lecturer, instructor, and assistant professor. Each semester a lecturer teaches 2 sections, an instructor teaches 3 sections, and an assistant professor teaches 4 sections. How many teachers should be hired at each rank?

33. *Scheduling.* In Problem 32 the department has been allocated $120,000 to pay the new teachers. The starting salary for a lecturer is $15,000, for an instructor is $20,000, and for an assistant professor is $22,000. How many teachers should be hired at each rank in order to teach the required number of sections and not exceed the salary allocation?

34. *Heat conduction.* Find the temperature at each intersection point in the grid in the figure. [*Hint:* Assume that the temperature at each intersection point is the average of the temperatures at the adjacent points (adjacent points are either other intersection points or ends of bars).]

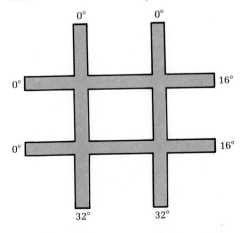

35. *Static equilibrium.* Determine the weights of the suspended objects that will place the mobile in the figure in static equilibrium.

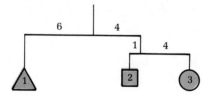

|2| Matrix Operations

<table>
<tr><td>| 2 |</td><td># Contents</td></tr>
</table>

In Chapter 1 we used matrices to make the process of solving linear systems more efficient. It turns out that matrices have many other important applications. In this chapter, we will examine some of the operations that can be performed with matrices and continue to investigate the relationship between matrices and systems of linear equations.

| 2-1| Matrix Addition and Scalar Multiplication

- Basic Definitions
- Matrix Addition
- Scalar Multiplication
- Application

▪ Basic Definitions

Recall that we defined a matrix as any rectangular array of numbers. The *size* or *dimension* of a matrix is important relative to operations on matrices. We define the **dimension** of a matrix to be $m \times n$ (read "*m* by *n*") if the matrix has m rows and n columns. It is important to note that the number of rows is always given first. A matrix that has the same number of rows and columns is called a **square matrix.** A matrix with only one column is called a **column matrix** and one with only one row is called a **row matrix.** These definitions are illustrated by the following:

$$
\begin{array}{cccc}
3 \times 2 & 3 \times 3 & 4 \times 1 & 1 \times 4 \\
\begin{bmatrix} -2 & 5 \\ 0 & -2 \\ 3 & 6 \end{bmatrix} &
\begin{bmatrix} 0.5 & 0.2 & 1.0 \\ 0.0 & 0.3 & 0.5 \\ 0.7 & 0.0 & 0.2 \end{bmatrix} &
\begin{bmatrix} 3 \\ -2 \\ 1 \\ 0 \end{bmatrix} &
\begin{bmatrix} 2 & \frac{1}{2} & 0 & -\frac{2}{3} \end{bmatrix} \\
& \text{Square matrix} & & \text{Row matrix} \\
& & \text{Column matrix} &
\end{array}
$$

Sometimes it will be necessary to write a matrix in general form. This is done by using double subscripts, as we did in writing the general form for linear systems. For example, the general form of a 3×4 matrix A can be written as

Column 3
$$\downarrow$$

$$A = [a_{ij}]_{3\times 4} = \begin{bmatrix} a_{11} & a_{12} & a_{13} & a_{14} \\ a_{21} & a_{22} & a_{23} & a_{24} \\ a_{31} & a_{32} & a_{33} & a_{34} \end{bmatrix} \leftarrow \text{Row 2}$$

In this notation, a_{ij} represents the element in row i and column j. Thus, a_{23} is the element in row 2 and column 3. In order to avoid confusion between letters representing matrices and those representing numbers, we will always use capital letters to represent matrices and small letters to represent numbers. In a row or column matrix, we will sometimes use just one subscript. Thus, we could write

$$X = \begin{bmatrix} x_{11} \\ x_{21} \\ x_{31} \end{bmatrix} \quad \text{or} \quad X = \begin{bmatrix} x_1 \\ x_2 \\ x_3 \end{bmatrix}$$

Two matrices are said to be **equal** if they have the same dimension and their corresponding elements are equal. For example,

$$\begin{bmatrix} a_{11} & a_{12} \\ a_{21} & a_{22} \\ a_{31} & a_{32} \end{bmatrix} = \begin{bmatrix} b_{11} & b_{12} \\ b_{21} & b_{22} \\ b_{31} & b_{32} \end{bmatrix} \quad \begin{array}{l} \text{if and} \\ \text{only if} \end{array} \quad \begin{array}{ll} a_{11} = b_{11} & a_{12} = b_{12} \\ a_{21} = b_{21} & a_{22} = b_{22} \\ a_{31} = b_{31} & a_{32} = b_{32} \end{array}$$

▪ Matrix Addition

The **sum of two matrices of the same dimension** is a matrix with elements that are the sum of the corresponding elements of the two given matrices. The dimension of the sum will be the same as the common dimension of the two given matrices. **Addition is not defined for matrices with different dimensions.**

Example 1 (A) $$\begin{bmatrix} a_{11} & a_{12} \\ a_{21} & a_{22} \\ a_{31} & a_{32} \end{bmatrix} + \begin{bmatrix} b_{11} & b_{12} \\ b_{21} & b_{22} \\ b_{31} & b_{32} \end{bmatrix} = \begin{bmatrix} a_{11}+b_{11} & a_{12}+b_{12} \\ a_{21}+b_{21} & a_{22}+b_{22} \\ a_{31}+b_{31} & a_{32}+b_{32} \end{bmatrix}$$

(B) $$\begin{bmatrix} 3 & 4 & 0 \\ 2 & -7 & 6 \end{bmatrix} + \begin{bmatrix} 1 & -2 & 2 \\ -1 & 4 & -3 \end{bmatrix} = \begin{bmatrix} 4 & 2 & 2 \\ 1 & -3 & 3 \end{bmatrix}$$

(C) $$\begin{bmatrix} \frac{1}{2} \\ 1 \\ 0 \\ -\frac{1}{4} \end{bmatrix} + \begin{bmatrix} \frac{3}{4} \\ -1 \\ 1 \\ \frac{1}{2} \end{bmatrix} = \begin{bmatrix} \frac{5}{4} \\ 0 \\ 1 \\ \frac{1}{4} \end{bmatrix}$$

Problem 1 Add:

(A) $$\begin{bmatrix} a_{11} & a_{12} & a_{13} \\ a_{21} & a_{22} & a_{23} \end{bmatrix} + \begin{bmatrix} b_{11} & b_{12} & b_{13} \\ b_{21} & b_{22} & b_{23} \end{bmatrix}$$

(B) $\begin{bmatrix} 2 & 1 \\ -1 & 0 \\ 1 & 4 \end{bmatrix} + \begin{bmatrix} 1 & -2 \\ 3 & 1 \\ 2 & -3 \end{bmatrix}$ (C) $[\frac{1}{10} \quad \frac{2}{5} \quad 2] + [\frac{2}{5} \quad \frac{1}{10} \quad 4]$ ▌▌

A matrix with elements that are all 0's is called a **zero matrix.** For example,

$$[0]_{2\times3} = \begin{bmatrix} 0 & 0 & 0 \\ 0 & 0 & 0 \end{bmatrix} \qquad [0]_{3\times1} = \begin{bmatrix} 0 \\ 0 \\ 0 \end{bmatrix} \qquad [0]_{4\times4} = \begin{bmatrix} 0 & 0 & 0 & 0 \\ 0 & 0 & 0 & 0 \\ 0 & 0 & 0 & 0 \\ 0 & 0 & 0 & 0 \end{bmatrix}$$

are all zero matrices. When the context makes it clear, we will use the number 0 to represent a zero matrix of the appropriate size. For example, in the equation

$$\begin{bmatrix} 1 & 2 \\ 0 & 1 \end{bmatrix} + \begin{bmatrix} -1 & -2 \\ 0 & -1 \end{bmatrix} = 0$$

the 0 on the right side of the equation represents the 2×2 zero matrix since this is the only one that will make the equation valid.

The **negative of a matrix** A, denoted by $-A$, is a matrix with elements that are the negatives of the elements in A. Thus, if

$$A = \begin{bmatrix} a_{11} & a_{12} & a_{13} & a_{14} \\ a_{21} & a_{22} & a_{23} & a_{24} \end{bmatrix}$$

then

$$-A = \begin{bmatrix} -a_{11} & -a_{12} & -a_{13} & -a_{14} \\ -a_{21} & -a_{22} & -a_{23} & -a_{24} \end{bmatrix}$$

If A and B are matrices of the same dimension, then we define **subtraction** as follows:

$$A - B = A + (-B)$$

Example 2 $\begin{bmatrix} 4 & -2 \\ 0 & 5 \end{bmatrix} - \begin{bmatrix} 6 & 3 \\ -1 & 5 \end{bmatrix} = \begin{bmatrix} 4 & -2 \\ 0 & 5 \end{bmatrix} + \begin{bmatrix} -6 & -3 \\ 1 & -5 \end{bmatrix} = \begin{bmatrix} -2 & -5 \\ 1 & 0 \end{bmatrix}$

Problem 2 Subtract: $\begin{bmatrix} 1 & 2 \\ -1 & 0 \\ 4 & -7 \end{bmatrix} - \begin{bmatrix} -1 & 2 \\ 1 & -3 \\ 2 & 5 \end{bmatrix}$ ▌▌

Matrix addition has many properties that are similar to ordinary number addition properties. Some of the most important properties are given in Theorem 1.

Theorem 1	If A, B, and C are matrices of the same dimension, then:

Theorem 1 | If A, B, and C are matrices of the same dimension, then:

(A) $A + B = B + A$ Commutative property
(B) $A + (B + C) = (A + B) + C$ Associative property
(C) $A + 0 = A$ Identity for addition
(D) $A + (-A) = 0$ Inverse for addition
(E) If $A = B$, then $A + C = B + C$.   Addition property

These properties all follow directly from the definition of matrix addition and the properties of numbers. To illustrate the techniques used in proving theorems involving matrices, we will prove part (E). Proofs of the remaining statements are left as exercises.

Proof   To prove part (E), that is, to show

$$A + C = B + C$$

when $A = B$, we must show that each element of $A + C$ is equal to the corresponding element of $B + C$. From the definition of matrix addition, the element in row i and column j of $A + C$ is $a_{ij} + c_{ij}$ and the corresponding element in $B + C$ is $b_{ij} + c_{ij}$. Thus, we must show that

$$a_{ij} + c_{ij} = b_{ij} + c_{ij}$$

Since we are given that $A = B$, we know that

$$a_{ij} = b_{ij}$$

The addition property for numbers then implies that

$$a_{ij} + c_{ij} = b_{ij} + c_{ij}$$

which completes the proof.   ▌▌

▪ Scalar Multiplication

The **product of a number k and a matrix A,** denoted by kA, is the matrix formed by multiplying each element of A by k. Traditionally, both the number k and the elements in the matrix A are called **scalars,** and the operation of multiplying a matrix by a number is referred to as **scalar multiplication.** In this text, the scalars will always be real numbers and we will use the terms "scalar" and "real number" interchangeably when dealing with scalar products.

Notice that if k is any positive integer, then the scalar product can be interpreted as repeated addition. For example, $3A = A + A + A$.

Example 3 $\dfrac{1}{2}\begin{bmatrix} 2 & 3 & -4 \\ 1 & 6 & 0 \\ -2 & 5 & -1 \end{bmatrix} = \begin{bmatrix} 1 & \frac{3}{2} & -2 \\ \frac{1}{2} & 3 & 0 \\ -1 & \frac{5}{2} & -\frac{1}{2} \end{bmatrix}$

Problem 3   Find: $10\begin{bmatrix} 2.4 & -1.3 \\ 5.2 & 1.0 \\ -4.5 & 9.7 \end{bmatrix}$ ∎

We will need the basic properties of scalar multiplication given in Theorem 2. The proof of Theorem 2 is left as an exercise.

Theorem 2   If A and B are matrices of the same dimension and k and ℓ are scalars, then:

(A) $k(\ell A) = (k\ell)A$ (B) $(k + \ell)A = kA + \ell A$
(C) $k(A + B) = kA + kB$ (D) If $A = B$, then $kA = kB$.

The following example illustrates the use of the properties listed in Theorems 1 and 2. Can you identify the properties used in each step?

Example 4   Solve for X:

$$5X - \begin{bmatrix} 7 & 2 & 3 \\ 9 & -4 & 1 \end{bmatrix} = 3X + \begin{bmatrix} 1 & -2 & 4 \\ 5 & 3 & -5 \end{bmatrix}$$

Solution

$$5X = 3X + \begin{bmatrix} 1 & -2 & 4 \\ 5 & 3 & -5 \end{bmatrix} + \begin{bmatrix} 7 & 2 & 3 \\ 9 & -4 & 1 \end{bmatrix}$$

$$= 3X + \begin{bmatrix} 8 & 0 & 7 \\ 14 & -1 & -4 \end{bmatrix}$$

$$5X - 3X = \begin{bmatrix} 8 & 0 & 7 \\ 14 & -1 & -4 \end{bmatrix}$$

$$2X = \begin{bmatrix} 8 & 0 & 7 \\ 14 & -1 & -4 \end{bmatrix}$$

$$X = \frac{1}{2}\begin{bmatrix} 8 & 0 & 7 \\ 14 & -1 & -4 \end{bmatrix}$$

$$= \begin{bmatrix} 4 & 0 & \frac{7}{2} \\ 7 & -\frac{1}{2} & -2 \end{bmatrix}$$

Problem 4   Solve for X:

$$2X + \begin{bmatrix} 1 & 2 \\ 0 & 1 \end{bmatrix} = \begin{bmatrix} 4 & -1 \\ 2 & 7 \end{bmatrix} - X$$ ∎

■ **Application**

Example 5
Sales Commissions

Mary Smith and Robert Jones are salespeople in a new car agency that sells only two models. August was the last month for this year's models, and next year's models were introduced in September. Gross dollar sales for each month are given in the following matrices:

August sales

$$\begin{matrix} & \text{Compact} & \text{Luxury} \\ \text{Smith} & \\ \text{Jones} \end{matrix} \begin{bmatrix} \$12{,}000 & \$24{,}000 \\ \$24{,}000 & 0 \end{bmatrix} = A$$

September sales

$$\begin{matrix} & \text{Compact} & \text{Luxury} \\ \text{Smith} & \\ \text{Jones} \end{matrix} \begin{bmatrix} \$48{,}000 & \$96{,}000 \\ \$60{,}000 & \$72{,}000 \end{bmatrix} = B$$

(For example, Smith had $12,000 in compact sales in August, and Jones had $72,000 in luxury car sales in September.)

(A) What was the combined dollar sales in August and September for each person and each model?
(B) What was the increase in dollar sales from August to September?
(C) If both salespeople receive 5% commissions on gross dollar sales, compute the commission for each person for each model sold in September.
(D) Compute the total commission for each person in September.

Solution (A) $A + B = \begin{matrix} \text{Compact} & \text{Luxury} \\ \begin{bmatrix} \$60{,}000 & \$120{,}000 \\ \$84{,}000 & \$72{,}000 \end{bmatrix} \end{matrix} \begin{matrix} \text{Smith} \\ \text{Jones} \end{matrix}$

(B) $B - A = \begin{matrix} \text{Compact} & \text{Luxury} \\ \begin{bmatrix} \$36{,}000 & \$72{,}000 \\ \$36{,}000 & \$72{,}000 \end{bmatrix} \end{matrix} \begin{matrix} \text{Smith} \\ \text{Jones} \end{matrix}$

(C) $0.05B = \begin{bmatrix} (0.05)(\$48{,}000) & (0.05)(\$96{,}000) \\ (0.05)(\$60{,}000) & (0.05)(\$72{,}000) \end{bmatrix}$

$= \begin{matrix} \text{Compact} & \text{Luxury} \\ \begin{bmatrix} \$2{,}400 & \$4{,}800 \\ \$3{,}000 & \$3{,}600 \end{bmatrix} \end{matrix} \begin{matrix} \text{Smith} \\ \text{Jones} \end{matrix}$

(D) $\begin{matrix} \text{Compact} & \text{Luxury} \\ \begin{bmatrix} \$2{,}400 \\ \$3{,}000 \end{bmatrix} + \begin{bmatrix} \$4{,}800 \\ \$3{,}600 \end{bmatrix} = \begin{bmatrix} \$7{,}200 \\ \$6{,}600 \end{bmatrix} \end{matrix} \begin{matrix} \text{Smith} \\ \text{Jones} \end{matrix}$

Problem 5   Repeat Example 5 with:

$$A = \begin{bmatrix} \$24{,}000 & \$24{,}000 \\ \$12{,}000 & \$24{,}000 \end{bmatrix} \qquad B = \begin{bmatrix} \$60{,}000 & \$72{,}000 \\ \$48{,}000 & \$72{,}000 \end{bmatrix}$$ ▮▮

Answers to   **1. (A)** $\begin{bmatrix} a_{11} + b_{11} & a_{12} + b_{12} & a_{13} + b_{13} \\ a_{21} + b_{21} & a_{22} + b_{22} & a_{23} + b_{23} \end{bmatrix}$
Matched Problems

(B) $\begin{bmatrix} 3 & -1 \\ 2 & 1 \\ 3 & 1 \end{bmatrix}$ **(C)** $[\frac{1}{2} \quad \frac{1}{2} \quad 6]$

2. $\begin{bmatrix} 2 & 0 \\ -2 & 3 \\ 2 & -12 \end{bmatrix}$ **3.** $\begin{bmatrix} 24 & -13 \\ 52 & 10 \\ -45 & 97 \end{bmatrix}$ **4.** $\begin{bmatrix} 1 & -1 \\ \frac{2}{3} & 2 \end{bmatrix}$

5. (A) $\begin{bmatrix} \$84{,}000 & \$96{,}000 \\ \$60{,}000 & \$96{,}000 \end{bmatrix}$ **(B)** $\begin{bmatrix} \$36{,}000 & \$48{,}000 \\ \$36{,}000 & \$48{,}000 \end{bmatrix}$

(C) $\begin{bmatrix} \$3{,}000 & \$3{,}600 \\ \$2{,}400 & \$3{,}600 \end{bmatrix}$ **(D)** $\begin{bmatrix} \$6{,}600 \\ \$6{,}000 \end{bmatrix}$

▮▮ Exercise 2-1

A   *Problems 1–20 refer to the following matrices:*

$$A = \begin{bmatrix} 2 & -1 \\ 3 & 0 \end{bmatrix} \qquad B = \begin{bmatrix} -3 & 1 \\ 2 & -3 \end{bmatrix} \qquad C = \begin{bmatrix} 2 \\ -3 \\ 0 \end{bmatrix}$$

$$D = \begin{bmatrix} 1 \\ 3 \\ 5 \end{bmatrix} \qquad E = [-4 \quad 1 \quad 0 \quad -2] \qquad F = \begin{bmatrix} 2 & -3 \\ -2 & 0 \\ 1 & 2 \\ 3 & 5 \end{bmatrix}$$

1. What are the dimensions of B? Of E?

2. What are the dimensions of F? Of D?

3. What element is in the third row and second column of matrix F?

4. What element is in the second row and first column of matrix F?

5. Write a zero matrix of the same dimension as B.

6. Write a zero matrix of the same dimension as E.

7. Identify all column matrices.

8. Identify all row matrices.

9. Identify all square matrices.

10. How many additional columns would F have to have to be a square matrix?

11. Find $A + B$.   **12.** Find $C + D$.

13. Write the negative of matrix C.   **14.** Write the negative of matrix B.

15. Find $D - C$.   16. Find $A - A$.
17. Find $5B$.   18. Find $-2E$.
19. Find $B + C$.   20. Find $D - E$.

B   *Problems 21–32 refer to the following matrices:*

$$A = \begin{bmatrix} 2 & 1 \\ -1 & 3 \end{bmatrix} \qquad B = \begin{bmatrix} -2 & 1 \\ 2 & -4 \end{bmatrix} \qquad C = \begin{bmatrix} 1 & 1 \\ 0 & 2 \end{bmatrix}$$

21. Compute $A + B$ and $B + A$.
22. Compute $A + C$ and $C + A$.
23. Compute $A + (B + C)$ and $(A + B) + C$.
24. Compute $B + (A + C)$ and $(B + A) + C$.
25. Compute $2A + 4B$.
26. Compute $-A + 3C$.
27. Compute $2(3A)$ and $6A$.
28. Compute $0.5(5B)$ and $2.5B$.
29. Compute $2(A + B)$ and $2A + 2B$.
30. Compute $3(B + C)$ and $3B + 3C$.
31. Compute $2(A - 3B) - 3(2B - C)$.
32. Compute $2(A + 2B) - 4(B - C)$.

In Problems 33–36, solve each equation for the matrix X.

33. $3X = \begin{bmatrix} 2 & 3 & 9 \\ -6 & 4 & -12 \end{bmatrix}$ 34. $2X + \begin{bmatrix} 1 & 2 \\ 1 & -1 \end{bmatrix} = X + \begin{bmatrix} 3 & 5 \\ 2 & 6 \end{bmatrix}$

35. $4\left(X + \begin{bmatrix} 1 & 1 \\ 0 & 1 \\ 1 & 0 \end{bmatrix} \right) = 2X + \begin{bmatrix} 2 & 4 \\ 6 & 8 \\ 10 & 12 \end{bmatrix}$

36. $5X - \begin{bmatrix} 1 & 0 & 1 \\ 0 & 1 & 0 \end{bmatrix} = 3\left(X + 2\begin{bmatrix} 0 & 1 & 1 \\ 1 & 1 & 1 \end{bmatrix} \right)$

C   *Problems 37–44 refer to statements from Theorems 1 and 2 in this section. Use the definitions of matrix operations and properties of real numbers to prove each statement.*

37. Theorem 1(A). $A + B = B + A$
38. Theorem 1(B). $A + (B + C) = (A + B) + C$
39. Theorem 1(C). $A + 0 = A$
40. Theorem 1(D). $A + (-A) = 0$
41. Theorem 2(A). $k(\ell A) = (k\ell)A$
42. Theorem 2(B). $(k + \ell)A = kA + \ell A$
43. Theorem 2(C). $k(A + B) = kA + kB$
44. Theorem 2(D). If $A = B$, then $kA = kB$.
45. A square matrix A is called an **upper triangular matrix** if all the elements below the diagonal are 0; that is, if

$$a_{ij} = 0 \quad \text{when} \quad i > j$$

If A and B are upper triangular matrices and k is a scalar, show that $A + B$ and kA are upper triangular matrices.

46. A square matrix A is called a **diagonal matrix** if all the elements off the diagonal are 0; that is, if

$$a_{ij} = 0 \quad \text{when} \quad i \neq j$$

If A and B are diagonal matrices and k is a scalar, show that $A + B$ and kA are diagonal matrices.

▌▫▌ Applications

47. *Cost analysis.* A company with two different plants manufactures guitars and banjos. Its production costs for each instrument are given in the following matrices:

	Plant X		Plant Y	
	Guitar	Banjo	Guitar	Banjo
Materials	$30	$25	$36	$27
Labor	$60	$80	$54	$74

$$\begin{bmatrix} \$30 & \$25 \\ \$60 & \$80 \end{bmatrix} = A \qquad \begin{bmatrix} \$36 & \$27 \\ \$54 & \$74 \end{bmatrix} = B$$

Find $\frac{1}{2}(A + B)$, the average cost of production for the two plants.

48. *Heredity.* Gregor Mendel (1822–1884), a Bavarian monk and botanist, made discoveries that revolutionized the science of genetics. In one experiment he crossed dihybrid yellow round peas (yellow and round are dominant characteristics; the peas also contained green and wrinkled as recessive genes) and obtained 560 peas of the types indicated in the matrix:

	Round	Wrinkled
Yellow	319	101
Green	108	32

$$\begin{bmatrix} 319 & 101 \\ 108 & 32 \end{bmatrix} = M$$

Suppose he carried out a second experiment of the same type and obtained 640 peas of the types indicated in this matrix:

	Round	Wrinkled
Yellow	370	124
Green	110	36

$$\begin{bmatrix} 370 & 124 \\ 110 & 36 \end{bmatrix} = N$$

If the results of the two experiments are combined, write the resulting matrix $M + N$. Compute the percentage of the total number of peas in each category of the combined results.

|2-2| Matrix Multiplication

- Dot Product
- Matrix Multiplication
- Properties of Matrix Multiplication
- Application

In this section we are going to introduce two types of matrix multiplication that will at first seem rather strange. In spite of this apparent strangeness, these operations are well founded on the general theory of matrices, and, as we will see, are extremely useful in practical problems.

■ Dot Product

We start by defining the **dot product** of two special matrices, a $1 \times n$ row matrix and an $n \times 1$ column matrix:

$$\underset{1 \times n}{[a_1 \quad a_2 \quad \cdots \quad a_n]} \cdot \overset{n \times 1}{\begin{bmatrix} b_1 \\ b_2 \\ \cdot \\ \cdot \\ \cdot \\ b_n \end{bmatrix}} = a_1 b_1 + a_2 b_2 + \cdots + a_n b_n \quad \text{A real number}$$

The dot between the two matrices is important. If the dot is omitted, the multiplication is of another type, which we will consider later.

Example 6 $[2 \quad -3 \quad 0] \cdot \begin{bmatrix} -5 \\ 2 \\ -2 \end{bmatrix} = (2)(-5) + (-3)(2) + (0)(-2)$

$$= -10 - 6 + 0 = -16$$

Problem 6 $[-1 \quad 0 \quad 3 \quad 2] \cdot \begin{bmatrix} 2 \\ 3 \\ 4 \\ -1 \end{bmatrix} = ?$ ∎

Example 7
Labor Costs   A factory produces a slalom water ski that requires 4 work-hours in the fabricating department and 1 work-hour in the finishing department. Fabricating personnel receive $8 per hour and finishing personnel receive $6 per hour. Total

labor cost per ski is given by the dot product:

$$[4 \quad 1] \cdot \begin{bmatrix} 8 \\ 6 \end{bmatrix} = (4)(8) + (1)(6) = 32 + 6 = \$38 \text{ per ski}$$

Problem 7   If the factory in Example 7 also produces a trick water ski that requires 6 work-hours in the fabricating department and 1.5 work-hours in the finishing department, write a dot product between appropriate row and column matrices that will give the total labor cost for this ski. Compute the cost.   ▮

▪ Matrix Multiplication

It is important to remember that the dot product of a row matrix and a column matrix (in that order) is a real number and not a matrix. We now define a *matrix product* for certain matrices. First, the product of two matrices A and B is defined only when the number of columns of A is equal to the number of rows of B. If A is an $m \times p$ matrix and B is a $p \times n$ matrix, then the **matrix product** of A and B (in that order), denoted by AB, is an $m \times n$ matrix whose element in the ith row and jth column is the dot product of the ith row matrix of A and the jth column matrix of B. Thus, if we let $C = AB$, then the element in the ith row and jth column of C is

$$c_{ij} = [a_{i1} \quad a_{i2} \quad \cdots \quad a_{ip}] \cdot \begin{bmatrix} b_{1j} \\ b_{2j} \\ \cdot \\ \cdot \\ \cdot \\ b_{pj} \end{bmatrix}$$

$$= a_{i1}b_{1j} + a_{i2}b_{2j} + \cdots + a_{ip}b_{pj}$$

Notice the pattern in the four subscripts of each term in this sum. The first subscript is the row number of the element being computed, the last subscript is the column number of the element being computed, and the inner two subscripts are always equal and assume all values from 1 to p.

It is important to check dimensions before starting the multiplication process. If matrix A has dimension $a \times b$ and matrix B has dimension $c \times d$, then if $b = c$, the product AB will exist and will have dimension $a \times d$. This is shown schematically in Figure 1. The definition is not as complicated as it might first seem. An example should help to clarify the process. For

$$A = \begin{bmatrix} 2 & 3 & -1 \\ -2 & 1 & 2 \end{bmatrix} \quad \text{and} \quad B = \begin{bmatrix} 1 & 3 \\ 2 & 0 \\ -1 & 2 \end{bmatrix}$$

A is 2×3, B is 3×2, and AB will be 2×2. The four dot products used to produce

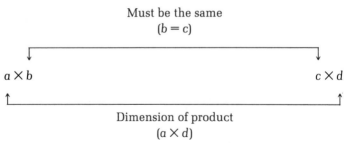

Must be the same
$(b = c)$

$a \times b$ $c \times d$

Dimension of product
$(a \times d)$

Figure 1

the four elements in AB (usually calculated mentally) are shown in the large matrix below:

$$2 \times 3 \quad 3 \times 2$$
$$\begin{bmatrix} 2 & 3 & -1 \\ -2 & 1 & 2 \end{bmatrix} \begin{bmatrix} 1 & 3 \\ 2 & 0 \\ -1 & 2 \end{bmatrix} = \begin{bmatrix} [\,2 \quad 3 \quad -1\,] \cdot \begin{bmatrix} 1 \\ 2 \\ -1 \end{bmatrix} & [\,2 \quad 3 \quad -1\,] \cdot \begin{bmatrix} 3 \\ 0 \\ 2 \end{bmatrix} \\ [-2 \quad 1 \quad 2\,] \cdot \begin{bmatrix} 1 \\ 2 \\ -1 \end{bmatrix} & [-2 \quad 1 \quad 2\,] \cdot \begin{bmatrix} 3 \\ 0 \\ 2 \end{bmatrix} \end{bmatrix} = \begin{bmatrix} 9 & 4 \\ -2 & -2 \end{bmatrix}$$

$$2 \times 2$$

The shaded portions highlight the steps involved in computing the element in the first row and second column of the product matrix.

Example 8   **(A)**
$$3 \times 2 \qquad 2 \times 4 \qquad\qquad 3 \times 4$$
$$\begin{bmatrix} 2 & 1 \\ 1 & 0 \\ -1 & 2 \end{bmatrix} \begin{bmatrix} 1 & -1 & 3 & 0 \\ 2 & 1 & 2 & 2 \end{bmatrix} = \begin{bmatrix} 4 & -1 & 8 & 2 \\ 1 & -1 & 3 & 0 \\ 3 & 3 & 1 & 4 \end{bmatrix}$$

(B)
$$2 \times 4 \qquad\qquad 3 \times 2$$
$$\begin{bmatrix} 1 & -1 & 3 & 0 \\ 2 & 1 & 2 & 2 \end{bmatrix} \begin{bmatrix} 2 & 1 \\ 1 & 0 \\ -1 & 2 \end{bmatrix} \quad \text{Not defined}$$

(C)
$$\begin{bmatrix} 2 & 2 \\ -1 & -1 \end{bmatrix} \begin{bmatrix} 1 & 2 \\ 1 & 2 \end{bmatrix} = \begin{bmatrix} 4 & 8 \\ -2 & -4 \end{bmatrix}$$

(D)
$$\begin{bmatrix} 1 & 2 \\ 1 & 2 \end{bmatrix} \begin{bmatrix} 2 & 2 \\ -1 & -1 \end{bmatrix} = \begin{bmatrix} 0 & 0 \\ 0 & 0 \end{bmatrix}$$

(E)
$$[\,2 \quad -3 \quad 0\,] \begin{bmatrix} -5 \\ 2 \\ -2 \end{bmatrix} = [-16]$$

(F)
$$\begin{bmatrix} -5 \\ 2 \\ -2 \end{bmatrix} [\,2 \quad -3 \quad 0\,] = \begin{bmatrix} -10 & 15 & 0 \\ 4 & -6 & 0 \\ -4 & 6 & 0 \end{bmatrix}$$

Notice that the matrix product of a $1 \times n$ row matrix and an $n \times 1$ column matrix is a 1×1 matrix [see Example 8(E)], whereas their dot product is a real number (see Example 6). This is a technical distinction and it is common to see 1×1 matrices written as real numbers.

Problem 8 Find the product:

(A) $\begin{bmatrix} 1 & 2 & 2 & 0 \\ -1 & 0 & 3 & -2 \end{bmatrix} \begin{bmatrix} 1 & -1 \\ 3 & 2 \\ 0 & 1 \end{bmatrix}$ (B) $\begin{bmatrix} 1 & -1 \\ 3 & 2 \\ 0 & 1 \end{bmatrix} \begin{bmatrix} 1 & 2 & 2 & 0 \\ -1 & 0 & 3 & -2 \end{bmatrix}$

(C) $\begin{bmatrix} 2 & 3 \\ 4 & 6 \end{bmatrix} \begin{bmatrix} 3 & 6 \\ -2 & -4 \end{bmatrix}$ (D) $\begin{bmatrix} 3 & 6 \\ -2 & -4 \end{bmatrix} \begin{bmatrix} 2 & 3 \\ 4 & 6 \end{bmatrix}$

(E) $\begin{bmatrix} -1 & 0 & 3 & 2 \end{bmatrix} \begin{bmatrix} 2 \\ 3 \\ 4 \\ -1 \end{bmatrix}$ (F) $\begin{bmatrix} 2 \\ 3 \\ 4 \\ -1 \end{bmatrix} \begin{bmatrix} -1 & 0 & 3 & 2 \end{bmatrix}$ ▮▮

▪ Properties of Matrix Multiplication

In the arithmetic of real numbers it does not matter in which order we multiply; for example, $5 \times 7 = 7 \times 5$. In matrix multiplication it does make a difference; that is, AB does not always equal BA, even if both multiplications are defined and result in products of the same size [see Examples 8(C) and (D)]. Thus, matrix multiplication is not commutative. Also, AB may be zero with neither A nor B zero [see Example 8(D)].

Matrix multiplication does have general properties, some of which are listed in Theorem 3. The proof of Theorem 3 is left as an exercise. Notice that since matrix multiplication is not commutative, there are two distributive laws and two multiplication properties.

Theorem 3

Assuming that all sums and products are defined, then:

(A) $A(BC) = (AB)C$ Associative property
(B) $A(B + C) = AB + AC$ Left distributive law
(C) $(B + C)A = BA + CA$ Right distributive law
(D) If $A = B$, then $CA = CB$.   Left multiplication property
(E) If $A = B$, then $AC = BC$.   Right multiplication property
(F) $k(AB) = (kA)B = A(kB)$

▮◻▮ **Example 9** Let us combine the time requirements discussed in Example 7 and Problem 7
Labor Costs into one matrix:

	Fabricating department	Finishing department
Trick ski	6 hr	1.5 hr
Slalom ski	4 hr	1 hr

$$\begin{bmatrix} 6 \text{ hr} & 1.5 \text{ hr} \\ 4 \text{ hr} & 1 \text{ hr} \end{bmatrix} = A$$

Now suppose the company has two manufacturing plants X and Y in different parts of the country and that their hourly rates for each department are given in the following matrix:

	Plant X	Plant Y
Fabricating department	$8	$7
Finishing department	$6	$4

$$\begin{bmatrix} \$8 & \$7 \\ \$6 & \$4 \end{bmatrix} = B$$

To find the total labor costs for each ski at each factory, we multiply A and B:

$$\begin{array}{cc} 2 \times 2 & 2 \times 2 \\ \end{array}$$

$$AB = \begin{bmatrix} 6 & 1.5 \\ 4 & 1 \end{bmatrix}\begin{bmatrix} 8 & 7 \\ 6 & 4 \end{bmatrix} = \begin{bmatrix} \$57 & \$48 \\ \$38 & \$32 \end{bmatrix} \quad \begin{array}{l} \text{Trick ski} \\ \text{Slalom ski} \end{array}$$

Notice that the dot product of the first row matrix of A and the first column matrix of B gives us the labor costs, $57, for a trick ski manufactured at plant X; the dot product of the second row matrix of A and the second column matrix of B gives us the labor costs, $32, for manufacturing a slalom ski at plant Y; and so on.

Example 9 is, of course, overly simplified. Companies manufacturing many different items in many different plants deal with matrices that have very large numbers of rows and columns.

Problem 9   Repeat Example 9 with

$$A = \begin{bmatrix} 7 \text{ hr} & 2 \text{ hr} \\ 5 \text{ hr} & 1.5 \text{ hr} \end{bmatrix} \quad \text{and} \quad B = \begin{bmatrix} \$10 & \$8 \\ \$6 & \$4 \end{bmatrix}$$ ▌▌

▪ Application

Example 10
Incidence Matrices

An airline serves five cities, but does not offer direct service between all five of the cities. Table 1 lists the cities and the direct flights offered by the airline. By combining direct flights, is it possible to fly from a given city to any other city? If so, what is the maximum number of flights required?

Table 1

Direct Flights			
Origin	Destination	Origin	Destination
Albany	Baltimore	Denver	Cleveland
Baltimore	Cleveland	Denver	Erie
Baltimore	Denver	Erie	Baltimore
Cleveland	Albany	Erie	Denver
Cleveland	Erie		

Solution   To begin, we draw a diagram illustrating the information listed in the table, as shown in Figure 2.

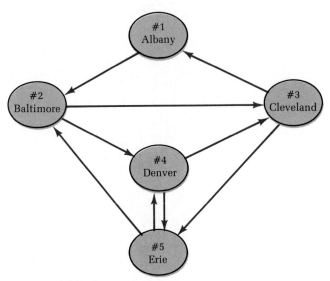

Figure 2   Flight diagram

In a problem of this size we could answer the questions by tracing routes on the diagram, but what would we do if there were 50 cities and 200 flights? Problems of this size require the use of a computer, and computers require numeric data rather than diagrams. The method we will use is easily adapted to large problems and computer solutions. First, using the numbers assigned to each city in the diagram, we will represent the flight information in a matrix $A = [a_{ij}]_{5 \times 5}$ defined by

$$a_{ij} = \begin{cases} 1 & \text{if there is a direct flight from city } \#i \text{ to city } \#j, \ i \neq j \\ 0 & \text{otherwise} \end{cases}$$

Thus,

$$
\begin{array}{c}
\text{To city \#} \\
\begin{array}{cc}
 & \begin{array}{ccccc} 1 & 2 & 3 & 4 & 5 \end{array} \\
\begin{array}{c} \text{From} \\ \text{city \#} \end{array}
\begin{array}{c} 1 \\ 2 \\ 3 \\ 4 \\ 5 \end{array}
&
\left[\begin{array}{ccccc}
0 & 1 & 0 & 0 & 0 \\
0 & 0 & 1 & 1 & 0 \\
1 & 0 & 0 & 0 & 1 \\
0 & 0 & 1 & 0 & 1 \\
0 & 1 & 0 & 1 & 0
\end{array}\right] = A
\end{array}
\end{array}
$$

Notice that all the diagonal elements in A are 0. This indicates that it is not possible to take a direct flight from city $\#i$ to city $\#i$.

Now we compute the matrix $A^2 = AA$:

$$A^2 = \begin{bmatrix} 0 & 1 & 0 & 0 & 0 \\ 0 & 0 & 1 & 1 & 0 \\ 1 & 0 & 0 & 0 & 1 \\ 0 & 0 & 1 & 0 & 1 \\ 0 & 1 & 0 & 1 & 0 \end{bmatrix} \begin{bmatrix} 0 & 1 & 0 & 0 & 0 \\ 0 & 0 & 1 & 1 & 0 \\ 1 & 0 & 0 & 0 & 1 \\ 0 & 0 & 1 & 0 & 1 \\ 0 & 1 & 0 & 1 & 0 \end{bmatrix} = \begin{bmatrix} 0 & 0 & 1 & 1 & 0 \\ 1 & 0 & 1 & 0 & 2 \\ 0 & 2 & 0 & 1 & 0 \\ 1 & 1 & 0 & 1 & 1 \\ 0 & 0 & 2 & 1 & 1 \end{bmatrix}$$

How can the elements in A^2 help us solve this problem? Let us consider how the 1 in position 1,3 was computed. Using the definition of matrix multiplication, we have

$$\begin{bmatrix} a_{11} & a_{12} & a_{13} & a_{14} & a_{15} \end{bmatrix} \cdot \begin{bmatrix} a_{13} \\ a_{23} \\ a_{33} \\ a_{43} \\ a_{53} \end{bmatrix} = a_{11}a_{13} + a_{12}a_{23} + a_{13}a_{33} + a_{14}a_{43} + a_{15}a_{53}$$

$$= (0)(0) + (1)(1) + (0)(0) + (0)(1) + (0)(0)$$
$$= 0 + 1 + 0 + 0 + 0 = 1$$

The result is 1 because a_{12} and a_{23} were both equal to 1. But $a_{12} = 1$ indicates that there is a direct flight from city #1 to city #2 and $a_{23} = 1$ indicates that there is a direct flight from city #2 to city #3. Thus, we can fly from city #1 to city #3 by going through city #2. In general, whenever the i,j entry in A^2 is nonzero, it is possible to go from city #i to city #j in two flights. The first flight will go from city #i to one of the other cities and the second flight will go from the destination of the first flight to city #j.

How can we interpret the 2 in position 3,2 in A^2? This 2 indicates that there are two *different* ways to take a two-flight trip from city #3 to city #2. Examining the flight diagram in Figure 2, we see that one of these two-flight trips would go through city #1 and the other through city #5. (Can you find the two different two-flight trips indicated by the 2 in position 2,5 in A^2?)

What information will $A + A^2$ give us? Let

$$S_2 = A + A^2 = \begin{bmatrix} 0 & 1 & 0 & 0 & 0 \\ 0 & 0 & 1 & 1 & 0 \\ 1 & 0 & 0 & 0 & 1 \\ 0 & 0 & 1 & 0 & 1 \\ 0 & 1 & 0 & 1 & 0 \end{bmatrix} + \begin{bmatrix} 0 & 0 & 1 & 1 & 0 \\ 1 & 0 & 1 & 0 & 2 \\ 0 & 2 & 0 & 1 & 0 \\ 1 & 1 & 0 & 1 & 1 \\ 0 & 0 & 2 & 1 & 1 \end{bmatrix}$$

$$= \begin{bmatrix} 0 & 1 & 1 & 1 & 0 \\ 1 & 0 & 2 & 1 & 2 \\ 1 & 2 & 0 & 1 & 1 \\ 1 & 1 & 1 & 1 & 2 \\ 0 & 1 & 2 & 2 & 1 \end{bmatrix}$$

A nonzero element in position i,j of S_2 indicates that we can go from city #i to city #j in one or two flights. A 0 entry indicates that this is not possible. For example, the 0 in position 1,5 indicates that we cannot fly from city #1 to city #5 in one or two flights. A nonzero diagonal entry indicates that it is possible to return to that city in two flights. Since we are concerned with flights between distinct cities, we will not be interested in the diagonal entries.

Reasoning as before, the nonzero entries in A^3 will tell us which cities can be reached in exactly three flights and the nonzero entries in $S_3 = A + A^2 + A^3$ will tell us which cities can be reached in three or fewer flights.

$$A^3 = AA^2 = \begin{bmatrix} 0 & 1 & 0 & 0 & 0 \\ 0 & 0 & 1 & 1 & 0 \\ 1 & 0 & 0 & 0 & 1 \\ 0 & 0 & 1 & 0 & 1 \\ 0 & 1 & 0 & 1 & 0 \end{bmatrix} \begin{bmatrix} 0 & 0 & 1 & 1 & 0 \\ 1 & 0 & 1 & 0 & 2 \\ 0 & 2 & 0 & 1 & 0 \\ 1 & 1 & 0 & 1 & 1 \\ 0 & 0 & 2 & 1 & 1 \end{bmatrix}$$

$$= \begin{bmatrix} 1 & 0 & 1 & 0 & 2 \\ 1 & 3 & 0 & 2 & 1 \\ 0 & 0 & 3 & 2 & 1 \\ 0 & 2 & 2 & 2 & 1 \\ 2 & 1 & 1 & 1 & 3 \end{bmatrix}$$

$$S_3 = A + A^2 + A^3 = S_2 + A^3 = \begin{bmatrix} 0 & 1 & 1 & 1 & 0 \\ 1 & 0 & 2 & 1 & 2 \\ 1 & 2 & 0 & 1 & 1 \\ 1 & 1 & 1 & 1 & 2 \\ 0 & 1 & 2 & 2 & 1 \end{bmatrix} + \begin{bmatrix} 1 & 0 & 1 & 0 & 2 \\ 1 & 3 & 0 & 2 & 1 \\ 0 & 0 & 3 & 2 & 1 \\ 0 & 2 & 2 & 2 & 1 \\ 2 & 1 & 1 & 1 & 3 \end{bmatrix}$$

$$= \begin{bmatrix} 1 & 1 & 2 & 1 & 2 \\ 2 & 3 & 2 & 3 & 3 \\ 1 & 2 & 3 & 3 & 2 \\ 1 & 3 & 3 & 3 & 3 \\ 2 & 2 & 3 & 3 & 4 \end{bmatrix}$$

Since S_3 has no 0 entries off the diagonal, we can conclude that it is possible to fly from any given city to any other city in three or fewer flights.

Problem 10 Repeat Example 10 if the flight from Erie to Baltimore and the flight from Baltimore to Denver are deleted from the flight schedule. ∎

Diagrams like the one used in Example 10 are called *directed graphs*. Each point in the graph (in Example 10, each city) is called a *vertex* and each arrow (direct flight) is called an *edge*. The matrix A used to represent this graph is called an *incidence matrix*. If the edges are not assigned directions, then the graph is called an *undirected graph*. Both types of graphs have a wide variety of applications. Matrices are very useful tools for analyzing graphs, especially when used in conjunction with a computer.

Answers to Matched Problems

6. 8   **7.** $\begin{bmatrix} 6 & 1.5 \end{bmatrix} \cdot \begin{bmatrix} 8 \\ 6 \end{bmatrix} = \$57$

8. (A) Not defined   (B) $\begin{bmatrix} 2 & 2 & -1 & 2 \\ 1 & 6 & 12 & -4 \\ -1 & 0 & 3 & -2 \end{bmatrix}$ (C) $\begin{bmatrix} 0 & 0 \\ 0 & 0 \end{bmatrix}$

(D) $\begin{bmatrix} 30 & 45 \\ -20 & -30 \end{bmatrix}$ (E) [8]   (F) $\begin{bmatrix} -2 & 0 & 6 & 4 \\ -3 & 0 & 9 & 6 \\ -4 & 0 & 12 & 8 \\ 1 & 0 & -3 & -2 \end{bmatrix}$

9. $\begin{matrix} & X & Y \\ & \begin{bmatrix} \$82 & \$64 \\ \$59 & \$46 \end{bmatrix} & \begin{matrix} \text{Trick} \\ \text{Slalom} \end{matrix} \end{matrix}$

10. All cities can be reached in four or fewer flights.

‖ Exercise 2-2

A   *In Problems 1–6, find the dot products.*

1. $\begin{bmatrix} 2 & 4 \end{bmatrix} \cdot \begin{bmatrix} 3 \\ 1 \end{bmatrix}$ **2.** $\begin{bmatrix} 3 & 2 \end{bmatrix} \cdot \begin{bmatrix} -1 \\ 2 \end{bmatrix}$

3. $\begin{bmatrix} 1 & 0 & 2 \end{bmatrix} \cdot \begin{bmatrix} -1 \\ 4 \\ 2 \end{bmatrix}$ **4.** $\begin{bmatrix} -1 & 2 & 3 \end{bmatrix} \cdot \begin{bmatrix} 2 \\ -2 \\ 2 \end{bmatrix}$

5. $\begin{bmatrix} 1 & -2 & 1 & 3 \end{bmatrix} \cdot \begin{bmatrix} 2 \\ 1 \\ -1 \\ 4 \end{bmatrix}$ **6.** $\begin{bmatrix} 1 & 2 & 3 & 4 \end{bmatrix} \cdot \begin{bmatrix} 4 \\ 3 \\ 2 \\ 1 \end{bmatrix}$

In Problems 7–16, find the matrix products.

7. $\begin{bmatrix} 2 & 1 \end{bmatrix} \begin{bmatrix} -1 & 2 \\ 3 & 1 \end{bmatrix}$ **8.** $\begin{bmatrix} -1 & 1 \end{bmatrix} \begin{bmatrix} 1 & 2 \\ 2 & 3 \end{bmatrix}$

9. $\begin{bmatrix} -1 & 2 \\ 3 & 1 \end{bmatrix} \begin{bmatrix} 2 \\ 1 \end{bmatrix}$ **10.** $\begin{bmatrix} 1 & 2 \\ 2 & 3 \end{bmatrix} \begin{bmatrix} -1 \\ 1 \end{bmatrix}$

11. $\begin{bmatrix} 2 & 1 \\ 1 & 3 \end{bmatrix} \begin{bmatrix} 1 & 3 \\ 4 & 5 \end{bmatrix}$ **12.** $\begin{bmatrix} 1 & -2 \\ -2 & 4 \end{bmatrix} \begin{bmatrix} 2 & 4 \\ 1 & 2 \end{bmatrix}$

13. $\begin{bmatrix} 1 & 3 \\ 4 & 5 \end{bmatrix} \begin{bmatrix} 2 & 1 \\ 1 & 3 \end{bmatrix}$ **14.** $\begin{bmatrix} 2 & 4 \\ 1 & 2 \end{bmatrix} \begin{bmatrix} 1 & -2 \\ -2 & 4 \end{bmatrix}$

15. $\begin{bmatrix} 3 & 2 \\ 1 & 1 \end{bmatrix} \begin{bmatrix} 2 & 2 \\ 1 & 0 \end{bmatrix}$ **16.** $\begin{bmatrix} 2 & 2 \\ 1 & 0 \end{bmatrix} \begin{bmatrix} 3 & 2 \\ 1 & 1 \end{bmatrix}$

B   In Problems 17–28, find the matrix products, if possible.

17. $\begin{bmatrix} 2 & -1 & 1 \\ 1 & 3 & -2 \end{bmatrix} \begin{bmatrix} 1 & 3 \\ 0 & -1 \\ -2 & 2 \end{bmatrix}$

18. $\begin{bmatrix} -1 & -4 & 3 \\ 2 & 0 & 1 \end{bmatrix} \begin{bmatrix} 2 & -3 \\ 1 & 2 \\ 0 & -1 \end{bmatrix}$

19. $\begin{bmatrix} 1 & 3 \\ 0 & -1 \\ -2 & 2 \end{bmatrix} \begin{bmatrix} 2 & -1 & 1 \\ 1 & 3 & -2 \end{bmatrix}$

20. $\begin{bmatrix} 2 & -3 \\ 1 & 2 \\ 0 & -1 \end{bmatrix} \begin{bmatrix} 1 & -4 & 3 \\ 2 & 0 & 1 \end{bmatrix}$

21. $\begin{bmatrix} 1 & -2 & 1 \\ 2 & 1 & 0 \\ -1 & 4 & -3 \end{bmatrix} \begin{bmatrix} 2 \\ 1 \\ -1 \end{bmatrix}$

22. $\begin{bmatrix} 1 \\ -3 \\ 2 \end{bmatrix} \begin{bmatrix} -1 & 3 & 0 \\ 4 & -2 & 5 \\ -1 & 1 & 2 \end{bmatrix}$

23. $\begin{bmatrix} -2 & 1 & 2 \\ 1 & -2 & 0 \\ 3 & 2 & -4 \\ 2 & 0 & 1 \end{bmatrix} \begin{bmatrix} 2 & 1 & 3 & 2 \\ 0 & -1 & 2 & -3 \end{bmatrix}$

24. $\begin{bmatrix} -1 & 4 & 2 & -1 \\ 1 & 0 & -1 & 0 \end{bmatrix} \begin{bmatrix} -1 & 0 & 2 \\ 0 & 2 & 1 \\ 1 & 2 & -1 \\ 3 & -2 & 0 \end{bmatrix}$

25. $[1 \quad 2 \quad 3] \begin{bmatrix} 4 \\ 5 \\ 6 \end{bmatrix}$

26. $[2 \quad -1 \quad -2] \begin{bmatrix} 3 \\ 1 \\ 2 \end{bmatrix}$

27. $\begin{bmatrix} 4 \\ 5 \\ 6 \end{bmatrix} [1 \quad 2 \quad 3]$

28. $\begin{bmatrix} 3 \\ 1 \\ 2 \end{bmatrix} [2 \quad -1 \quad 2]$

In Problems 29–32, verify each statement using the following matrices:

$$A = \begin{bmatrix} 1 & 2 \\ 0 & 1 \end{bmatrix} \qquad B = \begin{bmatrix} 1 & 1 \\ 2 & 3 \end{bmatrix} \qquad C = \begin{bmatrix} -3 & 1 \\ -1 & 2 \end{bmatrix}$$

29. $A(BC) = (AB)C$

30. $A(B + C) = AB + AC$

31. $(B + C)A = BA + CA$

32. $2(AB) = (2A)B = A(2B)$

C   33. If

$$A = \begin{bmatrix} 1 & 1 \\ 1 & 1 \end{bmatrix}$$

find all 2×2 matrices B that satisfy $AB = BA$.

34. If

$$A = \begin{bmatrix} 1 & 1 \\ 1 & 1 \end{bmatrix}$$

find all 2×2 matrices B that satisfy $AB = 0$.

35. Show that

$$X = \begin{bmatrix} 1 & 1 \\ 1 & 2 \end{bmatrix}$$

satisfies the equation

$$X^2 - 2X - \begin{bmatrix} 0 & 1 \\ 1 & 1 \end{bmatrix} = 0$$

36. Show that

$$X = \begin{bmatrix} 1 & -1 \\ -1 & 0 \end{bmatrix}$$

also satisfies the equation in Problem 35.

37. If A and B are $n \times n$ matrices, show that

$$(A + B)^2 = A^2 + 2AB + B^2$$

if and only if $AB = BA$.

38. Show that the matrix

$$A = \begin{bmatrix} ab & b^2 \\ -a^2 & -ab \end{bmatrix}$$

satisfies the equation $A^2 = 0$ for any scalars a and b.

39. Show that

$$A = \begin{bmatrix} a & a \\ 1-a & 1-a \end{bmatrix}$$

satisfies the equation $A^2 = A$ for any scalar a.

40. Show that

$$A = \begin{bmatrix} -a & a \\ 1-a & a-1 \end{bmatrix}$$

satisfies the equation $A^2 = -A$ for any scalar a.

Problems 41–46 are statements from Theorem 3 discussed in this section. Use the definitions of the matrix operations and properties of real numbers to prove each statement for the following matrices:

$$A = \begin{bmatrix} a_{11} & a_{12} \\ a_{21} & a_{22} \end{bmatrix} \qquad B = \begin{bmatrix} b_{11} & b_{12} \\ b_{21} & b_{22} \end{bmatrix} \qquad C = \begin{bmatrix} c_{11} & c_{12} \\ c_{21} & c_{22} \end{bmatrix}$$

Do not refer to Theorem 3 in your proof.

41. Theorem 3(A). $A(BC) = (AB)C$

42. Theorem 3(B). $A(B + C) = AB + AC$

43. Theorem 3(C). $(B + C)A = BA + CA$

44. *Theorem 3(D).* If $A = B$, then $CA = CB$.
45. *Theorem 3(E).* If $A = B$, then $AC = BC$.
46. *Theorem 3(F).* $k(AB) = (kA)B = A(kB)$

▮◻▮ Applications

47. *Labor costs.* A company with manufacturing plants located in different parts of the country has labor and wage requirements for the manufacturing of three types of inflatable boats as given in the following two matrices:

Work-hours per boat

	Cutting department	Assembly department	Packaging department	
$M = $	0.6 hr	0.6 hr	0.2 hr	One-person boat
	1.0 hr	0.9 hr	0.3 hr	Two-person boat
	1.5 hr	1.2 hr	0.4 hr	Four-person boat

Hourly wages

	Plant I	Plant II	
$N = $	\$6	\$7	Cutting department
	\$8	\$10	Assembly department
	\$3	\$4	Packaging department

(A) Use a dot product to find the labor costs for a one-person boat manufactured at plant I.
(B) Use a dot product to find the labor costs for a four-person boat manufactured at plant II.
(C) Find MN and interpret.

48. *Inventory value.* A retail company sells five different personal computer models through three stores located in a large metropolitan area. The inventory of each model on hand in each store is summarized in matrix M. Wholesale (W) and retail (R) values of each model computer are summarized in matrix N.

Model

	A	B	C	D	E	
$M = $	4	2	3	7	1	Store 1
	2	3	5	0	6	Store 2
	10	4	3	4	3	Store 3

	W	R	
$N = $	\$700	\$840	A
	\$1,400	\$1,800	B
	\$1,800	\$2,400	C
	\$2,700	\$3,300	D
	\$3,500	\$4,900	E

(A) What is the retail value of the inventory at store 2?

(B) What is the wholesale value of the inventory at store 3?

(C) Compute MN and interpret.

49. (A) Multiply M in Problem 48 by [1 1 1] and interpret. (The multiplication only makes sense in one direction.)

(B) Multiply MN in Problem 48 by [1 1 1] and interpret. (The multiplication only makes sense in one direction.)

50. *Nutrition.* A nutritionist for a cereal company blends two cereals in three different mixes. The amounts of protein, carbohydrate, and fat (in grams per ounce) in each cereal are given by matrix M. The amounts of each cereal used in the three mixes is given by matrix N.

$$M = \begin{matrix} & \text{Cereal } A & \text{Cereal } B & \\ & \begin{bmatrix} 4\text{ g} & 2\text{ g} \\ 20\text{ g} & 16\text{ g} \\ 3\text{ g} & 1\text{ g} \end{bmatrix} & & \begin{matrix} \text{Protein} \\ \text{Carbohydrate} \\ \text{Fat} \end{matrix} \end{matrix}$$

$$N = \begin{matrix} \text{Mix } X & \text{Mix } Y & \text{Mix } Z & \\ \begin{bmatrix} 15\text{ oz} & 10\text{ oz} & 5\text{ oz} \\ 5\text{ oz} & 10\text{ oz} & 15\text{ oz} \end{bmatrix} & & & \begin{matrix} \text{Cereal } A \\ \text{Cereal } B \end{matrix} \end{matrix}$$

(A) Use a dot product to find the amount of protein in mix X.

(B) Use a dot product to find the amount of fat in mix Z.

(C) Find MN and interpret.

(D) Find $\frac{1}{20}MN$ and interpret.

51. *Airline scheduling.* The table lists the direct flights offered by an airline serving the four cities.

Direct Flights	
Origin	Destination
Atlanta	Birmingham
Atlanta	Chattanooga
Birmingham	Atlanta
Birmingham	Durham
Chattanooga	Atlanta
Chattanooga	Durham
Durham	Atlanta
Durham	Birmingham

(A) Represent the flight information in a diagram.

(B) Find an incidence matrix A that also represents the flight information.

(C) Compute $S_2 = A + A^2$, $S_3 = A + A^2 + A^3$, . . . , until you obtain a matrix with no 0 entries off the diagonal. Use this matrix to determine the maximum number of flights required to fly between any two of these cities.

52. *Airline scheduling.* Repeat Problem 51 if the flight from Chattanooga to Durham and the flight from Durham to Atlanta are deleted from the flight schedule.

53. *Tournament ranking.* In order to rank players for an upcoming tennis tournament, a club decides to have each player play one set with every other player. The results are given in the table.

Player	Defeated
Ann	Barbara
Barbara	Carol
Carol	Ann, Diane
Diane	Ann, Barbara

(A) Express these results in an incidence matrix A where $a_{ij} = 1$ if player i defeated player j.

(B) Compute $A + A^2$, and use the sum of the entries in each row of $A + A^2$ to rank the players for the tournament.

54. *Tournament ranking.* Repeat Problem 53 for the results in the table below:

Player	Defeated
Ann	Barbara
Barbara	Carol, Diane
Carol	Ann
Diane	Ann, Carol

❚ 2-3 ❚ Inverse of a Square Matrix

- Identity Matrix for Multiplication
- Inverse of a Square Matrix
- Properties of Matrix Inverses
- Matrix Equations
- Application

▪ Identity Matrix for Multiplication

We know that

$$1a = a1 = a$$

for all real numbers a. The number 1 is called the **identity** for real number multiplication. Does the set of all matrices of a given dimension have an identity element for multiplication? The answer, in general, is no. However, the set of all **square matrices of order n** (dimension $n \times n$) does have an identity, and it is given as follows: The **identity element for multiplication** for the set of all square matrices of order n is the square matrix of order n, denoted by I, with 1's along the diagonal (from upper left to lower right) and 0's elsewhere. For example,

$$\begin{bmatrix} 1 & 0 \\ 0 & 1 \end{bmatrix} \quad \text{and} \quad \begin{bmatrix} 1 & 0 & 0 \\ 0 & 1 & 0 \\ 0 & 0 & 1 \end{bmatrix}$$

are the identity matrices for square matrices of order 2 and 3, respectively. In some cases, it will be convenient to use I_n to denote the identity matrix of order n. Thus,

$$I_2 = \begin{bmatrix} 1 & 0 \\ 0 & 1 \end{bmatrix} \qquad I_3 = \begin{bmatrix} 1 & 0 & 0 \\ 0 & 1 & 0 \\ 0 & 0 & 1 \end{bmatrix}$$

and so on.

Example 11

$$\begin{bmatrix} 1 & 0 & 0 \\ 0 & 1 & 0 \\ 0 & 0 & 1 \end{bmatrix} \begin{bmatrix} a_{11} & a_{12} & a_{13} \\ a_{21} & a_{22} & a_{23} \\ a_{31} & a_{32} & a_{33} \end{bmatrix} = \begin{bmatrix} a_{11} & a_{12} & a_{13} \\ a_{21} & a_{22} & a_{23} \\ a_{31} & a_{32} & a_{33} \end{bmatrix}$$

$$= \begin{bmatrix} a_{11} & a_{12} & a_{13} \\ a_{21} & a_{22} & a_{23} \\ a_{31} & a_{32} & a_{33} \end{bmatrix} \begin{bmatrix} 1 & 0 & 0 \\ 0 & 1 & 0 \\ 0 & 0 & 1 \end{bmatrix}$$

Problem 11   Multiply:

$$\begin{bmatrix} 1 & 0 \\ 0 & 1 \end{bmatrix} \begin{bmatrix} 2 & -3 \\ 5 & 7 \end{bmatrix} \quad \text{and} \quad \begin{bmatrix} 2 & -3 \\ 5 & 7 \end{bmatrix} \begin{bmatrix} 1 & 0 \\ 0 & 1 \end{bmatrix} \qquad \blacksquare$$

In general, we can show that if M is a square matrix of order n and I is the identity matrix of order n, then

$$IM = MI = M$$

▪ Inverse of a Square Matrix

In the set of real numbers we know that for each real number a (except 0) there exists a real number a^{-1} such that

$$a^{-1}a = 1$$

The number a^{-1} is called the **inverse** of the number a relative to multiplication, or the **multiplicative inverse** of a. For example, 2^{-1} is the multiplicative inverse of 2, since $2^{-1} \cdot 2 = 1$. For each square matrix M, does there exist an inverse matrix M^{-1} such that the following relation is true?

$$M^{-1}M = MM^{-1} = I$$

If M^{-1} does exist for a given matrix M, then M^{-1} is called the **inverse of M relative to multiplication.** Let us use this definition to find M^{-1} for

$$M = \begin{bmatrix} 2 & 3 \\ 1 & 2 \end{bmatrix}$$

We are looking for

$$M^{-1} = \begin{bmatrix} a_{11} & a_{12} \\ a_{21} & a_{22} \end{bmatrix}$$

such that

$$MM^{-1} = M^{-1}M = I$$

Thus, we write

$$\begin{array}{ccc} M & M^{-1} & I \end{array}$$
$$\begin{bmatrix} 2 & 3 \\ 1 & 2 \end{bmatrix} \begin{bmatrix} a_{11} & a_{12} \\ a_{21} & a_{22} \end{bmatrix} = \begin{bmatrix} 1 & 0 \\ 0 & 1 \end{bmatrix}$$

and try to find a_{11}, a_{12}, a_{21}, and a_{22} so that the product of M and M^{-1} is the identity matrix I. Multiplying M and M^{-1} on the left side, we obtain

$$\begin{bmatrix} 2a_{11} + 3a_{21} & 2a_{12} + 3a_{22} \\ a_{11} + 2a_{21} & a_{12} + 2a_{22} \end{bmatrix} = \begin{bmatrix} 1 & 0 \\ 0 & 1 \end{bmatrix}$$

which is true only if

$$2a_{11} + 3a_{21} = 1 \qquad 2a_{12} + 3a_{22} = 0$$
$$a_{11} + 2a_{21} = 0 \qquad a_{12} + 2a_{22} = 1$$

Solving these two systems, we find that $a_{11} = 2$, $a_{21} = -1$, $a_{12} = -3$, and $a_{22} = 2$. Thus,

$$M^{-1} = \begin{bmatrix} 2 & -3 \\ -1 & 2 \end{bmatrix}$$

as is easily checked:

$$\begin{array}{ccccc} M & M^{-1} & I & M^{-1} & M \end{array}$$
$$\begin{bmatrix} 2 & 3 \\ 1 & 2 \end{bmatrix} \begin{bmatrix} 2 & -3 \\ -1 & 2 \end{bmatrix} = \begin{bmatrix} 1 & 0 \\ 0 & 1 \end{bmatrix} = \begin{bmatrix} 2 & -3 \\ -1 & 2 \end{bmatrix} \begin{bmatrix} 2 & 3 \\ 1 & 2 \end{bmatrix}$$

Inverses do not always exist for square matrices. For example, if

$$M = \begin{bmatrix} 2 & 1 \\ 4 & 2 \end{bmatrix}$$

then, proceeding as above, we are led to the systems

$$2a_{11} + a_{21} = 1 \qquad 2a_{12} + a_{22} = 0$$
$$4a_{11} + 2a_{21} = 0 \qquad 4a_{12} + a_{22} = 1$$

These are both inconsistent and have no solution. Hence, M^{-1} does not exist.

If a square matrix M has an inverse, then M is called **invertible.** In some texts, matrices that do not have inverses are referred to as *singular matrices*; however, we will simply say that such matrices are not invertible. Referring to the previous two examples,

$$\begin{bmatrix} 2 & 3 \\ 1 & 2 \end{bmatrix} \text{ is invertible} \qquad \text{and} \qquad \begin{bmatrix} 2 & 1 \\ 4 & 2 \end{bmatrix} \text{ is not invertible}$$

Being able to find inverses, when they exist, leads to direct and simple solutions to many practical problems. Later in this section, for example, we will show how inverses can be used to solve systems of linear equations.

The method outlined above for finding M^{-1}, if it exists, gets very involved for matrices of order larger than 2. Now that we know what we are looking for, we can introduce the idea of the augmented matrix to make the process more efficient. For example, to find the inverse (if it exists) of

$$M = \begin{bmatrix} 1 & -1 & 1 \\ 0 & 2 & -1 \\ 2 & 3 & 0 \end{bmatrix}$$

we start as before and write

$$\overset{M}{\begin{bmatrix} 1 & -1 & 1 \\ 0 & 2 & -1 \\ 2 & 3 & 0 \end{bmatrix}} \overset{M^{-1}}{\begin{bmatrix} a_{11} & a_{12} & a_{13} \\ a_{21} & a_{22} & a_{23} \\ a_{31} & a_{32} & a_{33} \end{bmatrix}} = \overset{I}{\begin{bmatrix} 1 & 0 & 0 \\ 0 & 1 & 0 \\ 0 & 0 & 1 \end{bmatrix}}$$

which is true only if

$$a_{11} - a_{21} + a_{31} = 1 \qquad a_{12} - a_{22} + a_{32} = 0 \qquad a_{13} - a_{23} + a_{33} = 0$$
$$2a_{21} - a_{31} = 0 \qquad 2a_{22} - a_{32} = 1 \qquad 2a_{23} - a_{33} = 0$$
$$2a_{11} + 3a_{21} = 0 \qquad 2a_{12} + 3a_{22} = 0 \qquad 2a_{13} + 3a_{23} = 1$$

Now we write augmented matrices for each of the three systems:

$$\left[\begin{array}{ccc|c} 1 & -1 & 1 & 1 \\ 0 & 2 & -1 & 0 \\ 2 & 3 & 0 & 0 \end{array}\right] \quad \left[\begin{array}{ccc|c} 1 & -1 & 1 & 0 \\ 0 & 2 & -1 & 1 \\ 2 & 3 & 0 & 0 \end{array}\right] \quad \left[\begin{array}{ccc|c} 1 & -1 & 1 & 0 \\ 0 & 2 & -1 & 0 \\ 2 & 3 & 0 & 1 \end{array}\right]$$

Since each matrix to the left of the vertical bar is the same, exactly the same row operations can be used on each total matrix to transform it into a reduced form. We can speed up the process substantially by combining all three augmented matrices into the single augmented matrix form

$$\left[\begin{array}{ccc|ccc} 1 & -1 & 1 & 1 & 0 & 0 \\ 0 & 2 & -1 & 0 & 1 & 0 \\ 2 & 3 & 0 & 0 & 0 & 1 \end{array}\right] = [M|I] \tag{1}$$

We now try to perform row operations on matrix (1) until we obtain a row equivalent matrix that looks like matrix (2):

$$\left[\begin{array}{ccc|ccc} \multicolumn{3}{c}{I} & \multicolumn{3}{c}{B} \\ 1 & 0 & 0 & b_{11} & b_{12} & b_{13} \\ 0 & 1 & 0 & b_{21} & b_{22} & b_{23} \\ 0 & 0 & 1 & b_{31} & b_{32} & b_{33} \end{array}\right] \tag{2}$$

If this can be done, then the new matrix to the right of the vertical bar will be M^{-1}! Now let us try to transform (1) into (2):

$$
\begin{array}{ccc}
M & & I
\end{array}
$$

$$
\left[\begin{array}{ccc|ccc}
1 & -1 & 1 & 1 & 0 & 0 \\
0 & 2 & -1 & 0 & 1 & 0 \\
2 & 3 & 0 & 0 & 0 & 1
\end{array}\right] \quad R_3 + (-2)R_1 \to R_3
$$

$$
\sim \left[\begin{array}{ccc|ccc}
1 & -1 & 1 & 1 & 0 & 0 \\
0 & 2 & -1 & 0 & 1 & 0 \\
0 & 5 & -2 & -2 & 0 & 1
\end{array}\right] \quad \tfrac{1}{2}R_2 \to R_2
$$

$$
\sim \left[\begin{array}{ccc|ccc}
1 & -1 & 1 & 1 & 0 & 0 \\
0 & 1 & -\tfrac{1}{2} & 0 & \tfrac{1}{2} & 0 \\
0 & 5 & -2 & -2 & 0 & 1
\end{array}\right] \quad R_3 + (-5)R_2 \to R_3
$$

$$
\sim \left[\begin{array}{ccc|ccc}
1 & -1 & 1 & 1 & 0 & 0 \\
0 & 1 & -\tfrac{1}{2} & 0 & \tfrac{1}{2} & 0 \\
0 & 0 & \tfrac{1}{2} & -2 & -\tfrac{5}{2} & 1
\end{array}\right] \quad 2R_3 \to R_3
$$

$$
\sim \left[\begin{array}{ccc|ccc}
1 & -1 & 1 & 1 & 0 & 0 \\
0 & 1 & -\tfrac{1}{2} & 0 & \tfrac{1}{2} & 0 \\
0 & 0 & 1 & -4 & -5 & 2
\end{array}\right] \quad \begin{array}{l} R_1 + (-1)R_3 \to R_1 \\ R_2 + \tfrac{1}{2}R_3 \to R_2 \end{array}
$$

$$
\sim \left[\begin{array}{ccc|ccc}
1 & -1 & 0 & 5 & 5 & -2 \\
0 & 1 & 0 & -2 & -2 & 1 \\
0 & 0 & 1 & -4 & -5 & 2
\end{array}\right] \quad R_1 + R_2 \to R_1
$$

$$
\begin{array}{ccc}
I & & B
\end{array}
$$

$$
\sim \left[\begin{array}{ccc|ccc}
1 & 0 & 0 & 3 & 3 & -1 \\
0 & 1 & 0 & -2 & -2 & 1 \\
0 & 0 & 1 & -4 & -5 & 2
\end{array}\right]
$$

Converting back to systems of equations equivalent to our three original systems, we have

$$
\begin{array}{lll}
a_{11} = 3 & a_{12} = 3 & a_{13} = -1 \\
a_{21} = -2 & a_{22} = -2 & a_{23} = 1 \\
a_{31} = -4 & a_{32} = -5 & a_{33} = 2
\end{array}
$$

And these are just the elements of M^{-1} that we are looking for! Hence,

$$
M^{-1} = \left[\begin{array}{ccc}
3 & 3 & -1 \\
-2 & -2 & 1 \\
-4 & -5 & 2
\end{array}\right]
$$

Note that this is the matrix to the right of the vertical line in the last augmented matrix above. (You should check that $MM^{-1} = I$.*)

* According to the definition, to check that M^{-1} is correct, you should verify that $MM^{-1} = I$ and $M^{-1}M = I$. However, in Section 2-4 we will see that if one of these equations is valid, then the other is also valid. Thus, for checking purposes, it is sufficient to verify either that $MM^{-1} = I$ or that $M^{-1}M = I$.

The procedure just described for obtaining an inverse generalizes completely in the form of Theorem 4. (We delay proving Theorem 4 until the last part of Section 2-4.)

Theorem 4

Inverse of a Square Matrix M

If $[M|I]$ is transformed by row operations into $[I|B]$, then the resulting matrix B is M^{-1}. If, however, we obtain all 0's in one or more rows to the left of the vertical line, then M^{-1} does not exist.

Example 12 Find M^{-1}, given $M = \begin{bmatrix} 3 & -1 \\ -4 & 2 \end{bmatrix}$

Solution Can you identify the row operations used in the transformations?

$$\begin{bmatrix} 3 & -1 & \bigm| & 1 & 0 \\ -4 & 2 & \bigm| & 0 & 1 \end{bmatrix} \sim \begin{bmatrix} 1 & -\frac{1}{3} & \bigm| & \frac{1}{3} & 0 \\ -4 & 2 & \bigm| & 0 & 1 \end{bmatrix}$$

$$\sim \begin{bmatrix} 1 & -\frac{1}{3} & \bigm| & \frac{1}{3} & 0 \\ 0 & \frac{2}{3} & \bigm| & \frac{4}{3} & 1 \end{bmatrix}$$

$$\sim \begin{bmatrix} 1 & -\frac{1}{3} & \bigm| & \frac{1}{3} & 0 \\ 0 & 1 & \bigm| & 2 & \frac{3}{2} \end{bmatrix}$$

$$\sim \begin{bmatrix} 1 & 0 & \bigm| & 1 & \frac{1}{2} \\ 0 & 1 & \bigm| & 2 & \frac{3}{2} \end{bmatrix}$$

Thus,

$$M^{-1} = \begin{bmatrix} 1 & \frac{1}{2} \\ 2 & \frac{3}{2} \end{bmatrix} \qquad \begin{array}{l} \text{Check by showing that} \\ M^{-1}M = I. \end{array}$$

Problem 12 Find M^{-1}, given $M = \begin{bmatrix} 2 & -6 \\ 1 & -2 \end{bmatrix}$ ∎

Example 13 Find M^{-1}, given $M = \begin{bmatrix} 1 & -2 & 1 & 3 \\ 0 & 1 & -2 & 1 \\ 0 & 0 & 3 & -4 \\ 2 & -1 & -4 & 9 \end{bmatrix}$

Solution

$$\begin{bmatrix} 1 & -2 & 1 & 3 & \bigm| & 1 & 0 & 0 & 0 \\ 0 & 1 & -2 & 1 & \bigm| & 0 & 1 & 0 & 0 \\ 0 & 0 & 3 & -4 & \bigm| & 0 & 0 & 1 & 0 \\ 2 & -1 & -4 & 9 & \bigm| & 0 & 0 & 0 & 1 \end{bmatrix} \sim \begin{bmatrix} 1 & -2 & 1 & 3 & \bigm| & 1 & 0 & 0 & 0 \\ 0 & 1 & -2 & 1 & \bigm| & 0 & 1 & 0 & 0 \\ 0 & 0 & 3 & -4 & \bigm| & 0 & 0 & 1 & 0 \\ 0 & 3 & -6 & 3 & \bigm| & -2 & 0 & 0 & 1 \end{bmatrix}$$

$$\sim \begin{bmatrix} 1 & 2 & -3 & 5 & \bigm| & 1 & 2 & 0 & 0 \\ 0 & 1 & -2 & 1 & \bigm| & 0 & 1 & 0 & 0 \\ 0 & 0 & 3 & -4 & \bigm| & 0 & 0 & 3 & -4 \\ 0 & 0 & 0 & 0 & \bigm| & -2 & -3 & 0 & 1 \end{bmatrix}$$

Since this last matrix has a row of all 0's to the left of the vertical line, we can stop this process and conclude that M is not invertible.

Problem 13   Find M^{-1}, given $M = \begin{bmatrix} 1 & 2 & -4 \\ 2 & 5 & -6 \\ 0 & -2 & -4 \end{bmatrix}$ ∎

▪ Properties of Matrix Inverses

To complete our development of the basic properties of matrix arithmetic, we have listed some properties of matrix inverses in Theorem 5. We will prove parts of this theorem to illustrate some important ideas. [The proofs of parts (B), (D), and (E) are left as exercises.] Once again, notice that the noncommutativity of matrix multiplication requires two separate cancellation laws.

Theorem 5

> **Properties of Matrix Inverses**
>
> (A) If A is invertible, then A^{-1} is unique.
>
> (B) If A is invertible, then A^{-1} is invertible and
>
> $$(A^{-1})^{-1} = A$$
>
> (C) If A and B are invertible matrices of the same order, then AB is invertible and
>
> $$(AB)^{-1} = B^{-1}A^{-1}$$
>
> (D) If A is invertible and C and D are matrices satisfying $AD = AC$, then $D = C$.
>
> (E) If A is invertible and C and D are matrices satisfying $DA = CA$, then $D = C$.

Proof   (A) We readily accept the fact that the multiplicative inverse of a real number is unique; that is, each nonzero real number a has only one inverse, $a^{-1} = 1/a$. Since matrix multiplication is much more complicated than real number multiplication, we cannot just assume that matrix inverses are unique. Suppose that a matrix A has two inverses, B and C. Then by the definition of the inverse,

$$AB = BA = I \quad \text{and} \quad AC = CA = I$$

Now we want to show that $B = C$.

$$B = BI \qquad \text{Identity for multiplication}$$
$$\ = B(AC) \qquad \text{Substitution, using } AC = I$$

$$B = (BA)C \quad \text{Associative property}$$
$$= IC \quad \text{Substitution, using } BA = I$$
$$= C \quad \text{Identity for multiplication}$$

Thus, the inverse of A is unique.

(C) To show that AB is invertible, we must find a matrix C satisfying

$$C(AB) = I \quad \text{and} \quad (AB)C = I \tag{3}$$

Since A and B are invertible, we know that A^{-1} and B^{-1} exist. We will show that each equation in (3) is satisfied if we let $C = B^{-1}A^{-1}$.

$$C(AB) = (B^{-1}A^{-1})AB \quad \text{Substitution}$$
$$= B^{-1}(A^{-1}A)B \quad \text{Associative property}$$
$$= B^{-1}IB \quad A^{-1}A = I \text{ and } BB^{-1} = I$$
$$= B^{-1}B \quad \text{Identity}$$
$$= I \quad B^{-1}B = I \text{ and } AA^{-1} = I$$

$$(AB)C = (AB)(B^{-1}A^{-1})$$
$$= A(BB^{-1})A^{-1}$$
$$= AIA^{-1}$$
$$= AA^{-1}$$
$$= I$$

Thus, $C = B^{-1}A^{-1}$ satisfies the definition of the matrix inverse and must be the inverse of AB. Notice that the order of multiplication is critical. In general, $A^{-1}B^{-1}$ is not the inverse of AB. (What matrix would have $A^{-1}B^{-1}$ as its inverse?) ∎

Part (C) of Theorem 5 can be extended to an arbitrary product of invertible matrices. (The proof is omitted.)

Theorem 6

> If $A_1, A_2, \ldots, A_m$ are invertible matrices and
>
> $$A = A_1 A_2 \cdot \cdots \cdot A_m$$
>
> then A is invertible and A^{-1} is given by
>
> $$A^{-1} = A_m^{-1} A_{m-1}^{-1} \cdot \cdots \cdot A_1^{-1}$$

▪ Matrix Equations

We will now show how systems of equations can be solved using inverses of square matrices.

Example 14 Solve the system

$$3x_1 - x_2 = k_1 \tag{4}$$
$$-4x_1 + 2x_2 = k_2$$

given that:

(A) $k_1 = -5, k_2 = 8$ (B) $k_1 = 4, k_2 = -2$ (C) $k_1 = 0, k_2 = -4$

Solution   Once we obtain the inverse of the coefficient matrix,

$$A = \begin{bmatrix} 3 & -1 \\ -4 & 2 \end{bmatrix}$$

we will be able to solve parts (A)–(C) very easily. To see why, we convert system (4) into the following equivalent matrix equation:

$$\begin{array}{ccc} A & X & B \end{array}$$
$$\begin{bmatrix} 3 & -1 \\ -4 & 2 \end{bmatrix}\begin{bmatrix} x_1 \\ x_2 \end{bmatrix} = \begin{bmatrix} k_1 \\ k_2 \end{bmatrix} \tag{5}$$

You should check that matrix equation (5) is equivalent to system (4) by multiplying the left side and then equating corresponding elements on the left with those on the right.

We are now interested in finding a column matrix X that will satisfy the matrix equation

$$AX = B$$

To solve this equation, we multiply both sides by A^{-1} (if it exists) to isolate X. Since matrix multiplication is not commutative, it is important to perform the multiplication on both sides of the equation in the same order. We use the term **left-multiply** to indicate that each side of the equation should be multiplied on the left.

$$AX = B \qquad \text{Left-multiply both sides by } A^{-1}.$$
$$A^{-1}(AX) = A^{-1}B \qquad \text{Use the associative property.}$$
$$(A^{-1}A)X = A^{-1}B \qquad A^{-1}A = I$$
$$IX = A^{-1}B \qquad IX = X$$
$$X = A^{-1}B$$

The inverse of A was found in Example 12 to be

$$A^{-1} = \begin{bmatrix} 1 & \frac{1}{2} \\ 2 & \frac{3}{2} \end{bmatrix}$$

Thus,

$$\begin{array}{ccc} X & A^{-1} & B \end{array}$$
$$\begin{bmatrix} x_1 \\ x_2 \end{bmatrix} = \begin{bmatrix} 1 & \frac{1}{2} \\ 2 & \frac{3}{2} \end{bmatrix}\begin{bmatrix} k_1 \\ k_2 \end{bmatrix}$$

To solve parts (A)–(C), we simply replace k_1 and k_2 with appropriate values and multiply.

(A) $\begin{bmatrix} x_1 \\ x_2 \end{bmatrix} = \begin{bmatrix} 1 & \frac{1}{2} \\ 2 & \frac{3}{2} \end{bmatrix}\begin{bmatrix} -5 \\ 8 \end{bmatrix} = \begin{bmatrix} -1 \\ 2 \end{bmatrix}$ (B) $\begin{bmatrix} x_1 \\ x_2 \end{bmatrix} = \begin{bmatrix} 1 & \frac{1}{2} \\ 2 & \frac{3}{2} \end{bmatrix}\begin{bmatrix} 4 \\ -2 \end{bmatrix} = \begin{bmatrix} 3 \\ 5 \end{bmatrix}$

Thus, $x_1 = -1$ and $x_2 = 2$.   Thus, $x_1 = 3$ and $x_2 = 5$.

(C) $\begin{bmatrix} x_1 \\ x_2 \end{bmatrix} = \begin{bmatrix} 1 & \frac{1}{2} \\ 2 & \frac{3}{2} \end{bmatrix} \begin{bmatrix} 0 \\ -4 \end{bmatrix} = \begin{bmatrix} -2 \\ -6 \end{bmatrix}$

Thus, $x_1 = -2$ and $x_2 = -6$.

Problem 14   Solve the system

$$2x_1 - 3x_2 = k_1$$
$$-x_1 + 2x_2 = k_2$$

using the inverse of the coefficient matrix, given that:

(A) $k_1 = 10, k_2 = -6$ (B) $k_1 = 1, k_2 = 0$ (C) $k_1 = -8, k_2 = 5$ ▐▌

The process of using an inverse to solve a square system is completely generalized in Theorem 7. The proof of Theorem 7 was essentially given as part of
the solution to Example 14. Parts (A)–(C) of Example 14 illustrate the primary
advantage of using the matrix inverse method. That is, once the inverse of the
coefficient matrix is found, it can be used to solve any new system formed
through a change in the constant terms.

Theorem 7

Matrix Inverse Solution of Square Systems

The square linear system

$$a_{11}x_1 + a_{12}x_2 + \cdots + a_{1n}x_n = b_1$$
$$a_{21}x_1 + a_{22}x_2 + \cdots + a_{2n}x_n = b_2$$
$$\vdots \qquad \vdots \qquad \qquad \vdots \qquad \vdots$$
$$a_{n1}x_1 + a_{n2}x_2 + \cdots + a_{nn}x_n = b_n$$

is equivalent to the matrix equation

$$AX = B$$

where

$$A = \begin{bmatrix} a_{11} & a_{12} & \cdots & a_{1n} \\ a_{21} & a_{22} & \cdots & a_{2n} \\ \vdots & \vdots & & \vdots \\ a_{n1} & a_{n2} & \cdots & a_{nn} \end{bmatrix} \qquad X = \begin{bmatrix} x_1 \\ x_2 \\ \vdots \\ x_n \end{bmatrix} \qquad B = \begin{bmatrix} b_1 \\ b_2 \\ \vdots \\ b_n \end{bmatrix}$$

If A is invertible, then the unique solution of the linear system is given by

$$X = A^{-1}B$$

In the next section we will show that if A is not invertible, then the system $AX = B$ cannot have a unique solution. It may be dependent or it may be inconsistent. The only way to find out is to use Gauss–Jordan elimination.

If the number of variables and the number of equations in a system are not the same, then the inverse of the coefficient matrix is not defined. Thus, *the matrix inverse method can never be used to solve systems of equations that are not square. Gauss–Jordan elimination must always be used to solve nonsquare systems.*

▪ Application

The following application will illustrate the usefulness of the inverse method.

▌▫▌   **Example 15**   An investment advisor currently has two types of investments available for
Investment Mix   clients. A conservative investment A that pays 10% per year and an investment B of higher risk that pays 20% per year. Clients may divide their investments between the two to achieve any total return desired between 10% and 20%. However, the higher the desired return, the higher the risk. How should each client listed in Table 2 invest to achieve the indicated return?

Table 2

	Client 1	Client 2	Client 3	k
Total investment	$20,000	$50,000	$10,000	k_1
Annual return desired	$2,400 (12%)	$7,500 (15%)	$1,300 (13%)	k_2

Solution   We will solve the problem for an arbitrary client k, using inverses, and then apply the result to the three specific clients.

Let

$x_1 = $ Amount invested in A

$x_2 = $ Amount invested in B

Then

$$x_1 + x_2 = k_1 \quad \text{Total invested}$$
$$0.1x_1 + 0.2x_2 = k_2 \quad \text{Total annual return}$$

Write as a matrix equation:

$$\begin{array}{ccc} A & X & B \end{array}$$
$$\begin{bmatrix} 1 & 1 \\ 0.1 & 0.2 \end{bmatrix}\begin{bmatrix} x_1 \\ x_2 \end{bmatrix} = \begin{bmatrix} k_1 \\ k_2 \end{bmatrix}$$

If A^{-1} exists, then

$$X = A^{-1}B$$

We now find A^{-1} by starting with $[A|I]$ and proceeding as discussed earlier in this section:

$$\begin{bmatrix} 1 & 1 & | & 1 & 0 \\ 0.1 & 0.2 & | & 0 & 1 \end{bmatrix} \quad 10R_2 \to R_2$$

$$\sim \begin{bmatrix} 1 & 1 & | & 1 & 0 \\ 1 & 2 & | & 0 & 10 \end{bmatrix} \quad R_2 + (-1)R_1 \to R_2$$

$$\sim \begin{bmatrix} 1 & 1 & | & 1 & 0 \\ 0 & 1 & | & -1 & 10 \end{bmatrix} \quad R_1 + (-1)R_2 \to R_1$$

$$\sim \begin{bmatrix} 1 & 0 & | & 2 & -10 \\ 0 & 1 & | & -1 & 10 \end{bmatrix}$$

Thus,

$$A^{-1} = \begin{bmatrix} 2 & -10 \\ -1 & 10 \end{bmatrix} \quad \text{Check.} \quad \overset{A^{-1}}{\begin{bmatrix} 2 & -10 \\ -1 & 10 \end{bmatrix}} \overset{A}{\begin{bmatrix} 1 & 1 \\ 0.1 & 0.2 \end{bmatrix}} = \overset{I}{\begin{bmatrix} 1 & 0 \\ 0 & 1 \end{bmatrix}}$$

and

$$\overset{X}{\begin{bmatrix} x_1 \\ x_2 \end{bmatrix}} = \overset{A^{-1}}{\begin{bmatrix} 2 & -10 \\ -1 & 10 \end{bmatrix}} \overset{B}{\begin{bmatrix} k_1 \\ k_2 \end{bmatrix}}$$

To solve each client's investment problem, we replace k_1 and k_2 with appropriate values from Table 2 and multiply by A^{-1}:

Client 1

$$\begin{bmatrix} x_1 \\ x_2 \end{bmatrix} = \begin{bmatrix} 2 & -10 \\ -1 & 10 \end{bmatrix} \begin{bmatrix} 20{,}000 \\ 2{,}400 \end{bmatrix} = \begin{bmatrix} 16{,}000 \\ 4{,}000 \end{bmatrix}$$

Solution. $x_1 = \$16{,}000$ in A, $x_2 = \$4{,}000$ in B

Client 2

$$\begin{bmatrix} x_1 \\ x_2 \end{bmatrix} = \begin{bmatrix} 2 & -10 \\ -1 & 10 \end{bmatrix} \begin{bmatrix} 50{,}000 \\ 7{,}500 \end{bmatrix} = \begin{bmatrix} 25{,}000 \\ 25{,}000 \end{bmatrix}$$

Solution. $x_1 = \$25{,}000$ in A, $x_2 = \$25{,}000$ in B

Client 3

$$\begin{bmatrix} x_1 \\ x_2 \end{bmatrix} = \begin{bmatrix} 2 & -10 \\ -1 & 10 \end{bmatrix} \begin{bmatrix} 10{,}000 \\ 1{,}300 \end{bmatrix} = \begin{bmatrix} 7{,}000 \\ 3{,}000 \end{bmatrix}$$

Solution. $x_1 = \$7{,}000$ in A, $x_2 = \$3{,}000$ in B

Problem 15 Repeat Example 15 with investment A paying 8% and investment B paying 24%.

■

Answers to Matched Problems

11. $\begin{bmatrix} 2 & -3 \\ 5 & 7 \end{bmatrix}$ **12.** $\begin{bmatrix} -1 & 3 \\ -\frac{1}{2} & 1 \end{bmatrix}$ **13.** M is not invertible

14. (A) $x_1 = 2$, $x_2 = -2$ (B) $x_1 = 2$, $x_2 = 1$ (C) $x_1 = -1$, $x_2 = 2$

15. $A^{-1} = \begin{bmatrix} 1.5 & -6.25 \\ -0.5 & 6.25 \end{bmatrix}$

Client 1: $15,000 in A and $5,000 in B; client 2: $28,125 in A and $21,875 in B; client 3: $6,875 in A and $3,125 in B

■ Exercise 2-3

A In Problems 1–4, show that A and B are inverses of each other by showing that $AB = I$.

1. $A = \begin{bmatrix} 3 & -4 \\ -2 & 3 \end{bmatrix}$ $B = \begin{bmatrix} 3 & 4 \\ 2 & 3 \end{bmatrix}$

2. $A = \begin{bmatrix} 5 & -7 \\ -2 & 3 \end{bmatrix}$ $B = \begin{bmatrix} 3 & 7 \\ 2 & 5 \end{bmatrix}$

3. $A = \begin{bmatrix} 1 & -1 & 1 \\ 0 & 2 & -1 \\ 2 & 3 & 0 \end{bmatrix}$ $B = \begin{bmatrix} 3 & 3 & -1 \\ -2 & -2 & 1 \\ -4 & -5 & 2 \end{bmatrix}$

4. $A = \begin{bmatrix} 3 & 3 & -1 \\ -2 & -2 & 1 \\ -4 & -5 & 2 \end{bmatrix}$ $B = \begin{bmatrix} 1 & -1 & 1 \\ 0 & 2 & -1 \\ 2 & 3 & 0 \end{bmatrix}$

In Problems 5–8, find M^{-1} and show that $M^{-1}M = I$.

5. $M = \begin{bmatrix} 1 & 2 \\ 1 & 3 \end{bmatrix}$ **6.** $M = \begin{bmatrix} 2 & 1 \\ 5 & 3 \end{bmatrix}$

7. $M = \begin{bmatrix} 1 & 3 \\ 2 & 7 \end{bmatrix}$ **8.** $M = \begin{bmatrix} 2 & 1 \\ 1 & 1 \end{bmatrix}$

In Problems 9–12, write each system as a matrix equation and solve using inverses. [Note: The inverses were found in Problems 5–8.]

9. $x_1 + 2x_2 = k_1$
$x_1 + 3x_2 = k_2$
(A) $k_1 = 1$, $k_2 = 3$
(B) $k_1 = 3$, $k_2 = 5$
(C) $k_1 = -2$, $k_2 = 1$

10. $2x_1 + x_2 = k_1$
$5x_1 + 3x_2 = k_2$
(A) $k_1 = 2$, $k_2 = 13$
(B) $k_1 = -2$, $k_2 = 4$
(C) $k_1 = 1$, $k_2 = -3$

11. $x_1 + 3x_2 = k_1$
$\quad$ $2x_1 + 7x_2 = k_2$
$\quad$ (A) $k_1 = 2, k_2 = -1$
$\quad$ (B) $k_1 = 1, k_2 = 0$
$\quad$ (C) $k_1 = 3, k_2 = -1$

12. $2x_1 + x_2 = k_1$
$\quad$ $x_1 + x_2 = k_2$
$\quad$ (A) $k_1 = -1, k_2 = -2$
$\quad$ (B) $k_1 = 2, k_2 = 3$
$\quad$ (C) $k_1 = 2, k_2 = 0$

B   *In Problems 13–18, find M^{-1} and show that $M^{-1}M = I$.*

13. $M = \begin{bmatrix} 1 & -3 & 0 \\ 0 & 3 & 1 \\ 2 & -1 & 2 \end{bmatrix}$

14. $M = \begin{bmatrix} 2 & 9 & 0 \\ 1 & 2 & 3 \\ 0 & -1 & 1 \end{bmatrix}$

15. $M = \begin{bmatrix} 1 & 1 & 0 \\ 0 & 3 & -1 \\ 1 & 0 & 1 \end{bmatrix}$

16. $M = \begin{bmatrix} 1 & 0 & -1 \\ 2 & -1 & 0 \\ 1 & 1 & 1 \end{bmatrix}$

17. $M = \begin{bmatrix} 1 & 1 & 1 & 0 \\ 1 & 1 & 1 & 1 \\ 0 & 1 & 1 & 1 \\ 0 & 0 & 1 & 1 \end{bmatrix}$

18. $M = \begin{bmatrix} 1 & -1 & 1 & 0 \\ 0 & 1 & -1 & 1 \\ 1 & 0 & -1 & 1 \\ 1 & -1 & 0 & 1 \end{bmatrix}$

In Problems 19–24, write each system as a matrix equation and solve using inverses. [Note: The inverses were found in Problems 13–18.]

19. $x_1 - 3x_2 \qquad = k_1$
$\quad$ $3x_2 + x_3 = k_2$
$\quad$ $2x_1 - x_2 + 2x_3 = k_3$
$\quad$ (A) $k_1 = 1, k_2 = 0, k_3 = 2$
$\quad$ (B) $k_1 = -1, k_2 = 1, k_3 = 0$
$\quad$ (C) $k_1 = 2, k_2 = -2, k_3 = 1$

20. $2x_1 + 9x_2 \qquad = k_1$
$\quad$ $x_1 + 2x_2 + 3x_3 = k_2$
$\quad$ $-x_2 + x_3 = k_3$
$\quad$ (A) $k_1 = 0, k_2 = 2, k_3 = 1$
$\quad$ (B) $k_1 = -2, k_2 = 0, k_3 = 1$
$\quad$ (C) $k_1 = 3, k_2 = 1, k_3 = 0$

21. $x_1 + x_2 \qquad = k_1$
$\quad$ $3x_2 - x_3 = k_2$
$\quad$ $x_1 \qquad + x_3 = k_3$
$\quad$ (A) $k_1 = 2, k_2 = 0, k_3 = 4$
$\quad$ (B) $k_1 = 0, k_2 = 4, k_3 = -2$
$\quad$ (C) $k_1 = 4, k_2 = 2, k_3 = 0$

22. $x_1 \qquad - x_3 = k_1$
$\quad$ $2x_1 - x_2 \qquad = k_2$
$\quad$ $x_1 + x_2 + x_3 = k_3$
$\quad$ (A) $k_1 = 4, k_2 = 8, k_3 = 0$
$\quad$ (B) $k_1 = 4, k_2 = 0, k_3 = -4$
$\quad$ (C) $k_1 = 0, k_2 = 8, k_3 = -8$

23. $x_1 + x_2 + x_3 \qquad = k_1$
$\quad$ $x_1 + x_2 + x_3 + x_4 = k_2$
$\quad$ $x_2 + x_3 + x_4 = k_3$
$\quad$ $x_3 + x_4 = k_4$
$\quad$ (A) $k_1 = 1, k_2 = 1, k_3 = 1, k_4 = 1$
$\quad$ (B) $k_1 = 0, k_2 = 1, k_3 = 0, k_4 = 1$
$\quad$ (C) $k_1 = 1, k_2 = 2, k_3 = -1, k_4 = -2$

24. $\begin{aligned} x_1 - x_2 + x_3 \quad &= k_1 \\ x_2 - x_3 + x_4 &= k_2 \\ x_1 \quad - x_3 + x_4 &= k_3 \\ x_1 - x_2 \quad + x_4 &= k_4 \end{aligned}$
(A) $k_1 = 1,\ k_2 = -1,\ k_3 = 1,\ k_4 = -1$
(B) $k_1 = 1,\ k_2 = 2,\ k_3 = -1,\ k_4 = -2$
(C) $k_1 = -1,\ k_2 = 2,\ k_3 = -2,\ k_4 = 3$

In Problems 25–30, show that the inverse of the matrix does not exist.

25. $\begin{bmatrix} 3 & 9 \\ 2 & 6 \end{bmatrix}$

26. $\begin{bmatrix} 2 & -4 \\ -3 & 6 \end{bmatrix}$

27. $\begin{bmatrix} 2 & 1 & 1 \\ 1 & 1 & 0 \\ -1 & -1 & 0 \end{bmatrix}$

28. $\begin{bmatrix} 1 & -1 & 0 \\ 2 & -1 & 1 \\ 0 & 1 & 1 \end{bmatrix}$

29. $\begin{bmatrix} 1 & -1 & 1 & 2 \\ 0 & 1 & 2 & 1 \\ 2 & -3 & 2 & 2 \\ 1 & -2 & 1 & 0 \end{bmatrix}$

30. $\begin{bmatrix} 1 & 2 & -1 & 3 \\ -1 & 3 & -1 & 2 \\ -2 & 1 & 2 & -1 \\ -2 & 6 & 0 & 4 \end{bmatrix}$

31. Show that $(A^{-1})^{-1} = A$ for $A = \begin{bmatrix} 3 & 4 \\ 2 & 3 \end{bmatrix}$

32. Show that $(AB)^{-1} = B^{-1}A^{-1}$ for

$$A = \begin{bmatrix} 1 & 1 \\ -2 & -1 \end{bmatrix} \quad \text{and} \quad B = \begin{bmatrix} 2 & 3 \\ -1 & -1 \end{bmatrix}$$

C   **33.** Find the inverse of $A = \begin{bmatrix} 1 & a & b \\ 0 & 1 & c \\ 0 & 0 & 1 \end{bmatrix}$

34. Find the inverse of $A = \begin{bmatrix} 1 & 0 & 0 \\ a & 1 & 0 \\ b & c & 1 \end{bmatrix}$

35. If A is invertible and n is a positive integer, then we define A^{-n} by

$$A^{-n} = (A^{-1})^n$$

Find A^{-2} and show that $A^{-2}A^2 = I$ for

$$A = \begin{bmatrix} 2 & 1 \\ 1 & 1 \end{bmatrix}$$

36. Repeat Problem 35 for $A = \begin{bmatrix} 3 & 5 \\ 1 & 2 \end{bmatrix}$

37. If

$$A = \begin{bmatrix} a & b \\ c & d \end{bmatrix}$$

and $ad - bc \neq 0$, find A^{-1}. If $ad - bc = 0$, show that A is not invertible.

38. If

$$A = \begin{bmatrix} a_{11} & 0 & 0 \\ 0 & a_{22} & 0 \\ 0 & 0 & a_{33} \end{bmatrix}$$

and $a_{11} \neq 0$, $a_{22} \neq 0$, and $a_{33} \neq 0$, find A^{-1}.

39. If A, B, and C are $n \times n$ invertible matrices, show directly that $(ABC)^{-1} = C^{-1}B^{-1}A^{-1}$. (Do not refer to Theorem 6 in your proof.)

In Problems 40–42, prove the indicated statement from Theorem 5 in this section.

40. *Theorem 5(B).* If A is invertible, then A^{-1} is invertible and $(A^{-1})^{-1} = A$.

41. *Theorem 5(D).* If A is invertible and C and D are matrices satisfying $AD = AC$, then $D = C$.

42. *Theorem 5(E).* If A is invertible and C and D are matrices satisfying $DA = CA$, then $D = C$.

43. If P is an invertible $n \times n$ matrix and A and B are $n \times n$ matrices satisfying $A = P^{-1}BP$, show that $A^m = P^{-1}B^mP$ for any positive integer m.

44. Show that the statement in Problem 43 is also true for m a negative integer if A and B are invertible matrices.

▌▫▌ Applications

Guitar Model	Labor Cost	Material Cost
A	$30	$20
B	$40	$30

45. *Production scheduling.* Labor and material costs for manufacturing two guitar models are given in the table in the margin. If a total of $3,000 a week is allowed for labor and material, how many of each model should be produced each week to use exactly each of the allocations of the $3,000 indicated in the following table?

	Weekly Allocation		
	1	2	3
Labor	$1,800	$1,750	$1,720
Material	$1,200	$1,250	$1,280

Mix	Protein (%)	Fat (%)
A	20	2
B	10	6

46. *Diets.* A biologist has available two commercial food mixes with the percentages of protein and fat listed in the table in the margin. How many ounces of each mix should be used to prepare each of the diets listed in the following table?

	Diet		
	1	2	3
Protein	20 oz	10 oz	10 oz
Fat	6 oz	4 oz	6 oz

47. *Circuit analysis.* A direct current electrical circuit consisting of conductors (wires), resistors, and batteries is diagrammed below:

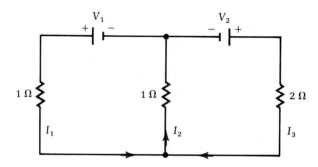

If I_1, I_2, and I_3 are the currents (in amperes) in the three branches of the circuit and V_1 and V_2 are the voltages (in volts) of the two batteries, then the currents can be shown to satisfy the following system of equations:

$$I_1 - I_2 + I_3 = 0$$
$$I_1 + I_2 \qquad = V_1$$
$$I_2 + 2I_3 = V_2$$

Use the inverse of the coefficient matrix to solve this system for the given battery voltages.

(A) $V_1 = 10$ volts, $V_2 = 10$ volts   (B) $V_1 = 10$ volts, $V_2 = 15$ volts
(C) $V_1 = 15$ volts, $V_2 = 10$ volts

48. *Circuit analysis.* Repeat Problem 47 for the following electrical circuit:

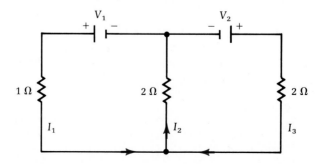

$$I_1 - I_2 + I_3 = 0$$
$$I_1 + 2I_2 \quad\quad = V_1$$
$$2I_2 + 2I_3 = V_2$$

49. *Heat conduction.* In the grid pictured here, let x_1, x_2, and x_3 represent the unknown temperatures at the intersection points, and let t_1, t_2, and t_3 represent the temperatures at the ends of the rods.

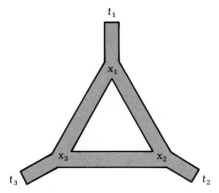

(A) If the temperature at each intersection is the average of the temperatures at the adjacent points (adjacent points are either other intersection points or ends of rods), show that x_1, x_2, and x_3 satisfy the following system of equations:

$$3x_1 - x_2 - x_3 = t_1$$
$$-x_1 + 3x_2 - x_3 = t_2$$
$$-x_1 - x_2 + 3x_3 = t_3$$

(B) Find the inverse of the coefficient matrix for the system in part (A).

(C) Use the inverse of the coefficient matrix to solve the system for the sets of temperatures given in the table.

	t_1	t_2	t_3
1	0°	0°	32°
2	0°	16°	32°
3	8°	16°	32°

50. *Heat conduction.* In the grid pictured at the top of the next page, let x_1, x_2, and x_3 represent the unknown temperatures at the intersection points, and let t_1, t_2, t_3, t_4, t_5, and t_6 represent the temperatures at the ends of the rods.

(A) Show that x_1, x_2, and x_3 satisfy the following system of equations:

$$-x_1 - x_2 + 4x_3 = t_1 + t_2$$
$$-x_1 + 4x_2 - x_3 = t_3 + t_4$$
$$4x_1 - x_2 - x_3 = t_5 + t_6$$

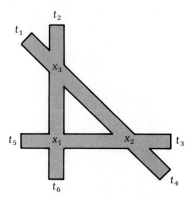

(B) Find the inverse of the coefficient matrix for the system in part (A).
(C) Use the inverse of the coefficient matrix to solve the system for the following sets of temperatures:

	t_1	t_2	t_3	t_4	t_5	t_6
1	0°	0°	0°	0°	10°	10°
2	0°	0°	0°	10°	10°	10°
3	5°	5°	10°	10°	15°	15°

2-4 Elementary Matrices and Row Equivalence

- Row Operations by Matrix Multiplication
- Elementary Matrices
- Inverse of an Elementary Matrix
- Fundamental Result

■ Row Operations by Matrix Multiplication

In the preceding section we saw that inverses can be found by using elementary row operations and that inverses can be used to solve certain systems of equations. In this section we present a new method for performing row operations. This method is introduced mainly to help us understand the relationship between invertible matrices and systems of equations. We begin by considering an example. Let

$$A = \begin{bmatrix} 0 & -2 & 1 \\ 2 & 4 & -2 \\ 3 & 5 & -2 \end{bmatrix}$$

Suppose we want to transform A to reduced form. We could begin by interchanging the first and second rows, producing a matrix A_1 which is row equivalent to A:

$$A = \begin{bmatrix} 0 & -2 & 1 \\ 2 & 4 & -2 \\ 3 & 5 & -2 \end{bmatrix} \overset{R_1 \leftrightarrow R_2}{\sim} \begin{bmatrix} 2 & 4 & -2 \\ 0 & -2 & 1 \\ 3 & 5 & -2 \end{bmatrix} = A_1$$

We will now show how to use matrix multiplication to accomplish this transformation. First, we perform the same row operation on the identity matrix of order 3:

$$I = \begin{bmatrix} 1 & 0 & 0 \\ 0 & 1 & 0 \\ 0 & 0 & 1 \end{bmatrix} \overset{R_1 \leftrightarrow R_2}{\sim} \begin{bmatrix} 0 & 1 & 0 \\ 1 & 0 & 0 \\ 0 & 0 & 1 \end{bmatrix} = E_1$$

Next we left-multiply A by E_1:

$$E_1 A = \begin{bmatrix} 0 & 1 & 0 \\ 1 & 0 & 0 \\ 0 & 0 & 1 \end{bmatrix} \begin{bmatrix} 0 & -2 & 1 \\ 2 & 4 & -2 \\ 3 & 5 & -2 \end{bmatrix} = \begin{bmatrix} 2 & 4 & -2 \\ 0 & -2 & 1 \\ 3 & 5 & -2 \end{bmatrix} = A_1$$

Thus, left-multiplying A by E_1 produces the same row equivalent matrix A_1 that we obtained by performing the row operation directly on A. The next step in transforming A to reduced form is to multiply the first row of A_1 by $\frac{1}{2}$:

$$E_1 A = A_1 = \begin{bmatrix} 2 & 4 & -2 \\ 0 & -2 & 1 \\ 3 & 5 & -2 \end{bmatrix} \overset{\frac{1}{2}R_1 \to R_1}{\sim} \begin{bmatrix} 1 & 2 & -1 \\ 0 & -2 & 1 \\ 3 & 5 & -2 \end{bmatrix} = A_2$$

We can also perform this operation by first operating on the identity to produce a matrix E_2 and then left-multiplying by this matrix:

$$I = \begin{bmatrix} 1 & 0 & 0 \\ 0 & 1 & 0 \\ 0 & 0 & 1 \end{bmatrix} \overset{\frac{1}{2}R_1 \to R_1}{\sim} \begin{bmatrix} \frac{1}{2} & 0 & 0 \\ 0 & 1 & 0 \\ 0 & 0 & 1 \end{bmatrix} = E_2$$

$$E_2(E_1 A) = E_2 A_1 = \begin{bmatrix} \frac{1}{2} & 0 & 0 \\ 0 & 1 & 0 \\ 0 & 0 & 1 \end{bmatrix} \begin{bmatrix} 2 & 4 & -2 \\ 0 & -2 & 1 \\ 3 & 5 & -2 \end{bmatrix}$$

$$= \begin{bmatrix} 1 & 2 & -1 \\ 0 & -2 & 1 \\ 3 & 5 & -2 \end{bmatrix} = A_2$$

Once again, left-multiplying A_1 by E_2 produces the same matrix that we obtained by performing the row operation on A_1. Furthermore, performing two successive row operations on A is equivalent to left-multiplying A by E_1 and then by E_2. We will do one more step to illustrate the third type of row operation:

$$E_2 E_1 A = A_2 = \begin{bmatrix} 1 & 2 & -1 \\ 0 & -2 & 1 \\ 3 & 5 & -2 \end{bmatrix} \overset{R_3 + (-3)R_1 \to R_3}{\sim} \begin{bmatrix} 1 & 2 & -1 \\ 0 & -2 & 1 \\ 0 & -1 & 1 \end{bmatrix} = A_3$$

Equivalently,

$$I = \begin{bmatrix} 1 & 0 & 0 \\ 0 & 1 & 0 \\ 0 & 0 & 1 \end{bmatrix} \underset{R_3 + (-3)R_1 \rightarrow R_3}{\sim} \begin{bmatrix} 1 & 0 & 0 \\ 0 & 1 & 0 \\ -3 & 0 & 1 \end{bmatrix} = E_3$$

$$E_3(E_2E_1A) = E_3A_2 = \begin{bmatrix} 1 & 0 & 0 \\ 0 & 1 & 0 \\ -3 & 0 & 1 \end{bmatrix} \begin{bmatrix} 1 & 2 & -1 \\ 0 & -2 & 1 \\ 3 & 5 & -2 \end{bmatrix}$$

$$= \begin{bmatrix} 1 & 2 & -1 \\ 0 & -2 & 1 \\ 0 & -1 & 1 \end{bmatrix} = A_3$$

Thus, performing three successive row operations on A produces the same matrix as does left-multiplying A by E_1, E_2, and E_3, in that order. It is essential that the order of multiplication agree with the order in which the row operations are performed. Since we are interested only in illustrating this new method, we will not continue the reduction of A. The two methods of row reduction are compared in Table 3.

Table 3
Comparison of Row Operation and Multiplication Methods

Row Operation Method	Multiplication Method	
	Row Operation on I	Left-Multiplication
1. $A \sim A_1$	1. $I \sim E_1$	$E_1A = A_1$
$R_1 \leftrightarrow R_2$	$R_1 \leftrightarrow R_2$	
2. $A_1 \sim A_2$	2. $I \sim E_2$	$E_2E_1A = E_2A_1 = A_2$
$\frac{1}{2}R_1 \rightarrow R_1$	$\frac{1}{2}R_1 \rightarrow R_1$	
3. $A_2 \sim A_3$	3. $I \sim E_3$	$E_3E_2E_1A = E_3A_2 = A_3$
$R_3 + (-3)R_1 \rightarrow R_3$	$R_3 + (-3)R_1 \rightarrow R_3$	

▪ Elementary Matrices

A matrix formed by performing exactly one elementary row operation on the identity matrix is called an **elementary matrix.** Matrices E_1, E_2, and E_3 in the preceding example illustrate the three different kinds of elementary matrices, corresponding to the three different row operations.

Example 16   Which of the following are elementary matrices?

$$\text{(A) } E_1 = \begin{bmatrix} 1 & 0 & 0 \\ 0 & 1 & -2 \\ 0 & 0 & 1 \end{bmatrix} \quad \text{(B) } E_2 = \begin{bmatrix} 0 & 0 & 1 \\ 0 & 1 & 0 \\ 1 & 0 & 0 \end{bmatrix} \quad \text{(C) } E_3 = \begin{bmatrix} 1 & 0 & 0 \\ 0 & 0 & 1 \\ 0 & 2 & 0 \end{bmatrix}$$

Solution   To show that a matrix is an elementary matrix, we must show that it can be obtained by performing exactly one row operation on I.

$$(A) \ I = \begin{bmatrix} 1 & 0 & 0 \\ 0 & 1 & 0 \\ 0 & 0 & 1 \end{bmatrix} \qquad R_2 + (-2)R_3 \rightarrow R_2$$

$$\sim \begin{bmatrix} 1 & 0 & 0 \\ 0 & 1 & -2 \\ 0 & 0 & 1 \end{bmatrix} = E_1$$

Thus, E_1 is an elementary matrix, since it took exactly one row operation to transform I into E_1.

$$(B) \ I = \begin{bmatrix} 1 & 0 & 0 \\ 0 & 1 & 0 \\ 0 & 0 & 1 \end{bmatrix} \qquad R_1 \leftrightarrow R_3$$

$$\sim \begin{bmatrix} 0 & 0 & 1 \\ 0 & 1 & 0 \\ 1 & 0 & 0 \end{bmatrix} = E_2$$

Thus, E_2 is an elementary matrix, since it took exactly one row operation to transform I into E_2.

$$(C) \ I = \begin{bmatrix} 1 & 0 & 0 \\ 0 & 1 & 0 \\ 0 & 0 & 1 \end{bmatrix} \qquad R_2 \leftrightarrow R_3$$

$$\sim \begin{bmatrix} 1 & 0 & 0 \\ 0 & 0 & 1 \\ 0 & 1 & 0 \end{bmatrix}$$

$$2R_3 \rightarrow R_3$$

$$\sim \begin{bmatrix} 1 & 0 & 0 \\ 0 & 0 & 1 \\ 0 & 2 & 0 \end{bmatrix} = E_3$$

Thus, E_3 is not an elementary matrix, since it took two row operations to transform I into E_3 (that is, E_3 cannot be obtained from I by performing exactly one row operation).

Problem 16   Which of the following are elementary matrices?

$$(A) \ E_1 = \begin{bmatrix} 1 & 3 & 0 \\ 0 & 1 & 0 \\ 0 & 0 & 1 \end{bmatrix} \qquad (B) \ E_2 = \begin{bmatrix} 1 & 2 & 3 \\ 0 & 1 & 0 \\ 0 & 0 & 1 \end{bmatrix} \qquad (C) \ E_3 = \begin{bmatrix} 1 & 0 & 0 \\ 0 & 0 & 1 \\ 0 & 1 & 0 \end{bmatrix} \qquad \blacksquare$$

If A and B are two row equivalent matrices, not necessarily square, then there must be a sequence of elementary row operations that transforms A into B. If we form the elementary matrices corresponding to these row operations and then left-multiply A by these matrices in exactly the same order as that in which the row operations were performed, we can conclude that

$$B = E_k E_{k-1} \cdot \cdots \cdot E_1 A$$

where E_k, E_{k-1}, . . . , E_1 are elementary matrices. This allows us to state the following relationship between row equivalent matrices (a formal proof is omitted):

Theorem 8

> Two $m \times n$ matrices A and B are row equivalent if and only if $B = PA$ where P is a product of elementary matrices.

The following example illustrates Theorem 8.

Example 17   Use elementary matrices to transform A to reduced form. If B is the reduced form of A, find P and verify that $B = PA$.

$$A = \begin{bmatrix} 0 & 1 & 3 \\ 3 & 6 & -9 \end{bmatrix}$$

Solution   To transform A to reduced form using elementary matrices, proceed as follows: start by looking at A to determine the first row operation required in the reduction process. Then, instead of applying the row operation to A (as we have before), apply it to I to obtain E_1; then left-multiply A by E_1 to achieve the transformation. This process is continued on each row equivalent version of A until a reduced form of A is obtained.

Looking back at A, we see that a nonzero element is needed in position 1,1. This can be achieved using $R_1 \leftrightarrow R_2$.

Step 1.   Perform the operation $R_1 \leftrightarrow R_2$:

$$I = \begin{bmatrix} 1 & 0 \\ 0 & 1 \end{bmatrix} \overset{R_1 \leftrightarrow R_2}{\sim} \begin{bmatrix} 0 & 1 \\ 1 & 0 \end{bmatrix} = E_1$$

$$E_1 A = \begin{bmatrix} 0 & 1 \\ 1 & 0 \end{bmatrix}\begin{bmatrix} 0 & 1 & 3 \\ 3 & 6 & -9 \end{bmatrix}$$

$$\left(\text{We now need a 1 here.}\right) = \begin{bmatrix} 3 & 6 & -9 \\ 0 & 1 & 3 \end{bmatrix}$$

Step 2.   Perform the operation $\tfrac{1}{3}R_1 \to R_1$.

$$I = \begin{bmatrix} 1 & 0 \\ 0 & 1 \end{bmatrix} \overset{\tfrac{1}{3}R_1 \to R_1}{\sim} \begin{bmatrix} \tfrac{1}{3} & 0 \\ 0 & 1 \end{bmatrix} = E_2$$

$$E_2 E_1 A = \begin{bmatrix} \tfrac{1}{3} & 0 \\ 0 & 1 \end{bmatrix}\begin{bmatrix} 3 & 6 & -9 \\ 0 & 1 & 3 \end{bmatrix}$$

$$\left(\begin{array}{c}\text{We now need}\\ \text{a 0 here.}\end{array}\right) = \begin{bmatrix} 1 & 2 & -3 \\ 0 & 1 & 3 \end{bmatrix}$$

Step 3. *Perform the operation* $R_1 + (-2)R_2 \rightarrow R_1$.

$$I = \begin{bmatrix} 1 & 0 \\ 0 & 1 \end{bmatrix} \begin{array}{c} R_1 + (-2)R_2 \rightarrow R_1 \\ \sim \end{array} \begin{bmatrix} 1 & -2 \\ 0 & 1 \end{bmatrix} = E_3$$

$$E_3 E_2 E_1 A = \begin{bmatrix} 1 & -2 \\ 0 & 1 \end{bmatrix}\begin{bmatrix} 1 & 2 & -3 \\ 0 & 1 & 3 \end{bmatrix}$$

$$= \begin{bmatrix} 1 & 0 & -9 \\ 0 & 1 & 3 \end{bmatrix} \qquad \text{Reduced form}$$

Thus,

$$B = \begin{bmatrix} 1 & 0 & -9 \\ 0 & 1 & 3 \end{bmatrix} = E_3 E_2 E_1 A = PA$$

where $P = E_3 E_2 E_1$.

We now multiply $E_3 E_2 E_1$ to find P and verify that $PA = B$.

$$P = \overset{E_3}{\begin{bmatrix} 1 & -2 \\ 0 & 1 \end{bmatrix}} \overset{E_2}{\begin{bmatrix} \frac{1}{3} & 0 \\ 0 & 1 \end{bmatrix}} \overset{E_1}{\begin{bmatrix} 0 & 1 \\ 1 & 0 \end{bmatrix}}$$

$$= \begin{bmatrix} \frac{1}{3} & -2 \\ 0 & 1 \end{bmatrix}\begin{bmatrix} 0 & 1 \\ 1 & 0 \end{bmatrix} = \begin{bmatrix} -2 & \frac{1}{3} \\ 1 & 0 \end{bmatrix}$$

$$PA = \begin{bmatrix} -2 & \frac{1}{3} \\ 1 & 0 \end{bmatrix}\begin{bmatrix} 0 & 1 & 3 \\ 3 & 6 & -9 \end{bmatrix} = \begin{bmatrix} 1 & 0 & -9 \\ 0 & 1 & 3 \end{bmatrix} = B$$

Problem 17   Repeat Example 17 for $A = \begin{bmatrix} 0 & 2 & -4 \\ 1 & 3 & 2 \end{bmatrix}$ ∎

The primary application of elementary matrices will be to help us understand the relationship between row equivalent matrices. Performing row operations directly on a matrix is still the most efficient procedure for finding the reduced form.

▪ Inverse of an Elementary Matrix

Since an elementary matrix is formed by performing one row operation on the identity, we can transform an elementary matrix back to the identity by reversing that operation. For example, if E is the elementary matrix formed by performing the operation $R_1 + 2R_2 \rightarrow R_1$ on the identity of order 3, then

$$I = \begin{bmatrix} 1 & 0 & 0 \\ 0 & 1 & 0 \\ 0 & 0 & 1 \end{bmatrix} \begin{array}{c} R_1 + 2R_2 \rightarrow R_1 \\ \sim \end{array} \begin{bmatrix} 1 & 2 & 0 \\ 0 & 1 & 0 \\ 0 & 0 & 1 \end{bmatrix} = E$$

The operation $R_1 + (-2)R_2 \rightarrow R_1$ transforms E back into the identity:

$$E = \begin{bmatrix} 1 & 2 & 0 \\ 0 & 1 & 0 \\ 0 & 0 & 1 \end{bmatrix} \begin{array}{c} R_1 + (-2)R_2 \rightarrow R_1 \\ \sim \end{array} \begin{bmatrix} 1 & 0 & 0 \\ 0 & 1 & 0 \\ 0 & 0 & 1 \end{bmatrix}$$

Now let F be the elementary matrix formed by applying this second operation to the identity:

$$I = \begin{bmatrix} 1 & 0 & 0 \\ 0 & 1 & 0 \\ 0 & 0 & 1 \end{bmatrix} \begin{array}{c} R_1 + (-2)R_2 \rightarrow R_1 \\ \sim \end{array} \begin{bmatrix} 1 & -2 & 0 \\ 0 & 1 & 0 \\ 0 & 0 & 1 \end{bmatrix} = F$$

We will call this second row operation the *reverse row operation*. What is the relationship between E and F? Left-multiplying E by F would be equivalent to performing the operation $R_1 + (-2)R_2 \rightarrow R_1$ directly on E. But this is the very operation that transforms E into the identity. That is,

$$FE = \begin{bmatrix} 1 & -2 & 0 \\ 0 & 1 & 0 \\ 0 & 0 & 1 \end{bmatrix} \begin{bmatrix} 1 & 2 & 0 \\ 0 & 1 & 0 \\ 0 & 0 & 1 \end{bmatrix} = \begin{bmatrix} 1 & 0 & 0 \\ 0 & 1 & 0 \\ 0 & 0 & 1 \end{bmatrix} = I$$

Furthermore,

$$EF = \begin{bmatrix} 1 & 2 & 0 \\ 0 & 1 & 0 \\ 0 & 0 & 1 \end{bmatrix} \begin{bmatrix} 1 & -2 & 0 \\ 0 & 1 & 0 \\ 0 & 0 & 1 \end{bmatrix} = \begin{bmatrix} 1 & 0 & 0 \\ 0 & 1 & 0 \\ 0 & 0 & 1 \end{bmatrix} = I$$

Thus, F must be the inverse of E. That is, *the inverse of the elementary matrix E is the elementary matrix formed by the reverse of the row operation that formed E*:

$$E = \begin{bmatrix} 1 & 2 & 0 \\ 0 & 1 & 0 \\ 0 & 0 & 1 \end{bmatrix} \qquad \text{Formed from } I \text{ using } R_1 + 2R_2 \rightarrow R_1$$

$$E^{-1} = \begin{bmatrix} 1 & -2 & 0 \\ 0 & 1 & 0 \\ 0 & 0 & 1 \end{bmatrix} \qquad \text{Formed from } I \text{ using } R_1 + (-2)R_2 \rightarrow R_1$$

Each of the three types of elementary row operations can be reversed by performing a second elementary row operation of the same type, as listed in the box.

Reverse Row Operations

Original Operation	**Reverse Operation**
1. $R_i + kR_j \rightarrow R_i$	1. $R_i + (-k)R_j \rightarrow R_i$
2. $kR_i \rightarrow R_i$ $\quad (k \neq 0)$	2. $\left(\dfrac{1}{k}\right)R_i \rightarrow R_i$ $\quad (k \neq 0)$
3. $R_i \leftrightarrow R_j$	3. $R_j \leftrightarrow R_i$

As indicated in Theorem 9, the reverse operations can be used to find the inverse of any elementary matrix. The proof will be left as an exercise.

Theorem 9

If E is an elementary matrix formed by any elementary row operation, then E is invertible and E^{-1} is the elementary matrix formed by the corresponding reverse row operation.

Example 18   Find the elementary matrix formed by performing each row operation on the identity of order 3, and then use Theorem 9 to find its inverse.

(A) $R_2 + (-3)R_3 \rightarrow R_2$ (B) $4R_1 \rightarrow R_1$ (C) $R_1 \leftrightarrow R_2$

Solution   (A) First, we use the given row operation to find the elementary matrix E:

$$I = \begin{bmatrix} 1 & 0 & 0 \\ 0 & 1 & 0 \\ 0 & 0 & 1 \end{bmatrix} \overset{R_2 + (-3)R_3 \rightarrow R_2}{\sim} \begin{bmatrix} 1 & 0 & 0 \\ 0 & 1 & -3 \\ 0 & 0 & 1 \end{bmatrix} = E$$

Now use the reverse row operation to find E^{-1}:

$$I = \begin{bmatrix} 1 & 0 & 0 \\ 0 & 1 & 0 \\ 0 & 0 & 1 \end{bmatrix} \overset{R_2 + 3R_3 \rightarrow R_2}{\sim} \begin{bmatrix} 1 & 0 & 0 \\ 0 & 1 & 3 \\ 0 & 0 & 1 \end{bmatrix} = E^{-1}$$

(B) $I = \begin{bmatrix} 1 & 0 & 0 \\ 0 & 1 & 0 \\ 0 & 0 & 1 \end{bmatrix} \overset{4R_1 \rightarrow R_1}{\sim} \begin{bmatrix} 4 & 0 & 0 \\ 0 & 1 & 0 \\ 0 & 0 & 1 \end{bmatrix} = E$

$$I = \begin{bmatrix} 1 & 0 & 0 \\ 0 & 1 & 0 \\ 0 & 0 & 1 \end{bmatrix} \overset{\frac{1}{4}R_1 \rightarrow R_1}{\sim} \begin{bmatrix} \frac{1}{4} & 0 & 0 \\ 0 & 1 & 0 \\ 0 & 0 & 1 \end{bmatrix} = E^{-1}$$

$$(C)\ I = \begin{bmatrix} 1 & 0 & 0 \\ 0 & 1 & 0 \\ 0 & 0 & 1 \end{bmatrix} \overset{R_1 \leftrightarrow R_2}{\sim} \begin{bmatrix} 0 & 1 & 0 \\ 1 & 0 & 0 \\ 0 & 1 & 0 \end{bmatrix} = E$$

$$I = \begin{bmatrix} 1 & 0 & 0 \\ 0 & 1 & 0 \\ 0 & 0 & 1 \end{bmatrix} \overset{R_2 \leftrightarrow R_1}{\sim} \begin{bmatrix} 0 & 1 & 0 \\ 1 & 0 & 0 \\ 0 & 1 & 0 \end{bmatrix} = E^{-1}$$

Problem 18　Repeat Example 18 for:

(A) $R_3 + 4R_2 \rightarrow R_3$　　(B) $\frac{1}{2}R_2 \rightarrow R_2$　　(C) $R_2 \leftrightarrow R_3$　　　■

▪ Fundamental Result

The following fundamental theorem ties together many of the ideas presented throughout this chapter and states some basic relationships between systems of n equations with n variables and square matrices of order n. Theorem 10 will have many applications in later chapters. In order to apply it correctly, it is important that you understand both its statement and its proof.

Theorem 10

> **Fundamental Theorem — Version 1**
>
> If A is a square matrix of order n, then all of the following statements are equivalent:
>
> (A)　A is invertible.
> (B)　$AX = B$ has a unique solution for any $n \times 1$ column matrix B.
> (C)　The only solution of $AX = 0$ is the trivial solution $X = 0$.
> (D)　A is row equivalent to I.
> (E)　A is a product of elementary matrices.

Proof　We will prove this theorem by establishing the following chain of implications:*

$$(A) \Rightarrow (B) \Rightarrow (C) \Rightarrow (D) \Rightarrow (E) \Rightarrow (A)$$

Once we have done this, if we know that any one of the five statements is true, then we can conclude that the other four are also true. That is what it means to say the five statements are equivalent.

(A) $\Rightarrow$ (B)　*Assume that A is invertible.* Then we can use A^{-1} to solve the system
$AX = B$ as follows:

$$\begin{aligned} AX &= B & &\text{Left-multiply by } A^{-1}. \\ A^{-1}AX &= A^{-1}B \\ X &= A^{-1}B \end{aligned}$$

* The expression "$p \Rightarrow q$" is read "p implies q" or "If p, then q."

Thus, the system $AX = B$ has the unique solution $X = A^{-1}B$ for any $n \times 1$ column matrix B.

(B) $\Rightarrow$ (C) *Assume that $AX = B$ always has exactly one solution for any $n \times 1$ column matrix B.* If we choose $B = 0$, then this implies that the homogeneous system $AX = 0$ has exactly one solution. But $X = 0$ is always a solution to any homogeneous system. By hypothesis, there can be no other solution, and thus, the only solution is the trivial solution.

(C) $\Rightarrow$ (D) *Assume that the only solution to the system $AX = 0$ is $X = 0$.* Now consider the system $IX = 0$, or

$$\begin{bmatrix} 1 & 0 & \cdots & 0 \\ 0 & 1 & \cdots & 0 \\ \cdot & \cdot & & \cdot \\ \cdot & \cdot & & \cdot \\ \cdot & \cdot & & \cdot \\ 0 & 0 & \cdots & 1 \end{bmatrix} \begin{bmatrix} x_1 \\ x_2 \\ \cdot \\ \cdot \\ \cdot \\ x_n \end{bmatrix} = \begin{bmatrix} 0 \\ 0 \\ \cdot \\ \cdot \\ \cdot \\ 0 \end{bmatrix}$$

The only solution to this system is $x_1 = 0$, $x_2 = 0$, $\ldots$, $x_n = 0$, or $X = 0$. This implies that the systems

$$AX = 0 \quad \text{and} \quad IX = 0$$

have the same solution set, and hence, they are equivalent. Applying Theorem 3 in Section 1-2, we can now conclude that the augmented coefficient matrices for these two systems are row equivalent. That is,

$$[A|0] \sim [I|0]$$

In other words, there is a sequence of elementary row operations that will transform $[A|0]$ into $[I|0]$. This same sequence will transform A into I. This shows that A is row equivalent to I.

(D) $\Rightarrow$ (E) *Assume A is row equivalent to I.* Since performing row operations is equivalent to multiplying by elementary matrices, there must be a sequence of elementary matrices $E_1, E_2, \ldots , E_k$ such that

$$E_k \cdot \cdots \cdot E_2 E_1 A = I \tag{1}$$

Since each of the elementary matrices $E_1, E_2, \ldots , E_k$ is invertible, Theorem 6 in Section 2-3 implies that their product is also invertible. Left-multiplying both sides of (1) by

$$(E_k \cdot \cdots \cdot E_2 E_1)^{-1} = E_1^{-1} E_2^{-1} \cdot \cdots \cdot E_k^{-1}$$

we have

$$(E_k \cdot \cdots \cdot E_2 E_1)^{-1} (E_k \cdot \cdots \cdot E_2 E_1) A = (E_k \cdot \cdots \cdot E_2 E_1)^{-1} I$$
$$A = E_1^{-1} E_2^{-1} \cdot \cdots \cdot E_k^{-1}$$

But the inverse of an elementary matrix is always another elementary matrix. Thus, we have expressed A as a product of elementary matrices.

$(E) \Rightarrow (A)$ *Assume A is a product of elementary matrices.* Since elementary matrices are invertible, Theorem 6 in Section 2-3 implies that A is invertible. ∎

We will now use Theorem 10 in proving Theorem 11 below. First, recall the definition of the inverse of a matrix: A square matrix A of order n is invertible if there exists another square matrix B of order n such that

$$AB = I \quad \text{and} \quad BA = I$$

Theorem 11 states that if either of these two equations is satisfied, then the other is also satisfied.

Theorem 11

> If A and B are square matrices of order n and either $AB = I$ or $BA = I$, then $B = A^{-1}$.

Proof   *Assume that A and B satisfy the equation $BA = I$.* First, we will use Theorem 10 to show that A is invertible. Let X be any solution of the homogeneous system $AX = 0$. Then

$AX = 0$	Left-multiply both sides by B.
$BAX = B0 = 0$	Substitute $BA = I$.
$IX = 0$	Property of I
$X = 0$	

Thus, the only solution to $AX = 0$ is the trivial one. Since this is one of the conditions listed in Theorem 10, we can conclude that the other four statements in Theorem 10 are also true. In particular, A is invertible and A^{-1} must exist. Now we must show that $B = A^{-1}$.

$BA = I$	Right-multiply by A^{-1}.
$BAA^{-1} = IA^{-1}$	$AA^{-1} = I$ and $IA^{-1} = A^{-1}$
$BI = A^{-1}$	$BI = B$
$B = A^{-1}$	

The proof in the case that $AB = I$ is obtained by interchanging the roles of A and B in the above proof. ∎

To see another application of Theorem 10, let's return to the method of computing inverses presented in Theorem 4, Section 2-3. Suppose we have performed row operations to show

$$[A|I] \sim [I|B]$$

How can we prove that B is actually A^{-1}? Since we were able to transform A into I, A must be row equivalent to I. Theorem 10 now implies that A is invertible. If $E_1, E_2, \ldots, E_k$ are the elementary matrices corresponding to the row operations that transform A into I, then

$$E_k \cdot \cdots \cdot E_2 E_1 A = I \tag{2}$$

Right-multiplying both sides of (2) by A^{-1}, we have

$$E_k \cdot \cdots \cdot E_2 E_1 A A^{-1} = I A^{-1}$$
$$E_k \cdot \cdots \cdot E_2 E_1 I = A^{-1}$$

Thus, the same row operations that transform A into I also transform I into A^{-1}. Since these are the very same operations we performed to transform $[A\,|\,I]$ into $[I\,|\,B]$, B must be A^{-1}.

Now suppose that it is not possible to transform $[A\,|\,I]$ into $[I\,|\,B]$. Then A must not be row equivalent to I. Theorem 10 now implies that A is not invertible. (If A were invertible, then it would be row equivalent to I.) Thus, the method of finding inverses presented in Theorem 4 of Section 2-3 always produces the inverse of an invertible matrix and always indicates when a matrix is not invertible. That is, we have proved Theorem 4.

Answers to Matched Problems

16. E_1 and E_3 are elementary matrices

17.

$$\begin{matrix} P & A & B \end{matrix}$$

$$\begin{bmatrix} -\frac{3}{2} & 1 \\ \frac{1}{2} & 0 \end{bmatrix} \begin{bmatrix} 0 & 2 & -4 \\ 1 & 3 & 2 \end{bmatrix} = \begin{bmatrix} 1 & 0 & 8 \\ 0 & 1 & -2 \end{bmatrix}$$

18. (A) $E = \begin{bmatrix} 1 & 0 & 0 \\ 0 & 1 & 0 \\ 0 & 4 & 1 \end{bmatrix}$ $E^{-1} = \begin{bmatrix} 1 & 0 & 0 \\ 0 & 1 & 0 \\ 0 & -4 & 1 \end{bmatrix}$

(B) $E = \begin{bmatrix} 1 & 0 & 0 \\ 0 & \frac{1}{2} & 0 \\ 0 & 0 & 1 \end{bmatrix}$ $E^{-1} = \begin{bmatrix} 1 & 0 & 0 \\ 0 & 2 & 0 \\ 0 & 0 & 1 \end{bmatrix}$ **(C)** $E = \begin{bmatrix} 1 & 0 & 0 \\ 0 & 0 & 1 \\ 0 & 1 & 0 \end{bmatrix} = E^{-1}$

‖ Exercise 2-4

A   *Which of the matrices in Problems 1–12 are elementary matrices?*

1. $\begin{bmatrix} 0 & 1 \\ 1 & 0 \end{bmatrix}$

2. $\begin{bmatrix} 1 & 0 \\ 0 & 2 \end{bmatrix}$

3. $\begin{bmatrix} 0 & 1 \\ 2 & 0 \end{bmatrix}$

4. $\begin{bmatrix} 1 & 2 \\ 0 & 1 \end{bmatrix}$

5. $\begin{bmatrix} 0 & 1 \\ 1 & 2 \end{bmatrix}$

6. $\begin{bmatrix} 1 & 0 \\ 0 & 0 \end{bmatrix}$

7. $\begin{bmatrix} 0 & 0 & 1 \\ 1 & 0 & 0 \\ 0 & 1 & 0 \end{bmatrix}$

8. $\begin{bmatrix} 1 & 0 & 2 \\ 0 & 0 & 1 \\ 1 & 0 & 0 \end{bmatrix}$

9. $\begin{bmatrix} 1 & 0 & 0 \\ 0 & 2 & 0 \\ 0 & 0 & 1 \end{bmatrix}$

10. $\begin{bmatrix} 0 & 2 & 0 \\ 1 & 0 & 0 \\ 0 & 0 & 1 \end{bmatrix}$ **11.** $\begin{bmatrix} 1 & 0 & 0 \\ 0 & 0 & 1 \\ 0 & 1 & 0 \end{bmatrix}$ **12.** $\begin{bmatrix} 1 & 0 & 0 \\ 2 & 1 & 0 \\ 0 & 0 & 1 \end{bmatrix}$

In Problems 13–16, use an elementary matrix to perform the indicated row operation on the matrix

$$A = \begin{bmatrix} 1 & -4 & -1 \\ 3 & 6 & 9 \end{bmatrix}$$

13. $R_1 \leftrightarrow R_2$ **14.** $\frac{1}{3}R_2 \rightarrow R_2$

15. $R_2 + (-3)R_1 \rightarrow R_2$ **16.** $R_1 + (-\frac{1}{3})R_2 \rightarrow R_1$

B   In Problems 17–22, use elementary matrices to transform A to reduced form. If B is the reduced form of A, find a matrix P so that $B = PA$.

17. $A = \begin{bmatrix} 1 & 2 \\ 2 & 4 \end{bmatrix}$ **18.** $A = \begin{bmatrix} 1 & -1 \\ 3 & -3 \end{bmatrix}$

19. $A = \begin{bmatrix} 2 & 4 & 6 \\ -1 & -1 & -1 \end{bmatrix}$ **20.** $A = \begin{bmatrix} 1 & -2 & 1 \\ 4 & -5 & 1 \end{bmatrix}$

21. $A = \begin{bmatrix} 1 & 0 & -1 \\ 1 & 1 & 1 \\ 1 & -2 & -5 \end{bmatrix}$ **22.** $A = \begin{bmatrix} 1 & 1 & 0 \\ 2 & 2 & 1 \\ 2 & 2 & -2 \end{bmatrix}$

In Problems 23–30, find the inverse of each elementary matrix.

23. $\begin{bmatrix} 0 & 1 \\ 1 & 0 \end{bmatrix}$ **24.** $\begin{bmatrix} 1 & -2 \\ 0 & 1 \end{bmatrix}$ **25.** $\begin{bmatrix} 1 & 0 \\ 3 & 1 \end{bmatrix}$

26. $\begin{bmatrix} 2 & 0 \\ 0 & 1 \end{bmatrix}$ **27.** $\begin{bmatrix} 1 & 0 & 0 \\ 0 & 0 & 1 \\ 0 & 1 & 0 \end{bmatrix}$ **28.** $\begin{bmatrix} 1 & 0 & 2 \\ 0 & 1 & 0 \\ 0 & 0 & 1 \end{bmatrix}$

29. $\begin{bmatrix} 1 & 0 & 0 \\ -5 & 1 & 0 \\ 0 & 0 & 1 \end{bmatrix}$ **30.** $\begin{bmatrix} 1 & 0 & 0 \\ 0 & 1 & 0 \\ 0 & 0 & \frac{1}{4} \end{bmatrix}$

In Problems 31–36, use the inverses of the elementary matrices in each product to find the inverse of the product.

31. $\begin{bmatrix} 16 & 3 \\ 5 & 1 \end{bmatrix} = \begin{bmatrix} 1 & 3 \\ 0 & 1 \end{bmatrix}\begin{bmatrix} 1 & 0 \\ 5 & 1 \end{bmatrix}$

32. $\begin{bmatrix} 1 & 4 \\ -2 & -7 \end{bmatrix} = \begin{bmatrix} 1 & 0 \\ -2 & 1 \end{bmatrix}\begin{bmatrix} 1 & 4 \\ 0 & 1 \end{bmatrix}$

33. $\begin{bmatrix} 1 & 2 \\ -1 & 2 \end{bmatrix} = \begin{bmatrix} 1 & 0 \\ -1 & 1 \end{bmatrix}\begin{bmatrix} 1 & 0 \\ 0 & 4 \end{bmatrix}\begin{bmatrix} 1 & 2 \\ 0 & 1 \end{bmatrix}$

34. $\begin{bmatrix} 1 & 3 \\ -1 & -1 \end{bmatrix} = \begin{bmatrix} 1 & 0 \\ 0 & 2 \end{bmatrix}\begin{bmatrix} 1 & 0 \\ -\frac{1}{2} & 1 \end{bmatrix}\begin{bmatrix} 1 & 3 \\ 0 & 1 \end{bmatrix}$

35. $\begin{bmatrix} 0 & 1 & 0 \\ 1 & 2 & 0 \\ -3 & 0 & 1 \end{bmatrix} = \begin{bmatrix} 0 & 1 & 0 \\ 1 & 0 & 0 \\ 0 & 0 & 1 \end{bmatrix} \begin{bmatrix} 1 & 2 & 0 \\ 0 & 1 & 0 \\ 0 & 0 & 1 \end{bmatrix} \begin{bmatrix} 1 & 0 & 0 \\ 0 & 1 & 0 \\ -3 & 0 & 1 \end{bmatrix}$

36. $\begin{bmatrix} 1 & 4 & 0 \\ 0 & -2 & 1 \\ 0 & 1 & 0 \end{bmatrix} = \begin{bmatrix} 1 & 0 & 0 \\ 0 & 1 & -2 \\ 0 & 0 & 1 \end{bmatrix} \begin{bmatrix} 1 & 0 & 0 \\ 0 & 0 & 1 \\ 0 & 1 & 0 \end{bmatrix} \begin{bmatrix} 1 & 4 & 0 \\ 0 & 1 & 0 \\ 0 & 0 & 1 \end{bmatrix}$

In Problems 37–40, express each matrix as a product of elementary matrices.

37. $\begin{bmatrix} 1 & 2 \\ 3 & 7 \end{bmatrix}$

38. $\begin{bmatrix} 1 & -4 \\ -2 & 9 \end{bmatrix}$

39. $\begin{bmatrix} 1 & 0 & -2 \\ 0 & 0 & 3 \\ 2 & 1 & -4 \end{bmatrix}$

40. $\begin{bmatrix} 0 & 0 & 2 \\ -1 & 1 & -3 \\ 1 & 0 & 3 \end{bmatrix}$

C **41.** Find all possible 2×2 elementary matrices corresponding to each of the three elementary row operations.

42. Find all possible 3×3 elementary matrices corresponding to each of the three elementary row operations.

43. If E is a 3×3 elementary matrix formed by the operation $R_i \leftrightarrow R_j$, show that $E^2 = I$.

44. If E is a 3×3 elementary matrix formed by the operation $kR_i \rightarrow R_i$, show that E^2 is also an elementary matrix. What row operation can be used to form E^2?

45. If E is a 3×3 elementary matrix formed by the operation $R_i + kR_j \rightarrow R_i$, show that E^2 is also an elementary matrix. What row operation can be used to form E^2?

46. Prove Theorem 9.

47. Use Theorem 8 to show that if A is row equivalent to B and B is row equivalent to C, then A is row equivalent to C.

48. If A is an $n \times n$ matrix with a row of 0's, show that A is not invertible.

49. If A is an $n \times n$ matrix with two identical rows, show that A is not invertible.

50. If A is row equivalent to B and A is invertible, show that B is also invertible.

51. If A and B are $n \times n$ matrices and AB is invertible, then show that A and B are both invertible.

▌2-5▐ Chapter Review

Important Terms and Symbols

2-1. *Matrix addition and scalar multiplication.* Size or dimension of a matrix, $m \times n$ matrix, square matrix, column matrix, row matrix, equal matrices, sum of two matrices, zero matrix, negative of a matrix, matrix subtraction, properties of matrix addition, product of a number k and a matrix A, scalar, scalar multiplication, properties of scalar multiplication, $[a_{ij}]_{m \times n}$

2-2. *Matrix multiplication.* Dot product, matrix product, properties of matrix multiplication, incidence matrices

2-3. *Inverse of a square matrix.* Identity matrix, multiplicative inverse, invertible matrix, properties of matrix inverses, matrix equations, left-multiply, right-multiply

2-4. *Elementary matrices and row equivalence.* Row operations by multiplication, elementary matrices, inverses of elementary matrices, reverse operations, fundamental theorem

▐ Exercise 2-5 Chapter Review

Work through all the problems in this chapter review and check your answers in the back of the book. (Answers to most review problems are there.) Where weaknesses show up, review appropriate sections in the text.

A   *In Problems 1–14, perform the operations that are defined, given the following matrices:*

$$A = \begin{bmatrix} 2 & 1 \\ 5 & 3 \end{bmatrix} \qquad B = \begin{bmatrix} 1 & -1 \\ 2 & 1 \end{bmatrix} \qquad C = \begin{bmatrix} 1 & 2 \end{bmatrix}$$

$$D = \begin{bmatrix} 2 \\ 3 \end{bmatrix} \qquad E = \begin{bmatrix} 1 & 2 & -1 \\ 1 & 0 & -2 \end{bmatrix} \qquad F = \begin{bmatrix} 1 & -1 \\ 2 & -1 \\ -1 & 0 \end{bmatrix}$$

1. $C \cdot D$ **2.** CE **3.** FD **4.** FE

5. EC **6.** AEF **7.** AFE **8.** $2A + 3B$

9. $B + 3FE$ **10.** $BE - 2E$ **11.** $AB + BA$ **12.** $CD + EA$

13. $A - DC$ **14.** $DC - EF$

15. Find the inverse of $A = \begin{bmatrix} 3 & 2 \\ 4 & 3 \end{bmatrix}$

16. Write the system given below as a matrix equation, and then solve using an inverse. [Note: The inverse was found in Problem 15.]

$$3x_1 + 2x_2 = k_1$$
$$4x_1 + 3x_2 = k_2$$

(A) $k_1 = 3, k_2 = 5$ (B) $k_1 = 7, k_2 = 10$ (C) $k_1 = 4, k_2 = 2$

In Problems 17–19, use elementary matrices to perform each row operation on the matrix below:

$$\begin{bmatrix} 2 & -4 & 6 \\ 0 & 2 & 5 \end{bmatrix}$$

17. $\frac{1}{2}R_1 \to R_1$ **18.** $R_1 \leftrightarrow R_2$ **19.** $R_1 + 2R_2 \to R_1$

B In Problems 20–25, A, B, and X are all $n \times n$ invertible matrices. Solve each equation for X.

20. $AX = AB$

21. $AX = BA$

22. $AX = A + B$

23. $ABX = A + B$

24. $AXB = A + B$

25. $XAB = A + B$

In Problems 26 and 27, find A^{-1} and show that $A^{-1}A = I$.

26. $A = \begin{bmatrix} 1 & 2 & 3 \\ 2 & 3 & 4 \\ 1 & 2 & 1 \end{bmatrix}$

27. $A = \begin{bmatrix} 4 & 5 & 6 \\ 4 & 5 & -6 \\ 1 & 1 & 1 \end{bmatrix}$

In Problems 28 and 29, write each system as a matrix equation and solve using inverses. [Note: The inverses were found in Problems 26 and 27.]

28. $\begin{aligned} x_1 + 2x_2 + 3x_3 &= k_1 \\ 2x_1 + 3x_2 + 4x_3 &= k_2 \\ x_1 + 2x_2 + x_3 &= k_3 \end{aligned}$
(A) $k_1 = 1, k_2 = 3, k_3 = 3$
(B) $k_1 = 0, k_2 = 0, k_3 = -2$
(C) $k_1 = -3, k_2 = -4, k_3 = 1$

29. $\begin{aligned} 4x_1 + 5x_2 + 6x_3 &= k_1 \\ 4x_1 + 5x_2 - 6x_3 &= k_2 \\ x_1 + x_2 + x_3 &= k_3 \end{aligned}$
(A) $k_1 = 3, k_2 = -1, k_3 = 0$
(B) $k_1 = 1, k_2 = 2, k_3 = -4$
(C) $k_1 = 0, k_2 = 1, k_3 = -1$

In Problems 30 and 31, use elementary matrices to transform each matrix A to its reduced form B, and then find a matrix P satisfying $PA = B$.

30. $A = \begin{bmatrix} 1 & 2 & 3 \\ -1 & -1 & 1 \end{bmatrix}$

31. $A = \begin{bmatrix} 2 & 0 & 0 \\ -2 & 1 & 0 \\ -4 & 2 & 1 \end{bmatrix}$

32. Find the inverse of $\begin{bmatrix} 1 & k & k \\ 0 & 1 & k \\ 0 & 0 & 1 \end{bmatrix}$

33. Find the inverse of $\begin{bmatrix} 1 & -k & 0 & 0 \\ 0 & 1 & -k & 0 \\ 0 & 0 & 1 & -k \\ 0 & 0 & 0 & 1 \end{bmatrix}$

C **34.** If A is an $n \times n$ matrix satisfying

$$A^2 - kA + I = 0$$

for some constant k, show that $A^{-1} = kI - A$.

35. (A) If A, B, and P are $n \times n$ matrices satisfying $B = PAP^{-1}$, show that $B^2 = PA^2P^{-1}$.

(B) If A is invertible, show that $B^{-1} = PA^{-1}P^{-1}$.

36. If X_1 is a solution to the system $AX = B$ and X_0 is a solution to the homogeneous system $AX = 0$, show that $X_1 + aX_0$ is also a solution to the system $AX = B$ for any real number a.

37. If X_1 and X_2 are solutions of the homogeneous system $AX = 0$, show that $aX_1 + bX_2$ is also a solution for any real numbers a and b.

38. If A and B are $n \times n$ matrices and $AB = 0$, show that either $A = 0$ or $B = 0$, or both A and B are not invertible.

39. If A and B are $n \times n$ invertible matrices, show that $AB \neq 0$.

40. If A is row equivalent to I, show that A^{-1} is also row equivalent to I.

▮□▮ Applications

Ore	Nickel (%)	Copper (%)
A	1	2
B	2	5

41. *Resource allocation.* A mining company has two mines with ore compositions as given in the table. How many tons of each ore should be used to obtain 4.5 tons of nickel and 10 tons of copper? Set up a system of equations and solve using Gauss–Jordan elimination.

42. (A) Set up Problem 41 as a matrix equation and solve using the inverse of the coefficient matrix.

(B) Solve Problem 41 as in part (A) if 2.3 tons of nickel and 5 tons of copper are needed.

43. *Airline scheduling.* Use an incidence matrix and matrix multiplication to determine the maximum number of flights necessary to fly between any two cities in the schedule given in the table below:

Origin	Destination
Akron	Dayton
Bay City	Akron
Canton	Dayton
Dayton	Bay City
Dayton	Canton

44. *Heat conduction.* In the grid pictured here let x_1 and x_2 represent the unknown temperatures at the interior points and let t_1, t_2, t_3, and t_4 represent the temperatures at the ends of the rods.

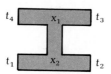

(A) If the temperature at each intersection point is the average of the temperatures at the adjacent points in the grid (adjacent points are either other intersection points or ends of bars), show that x_1 and x_2 satisfy the following system of equations:

$$-x_1 + 3x_2 = t_1 + t_2$$
$$3x_1 - x_2 = t_3 + t_4$$

(B) Find the inverse of the coefficient matrix for the system in part (A).

(C) Use the inverse of the coefficient matrix to solve the system for the following sets of temperatures:

	t_1	t_2	t_3	t_4
1	0°	0°	0°	16°
2	0°	0°	16°	16°
3	0°	8°	8°	16°

|3| Determinants

| 3 | Contents

In this chapter we will associate with each square matrix a number called the *determinant* of the matrix. It turns out that this number will tell us whether or not a matrix A is invertible. Furthermore, the determinant will provide us with a formula for A^{-1} and a formula for the solutions to the linear system $AX = B$. The definition of the determinant given in Section 3-1 is rather complicated, and the evaluation of determinants of square matrices of order 4 or greater is a tedious calculation. The properties of determinants discussed in Sections 3-2 and 3-3 will simplify the evaluation of determinants.

   Although determinants will provide alternate methods for finding inverses and solving systems of equations, these methods are very inefficient when compared to the methods presented in Chapters 1 and 2. Thus, we will be much more interested in the theoretical implications of determinants than in their application to problems involving large matrices or systems.

| 3-1 | Definition of the Determinant

- Determinants of 2×2 Matrices
- Cofactor Expansion
- Determinant of an $n \times n$ Matrix
- Application

■ Determinants of 2×2 Matrices

Since a matrix is an array of numbers, it seems unlikely that one number could provide important information about the matrix. However, there is a particular number called the *determinant* associated with each square matrix that does provide some useful information about the matrix. To develop the definition of the determinant of a 2×2 matrix, it is helpful to begin by considering the problem of finding the inverse of such a matrix. It can be shown that if

$$A = \begin{bmatrix} a_{11} & a_{12} \\ a_{21} & a_{22} \end{bmatrix}$$

is any 2×2 matrix and if

$$a_{11}a_{22} - a_{21}a_{12} \neq 0$$

then A is invertible and A^{-1} is given by

$$A^{-1} = \frac{1}{a_{11}a_{22} - a_{21}a_{12}} \begin{bmatrix} a_{22} & -a_{12} \\ -a_{21} & a_{11} \end{bmatrix} \qquad (1)$$

If $a_{11}a_{22} - a_{21}a_{12} = 0$, then A is not invertible. This formula can be derived by the method used in Section 2-3. We will omit the calculations and simply verify the formula by multiplication:

$$A^{-1}A = \frac{1}{a_{11}a_{22} - a_{21}a_{12}} \begin{bmatrix} a_{22} & -a_{12} \\ -a_{21} & a_{11} \end{bmatrix}\begin{bmatrix} a_{11} & a_{12} \\ a_{21} & a_{22} \end{bmatrix}$$

$$= \frac{1}{a_{11}a_{22} - a_{21}a_{12}} \begin{bmatrix} a_{11}a_{22} - a_{21}a_{12} & 0 \\ 0 & a_{11}a_{22} - a_{21}a_{12} \end{bmatrix} = I$$

If $a_{11}a_{22} - a_{21}a_{12} = 0$, then the reduced form of $[A|I]$ will always contain a row of 0's to the left of the vertical line and A^{-1} will not exist.

Thus, the number $a_{11}a_{22} - a_{21}a_{12}$ gives us some very important information about A: A is invertible if and only if $a_{11}a_{22} - a_{21}a_{12} \neq 0$.

The number $a_{11}a_{22} - a_{21}a_{12}$ is called the **determinant** of A and is denoted by any of the following:

$$\det(A) = \begin{vmatrix} a_{11} & a_{12} \\ a_{21} & a_{22} \end{vmatrix} = a_{11}a_{22} - a_{21}a_{12}$$

Example 1 Find the determinant of each matrix. If the matrix is invertible, find the inverse.

(A) $A = \begin{bmatrix} 3 & 4 \\ 1 & 2 \end{bmatrix}$ (B) $B = \begin{bmatrix} 2 & -4 \\ 1 & -2 \end{bmatrix}$

Solution (A) $\det(A) = \begin{vmatrix} 3 & 4 \\ 1 & 2 \end{vmatrix} = (3)(2) - (1)(4) = 6 - 4 = 2$

Thus, A is invertible and A^{-1} is found using (1):

$$A^{-1} = \begin{bmatrix} 3 & 4 \\ 1 & 2 \end{bmatrix}^{-1} = \frac{1}{2}\begin{bmatrix} 2 & -4 \\ -1 & 3 \end{bmatrix} = \begin{bmatrix} 1 & -2 \\ -\frac{1}{2} & \frac{3}{2} \end{bmatrix}$$

Check

$$\begin{bmatrix} 1 & -2 \\ -\frac{1}{2} & \frac{3}{2} \end{bmatrix}\begin{bmatrix} 3 & 4 \\ 1 & 2 \end{bmatrix} = \begin{bmatrix} 3 - 2 & 4 - 4 \\ -\frac{3}{2} + \frac{3}{2} & -2 + 3 \end{bmatrix} = \begin{bmatrix} 1 & 0 \\ 0 & 1 \end{bmatrix} = I$$

(B) $\det(B) = \begin{vmatrix} 2 & -4 \\ 1 & -2 \end{vmatrix} = (2)(-2) - (1)(-4) = -4 + 4 = 0$

Thus, B is not invertible.

Problem 1 Find the determinant of each matrix. If the matrix is invertible, find the inverse.

(A) $\begin{bmatrix} 3 & -6 \\ 2 & -4 \end{bmatrix}$ (B) $\begin{bmatrix} 3 & 7 \\ 2 & 5 \end{bmatrix}$ ∎

Now let us consider the problem of solving the system

$$a_{11}x_1 + a_{12}x_2 = b_1$$
$$a_{21}x_1 + a_{22}x_2 = b_2$$

or

$$AX = B \tag{2}$$

If $\det(A) \neq 0$, then A^{-1} is given by (1). Left-multiplying both sides of (2) by A^{-1}, we have

$$X = A^{-1}B = \frac{1}{a_{11}a_{22} - a_{21}a_{12}} \begin{bmatrix} a_{22} & -a_{12} \\ -a_{21} & a_{11} \end{bmatrix} \begin{bmatrix} b_1 \\ b_2 \end{bmatrix}$$

$$= \begin{bmatrix} \dfrac{b_1 a_{22} - b_2 a_{12}}{a_{11}a_{22} - a_{21}a_{12}} \\ \\ \dfrac{a_{11}b_2 - a_{21}b_1}{a_{11}a_{22} - a_{21}a_{12}} \end{bmatrix}$$

Thus, the unique solution is given by

$$x_1 = \frac{b_1 a_{22} - b_2 a_{12}}{a_{11}a_{22} - a_{21}a_{12}} \quad \text{and} \quad x_2 = \frac{a_{11}b_2 - a_{21}b_1}{a_{11}a_{22} - a_{21}a_{12}}$$

Notice the pattern in the numerators of the expressions for x_1 and x_2. These look like determinants. In fact, we can use determinant notation to express these solutions as

$$x_1 = \frac{\begin{vmatrix} b_1 & a_{12} \\ b_2 & a_{22} \end{vmatrix}}{\begin{vmatrix} a_{11} & a_{12} \\ a_{21} & a_{22} \end{vmatrix}} \quad \text{and} \quad x_2 = \frac{\begin{vmatrix} a_{11} & b_1 \\ a_{21} & b_2 \end{vmatrix}}{\begin{vmatrix} a_{11} & a_{12} \\ a_{21} & a_{22} \end{vmatrix}}$$

These formulas are very easy to remember. The denominator of each fraction is $\det(A)$. The numerator for x_1 is formed from $\det(A)$ by replacing the coefficients of x_1 with the constant terms. Similarly, the numerator for x_2 is formed from $\det(A)$ by replacing the coefficients of x_2 with the constant terms. Thus, determinant notation gives us a general formula for any 2×2 system of linear equations that has a unique solution. These formulas, known as *Cramer's Rule* for a 2×2 system, will be extended to $n \times n$ systems in Section 3-4.

We summarize our discussion of second-order determinants in the box for convenient reference.

Determinant of a 2 × 2 Matrix

1. If $A = \begin{bmatrix} a_{11} & a_{12} \\ a_{21} & a_{22} \end{bmatrix}$, then

$$\det(A) = \begin{vmatrix} a_{11} & a_{12} \\ a_{21} & a_{22} \end{vmatrix} = a_{11}a_{22} - a_{21}a_{12}$$

2. If $\det(A) \neq 0$, then A is invertible and

$$A^{-1} = \frac{1}{\det(A)} \begin{bmatrix} a_{22} & -a_{12} \\ -a_{21} & a_{11} \end{bmatrix}$$

3. If $\det(A) \neq 0$, then the linear system

$$AX = B \qquad \text{where} \qquad X = \begin{bmatrix} x_1 \\ x_2 \end{bmatrix} \qquad \text{and} \qquad B = \begin{bmatrix} b_1 \\ b_2 \end{bmatrix}$$

has the unique solution given by

$$x_1 = \frac{\begin{vmatrix} b_1 & a_{12} \\ b_2 & a_{22} \end{vmatrix}}{\det(A)} \qquad x_2 = \frac{\begin{vmatrix} a_{11} & b_1 \\ a_{21} & b_2 \end{vmatrix}}{\det(A)}$$

The last two formulas are known as **Cramer's Rule** for a 2 × 2 linear system.

Example 2 Use Cramer's Rule to solve the system: $2x_1 + 3x_2 = 17$
$$3x_1 + 6x_2 = 18$$

Solution $\quad x_1 = \dfrac{\begin{vmatrix} 17 & 3 \\ 18 & 6 \end{vmatrix}}{\begin{vmatrix} 2 & 3 \\ 3 & 6 \end{vmatrix}} = \dfrac{(17)(6) - (18)(3)}{(2)(6) - (3)(3)} = \dfrac{48}{3} = 16$

$\quad x_2 = \dfrac{\begin{vmatrix} 2 & 17 \\ 3 & 18 \end{vmatrix}}{\begin{vmatrix} 2 & 3 \\ 3 & 6 \end{vmatrix}} = \dfrac{(2)(18) - (3)(17)}{3} = \dfrac{-15}{3} = -5$

Problem 2 Use Cramer's Rule to solve the system: $3x_1 + 4x_2 = 15$
$$x_1 + 3x_2 = 20 \qquad ∎$$

To summarize what we have done so far, the determinant of a 2 × 2 matrix A tells us whether A is invertible and can be used to state explicit formulas for both A^{-1} and the solution of $AX = B$. In Sections 3-3 and 3-4 we will see that the determinant of an $n \times n$ matrix can be used in exactly the same fashion.

▪ Cofactor Expansion

The definition of the determinant of an $n \times n$ matrix is considerably more complicated when n is greater than 2. Preliminary to defining the determinant of an $n \times n$ matrix, we will consider several related topics.

First, given an $n \times n$ matrix A, the $(n - 1) \times (n - 1)$ submatrix M_{ij} formed by deleting the ith row and jth column of A is called the ***ij*th minor** of A or the **minor of element a_{ij}.**

A 2×2 matrix has four elements, and hence, has four minors. Each minor is a 1×1 matrix. A 3×3 matrix has nine elements, and hence, has nine minors. Each minor is a 2×2 matrix. For example, for

$$A = \begin{bmatrix} a_{11} & a_{12} & a_{13} \\ a_{21} & a_{22} & a_{23} \\ a_{31} & a_{32} & a_{33} \end{bmatrix}$$

the minor of a_{12}, denoted by M_{12}, is found by deleting the first row and second column in A (since a_{12} is in the first row and second column):

$$\begin{bmatrix} a_{11} & a_{12} & a_{13} \\ a_{21} & a_{22} & a_{23} \\ a_{31} & a_{32} & a_{33} \end{bmatrix} \qquad M_{12} = \begin{bmatrix} a_{21} & a_{23} \\ a_{31} & a_{33} \end{bmatrix}$$

Similarly, M_{23} is found by deleting the second row and third column of A:

$$\begin{bmatrix} a_{11} & a_{12} & a_{13} \\ a_{21} & a_{22} & a_{23} \\ a_{31} & a_{32} & a_{33} \end{bmatrix} \qquad M_{23} = \begin{bmatrix} a_{11} & a_{12} \\ a_{31} & a_{32} \end{bmatrix}$$

The actual crossouts of rows and columns are usually done mentally. Can you find the remaining seven minors for A?

Example 3   Find M_{22} and M_{44} for the matrix:

$$A = \begin{bmatrix} 1 & 2 & -1 & 0 \\ 3 & 1 & 5 & -2 \\ -4 & 1 & 6 & -7 \\ 9 & -3 & 4 & 2 \end{bmatrix}$$

Solution   The submatrix M_{22} is the minor of the element in the second row and second column. Deleting the second row and second column, we obtain

$$\begin{bmatrix} 1 & 2 & -1 & 0 \\ 3 & 1 & 5 & -2 \\ -4 & 1 & 6 & -7 \\ 9 & -3 & 4 & 2 \end{bmatrix} \qquad M_{22} = \begin{bmatrix} 1 & -1 & 0 \\ -4 & 6 & -7 \\ 9 & 4 & 2 \end{bmatrix}$$

└── Done mentally

The submatrix M_{44} is the minor of the element in the fourth row and fourth column. Deleting the fourth row and fourth column, we obtain

$$
\begin{bmatrix}
1 & 2 & -1 & 0 \\
3 & 1 & 5 & -2 \\
-4 & 1 & 6 & -7 \\
9 & -3 & 4 & 2
\end{bmatrix}
\qquad
M_{44} =
\begin{bmatrix}
1 & 2 & -1 \\
3 & 1 & 5 \\
-4 & 1 & 6
\end{bmatrix}
$$

$\llcorner$ Done mentally

Problem 3   Find M_{14} and M_{32} for the matrix in Example 3.   ▐▌

To continue the development of the concepts to be used in the definition of the determinant of an $n \times n$ matrix, we define the **ijth cofactor,** A_{ij}, of an $n \times n$ matrix to be

$$A_{ij} = (-1)^{i+j} \det(M_{ij})$$

Notice that the ijth minor is a *matrix*, while the ijth cofactor is a *number*.

Example 4   Find A_{11}, A_{12}, and A_{13} for

$$
A =
\begin{bmatrix}
1 & -1 & -2 \\
-4 & -3 & 3 \\
2 & 0 & 4
\end{bmatrix}
$$

Solution $A_{11} = (-1)^{1+1} \det(M_{11}) = (-1)^2 \begin{vmatrix} -3 & 3 \\ 0 & 4 \end{vmatrix} = -12$

$A_{12} = (-1)^{1+2} \det(M_{12}) = (-1)^3 \begin{vmatrix} -4 & 3 \\ 2 & 4 \end{vmatrix} = 22$

$A_{13} = (-1)^{1+3} \det(M_{13}) = (-1)^4 \begin{vmatrix} -4 & -3 \\ 2 & 0 \end{vmatrix} = 6$

Problem 4   Find A_{31}, A_{32}, and A_{33} for the matrix in Example 4.   ▐▌

The expression $(-1)^{i+j}$ which is used in the definition of the cofactor can be evaluated by the following rule:

$$(-1)^{i+j} = \begin{cases} 1 & \text{if } i+j \text{ is even} \\ -1 & \text{if } i+j \text{ is odd} \end{cases}$$

This rule determines the sign in front of the determinant of the minor M_{ij}. The proper sign also can be determined by using an alternating pattern of $+$ and $-$ signs, starting with $+$ in the upper left-hand corner of the matrix. For example, the sign pattern for a 4×4 matrix would be

$$\begin{bmatrix} + & - & + & - \\ - & + & - & + \\ + & - & + & - \\ - & + & - & + \end{bmatrix}$$

The **cofactor expansion** of an $n \times n$ matrix A along a given row (or column) is formed by multiplying each element in the row (or column) by its cofactor and summing. Thus, the **cofactor expansion along the ith row is**

$$a_{i1}A_{i1} + a_{i2}A_{i2} + \cdots + a_{in}A_{in}$$

and the **cofactor expansion along the jth column is**

$$a_{1j}A_{1j} + a_{2j}A_{2j} + \cdots + a_{nj}A_{nj}$$

Example 5   Find the cofactor expansion along the second row and also along the third column:

$$A = \begin{bmatrix} 1 & -1 & 0 \\ -6 & 3 & 5 \\ 2 & -3 & -4 \end{bmatrix}$$

Solution   The cofactor expansion along the second row is

$$a_{21}A_{21} + a_{22}A_{22} + a_{23}A_{23} \qquad \text{Row 2}$$

$$= (-6)(-1)^{2+1}\begin{vmatrix} -1 & 0 \\ -3 & -4 \end{vmatrix} + (3)(-1)^{2+2}\begin{vmatrix} 1 & 0 \\ 2 & -4 \end{vmatrix} + (5)(-1)^{2+3}\begin{vmatrix} 1 & -1 \\ 2 & -3 \end{vmatrix}$$

$$= (-6)(-1)(4) + (3)(1)(-4) + (5)(-1)(-1) = 17$$

The cofactor expansion along the third column is

$$a_{13}A_{13} + a_{23}A_{23} + a_{33}A_{33} \qquad \text{Column 3}$$

$$= (0)(-1)^{1+3}\begin{vmatrix} -6 & 3 \\ 2 & -3 \end{vmatrix} + (5)(-1)^{2+3}\begin{vmatrix} 1 & -1 \\ 2 & -3 \end{vmatrix} + (-4)(-1)^{3+3}\begin{vmatrix} 1 & -1 \\ -6 & 3 \end{vmatrix}$$

$$= (0)(1)(12) + (5)(-1)(-1) + (-4)(1)(-3) = 17$$

Problem 5   Find the cofactor expansion along the first row and also along the second column for the matrix in Example 5.   **∎**

In Example 5, both cofactor expansions had the same value. Surprisingly, this is not just a coincidence. Theorem 1 states that all cofactor expansions (for all rows and columns) for a given matrix have the same value. (We omit the proof.)

Theorem 1 | All cofactor expansions for a given $n \times n$ matrix A have the same value.

▪ Determinant of an $n \times n$ Matrix

We are now ready to define the *determinant* of an $n \times n$ matrix A. The **determinant** of an $n \times n$ matrix A is the common value of all cofactor expansions for A. That is, $\det(A)$ is given by any of the formulas in the box.

Determinant of an $n \times n$ Matrix

$\det(A) = a_{i_1} A_{i_1} + a_{i_2} A_{i_2} + \cdots + a_{in} A_{in}$ Expansion along
for any i, $1 \le i \le n$ the ith row

$\det(A) = a_{1j} A_{1j} + a_{2j} A_{2j} + \cdots + a_{nj} A_{nj}$ Expansion along
for any j, $1 \le j \le n$ the jth column

Example 6   Find the determinant of each matrix.

$$\text{(A)}\ A = \begin{bmatrix} 0 & 1 & 2 \\ -1 & 3 & -2 \\ 5 & -4 & -3 \end{bmatrix} \qquad \text{(B)}\ A = \begin{bmatrix} 1 & 0 & -2 & 0 \\ 2 & 0 & 7 & 5 \\ -1 & 2 & 3 & -4 \\ -6 & 0 & 4 & 0 \end{bmatrix}$$

Solution   (A) Expanding along the first row, we have

$$\det(A) = a_{11}A_{11} + a_{12}A_{12} + a_{13}A_{13}$$

$$= (0)(-1)^{1+1}\begin{vmatrix} 3 & -2 \\ -4 & -3 \end{vmatrix} + (1)(-1)^{1+2}\begin{vmatrix} -1 & -2 \\ 5 & -3 \end{vmatrix} + (2)(-1)^{1+3}\begin{vmatrix} -1 & 3 \\ 5 & -4 \end{vmatrix}$$

$$= 0 + (1)(-1)(13) + (2)(1)(-11) = -35$$

Notice that if $a_{ij} = 0$, then it is not necessary to evaluate A_{ij}. Selecting the row or column with the greatest number of zero elements will simplify the calculations necessary to evaluate a determinant.

(B) Since column 2 has three 0's, we expand along this column:

$$\det(A) = a_{12}A_{12} + a_{22}A_{22} + a_{32}A_{32} + a_{42}A_{42}$$

$$= (0)A_{12} + (0)A_{22} + (2)(-1)^{3+2}\begin{vmatrix} 1 & -2 & 0 \\ 2 & 7 & 5 \\ -6 & 4 & 0 \end{vmatrix} + (0)A_{42}$$

$$= (-2)\begin{vmatrix} 1 & -2 & 0 \\ 2 & 7 & 5 \\ -6 & 4 & 0 \end{vmatrix} \qquad \begin{array}{l}\text{Expand this determinant} \\ \text{along the third column.}\end{array}$$

$$= (-2)\left(0 + (5)(-1)^{2+3}\begin{vmatrix} 1 & -2 \\ -6 & 4 \end{vmatrix} + 0\right)$$

$$= (-2)(5)(-1)(-8) = -80$$

Problem 6   Find the determinant of each matrix.

$$\text{(A) } A = \begin{bmatrix} 1 & -1 & 1 \\ 2 & 0 & 3 \\ -2 & 1 & 2 \end{bmatrix} \quad \text{(B) } A = \begin{bmatrix} 0 & -2 & 7 & 0 \\ 1 & -1 & 4 & 5 \\ 0 & 0 & 3 & 0 \\ 2 & -2 & 1 & 3 \end{bmatrix}$$

$\blacksquare\blacksquare$

The definition of the determinant is an example of a *recursive definition*. The determinant of an $n \times n$ matrix is defined in terms of $(n-1) \times (n-1)$ determinants, which in turn are defined in terms of $(n-2) \times (n-2)$ determinants, and so on, as illustrated in Example 6(B). This process must be repeated until we reach 2×2 determinants, which can be evaluated directly. For example, the evaluation of a 5×5 determinant requires the evaluation of five 4×4 determinants; each of these requires the evaluation of four 3×3 determinants; and each of these requires the evaluation of three 2×2 determinants. Thus, there are $5 \cdot 4 \cdot 3 = 60$ total 2×2 determinants involved in the evaluation of one 5×5 determinant. (How many 2×2 determinants would be involved in the evaluation of a 6×6 determinant?) The direct evaluation of determinants is a formidable task, even if a computer is used. Fortunately, the calculations can be reduced to a reasonable level if the matrix has many zero elements, as Example 6(B) illustrates. In later sections, we will learn techniques that will enable us to introduce zero elements in desirable locations in any matrix, making the evaluation of determinants of large matrices a more reasonable task.

▪ Application

$\blacksquare\square\blacksquare$

Example 7
Biology

A biologist is conducting an experiment with two species of bacteria. If x_1 and x_2 are the populations of the species, empirical evidence suggests that x_1 and x_2 must satisfy the system

$$3x_1 + x_2 = \lambda x_1$$
$$-4x_1 + 7x_2 = \lambda x_2$$

for some unknown number λ. (The Greek letter lambda is traditionally used in problems of this type.) Find the value of the number λ for which this system has nontrivial solutions, and then find those solutions.

Solution   The original system can be written as a homogeneous system:

$$(3 - \lambda)x_1 + x_2 = 0$$
$$-4x_1 + (7 - \lambda)x_2 = 0 \tag{3}$$

or

$$\begin{bmatrix} 3 - \lambda & 1 \\ -4 & 7 - \lambda \end{bmatrix} \begin{bmatrix} x_1 \\ x_2 \end{bmatrix} = 0 \tag{4}$$

According to Theorem 10 in Section 2-4, the homogeneous system $AX = 0$ has nontrivial solutions if and only if A is not invertible. Earlier in this section we saw that a 2×2 matrix is invertible if and only if $\det(A) \neq 0$; that is, a 2×2 matrix is not invertible if and only if $\det(A) = 0$. Thus, for system (3) to have nontrivial solutions, the determinant of the coefficient matrix in (4) must be 0. Consequently, system (3) will have nontrivial solutions if λ satisfies

$$\begin{vmatrix} 3 - \lambda & 1 \\ -4 & 7 - \lambda \end{vmatrix} = 0$$

$$(3 - \lambda)(7 - \lambda) + 4 = 0$$

$$\lambda^2 - 10\lambda + 25 = 0$$

$$(\lambda - 5)^2 = 0$$

$$\lambda = 5$$

Substituting $\lambda = 5$ in (3) produces the system

$$-2x_1 + x_2 = 0$$
$$-4x_1 + 2x_2 = 0$$

The nontrivial solutions of this system are given by $x_1 = t/2$ and $x_2 = t$ for any nonzero real number t.

Problem 7 Repeat Example 7 for the system

$$5x_1 - 4x_2 = \lambda x_1$$
$$x_1 + x_2 = \lambda x_2$$

∎

Answers to Matched Problems

1. (A) $\det(A) = 0$, A^{-1} does not exist   (B) $\det(A) = 1$, $A^{-1} = \begin{bmatrix} 5 & -7 \\ -2 & 3 \end{bmatrix}$

2. $x_1 = \dfrac{\begin{vmatrix} 15 & 4 \\ 20 & 3 \end{vmatrix}}{\begin{vmatrix} 3 & 4 \\ 1 & 3 \end{vmatrix}} = -7$, $x_2 = \dfrac{\begin{vmatrix} 3 & 15 \\ 1 & 20 \end{vmatrix}}{\begin{vmatrix} 3 & 4 \\ 1 & 3 \end{vmatrix}} = 9$

3. $M_{14} = \begin{bmatrix} 3 & 1 & 5 \\ -4 & 1 & 6 \\ 9 & -3 & 4 \end{bmatrix}$, $M_{32} = \begin{bmatrix} 1 & -1 & 0 \\ 3 & 5 & -2 \\ 9 & 4 & 2 \end{bmatrix}$

4. $A_{31} = -9$, $A_{32} = 5$, $A_{33} = -7$ 5. 17   6. (A) 9   (B) -42

7. $\lambda = 3$; $x_1 = 2t$, $x_2 = t$, t any nonzero real number

∎ Exercise 3-1

A *Evaluate the determinant of each matrix.*

1. $\begin{bmatrix} 2 & 1 \\ 1 & 1 \end{bmatrix}$ 2. $\begin{bmatrix} 4 & 2 \\ 2 & 1 \end{bmatrix}$ 3. $\begin{bmatrix} -2 & -3 \\ 2 & 4 \end{bmatrix}$ 4. $\begin{bmatrix} -4 & 3 \\ 2 & 0 \end{bmatrix}$

Use Equation (1) to find the inverse of each matrix, if it exists.

5. $\begin{bmatrix} 5 & 2 \\ 2 & 1 \end{bmatrix}$
6. $\begin{bmatrix} 2 & 3 \\ -3 & -4 \end{bmatrix}$
7. $\begin{bmatrix} 2 & 0 \\ 4 & 1 \end{bmatrix}$

8. $\begin{bmatrix} 3 & -2 \\ -6 & 4 \end{bmatrix}$
9. $\begin{bmatrix} 4 & -2 \\ 3 & 1 \end{bmatrix}$
10. $\begin{bmatrix} -4 & 4 \\ 2 & 3 \end{bmatrix}$

Use Cramer's Rule to solve each system.

11. $3x_1 + 7x_2 = 15$
$\quad\;\; x_1 + 3x_2 = \;\; 7$

12. $5x_1 + 8x_2 = 1$
$\quad\;\; 2x_1 + 3x_2 = 1$

13. $7x_1 - 6x_2 = 28$
$\quad\;\; 4x_1 - 3x_2 = 16$

14. $\quad\; 7x_1 - 8x_2 = -4$
$\quad\; -2x_1 + 3x_2 = \;\; 2$

Problems 15–20 refer to the matrix

$$A = \begin{bmatrix} 1 & 2 & 3 \\ 4 & 5 & 6 \\ 7 & 8 & 9 \end{bmatrix}$$

15. Find M_{11}, M_{12}, and M_{13}.
16. Find A_{11}, A_{12}, and A_{13}.
17. Evaluate the cofactor expansion along the first row of A.
18. Find M_{22} and M_{32}.
19. Find A_{22} and A_{32}.
20. Evaluate the cofactor expansion along the second column of A.

B *Problems 21–26 refer to the matrix*

$$A = \begin{bmatrix} 2 & 1 & 4 & 7 \\ -1 & 0 & -2 & 5 \\ 6 & -3 & -4 & 3 \\ 8 & -6 & -5 & -7 \end{bmatrix}$$

21. Find M_{31}, M_{32}, M_{33}, and M_{34}.
22. Expand M_{31} along the first column in order to find A_{31}.
23. Expand M_{32} along the first row in order to find A_{32}.
24. Expand M_{33} along the second column in order to find A_{33}.
25. Expand M_{34} along the second row in order to find A_{34}.
26. Evaluate the cofactor expansion along the third row of A.

In Problems 27–36, find the determinant.

27. $\begin{bmatrix} 2 & 0 & 0 \\ 3 & 1 & -1 \\ 4 & 2 & -1 \end{bmatrix}$
28. $\begin{bmatrix} 3 & 4 & -4 \\ 2 & 0 & -2 \\ 5 & 0 & 1 \end{bmatrix}$

29. $\begin{bmatrix} 4 & -2 & 1 \\ 4 & 2 & -1 \\ 8 & -4 & 2 \end{bmatrix}$
30. $\begin{bmatrix} 2 & -5 & 7 \\ -1 & 6 & -6 \\ 1 & 1 & 1 \end{bmatrix}$

31. $\begin{bmatrix} 0 & 1 & 0 & 3 \\ -1 & 2 & -3 & 1 \\ 0 & 0 & 0 & -4 \\ -1 & 2 & 1 & -3 \end{bmatrix}$
32. $\begin{bmatrix} -1 & 0 & 2 & 0 \\ 4 & 0 & -3 & 2 \\ 1 & -2 & 3 & 2 \\ 5 & 0 & -2 & 0 \end{bmatrix}$

33. $\begin{bmatrix} 1 & 0 & 2 & 0 \\ 0 & 1 & 0 & 2 \\ 2 & 0 & 1 & 0 \\ 0 & 2 & 0 & 1 \end{bmatrix}$
34. $\begin{bmatrix} 1 & 4 & 7 & 3 \\ 0 & 2 & -1 & 5 \\ 0 & 0 & 3 & 9 \\ 0 & 0 & 0 & 4 \end{bmatrix}$

35. $\begin{bmatrix} 2 & -1 & 7 & -4 & 5 \\ 0 & 3 & 0 & 0 & 0 \\ 0 & -4 & 5 & 6 & -9 \\ 0 & -7 & 0 & 4 & 2 \\ 0 & 5 & 0 & 2 & 6 \end{bmatrix}$
36. $\begin{bmatrix} 2 & 1 & 0 & 0 & 0 \\ 1 & 2 & 1 & 0 & 0 \\ 0 & 1 & 2 & 1 & 0 \\ 0 & 0 & 1 & 2 & 1 \\ 0 & 0 & 0 & 1 & 2 \end{bmatrix}$

In Problems 37–40, find the value(s) of λ for which the given system has nontrivial solutions, and then find those solutions.

37. $6x_1 - 4x_2 = \lambda x_1$
    $x_1 + 2x_2 = \lambda x_2$

38. $3x_1 - x_2 = \lambda x_1$
    $x_1 + x_2 = \lambda x_2$

39. $7x_1 + 3x_2 = \lambda x_1$
    $x_1 + 5x_2 = \lambda x_2$

40. $x_1 + 2x_2 = \lambda x_1$
    $2x_1 - 2x_2 = \lambda x_2$

In Problems 41–44, find all values of λ that satisfy the equation.

41. $\begin{vmatrix} 1-\lambda & 0 & -1 \\ 0 & 2-\lambda & 0 \\ -1 & 0 & 1-\lambda \end{vmatrix} = 0$

42. $\begin{vmatrix} 2-\lambda & 1 & -2 \\ 1 & 3-\lambda & -1 \\ -2 & -1 & 2-\lambda \end{vmatrix} = 0$

43. $\begin{vmatrix} 3-\lambda & -2 & 1 \\ -2 & -\lambda & 2 \\ 1 & 2 & -5-\lambda \end{vmatrix} = 0$

44. $\begin{vmatrix} 3-\lambda & 2 & -2 \\ 2 & 1-\lambda & 0 \\ -2 & 0 & -4-\lambda \end{vmatrix} = 0$

C   In Problems 45 and 46, let

$$A = \begin{bmatrix} a_{11} & a_{12} & a_{13} \\ a_{21} & a_{22} & a_{23} \\ a_{31} & a_{32} & a_{33} \end{bmatrix}$$

45. Show that the cofactor expansions along the first row of A and along the second column of A are equal.

46. Show that the cofactor expansions along the second row of A and along the third row of A are equal.

47. *Diagonal expansion.* If A is a 3×3 matrix, det(A) can be evaluated by the following **diagonal expansion.** Form a 3×5 array by augmenting A with its first two columns, and compute the diagonal products $p_1, p_2, \ldots, p_6$ indicated by the arrows:

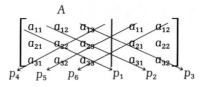

The determinant of A is given by

$$\det(A) = p_1 + p_2 + p_3 - p_4 - p_5 - p_6$$
$$= a_{11}a_{22}a_{33} + a_{12}a_{23}a_{31} + a_{13}a_{21}a_{32} - a_{13}a_{22}a_{31} - a_{11}a_{23}a_{32} - a_{12}a_{21}a_{33}$$

[*Caution:* The diagonal expansion procedure works only for 3×3 matrices. Do not apply it to matrices of any other size.] Use any cofactor expansion of A to verify the diagonal expansion formula.

48. Use the diagonal expansion formula in Problem 47 to evaluate the following determinants:

$$\text{(A)} \quad \begin{vmatrix} 1 & -2 & 1 \\ 3 & 2 & -3 \\ 1 & -1 & 1 \end{vmatrix} \qquad \text{(B)} \quad \begin{vmatrix} 2 & -1 & 1 \\ -1 & 3 & -4 \\ 1 & 2 & -3 \end{vmatrix}$$

In Problems 49–52, find the determinant of the indicated matrix.

49. $\begin{bmatrix} a & 0 & 0 \\ 0 & b & 0 \\ 0 & 0 & c \end{bmatrix}$ 　　　　　　　　 **50.** $\begin{bmatrix} a & d & e \\ 0 & b & f \\ 0 & 0 & c \end{bmatrix}$

51. $\begin{bmatrix} a & b & c \\ a & b & c \\ d & e & f \end{bmatrix}$ 　　　　　　　　 **52.** $\begin{bmatrix} a & b & c \\ d & e & f \\ a+d & b+e & c+f \end{bmatrix}$

53. Show that

$$\begin{vmatrix} 1 & 1 & 1 \\ a & b & c \\ a^2 & b^2 & c^2 \end{vmatrix} = (a - b)(b - c)(c - a)$$

54. Show that

$$\begin{vmatrix} 1+a & b & c \\ a & 1+b & c \\ a & b & 1+c \end{vmatrix} = 1 + a + b + c$$

55. Show that

$$\begin{vmatrix} a & b & c \\ b & c & a \\ c & a & b \end{vmatrix} = 3abc - (a^3 + b^3 + c^3)$$

56. If the determinant of the 1×1 matrix $[a_{11}]$ is defined by $\det([a_{11}]) = a_{11}$, then the cofactor expansion formula can be applied to 2×2 matrices. Show that the cofactor expansion along any row or column of

$$A = \begin{bmatrix} a_{11} & a_{12} \\ a_{21} & a_{22} \end{bmatrix}$$

is $a_{11}a_{22} - a_{21}a_{12}$.

▮▢▮ Applications

57. *Mechanics.* The study of the vibrations of a mechanical system consisting of three interconnected springs (see the figure) leads to the system of equations

$$-kx_1 + k(x_2 - x_1) = \lambda x_1$$
$$-k(x_2 - x_1) - kx_2 = \lambda x_2$$

where k is a known positive constant associated with the spring and λ is an unknown parameter. Find the values of λ for which this system will have nontrivial solutions, and then find those solutions.

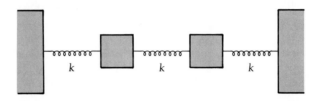

58. *Mechanics.* Repeat Problem 57 for the following mechanical system:

$$-kx_1 + \tfrac{1}{2}k(x_2 - x_1) = \lambda x_1$$
$$-\tfrac{1}{2}k(x_2 - x_1) - kx_2 = \lambda x_2$$

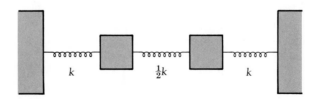

▮3-2▮ Determinants and Row Operations

- ■ Determinants of Triangular Matrices
- ■ Determinants of Elementary Matrices
- ■ Determinant of the Product of Two Matrices
- ■ Determinants and Row Operations
- ■ Application

▪ Determinants of Triangular Matrices

In the preceding section we saw that the evaluation of determinants is greatly simplified if a matrix has a large number of elements that are 0's. Now we want to apply this idea to some matrices with 0's that form specific patterns.

Example 8   Find the determinant of each matrix:

$$(A) \begin{bmatrix} 2 & 0 & 0 \\ 1 & 3 & 0 \\ -4 & 5 & 4 \end{bmatrix} \qquad (B) \begin{bmatrix} 5 & -1 & 7 & 9 \\ 0 & 6 & -6 & 1 \\ 0 & 0 & 7 & 2 \\ 0 & 0 & 0 & 8 \end{bmatrix}$$

Solution   (A)

$$\begin{vmatrix} 2 & 0 & 0 \\ 1 & 3 & 0 \\ -4 & 5 & 4 \end{vmatrix} = (2)\begin{vmatrix} 3 & 0 \\ 5 & 4 \end{vmatrix}$$ If we expand along the first row, the second and third terms will be 0.

$$= (2)(3)(4)$$ Notice that the determinant is just the product of the diagonal entries.

$$= 24$$

(B)

$$\begin{vmatrix} 5 & -1 & 7 & 9 \\ 0 & 6 & -6 & 1 \\ 0 & 0 & 7 & 2 \\ 0 & 0 & 0 & 8 \end{vmatrix} = (5)\begin{vmatrix} 6 & -6 & 1 \\ 0 & 7 & 2 \\ 0 & 0 & 8 \end{vmatrix}$$ Expansion along the first column

$$= (5)(6)\begin{vmatrix} 7 & 2 \\ 0 & 8 \end{vmatrix}$$ Expansion along the first column

$$= (5)(6)(7)(8)$$ Notice that the determinant is just the product of the diagonal entries.

$$= 1{,}680$$

Problem 8   Find the determinant of each matrix:

$$(A) \begin{bmatrix} 4 & -2 & 1 \\ 0 & 3 & -4 \\ 0 & 0 & 2 \end{bmatrix} \qquad (B) \begin{bmatrix} 8 & 0 & 0 & 0 \\ 2 & 7 & 0 & 0 \\ -2 & 3 & 6 & 0 \\ -4 & -3 & 1 & 5 \end{bmatrix}$$ ▮▮

The determinants of the matrices in Example 8 were easy to evaluate because each one had all zero elements either above or below the diagonal. Matrices with this property are called **triangular matrices** (Figure 1). An n × n matrix is an **upper triangular matrix** if all the elements below the diagonal are 0's, a **lower triangular matrix** if all the elements above the diagonal are 0's, and a **diagonal matrix** if the only nonzero elements are on the diagonal. A diagonal matrix is both upper triangular and lower triangular.

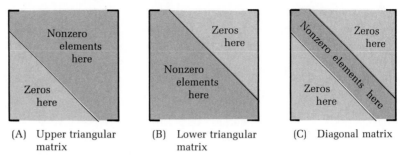

(A)   Upper triangular
      matrix

(B)   Lower triangular
      matrix

(C)   Diagonal matrix

Figure 1   Triangular matrices

Example 9   (A) $\begin{bmatrix} 1 & 2 & 3 & 4 \\ 0 & 2 & 3 & 4 \\ 0 & 0 & 3 & 4 \\ 0 & 0 & 0 & 4 \end{bmatrix}$ This is an upper triangular matrix.

(B) $\begin{bmatrix} 1 & 0 & 0 & 0 \\ 1 & 2 & 0 & 0 \\ 1 & 2 & 3 & 0 \\ 1 & 2 & 3 & 4 \end{bmatrix}$ This is a lower triangular matrix.

(C) $\begin{bmatrix} 1 & 0 & 0 & 0 \\ 0 & 2 & 0 & 0 \\ 0 & 0 & 3 & 0 \\ 0 & 0 & 0 & 4 \end{bmatrix}$ This is a diagonal matrix (and also an upper triangular matrix and a lower triangular matrix).

(D) $\begin{bmatrix} 1 & 0 & 0 & 0 \\ 0 & 0 & 2 & 0 \\ 0 & 3 & 0 & 0 \\ 0 & 0 & 0 & 4 \end{bmatrix}$ This matrix is none of these types.

Problem 9   Classify each of the following as upper triangular, lower triangular, diagonal, or none of these:

(A) $\begin{bmatrix} 1 & 0 & 0 & 0 \\ 0 & 2 & 0 & 0 \\ 0 & 0 & 0 & 0 \\ 1 & 0 & 0 & 3 \end{bmatrix}$ (B) $\begin{bmatrix} 1 & 0 & 0 & 0 \\ 0 & 0 & 0 & 0 \\ 0 & 0 & 0 & 0 \\ 0 & 0 & 0 & 0 \end{bmatrix}$

(C) $\begin{bmatrix} 0 & 0 & 0 & 1 \\ 0 & 1 & 0 & 0 \\ 0 & 0 & 1 & 0 \\ 1 & 0 & 0 & 0 \end{bmatrix}$ (D) $\begin{bmatrix} 0 & 0 & 0 & 0 \\ 0 & 0 & 0 & 0 \\ 0 & 0 & 0 & 0 \\ 1 & 0 & 0 & 0 \end{bmatrix}$ ∎

In Example 8, the determinant of each triangular matrix turned out to be the product of the diagonal elements. Theorem 2 asserts that this is always the case.

Theorem 2

> If A is an $n \times n$ triangular matrix, then the determinant of A is the product of the diagonal elements; that is,
>
> $$\det(A) = (a_{11})(a_{22}) \cdot \cdots \cdot (a_{nn})$$

This theorem can be proved by repeatedly expanding each determinant along the first row in the lower triangular case and along the first column in the upper triangular case. Several special cases will be given as exercises.

Example 10   Find the determinant of each matrix:

(A) $\begin{bmatrix} 2 & -1 & 3 \\ 0 & 4 & 2 \\ 0 & 0 & 6 \end{bmatrix}$ (B) $\begin{bmatrix} 1 & 0 & 0 \\ 2 & 0 & 0 \\ 1 & 2 & 1 \end{bmatrix}$ (C) $\begin{bmatrix} 1 & 0 & 0 \\ 0 & -1 & 0 \\ 0 & 0 & 1 \end{bmatrix}$

Solution

(A) $\begin{vmatrix} 2 & -1 & 3 \\ 0 & 4 & 2 \\ 0 & 0 & 6 \end{vmatrix} = (2)(4)(6) = 48$

(B) $\begin{vmatrix} 1 & 0 & 0 \\ 2 & 0 & 0 \\ 1 & 2 & 1 \end{vmatrix} = (1)(0)(1) = 0$

(C) $\begin{vmatrix} 1 & 0 & 0 \\ 0 & -1 & 0 \\ 0 & 0 & 1 \end{vmatrix} = (1)(-1)(1) = -1$

Problem 10   Find the determinant of each matrix:

(A) $\begin{bmatrix} 1 & 0 & 0 \\ 0 & 2 & 0 \\ 1 & 2 & 3 \end{bmatrix}$ (B) $\begin{bmatrix} 4 & 1 & 2 \\ 0 & 8 & 1 \\ 0 & 0 & 1 \end{bmatrix}$ (C) $\begin{bmatrix} 1 & 0 & 0 \\ 0 & 0 & 0 \\ 0 & 0 & 0 \end{bmatrix}$ ∎

▪ Determinants of Elementary Matrices

In Section 2-4, we found that elementary matrices were useful in studying the relationship between matrices and linear systems. Elementary matrices are also useful in studying the properties of determinants (as we will see later in this section). The values of the determinants of the three types of elementary matrices are given in Theorem 3.

Theorem 3	(A) If E is an elementary matrix formed by performing the operation $R_i \leftrightarrow R_j$ on I, then $\det(E) = -1$.
	(B) If E is an elementary matrix formed by performing the operation $kR_i \rightarrow R_i$ on I, then $\det(E) = k$.
	(C) If E is an elementary matrix formed by performing the operation $R_i + kR_j \rightarrow R_i$ on I, then $\det(E) = 1$.

Parts (B) and (C) of Theorem 3 follow directly from Theorem 2 and the fact that the elementary matrices formed by these operations are always triangular matrices. But an elementary matrix formed by interchanging two rows of the identity matrix is not a triangular matrix, and so Theorem 2 does not apply to part (A). An actual proof of part (A) is rather involved and we will not include it here. All three cases of Theorem 3 are illustrated in Example 11.

Example 11   (A) $E = \begin{bmatrix} 1 & 0 & 0 \\ 0 & 0 & 1 \\ 0 & 1 & 0 \end{bmatrix}$ E is formed by performing the operation $R_2 \leftrightarrow R_3$ on I.

$$\det(E) = (1)\begin{vmatrix} 0 & 1 \\ 1 & 0 \end{vmatrix}$$ Expansion along first row

$$= (1)(-1) = -1$$

(B) $E = \begin{bmatrix} k & 0 & 0 \\ 0 & 1 & 0 \\ 0 & 0 & 1 \end{bmatrix}$ E is formed by performing the operation $kR_1 \rightarrow R_1$ on I.

$$\det(E) = (k)(1)(1) = k$$ Theorem 2 applies.

(C) $E = \begin{bmatrix} 1 & 0 & k \\ 0 & 1 & 0 \\ 0 & 0 & 1 \end{bmatrix}$ E is formed by performing the operation $R_1 + kR_3 \rightarrow R_1$ on I.

$$\det(E) = (1)(1)(1) = 1$$ Theorem 2 applies.

Problem 11   Evaluate the determinant of each matrix:

(A) $\begin{bmatrix} 1 & 0 & 0 \\ 0 & k & 0 \\ 0 & 0 & 1 \end{bmatrix}$ (B) $\begin{bmatrix} 1 & 0 & 0 \\ k & 1 & 0 \\ 0 & 0 & 1 \end{bmatrix}$ (C) $\begin{bmatrix} 0 & 0 & 1 \\ 0 & 1 & 0 \\ 1 & 0 & 0 \end{bmatrix}$ ∎

▪ Determinant of the Product of Two Matrices

Is there a relationship between the determinant of the product of two matrices and the determinants of each of the two separate matrices? Since the definition of matrix multiplication and the definition of the determinant of a matrix are both rather complicated, you might expect any such relationship (if it exists) to be complicated. Fortunately, this is not the case, as you can see by Theorem 4 (which we state without proof).

Theorem 4

If A and B are square matrices of order n, then

$$\det(AB) = \det(A)\,\det(B)$$

In other words, the determinant of a product is equal to the product of the determinants. Theorem 4 can easily be extended to products of three or more matrices.

Example 12 Compute $\det(A)$, $\det(B)$, and $\det(AB)$ for the following:

$$A = \begin{bmatrix} 2 & 3 \\ -1 & 1 \end{bmatrix} \qquad B = \begin{bmatrix} 4 & 2 \\ 3 & 2 \end{bmatrix}$$

Solution

$$\det(A) = \begin{vmatrix} 2 & 3 \\ -1 & 1 \end{vmatrix} = (2)(1) - (-1)(3) = 5$$

$$\det(B) = \begin{vmatrix} 4 & 2 \\ 3 & 2 \end{vmatrix} = (4)(2) - (3)(2) = 2$$

$$AB = \begin{bmatrix} 2 & 3 \\ -1 & 1 \end{bmatrix}\begin{bmatrix} 4 & 2 \\ 3 & 2 \end{bmatrix} = \begin{bmatrix} 17 & 10 \\ -1 & 0 \end{bmatrix}$$

$$\det(AB) = (17)(0) - (-1)(10) = 10$$

As Theorem 4 guarantees,

$$\det(A)\,\det(B) = (5)(2) = 10 = \det(AB)$$

Problem 12 Compute $\det(A)$, $\det(B)$, and $\det(AB)$ for the following:

$$A = \begin{bmatrix} 7 & 4 \\ 3 & 2 \end{bmatrix} \qquad B = \begin{bmatrix} 3 & 3 \\ 5 & 6 \end{bmatrix} \qquad \blacksquare\blacksquare$$

■ Determinants and Row Operations

Performing a row operation on a matrix A always produces an equivalent matrix B. What is the relationship between $\det(A)$ and $\det(B)$? We can answer this question by applying Theorems 3 and 4 and the fact that row operations on a matrix can be performed by multiplying the matrix by an appropriate elementary matrix (as was seen in Section 2-4). Let's look at three specific examples, each involving the use of one of the three elementary row operations.

If

$$A = \begin{bmatrix} 0 & 4 & -8 \\ 1 & 2 & -3 \\ 3 & 9 & 2 \end{bmatrix}$$

and we interchange the first and second rows, then

$$A = \begin{bmatrix} 0 & 4 & -8 \\ 1 & 2 & -3 \\ 3 & 9 & 2 \end{bmatrix} \overset{R_1 \leftrightarrow R_2}{\sim} \begin{bmatrix} 1 & 2 & -3 \\ 0 & 4 & -8 \\ 3 & 9 & 2 \end{bmatrix} = B$$

If E_1 is the elementary matrix corresponding to this row operation, then $E_1 A = B$ and

$$\begin{aligned} \det(B) &= \det(E_1 A) \\ &= \det(E_1) \det(A) && \text{Theorem 4} \\ &= (-1) \det(A) && \text{Theorem 3(A)} \end{aligned}$$

Thus, $\det(A) = -\det(B)$; that is,

$$\overset{\det(A)}{\begin{vmatrix} 0 & 4 & -8 \\ 1 & 2 & -3 \\ 3 & 9 & 2 \end{vmatrix}} \overset{R_1 \leftrightarrow R_2}{=} (-1) \overset{\det(B)}{\begin{vmatrix} 1 & 2 & -3 \\ 0 & 4 & -8 \\ 3 & 9 & 2 \end{vmatrix}} \tag{1}$$

In other words, **interchanging two rows in a determinant changes the sign of the determinant.**

Now suppose we perform the operation $R_3 + (-3)R_1 \to R_3$ on the matrix B:

$$B = \begin{bmatrix} 1 & 2 & -3 \\ 0 & 4 & -8 \\ 3 & 9 & 2 \end{bmatrix} \underset{R_3 + (-3)R_1 \to R_3}{\sim} \begin{bmatrix} 1 & 2 & -3 \\ 0 & 4 & -8 \\ 0 & 3 & 11 \end{bmatrix} = C$$

If E_2 is the elementary matrix corresponding to the row operation $R_3 + (-3)R_1 \to R_3$, then $E_2 B = C$ and

$$\begin{aligned} \det(C) &= \det(E_2 B) \\ &= \det(E_2) \det(B) && \text{Theorem 4} \\ &= (1) \det(B) && \text{Theorem 3(C)} \\ &= \det(B) \end{aligned}$$

or

$$\overset{\det(B)}{\begin{vmatrix} 1 & 2 & -3 \\ 0 & 4 & -8 \\ 3 & 9 & 2 \end{vmatrix}} \underset{R_3 + (-3)R_1 \to R_3}{=} \overset{\det(C)}{\begin{vmatrix} 1 & 2 & -3 \\ 0 & 4 & -8 \\ 0 & 3 & 11 \end{vmatrix}} \tag{2}$$

Thus, **performing row operations of the form $R_i + kR_j \rightarrow R_i$ does not change the value of the determinant.**

To illustrate the effect of the remaining type of row operation, we will multiply the second row of C by $\frac{1}{4}$:

$$C = \begin{bmatrix} 1 & 2 & -3 \\ 0 & 4 & -8 \\ 0 & 3 & 11 \end{bmatrix} \overset{\frac{1}{4}R_2 \rightarrow R_2}{\sim} \begin{bmatrix} 1 & 2 & -3 \\ 0 & 1 & -2 \\ 0 & 3 & 11 \end{bmatrix} = D$$

If E_3 is the elementary matrix corresponding to the row operation $\frac{1}{4}R_2 \rightarrow R_2$, then $D = E_3 C$ and

$$\begin{aligned} \det(D) &= \det(E_3 C) \\ &= \det(E_3)\det(C) & \text{Theorem 4} \\ &= \frac{1}{4}\det(C) & \text{Theorem 3(B)} \end{aligned}$$

or

$$\det(C) = 4\det(D)$$

Thus,

$$\underset{\det(C)}{\begin{vmatrix} 1 & 2 & -3 \\ 0 & 4 & -8 \\ 0 & 3 & 11 \end{vmatrix}} \overset{\frac{1}{4}R_2 \rightarrow R_2}{=} (4)\underset{\det(D)}{\begin{vmatrix} 1 & 2 & -3 \\ 0 & 1 & -2 \\ 0 & 3 & 11 \end{vmatrix}} \tag{3}$$

In this case, **performing the operation $\frac{1}{4}R_2 \rightarrow R_2$ has the effect of factoring 4 out of the second row and then multiplying the resulting determinant by 4.**

We have illustrated the effect of each type of row operation on a determinant. How will this help us evaluate determinants? Using back substitution from (3) to (2) to (1), we see that

$$\det(A) = (-1)(4)\det(D)$$

That is,

$$\underset{\det(A)}{\begin{vmatrix} 0 & 4 & -8 \\ 1 & 2 & -3 \\ 3 & 9 & 2 \end{vmatrix}} = (-1)(4)\underset{\det(D)}{\begin{vmatrix} 1 & 2 & -3 \\ 0 & 1 & -2 \\ 0 & 3 & 11 \end{vmatrix}}$$

One more row operation will transform matrix D into a triangular matrix whose determinant is easy to evaluate. We need a 0 in position 3,2:

$$D = \begin{bmatrix} 1 & 2 & -3 \\ 0 & 1 & -2 \\ 0 & 3 & 11 \end{bmatrix} \overset{R_3 + (-3)R_2 \rightarrow R_3}{\sim} \begin{bmatrix} 1 & 2 & -3 \\ 0 & 1 & -2 \\ 0 & 0 & 17 \end{bmatrix}$$

Since the row operation $R_3 + (-3)R_2 \rightarrow R_3$ does not change the value of the determinant, we have

$$\begin{vmatrix} 0 & 4 & -8 \\ 1 & 2 & -3 \\ 3 & 9 & 2 \end{vmatrix} = (-1)(4) \begin{vmatrix} 1 & 2 & -3 \\ 0 & 1 & -2 \\ 0 & 3 & 11 \end{vmatrix} \qquad R_3 + (-3)R_2 \rightarrow R_3$$

$$= (-1)(4) \begin{vmatrix} 1 & 2 & -3 \\ 0 & 1 & -2 \\ 0 & 0 & 17 \end{vmatrix} \qquad \begin{array}{l} \text{Multiply the diagonal} \\ \text{elements (Theorem 2).} \end{array}$$

$$= (-1)(4)(1)(1)(17) = -68$$

Since any square matrix can be transformed into a triangular matrix by performing a sequence of row operations, we can always evaluate the determinant by keeping track of the effects of the row operations used. This is the most efficient method for calculating determinants, either by hand or on a computer.* The effect of row operations on determinants is summarized in the box for convenient reference.

Row Operations and Determinants

Row Operation	Effect on Determinant	Example
$R_i \leftrightarrow R_j$	Change sign of determinant	$\begin{vmatrix} 2 & 4 \\ 3 & 5 \end{vmatrix} \overset{R_1 \leftrightarrow R_2}{=} - \begin{vmatrix} 3 & 5 \\ 2 & 4 \end{vmatrix}$
$R_i + kR_j \rightarrow R_i$	No change	$\begin{vmatrix} 2 & 4 \\ 3 & 5 \end{vmatrix} \underset{R_2 + (-1)R_1 \rightarrow R_2}{=} \begin{vmatrix} 2 & 4 \\ 1 & 1 \end{vmatrix}$
$\dfrac{1}{k}R_i \rightarrow R_i$	Factor k out of row i and multiply determinant by k	$\begin{vmatrix} 2 & 4 \\ 3 & 5 \end{vmatrix} \overset{\frac{1}{2}R_1 \rightarrow R_1}{=} 2 \begin{vmatrix} 1 & 2 \\ 3 & 5 \end{vmatrix}$

Example 13 Use row operations to evaluate:

$$\begin{vmatrix} 1 & -2 & 3 & -1 \\ 2 & -4 & 9 & 1 \\ 0 & 2 & -6 & 4 \\ -2 & 1 & 4 & 6 \end{vmatrix}$$

---

* The computer supplement to this book (see the Preface) contains a program that uses this procedure to evaluate determinants.

Solution

$$
\begin{vmatrix} 1 & -2 & 3 & -1 \\ 2 & -4 & 9 & 1 \\ 0 & 2 & -6 & 4 \\ -2 & 1 & 4 & 6 \end{vmatrix}
\quad
\begin{aligned}
&R_2 + (-2)R_1 \to R_2 \text{ (No change)} \\[2ex]
&R_4 + 2R_1 \to R_4 \text{ (No change)}
\end{aligned}
$$

$$
= \begin{vmatrix} 1 & -2 & 3 & -1 \\ 0 & 0 & 3 & 3 \\ 0 & 2 & -6 & 4 \\ 0 & -3 & 10 & 4 \end{vmatrix}
\quad R_2 \leftrightarrow R_3 \text{ (Sign change)}
$$

$$
= (-1) \begin{vmatrix} 1 & -2 & 3 & -1 \\ 0 & 2 & -6 & 4 \\ 0 & 0 & 3 & 3 \\ 0 & -3 & 10 & 4 \end{vmatrix}
\quad \tfrac{1}{2}R_2 \to R_2 \text{ (Factor 2 out of row 2.)}
$$

$$
= (-1)(2) \begin{vmatrix} 1 & -2 & 3 & -1 \\ 0 & 1 & -3 & 2 \\ 0 & 0 & 3 & 3 \\ 0 & -3 & 10 & 4 \end{vmatrix}
\quad R_4 + 3R_2 \to R_4 \text{ (No change)}
$$

$$
= (-1)(2) \begin{vmatrix} 1 & -2 & 3 & -1 \\ 0 & 1 & -3 & 2 \\ 0 & 0 & 3 & 3 \\ 0 & 0 & 1 & 10 \end{vmatrix}
\quad R_4 + (-\tfrac{1}{3})R_3 \to R_4 \text{ (No change)}
$$

$$
= (-1)(2) \begin{vmatrix} 1 & -2 & 3 & -1 \\ 0 & 1 & -3 & 2 \\ 0 & 0 & 3 & 3 \\ 0 & 0 & 0 & 9 \end{vmatrix}
\quad \begin{aligned}&\text{Triangular form;} \\ &\text{apply Theorem 2.}\end{aligned}
$$

$$
= (-1)(2)(1)(1)(3)(9) = -54
$$

Problem 13   Use row operations to evaluate:

$$
\begin{vmatrix} 2 & 6 & -4 & 2 \\ -1 & -3 & 5 & 9 \\ 0 & 0 & 3 & -12 \\ 0 & 1 & 4 & -7 \end{vmatrix}
$$

In some cases we can recognize that a determinant is 0, even if the matrix is not a triangular matrix. Two of these cases are listed in Theorem 5. If either case occurs while you are performing row operations to evaluate a determinant, you can stop at that point and conclude that the value of the determinant is 0. The proof is left as an exercise.

Theorem 5

(A) If a matrix has a row of 0's, then its determinant is 0.
(B) If one row of a matrix is a multiple of another row in the matrix, then its determinant is 0.

Example 14 Evaluate each determinant:

$$\text{(A)} \begin{vmatrix} 0 & 0 & 0 \\ 2 & 1 & 3 \\ 4 & 5 & 6 \end{vmatrix} \quad \text{(B)} \begin{vmatrix} 1 & -1 & 2 \\ 2 & -3 & 4 \\ -2 & 2 & -4 \end{vmatrix} \quad \text{(C)} \begin{vmatrix} 1 & -1 & 2 \\ 2 & -4 & 5 \\ -1 & -1 & -1 \end{vmatrix}$$

Solution   (A) $\begin{vmatrix} 0 & 0 & 0 \\ 2 & 1 & 3 \\ 4 & 5 & 6 \end{vmatrix} = 0$ Since every element in R_1 is 0

(B) $\begin{vmatrix} 1 & -1 & 2 \\ 2 & -3 & 4 \\ -2 & 2 & -4 \end{vmatrix} = 0$ Since $(-2)R_1 = R_3$

(C) $\begin{vmatrix} 1 & -1 & 2 \\ 2 & -4 & 5 \\ -1 & -1 & -1 \end{vmatrix} \quad \begin{matrix} R_2 + (-2)R_1 \to R_2 = \\ R_3 + R_1 \to R_3 \end{matrix} \begin{vmatrix} 1 & -1 & 2 \\ 0 & -2 & 1 \\ 0 & -2 & 1 \end{vmatrix} = 0$ Since $R_2 = R_3$

Problem 14 Evaluate each determinant:

$$\text{(A)} \begin{vmatrix} 1 & 2 & 4 \\ 0 & 0 & 0 \\ 2 & 1 & 3 \end{vmatrix} \quad \text{(B)} \begin{vmatrix} 1 & -2 & 3 \\ 2 & -3 & 5 \\ -1 & 3 & -4 \end{vmatrix} \quad \text{(C)} \begin{vmatrix} 1 & 2 & 4 \\ 3 & 6 & 9 \\ 1 & 2 & 3 \end{vmatrix} \qquad \blacksquare\blacksquare$$

■ **Application**

Determinants are often used as a convenient method for expressing formulas. For example, if (x_1, y_1) and (x_2, y_2) are any two distinct points in the plane, then the equation of the line through these points can be expressed as

$$\begin{vmatrix} x & y & 1 \\ x_1 & y_1 & 1 \\ x_2 & y_2 & 1 \end{vmatrix} = 0 \tag{4}$$

To verify this, we begin by evaluating the determinant in (4) by expanding it along the first row:

$$\begin{vmatrix} x & y & 1 \\ x_1 & y_1 & 1 \\ x_2 & y_2 & 1 \end{vmatrix} = x \begin{vmatrix} y_1 & 1 \\ y_2 & 1 \end{vmatrix} - y \begin{vmatrix} x_1 & 1 \\ x_2 & 1 \end{vmatrix} + \begin{vmatrix} x_1 & y_1 \\ x_2 & y_2 \end{vmatrix}$$

$$= x(y_1 - y_2) - y(x_1 - x_2) + (x_1 y_2 - x_2 y_1)$$

Thus, equation (4) is equivalent to the standard equation of a line,

$$ax + by + c = 0 \qquad a = y_1 - y_2, \; b = x_2 - x_1, \; c = x_1 y_2 - x_2 y_1$$

This shows that equation (4) is the equation of a line in the xy plane.

To show that this is the line through the given points, we must show that each given point satisfies the equation. Substituting $x = x_1$, and $y = y_1$ in (4), we have

$$\begin{vmatrix} x_1 & y_1 & 1 \\ x_1 & y_1 & 1 \\ x_2 & y_2 & 1 \end{vmatrix} \stackrel{?}{=} 0$$

Since this matrix has two equal rows, the determinant is 0 and the equation is satisfied. The same thing will happen if we substitute $x = x_2$ and $y = y_2$ in (4). Thus, (4) is the equation of the line through the points (x_1, y_1) and (x_2, y_2).

Example 15 Use equation (4) to find the equation of the line through the points (1, 2) and (4, 9).

Solution Let $(x_1, y_1) = (1, 2)$ and $(x_2, y_2) = (4, 9)$ and substitute in (4):

$$\begin{vmatrix} x & y & 1 \\ 1 & 2 & 1 \\ 4 & 9 & 1 \end{vmatrix} = 0$$

Expand this determinant along the first row:

$$x\begin{vmatrix} 2 & 1 \\ 9 & 1 \end{vmatrix} - y\begin{vmatrix} 1 & 1 \\ 4 & 1 \end{vmatrix} + \begin{vmatrix} 1 & 2 \\ 4 & 9 \end{vmatrix} = 0$$

$$-7x + 3y + 1 = 0$$

You should check that (1, 2) and (4, 9) each satisfy this equation.

Problem 15 Use equation (4) to find the equation of the line through the points $(-1, 4)$ and $(2, -3)$.   ▋▋

Answers to Matched Problems

8. (A) 24   (B) 1,680

9. (A) Lower triangular   (B) Diagonal   (C) None of these
   (D) Lower triangular

10. (A) 6   (B) 32   (C) 0   11. (A) k (B) 1   (C) -1

12. $\det(A) = 2$, $\det(B) = 3$, $\det(AB) = 6$ 13. -132

14. (A) 0   (B) 0   (C) 0   15. $7x + 3y - 5 = 0$

▋▋ Exercise 3-2

A *In Problems 1–8, classify each matrix as upper triangular, lower triangular, diagonal, or none of these. Then use the theorems in this section to evaluate the determinant of each matrix.*

1. $\begin{bmatrix} 2 & -1 & 3 \\ 0 & -4 & 7 \\ 0 & 0 & 5 \end{bmatrix}$
2. $\begin{bmatrix} 0 & 1 & 0 \\ 1 & 1 & 1 \\ 0 & -3 & 0 \end{bmatrix}$

3. $\begin{bmatrix} 5 & 0 & 0 \\ 7 & 6 & 0 \\ 9 & 8 & 7 \end{bmatrix}$
4. $\begin{bmatrix} 2 & -1 & 0 \\ 1 & 0 & 0 \\ 0 & 0 & 0 \end{bmatrix}$

5. $\begin{bmatrix} 1 & 0 & 0 & 0 \\ 0 & -1 & 0 & 0 \\ 0 & 0 & 1 & 0 \\ 0 & 0 & 0 & -1 \end{bmatrix}$
6. $\begin{bmatrix} 4 & -2 & 0 & 0 \\ 0 & 1 & 0 & 0 \\ 0 & 0 & 3 & 0 \\ 0 & 0 & 0 & 5 \end{bmatrix}$

7. $\begin{bmatrix} 4 & -2 & 0 & 0 \\ -2 & 1 & 0 & 0 \\ 0 & 0 & 3 & 0 \\ 0 & 0 & 0 & 5 \end{bmatrix}$
8. $\begin{bmatrix} 4 & 0 & 0 & 0 \\ -2 & 1 & 0 & 0 \\ 0 & 0 & 3 & 0 \\ 0 & 0 & 0 & 5 \end{bmatrix}$

Given that

$$\begin{vmatrix} a_{11} & a_{12} \\ a_{21} & a_{22} \end{vmatrix} = 10$$

use the properties of determinants discussed in this section to evaluate each determinant given below.

9. $\begin{vmatrix} a_{21} & a_{22} \\ a_{11} & a_{12} \end{vmatrix}$
10. $\begin{vmatrix} 2a_{11} & 2a_{12} \\ a_{21} & a_{22} \end{vmatrix}$

11. $\begin{vmatrix} a_{11} + a_{21} & a_{12} + a_{22} \\ a_{21} & a_{22} \end{vmatrix}$
12. $\begin{vmatrix} a_{11} & a_{12} \\ \frac{1}{2}a_{21} & \frac{1}{2}a_{22} \end{vmatrix}$

13. $\begin{vmatrix} a_{11} & a_{12} \\ a_{11} & a_{12} \end{vmatrix}$
14. $\begin{vmatrix} a_{11} + a_{21} & a_{12} + a_{22} \\ -a_{11} & -a_{12} \end{vmatrix}$

In Problems 15 and 16, the matrix A has been expressed as a product of elementary and triangular matrices. Find the determinant of each matrix in the product and then use Theorem 4 to find det(A).

15. $A = \begin{bmatrix} 0 & 4 & 3 \\ 10 & -5 & 5 \\ 4 & -2 & -3 \end{bmatrix} = \begin{bmatrix} 1 & 0 & 0 \\ 0 & 5 & 0 \\ 0 & 0 & 1 \end{bmatrix} \begin{bmatrix} 0 & 1 & 0 \\ 1 & 0 & 0 \\ 0 & 0 & 1 \end{bmatrix} \begin{bmatrix} 1 & 0 & 0 \\ 0 & 1 & 0 \\ 2 & 0 & 1 \end{bmatrix} \begin{bmatrix} 2 & -1 & 1 \\ 0 & 4 & 3 \\ 0 & 0 & -5 \end{bmatrix}$

16. $A = \begin{bmatrix} 1 & -2 & 2 \\ -3 & -3 & 0 \\ 1 & 2 & -2 \end{bmatrix}$

$$= \begin{bmatrix} 1 & 0 & 0 \\ 0 & 1 & 0 \\ -1 & 0 & 1 \end{bmatrix} \begin{bmatrix} 0 & 0 & 1 \\ 0 & 1 & 0 \\ 1 & 0 & 0 \end{bmatrix} \begin{bmatrix} 1 & 0 & 0 \\ 0 & 3 & 0 \\ 0 & 0 & 1 \end{bmatrix} \begin{bmatrix} 2 & 0 & 0 \\ -1 & -1 & 0 \\ 1 & -2 & 2 \end{bmatrix}$$

B *Use row operations to evaluate each determinant.*

17. $\begin{vmatrix} -2 & 4 & 4 \\ 2 & -4 & -3 \\ 5 & -5 & 4 \end{vmatrix}$

18. $\begin{vmatrix} 3 & -1 & 5 \\ -6 & 2 & -7 \\ 3 & 2 & 0 \end{vmatrix}$

19. $\begin{vmatrix} 1 & -1 & 2 & 0 \\ -2 & 3 & -1 & 3 \\ 1 & -1 & -1 & 6 \\ -1 & 2 & 0 & -2 \end{vmatrix}$

20. $\begin{vmatrix} 0 & 1 & 0 & 2 \\ 2 & 0 & 4 & 0 \\ 2 & 1 & 4 & 2 \\ -2 & 1 & -4 & 2 \end{vmatrix}$

21. $\begin{vmatrix} -1 & 0 & 2 & -1 \\ 2 & -2 & 0 & 3 \\ 1 & -1 & 4 & 0 \\ 0 & -1 & 2 & -1 \end{vmatrix}$

22. $\begin{vmatrix} 0 & -3 & 2 & -1 \\ 2 & -2 & 4 & -4 \\ 4 & -3 & 2 & -5 \\ -2 & 3 & -4 & 5 \end{vmatrix}$

23. $\begin{vmatrix} 2 & 3 & 0 & 0 & 0 \\ 1 & 2 & 3 & 0 & 0 \\ 0 & 1 & 2 & 3 & 0 \\ 0 & 0 & 1 & 2 & 3 \\ 0 & 0 & 0 & 1 & 2 \end{vmatrix}$

24. $\begin{vmatrix} 1 & 2 & 3 & 4 & 5 \\ 2 & 3 & 4 & 5 & 6 \\ 3 & 4 & 5 & 6 & 7 \\ 4 & 5 & 6 & 7 & 8 \\ 5 & 6 & 7 & 8 & 9 \end{vmatrix}$

25. $\begin{vmatrix} 1 & 1 & 1 & 1 & 1 \\ 1 & 2 & 4 & 8 & 16 \\ 1 & 3 & 9 & 27 & 81 \\ 1 & 4 & 16 & 64 & 256 \\ 1 & 5 & 25 & 125 & 625 \end{vmatrix}$

26. $\begin{vmatrix} 0 & 1 & 2 & 1 & 0 \\ 0 & 0 & 1 & 2 & 1 \\ 1 & 0 & 0 & 1 & 2 \\ 2 & 1 & 0 & 0 & 1 \\ 1 & 2 & 1 & 0 & 0 \end{vmatrix}$

27. $\begin{vmatrix} 1 & 1 & 1 & 1 & 1 & 1 \\ 1 & 2 & 2 & 2 & 2 & 2 \\ 1 & 1 & 3 & 3 & 3 & 3 \\ 1 & 1 & 1 & 4 & 4 & 4 \\ 1 & 1 & 1 & 1 & 5 & 5 \\ 1 & 1 & 1 & 1 & 1 & 6 \end{vmatrix}$

28. $\begin{vmatrix} 1 & 1 & 1 & 1 & 1 & 1 \\ 1 & 2 & 2 & 2 & 2 & 2 \\ 1 & 2 & 3 & 3 & 3 & 3 \\ 1 & 2 & 3 & 4 & 4 & 4 \\ 1 & 2 & 3 & 4 & 5 & 5 \\ 1 & 2 & 3 & 4 & 5 & 6 \end{vmatrix}$

C *Problems 29–34 are representative cases of theorems discussed in this section. Prove each statement directly, without reference to the theorem it represents.*

29. *Theorem 2.* Show that

$$\begin{vmatrix} a_{11} & a_{12} & a_{13} & a_{14} \\ 0 & a_{22} & a_{23} & a_{24} \\ 0 & 0 & a_{33} & a_{34} \\ 0 & 0 & 0 & a_{44} \end{vmatrix} = a_{11}a_{22}a_{33}a_{44}$$

30. *Theorem 2.* Show that

$$\begin{vmatrix} a_{11} & 0 & 0 & 0 \\ a_{21} & a_{22} & 0 & 0 \\ a_{31} & a_{32} & a_{33} & 0 \\ a_{41} & a_{42} & a_{43} & a_{44} \end{vmatrix} = a_{11}a_{22}a_{33}a_{44}$$

31. Theorem 3(A). Show that

$$\begin{vmatrix} 0 & 0 & 1 & 0 \\ 0 & 1 & 0 & 0 \\ 1 & 0 & 0 & 0 \\ 0 & 0 & 0 & 1 \end{vmatrix} = -1$$

32. Theorem 4. If

$$A = \begin{bmatrix} a_{11} & a_{12} \\ a_{21} & a_{22} \end{bmatrix} \quad \text{and} \quad B = \begin{bmatrix} b_{11} & b_{12} \\ b_{21} & b_{22} \end{bmatrix}$$

find AB, and show that $\det(AB) = \det(A)\det(B)$.

33. Theorem 5(A). Show that

$$\begin{vmatrix} a_{11} & a_{12} & a_{13} \\ 0 & 0 & 0 \\ a_{31} & a_{32} & a_{33} \end{vmatrix} = 0$$

34. Theorem 5(B). Show that

$$\begin{vmatrix} a_{11} & a_{12} & a_{13} \\ a_{21} & a_{22} & a_{23} \\ a_{21} & a_{22} & a_{23} \end{vmatrix} = 0$$

35. Show that

$$\begin{vmatrix} a_{11} & a_{12} & a_{13} \\ a_{21} & a_{22} & 0 \\ a_{31} & 0 & 0 \end{vmatrix} = -a_{13}a_{22}a_{31}$$

36. Show that

$$\begin{vmatrix} 0 & 0 & a_{13} \\ 0 & a_{22} & a_{23} \\ a_{31} & a_{32} & a_{33} \end{vmatrix} = -a_{13}a_{22}a_{31}$$

37. Show that

$$\begin{vmatrix} 1+a & b & c & d \\ a & 1+b & c & d \\ a & b & 1+c & d \\ a & b & c & 1+d \end{vmatrix} = 1+a+b+c+d$$

38. Show that

$$\begin{vmatrix} 1 & a & a^2 & a^3 \\ 1 & b & b^2 & b^3 \\ 1 & c & c^2 & c^3 \\ 1 & d & d^2 & d^3 \end{vmatrix} = (b-a)(c-a)(d-a)(c-b)(d-b)(d-c)$$

39. If A is a square matrix satisfying $A^2 = A$, what are the possible values of $\det(A)$?

40. If A is an $n \times n$ matrix and k is a scalar, show that

$$\det(kA) = k^n \det(A)$$

41. If A is a square matrix and m is a positive integer, show that

$$\det(A^m) = [\det(A)]^m$$

42. If A is an invertible matrix, show that

$$\det(A) \det(A^{-1}) = 1$$

43. If A is an $n \times n$ matrix and P is an invertible $n \times n$ matrix, show that

$$\det(P^{-1}AP) = \det(A)$$

44. If A and B are row equivalent square matrices, show that $\det(A) = 0$ if and only if $\det(B) = 0$.

▮▢▮ Applications

45. *Equation of a line.* Find the equation of the line through the points $(1, 5)$ and $(4, -7)$.

46. *Equation of a line.* Find the equation of the line through the points $(3, 4)$ and $(-1, 2)$.

47. *Collinear points.* Show that if

$$\begin{vmatrix} x_1 & y_1 & 1 \\ x_2 & y_2 & 1 \\ x_3 & y_3 & 1 \end{vmatrix} = 0$$

then (x_1, y_1), (x_2, y_2), and (x_3, y_3) are collinear (all lie on the same line).

48. *Collinear points.* Show that if (x_1, y_1), (x_2, y_2), and (x_3, y_3) are collinear, then

$$\begin{vmatrix} x_1 & y_1 & 1 \\ x_2 & y_2 & 1 \\ x_3 & y_3 & 1 \end{vmatrix} = 0$$

49. *Interpolating polynomial.* Given three noncollinear points (x_1, y_1), (x_2, y_2), and (x_3, y_3) with distinct x coordinates, there is a unique second-degree polynomial whose graph will pass through these points (see Section 1-2). Show that the equation of this polynomial is given by

$$\begin{vmatrix} y & 1 & x & x^2 \\ y_1 & 1 & x_1 & x_1^2 \\ y_2 & 1 & x_2 & x_2^2 \\ y_3 & 1 & x_3 & x_3^2 \end{vmatrix} = 0$$

50. *Interpolating polynomial.* Use the equation in Problem 49 to find the equation of the second-degree polynomial whose graph passes through the points $(-1, 2)$, $(1, 1)$, and $(3, 4)$.

51. *Interpolating polynomial.* Use the equation in Problem 49 to find the equation of the second-degree polynomial whose graph passes through the points $(1, 6)$, $(2, 7)$, and $(4, 3)$.

52. *Interpolating polynomial.* Given four noncollinear points (x_1, y_1), (x_2, y_2), (x_3, y_3), and (x_4, y_4) with distinct x coordinates, show that the third-degree polynomial whose graph passes through these four points is given by

$$\begin{vmatrix} y & 1 & x & x^2 & x^3 \\ y_1 & 1 & x_1 & x_1^2 & x_1^3 \\ y_2 & 1 & x_2 & x_2^2 & x_2^3 \\ y_3 & 1 & x_3 & x_3^2 & x_3^3 \\ y_4 & 1 & x_4 & x_4^2 & x_4^3 \end{vmatrix} = 0$$

53. *Interpolating polynomial.* Use the equation in Problem 52 to find the equation of the third-degree polynomial whose graph passes through the points $(-1, 3)$, $(0, 0)$, $(1, -3)$, and $(2, 0)$.

54. *Interpolating polynomial.* Use the equation in Problem 52 to find the equation of the third-degree polynomial whose graph passes through the points $(-1, 0)$, $(0, 0)$, $(1, -2)$, and $(2, 0)$.

▌3-3▏ Determinants and Matrix Inversion

- Transpose of a Matrix
- Cofactor and Adjoint Matrices
- The Adjoint Method for Finding A^{-1}
- Application

In this section we will develop a formula for the inverse of a matrix. We begin this development by introducing a new operation called *transposition*.

▪ Transpose of a Matrix

If A is an $n \times m$ matrix, then the **transpose** of A is an $m \times n$ matrix denoted by A^T and formed by interchanging the rows and columns of A. Thus, the first row of A becomes the first column of A^T, the second row of A becomes the second column of A^T, and so on:

$$A = \begin{bmatrix} a_{11} & a_{12} & a_{13} \\ a_{21} & a_{22} & a_{23} \end{bmatrix}$$

R_1 in $A = C_1$ in A^T

R_2 in $A = C_2$ in A^T

Transpose of A

$$\begin{bmatrix} a_{11} & a_{21} \\ a_{12} & a_{22} \\ a_{13} & a_{23} \end{bmatrix} = A^\mathsf{T}$$

Notice that the element in position *ij* of *A* is equal to the element in position *ji* in *A*ᵀ.

Same element

$$A = \begin{bmatrix} a_{11} & a_{12} & a_{13} \\ a_{21} & a_{22} & a_{23} \end{bmatrix} \qquad A^T = \begin{bmatrix} a_{11} & a_{21} \\ a_{12} & a_{22} \\ a_{13} & a_{23} \end{bmatrix}$$

Example 16   Find the transpose of each matrix:

(A) $\begin{bmatrix} 1 & 2 & 3 \\ 4 & 5 & 6 \end{bmatrix}$ (B) [1 2 3]   (C) $\begin{bmatrix} 1 & 2 & 3 \\ 0 & 0 & 0 \\ 3 & 2 & 1 \end{bmatrix}$

Solution   (A) $\begin{bmatrix} 1 & 2 & 3 \\ 4 & 5 & 6 \end{bmatrix}^T = \begin{bmatrix} 1 & 4 \\ 2 & 5 \\ 3 & 6 \end{bmatrix}$ (B) $[1 \quad 2 \quad 3]^T = \begin{bmatrix} 1 \\ 2 \\ 3 \end{bmatrix}$

(C) $\begin{bmatrix} 1 & 2 & 3 \\ 0 & 0 & 0 \\ 3 & 2 & 1 \end{bmatrix}^T = \begin{bmatrix} 1 & 0 & 3 \\ 2 & 0 & 2 \\ 3 & 0 & 1 \end{bmatrix}$

Problem 16   Find the transpose of each matrix:

(A) $\begin{bmatrix} 1 & 2 \\ 3 & 4 \\ 5 & 6 \end{bmatrix}$ (B) $\begin{bmatrix} 1 \\ 2 \\ 3 \\ 4 \end{bmatrix}$ (C) $\begin{bmatrix} 1 & 1 & 1 \\ 2 & 2 & 2 \\ 3 & 3 & 3 \end{bmatrix}$ ▮▮

Transposition is an important operation that we will use frequently. Theorem 6 lists some basic properties of the transpose. The proof is omitted, but some special cases of the proof will be considered in the exercises.

Theorem 6

(A) $(A^T)^T = A$	
(B) $(A + B)^T = A^T + B^T$	Assuming $A + B$ is defined
(C) $(AB)^T = B^T A^T$	Assuming AB is defined
(D) $(kA)^T = kA^T$	
(E) $\det(A^T) = \det(A)$	Assuming A is a square matrix

■ **Cofactor and Adjoint Matrices**

In Section 3-1 we defined the *ij*th cofactor of an $n \times n$ matrix A to be the number

$$A_{ij} = (-1)^{i+j} \det(M_{ij})$$

where M_{ij} is the $(n-1) \times (n-1)$ submatrix formed by deleting the ith row and jth column of A. The **cofactor matrix** of A, denoted cof(A), is the $n \times n$ matrix whose ijth element is A_{ij}. Thus, if

$$A = \begin{bmatrix} a_{11} & a_{12} & a_{13} \\ a_{21} & a_{22} & a_{23} \\ a_{31} & a_{32} & a_{33} \end{bmatrix}$$

then

$$\text{cof}(A) = \begin{bmatrix} A_{11} & A_{12} & A_{13} \\ A_{21} & A_{22} & A_{23} \\ A_{31} & A_{32} & A_{33} \end{bmatrix}$$

Example 17   Find the cofactor matrix. Also, find det(A).

$$A = \begin{bmatrix} 1 & -2 & 0 \\ 3 & -1 & 2 \\ 2 & 0 & 2 \end{bmatrix}$$

Solution $A_{11} = \begin{vmatrix} -1 & 2 \\ 0 & 2 \end{vmatrix} = -2 \qquad A_{12} = -\begin{vmatrix} 3 & 2 \\ 2 & 2 \end{vmatrix} = -2 \qquad A_{13} = \begin{vmatrix} 3 & -1 \\ 2 & 0 \end{vmatrix} = 2$

$A_{21} = -\begin{vmatrix} -2 & 0 \\ 0 & 2 \end{vmatrix} = 4 \qquad A_{22} = \begin{vmatrix} 1 & 0 \\ 2 & 2 \end{vmatrix} = 2 \qquad A_{23} = -\begin{vmatrix} 1 & -2 \\ 2 & 0 \end{vmatrix} = -4$

$A_{31} = \begin{vmatrix} -2 & 0 \\ -1 & 2 \end{vmatrix} = -4 \qquad A_{32} = -\begin{vmatrix} 1 & 0 \\ 3 & 2 \end{vmatrix} = -2 \qquad A_{33} = \begin{vmatrix} 1 & -2 \\ 3 & -1 \end{vmatrix} = 5$

$$\text{cof}(A) = \begin{bmatrix} -2 & -2 & 2 \\ 4 & 2 & -4 \\ -4 & -2 & 5 \end{bmatrix}$$

Expanding along the first row of A,

$$\begin{aligned} \det(A) &= a_{11}A_{11} + a_{12}A_{12} + a_{13}A_{13} \\ &= (1)(-2) + (-2)(-2) + (0)(2) = 2 \end{aligned}$$

Problem 17   Find the cofactor matrix. Also, find det(A).

$$A = \begin{bmatrix} 2 & -1 & 1 \\ 0 & 1 & -2 \\ 3 & -1 & 2 \end{bmatrix}$$

▐▐

The transpose of the cofactor matrix is called the **adjoint** of A and is denoted by adj(A). That is,

adj(A) = [cof(A)]ᵀ

If

$$A = \begin{bmatrix} a_{11} & a_{12} & a_{13} \\ a_{21} & a_{22} & a_{23} \\ a_{31} & a_{32} & a_{33} \end{bmatrix}$$

then

$$\text{adj}(A) = \begin{bmatrix} A_{11} & A_{12} & A_{13} \\ A_{21} & A_{22} & A_{23} \\ A_{31} & A_{32} & A_{33} \end{bmatrix}^{T} = \begin{bmatrix} A_{11} & A_{21} & A_{31} \\ A_{12} & A_{22} & A_{32} \\ A_{13} & A_{23} & A_{33} \end{bmatrix}$$

Example 18   Find the adjoint of the matrix A in Example 17. Also, compute $A \text{ adj}(A)$.

Solution $\text{adj}(A) = [\text{cof}(A)]^{T} = \begin{bmatrix} -2 & -2 & 2 \\ 4 & 2 & -4 \\ -4 & -2 & 5 \end{bmatrix}^{T}$

$$= \begin{bmatrix} -2 & 4 & -4 \\ -2 & 2 & -2 \\ 2 & -4 & 5 \end{bmatrix}$$

$$A \text{ adj}(A) = \begin{bmatrix} 1 & -2 & 0 \\ 3 & -1 & 2 \\ 2 & 0 & 2 \end{bmatrix} \begin{bmatrix} -2 & 4 & -4 \\ -2 & 2 & -2 \\ 2 & -4 & 5 \end{bmatrix}$$

$$= \begin{bmatrix} 2 & 0 & 0 \\ 0 & 2 & 0 \\ 0 & 0 & 2 \end{bmatrix}$$

Problem 18   Find the adjoint of the matrix A in Problem 17. Also, compute $A \text{ adj}(A)$.   ▌▌

▪ The Adjoint Method for Finding A^{-1}

The calculations in Examples 17 and 18 show that for this particular matrix A,

$$A \text{ adj}(A) = \begin{bmatrix} \det(A) & 0 & 0 \\ 0 & \det(A) & 0 \\ 0 & 0 & \det(A) \end{bmatrix} = \det(A)I$$

Theorem 7 states that this equation holds for any square matrix A.

Theorem 7   If A is an $n \times n$ matrix, then

$$A \text{ adj}(A) = \det(A)I$$

Proof   To keep the proof simple, we will prove this theorem only for 3×3 matrices.

The proof for $n \times n$ matrices is similar. Let

$$A = \begin{bmatrix} a_{11} & a_{12} & a_{13} \\ a_{21} & a_{22} & a_{23} \\ a_{31} & a_{32} & a_{33} \end{bmatrix}$$

and let

$$B = A \, \mathrm{adj}(A) = \begin{bmatrix} a_{11} & a_{12} & a_{13} \\ a_{21} & a_{22} & a_{23} \\ a_{31} & a_{32} & a_{33} \end{bmatrix} \begin{bmatrix} A_{11} & A_{21} & A_{31} \\ A_{12} & A_{22} & A_{32} \\ A_{13} & A_{23} & A_{33} \end{bmatrix}$$

We need to examine each element b_{ij} in this product. To begin, we find the diagonal elements b_{11}, b_{22}, and b_{33} in B:

$$b_{11} = \begin{bmatrix} a_{11} & a_{12} & a_{13} \end{bmatrix} \cdot \begin{bmatrix} A_{11} \\ A_{12} \\ A_{13} \end{bmatrix}$$

$$= a_{11}A_{11} + a_{12}A_{12} + a_{13}A_{13}$$

You should recognize this as the cofactor expansion along the first row of A. Since $\det(A)$ is given by the cofactor expansion along any row (or column), we can conclude that

$$b_{11} = \det(A)$$

In the same way,

$$b_{22} = \begin{bmatrix} a_{21} & a_{22} & a_{23} \end{bmatrix} \cdot \begin{bmatrix} A_{21} \\ A_{22} \\ A_{23} \end{bmatrix}$$

$$= a_{21}A_{21} + a_{22}A_{22} + a_{23}A_{23} = \det(A)$$

and

$$b_{33} = \begin{bmatrix} a_{31} & a_{32} & a_{33} \end{bmatrix} \cdot \begin{bmatrix} A_{31} \\ A_{32} \\ A_{33} \end{bmatrix}$$

$$= a_{31}A_{31} + a_{32}A_{32} + a_{33}A_{33} = \det(A)$$

Thus, all the diagonal elements of B are equal to $\det(A)$. Now let's examine one of the nondiagonal elements in B:

$$b_{12} = \begin{bmatrix} a_{11} & a_{12} & a_{13} \end{bmatrix} \cdot \begin{bmatrix} A_{21} \\ A_{22} \\ A_{23} \end{bmatrix}$$

$$= a_{11}A_{21} + a_{12}A_{22} + a_{13}A_{23} \tag{1}$$

This is not a cofactor expansion of A since the elements a_{11}, a_{12}, and a_{13} come from the first row of A while the cofactors A_{21}, A_{22}, and A_{23} correspond to the elements in the second row of A. To evaluate this expression, we want to define a new matrix D with the property that (1) is a cofactor expansion for D. We claim that the following matrix has this property:

$$D = \begin{bmatrix} a_{11} & a_{12} & a_{13} \\ a_{11} & a_{12} & a_{13} \\ a_{31} & a_{32} & a_{33} \end{bmatrix} \quad \leftarrow \text{Elements in second row of } D$$

The cofactor expansion along the second row of D is

$$\overset{\text{Elements in second row of } D}{\underset{\downarrow \qquad \downarrow \qquad \downarrow}{}}$$

$$\det(D) = a_{11}D_{21} + a_{12}D_{22} + a_{13}D_{23} \tag{2}$$

Now each of the cofactors D_{21}, D_{22}, and D_{23} is computed by blocking out row 2 of D. But rows 1 and 3 of D agree with rows 1 and 3 of A:

$$D = \begin{bmatrix} a_{11} & a_{12} & a_{13} \\ \cancel{a_{11}} & \cancel{a_{12}} & \cancel{a_{13}} \\ a_{31} & a_{32} & a_{33} \end{bmatrix} \qquad A = \begin{bmatrix} a_{11} & a_{12} & a_{13} \\ \cancel{a_{21}} & \cancel{a_{22}} & \cancel{a_{23}} \\ a_{31} & a_{32} & a_{33} \end{bmatrix}$$

Thus, $D_{21} = A_{21}$, $D_{22} = A_{22}$, and $D_{23} = A_{23}$. Substituting in (2) and comparing with (1), we now have

$$b_{12} = a_{11}A_{21} + a_{12}A_{22} + a_{13}A_{23} = \det(D)$$

But D is a matrix with two identical rows, so its determinant must be 0. That is,

$$b_{12} = \det(D) = 0$$

A similar argument can be used to show that all the nondiagonal elements are 0's. Thus,

$$A \text{ adj}(A) = \begin{bmatrix} \det(A) & 0 & 0 \\ 0 & \det(A) & 0 \\ 0 & 0 & \det(A) \end{bmatrix}$$

$$= \det(A)I \qquad\qquad\qquad \blacksquare$$

We are now able to establish the relationship between the inverse of a matrix and its determinant.

Theorem 8

> If A is invertible, then $\det(A) \neq 0$.

Proof   If A is invertible, then A^{-1} exists and $AA^{-1} = I$. Thus,

$$\begin{aligned} 1 &= \det(I) & &\text{Theorem 2} \\ &= \det(AA^{-1}) & &AA^{-1} = I \\ &= \det(A)\det(A^{-1}) & &\text{Theorem 4} \end{aligned}$$

If the product of two numbers is nonzero, then each number must be nonzero. Thus, we have shown that $\det(A) \neq 0$.   ∎

Theorem 9

If $\det(A) \neq 0$, then A is invertible and

$$A^{-1} = \frac{1}{\det(A)} \text{ adj}(A)$$

Proof   Since $\det(A) \neq 0$, the matrix

$$\frac{1}{\det(A)} \text{ adj}(A)$$

certainly exists. Now we must show that it satisfies the definition of the inverse.

$$A\left(\frac{1}{\det(A)} \text{ adj}(A)\right) = \frac{1}{\det(A)} (A \text{ adj}(A)) \qquad \text{Theorem 3(F) in Section 2-2}$$

$$= \frac{1}{\det(A)} (\det(A)I) \qquad \text{Theorem 7}$$

$$= I \qquad \text{Theorem 2(A) in Section 2-1}$$

This shows that

$$A^{-1} = \frac{1}{\det(A)} \text{ adj}(A)$$ ∎

Combining Theorems 8 and 9, we can conclude that A is invertible if and only if $\det(A) \neq 0$. We can now add this to the five equivalent properties listed in Theorem 10 in Section 2-4.

Theorem 10

Fundamental Theorem — Version 2

If A is a square matrix of order n, then all of the following statements are equivalent:

(A) A is invertible.
(B) $AX = B$ has a unique solution for any $n \times 1$ column matrix B.
(C) The only solution of $AX = 0$ is the trivial solution $X = 0$.
(D) A is row equivalent to I.
(E) A is a product of elementary matrices.
(F) $\det(A) \neq 0$.

Example 19   Determine which of the following matrices are invertible without finding the inverse:

$$(A) \ A = \begin{bmatrix} 5 & 0 & 0 \\ 0 & 2 & 1 \\ 0 & 1 & 1 \end{bmatrix} \qquad (B) \ B = \begin{bmatrix} 1 & 2 & 3 \\ -1 & 0 & 1 \\ 1 & 2 & 3 \end{bmatrix}$$

Solution   (A) $\det(A) = 5 \begin{vmatrix} 2 & 1 \\ 1 & 1 \end{vmatrix} = 5 \neq 0$ Thus, A is invertible.

(B) Since the first and third rows of B are identical, $\det(B) = 0$ and B is not invertible.

Problem 19   Repeat Example 19 for:

$$(A) \ A = \begin{bmatrix} 1 & 0 & -1 \\ -2 & 0 & 2 \\ 1 & 1 & 2 \end{bmatrix} \qquad (B) \ B = \begin{bmatrix} 2 & 1 & 0 \\ 0 & -1 & 1 \\ 0 & 0 & 2 \end{bmatrix}$$ ▮

The formula

$$A^{-1} = \frac{1}{\det(A)} \ \text{adj}(A)$$

in Theorem 9 also provides an alternate method for computing A^{-1}. This formula has some important theoretical implications; however, it is not a very practical method for computing inverses, as the following example illustrates. Using this formula even for a 3×3 matrix involves some lengthy calculations. In general, it is much more efficient to use row operations to find A^{-1} than it is to use this formula.

Example 20   Use the formula

$$A^{-1} = \frac{1}{\det(A)} \ \text{adj}(A)$$

to find the inverse of

$$A = \begin{bmatrix} 1 & 0 & 1 \\ 0 & -1 & 0 \\ 1 & 1 & 2 \end{bmatrix}$$

Solution   First, we must find cof(A):

$$A_{11} = \begin{vmatrix} -1 & 0 \\ 1 & 2 \end{vmatrix} = -2 \qquad A_{12} = -\begin{vmatrix} 0 & 0 \\ 1 & 2 \end{vmatrix} = 0 \qquad A_{13} = \begin{vmatrix} 0 & -1 \\ 1 & 1 \end{vmatrix} = 1$$

$$A_{21} = -\begin{vmatrix} 0 & 1 \\ 1 & 2 \end{vmatrix} = 1 \qquad A_{22} = \begin{vmatrix} 1 & 1 \\ 1 & 2 \end{vmatrix} = 1 \qquad A_{23} = -\begin{vmatrix} 1 & 0 \\ 1 & 1 \end{vmatrix} = -1$$

$$A_{31} = \begin{vmatrix} 0 & 1 \\ -1 & 0 \end{vmatrix} = 1 \qquad A_{32} = -\begin{vmatrix} 1 & 1 \\ 0 & 0 \end{vmatrix} = 0 \qquad A_{33} = \begin{vmatrix} 1 & 0 \\ 0 & -1 \end{vmatrix} = -1$$

$$\text{cof}(A) = \begin{bmatrix} -2 & 0 & 1 \\ 1 & 1 & -1 \\ 1 & 0 & -1 \end{bmatrix}$$

$$\text{adj}(A) = \begin{bmatrix} -2 & 0 & 1 \\ 1 & 1 & -1 \\ 1 & 0 & -1 \end{bmatrix}^T = \begin{bmatrix} -2 & 1 & 1 \\ 0 & 1 & 0 \\ 1 & -1 & -1 \end{bmatrix}$$

$$\det(A) = 0 + (-1)(1) + 0 \qquad \text{Expansion along}$$
$$= -1 \qquad\qquad\qquad \text{second row}$$

Thus,

$$A^{-1} = \frac{1}{-1} \begin{bmatrix} -2 & 1 & 1 \\ 0 & 1 & 0 \\ 1 & -1 & -1 \end{bmatrix} = \begin{bmatrix} 2 & -1 & -1 \\ 0 & -1 & 0 \\ -1 & 1 & 1 \end{bmatrix}$$

You should check that $AA^{-1} = I$.

Problem 20   Repeat Example 20 for $A = \begin{bmatrix} 0 & -1 & -2 \\ 3 & 2 & 3 \\ 5 & 3 & 5 \end{bmatrix}$ ∎

- ## Application

The adjoint method for finding A^{-1} is useful if some of the elements in A are parameters.

Example 21   In the income determination model in economics, the equilibrium level of
Economics   income is determined by the system

$$y = c + i$$
$$c = c_0 + by$$
$$i = i_0 + ay$$

where y is income, c is consumption, i is investment, and a, b, c_0, and i_0 are parameters whose values depend on the particular economy under consideration. Express y, c, and i in terms of the parameters a, b, c_0, and i_0.

Solution   The original system can be written as

$$y - c - i = 0$$
$$-by + c \qquad = c_0$$
$$-ay \qquad + i = i_0$$

or

$$AX = B$$

where

$$A = \begin{bmatrix} 1 & -1 & -1 \\ -b & 1 & 0 \\ -a & 0 & 1 \end{bmatrix} \qquad X = \begin{bmatrix} y \\ c \\ i \end{bmatrix} \qquad B = \begin{bmatrix} 0 \\ c_0 \\ i_0 \end{bmatrix}$$

We could solve this system by using Gauss–Jordan elimination; however, the presence of the parameters would make the calculations somewhat complicated. Instead, we will use the adjoint method to find A^{-1}.

$$A_{11} = \begin{vmatrix} 1 & 0 \\ 0 & 1 \end{vmatrix} = 1 \qquad A_{12} = -\begin{vmatrix} -b & 0 \\ -a & 1 \end{vmatrix} = b \qquad A_{13} = \begin{vmatrix} -b & 1 \\ -a & 0 \end{vmatrix} = a$$

$$A_{21} = -\begin{vmatrix} -1 & -1 \\ 0 & 1 \end{vmatrix} = 1 \qquad A_{22} = \begin{vmatrix} 1 & -1 \\ -a & 1 \end{vmatrix} = 1-a \qquad A_{23} = -\begin{vmatrix} 1 & -1 \\ -a & 0 \end{vmatrix} = a$$

$$A_{31} = \begin{vmatrix} -1 & -1 \\ 1 & 0 \end{vmatrix} = 1 \qquad A_{32} = -\begin{vmatrix} 1 & -1 \\ -b & 0 \end{vmatrix} = b \qquad A_{33} = \begin{vmatrix} 1 & -1 \\ -b & 1 \end{vmatrix} = 1-b$$

$$\begin{aligned} \det(A) &= (1)(1) + (-1)(b) + (-1)(a) \qquad \text{Expansion along the} \\ &= 1 - b - a \qquad\qquad\qquad\qquad\quad \text{first row of } A \end{aligned}$$

To ensure that the system has a unique solution, we will assume that $a + b \neq 1$, so that $\det(A) \neq 0$. Then A^{-1} is given by

$$A^{-1} = \frac{1}{\det(A)} \, \mathrm{adj}(A)$$

$$= \frac{1}{1-b-a} \begin{bmatrix} 1 & 1 & 1 \\ b & 1-a & b \\ a & a & 1-b \end{bmatrix}$$

and the solution of the system is

$$X = A^{-1}B$$

$$\begin{bmatrix} y \\ c \\ i \end{bmatrix} = \frac{1}{1-b-a} \begin{bmatrix} 1 & 1 & 1 \\ b & 1-a & b \\ a & a & 1-b \end{bmatrix} \begin{bmatrix} 0 \\ c_0 \\ i_0 \end{bmatrix}$$

$$= \frac{1}{1-b-a} \begin{bmatrix} c_0 + i_0 \\ (1-a)c_0 + bi_0 \\ ac_0 + (1-b)i_0 \end{bmatrix}$$

or

$$y = \frac{c_0 + i_0}{1-b-a}$$

$$c = \frac{(1-a)c_0 + bi_0}{1-b-a}$$

$$i = \frac{ac_0 + (1-b)i_0}{1-b-a} \qquad a+b \neq 1$$

Problem 21   If a lump sum tax is added to the model in Example 21, the equilibrium level is determined by the system

$$y = c + i_0$$
$$c = c_0 + bd$$
$$d = y - t_0$$

where y is income, c is consumption, d is disposable income, and i_0, c_0, b, and t_0 are parameters. Express y, c, and d in terms of the parameters i_0, c_0, b, and t_0. ▌▌

Answers to Matched Problems

16. (A) $\begin{bmatrix} 1 & 3 & 5 \\ 2 & 4 & 6 \end{bmatrix}$ **(B)** $\begin{bmatrix} 1 & 2 & 3 & 4 \end{bmatrix}$ **(C)** $\begin{bmatrix} 1 & 2 & 3 \\ 1 & 2 & 3 \\ 1 & 2 & 3 \end{bmatrix}$

17. $\operatorname{cof}(A) = \begin{bmatrix} 0 & -6 & -3 \\ 1 & 1 & -1 \\ 1 & 4 & 2 \end{bmatrix}$, $\det(A) = 3$

18. $\operatorname{adj}(A) = \begin{bmatrix} 0 & 1 & 1 \\ -6 & 1 & 4 \\ -3 & -1 & 2 \end{bmatrix}$, $A \operatorname{adj}(A) = \begin{bmatrix} 3 & 0 & 0 \\ 0 & 3 & 0 \\ 0 & 0 & 3 \end{bmatrix}$

19. (A) $\det(A) = 0$, A is not invertible   **(B)** $\det(B) = -4$, B is invertible

20. $A^{-1} = \dfrac{1}{2} \begin{bmatrix} 1 & -1 & 1 \\ 0 & 10 & -6 \\ -1 & -5 & 3 \end{bmatrix}$

21. $y = \dfrac{i_0 + c_0 - bt_0}{1 - b}$, $c = \dfrac{bi_0 + c_0 - bt_0}{1 - b}$, $d = \dfrac{i_0 + c_0 - t_0}{1 - b}$, $b \neq 1$

▌▌ Exercise 3-3

Use the adjoint method to find any inverses required in the solution of the problems in this exercise set.

A   *Find the transpose of each matrix.*

1. $\begin{bmatrix} 1 & 2 & 3 \\ 4 & 5 & 6 \end{bmatrix}$

2. $\begin{bmatrix} 1 & 2 \\ 3 & 4 \\ 5 & 6 \\ 7 & 8 \end{bmatrix}$

3. $\begin{bmatrix} 1 & -1 & 0 & -4 \end{bmatrix}$

4. $\begin{bmatrix} -6 \\ 4 \\ 3 \end{bmatrix}$

5. $\begin{bmatrix} 1 & 2 & 3 \\ 2 & 0 & 4 \\ 3 & 4 & -1 \end{bmatrix}$

6. $\begin{bmatrix} 0 & 1 & -2 \\ -1 & 0 & 3 \\ 2 & -3 & 0 \end{bmatrix}$

Problems 7–14 pertain to the following matrices:

$$A = \begin{bmatrix} 1 & -2 & 1 \\ 2 & -3 & 4 \\ -1 & -1 & 1 \end{bmatrix} \qquad B = \begin{bmatrix} 2 & 1 & 0 \\ -1 & 1 & 2 \\ 1 & 1 & 2 \end{bmatrix}$$

7. Find cof(A).

9. Find det(A).

11. Find adj(A).

13. Find A adj(A).

8. Find cof(B).

10. Find det(B).

12. Find adj(B).

14. Find B adj(B).

B In Problems 15–22, find the inverse, if it exists.

15. $\begin{bmatrix} 1 & 0 & -1 \\ 2 & 1 & 0 \\ 2 & 0 & -1 \end{bmatrix}$

16. $\begin{bmatrix} 0 & 0 & 1 \\ 3 & 1 & 0 \\ 5 & 2 & 0 \end{bmatrix}$

17. $\begin{bmatrix} 3 & -5 & 2 \\ 4 & -1 & 0 \\ 4 & -5 & 2 \end{bmatrix}$

18. $\begin{bmatrix} 2 & -1 & 3 \\ -4 & 2 & -2 \\ 1 & -1 & 5 \end{bmatrix}$

19. $\begin{bmatrix} 1 & 0 & 0 & 0 \\ 2 & -1 & 0 & 0 \\ 0 & -2 & 1 & 0 \\ 0 & 0 & 2 & -1 \end{bmatrix}$

20. $\begin{bmatrix} 1 & -1 & 0 & 0 \\ 0 & 1 & -1 & 0 \\ 0 & 0 & 1 & -1 \\ 0 & 0 & 0 & 1 \end{bmatrix}$

21. $\begin{bmatrix} 2 & 0 & 0 & 0 \\ 1 & 1 & 1 & 1 \\ 2 & 0 & 2 & 0 \\ 1 & 0 & 1 & 1 \end{bmatrix}$

22. $\begin{bmatrix} 1 & 0 & -2 & 0 \\ 0 & 1 & 0 & -2 \\ -1 & 0 & 2 & 0 \\ 0 & -1 & 0 & 2 \end{bmatrix}$

In Problems 23 and 24, perform the indicated multiplication.

23. $\begin{bmatrix} x_1 \\ x_2 \end{bmatrix}^T \begin{bmatrix} 1 & 2 \\ 2 & 3 \end{bmatrix} \begin{bmatrix} x_1 \\ x_2 \end{bmatrix}$

24. $\begin{bmatrix} x_1 \\ x_2 \end{bmatrix}^T \begin{bmatrix} 1 & -1 \\ -1 & 1 \end{bmatrix} \begin{bmatrix} x_1 \\ x_2 \end{bmatrix}$

In Problems 25–30, find the inverse of each matrix. State any restrictions that must be made on the elements of the matrix in order for the inverse to exist.

25. $\begin{bmatrix} a_{11} & 0 & 0 \\ 0 & a_{22} & 0 \\ 0 & 0 & a_{33} \end{bmatrix}$

26. $\begin{bmatrix} a_{11} & a_{12} & 0 \\ a_{21} & a_{22} & 0 \\ 0 & 0 & 1 \end{bmatrix}$

27. $\begin{bmatrix} a_{11} & a_{12} & a_{13} \\ 0 & a_{22} & a_{23} \\ 0 & 0 & a_{33} \end{bmatrix}$

28. $\begin{bmatrix} a_{11} & a_{12} & a_{13} \\ a_{21} & a_{22} & 0 \\ a_{31} & 0 & 0 \end{bmatrix}$

29. $\begin{bmatrix} 1 & 0 & a \\ 0 & 1 & 0 \\ a & 0 & 1 \end{bmatrix}$

30. $\begin{bmatrix} a & 0 & 1 \\ 0 & a & 0 \\ 1 & 0 & 1 \end{bmatrix}$

C *Problems 31–34 are representative cases of Theorem 6. Prove each statement directly, without reference to the theorem it represents.*

31. Theorem 6(A). If A is a 3×3 matrix, show that $(A^T)^T = A$.

32. Theorem 6(B). If A and B are 2×2 matrices, show that $(A + B)^T = A^T + B^T$.

33. Theorem 6(C). If A and B are 2×2 matrices, show that $(AB)^T = B^T A^T$.

34. Theorem 6(E). If A is a 3×3 matrix, show that $\det(A^T) = \det(A)$.

35. If A is an invertible matrix, show that $(A^{-1})^T = (A^T)^{-1}$.

36. A matrix is said to be **symmetric** if $A^T = A$. If A is an invertible symmetric matrix, show that A^{-1} is also symmetric.

37. A matrix is said to be **skew-symmetric** if $A^T = -A$. If A is an invertible skew-symmetric matrix, show that A^{-1} is also skew-symmetric.

38. A matrix A is said to be **orthogonal** if A is invertible and $A^{-1} = A^T$. If A is an orthogonal matrix, show that $\det(A) = \pm 1$.

39. Find all 2×2 matrices satisfying $A = A^{-1}$.

40. Find all 2×2 matrices satisfying $A^{-1} = A^T$.

41. If A is an $n \times n$ matrix, show that $\det[\operatorname{adj}(A)] = [\det(A)]^{n-1}$.

42. If A and B are $n \times n$ matrices, show that $\operatorname{adj}(AB) = \operatorname{adj}(B) \operatorname{adj}(A)$.

▮◻▮ Applications

43. *Circuit analysis.* A direct current electrical circuit consists of a combination of conductors, resistors, and batteries (see the figure). The currents I_1, I_2, and I_3, the voltages V_1 and V_2, and the resistances R_1, R_2, and R_3 can be shown to satisfy

$$
\begin{aligned}
I_1 - \ I_2 - \ I_3 &= 0 \\
R_1 I_1 + R_2 I_2 \qquad &= V_1 \\
- R_2 I_2 + R_3 I_3 &= V_2
\end{aligned}
$$

where it is assumed that V_1, V_2, R_1, R_2, and R_3 are positive constants.

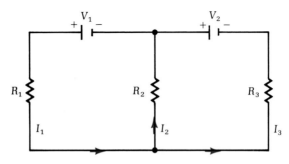

(A) Write this system as a matrix equation of the form $AX = B$ where

$$
X = \begin{bmatrix} I_1 \\ I_2 \\ I_3 \end{bmatrix}
$$

(B) Find A^{-1}.

(C) Use A^{-1} to solve for I_1, I_2, and I_3 in terms of V_1, V_2, R_1, R_2, and R_3.

44. *Circuit analysis.* Repeat Problem 43 for the following electrical circuit:

$$I_1 - \quad I_2 - \quad I_3 = 0$$
$$R_1 I_1 \qquad + R_3 I_3 = V_1$$
$$R_2 I_2 - R_3 I_3 = V_2$$

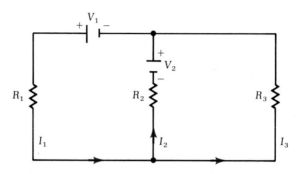

45. *Demand functions.* In order to construct *Marshallian demand functions*, which maximize consumer satisfaction subject to a budgetary constraint, economists use a constrained optimizaton technique called the **method of Lagrange multipliers.** In one particular model, the Marshallian demand functions for consumer income x and product price y are given by

$$ay + p\lambda = -b$$
$$ax \quad + q\lambda = -c$$
$$px + qy \quad = d$$

where a, b, c, d, p, and q are positive constants and λ is an auxiliary variable introduced as part of the optimization process.

(A) Write this system as a matrix equation of the form $AX = B$ where

$$X = \begin{bmatrix} x \\ y \\ \lambda \end{bmatrix}$$

(B) Find A^{-1}.

(C) Use A^{-1} to solve for x, y, and λ in terms of a, b, c, d, p, and q.

46. *Demand functions.* Repeat Problem 45 if consumer income x and product price y are given by

$$ax \quad + p\lambda = 0$$
$$by + q\lambda = 0$$
$$px + qy \quad = c$$

where a, b, c, p, and q are positive constants and λ is an auxiliary variable.

3-4 Cramer's Rule

- Cramer's Rule
- Application

■ Cramer's Rule

In this section we will see that the solution to a consistent system of n equations in n variables can be expressed in terms of determinants. The formula for the solution is generally referred to as *Cramer's Rule*.

Theorem 11

Cramer's Rule

If A is an $n \times n$ matrix and $\det(A) \neq 0$, then the solution to the system

$$AX = B$$

is given by

$$x_i = \frac{\det(B_i)}{\det(A)} \qquad 1 \leq i \leq n$$

where B_i is the matrix formed by replacing the ith column of A with the column of constants B.

Proof Once again, we will restrict our proof to the 3×3 case. Consider the system

$$
\begin{matrix} A & X & B \end{matrix}
$$

$$
\begin{bmatrix} a_{11} & a_{12} & a_{13} \\ a_{21} & a_{22} & a_{23} \\ a_{31} & a_{32} & a_{33} \end{bmatrix}
\begin{bmatrix} x_1 \\ x_2 \\ x_3 \end{bmatrix}
=
\begin{bmatrix} b_1 \\ b_2 \\ b_3 \end{bmatrix}
$$

Since $\det(A) \neq 0$, A is invertible and the unique solution to the system is

$$X = A^{-1}B$$

Furthermore,

$$A^{-1} = \frac{1}{\det(A)} \, \text{adj}(A)$$

Thus,

$$
\begin{bmatrix} x_1 \\ x_2 \\ x_3 \end{bmatrix}
= \frac{1}{\det(A)} \, \text{adj}(A)B = \frac{1}{\det(A)}
\begin{bmatrix} A_{11} & A_{21} & A_{31} \\ A_{12} & A_{22} & A_{32} \\ A_{13} & A_{23} & A_{33} \end{bmatrix}
\begin{bmatrix} b_1 \\ b_2 \\ b_3 \end{bmatrix}
$$

$$
= \frac{1}{\det(A)}
\begin{bmatrix} b_1A_{11} + b_2A_{21} + b_3A_{31} \\ b_1A_{12} + b_2A_{22} + b_3A_{32} \\ b_1A_{13} + b_2A_{23} + b_3A_{33} \end{bmatrix}
$$

and

$$x_1 = \frac{b_1 A_{11} + b_2 A_{21} + b_3 A_{31}}{\det(A)} \tag{1}$$

$$x_2 = \frac{b_1 A_{12} + b_2 A_{22} + b_3 A_{32}}{\det(A)} \tag{2}$$

$$x_3 = \frac{b_1 A_{13} + b_2 A_{23} + b_3 A_{33}}{\det(A)} \tag{3}$$

Now consider the numerator of x_1. We want to express this as a determinant. Compare the matrix

$$B_1 = \begin{bmatrix} b_1 & a_{12} & a_{13} \\ b_2 & a_{22} & a_{23} \\ b_3 & a_{32} & a_{33} \end{bmatrix}$$

with

$$A = \begin{bmatrix} a_{11} & a_{12} & a_{13} \\ a_{21} & a_{22} & a_{23} \\ a_{31} & a_{32} & a_{33} \end{bmatrix}$$

Remember that when you compute the cofactor of an element you delete the row and column containing that element. If we use the cofactor expansion along the first column of B_1 to compute $\det(B_1)$, we will be using the second and third columns of B_1, which are identical to the second and third columns of A. Thus, we can conclude that

$$\det(B_1) = b_1 A_{11} + b_2 A_{21} + b_3 A_{31}$$

Substituting in (1), we have

$$x_1 = \frac{\det(B_1)}{\det(A)}$$

Formulas (2) and (3) can be established by similar arguments.   ▮

A convenient way to remember Cramer's Rule in the 3×3 case is to state it as follows:

$$x_1 = \frac{\begin{vmatrix} b_1 & a_{12} & a_{13} \\ b_2 & a_{22} & a_{23} \\ b_3 & a_{32} & a_{33} \end{vmatrix}}{\begin{vmatrix} a_{11} & a_{12} & a_{13} \\ a_{21} & a_{22} & a_{23} \\ a_{31} & a_{32} & a_{33} \end{vmatrix}} \qquad x_2 = \frac{\begin{vmatrix} a_{11} & b_1 & a_{13} \\ a_{21} & b_2 & a_{23} \\ a_{31} & b_3 & a_{33} \end{vmatrix}}{\begin{vmatrix} a_{11} & a_{12} & a_{13} \\ a_{21} & a_{22} & a_{23} \\ a_{31} & a_{32} & a_{33} \end{vmatrix}} \qquad x_3 = \frac{\begin{vmatrix} a_{11} & a_{12} & b_1 \\ a_{21} & a_{22} & b_2 \\ a_{31} & a_{32} & b_3 \end{vmatrix}}{\begin{vmatrix} a_{11} & a_{12} & a_{13} \\ a_{21} & a_{22} & a_{23} \\ a_{31} & a_{32} & a_{33} \end{vmatrix}}$$

Example 22 Use Cramer's Rule to solve the system:

$$x_1 - 2x_2 + x_3 = 3$$
$$x_1 + x_2 - x_3 = 4$$
$$2x_1 - x_2 + x_3 = 5$$

Solution $A = \begin{bmatrix} 1 & -2 & 1 \\ 1 & 1 & -1 \\ 2 & -1 & 1 \end{bmatrix} \qquad B = \begin{bmatrix} 3 \\ 4 \\ 5 \end{bmatrix}$

Since Cramer's Rule cannot be used if $\det(A) = 0$, we first compute $\det(A)$:

$$\det(A) = \begin{vmatrix} 1 & -1 \\ -1 & 1 \end{vmatrix} + 2 \begin{vmatrix} 1 & -1 \\ 2 & 1 \end{vmatrix} + \begin{vmatrix} 1 & 1 \\ 2 & -1 \end{vmatrix} \qquad \text{Expansion along the first row of } A$$

$$= 0 + 6 - 3 = 3$$

Since $\det(A) \neq 0$, Cramer's Rule can be used.

Step 1. *Compute* x_1:

$$\begin{array}{c} B \\ B_1 = \begin{bmatrix} 3 & -2 & 1 \\ 4 & 1 & -1 \\ 5 & -1 & 1 \end{bmatrix} \end{array} \qquad \begin{array}{l} \text{Replace the first column of } A \\ \text{with the constant terms from } B. \end{array}$$

$$\det(B_1) = 3 \begin{vmatrix} 1 & -1 \\ -1 & 1 \end{vmatrix} + 2 \begin{vmatrix} 4 & -1 \\ 5 & 1 \end{vmatrix} + \begin{vmatrix} 4 & 1 \\ 5 & -1 \end{vmatrix} \qquad \text{Expansion along the first row of } B_1$$

$$= 0 + 18 - 9 = 9$$

$$x_1 = \frac{\det(B_1)}{\det(A)} = \frac{9}{3} = 3 \qquad \text{Cramer's Rule}$$

Step 2. *Compute* x_2:

$$\begin{array}{c} B \\ B_2 = \begin{bmatrix} 1 & 3 & 1 \\ 1 & 4 & -1 \\ 2 & 5 & 1 \end{bmatrix} \end{array} \qquad \begin{array}{l} \text{Replace the second column of } \\ A \text{ with the constant terms} \\ \text{from } B. \end{array}$$

$$\det(B_2) = \begin{vmatrix} 4 & -1 \\ 5 & 1 \end{vmatrix} - 3 \begin{vmatrix} 1 & -1 \\ 2 & 1 \end{vmatrix} + \begin{vmatrix} 1 & 4 \\ 2 & 5 \end{vmatrix} \qquad \text{Expansion along the first row of } B_2$$

$$= 9 - 9 - 3 = -3$$

$$x_2 = \frac{\det(B_2)}{\det(A)} = \frac{-3}{3} = -1 \qquad \text{Cramer's Rule}$$

Step 3.　*Compute* x_3:

$$B$$

$$B_3 = \begin{bmatrix} 1 & -2 & 3 \\ 1 & 1 & 4 \\ 2 & -1 & 5 \end{bmatrix}$$

Replace the third column of A with the constant terms from B.

$$\det(B_3) = \begin{vmatrix} 1 & 4 \\ -1 & 5 \end{vmatrix} + 2\begin{vmatrix} 1 & 4 \\ 2 & 5 \end{vmatrix} + 3\begin{vmatrix} 1 & 1 \\ 2 & -1 \end{vmatrix}$$

Expansion along the first row of B_3

$$= 9 - 6 - 9 = -6$$

$$x_3 = \frac{\det(B_3)}{\det(A)} = \frac{-6}{3} = -2 \qquad \text{Cramer's Rule}$$

Thus, the solution is $x_1 = 3$, $x_2 = -1$, and $x_3 = -2$. You should check this solution.

Problem 22　Use Cramer's Rule to solve the system:

$$2x_1 - x_2 + 3x_3 = -3$$
$$-x_1 + 2x_2 - 2x_3 = 8$$
$$3x_1 + x_2 + 3x_3 = 7$$

∎

In terms of the number of calculations involved, Cramer's Rule and Gauss–Jordan elimination are roughly equivalent for 3 × 3 systems. For larger systems, Gauss–Jordan elimination is much more efficient. Furthermore, Gauss–Jordan elimination can be used to find the solutions to dependent systems. Cramer's Rule cannot be used at all for such systems. On the other hand, Cramer's Rule is useful if the coefficients of the system involve parameters.

▪ Application

Many mathematical models of real-world phenomena involve systems of differential equations (systems of equations involving calculus forms). One way to solve certain types of these systems is to use the *method of Laplace transforms*. This method transforms the original system of differential equations into a system of linear equations whose coefficients involve a parameter. The following example illustrates the type of systems that arise in applications of this method.

Example 23　In the following system, s is a parameter. Solve the system:

$$2sx_1 + sx_2 - x_3 = 2$$
$$sx_1 + sx_2 - x_3 = 1$$
$$sx_1 + x_2 + sx_3 = 2$$

Solution $\quad A = \begin{bmatrix} 2s & s & -1 \\ s & s & -1 \\ s & 1 & s \end{bmatrix} \qquad B = \begin{bmatrix} 2 \\ 1 \\ 2 \end{bmatrix}$

$$\det(A) = 2s \begin{vmatrix} s & -1 \\ 1 & s \end{vmatrix} - s \begin{vmatrix} s & -1 \\ s & s \end{vmatrix} - \begin{vmatrix} s & s \\ s & 1 \end{vmatrix}$$

$$= 2s(s^2 + 1) - s(s^2 + s) - (s - s^2)$$

$$= 2s^3 + 2s - s^3 - s^2 - s + s^2$$

$$= s^3 + s$$

$$= s(s^2 + 1)$$

If $s = 0$, then $\det(A) = 0$ and the system does not have a unique solution. In applications of this type we are only interested in unique solutions, so we assume that $s \neq 0$.

Step 1. Compute x_1:

$$B$$

$$\det(B_1) = \begin{vmatrix} 2 & s & -1 \\ 1 & s & -1 \\ 2 & 1 & s \end{vmatrix} = 2 \begin{vmatrix} s & -1 \\ 1 & s \end{vmatrix} - s \begin{vmatrix} 1 & -1 \\ 2 & s \end{vmatrix} - \begin{vmatrix} 1 & s \\ 2 & 1 \end{vmatrix}$$

$$= 2(s^2 + 1) - s(s + 2) - (1 - 2s) = s^2 + 1$$

$$x_1 = \frac{\det(B_1)}{\det(A)} = \frac{s^2 + 1}{s(s^2 + 1)} = \frac{1}{s}$$

Step 2. Compute x_2:

$$B$$

$$\det(B_2) = \begin{vmatrix} 2s & 2 & -1 \\ s & 1 & -1 \\ s & 2 & s \end{vmatrix} = 2s \begin{vmatrix} 1 & -1 \\ 2 & s \end{vmatrix} - 2 \begin{vmatrix} s & -1 \\ s & s \end{vmatrix} - \begin{vmatrix} s & 1 \\ s & 2 \end{vmatrix}$$

$$= 2s(s + 2) - 2(s^2 + s) - (2s - s) = s$$

$$x_2 = \frac{\det(B_2)}{\det(A)} = \frac{s}{s(s^2 + 1)} = \frac{1}{s^2 + 1}$$

Step 3. Compute x_3:

$$B$$

$$\det(B_3) = \begin{vmatrix} 2s & s & 2 \\ s & s & 1 \\ s & 1 & 2 \end{vmatrix} = 2s \begin{vmatrix} s & 1 \\ 1 & 2 \end{vmatrix} - s \begin{vmatrix} s & 1 \\ s & 2 \end{vmatrix} + 2 \begin{vmatrix} s & s \\ s & 1 \end{vmatrix}$$

$$= 2s(2s - 1) - s(2s - s) + 2(s - s^2) = s^2$$

$$x_3 = \frac{\det(B_3)}{\det(A)} = \frac{s^2}{s(s^2 + 1)} = \frac{s}{s^2 + 1}$$

Thus, the solution is

$$x_1 = \frac{1}{s}, \qquad x_2 = \frac{1}{s^2 + 1}, \qquad x_3 = \frac{s}{s^2 + 1}, \qquad s \neq 0$$

You should check this solution.

Problem 23   Solve the system:

$$\begin{aligned} sx_1 + \ sx_2 + \ x_3 &= \ \ 1 \\ sx_1 + 2sx_2 + 2x_3 &= \ \ 1 \\ sx_1 + \ \ x_2 + sx_3 &= -1 \end{aligned}$$

II

Answers to   22. $x_1 = 2, \quad x_2 = 4, \quad x_3 = -1$

Matched Problems

23. $x_1 = \dfrac{1}{s}, \quad x_2 = \dfrac{2}{s^2 - 1}, \quad x_3 = \dfrac{-2s}{s^2 - 1}, \quad s \neq 1, 0, 1$

II   Exercise 3-4

A   *Use Cramer's Rule, if it applies, to solve the following systems. If Cramer's Rule cannot be used, do not solve the system.*

1. $\begin{aligned} 5x_1 + 7x_2 &= 16 \\ 2x_1 + 3x_2 &= \ 6 \end{aligned}$

2. $\begin{aligned} 4x_1 + 2x_2 &= 5 \\ 6x_1 + 3x_2 &= 7 \end{aligned}$

3. $\begin{aligned} 3x_1 - 6x_2 &= 11 \\ 4x_1 - 8x_2 &= 17 \end{aligned}$

4. $\begin{aligned} 7x_1 + 4x_2 &= 2 \\ 5x_1 + 3x_2 &= 2 \end{aligned}$

B   5. $\begin{aligned} 2x_1 - 3x_2 + \ x_3 &= \ -3 \\ -4x_1 + 3x_2 + 2x_3 &= -11 \\ x_1 - \ x_2 - \ x_3 &= \ \ \ 3 \end{aligned}$

6. $\begin{aligned} x_1 + 4x_2 - 3x_3 &= 25 \\ 3x_1 + \ x_2 - \ x_3 &= \ 2 \\ -4x_1 + \ x_2 + 2x_3 &= \ 1 \end{aligned}$

7. $\begin{aligned} 2x_1 + 3x_2 - x_3 &= 9 \\ x_1 - 2x_2 + x_3 &= 7 \end{aligned}$

8. $\begin{aligned} 2x_1 + \ x_2 &= \ \ 5 \\ -3x_1 + 4x_2 &= \ \ 6 \\ x_1 - \ x_2 &= -4 \end{aligned}$

9. $\begin{aligned} 12x_1 - 14x_2 + 11x_3 &= \ \ \ 5 \\ 15x_1 + \ 7x_2 - \ 9x_3 &= -13 \\ 5x_1 - \ 3x_2 + \ 2x_3 &= \ \ \ 0 \end{aligned}$

10. $\begin{aligned} 2x_1 - \ x_2 + 4x_3 &= 15 \\ -x_1 + \ x_2 + 2x_3 &= \ 5 \\ 3x_1 + 4x_2 - 2x_3 &= \ 4 \end{aligned}$

11. $\begin{aligned} x_1 + \ x_2 + \ x_3 &= 0 \\ x_1 + 2x_2 + 2x_3 &= 0 \\ 2x_1 + \ x_2 + \ x_3 &= 0 \end{aligned}$

12. $\begin{aligned} x_1 + \ x_2 + \ x_3 &= 0 \\ x_1 + 2x_2 + 2x_3 &= 0 \\ x_1 + 2x_2 + \ x_3 &= 0 \end{aligned}$

13. $\begin{aligned} 10x_1 + 10x_2 + 20x_3 &= \ \ \ 300 \\ 5x_1 - 15x_2 + 10x_3 &= -250 \\ 10x_1 + 10x_2 + 15x_3 &= \ \ \ 350 \end{aligned}$

14. $\begin{aligned} 5x_1 - 10x_2 + 15x_3 &= 125 \\ 10x_1 + 20x_2 - \ 5x_3 &= 250 \\ 15x_1 + 10x_2 + 10x_3 &= 200 \end{aligned}$

15. $\begin{aligned} 20x_1 - 10x_2 + \ 5x_3 &= 300 \\ 10x_1 + 10x_2 - 20x_3 &= 200 \\ 10x_1 - 10x_2 + 10x_3 &= 250 \end{aligned}$

16. $\begin{aligned} 10x_1 - 10x_2 + 15x_3 &= 100 \\ 20x_1 - 15x_2 + \ 5x_3 &= 450 \\ 10x_1 - 10x_2 + 20x_3 &= 100 \end{aligned}$

17. $5x_1 - 4x_2 + 3x_3 = 18$
$7x_1 + 8x_2 - 9x_3 = -13$
$10x_1 - 7x_2 + 5x_3 = 33$

18. $10x_1 + 11x_2 + 13x_3 = 2$
$7x_1 + 8x_2 + 10x_3 = 1$
$4x_1 + 5x_2 + 8x_3 = 4$

In Problems 19–24, use Cramer's Rule to solve for x_1 only.

19. $2x_1 - 3x_2 + 4x_3 = 5$
$-x_1 + 2x_2 - 5x_3 = 2$
$4x_1 - 5x_2 + 7x_3 = 1$

20. $2x_1 - x_2 + 7x_3 = 7$
$x_1 - 5x_2 + 4x_3 = 4$
$-4x_1 + 2x_2 + 10x_3 = 10$

21. $x_1 + x_2 + x_3 + x_4 = 1$
$x_1 - x_2 = 1$
$x_1 - x_3 = 1$
$x_1 - x_4 = 1$

22. $x_1 + x_2 + x_3 + x_4 = 1$
$x_1 - x_2 = 0$
$x_1 - x_3 = 0$
$x_1 - x_4 = 0$

23. $x_1 + 2x_2 = 2$
$x_2 + 3x_3 = 3$
$x_3 + 4x_4 = 4$
$-5x_1 + x_4 = 5$

24. $x_1 + 2x_2 = 1$
$x_2 + 3x_3 = 2$
$x_3 + 4x_4 = 3$
$-5x_1 + x_4 = 4$

C In Problems 25 and 26, find the values of s for which the system has a unique solution, and then find the solution.

25. $2sx_1 + sx_2 + sx_3 = -1$
$-3sx_1 + sx_2 + sx_3 = 4$
$x_1 - sx_2 + x_3 = -1$

26. $sx_1 + sx_2 + x_3 = 2$
$sx_1 + 2sx_2 + x_3 = 4$
$-x_1 + sx_2 + x_3 = 1$

27. In the system of equations given below, θ, ϕ, and r are constants and u, v, w, x, y, and z are variables. Use Cramer's Rule to express z in terms of u, v, and w.

$$u = (\sin \theta \cos \phi)x + (\sin \theta \sin \phi)y + (\cos \theta)z$$
$$v = (r \cos \theta \cos \phi)x + (r \cos \theta \sin \phi)y - (r \sin \theta)z$$
$$w = (-r \sin \theta \sin \phi)x + (r \sin \theta \cos \phi)y$$

28. Using elementary trigonometry (see the figure), it can be shown that the angles A, B, and C and the sides a, b, and c of a triangle satisfy

$$c = b \cos A + a \cos B$$
$$b = c \cos A + a \cos C$$
$$a = b \cos C + c \cos B$$

Use Cramer's Rule to express $\cos A$ in terms of a, b, and c, thereby deriving the **law of cosines:**

$$\cos A = \frac{b^2 + c^2 - a^2}{2bc}$$

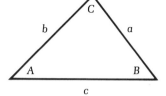

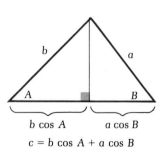

$c = b \cos A + a \cos B$

Use Cramer's Rule to prove the following statements:

29. If A is a square matrix and $\det(A) \neq 0$, then the only solution of the system $AX = 0$ is the trivial solution.

30. If A is a square matrix and $\det(A) \neq 0$, then the system $AX = B$ has a unique solution for any $n \times 1$ column matrix B.

| 3-5 | Chapter Review

Important Terms
and Symbols

3-1. *Definition of the determinant.* Determinant of a 2×2 matrix, formula for A^{-1}, Cramer's Rule, minor, cofactor, cofactor expansion, determinant of an $n \times n$ matrix, $\det(A)$

3-2. *Determinants and row operations.* Triangular matrix, upper triangular matrix, lower triangular matrix, diagonal matrix, determinants of elementary matrices, row operations and determinants

3-3. *Determinants and matrix inverson.* Transpose of a matrix, cofactor matrix, adjoint of a matrix, adjoint method for finding A^{-1}, adj(A)

3-4. *Cramer's Rule.* Cramer's Rule

‖ Exercise 3-5   Chapter Review

Work through all the problems in this chapter review and check your answers in the back of the book. (Answers to most review problems are there.) Where weaknesses show up, review appropriate sections in the text.

A   *Problems 1–3 refer to the matrix below:*

$$A = \begin{bmatrix} 4 & -3 \\ -2 & 2 \end{bmatrix}$$

1. Find $\det(A)$.

2. Find A^{-1}.

3. Find A^{T}.

4. Use Cramer's Rule to solve the system:

$$3x_1 + 5x_2 = 9$$
$$x_1 + 3x_2 = 7$$

Problems 5–12 refer to the matrix below:

$$A = \begin{bmatrix} 1 & 2 & -3 \\ 3 & 0 & 5 \\ 2 & 2 & -2 \end{bmatrix}$$

5. Find A^T.   **6.** Find M_{11}, M_{12}, and M_{13}.

7. Find A_{11}, A_{12}, and A_{13}.

8. Use the cofactor expansion along the first row of A to find $\det(A)$.

9. Use elementary row operations to find $\det(A)$.

10. Find $\text{cof}(A)$.   **11.** Find $\text{adj}(A)$.   **12.** Find A^{-1}.

13. If

$$\begin{vmatrix} a_{11} & a_{12} & a_{13} \\ a_{21} & a_{22} & a_{23} \\ a_{31} & a_{32} & a_{33} \end{vmatrix} = 5$$

find:

$$(A) \begin{vmatrix} a_{11} & a_{12} & a_{13} \\ 3a_{21} & 3a_{22} & 3a_{23} \\ a_{31} & a_{32} & a_{33} \end{vmatrix} \qquad (B) \begin{vmatrix} a_{31} & a_{32} & a_{33} \\ a_{11} & a_{12} & a_{13} \\ a_{21} & a_{22} & a_{23} \end{vmatrix}$$

$$(C) \begin{vmatrix} a_{11} + 3a_{21} & a_{12} + 3a_{22} & a_{13} + 3a_{23} \\ a_{21} & a_{22} & a_{23} \\ a_{31} & a_{32} & a_{33} \end{vmatrix}$$

B   *In Problems 14–28, find the determinant of each matrix.*

14. $\begin{bmatrix} 2 & 0 & 0 \\ 0 & 3 & 0 \\ 0 & 0 & 4 \end{bmatrix}$ **15.** $\begin{bmatrix} 1 & 0 & 2 \\ 0 & 1 & 0 \\ 0 & 0 & 1 \end{bmatrix}$ **16.** $\begin{bmatrix} -1 & 0 & 0 \\ 7 & 1 & 0 \\ -9 & 4 & -3 \end{bmatrix}$

17. $\begin{bmatrix} 0 & 1 & 0 \\ 1 & 0 & 0 \\ 0 & 0 & 1 \end{bmatrix}$ **18.** $\begin{bmatrix} 2 & 0 & 0 \\ 3 & 4 & 5 \\ 6 & 7 & 8 \end{bmatrix}$ **19.** $\begin{bmatrix} 2 & 1 & -1 \\ 4 & -2 & 2 \\ 1 & 2 & -7 \end{bmatrix}$

20. $\begin{bmatrix} 2 & 1 & -1 \\ -4 & -2 & 2 \\ 1 & 2 & -7 \end{bmatrix}$ **21.** $\begin{bmatrix} 3 & 1 & -1 \\ 0 & 6 & 2 \\ 0 & 0 & 5 \end{bmatrix}$ **22.** $\begin{bmatrix} 3 & 0 & 0 \\ 1 & 6 & 0 \\ -1 & 2 & 5 \end{bmatrix}$

23. $\begin{bmatrix} 2 & -1 & 1 & 2 \\ 3 & 1 & 4 & 2 \\ 0 & 0 & 0 & 0 \\ 5 & 2 & 1 & 6 \end{bmatrix}$ **24.** $\begin{bmatrix} 9 & 2 & -1 & 3 \\ -14 & 2 & 4 & 0 \\ 12 & 1 & 5 & 0 \\ 7 & 0 & 0 & 0 \end{bmatrix}$

25. $\begin{bmatrix} 1 & 1 & 1 & 1 \\ 1 & 2 & 3 & 4 \\ 0 & 2 & 5 & 8 \\ 0 & 0 & 4 & 5 \end{bmatrix}$ **26.** $\begin{bmatrix} 1 & 2 & 3 & 4 \\ 2 & 3 & 4 & 5 \\ 3 & 4 & 5 & 6 \\ 4 & 5 & 6 & 7 \end{bmatrix}$

$$
\textbf{27.} \begin{bmatrix} 1 & -1 & 1 & -1 & 1 \\ 3 & -1 & 1 & -1 & 1 \\ 0 & 4 & -1 & 1 & -1 \\ 0 & 0 & 3 & 1 & -1 \\ 0 & 0 & 0 & 4 & 1 \end{bmatrix} \qquad \textbf{28.} \begin{bmatrix} 1 & -1 & 0 & 1 & -1 \\ 0 & 1 & -1 & 1 & -1 \\ -1 & 1 & 0 & 1 & -1 \\ 1 & 0 & 1 & -1 & 1 \\ 1 & -1 & -1 & 1 & 0 \end{bmatrix}
$$

In Problems 29 and 30, solve for x_2 only.

29.
$$
\begin{aligned}
3x_1 - 3x_2 + 5x_3 &= -4 \\
7x_1 + 3x_2 - 4x_3 &= 5 \\
-6x_1 - 4x_2 + 9x_3 &= 7
\end{aligned}
$$

30.
$$
\begin{aligned}
x_1 + x_2 + x_3 + x_4 &= 1 \\
x_1 - x_2 &= 0 \\
x_2 + x_3 &= \tfrac{1}{2} \\
x_3 - 2x_4 &= 0
\end{aligned}
$$

In Problems 31–34, use the equation $A^{-1} = [1/\det(A)]\, adj(A)$ to find the inverse of each matrix, if it exists.

$$
\textbf{31.} \begin{bmatrix} 4 & -3 & 7 \\ 2 & 5 & -4 \\ 3 & -2 & 5 \end{bmatrix} \qquad \textbf{32.} \begin{bmatrix} 2 & -2 & 3 \\ 4 & 2 & -5 \\ 1 & 1 & -2 \end{bmatrix}
$$

$$
\textbf{33.} \begin{bmatrix} 5 & -2 & 3 \\ 1 & 4 & 7 \\ -2 & 3 & 2 \end{bmatrix} \qquad \textbf{34.} \begin{bmatrix} 1 & -1 & 0 \\ 2 & 0 & -1 \\ 2 & 2 & 3 \end{bmatrix}
$$

C   **35.** If $abc \neq 0$, show that

$$
\begin{vmatrix} 1+a & 1 & 1 \\ 1 & 1+b & 1 \\ 1 & 1 & 1+c \end{vmatrix} = abc\left(1 + \frac{1}{a} + \frac{1}{b} + \frac{1}{c}\right)
$$

36. Show that

$$
\begin{vmatrix} a & b & b & b \\ b & a & b & b \\ b & b & a & b \\ b & b & b & a \end{vmatrix} = (a - b)^3 (a + 3b)
$$

37. Show that

$$
\begin{vmatrix} a & 0 & 0 & 0 & -1 \\ 0 & a & 0 & 0 & 0 \\ 0 & 0 & a & 0 & 0 \\ 0 & 0 & 0 & a & 0 \\ -1 & 0 & 0 & 0 & a \end{vmatrix} = a^3(a^2 - 1)
$$

38. Show that

$$\begin{vmatrix} x & 0 & 0 & 0 & a_0 \\ -1 & x & 0 & 0 & a_1 \\ 0 & -1 & x & 0 & a_2 \\ 0 & 0 & -1 & x & a_3 \\ 0 & 0 & 0 & -1 & x+a_4 \end{vmatrix} = x^5 + a_4x^4 + a_3x^3 + a_2x^2 + a_1x + a_0$$

In Problems 39 and 40, use the equation $A^{-1} = [1/\det(A)]\,adj(A)$ to find the inverse of the given matrix.

39. $\begin{bmatrix} a & 0 & b \\ 0 & 1 & 0 \\ c & 0 & d \end{bmatrix}$
 40. $\begin{bmatrix} 1 & 0 & 0 \\ a & 1 & 0 \\ 0 & b & 1 \end{bmatrix}$

41. Find the values of λ for which the following system has nontrivial solutions, and then find those solutions:

$$2x_1 - 3x_2 = \lambda x_1$$
$$x_1 - 2x_2 = \lambda x_2$$

42. Find the values of λ that satisfy the equation

$$\begin{vmatrix} 2-\lambda & -2 & 2 \\ -2 & 3-\lambda & -1 \\ 2 & -1 & 3-\lambda \end{vmatrix} = 0$$

43. Find the values of s for which the following system has a unique solution, and then find the solution:

$$sx_1 + sx_2 + x_3 = 2$$
$$4x_1 + sx_2 - sx_3 = 3$$
$$sx_1 + 2sx_2 + x_3 = 4$$

44. In the system of equations given below, r and θ are constants and u, v, x, and y are variables. Use Cramer's Rule to express u and v in terms of x and y.

$$x = (\cos\theta)u + (\sin\theta)v$$
$$y = (-r\sin\theta)u + (r\cos\theta)v$$

45. If A and B are invertible matrices with $\det(A) = 2$ and $\det(B) = 5$, evaluate each of the following:

(A) $\det(AB)$ (B) $\det(A^2)$ (C) $\det(A^T)$
(D) $\det[(B^T)^2]$ (E) $\det(A^{-1})$ (F) $\det[(A^{-1}B^{-1})^T]$

46. Show that for $n \times n$ matrices A and B

$$\det(AB) = \det(BA)$$

even though $AB \ne BA$.

47. If $A^p = 0$ for some positive integer p, show that A is not invertible.

48. A matrix A is symmetric if $A = A^T$ and skew-symmetric if $A = -A^T$. If A is any square matrix, show that the matrix $A + A^T$ is symmetric and the matrix $A - A^T$ is skew-symmetric. Then show that A can be expressed as a sum of a symmetric matrix and a skew-symmetric matrix.

▮◻▮ Applications

49. *Equation of a line.* Use the equation

$$\begin{vmatrix} x & y & 1 \\ x_1 & y_1 & 1 \\ x_2 & y_2 & 1 \end{vmatrix} = 0$$

to find the equation of the line that passes through the points $(1, -3)$ and $(4, 5)$.

50. *Interpolating polynomial.* Use the equation

$$\begin{vmatrix} y & 1 & x & x^2 \\ y_1 & 1 & x_1 & x_1^2 \\ y_2 & 1 & x_2 & x_2^2 \\ y_3 & 1 & x_3 & x_3^2 \end{vmatrix} = 0$$

to find the equation of the second-degree polynomial whose graph passes through the points $(1, 1)$, $(2, 0)$, and $(4, 4)$.

51. *Equation of a circle.* If (x_1, y_1), (x_2, y_2), and (x_3, y_3) are distinct noncollinear points, show that the equation of the circle through these points is given by

$$\begin{vmatrix} x^2 + y^2 & x & y & 1 \\ x_1^2 + y_1^2 & x_1 & y_1 & 1 \\ x_2^2 + y_2^2 & x_2 & y_2 & 1 \\ x_3^2 + y_3^2 & x_3 & y_3 & 1 \end{vmatrix} = 0$$

Use this equation to find the equation of the circle through each set of points.

(A) $(5, 0)$, $(4, 3)$, $(-3, 4)$ (B) $(6, -2)$, $(5, 1)$, $(-2, 2)$

52. *Mechanics.* The study of the vibrations of a mechanical system consisting of two interconnected springs (see the figure) leads to the system of equations

$$-\tfrac{3}{2}kx_1 + k(x_2 - x_1) = \lambda x_1$$
$$kx_1 - kx_2 \quad\quad\; = \lambda x_2$$

where k is a known positive constant associated with the spring and λ is an unknown parameter. Find the values of λ for which this system will have nontrivial solutions, and then find those solutions.

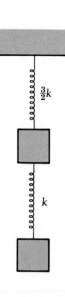

53. *Circuit analysis.* The currents I_1, I_2, and I_3, the resistances R_1, R_2, and R_3, and the voltage V in an electrical circuit (see the figure) can be shown to satisfy

$$\begin{aligned}
I_1 - I_2 - I_3 &= 0 \\
R_1I_1 + R_2I_2 &= V \\
R_1I_1 + R_3I_3 &= V
\end{aligned}$$

where R_1, R_2, R_3, and V are positive constants.

(A) Use the adjoint method to find the inverse of the coefficient matrix for this system.

(B) Use the inverse of the coefficient matrix to solve for I_1, I_2, and I_3.

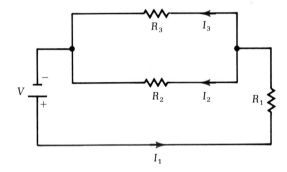

|4| Vectors in Two and Three Dimensions

| 4 | **Contents**

In the first three chapters our work with matrices was essentially algebraic in nature. In this chapter we will introduce the concept of vector, and we will explore this concept both geometrically in two and three dimensions and algebraically. Two algebraic operations, the dot product and cross product, are introduced. These operations have important geometric applications. We will also formulate equations for lines and planes in two and three dimensions that can be generalized easily to more abstract higher-dimensional spaces. A firm understanding of the interplay between algebra and geometry in two and three dimensions is essential to the understanding of the interplay between algebra and geometry in more abstract spaces.

| 4-1 | Vectors in the Plane

- Vectors: Geometrically Defined
- Vectors: Algebraically Defined
- Vector Addition and Scalar Multiplication
- Unit Vectors
- Properties of Vector Operations

▪ Vectors: Geometrically Defined

A professional bowler usually rolls a bowling ball so that it follows a curved path down the bowling lane. How can we describe the velocity of the ball at any given point in its path, the force with which it will strike a pin, and the resulting displacement of that pin? In Figure 1 we have used arrows to represent these three quantities. The *length* of the arrow in part (A) tells us how fast the ball is traveling, in part (B) how hard the ball strikes the pin, and in part (C) how far the pin moves. The *direction* of the arrow in part (A) tells us the direction in which the ball is rolling, in part (B) the direction in which the force of the impact is applied, and in part (C) the direction in which the pin moves after impact.

In general, a quantity that can be completely described by a direction and a length is called a **vector.** Vectors are often denoted by boldface lowercase letters, such as **v**, or lowercase letters with an arrow over them, such as $\vec{v}$. The latter is easier to write, but the former is easier to print, so we will use boldface

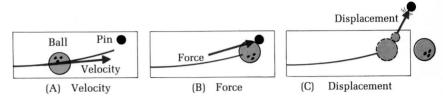

(A)   Velocity   (B)   Force   (C)   Displacement

Figure 1

letters to indicate vectors in this book. When vectors are represented geometrically in two and three dimensions, *directed line segments* (arrows) may be used. If A and B are two points, then the **directed line segment from A to B,** denoted $\overrightarrow{AB}$, is the line segment joining A to B with an arrowhead placed at B to indicate the direction. Point A is called the **initial point** and point B is called the **terminal point.** The absolute value notation $|\overrightarrow{AB}|$ is used to denote the **length** of the directed line segment $\overrightarrow{AB}$ (see Figure 2).

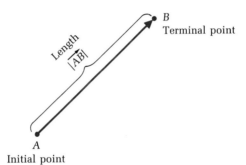

Figure 2   Directed line segment

Example 1   Graph each directed line segment and find its length.

(A) $\overrightarrow{AB}$ where $A = (1, 2)$ and $B = (5, 5)$
(B) $\overrightarrow{CD}$ where $C = (3, -1)$ and $D = (7, 2)$

Solution   Here, we must use the formula for the distance between two points (see the figure).

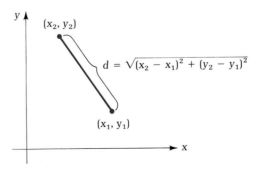

(A) $|\overrightarrow{AB}| = \sqrt{(5-1)^2 + (5-2)^2}$ (B) $|\overrightarrow{CD}| = \sqrt{(7-3)^2 + [2-(-1)]^2}$
$= \sqrt{25} = 5$ $= \sqrt{25} = 5$

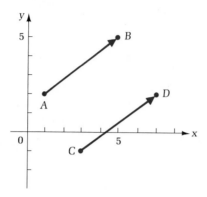

Problem 1   Graph each directed line segment and find its length.

(A) $\overrightarrow{AB}$ where $A = (2, 2)$ and $B = (7, 14)$
(B) $\overrightarrow{CD}$ where $C = (4, -3)$ and $D = (9, 9)$ **▌▌**

   How are vectors and directed line segments related? Technically speaking, they are not quite the same thing. A directed line segment has a direction and a length like a vector, but it also has a *fixed position.* A vector is completely determined by its direction and length and is considered to be *position-free.* Any directed line segment with the same direction and length as a given vector is called a **representative** of that vector. The directed line segments $\overrightarrow{AB}$ and $\overrightarrow{CD}$ in Example 1 have the same length and direction; thus, they represent the same vector. It is perfectly proper to label both of these directed line segments with the same symbol **v**, as in Figure 3.

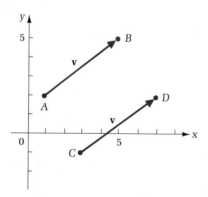

Figure 3

Every vector $\mathbf{v}$ has infinitely many representatives. In order to avoid constant repetition of the phrase "let $\overrightarrow{AB}$ be a representative of the vector $\mathbf{v}$," it is customary to write $\mathbf{v} = \overrightarrow{AB}$ where it is understood that $\overrightarrow{AB}$ is a particular representative of $\mathbf{v}$.

Since vectors are position-free, a representative can be translated to any position in the plane. In particular, it is always possible to find a representative of a vector with its initial point at the origin. This representative is called the **position vector** for $\mathbf{v}$, and selecting this representative is referred to as placing $\mathbf{v}$ in **standard position.**

Example 2   If $\mathbf{v} = \overrightarrow{AB}$ where $A = (2, 5)$ and $B = (7, 8)$, find a representative of $\mathbf{v}$ with its initial point at the origin (that is, find the position vector for $\mathbf{v}$).

Solution

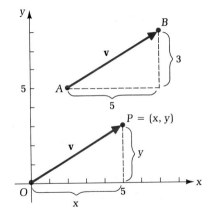

If $P = (x, y)$ is the terminal point of the representative of $\mathbf{v}$ with initial point at the origin $O = (0, 0)$, then the two right triangles in the figure must be congruent. This implies that $x = 5$ and $y = 3$. Thus, $\mathbf{v} = \overrightarrow{OP}$ where $P = (5, 3)$.

Problem 2   If $\mathbf{v} = \overrightarrow{AB}$ where $A = (1, 4)$ and $B = (5, 2)$, find a representative of $\mathbf{v}$ with its initial point at the origin. ∎

▪ Vectors: Algebraically Defined

Example 2 suggests another way to look at vectors. Given any vector $\mathbf{v}$, we can always find a point $P = (x, y)$ in the plane so that $\mathbf{v} = \overrightarrow{OP}$. The point (x, y) completely specifies the vector $\mathbf{v}$. Conversely, given any point $P = (x, y)$ in the plane, the directed line segment $\overrightarrow{OP}$ defines a vector $\mathbf{v}$. This leads to the following *algebraic definition* of vectors:

Algebraic Definition of Vectors

A **vector v** in the plane is an ordered pair of real numbers. If $\mathbf{v} = (x, y)$, then x and y are called the **components** of $\mathbf{v}$. The symbol R^2 is used to denote the set of all vectors in the plane.* That is,

$R^2 = \{(x, y) \mid x$ and y any real numbers$\}$

The vectors $\mathbf{u} = (x_1, y_1)$ and $\mathbf{v} = (x_2, y_2)$ in R^2 are said to be **equal** if their components are equal; that is, if

$$x_1 = x_2 \quad \text{and} \quad y_1 = y_2$$

We now have two ways to view a vector: *geometrically* as a directed line segment and *algebraically* as an ordered pair of real numbers. The interplay between the geometric and algebraic interpretations is extremely important. Theorem 1 connects these two viewpoints.

Theorem 1

If $\mathbf{v} = (v_1, v_2)$, $A = (a_1, a_2)$, and $B = (b_1, b_2)$, then

$$\mathbf{v} = \overrightarrow{AB}$$

if and only if

$$v_1 = b_1 - a_1 \quad \text{and} \quad v_2 = b_2 - a_2$$

Example 3   Represent $\mathbf{v}$ as an ordered pair if $\mathbf{v} = \overrightarrow{AB}$ where $A = (2, -3)$ and $B = (4, 5)$.

Solution   According to Theorem 1, $\mathbf{v} = (v_1, v_2)$ where

$$v_1 = b_1 - a_1 = 4 - 2 = 2$$

and

$$v_2 = b_2 - a_2 = 5 - (-3) = 8$$

Thus, $\mathbf{v} = (2, 8)$, as shown in the figure.

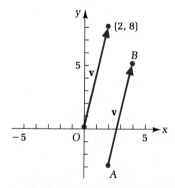

* R^2 is also referred to as **two-dimensional Euclidean space**.

Problem 3   Represent **v** as an ordered pair if $\mathbf{v} = \overrightarrow{AB}$ where $A = (4, 5)$ and $B = (6, -2)$. ▌▌

Example 4   Find a representative of $\mathbf{v} = (3, -2)$ with initial point at $A = (-4, 5)$.

Solution   If $B = (b_1, b_2)$, then according to Theorem 1,

$$3 = b_1 - (-4) \qquad \text{and} \qquad -2 = b_2 - 5$$
$$b_1 = -1 \qquad\qquad\qquad b_2 = 3$$

Thus, $\mathbf{v} = \overrightarrow{AB}$ where $B = (-1, 3)$, as shown in the figure.

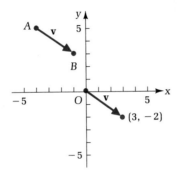

Problem 4   Find a representative of $\mathbf{v} = (-3, -1)$ with initial point at $A = (4, 6)$. ▌▌

The **length** of a vector **v**, denoted by $\|\mathbf{v}\|$, is easily found using the algebraic definition of a vector. This quantity is sometimes called the **magnitude** or **norm** of the vector.

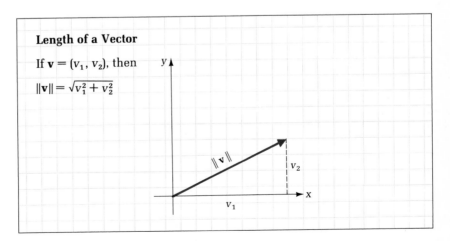

Length of a Vector

If $\mathbf{v} = (v_1, v_2)$, then

$$\|\mathbf{v}\| = \sqrt{v_1^2 + v_2^2}$$

Note that if **v** is represented by a directed line segment $\overrightarrow{AB}$, then the length of **v** is the length of the directed line segment; that is, $\|\mathbf{v}\| = |\overrightarrow{AB}|$.

Example 5   Find the length of each vector.

(A) $\mathbf{v} = (5, 12)$ (B) $\mathbf{u} = (-1, 2)$ (C) $\mathbf{0} = (0, 0)$

Solution   (A) $\|\mathbf{v}\| = \sqrt{5^2 + 12^2} = \sqrt{25 + 144} = \sqrt{169} = 13$
(B) $\|\mathbf{u}\| = \sqrt{(-1)^2 + 2^2} = \sqrt{1 + 4} = \sqrt{5}$
(C) $\|\mathbf{0}\| = \sqrt{0^2 + 0^2} = 0$

Problem 5   Find the length of each vector.

(A) $\mathbf{v} = (-3, 4)$ (B) $\mathbf{u} = (1, 1)$ (C) $\mathbf{w} = (0, -2)$ ▮

The vector $\mathbf{0} = (0, 0)$ is called the **zero vector.** It is the only vector with length equal to 0. What is the direction of the zero vector? If $(v_1, v_2) \neq (0, 0)$, then the direction of $\mathbf{v} = (v_1, v_2)$ is determined by the directed line segment from $(0, 0)$ to (v_1, v_2). But since the directed line segment from $(0, 0)$ to $(0, 0)$ does not determine a unique direction, *the zero vector is assumed to have arbitrary direction.* That is, the zero vector can be assigned any direction that is convenient for the solution of a particular problem.

▪ Vector Addition and Scalar Multiplication

We now want to consider two basic operations on vectors: *addition and multiplication by a number.* The algebraic definitions of these vector operations are the same as the definitions of the corresponding matrix operations we considered in Chapter 2. However, the vector operations also have geometric interpretations. Matrix operations do not.

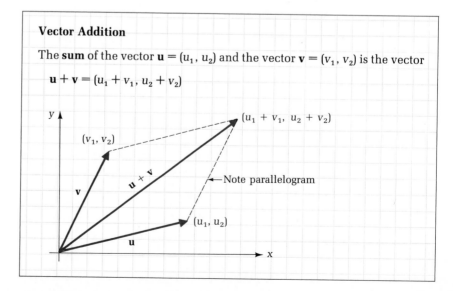

Vector Addition

The **sum** of the vector $\mathbf{u} = (u_1, u_2)$ and the vector $\mathbf{v} = (v_1, v_2)$ is the vector

$$\mathbf{u} + \mathbf{v} = (u_1 + v_1, u_2 + v_2)$$

Geometrically, $\mathbf{u} + \mathbf{v}$ can be formed by positioning $\mathbf{v}$ so that its initial point coincides with the terminal point of $\mathbf{u}$. Then $\mathbf{u} + \mathbf{v}$ is the vector from the initial point of $\mathbf{u}$ to the terminal point of $\mathbf{v}$ (see Figure 4).

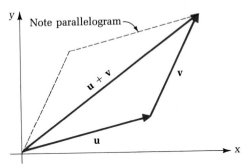

Figure 4   Geometric definition of addition

Example 6   For each pair $\mathbf{u}$ and $\mathbf{v}$, find and graph $\mathbf{u} + \mathbf{v}$.

(A) $\mathbf{u} = (1, 4)$ and $\mathbf{v} = (3, 2)$ (B) $\mathbf{u} = (5, -1)$ and $\mathbf{v} = (-2, 3)$

Solution   (A) $\mathbf{u} + \mathbf{v} = (1, 4) + (3, 2)$ (B) $\mathbf{u} + \mathbf{v} = (5, -1) + (-2, 3)$

$= (1 + 3, 4 + 2) = (4, 6)$ $= (5 - 2, -1 + 3) = (3, 2)$

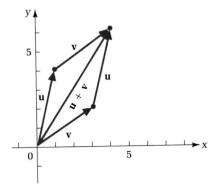
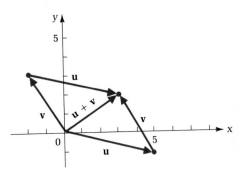

Problem 6   For each pair $\mathbf{u}$ and $\mathbf{v}$, find and graph $\mathbf{u} + \mathbf{v}$.

(A) $\mathbf{u} = (2, 5)$ and $\mathbf{v} = (4, 1)$ (B) $\mathbf{u} = (1, 6)$ and $\mathbf{v} = (5, -4)$ ▐▐

The **negative** of the vector $\mathbf{v} = (v_1, v_2)$ is the vector $-\mathbf{v} = (-v_1, -v_2)$. Geometrically, $-\mathbf{v}$ is a vector with the same length as $\mathbf{v}$ but pointing in the opposite direction, as shown in Figure 5 on page 194.

If $\mathbf{u} = (u_1, u_2)$ and $\mathbf{v} = (v_1, v_2)$, then we define **subtraction** as

$$\mathbf{u} - \mathbf{v} = \mathbf{u} + (-\mathbf{v}) = (u_1 - v_1, u_2 - v_2)$$

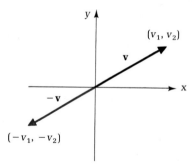

Figure 5   Negative of a vector

In the arithmetic of real numbers, subtraction is defined so that $a - b = c$ if and only if $b + c = a$. This same property holds for vector subtraction. Note what happens when we add $\mathbf{u} - \mathbf{v}$ to $\mathbf{v}$:

$$\mathbf{v} + (\mathbf{u} - \mathbf{v}) = (v_1, v_2) + (u_1 - v_1, u_2 - v_2)$$
$$= (v_1 + u_1 - v_1, v_2 + u_2 - v_2)$$
$$= (u_1, u_2) = \mathbf{u}$$

In other words, when the vector $\mathbf{u} - \mathbf{v}$ is added to $\mathbf{v}$, the sum is $\mathbf{u}$.

Given $\mathbf{u}$ and $\mathbf{v}$, we can easily find $\mathbf{u} - \mathbf{v}$ geometrically. We draw $\mathbf{u}$ and $\mathbf{v}$ in standard position. Then we join their terminal points with an arrow (vector) pointing in the direction so that when the arrow (vector) is added to $\mathbf{v}$ the sum is $\mathbf{u}$ (see Figure 6). Vector $\mathbf{u} - \mathbf{v}$ in standard position is found by translating the arrow (vector) joining the terminal points of $\mathbf{u}$ and $\mathbf{v}$ to the origin (see Figure 6).

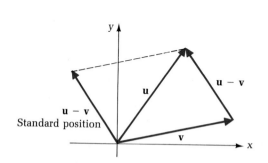

Figure 6   Vector subtraction shown geometrically

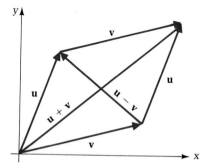

Figure 7   Parallelogram law

A convenient way to remember the geometric definitions of addition and subtraction is by use of the *parallelogram law*, as illustrated in Figure 7.

Example 7   If $\mathbf{u} = (2, 6)$ and $\mathbf{v} = (7, 3)$, compute $\mathbf{u} - \mathbf{v}$ and illustrate the difference geometrically.

Solution $\mathbf{u} - \mathbf{v} = (2, 6) - (7, 3) = (2 - 7, 6 - 3) = (-5, 3)$

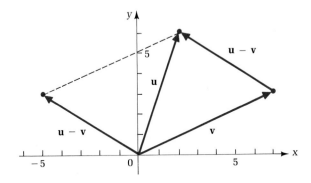

Problem 7   If $\mathbf{u} = (6, 1)$ and $\mathbf{v} = (2, 4)$, compute $\mathbf{u} - \mathbf{v}$ and illustrate the difference geometrically. ▌▌

In Section 2-1 we formed the product of a matrix and a scalar (real number) by multiplying each element in the matrix by the scalar. The *scalar product for vectors* is defined in the same manner.

Scalar Product

The **scalar product*** of a vector $\mathbf{v} = (v_1, v_2)$ and a real number k is

$k\mathbf{v} = (kv_1, kv_2)$

Theorem 2 relates the length of the scalar product $k\mathbf{v}$ to k and $\|\mathbf{v}\|$.

Theorem 2   If $\mathbf{v} = (v_1, v_2)$ and k is a real number, then

$\|k\mathbf{v}\| = |k|\, \|\mathbf{v}\|$

---

* The term "scalar product" is used interchangeably with "dot product" by some. We will always use it as defined above.

Proof $\|k\mathbf{v}\| = \sqrt{(kv_1)^2 + (kv_2)^2}$

$\qquad\quad = \sqrt{k^2v_1^2 + k^2v_2^2}$

$\qquad\quad = \sqrt{k^2(v_1^2 + v_2^2)}$

$\qquad\quad = |k|\sqrt{v_1^2 + v_2^2}$

$\qquad\quad = |k|\,\|\mathbf{v}\|$ Remember, $\sqrt{x^2} = |x|$ for any real number x (we write $\sqrt{x^2} = x$ only if x is nonnegative).   ∎

It can also be shown that $k\mathbf{v}$ has the same direction as $\mathbf{v}$ if $k > 0$ and the opposite direction if $k < 0$. Of course, if $k = 0$, then $k\mathbf{v} = 0$ and its direction is arbitrary. Thus, multiplying a vector by a positive scalar k multiplies the length by k and leaves the direction unchanged, while multiplying by a negative scalar k multiplies the length by $|k|$ and reverses the direction. If $|k| > 1$, then $k\mathbf{v}$ is longer than $\mathbf{v}$, while if $|k| < 1$, then $k\mathbf{v}$ is shorter than $\mathbf{v}$ (see Figure 8).

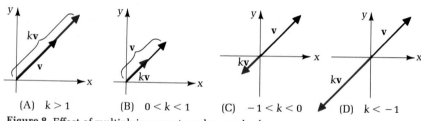

(A) $k > 1$ (B) $0 < k < 1$ (C) $-1 < k < 0$ (D) $k < -1$

Figure 8 Effect of multiplying a vector $\mathbf{v}$ by a scalar k

▪ Unit Vectors

A vector $\mathbf{v}$ is called a **unit vector** if its length is 1; that is, if $\|\mathbf{v}\| = 1$. Special unit vectors play an important role in much of the work that follows. Unit vectors can be formed from arbitrary nonzero vectors as described below:

Forming Unit Vectors

If $\mathbf{v}$ is a nonzero vector, then

$$\mathbf{u} = \frac{1}{\|\mathbf{v}\|}\mathbf{v}$$

is a unit vector with the same direction as $\mathbf{v}$.

Example 8 For each vector, find a unit vector with the same direction as the given vector.

(A) $\mathbf{v} = (1, 3)$ (B) $\mathbf{v} = \left(\dfrac{1}{\sqrt{2}}, -\dfrac{1}{\sqrt{2}}\right)$

Solution   (A) $\|\mathbf{v}\| = \sqrt{1^2 + 3^2} = \sqrt{1 + 9} = \sqrt{10}$

Thus,

$$\mathbf{u} = \frac{1}{\|\mathbf{v}\|}\,\mathbf{v} = \frac{1}{\sqrt{10}}\,(1, 3) = \left(\frac{1}{\sqrt{10}}, \frac{3}{\sqrt{10}}\right)$$

Check:

$$\|\mathbf{u}\| = \sqrt{\left(\frac{1}{\sqrt{10}}\right)^2 + \left(\frac{3}{\sqrt{10}}\right)^2} = \sqrt{\frac{1}{10} + \frac{9}{10}} = \sqrt{1} = 1$$

Thus, $\mathbf{u}$ is a unit vector with the same direction as $\mathbf{v}$.

(B) $\|\mathbf{v}\| = \sqrt{\left(\dfrac{1}{\sqrt{2}}\right)^2 + \left(-\dfrac{1}{\sqrt{2}}\right)^2} = \sqrt{\dfrac{1}{2} + \dfrac{1}{2}} = \sqrt{1} = 1$

Thus, $\mathbf{v}$ is a unit vector.

Problem 8 For each vector, find a unit vector with the same direction as the given vector.

(A) $\mathbf{v} = (-2, 1)$ (B) $\mathbf{v} = \left(\tfrac{3}{5}, -\tfrac{4}{5}\right)$ ▌▌

There are two special unit vectors, called the *standard unit vectors*, which can be used to provide a very useful representation of an arbitrary vector.

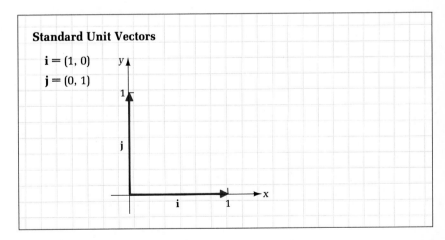

Standard Unit Vectors

$\mathbf{i} = (1, 0)$

$\mathbf{j} = (0, 1)$

Any vector $\mathbf{v}$ can be expressed in terms of the standard unit vectors, as follows:

$$\mathbf{v} = (v_1, v_2) = (v_1, 0) + (0, v_2)$$
$$= v_1\,(1, 0) + v_2\,(0, 1) = v_1\mathbf{i} + v_2\mathbf{j}$$

Thus,

$$\mathbf{v} = (v_1, v_2) = v_1\mathbf{i} + v_2\mathbf{j}$$

Example 9   Express each vector in terms of the standard unit vectors $\mathbf{i}$ and $\mathbf{j}$.

(A) $(2, -5)$ (B) $(3, 0)$ (C) $(0, -2)$

Solution   (A) $(2, -5) = 2\mathbf{i} - 5\mathbf{j}$ (B) $(3, 0) = 3\mathbf{i} + 0\mathbf{j} = 3\mathbf{i}$ (C) $(0, -2) = 0\mathbf{i} - 2\mathbf{j} = -2\mathbf{j}$

Problem 9   Express each vector in terms of the standard unit vectors $\mathbf{i}$ and $\mathbf{j}$.

(A) $(-3, 4)$ (B) $(0, 5)$ (C) $(-6, 0)$ ▌▌

▪ Properties of Vector Operations

Vector addition and scalar multiplication possess properties similar to the properties of matrix addition and scalar multiplication in Theorems 1 and 2 in Section 2-1. All the properties of vector arithmetic listed below in Theorem 3 follow directly from the algebraic definitions of addition and scalar multiplication and the properties of real numbers. The proof is left as an exercise.

Theorem 3

Algebraic Properties of R^2

(A) The following **addition properties** are satisfied for all vectors $\mathbf{u}, \mathbf{v}$, and $\mathbf{w}$ in R^2:

(1) R^2 is closed under addition; that is,   Closure property
$\mathbf{u} + \mathbf{v}$ is in R^2

(2) $\mathbf{u} + \mathbf{v} = \mathbf{v} + \mathbf{u}$ Commutative property

(3) $\mathbf{u} + (\mathbf{v} + \mathbf{w}) = (\mathbf{u} + \mathbf{v}) + \mathbf{w}$ Associative property

(4) $\mathbf{u} + \mathbf{0} = \mathbf{0} + \mathbf{u} = \mathbf{u}$ Additive identity

(5) $\mathbf{u} + (-\mathbf{u}) = (-\mathbf{u}) + \mathbf{u} = \mathbf{0}$ Additive inverse

(B) The following **scalar multiplication properties** are satisfied for all vectors $\mathbf{u}$ and $\mathbf{v}$ in R^2 and all scalars k and ℓ:

(1) R^2 is closed under scalar multiplication; that is, $k\mathbf{u}$ is in R^2 Closure property

(2) $k(\ell\mathbf{u}) = (k\ell)\mathbf{u}$ Associative property

(3) $k(\mathbf{u} + \mathbf{v}) = k\mathbf{u} + k\mathbf{v}$ Distributive property

(4) $(k + \ell)\mathbf{u} = k\mathbf{u} + \ell\mathbf{u}$ Distributive property

(5) $1\mathbf{u} = \mathbf{u}$ Multiplicative identity

When vectors are represented in terms of the standard unit vectors and the properties listed in Theorem 3 are applied, we have a very efficient procedure for performing algebraic operations with vectors.

Example 10   If $\mathbf{u} = 2\mathbf{i} - \mathbf{j}$ and $\mathbf{v} = 4\mathbf{i} + 5\mathbf{j}$, compute each of the following:

(A) $\mathbf{u} + \mathbf{v}$ (B) $\mathbf{u} - \mathbf{v}$ (C) $4\mathbf{u}$ (D) $3\mathbf{u} + 2\mathbf{v}$

Solution   (A) $\mathbf{u} + \mathbf{v} = (2\mathbf{i} - \mathbf{j}) + (4\mathbf{i} + 5\mathbf{j})$ (B) $\mathbf{u} - \mathbf{v} = (2\mathbf{i} - \mathbf{j}) - (4\mathbf{i} + 5\mathbf{j})$

$\qquad\qquad = 2\mathbf{i} + 4\mathbf{i} - \mathbf{j} + 5\mathbf{j}$ $\qquad\qquad = 2\mathbf{i} - \mathbf{j} - 4\mathbf{i} - 5\mathbf{j}$

$\qquad\qquad = (2 + 4)\mathbf{i} + (-1 + 5)\mathbf{j}$ $\qquad\qquad = (2 - 4)\mathbf{i} + (-1 - 5)\mathbf{j}$

$\qquad\qquad = 6\mathbf{i} + 4\mathbf{j}$ $\qquad\qquad = -2\mathbf{i} - 6\mathbf{j}$

(C) $4\mathbf{u} = 4(2\mathbf{i} - \mathbf{j})$ (D) $3\mathbf{u} + 2\mathbf{v} = 3(2\mathbf{i} - \mathbf{j}) + 2(4\mathbf{i} + 5\mathbf{j})$

$\qquad\quad = 8\mathbf{i} - 4\mathbf{j}$ $\qquad\qquad\qquad = 6\mathbf{i} - 3\mathbf{j} + 8\mathbf{i} + 10\mathbf{j}$

$\qquad\qquad\qquad = (6 + 8)\mathbf{i} + (-3 + 10)\mathbf{j}$

$\qquad\qquad\qquad = 14\mathbf{i} + 7\mathbf{j}$

Problem 10   If $\mathbf{u} = 3\mathbf{i} + 2\mathbf{j}$ and $\mathbf{v} = 4\mathbf{i} - \mathbf{j}$, compute each of the following:

(A) $\mathbf{u} + \mathbf{v}$ (B) $\mathbf{u} - \mathbf{v}$ (C) $3\mathbf{v}$ (D) $2\mathbf{u} + 5\mathbf{v}$ ▐▌

Answers to   **1.** (A) $|\overrightarrow{AB}| = 13$
Matched Problems   (B) $|\overrightarrow{CD}| = 13$

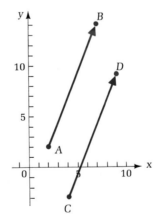

2. $\mathbf{v} = \overrightarrow{OP}$ where $P = (4, -2)$ **3.** $(2, -7)$

4. $\overrightarrow{AB}$ where $B = (1, 5)$ **5.** (A) 5   (B) $\sqrt{2}$ (C) 2

6. (A) $(6, 6)$ (B) $(6, 2)$

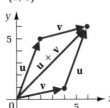

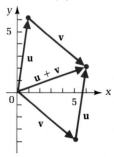

7. $(4, -3)$

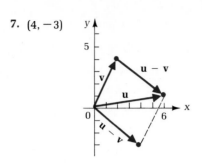

8. (A) $(-2/\sqrt{5}, 1/\sqrt{5})$ (B) $(\frac{3}{5}, -\frac{4}{5})$; that is, $\mathbf{v}$ is a unit vector

9. (A) $-3\mathbf{i} + 4\mathbf{j}$ (B) $5\mathbf{j}$ (C) $-6\mathbf{i}$

10. (A) $7\mathbf{i} + \mathbf{j}$ (B) $-\mathbf{i} + 3\mathbf{j}$ (C) $12\mathbf{i} - 3\mathbf{j}$ (D) $26\mathbf{i} - \mathbf{j}$

▌ Exercise 4-1

A   In Problems 1–4, $\overrightarrow{AB}$ is a representative of the vector $\mathbf{v}$. Find the length of $\mathbf{v}$. Find a representative of $\mathbf{v}$ with its initial point at the origin. Graph this representative and $\overrightarrow{AB}$ on the same set of axes.

1. $A = (-6, 1)$; $B = (6, 6)$ 　　　　　　 **2.** $A = (2, 5)$; $B = (8, -3)$

3. $A = (6, -2)$; $B = (3, 4)$ 　　　　　　 **4.** $A = (4, 4)$; $B = (0, 0)$

In Problems 5–8, represent the vector $\overrightarrow{AB}$ as an ordered pair.

5. $A = (-2, 2)$; $B = (3, -5)$ 　　　　　 **6.** $A = (4, -2)$; $B = (1, 3)$

7. $A = (2, 3)$; $B = (5, 3)$ 　　　　　　 **8.** $A = (7, -7)$; $B = (7, 7)$

In Problems 9–12, find $\mathbf{u} + \mathbf{v}$ and $\mathbf{u} - \mathbf{v}$. Graph $\mathbf{u}$, $\mathbf{v}$, $\mathbf{u} + \mathbf{v}$, and $\mathbf{u} - \mathbf{v}$ on the same set of axes.

9. $\mathbf{u} = (5, 2)$; $\mathbf{v} = (1, 4)$ 　　　　　 **10.** $\mathbf{u} = (7, 3)$; $\mathbf{v} = (-4, 8)$

11. $\mathbf{u} = (2, 3)$; $\mathbf{v} = (1, -4)$ 　　　　 **12.** $\mathbf{u} = (-2, 4)$; $\mathbf{v} = (5, -1)$

B   In Problems 13–16, find a representative of $\mathbf{v}$ with its initial point at A.

13. $\mathbf{v} = (-3, 2)$; $A = (4, 1)$ 　　　　 **14.** $\mathbf{v} = (1, 1)$; $A = (1, 1)$

15. $\mathbf{v} = (2, 3)$; $A = (-2, -3)$ 　　　 **16.** $\mathbf{v} = (4, -5)$; $A = (-3, 7)$

In Problems 17–20, find a representative of $\mathbf{v}$ with its terminal point at B.

17. $\mathbf{v} = (2, 4)$; $B = (9, 5)$ 　　　　　 **18.** $\mathbf{v} = (-1, -3)$; $B = (-4, 2)$

19. $\mathbf{v} = (2, 3)$; $B = (-2, -3)$ 　　　 **20.** $\mathbf{v} = (1, 1)$; $B = (1, 1)$

In Problems 21 and 22, find $\|\mathbf{v}\|$. For each value of k, find $k\mathbf{v}$ and the length of $k\mathbf{v}$. Graph $\mathbf{v}$ and $k\mathbf{v}$ on the same set of axes.

21. $\mathbf{v} = (6, -8)$
    (A) $k = 2$ (B) $k = -\frac{1}{2}$

22. $\mathbf{v} = (-3, -1)$
    (A) $k = -3$ (B) $k = \frac{2}{3}$

In Problems 23–26, express **v** in terms of the standard unit vectors **i** and **j**.

23. $\mathbf{v} = (4, -5)$

24. $\mathbf{v} = (-3, 7)$

25. $\mathbf{v} = \overrightarrow{AB};$ $A = (-1, 2),$
    $B = (4, -3)$

26. $\mathbf{v} = \overrightarrow{AB};$ $A = (3, -2),$ $B = (3, 5)$

In Problems 27–30, find a unit vector in the direction of **v**.

27. $\mathbf{v} = (-3, 4)$ **28.** $\mathbf{v} = (1, -2)$ **29.** $\mathbf{v} = 2\mathbf{i} - 4\mathbf{j}$ **30.** $\mathbf{v} = 5\mathbf{i} + 12\mathbf{j}$

In Problems 31–34, let $\mathbf{u} = 3\mathbf{i} - 4\mathbf{j}$ and $\mathbf{v} = 2\mathbf{i} + 5\mathbf{j}$, and perform the indicated operations.

31. $2\mathbf{u} + 3\mathbf{v}$ **32.** $-\mathbf{u} + 2\mathbf{v}$ **33.** $4\mathbf{u} - 6\mathbf{v}$ **34.** $5\mathbf{u} + 4\mathbf{v}$

35. If $\mathbf{v} = (1, -3)$, find a vector of length 20 that has the same direction as **v**.

36. If $\mathbf{v} = (-1, 1)$, find a vector of length 10 that has the opposite direction as **v**.

C   Problems 37–40 refer to statements from Theorem 3 in this section. Use the ordered pair representation of vectors and the properties of real numbers to prove each statement. Do not refer to Theorem 3 in your proof.

37. Theorem 3(A), part (2). $\mathbf{u} + \mathbf{v} = \mathbf{v} + \mathbf{u}$

38. Theorem 3 (A), part (5). $\mathbf{u} + (-\mathbf{u}) = \mathbf{0}$

39. Theorem 3(B), part (2). $k(\ell \mathbf{u}) = (k\ell)\mathbf{u}$

40. Theorem 3(B), part (4). $(k + \ell)\mathbf{u} = k\mathbf{u} + \ell \mathbf{u}$

41. If **u**, **v**, and **w** are vectors in R^2 and $\mathbf{u} = \mathbf{v}$, show that $\mathbf{u} + \mathbf{w} = \mathbf{v} + \mathbf{w}$.

42. If **u** and **v** are vectors in R^2, k is a scalar, and $\mathbf{u} = \mathbf{v}$, show that $k\mathbf{u} = k\mathbf{v}$.

In Problems 43 and 44, solve each system and express the solution set in the form

    $\{t\mathbf{v}|t \text{ any scalar}\}$

where **v** is an appropriately chosen vector.

43. $x_1 + x_2 = 0$
      $2x_1 + 2x_2 = 0$

44. $x_1 - 2x_2 = 0$
     $-2x_1 + 4x_2 = 0$

In Problems 45 and 46, solve each system and express the solution set in the form

    $\{\mathbf{u} + t\mathbf{v}|t \text{ any scalar}\}$

where **u** and **v** are appropriately chosen vectors.

45. $x_1 + 2x_2 = 5$
   $-2x_1 - 4x_2 = -10$

46. $x_1 - 3x_2 = -1$
   $-3x_1 + 9x_2 = 3$

47. Prove or disprove: If $\mathbf{a} = (1, 1)$, $\mathbf{b} = (1, -1)$, and **u** is any vector, then there exist scalars k_1 and k_2 satisfying

    $\mathbf{u} = k_1\mathbf{a} + k_2\mathbf{b}$

48. Rework Problem 47 for **a** = (1, 1) and **b** = (−1, −1).

In Problems 49 and 50, solve each equation for **x**.

49. $3\mathbf{x} + 2(\mathbf{i} + 2\mathbf{j}) = 2\mathbf{x} + 4(2\mathbf{i} − 3\mathbf{j})$ **50.** $2\mathbf{x} + 3(2\mathbf{i} − \mathbf{j}) = 4\mathbf{x} − 2\mathbf{i} − 3\mathbf{j}$

51. If **v** is a vector, show that **v** = 0 if and only if $\|\mathbf{v}\| = 0$.

52. Show that $\|\mathbf{v}\| = \|−\mathbf{v}\|$ for any vector **v**.

53. Use the parallelogram law to prove the triangle inequality for vectors:

$$\|\mathbf{u} + \mathbf{v}\| \leq \|\mathbf{u}\| + \|\mathbf{v}\|$$

54. Refer to Problem 53. Under what conditions will $\|\mathbf{u} + \mathbf{v}\| = \|\mathbf{u}\| + \|\mathbf{v}\|$?

|4-2| Vectors in Three Dimensions

- ■ Three-Dimensional Coordinate Systems
- ■ Distance in Three Dimensions
- ■ Vectors in Three Dimensions
- ■ Vector Operations

■ Three-Dimensional Coordinate Systems

Many situations (real-world and mathematical) call for the use of vectors which do not all lie in a single plane. In order to describe such vectors, it is necessary to consider a three-dimensional coordinate system. We select a point O, called the *origin*, and construct three mutually perpendicular number lines intersecting at the point O. The number lines are called the *x axis*, *y axis*, and *z axis*. The plane determined by the x and y axes is called the *xy coordinate plane*. Similarly, the x and z axes determine the *xz coordinate plane*, and the y and z axes determine the *yz coordinate plane* (see Figure 9).

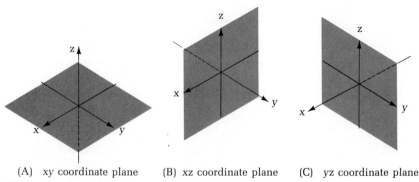

(A) xy coordinate plane   (B) xz coordinate plane   (C) yz coordinate plane

Figure 9

Three-dimensional coordinate systems all have one of two possible orientations. If we draw the *y* and *z* axes on a piece of paper with the positive *y* axis pointing to the right and the positive *z* axis pointing to the top of the page, then there are two possible directions for the positive *x* axis, toward us or away from us. The system is called a *right-hand coordinate system* if the positive *x* axis points toward us and a *left-hand coordinate system* if the positive *x* axis points away from us (see Figure 10).

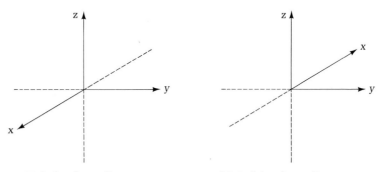

(A) Right-hand coordinate system   (B) Left-hand coordinate system
Figure 10

To determine whether a given coordinate system is a right-hand system, point the forefinger of your right hand in the direction of the positive *x* axis and the second finger in the direction of the positive *y* axis. If your thumb now points in the direction of the positive *z* axis, then the coordinate system is right-handed (see Figure 11). All the coordinate systems we will consider will be right-hand systems.

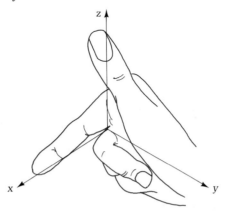

Figure 11   Test for a right-hand coordinate system

A point *P* in a three-dimensional coordinate system is described by an ordered triple of numbers (x, y, z), where x is the directed distance from *P* to the yz plane, y is the directed distance from P to the xz plane, and z is the directed distance from P to the xy plane (see Figure 12).

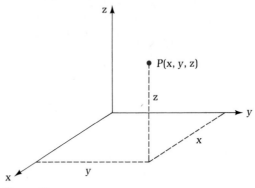

Figure 12

Example 11   Locate all the following points in a three-dimensional coordinate system:

(A) $(3, 5, 4)$ (B) $(2, -1, 8)$ (C) $(-1, 3, -5)$

Solution

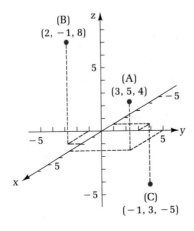

Problem 11   Locate all the following points in a three-dimensional coordinate system:

(A) $(2, 4, 3)$ (B) $(2, 1, -3)$ (C) $(-3, -5, 2)$ ▌▌

▪ Distance in Three Dimensions

Some of the familiar concepts from two-dimensional coordinate systems can be easily extended to three dimensions. One of the most important of these concepts is the *distance formula*.

Theorem 4

> The distance between the points $P_1 = (x_1, y_1, z_1)$ and $P_2 = (x_2, y_2, z_2)$ is
>
> $$d = \sqrt{(x_2 - x_1)^2 + (y_2 - y_1)^2 + (z_2 - z_1)^2}$$

Proof   Let a, b, and c be the lengths of the sides of the box shown in Figure 13, and let d be the length of the diagonal.

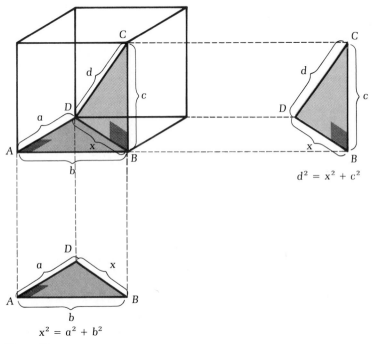

Figure 13

Applying the Pythagorean theorem to right triangles DBC and DAB, we have

$$d^2 = x^2 + c^2 \qquad \text{Triangle } DBC$$
$$= a^2 + b^2 + c^2 \qquad \text{Triangle } DAB$$

Thus,

$$d = \sqrt{a^2 + b^2 + c^2} \tag{1}$$

Now, given two points P_1 and P_2, we construct a box with D at P_1, C at P_2, and sides parallel to the coordinate axes (Figure 14). The lengths of the sides of this box are $a = |x_2 - x_1|$, $b = |y_2 - y_1|$, and $c = |z_2 - z_1|$. Absolute values are used since some of the differences may be negative. Substituting in (1) and remem-

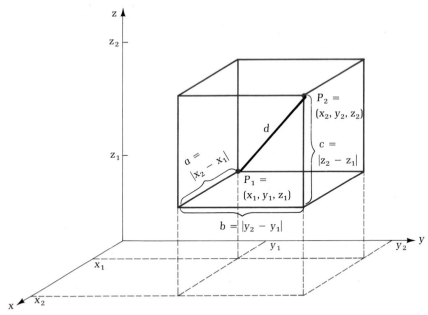

Figure 14

bering that $|w|^2 = w^2$ for any real number w, we have

$$d = \sqrt{(x_2 - x_1)^2 + (y_2 - y_1)^2 + (z_2 - z_1)^2} \qquad (2)$$

This completes the proof.   ∎

Example 12   Find the distance between $P_1 = (1, -1, 2)$ and $P_2 = (3, 2, 4)$.

Solution $d = \sqrt{(3 - 1)^2 + [2 - (-1)]^2 + (4 - 2)^2}$

             $= \sqrt{4 + 9 + 4} = \sqrt{17}$

Problem 12   Find the distance between $P_1 = (2, 1, -3)$ and $P_2 = (-1, 2, 2)$.   ∎

▪ Vectors in Three Dimensions

If A and B are two points in a three-dimensional coordinate system, then the directed line segment $\overrightarrow{AB}$ represents a three-dimensional vector. If $\mathbf{v}$ is the vector represented by $\overrightarrow{AB}$, then we can find a point $P = (v_1, v_2, v_3)$ so that $\mathbf{v} = \overrightarrow{OP}$ where $O = (0, 0, 0)$ is the origin of the coordinate system (see Figure 15). This leads to a natural extension of the algebraic definition of a vector from two dimensions to three dimensions.

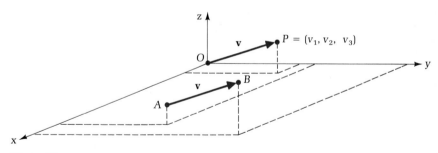

Figure 15

Three-Dimensional Vectors

A **vector v** in three dimensions is an ordered triple of real numbers. If $\mathbf{v} = (v_1, v_2, v_3)$, then v_1, v_2, and v_3 are called the **components** of **v**. The set of all three-dimensional vectors is*

$$R^3 = \{(v_1, v_2, v_3)|v_1, v_2, \text{ and } v_3 \text{ any real numbers}\}$$

The vectors $\mathbf{u} = (u_1, u_2, u_3)$ and $\mathbf{v} = (v_1, v_2, v_3)$ in R^3 are said to be **equal** if their components are equal; that is, if

$$u_1 = v_1 \qquad u_2 = v_2 \qquad u_3 = v_3$$

Most of the properties of two-dimensional vectors studied in Section 4-1 can be extended to three-dimensional vectors.

Theorem 5

If $\mathbf{v} = (v_1, v_2, v_3)$, $A = (a_1, a_2, a_3)$, and $B = (b_1, b_2, b_3)$, then

$$\mathbf{v} = \overrightarrow{AB}$$

if and only if

$$v_1 = b_1 - a_1, \qquad v_2 = b_2 - a_2, \qquad \text{and} \qquad v_3 = b_3 - a_3$$

Example 13   If $\mathbf{v} = \overrightarrow{AB}$ where $A = (2, 12, 2)$ and $B = (6, 8, 7)$, represent **v** as an ordered triple.

Solution $v_1 = 6 - 2 = 4 \qquad v_2 = 8 - 12 = -4 \qquad v_3 = 7 - 2 = 5$

Thus, $\mathbf{v} = (4, -4, 5)$, as shown in the figure at the top of the next page.

---

* R^3 is also called **three-dimensional Euclidean space.**

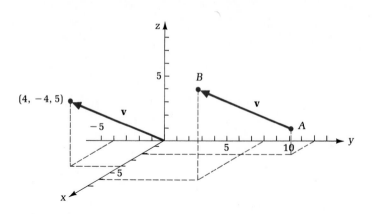

Problem 13   If $\mathbf{v} = \overrightarrow{AB}$ where $A = (-1, 2, 0)$ and $B = (1, -1, 3)$, represent $\mathbf{v}$ as an ordered triple.   ▐▌

Example 14   Find a representative of $\mathbf{v} = (-3, 5, 4)$ with initial point at $A = (7, 12, 1)$.

Solution   If $B = (b_1, b_2, b_3)$, then

$$-3 = b_1 - 7 \qquad 5 = b_2 - 12 \qquad 4 = b_3 - 1$$
$$b_1 = 4 \qquad\qquad b_2 = 17 \qquad\qquad b_3 = 5$$

Thus, $B = (4, 17, 5)$, as shown in the figure.

Problem 14   Find a representative of $\mathbf{v} = (2, -4, 3)$ with initial point at $A = (4, 2, 1)$.   ▐▌

The length of a vector $\mathbf{v}$ is easily found using Theorem 4 and $\mathbf{v}$ in its standard position.

Length of a Vector

If $\mathbf{v} = (v_1, v_2, v_3)$, then the **length (magnitude** or **norm)** of $\mathbf{v}$ is

$$\|\mathbf{v}\| = \sqrt{v_1^2 + v_2^2 + v_3^2}$$

If $\|\mathbf{v}\| = 1$, then $\mathbf{v}$ is called a **unit vector.**

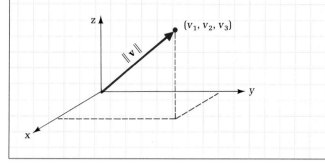

Example 15   Find the length of $\mathbf{v} = (4, -2, 5)$.

Solution $\|\mathbf{v}\| = \sqrt{4^2 + (-2)^2 + 5^2}$

$= \sqrt{16 + 4 + 25}$

$= \sqrt{45} = 3\sqrt{5}$

Problem 15   Find the length of $\mathbf{v} = (14, 4, 2)$.   ▌▌

▪ Vector Operations

The geometric definitions of addition and subtraction for three-dimensional vectors are exactly the same as in two dimensions. The algebraic definitions are easily extended from two to three dimensions by including the third component.

Addition and Subtraction

If $\mathbf{u} = (u_1, u_2, u_3)$ and $\mathbf{v} = (v_1, v_2, v_3)$, then:

(A) $\mathbf{u} + \mathbf{v} = (u_1 + v_1, u_2 + v_2, u_3 + v_3)$
(B) $-\mathbf{v} = (-v_1, -v_2, -v_3)$
(C) $\mathbf{u} - \mathbf{v} = \mathbf{u} + (-\mathbf{v}) = (u_1 - v_1, u_2 - v_2, u_3 - v_3)$

Notice that $\mathbf{u} + \mathbf{v}$ and $\mathbf{u} - \mathbf{v}$ are the diagonals of the parallelogram determined by $\mathbf{u}$ and $\mathbf{v}$, just as in the two-dimensional case.

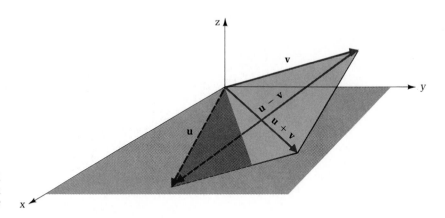

Vector addition and
subtraction shown
geometrically

Example 16   If $\mathbf{u} = (7, 3, -2)$ and $\mathbf{v} = (2, 9, 3)$, find: (A) $\mathbf{u} + \mathbf{v}$ (B) $\mathbf{u} - \mathbf{v}$

Solution   (A) $\mathbf{u} + \mathbf{v} = (7, 3, -2) + (2, 9, 3) = (9, 12, 1)$
(B) $\mathbf{u} - \mathbf{v} = \mathbf{u} + (-\mathbf{v}) = (7, 3, -2) + (-2, -9, -3) = (5, -6, -5)$

Problem 16   If $\mathbf{u} = (-2, 3, 1)$ and $\mathbf{v} = (4, 3, 3)$, find: (A) $\mathbf{u} + \mathbf{v}$ (B) $\mathbf{u} - \mathbf{v}$ ▌▌

Scalar Product

The **scalar product** of the vector $\mathbf{v} = (v_1, v_2, v_3)$ and the real number k is

$$k\mathbf{v} = (kv_1, kv_2, kv_3)$$

The length of $k\mathbf{v}$ satisfies

$$\|k\mathbf{v}\| = |k|\, \|\mathbf{v}\|$$

If $k > 0$, $k\mathbf{v}$ and $\mathbf{v}$ have the same direction.

If $k < 0$, $k\mathbf{v}$ and $\mathbf{v}$ have opposite directions.

If $\mathbf{v}$ is a nonzero vector, then

$$\mathbf{u} = \frac{1}{\|\mathbf{v}\|}\mathbf{v}$$

is a unit vector with the same direction as $\mathbf{v}$.

Example 17   If $\mathbf{v} = (2, 6, 5)$, find: (A) $2\mathbf{v}$ (B) $-\frac{1}{2}\mathbf{v}$

Solution   (A) $2\mathbf{v} = 2(2, 6, 5) = (4, 12, 10)$ (B) $-\frac{1}{2}\mathbf{v} = -\frac{1}{2}(2, 6, 5) = (-1, -3, -\frac{5}{2})$

Problem 17   If $\mathbf{v} = (3, -2, 6)$, find: (A) $4\mathbf{v}$ (B) $-\frac{1}{3}\mathbf{v}$ ▌▌

Example 18   Find a unit vector with the same direction as $\mathbf{v} = (2, 4, 4)$.

Solution $\|\mathbf{v}\| = \sqrt{2^2 + 4^2 + 4^2} = \sqrt{4 + 16 + 16} = \sqrt{36} = 6$

Thus,

$$\mathbf{u} = \frac{1}{\|\mathbf{v}\|}\,\mathbf{v}$$

$$= \tfrac{1}{6}(2, 4, 4) = (\tfrac{1}{3}, \tfrac{2}{3}, \tfrac{2}{3})$$

Check:

$$\|\mathbf{u}\| = \sqrt{(\tfrac{1}{3})^2 + (\tfrac{2}{3})^2 + (\tfrac{2}{3})^2} = \sqrt{\tfrac{1}{9} + \tfrac{4}{9} + \tfrac{4}{9}} = \sqrt{1} = 1$$

Problem 18   Find a unit vector with the same direction as $\mathbf{v} = (-3, 2, 6)$.   ▌▌

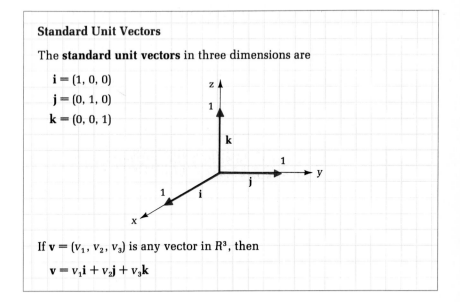

Standard Unit Vectors

The **standard unit vectors** in three dimensions are

$\mathbf{i} = (1, 0, 0)$

$\mathbf{j} = (0, 1, 0)$

$\mathbf{k} = (0, 0, 1)$

If $\mathbf{v} = (v_1, v_2, v_3)$ is any vector in R^3, then

$\mathbf{v} = v_1\mathbf{i} + v_2\mathbf{j} + v_3\mathbf{k}$

Example 19   Express each vector in terms of $\mathbf{i}$, $\mathbf{j}$, and $\mathbf{k}$.

(A) $\mathbf{u} = (2, -3, 1)$ (B) $\mathbf{v} = (3, 4, 0)$

Solution   (A) $\mathbf{u} = (2, -3, 1) = 2\mathbf{i} - 3\mathbf{j} + \mathbf{k}$
(B) $\mathbf{v} = (3, 4, 0) = 3\mathbf{i} + 4\mathbf{j} + 0\mathbf{k} = 3\mathbf{i} + 4\mathbf{j}$

Problem 19   Express each vector in terms of **i**, **j**, and **k**.

(A) **u** $= (-1, 2, 4)$ (B) **v** $= (0, 2, -5)$ ▌▌

The basic algebraic properties of vectors in R^3 are identical to those of vectors in R^2. Compare Theorem 6 with Theorem 3 in Section 4-1.

Theorem 6

> **Algebraic Properties of R^3**
>
> (A) The following **addition properties** are satisfied for all vectors **u**, **v**, and **w** in R^3:
>
> | | |
> |---|---|
> | (1) R^3 is closed under addition; that is, **u** + **v** is in R^3 | Closure property |
> | (2) **u** + **v** = **v** + **u** | Commutative property |
> | (3) **u** + (**v** + **w**) = (**u** + **v**) + **w** | Associative property |
> | (4) **u** + **0** = **0** + **u** = **u** | Additive identity |
> | (5) **u** + (−**u**) = (−**u**) + **u** = **0** | Additive inverse |
>
> (B) The following **scalar multiplication properties** are satisfied for all vectors **u** and **v** in R^3 and all scalars k and ℓ:
>
> | | |
> |---|---|
> | (1) R^3 is closed under scalar multiplication; that is, k**u** is in R^3 | Closure property |
> | (2) $k(\ell$**u**$) = (k\ell)$**u** | Associative property |
> | (3) $k($**u** + **v**$) = k$**u** + k**v** | Distributive property |
> | (4) $(k + \ell)$**u** = k**u** + ℓ**u** | Distributive property |
> | (5) 1**u** = **u** | Multiplicative identity |

Example 20   If **u** $= $ **i** + 2**j** − **k** and **v** $= $ 2**i** − **j** + 3**k**, find each of the following:

(A) **u** + **v**   (B) **u** − **v**   (C) 2**u** + 3**v**   (D) ‖2**u** + 3**v**‖

Solution   (A) **u** + **v** = (**i** + 2**j** − **k**) + (2**i** − **j** + 3**k**)
$\qquad\qquad$ = (1 + 2)**i** + (2 − 1)**j** + (−1 + 3)**k**
$\qquad\qquad$ = 3**i** + **j** + 2**k**

$\qquad$ (B) **u** − **v** = (**i** + 2**j** − **k**) − (2**i** − **j** + 3**k**)
$\qquad\qquad$ = **i** + 2**j** − **k** − 2**i** + **j** − 3**k**
$\qquad\qquad$ = (1 − 2)**i** + (2 + 1)**j** + (−1 − 3)**k**
$\qquad\qquad$ = −**i** + 3**j** − 4**k**

$\qquad$ (C) 2**u** + 3**v** = 2(**i** + 2**j** − **k**) + 3(2**i** − **j** + 3**k**)
$\qquad\qquad$ = 2**i** + 4**j** − 2**k** + 6**i** − 3**j** + 9**k**
$\qquad\qquad$ = (2 + 6)**i** + (4 − 3)**j** + (−2 + 9)**k**
$\qquad\qquad$ = 8**i** + **j** + 7**k**

$\qquad$ (D) ‖2**u** + 3**v**‖ = ‖8**i** + **j** + 7**k**‖
$\qquad\qquad$ = $\sqrt{8^2 + 1^2 + 7^2}$
$\qquad\qquad$ = $\sqrt{64 + 1 + 49} = \sqrt{114}$

Problem 20   If $\mathbf{u} = 2\mathbf{i} - \mathbf{j} + 4\mathbf{k}$ and $\mathbf{v} = -\mathbf{i} + 2\mathbf{j} - 3\mathbf{k}$, find each of the following:

(A) $\mathbf{u} + \mathbf{v}$ (B) $\mathbf{u} - \mathbf{v}$ (C) $3\mathbf{u} + 4\mathbf{v}$ (D) $\|3\mathbf{u} + 4\mathbf{v}\|$ ▌▌

Answers to   **11.**
Matched Problems

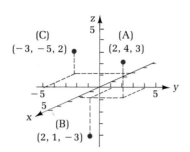

12. $\sqrt{35}$ **13.** $(2, -3, 3)$ **14.** $\mathbf{v} = \overrightarrow{AB}$ where $B = (6, -2, 4)$

15. $6\sqrt{6}$ **16.** (A) $(2, 6, 4)$ (B) $(-6, 0, -2)$

17. (A) $(12, -8, 24)$ (B) $(-1, \frac{2}{3}, -2)$

18. $(-\frac{3}{7}, \frac{2}{7}, \frac{6}{7})$ **19.** (A) $-\mathbf{i} + 2\mathbf{j} + 4\mathbf{k}$ (B) $2\mathbf{j} - 5\mathbf{k}$

20. (A) $\mathbf{i} + \mathbf{j} + \mathbf{k}$ (B) $3\mathbf{i} - 3\mathbf{j} + 7\mathbf{k}$ (C) $2\mathbf{i} + 5\mathbf{j}$ (D) $\sqrt{29}$

▌▌ Exercise 4-2

A   In Problems 1–4, locate each pair of points in a three-dimensional coordinate system and find the distance between the points.

1. $(2, 3, 8)$ and $(8, 9, 5)$ **2.** $(2, -2, 5)$ and $(6, 2, -2)$

3. $(3, -3, -1)$ and $(-4, 2, 5)$ **4.** $(-3, -4, 2)$ and $(-4, 3, -6)$

In Problems 5–8, $\overrightarrow{AB}$ is a representative of a vector $\mathbf{v}$. Find a representative of $\mathbf{v}$ with its initial point at the origin. Find the length of $\mathbf{v}$. Graph both representatives on the same set of axes.

5. $A = (2, 4, 1);$ $B = (4, 7, 7)$ **6.** $A = (2, 8, 3);$ $B = (4, 4, 7)$

7. $A = (3, -3, 2);$ $B = (5, 2, -2)$ **8.** $A = (4, 0, 0);$ $B = (0, 4, 0)$

In Problems 9–12, find $\mathbf{u} + \mathbf{v}$ and $\mathbf{u} - \mathbf{v}$.

9. $\mathbf{u} = (5, -2, 3);$ $\mathbf{v} = (4, 6, -2)$ **10.** $\mathbf{u} = (6, 3, 7);$ $\mathbf{v} = (2, 8, 3)$

11. $\mathbf{u} = (5, 1, 0);$ $\mathbf{v} = (0, 6, 2)$ **12.** $\mathbf{u} = (0, -4, 0);$ $\mathbf{v} = (-5, 0, 0)$

B   In Problems 13–16, find a representative of $\mathbf{v}$ with its initial point at A.

13. $\mathbf{v} = (2, 1, -4);$ $A = (1, -2, 3)$ **14.** $\mathbf{v} = (-1, 3, -4);$ $A = (5, -6, 2)$

15. $\mathbf{v} = (2, 3, 1);$ $A = (-2, -3, -1)$ **16.** $\mathbf{v} = (-4, -2, -3);$ $A = (4, 2, -3)$

In Problems 17–20, find a representative of **v** *with its terminal point at B.*

17. $\mathbf{v} = (1, 2, 1); \quad B = (5, 4, 6)$ **18.** $\mathbf{v} = (-2, 0, 3); \quad B = (5, -2, -6)$

19. $\mathbf{v} = (0, 1, -3); \quad B = (2, 0, -4)$ **20.** $\mathbf{v} = (1, -1, 1); \quad B = (2, -2, 2)$

*In Problems 21 and 22, find $\|\mathbf{v}\|$. For each value of k, find k**v** and the length of k**v**.*

21. $\mathbf{v} = (8, -8, 4)$ **22.** $\mathbf{v} = (1, 1, 1)$
     (A) $k = 2$ (B) $k = -\frac{1}{4}$ (A) $k = -5$ (B) $k = \frac{1}{2}$

In Problems 23–26, express **v** *in terms of the standard unit vectors* **i**, **j**, *and* **k**.

23. $\mathbf{v} = (-1, 2, -4)$ **24.** $\mathbf{v} = (0, 1, -2)$
25. $\mathbf{v} = \overrightarrow{AB}; \quad A = (2, 4, 2), \quad B = (2, 0, 2)$
26. $\mathbf{v} = \overrightarrow{AB}; \quad A = (-1, 1, -2), \quad B = (1, 1, 2)$

In Problems 27–30, find a unit vector in the direction of **v**.

27. $\mathbf{v} = (10, 10, -5)$ **28.** $\mathbf{v} = (-9, 6, 2)$

29. $\mathbf{v} = \mathbf{i} - \mathbf{j} - 5\mathbf{k}$ **30.** $\mathbf{v} = \frac{1}{4}\mathbf{i} + \frac{1}{2}\mathbf{j} + \frac{1}{2}\mathbf{k}$

31. If $\mathbf{v} = (11, 10, 2)$, find a vector of length 5 with the same direction as **v**.

32. If $\mathbf{v} = -8\mathbf{i} - 4\mathbf{j} + \mathbf{k}$, find a vector of length 3 with the opposite direction.

In Problems 33–36, let $\mathbf{u} = \mathbf{i} - 2\mathbf{j} + \mathbf{k}$, $\mathbf{v} = 2\mathbf{i} + \mathbf{j} + 2\mathbf{k}$, *and* $\mathbf{w} = \mathbf{i} - \mathbf{j} + \mathbf{k}$, *and perform the indicated operations.*

33. $\mathbf{u} + \mathbf{v} + 2\mathbf{w}$ **34.** $\mathbf{u} + 4\mathbf{v} + 2\mathbf{w}$

35. $3\mathbf{u} + \mathbf{v} - 5\mathbf{w}$ **36.** $\mathbf{u} - 2\mathbf{v} + 3\mathbf{w}$

C   *In Problems 37 and 38, solve each equation for* **x**.

37. $2\mathbf{x} - 2\mathbf{i} - 3\mathbf{j} + \mathbf{k} = \mathbf{x} + 3\mathbf{i} + 4\mathbf{j} + 2\mathbf{k}$
38. $5\mathbf{x} - 3(\mathbf{i} + \mathbf{j} + 2\mathbf{k}) = 3(\mathbf{x} + 2\mathbf{i} + 3\mathbf{j} - 5\mathbf{k})$

In Problems 39 and 40, solve each system and express the solution set in the form

     $\{t\mathbf{v} | t \text{ any scalar}\}$

where **v** *is an appropriately chosen vector.*

39. $\begin{aligned} x + y - z &= 0 \\ x - y - 3z &= 0 \\ 2x + y - 3z &= 0 \end{aligned}$ **40.** $\begin{aligned} x + y - 2z &= 0 \\ 2x - y + 5z &= 0 \\ 4x + y + z &= 0 \end{aligned}$

In Problems 41 and 42, solve each system and express the solution set in the form

     $\{s\mathbf{u} + t\mathbf{v} | s \text{ and } t \text{ any scalars}\}$

where **u** *and* **v** *are appropriately chosen vectors.*

41. $\begin{aligned} x - 2y + 3z &= 0 \\ -2x + 4y - 6z &= 0 \end{aligned}$ **42.** $\begin{aligned} x + y - 2z &= 0 \\ -2x - 2y + 4z &= 0 \end{aligned}$

In Problems 43 and 44, solve each system and express the solution set in the form

$$\{\mathbf{u} + t\mathbf{v} | t \text{ any scalar}\}$$

where **u** *and* **v** *are appropriately chosen vectors.*

43. $\begin{aligned} x + y - z &= 6 \\ 2x + 3y &= 16 \\ x + 2y + z &= 10 \end{aligned}$ **44.** $\begin{aligned} x + y - z &= 3 \\ 2x - y - 2z &= 0 \\ -x + 2y + z &= 3 \end{aligned}$

45. Prove or disprove: If $\mathbf{a} = (1, 1, 2)$, $\mathbf{b} = (1, 2, 3)$, $\mathbf{c} = (-1, 2, 1)$, and x_1, x_2, and x_3 are scalars, then the vector equation

$$x_1\mathbf{a} + x_2\mathbf{b} + x_3\mathbf{c} = \mathbf{0}$$

has nontrivial solutions.

46. Rework Problem 45 for $\mathbf{a} = (1, 1, 1)$, $\mathbf{b} = (1, 2, 3)$, and $\mathbf{c} = (-1, 2, 1)$.

47. Let $\mathbf{a} = (a_1, a_2, a_3)$, $\mathbf{b} = (b_1, b_2, b_3)$, and $\mathbf{c} = (c_1, c_2, c_3)$ be vectors, and let x_1, x_2, x_3 be scalars. Show that the vector equation

$$x_1\mathbf{a} + x_2\mathbf{b} + x_3\mathbf{c} = \mathbf{0}$$

can be written as the matrix equation

$$\begin{bmatrix} a_1 & b_1 & c_1 \\ a_2 & b_2 & c_2 \\ a_3 & b_3 & c_3 \end{bmatrix} \begin{bmatrix} x_1 \\ x_2 \\ x_3 \end{bmatrix} = \begin{bmatrix} 0 \\ 0 \\ 0 \end{bmatrix}$$

48. Prove that the vector equation in Problem 47 has nontrivial solutions if and only if

$$\begin{vmatrix} a_1 & b_1 & c_1 \\ a_2 & b_2 & c_2 \\ a_3 & b_3 & c_3 \end{vmatrix} = 0$$

|4-3| The Dot Product

- Dot Product of Two Vectors
- Angle Between Two Vectors
- Direction Angles and Direction Cosines
- Projection

■ Dot Product of Two Vectors

In Section 2-2 we defined the dot product of a $1 \times n$ row matrix and an $n \times 1$ column matrix as

$$\underset{1 \times n}{[a_1 \quad a_2 \quad \cdots \quad a_n]} \cdot \underset{n \times 1}{\begin{bmatrix} b_1 \\ b_2 \\ \cdot \\ \cdot \\ \cdot \\ b_n \end{bmatrix}} = a_1 b_1 + a_2 b_2 + \cdots + a_n b_n \qquad \text{Real number}$$

In this section we will define the dot product of two vectors and study some of its geometric properties.

Dot Product

The **dot product** of the vectors **u** and **v** is a number denoted by **u** · **v** and defined as follows:

1. If $\mathbf{u} = (u_1, u_2)$ and $\mathbf{v} = (v_1, v_2)$ are vectors in R^2, then

   $$\mathbf{u} \cdot \mathbf{v} = u_1 v_1 + u_2 v_2 \qquad \text{Dot product in } R^2$$

2. If $\mathbf{u} = (u_1, u_2, u_3)$ and $\mathbf{v} = (v_1, v_2, v_3)$ are vectors in R^3, then

   $$\mathbf{u} \cdot \mathbf{v} = u_1 v_1 + u_2 v_2 + u_3 v_3 \qquad \text{Dot product in } R^3$$

Example 21   Find each dot product.

(A) $(2, 1) \cdot (-1, 3)$ (B) $(2, -1, 3) \cdot (4, 5, -1)$
(C) $(3\mathbf{i} - \mathbf{j} + \mathbf{k}) \cdot (2\mathbf{i} + \mathbf{j} + 2\mathbf{k})$ (D) $\mathbf{i} \cdot \mathbf{k}$

Solution   (A) $(2, 1) \cdot (-1, 3) = (2)(-1) + (1)(3) = -2 + 3 = 1$
(B) $(2, -1, 3) \cdot (4, 5, -1) = (2)(4) + (-1)(5) + (3)(-1)$
$$= 8 - 5 - 3 = 0$$
(C) $(3\mathbf{i} - \mathbf{j} + \mathbf{k}) \cdot (2\mathbf{i} + \mathbf{j} + 2\mathbf{k}) = (3, -1, 1) \cdot (2, 1, 2)$
$$= (3)(2) + (-1)(1) + (1)(2)$$
$$= 6 - 1 + 2 = 7$$
(D) $\mathbf{i} \cdot \mathbf{k} = (1, 0, 0) \cdot (0, 0, 1) = (1)(0) + (0)(0) + (0)(1) = 0$

Problem 21   Find each dot product.

(A) $(3, 6) \cdot (2, -1)$ (B) $(2, -1, 0) \cdot (3, -2, 5)$
(C) $(2\mathbf{i} + 3\mathbf{j} - 4\mathbf{k}) \cdot (-\mathbf{i} + 2\mathbf{j} + 2\mathbf{k})$ (D) $\mathbf{k} \cdot \mathbf{j}$ ∎

Some important properties of the dot product are listed in Theorem 7. All these properties follow directly from the definitions of vector operations and the properties of real numbers. We will prove part (G) and leave the rest for exercises.

Theorem 7

Properties of the Dot Product in R^2 and R^3

If $\mathbf{u}$, $\mathbf{v}$, and $\mathbf{w}$ are all vectors in R^3 (or all in R^2) and a is a real number, then

(A) $\mathbf{u} \cdot \mathbf{u} = \|\mathbf{u}\|^2 \geq 0$
(B) $\mathbf{u} \cdot \mathbf{u} = 0$ if and only if $\mathbf{u} = \mathbf{0}$
(C) $\mathbf{u} \cdot \mathbf{0} = 0$
(D) $\mathbf{u} \cdot \mathbf{v} = \mathbf{v} \cdot \mathbf{u}$
(E) $\mathbf{u} \cdot (\mathbf{v} + \mathbf{w}) = \mathbf{u} \cdot \mathbf{v} + \mathbf{u} \cdot \mathbf{w}$
(F) $(\mathbf{u} + \mathbf{v}) \cdot \mathbf{w} = \mathbf{u} \cdot \mathbf{w} + \mathbf{v} \cdot \mathbf{w}$
(G) $(a\mathbf{u}) \cdot \mathbf{v} = a(\mathbf{u} \cdot \mathbf{v})$
(H) $\mathbf{u} \cdot (a\mathbf{v}) = a(\mathbf{u} \cdot \mathbf{v})$

Proof   (G) Let $\mathbf{u} = (u_1, u_2, u_3)$ and $\mathbf{v} = (v_1, v_2, v_3)$. Then

$$(a\mathbf{u}) \cdot \mathbf{v} = (au_1, au_2, au_3) \cdot (v_1, v_2, v_3)$$
$$= au_1v_1 + au_2v_2 + au_3v_3 \tag{1}$$

$$a(\mathbf{u} \cdot \mathbf{v}) = a[(u_1, u_2, u_3) \cdot (v_1, v_2, v_3)]$$
$$= a[u_1v_1 + u_2v_2 + u_3v_3]$$
$$= au_1v_1 + au_2v_2 + au_3v_3 \tag{2}$$

Comparing (1) and (2), we see that

$$(a\mathbf{u}) \cdot \mathbf{v} = a(\mathbf{u} \cdot \mathbf{v})$$ ∎

Part (A) of Theorem 7 provides us with a geometric application of the dot product of a vector with itself.

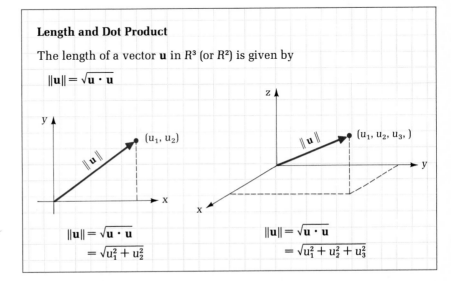

Length and Dot Product

The length of a vector $\mathbf{u}$ in R^3 (or R^2) is given by

$$\|\mathbf{u}\| = \sqrt{\mathbf{u} \cdot \mathbf{u}}$$

$$\|\mathbf{u}\| = \sqrt{\mathbf{u} \cdot \mathbf{u}}$$
$$= \sqrt{u_1^2 + u_2^2}$$

$$\|\mathbf{u}\| = \sqrt{\mathbf{u} \cdot \mathbf{u}}$$
$$= \sqrt{u_1^2 + u_2^2 + u_3^2}$$

▪ Angle Between Two Vectors

Another important geometric application of the dot product is related to the angle between two vectors. If **u** and **v** are two nonzero vectors, we can position **u** and **v** so that their initial points coincide. This forms two angles, θ and ϕ, which satisfy $\theta + \phi = 2\pi$ (Figure 16). The angle between the vectors **u** and **v** is defined to be the angle θ formed by the vectors which satisfies $0 \le \theta \le \pi$ (Figure 17).

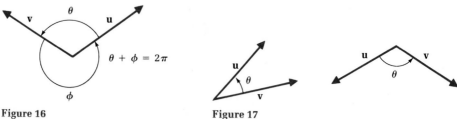

Figure 16   **Figure 17**

If $\theta = \pi/2$, the vectors **u** and **v** are said to be **orthogonal** or **perpendicular.** If $\theta = 0$ or $\theta = \pi$, then each vector is a scalar multiple of the other and the vectors are said to be **parallel** (Figure 18).

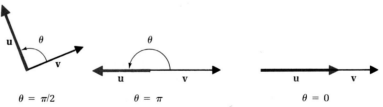

$\theta = \pi/2$ $\theta = \pi$ $\theta = 0$

Figure 18

Theorem 8 establishes the relationship between the dot product and the angle between nonzero vectors. Before we state and prove Theorem 8, let us recall the law of cosines from trigonometry.

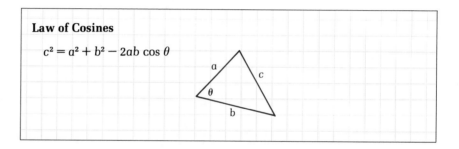

Law of Cosines

$$c^2 = a^2 + b^2 - 2ab \cos \theta$$

Theorem 8

> If $\mathbf{u}$ and $\mathbf{v}$ are two nonzero vectors in R^3 (or in R^2) and θ is the angle between $\mathbf{u}$ and $\mathbf{v}$, then
>
> $$\cos\theta = \frac{\mathbf{u}\cdot\mathbf{v}}{\|\mathbf{u}\|\,\|\mathbf{v}\|}$$

Proof   Given two nonzero vectors $\mathbf{u}$ and $\mathbf{v}$, we can position $\mathbf{u}$, $\mathbf{v}$, and $\mathbf{u}-\mathbf{v}$ so that they form a triangle where θ is the angle between $\mathbf{u}$ and $\mathbf{v}$ [see Figure 19(A)].

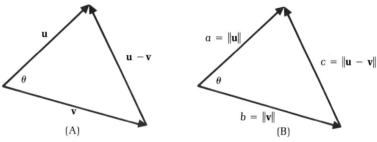

Figure 19

Applying the law of cosines with $a=\|\mathbf{u}\|$, $b=\|\mathbf{v}\|$, and $c=\|\mathbf{u}-\mathbf{v}\|$ gives

$$\|\mathbf{u}-\mathbf{v}\|^2 = \|\mathbf{u}\|^2 + \|\mathbf{v}\|^2 - 2\|\mathbf{u}\|\,\|\mathbf{v}\|\cos\theta \tag{3}$$

We now apply some of the properties listed in Theorem 7 to $\|\mathbf{u}-\mathbf{v}\|^2$:

$$
\begin{aligned}
\|\mathbf{u}-\mathbf{v}\|^2 &= (\mathbf{u}-\mathbf{v})\cdot(\mathbf{u}-\mathbf{v}) && \text{Theorem 7(A)}\\
&= (\mathbf{u}-\mathbf{v})\cdot\mathbf{u} - (\mathbf{u}-\mathbf{v})\cdot\mathbf{v} && \text{Theorem 7(E)}\\
&= \mathbf{u}\cdot(\mathbf{u}-\mathbf{v}) - \mathbf{v}\cdot(\mathbf{u}-\mathbf{v}) && \text{Theorem 7(D)}\\
&= \mathbf{u}\cdot\mathbf{u} - \mathbf{u}\cdot\mathbf{v} - \mathbf{v}\cdot\mathbf{u} + \mathbf{v}\cdot\mathbf{v} && \text{Theorem 7(E)}\\
&= \|\mathbf{u}\|^2 - 2(\mathbf{u}\cdot\mathbf{v}) + \|\mathbf{v}\|^2 && \text{Theorem 7(A) and (D)}
\end{aligned}
$$

Now we substitute this last expression in (3):

$$\|\mathbf{u}\|^2 - 2(\mathbf{u}\cdot\mathbf{v}) + \|\mathbf{v}\|^2 = \|\mathbf{u}\|^2 + \|\mathbf{v}\|^2 - 2\|\mathbf{u}\|\,\|\mathbf{v}\|\cos\theta$$

This equation simplifies to

$$2\|\mathbf{u}\|\,\|\mathbf{v}\|\cos\theta = 2(\mathbf{u}\cdot\mathbf{v})$$

Since $\mathbf{u}$ and $\mathbf{v}$ are nonzero, we can solve for $\cos\theta$:

$$\cos\theta = \frac{\mathbf{u}\cdot\mathbf{v}}{\|\mathbf{u}\|\,\|\mathbf{v}\|}$$

∎

Example 22   Find the angle between each pair of vectors.

(A) $\mathbf{u} = (2, 0, 2);$ $\mathbf{v} = (1, 1, 0)$ (B) $\mathbf{u} = (-1, 7);$ $\mathbf{v} = (-3, -4)$
(C) $\mathbf{u} = 3\mathbf{i} + \mathbf{j} - \mathbf{k};$ $\mathbf{v} = \mathbf{i} - 2\mathbf{j} + 4\mathbf{k}$

Solution   (A) $\|\mathbf{u}\| = \sqrt{(2, 0, 2) \cdot (2, 0, 2)} = \sqrt{4 + 0 + 4} = \sqrt{8}$
$\|\mathbf{v}\| = \sqrt{(1, 1, 0) \cdot (1, 1, 0)} = \sqrt{1 + 1 + 0} = \sqrt{2}$
$\mathbf{u} \cdot \mathbf{v} = (2, 0, 2) \cdot (1, 1, 0) = 2 + 0 + 0 = 2$
Thus,

$$\cos \theta = \frac{\mathbf{u} \cdot \mathbf{v}}{\|\mathbf{u}\| \, \|\mathbf{v}\|} = \frac{2}{\sqrt{8} \, \sqrt{2}} = \frac{1}{2}$$

and $\theta = \pi/3$ or $60°$.

(B) $\|\mathbf{u}\| = \sqrt{(-1, 7) \cdot (-1, 7)} = \sqrt{1 + 49} = \sqrt{50} = 5\sqrt{2}$
$\|\mathbf{v}\| = \sqrt{(-3, -4) \cdot (-3, -4)} = \sqrt{9 + 16} = \sqrt{25} = 5$
$\mathbf{u} \cdot \mathbf{v} = (-1, 7) \cdot (-3, -4) = 3 - 28 = -25$
Thus,

$$\cos \theta = \frac{\mathbf{u} \cdot \mathbf{v}}{\|\mathbf{u}\| \, \|\mathbf{v}\|} = \frac{-25}{(5\sqrt{2})(5)} = -\frac{1}{\sqrt{2}}$$

and $\theta = 3\pi/4$ or $135°$.

(C) $\|\mathbf{u}\| = \sqrt{(3\mathbf{i} + \mathbf{j} - \mathbf{k}) \cdot (3\mathbf{i} + \mathbf{j} - \mathbf{k})} = \sqrt{9 + 1 + 1} = \sqrt{11}$
$\|\mathbf{v}\| = \sqrt{(\mathbf{i} - 2\mathbf{j} + 4\mathbf{k}) \cdot (\mathbf{i} - 2\mathbf{j} + 4\mathbf{k})} = \sqrt{1 + 4 + 16} = \sqrt{21}$
$\mathbf{u} \cdot \mathbf{v} = (3\mathbf{i} + \mathbf{j} - \mathbf{k}) \cdot (\mathbf{i} - 2\mathbf{j} + 4\mathbf{k}) = 3 - 2 - 4 = -3$
Thus,

$$\cos \theta = \frac{\mathbf{u} \cdot \mathbf{v}}{\|\mathbf{u}\| \, \|\mathbf{v}\|} = \frac{-3}{\sqrt{11} \, \sqrt{21}} \approx -0.19739$$

Using a calculator (or a table), we find that $\theta \approx 101.4°$.

Problem 22   Find the angle between each pair of vectors.

(A) $\mathbf{u} = (-1, 0, 1);$ $\mathbf{v} = (3, 3, 0)$ (B) $\mathbf{u} = (1, 2);$ $\mathbf{v} = (-1, 3)$
(C) $\mathbf{u} = \mathbf{i} + 2\mathbf{j} + 2\mathbf{k};$ $\mathbf{v} = 2\mathbf{i} - \mathbf{j} + 5\mathbf{k}$ ▌▌

If $\theta = \pi/2$, then $\cos \theta = 0$ and it follows that $\mathbf{u} \cdot \mathbf{v} = 0$. Conversely, if $\mathbf{u} \cdot \mathbf{v} = 0$, then $\cos \theta = 0$ and $\theta = \pi/2$. Thus, to test for orthogonality, we need only compute $\mathbf{u} \cdot \mathbf{v}$.

Test for Orthogonal Vectors

Two vectors $\mathbf{u}$ and $\mathbf{v}$ are orthogonal if and only if

  $\mathbf{u} \cdot \mathbf{v} = 0$

Note: The zero vector is orthogonal to every vector.

Example 23   Determine which of the following pairs of vectors are orthogonal:

(A) $\mathbf{u} = (1, 3, -2)$; $\mathbf{v} = (4, -2, -1)$ (B) $\mathbf{u} = \mathbf{i} + \mathbf{j}$; $\mathbf{v} = \mathbf{j} + \mathbf{k}$

Solution   (A) $\mathbf{u} \cdot \mathbf{v} = (1, 3, -2) \cdot (4, -2, -1) = 4 - 6 + 2 = 0$
Thus, $\mathbf{u}$ and $\mathbf{v}$ are orthogonal.
(B) $\mathbf{u} \cdot \mathbf{v} = (\mathbf{i} + \mathbf{j}) \cdot (\mathbf{j} + \mathbf{k}) = 0 + 1 + 0 = 1$
Thus, $\mathbf{u}$ and $\mathbf{v}$ are not orthogonal.

Problem 23   Determine which of the following pairs of vectors are orthogonal:

(A) $\mathbf{u} = (2, -1, 1)$; $\mathbf{v} = (1, -1, 3)$ (B) $\mathbf{u} = \mathbf{i} + \mathbf{k}$; $\mathbf{v} = \mathbf{i} - \mathbf{k}$ ▐▐

Example 24   Find all vectors in R^2 that are orthogonal to the vector $\mathbf{u} = (1, 2)$.

Solution   Let $\mathbf{v} = (v_1, v_2)$ be any vector in R^2. In order for $\mathbf{u}$ and $\mathbf{v}$ to be orthogonal, we must have

$$\mathbf{u} \cdot \mathbf{v} = (1, 2) \cdot (v_1, v_2) = v_1 + 2v_2 = 0$$

or

$$v_1 = -2v_2$$

If we let $v_2 = t$, then $v_1 = -2t$ and $\mathbf{v} = (-2t, t)$ is orthogonal to $\mathbf{u}$ for any real number t (including $t = 0$).
The graph of the set of points

$$\{(-2t, t) | t \text{ any real number}\}$$

is the line $y = -\tfrac{1}{2}x$ (see the figure). Thus, all the vectors orthogonal to $\mathbf{u}$ lie on this line.

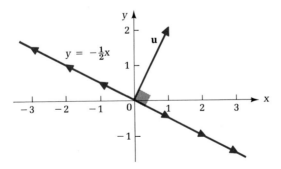

Problem 24   Find all vectors in R^2 that are orthogonal to the vector $\mathbf{u} = (-1, 3)$.   ▐▐

▪ Direction Angles and Direction Cosines

The angles α, β, and γ which a nonzero vector $\mathbf{v} = (v_1, v_2, v_3)$ forms with the positive x, y, and z axes, respectively, are called the **direction angles** for $\mathbf{v}$ (Figure 20).

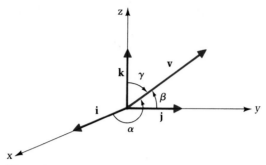

Figure 20   Direction angles

Theorem 8 can be used to find the cosine of each direction angle. Since α is the angle between **v** and the positive x axis, α is also the angle between **v** and the standard unit vector **i**. Thus,

$$\cos \alpha = \frac{\mathbf{v} \cdot \mathbf{i}}{\|\mathbf{v}\|\,\|\mathbf{i}\|} = \frac{(v_1, v_2, v_3) \cdot (1, 0, 0)}{\sqrt{v_1^2 + v_2^2 + v_3^2}\,\sqrt{1 + 0 + 0}}$$

$$= \frac{v_1}{\sqrt{v_1^2 + v_2^2 + v_3^2}}$$

In the same manner,

$$\cos \beta = \frac{\mathbf{v} \cdot \mathbf{j}}{\|\mathbf{v}\|\,\|\mathbf{j}\|} = \frac{v_2}{\sqrt{v_1^2 + v_2^2 + v_3^2}}$$

and

$$\cos \gamma = \frac{\mathbf{v} \cdot \mathbf{k}}{\|\mathbf{v}\|\,\|\mathbf{k}\|} = \frac{v_3}{\sqrt{v_1^2 + v_2^2 + v_3^2}}$$

The numbers $\cos \alpha$, $\cos \beta$, $\cos \gamma$ are called the *direction cosines* for **v**, and the vector

$$\mathbf{u} = \cos \alpha\,\mathbf{i} + \cos \beta\,\mathbf{j} + \cos \gamma\,\mathbf{k}$$

$$= \frac{v_1}{\sqrt{v_1^2 + v_2^2 + v_3^2}}\,\mathbf{i} + \frac{v_2}{\sqrt{v_1^2 + v_2^2 + v_3^2}}\,\mathbf{j} + \frac{v_3}{\sqrt{v_1^2 + v_2^2 + v_3^2}}\,\mathbf{k}$$

$$= \frac{1}{\|\mathbf{v}\|}\,\mathbf{v}$$

is a unit vector in the direction of **v**. We have just sketched the proof for Theorem 9.

| Theorem 9 | **Direction Cosines** |

The **direction cosines** for the nonzero vector

$$\mathbf{v} = (v_1, v_2, v_3)$$

are given by

$$\cos \alpha = \frac{v_1}{\|\mathbf{v}\|} = \frac{v_1}{\sqrt{v_1^2 + v_2^2 + v_3^2}}$$

$$\cos \beta = \frac{v_2}{\|\mathbf{v}\|} = \frac{v_2}{\sqrt{v_1^2 + v_2^2 + v_3^2}}$$

$$\cos \gamma = \frac{v_3}{\|\mathbf{v}\|} = \frac{v_3}{\sqrt{v_1^2 + v_2^2 + v_3^2}}$$

The vector

$$\mathbf{u} = \frac{1}{\|\mathbf{v}\|}\,\mathbf{v} = \cos \alpha\,\mathbf{i} + \cos \beta\,\mathbf{j} + \cos \gamma\,\mathbf{k}$$

is a unit vector in the direction of **v**.

Example 25   Find a unit vector in the direction of $\mathbf{v} = (3, -2, 6)$ and find the direction cosines for **v**.

Solution $\|\mathbf{v}\| = \sqrt{3^2 + (-2)^2 + 6^2} = \sqrt{9 + 4 + 36} = \sqrt{49} = 7$

A unit vector in the direction of **v** is therefore

$$\mathbf{u} = \frac{1}{\|\mathbf{v}\|}\,\mathbf{v} = \frac{1}{7}(3, -2, 6) = \left(\frac{3}{7}, -\frac{2}{7}, \frac{6}{7}\right)$$

The direction cosines are

$$\cos \alpha = \frac{3}{7} \qquad \cos \beta = -\frac{2}{7} \qquad \cos \gamma = \frac{6}{7}$$

Problem 25   Find a unit vector in the direction of $\mathbf{v} = (8, 4, -1)$ and find the direction cosines for **v**.   ∎

▪ Projection

If $\mathbf{u} = (u_1, u_2)$ is a vector in R^2, then we have already seen that **u** can be expressed in terms of the standard unit vectors as $\mathbf{u} = u_1\mathbf{i} + u_2\mathbf{j}$. We can now consider **u** as a sum of two orthogonal vectors, $\mathbf{p} = u_1\mathbf{i}$ and $\mathbf{q} = u_2\mathbf{j}$ (Figure 21).

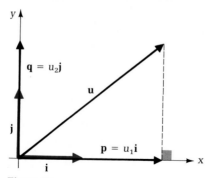

Figure 21

The vector **p** is called the projection of **u** onto **i**. Notice that the line through the terminal points of **u** and **p** is perpendicular to the x axis. Now suppose that we want to project a vector **u** (in R^2 or R^3) onto an arbitrary nonzero vector **v**. This can be accomplished by dropping a perpendicular line from the terminal point of **u** to a line that contains **v** (see Figure 22).

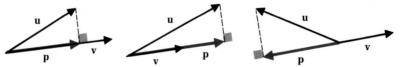

Figure 22

Since the projection **p** also lies on the line containing **v**, **p** must be a scalar multiple of **v**. Thus, there must be a scalar a so that $\mathbf{p} = a\mathbf{v}$. In order to determine **p**, we must find the appropriate value of a. If we form the vector $\mathbf{q} = \mathbf{u} - \mathbf{p}$ which goes from the terminal point of **p** to the terminal point of **u**, then **q** must be orthogonal to both **v** and **p** (see Figure 23).

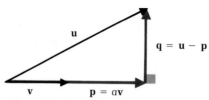

Figure 23

We can use this fact to obtain the value of a, as follows:

$$\mathbf{q} \cdot \mathbf{v} = 0 \qquad \text{**q** and **v** are orthogonal}$$

$$(\mathbf{u} - \mathbf{p}) \cdot \mathbf{v} = 0 \qquad \mathbf{q} = \mathbf{u} - \mathbf{p}$$

$$\mathbf{u} \cdot \mathbf{v} - \mathbf{p} \cdot \mathbf{v} = 0 \qquad \text{Theorem 7(F)}$$

$$\mathbf{u} \cdot \mathbf{v} - a\mathbf{v} \cdot \mathbf{v} = 0 \qquad \mathbf{p} = a\mathbf{v}$$

$$a = \frac{\mathbf{u} \cdot \mathbf{v}}{\|\mathbf{v}\|^2} \qquad \mathbf{v} \cdot \mathbf{v} = \|\mathbf{v}\|^2$$

Thus, the *projection of* **u** *onto* **v** is

$$\mathbf{p} = \frac{\mathbf{u} \cdot \mathbf{v}}{\|\mathbf{v}\|^2} \mathbf{v}$$

Furthermore, since $\mathbf{q} = \mathbf{u} - \mathbf{p}$, we can write

$$\mathbf{u} = \mathbf{p} + \mathbf{q}$$

which expresses **u** as the sum of two orthogonal vectors. The vector **q** is the *component of* **u** *orthogonal to* **v**. These facts are summarized in the box.

Projection

The **projection** of a vector **u** onto a nonzero vector **v** is

$$\mathbf{p} = \frac{\mathbf{u} \cdot \mathbf{v}}{\|\mathbf{v}\|^2} \mathbf{v}$$

The **component of u orthogonal to v** is

$$\mathbf{q} = \mathbf{u} - \mathbf{p}$$

The **orthogonal decomposition** of **u** relative to **v** is

$$\mathbf{u} = \mathbf{p} + \mathbf{q}$$

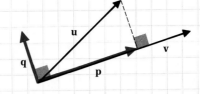

Example 26 For each pair of vectors **u** and **v**, find the projection of **u** onto **v**, the component of **u** orthogonal to **v**, and the orthogonal decomposition of **u** relative to **v**.

(A) $\mathbf{u} = (3, 6)$; $\mathbf{v} = (3, 1)$ (B) $\mathbf{u} = 6\mathbf{i} + 8\mathbf{j} + 7\mathbf{k}$; $\mathbf{v} = 2\mathbf{i} - \mathbf{j} + 2\mathbf{k}$

Solution (A) $\mathbf{u} \cdot \mathbf{v} = (3, 6) \cdot (3, 1) = 9 + 6 = 15$

$\|\mathbf{v}\|^2 = \mathbf{v} \cdot \mathbf{v} = (3, 1) \cdot (3, 1) = 9 + 1 = 10$

Thus,

$$\mathbf{p} = \frac{\mathbf{u} \cdot \mathbf{v}}{\|\mathbf{v}\|^2}\mathbf{v} = \frac{15}{10}(3, 1) = \left(\frac{9}{2}, \frac{3}{2}\right)$$

and

$$\mathbf{q} = \mathbf{u} - \mathbf{p} = (3, 6) - \left(\frac{9}{2}, \frac{3}{2}\right) = \left(-\frac{3}{2}, \frac{9}{2}\right)$$

The orthogonal decomposition of $\mathbf{u}$ relative to $\mathbf{v}$ is

$$\mathbf{u} = \left(\frac{9}{2}, \frac{3}{2}\right) + \left(-\frac{3}{2}, \frac{9}{2}\right)$$

Check: $\mathbf{u} \stackrel{?}{=} \mathbf{p} + \mathbf{q}$ $\mathbf{p} \cdot \mathbf{q} \stackrel{?}{=} 0$

$$(3, 6) = \left(\frac{9}{2}, \frac{3}{2}\right) + \left(-\frac{3}{2}, \frac{9}{2}\right) \qquad \left(\frac{9}{2}, \frac{3}{2}\right) \cdot \left(-\frac{3}{2}, \frac{9}{2}\right) = 0$$

$$(3, 6) \stackrel{\checkmark}{=} (3, 6) \qquad\qquad\qquad -\frac{27}{4} + \frac{27}{4} = 0$$

$$0 \stackrel{\checkmark}{=} 0$$

The figure shows the geometric relationships of all the vectors in this example.

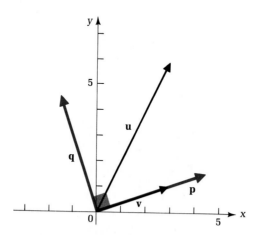

(B) $\mathbf{u} \cdot \mathbf{v} = (6\mathbf{i} + 8\mathbf{j} + 7\mathbf{k}) \cdot (2\mathbf{i} - \mathbf{j} + 2\mathbf{k})$

$\qquad = 12 - 8 + 14 = 18$

$\|\mathbf{v}\|^2 = (2\mathbf{i} - \mathbf{j} + 2\mathbf{k}) \cdot (2\mathbf{i} - \mathbf{j} + 2\mathbf{k})$

$\qquad = 4 + 1 + 4 = 9$

Thus,

$$\mathbf{p} = \frac{\mathbf{u} \cdot \mathbf{v}}{\|\mathbf{v}\|^2} \mathbf{v} = \frac{18}{9} (2\mathbf{i} - \mathbf{j} + 2\mathbf{k})$$

$$= 4\mathbf{i} - 2\mathbf{j} + 4\mathbf{k}$$

and

$$\mathbf{q} = \mathbf{u} - \mathbf{p} = (6\mathbf{i} + 8\mathbf{j} + 7\mathbf{k}) - (4\mathbf{i} - 2\mathbf{j} + 4\mathbf{k})$$

$$= 2\mathbf{i} + 10\mathbf{j} + 3\mathbf{k}$$

The orthogonal decomposition of $\mathbf{u}$ relative to $\mathbf{v}$ is

$$\mathbf{u} = (4\mathbf{i} - 2\mathbf{j} + 4\mathbf{k}) + (2\mathbf{i} + 10\mathbf{j} + 3\mathbf{k})$$

Check:
$$\mathbf{u} \overset{?}{=} \mathbf{p} + \mathbf{q}$$
$$6\mathbf{i} + 8\mathbf{j} + 7\mathbf{k} = (4\mathbf{i} - 2\mathbf{j} + 4\mathbf{k}) + (2\mathbf{i} + 10\mathbf{j} + 3\mathbf{k})$$
$$6\mathbf{i} + 8\mathbf{j} + 7\mathbf{k} \overset{\checkmark}{=} 6\mathbf{i} + 8\mathbf{j} + 7\mathbf{k}$$

$$\mathbf{p} \cdot \mathbf{q} \overset{?}{=} 0$$
$$(4\mathbf{i} - 2\mathbf{j} + 4\mathbf{k}) \cdot (2\mathbf{i} + 10\mathbf{j} + 3\mathbf{k}) = 0$$
$$8 - 20 + 12 = 0$$
$$0 \overset{\checkmark}{=} 0$$

The figure shows the geometric relationships of all the vectors in this example.

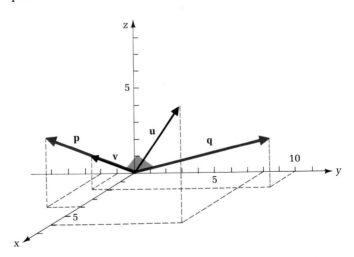

Problem 26 Repeat Example 26 for the following pairs of vectors.

(A) $\mathbf{u} = (5, 1)$; $\mathbf{v} = (4, 6)$ (B) $\mathbf{u} = 3\mathbf{i} + 8\mathbf{j} - 2\mathbf{k}$; $\mathbf{v} = \mathbf{i} + 2\mathbf{j} + 2\mathbf{k}$

Answers to
Matched Problems

21. (A) 0   (B) 8   (C) −4   (D) 0

22. (A) $\cos \theta = -\frac{1}{2}$, $\theta = 2\pi/3$ or $120°$
(B) $\cos \theta = 1/\sqrt{2}$, $\theta = \pi/4$ or $45°$
(C) $\cos \theta = \sqrt{30}/9 \approx 0.60858$, $\theta \approx 52.5°$

23. (A) $\mathbf{u} \cdot \mathbf{v} = 6$, not orthogonal   (B) $\mathbf{u} \cdot \mathbf{v} = 0$, orthogonal

24. $\{(3t, t)|t$ any real number$\}$

25. $\mathbf{u} = (\frac{8}{9}, \frac{4}{9}, -\frac{1}{9})$; $\cos \alpha = \frac{8}{9}$, $\cos \beta = \frac{4}{9}$, $\cos \gamma = -\frac{1}{9}$

26. (A) $\mathbf{p} = (2, 3)$, $\mathbf{q} = (3, -2)$, $\mathbf{u} = (2, 3) + (3, -2)$
(B) $\mathbf{p} = \frac{5}{3}\mathbf{i} + \frac{10}{3}\mathbf{j} + \frac{10}{3}\mathbf{k}$, $\mathbf{q} = \frac{4}{3}\mathbf{i} + \frac{14}{3}\mathbf{j} - \frac{16}{3}\mathbf{k}$,
$\mathbf{u} = (\frac{5}{3}\mathbf{i} + \frac{10}{3}\mathbf{j} + \frac{10}{3}\mathbf{k}) + (\frac{4}{3}\mathbf{i} + \frac{14}{3}\mathbf{j} - \frac{16}{3}\mathbf{k})$

▌ Exercise 4-3

A   **1.** If $\mathbf{i}$ and $\mathbf{j}$ are the standard unit vectors in R^2, find:
(A) $\mathbf{i} \cdot \mathbf{i}$ (B) $\mathbf{i} \cdot \mathbf{j}$ (C) $\mathbf{j} \cdot \mathbf{j}$

2. If $\mathbf{i}$, $\mathbf{j}$, and $\mathbf{k}$ are the standard unit vectors in R^3, find:
(A) $\mathbf{i} \cdot \mathbf{i}$ (B) $\mathbf{i} \cdot \mathbf{j}$ (C) $\mathbf{i} \cdot \mathbf{k}$
(D) $\mathbf{j} \cdot \mathbf{j}$ (E) $\mathbf{j} \cdot \mathbf{k}$ (F) $\mathbf{k} \cdot \mathbf{k}$

In Problems 3–10, find the indicated dot product.

3. $(1, 2) \cdot (3, 4)$ **4.** $(3, 4) \cdot (5, -2)$

5. $(3, 2, -1) \cdot (1, -1, 1)$ **6.** $(2, 1, -1) \cdot (3, -4, 3)$

7. $(2\mathbf{i} + 3\mathbf{j}) \cdot (4\mathbf{i} - 2\mathbf{j})$ **8.** $(5\mathbf{i} + 6\mathbf{j}) \cdot (6\mathbf{i} - 5\mathbf{j})$

9. $(2\mathbf{i} + \mathbf{j} + 2\mathbf{k}) \cdot (\mathbf{i} - \mathbf{j} + \mathbf{k})$ **10.** $(\mathbf{i} + 2\mathbf{j}) \cdot (3\mathbf{j} - 4\mathbf{k})$

B   *In Problems 11–18, find the cosine of the angle between the indicated vectors. Then find the angle itself. (This may require the use of a calculator or a table of values of the cosine function.)*

11. $\mathbf{u} = (5, 2)$; $\mathbf{v} = (3, 7)$ **12.** $\mathbf{u} = (-1, 2)$; $\mathbf{v} = (3, -1)$

13. $\mathbf{u} = \mathbf{i} + \mathbf{j}$; $\mathbf{v} = \mathbf{i} - \mathbf{j}$ **14.** $\mathbf{u} = \mathbf{i} - 2\mathbf{j}$; $\mathbf{v} = -2\mathbf{i} + 4\mathbf{j}$

15. $\mathbf{u} = (-1, 4, 3)$; $\mathbf{v} = (2, 5, 7)$ **16.** $\mathbf{u} = (3, -8, -5)$; $\mathbf{v} = (1, 9, -4)$

17. $\mathbf{u} = \mathbf{i} + 2\mathbf{k}$; $\mathbf{v} = \mathbf{j} + \mathbf{k}$

18. $\mathbf{u} = -6\mathbf{i} + 3\mathbf{j} + 2\mathbf{k}$; $\mathbf{v} = 8\mathbf{i} + 4\mathbf{j} + \mathbf{k}$

In Problems 19–22, determine whether $\mathbf{u}$ and $\mathbf{v}$ are orthogonal.

19. $\mathbf{u} = (1, -2)$; $\mathbf{v} = (4, 2)$ **20.** $\mathbf{u} = (1, 3)$; $\mathbf{v} = (4, -1)$

21. $\mathbf{u} = \mathbf{i} + 2\mathbf{j} - \mathbf{k}$; $\mathbf{v} = 3\mathbf{i} + 2\mathbf{j} + 5\mathbf{k}$ **22.** $\mathbf{u} = \mathbf{i} + 2\mathbf{j} - \mathbf{k}$; $\mathbf{v} = \mathbf{i} + 2\mathbf{j} + 5\mathbf{k}$

In Problems 23–26, find a unit vector in the direction of the given vector and find the direction cosines.

23. $\mathbf{v} = (7, -4, 4)$ **24.** $\mathbf{v} = (-10, 6, 8)$

25. $v = i + k$ **26.** $v = 3i + 4j$

In Problems 27 – 32, determine the value(s) of t that will make u and v orthogonal.

27. $u = (1, -3);\quad v = (2, t)$ **28.** $u = (2, -1);\quad v = (t, 3t)$

29. $u = ti + j;\quad v = i - tj$ **30.** $u = ti + j;\quad v = ti + 4j$

31. $u = ti + j - k;\quad v = ti - j + 3k$ **32.** $u = ti + tj + 3k;\quad v = ti + j - 4k$

In Problems 33 and 34, find all vectors in R^2 that are orthogonal to the given vector.

33. $u = (1, -5)$ **34.** $u = (4, -1)$

In Problems 35 – 38, find p, the projection of u onto v, and q, the component of u orthogonal to v. Graph u, v, p, and q on the same set of axes.

35. $u = (9, 8);\quad v = (1, 2)$ **36.** $u = (7, -6);\quad v = (-1, 3)$

37. $u = i + k;\quad v = i - j$ **38.** $u = i + 2j + 2k;\quad v = i - j + 2k$

C   A set of three vectors, u, v, and w, is said to be mutually orthogonal if $u \cdot v = u \cdot w = v \cdot w = 0$. For example, i, j, and k are mutually orthogonal. In Problems 39 and 40 determine whether the indicated set of vectors is mutually orthogonal.

39. $u = (2, 3, -4);\quad v = (1, 2, 2);\quad w = (14, -8, 1)$

40. $u = (1, 0, 1);\quad v = (0, 1, 0);\quad w = (1, 1, -1)$

41. Let $x = x_1 i + x_2 j + x_3 k$ be orthogonal to both $u = i + j - 2k$ and $v = 2i + j + k$.

   (A) Find a system of equations that the components of x must satisfy.

   (B) Express the solution set of this system in the form

   $\{tw | t \text{ any scalar}\}$

   for an appropriately chosen vector w.

42. Rework Problem 41 if $u = i + j - 2k$ and $v = 5i + 3j + 4k$.

Problems 43 and 44 refer to statements from Theorem 7 in this section. Use the definition of the dot product and properties of real numbers to prove each statement for u, v, and w any vectors in R^3. Do not refer to Theorem 7 in your proof.

43. Theorem 7(D). $u \cdot v = v \cdot u$

44. Theorem 7(E). $u \cdot (v + w) = u \cdot v + u \cdot w$

In Problems 45 and 46, u and v are both vectors in R^3 (or both in R^2). Prove that each statement is true.

45. $\|u + v\|^2 + \|u - v\|^2 = 2\|u\|^2 + 2\|v\|^2$

46. $u \cdot v = \frac{1}{4}\|u + v\|^2 - \frac{1}{4}\|u - v\|^2$

47. If u, v, and w are vectors in R^3 and w is orthogonal to both u and v, then show that **w** is orthogonal to all vectors of the form $k\mathbf{u} + \ell\mathbf{v}$ where k and ℓ are scalars.

48. If **u**, **v**, and **w** are mutually orthogonal nonzero vectors in R^3 (that is, $\mathbf{u} \cdot \mathbf{v} = \mathbf{u} \cdot \mathbf{w} = \mathbf{v} \cdot \mathbf{w} = 0$) and a, b, and c are scalars satisfying $a\mathbf{u} + b\mathbf{v} + c\mathbf{w} = 0$, then show that $a = b = c = 0$.

49. If **p** is the projection of **u** onto **v** and θ is the angle between **u** and **v**, show that $\|\mathbf{p}\| = \|\mathbf{u}\| \, |\cos \theta|$.

50. Show that the direction cosines of any nonzero vector satisfy $\cos^2\alpha + \cos^2\beta + \cos^2\gamma = 1$.

▌▫▌ Applications

51. *Geometry.* Use the accompanying figure with the indicated vector assignments and appropriate properties of the dot product to prove that an angle inscribed in a semicircle is a right angle. (*Note:* $\|\mathbf{a}\| = \|\mathbf{c}\| =$ Radius)

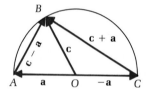

52. *Geometry.* Use the accompanying figure with the indicated vector assignments and appropriate properties of the dot product to prove that the diagonals of a rhombus are perpendicular. (*Note:* $\|\mathbf{a}\| = \|\mathbf{b}\|$)

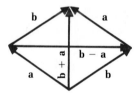

▌4-4▌ The Cross Product

- Definition of the Cross Product
- Geometric Properties
- Algebraic Properties
- Scalar Triple Product

▪ Definition of the Cross Product

In Section 4-3 we saw that the dot product of two vectors is a number with important geometric significance. In this section we will study another type of vector multiplication, called the *cross product*. The cross product of two vectors in R^3 is a vector—not a number—and we will see that it also has important geometric significance.

The Cross Product

The **cross product** of $\mathbf{u} = (u_1, u_2, u_3)$ and $\mathbf{v} = (v_1, v_2, v_3)$, denoted $\mathbf{u} \times \mathbf{v}$, is

$$\mathbf{u} \times \mathbf{v} = \begin{vmatrix} u_2 & u_3 \\ v_2 & v_3 \end{vmatrix} \mathbf{i} - \begin{vmatrix} u_1 & u_3 \\ v_1 & v_3 \end{vmatrix} \mathbf{j} + \begin{vmatrix} u_1 & u_2 \\ v_1 & v_2 \end{vmatrix} \mathbf{k} \qquad (1)$$

$$= (u_2 v_3 - v_2 u_3)\mathbf{i} - (u_1 v_3 - v_1 u_3)\mathbf{j} + (u_1 v_2 - v_1 u_2)\mathbf{k} \qquad (2)$$

Note: The cross product is defined only for vectors in R^3.

Determinant notation can be used to make the formulas in (2) easier to remember. In fact, (1) appears to be the cofactor expansion of the following 3×3 determinant along the top row:

$$\mathbf{u} \times \mathbf{v} = \begin{vmatrix} \mathbf{i} & \mathbf{j} & \mathbf{k} \\ u_1 & u_2 & u_3 \\ v_1 & v_2 & v_3 \end{vmatrix} \qquad (3)$$

Since $\mathbf{i}$, $\mathbf{j}$, and $\mathbf{k}$ are vectors and the entries in a determinant are supposed to be numbers, we will refer to (3) as a *symbolic determinant*. Expanding this symbolic determinant along the first row as though it were an ordinary determinant provides an easy way to remember and calculate the cross product. Furthermore, we will be able to use some of the formal properties of determinants that we studied in Chapter 3 as we investigate the properties of the cross product (as long as we continue to interpret $\mathbf{i}$, $\mathbf{j}$, and $\mathbf{k}$ as vectors). For example, if two rows of a symbolic determinant are interchanged, the new determinant is the negative of the original; if one row is zero, then the value of the determinant (a vector) is zero; and so on.

Example 27   If $\mathbf{u} = \mathbf{i} + 2\mathbf{j} + 3\mathbf{k}$, $\mathbf{v} = -\mathbf{i} + 4\mathbf{j} - 2\mathbf{k}$, and $\mathbf{w} = 2\mathbf{i} - \mathbf{j} + \mathbf{k}$, find:

(A) $\mathbf{u} \times \mathbf{v}$ (B) $(\mathbf{u} \times \mathbf{v}) \times \mathbf{w}$

Solution   (A) $\mathbf{u} \times \mathbf{v} = \begin{vmatrix} \mathbf{i} & \mathbf{j} & \mathbf{k} \\ 1 & 2 & 3 \\ -1 & 4 & -2 \end{vmatrix}$ Expand along the first row.

$$= \begin{vmatrix} 2 & 3 \\ 4 & -2 \end{vmatrix} \mathbf{i} - \begin{vmatrix} 1 & 3 \\ -1 & -2 \end{vmatrix} \mathbf{j} + \begin{vmatrix} 1 & 2 \\ -1 & 4 \end{vmatrix} \mathbf{k} \qquad \text{Evaluate each } 2 \times 2 \text{ determinant.}$$

$$= [2(-2) - 3(4)]\mathbf{i} - [1(-2) - (-1)(3)]\mathbf{j} + [1(4) - (-1)(2)]\mathbf{k}$$

$$= -16\mathbf{i} - \mathbf{j} + 6\mathbf{k}$$

(B) From part (A), $\mathbf{u} \times \mathbf{v} = -16\mathbf{i} - \mathbf{j} + 6\mathbf{k}$. Thus,

$$(\mathbf{u} \times \mathbf{v}) \times \mathbf{w} = \begin{vmatrix} \mathbf{i} & \mathbf{j} & \mathbf{k} \\ -16 & -1 & 6 \\ 2 & -1 & 1 \end{vmatrix}$$

$$= \begin{vmatrix} -1 & 6 \\ -1 & 1 \end{vmatrix} \mathbf{i} - \begin{vmatrix} -16 & 6 \\ 2 & 1 \end{vmatrix} \mathbf{j} + \begin{vmatrix} -16 & -1 \\ 2 & -1 \end{vmatrix} \mathbf{k}$$

$$= 5\mathbf{i} + 28\mathbf{j} + 18\mathbf{k}$$

Problem 27   Using $\mathbf{u}$, $\mathbf{v}$, and $\mathbf{w}$ in Example 27, find:

(A) $\mathbf{v} \times \mathbf{u}$ [Compare the result with that of Example 27(A).]

(B) $\mathbf{v} \times \mathbf{w}$

(C) $\mathbf{u} \times (\mathbf{v} \times \mathbf{w})$ [Compare the result with that of Example 27(B).]   ∎

The cross products computed in Example 27 and Problem 27 illustrate several important points about this operation. First, comparing the cross products computed in Example 27(A) and Problem 27(A), we see that $\mathbf{u} \times \mathbf{v}$ is not equal to $\mathbf{v} \times \mathbf{u}$. (Later we will prove that $\mathbf{u} \times \mathbf{v} = -\mathbf{v} \times \mathbf{u}$ for any vectors $\mathbf{u}$ and $\mathbf{v}$.) Thus, *the cross product is not a commutative operation.* Comparing Example 27(B) and Problem 27(C), we see that $(\mathbf{u} \times \mathbf{v}) \times \mathbf{w}$ and $\mathbf{u} \times (\mathbf{v} \times \mathbf{w})$ are not equal. This means that *the cross product is not an associative operation.* In general, it is incorrect to write the expression $\mathbf{u} \times \mathbf{v} \times \mathbf{w}$ without including parentheses to indicate the order in which the operations are to be performed.

▪ Geometric Properties

Let us look at an important geometric property of the cross product. If

$$\mathbf{u} = \mathbf{i} + 2\mathbf{j} + 3\mathbf{k} \qquad \text{and} \qquad \mathbf{v} = -\mathbf{i} + 4\mathbf{j} - 2\mathbf{k}$$

then in Example 27(A) we saw that

$$\mathbf{u} \times \mathbf{v} = -16\mathbf{i} - \mathbf{j} + 6\mathbf{k}$$

Now let us compute $\mathbf{u} \cdot (\mathbf{u} \times \mathbf{v})$ and $\mathbf{v} \cdot (\mathbf{u} \times \mathbf{v})$:

$$\mathbf{u} \cdot (\mathbf{u} \times \mathbf{v}) = (\mathbf{i} + 2\mathbf{j} + 3\mathbf{k}) \cdot (-16\mathbf{i} - \mathbf{j} + 6\mathbf{k})$$

$$= -16 - 2 + 18 = 0$$

$$\mathbf{v} \cdot (\mathbf{u} \times \mathbf{v}) = (-\mathbf{i} + 4\mathbf{j} - 2\mathbf{k}) \cdot (-16\mathbf{i} - \mathbf{j} + 6\mathbf{k})$$

$$= 16 - 4 - 12 = 0$$

Thus, $\mathbf{u} \times \mathbf{v}$ is orthogonal to both $\mathbf{u}$ and $\mathbf{v}$. Theorem 10 states that this is always the case for any vectors $\mathbf{u}$ and $\mathbf{v}$.

Theorem 10	If $\mathbf{u}$ and $\mathbf{v}$ are vectors in R^3, then
	$$\mathbf{u} \cdot (\mathbf{u} \times \mathbf{v}) = 0 \qquad \text{and} \qquad \mathbf{v} \cdot (\mathbf{u} \times \mathbf{v}) = 0$$

Proof   If $\mathbf{u} = u_1\mathbf{i} + u_2\mathbf{j} + u_3\mathbf{k}$ and $\mathbf{v} = v_1\mathbf{i} + v_2\mathbf{j} + v_3\mathbf{k}$, then

$$\mathbf{u} \cdot (\mathbf{u} \times \mathbf{v}) = (u_1\mathbf{i} + u_2\mathbf{j} + u_3\mathbf{k}) \cdot \left(\begin{vmatrix} u_2 & u_3 \\ v_2 & v_3 \end{vmatrix} \mathbf{i} - \begin{vmatrix} u_1 & u_3 \\ v_1 & v_3 \end{vmatrix} \mathbf{j} + \begin{vmatrix} u_1 & u_2 \\ v_1 & v_2 \end{vmatrix} \mathbf{k} \right)$$

$$= u_1 \begin{vmatrix} u_2 & u_3 \\ v_2 & v_3 \end{vmatrix} - u_2 \begin{vmatrix} u_1 & u_3 \\ v_1 & v_3 \end{vmatrix} + u_3 \begin{vmatrix} u_1 & u_2 \\ v_1 & v_2 \end{vmatrix}$$

Cofactor expansion
along the first
row of the matrix

$$\begin{bmatrix} u_1 & u_2 & u_3 \\ u_1 & u_2 & u_3 \\ v_1 & v_2 & v_3 \end{bmatrix}$$

$$= \begin{vmatrix} u_1 & u_2 & u_3 \\ u_1 & u_2 & u_3 \\ v_1 & v_2 & v_3 \end{vmatrix}$$

Two rows are identical

$$= 0$$

The proof that $\mathbf{v} \cdot (\mathbf{u} \times \mathbf{v}) = 0$ is similar.   ∥

Example 28   Find a vector orthogonal to both

$$\mathbf{u} = 2\mathbf{i} - \mathbf{j} + 2\mathbf{k} \qquad \text{and} \qquad \mathbf{v} = \mathbf{i} + 2\mathbf{j} - \mathbf{k}$$

Solution $\mathbf{u} \times \mathbf{v} = \begin{vmatrix} \mathbf{i} & \mathbf{j} & \mathbf{k} \\ 2 & -1 & 2 \\ 1 & 2 & -1 \end{vmatrix}$

$$= \begin{vmatrix} -1 & 2 \\ 2 & -1 \end{vmatrix} \mathbf{i} - \begin{vmatrix} 2 & 2 \\ 1 & -1 \end{vmatrix} \mathbf{j} + \begin{vmatrix} 2 & -1 \\ 1 & 2 \end{vmatrix} \mathbf{k}$$

$$= -3\mathbf{i} + 4\mathbf{j} + 5\mathbf{k}$$

Check:

$$\mathbf{u} \cdot (\mathbf{u} \times \mathbf{v}) = (2\mathbf{i} - \mathbf{j} + 2\mathbf{k}) \cdot (-3\mathbf{i} + 4\mathbf{j} + 5\mathbf{k})$$

$$= -6 - 4 + 10 = 0$$

$$\mathbf{v} \cdot (\mathbf{u} \times \mathbf{v}) = (\mathbf{i} + 2\mathbf{j} - \mathbf{k}) \cdot (-3\mathbf{i} + 4\mathbf{j} + 5\mathbf{k})$$

$$= -3 + 8 - 5 = 0$$

The figure shows the geometric relationship of the vectors in this example.

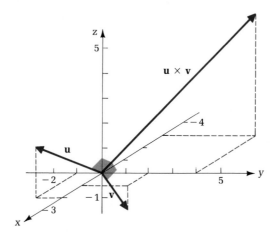

Each time you compute a cross product, you should mentally compute $\mathbf{u} \cdot (\mathbf{u} \times \mathbf{v})$ and $\mathbf{v} \cdot (\mathbf{u} \times \mathbf{v})$ as a check. If either of these dot products is nonzero, then you know that you have made an error in computing the cross product.

Problem 28   Find a vector orthogonal to both

$$\mathbf{u} = \mathbf{i} + \mathbf{j} + \mathbf{k} \qquad \text{and} \qquad \mathbf{v} = 2\mathbf{i} - \mathbf{j} + 3\mathbf{k}$$ ▐▌

If $\mathbf{u}$ and $\mathbf{v}$ are two nonparallel vectors in R^2, then it is impossible to find a third vector in R^2 that is orthogonal to both of them. However, we can always consider $\mathbf{u}$ and $\mathbf{v}$ to be vectors lying in the xy plane of R^3. The cross product will then be a vector perpendicular to the xy plane.

Example 29   Find the cross product of

$$\mathbf{u} = 2\mathbf{i} + \mathbf{j} \qquad \text{and} \qquad \mathbf{v} = \mathbf{i} + 4\mathbf{j}$$

Solution   Assuming that $\mathbf{u}$ and $\mathbf{v}$ are vectors in R^3—that is, assuming $\mathbf{u} = 2\mathbf{i} + \mathbf{j} + 0\mathbf{k}$ and $\mathbf{v} = \mathbf{i} + 4\mathbf{j} + 0\mathbf{k}$—then

$$\mathbf{u} \times \mathbf{v} = \begin{vmatrix} \mathbf{i} & \mathbf{j} & \mathbf{k} \\ 2 & 1 & 0 \\ 1 & 4 & 0 \end{vmatrix}$$

$$= \begin{vmatrix} 1 & 0 \\ 4 & 0 \end{vmatrix} \mathbf{i} - \begin{vmatrix} 2 & 0 \\ 1 & 0 \end{vmatrix} \mathbf{j} + \begin{vmatrix} 2 & 1 \\ 1 & 4 \end{vmatrix} \mathbf{k}$$

$$= 0\mathbf{i} - 0\mathbf{j} + 7\mathbf{k}$$

$$= 7\mathbf{k}$$

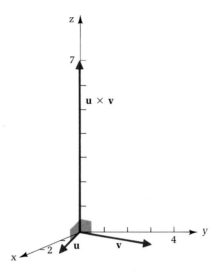

Problem 29   Find the cross product of

$$\mathbf{u} = 3\mathbf{i} + 4\mathbf{j} \qquad \text{and} \qquad \mathbf{v} = 2\mathbf{i} - \mathbf{j}$$

Can we determine the direction of $\mathbf{u} \times \mathbf{v}$ geometrically without actually computing $\mathbf{u} \times \mathbf{v}$ algebraically? Since $\mathbf{u} \times \mathbf{v}$ is orthogonal to both $\mathbf{u}$ and $\mathbf{v}$, there are only two possible directions for $\mathbf{u} \times \mathbf{v}$ (see Figure 24). It can be shown that $\mathbf{u}$, $\mathbf{v}$, and $\mathbf{u} \times \mathbf{v}$, in this order, always form a right-handed system (see Figure 25). Thus, Figure 24(A) is correct.

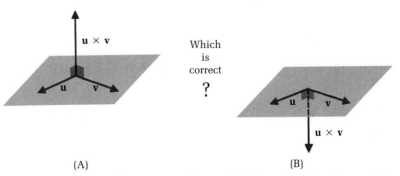

(A)   (B)

Figure 24   Vectors $\mathbf{u}$ and $\mathbf{v}$ point toward you, and the plane determined by $\mathbf{u}$ and $\mathbf{v}$ is seen from above

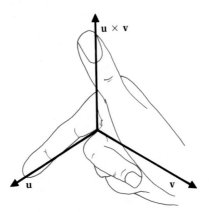

Figure 25 Right-hand system formed by **u**, **v**, and **u** $\times$ **v**

Now we want to find a geometric interpretation for the length of the cross product. First we recall that the area of a parallelogram with base b and altitude h is $A = bh$ (see Figure 26). If **u** and **v** are two nonzero vectors that are not parallel, then they can be positioned so that they form adjacent sides of a parallelogram (see Figure 27).

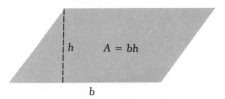

Figure 26 Area of a parallelogram

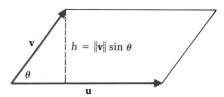

Figure 27 Parallelogram determined by **u** and **v**

If θ is the angle between **u** and **v**, then

$$\sin \theta = \frac{h}{\|\mathbf{v}\|} \qquad \text{or} \qquad h = \|\mathbf{v}\| \sin \theta$$

Using $b = \|\mathbf{u}\|$ and $h = \|\mathbf{v}\| \sin \theta$ in the area formula, $A = bh$, the area of the parallelogram determined by **u** and **v** is

$$A = \|\mathbf{u}\| \, \|\mathbf{v}\| \sin \theta \tag{4}$$

Theorem 11(B) relates this area to the length of the cross product.

Theorem 11

> If $\mathbf{u}$ and $\mathbf{v}$ are vectors in R^3 and θ is the angle between $\mathbf{u}$ and $\mathbf{v}$, then
>
> (A) $\|\mathbf{u} \times \mathbf{v}\|^2 = \|\mathbf{u}\|^2\|\mathbf{v}\|^2 - (\mathbf{u} \cdot \mathbf{v})^2$
> (B) $\|\mathbf{u} \times \mathbf{v}\| = \|\mathbf{u}\|\,\|\mathbf{v}\|\sin\theta$ Area of a parallelogram with adjacent sides $\mathbf{u}$ and $\mathbf{v}$

Proof (A) If $\mathbf{u} = u_1\mathbf{i} + u_2\mathbf{j} + u_3\mathbf{k}$ and $\mathbf{v} = v_1\mathbf{i} + v_2\mathbf{j} + v_3\mathbf{k}$, then

$$\|\mathbf{u} \times \mathbf{v}\|^2 = (u_2v_3 - v_2u_3)^2 + (u_1v_3 - v_1u_3)^2 + (u_1v_2 - v_1u_2)^2$$

and

$$\|\mathbf{u}\|^2\|\mathbf{v}\|^2 - (\mathbf{u} \cdot \mathbf{v})^2 = (u_1^2 + u_2^2 + u_3^2)(v_1^2 + v_2^2 + v_3^2) - (u_1v_1 + u_2v_2 + u_3v_3)^2$$

A lengthy algebraic computation then shows that these two expressions are equal.

(B) If $\mathbf{u}$ and $\mathbf{v}$ are both nonzero vectors, then Theorem 8 in Section 4-3 states that

$$\cos\theta = \frac{\mathbf{u} \cdot \mathbf{v}}{\|\mathbf{u}\|\,\|\mathbf{v}\|}$$

or, equivalently,

$$\mathbf{u} \cdot \mathbf{v} = \|\mathbf{u}\|\,\|\mathbf{v}\|\cos\theta$$

where θ is the angle between $\mathbf{u}$ and $\mathbf{v}$, $0 \leq \theta \leq \pi$. Substituting in Theorem 11(A), we have

$$\begin{aligned}
\|\mathbf{u} \times \mathbf{v}\|^2 &= \|\mathbf{u}\|^2\|\mathbf{v}\|^2 - (\mathbf{u} \cdot \mathbf{v})^2 \\
&= \|\mathbf{u}\|^2\|\mathbf{v}\|^2 - \|\mathbf{u}\|^2\|\mathbf{v}\|^2\cos^2\theta \\
&= \|\mathbf{u}\|^2\|\mathbf{v}\|^2(1 - \cos^2\theta) \qquad 1 - \cos^2\theta = \sin^2\theta\\
&= \|\mathbf{u}\|^2\|\mathbf{v}\|^2\sin^2\theta
\end{aligned}$$

Since the length of a vector is nonnegative, $\|\mathbf{u} \times \mathbf{v}\|$ must be the positive square root of the right side.

$$\begin{aligned}
\|\mathbf{u} \times \mathbf{v}\| &= \|\mathbf{u}\|\,\|\mathbf{v}\|\,|\sin\theta| \qquad |\sin\theta| = \sin\theta, \text{ since } \sin\theta \geq 0 \text{ for } 0 \leq \theta \leq \pi \\
&= \|\mathbf{u}\|\,\|\mathbf{v}\|\sin\theta
\end{aligned}$$

which completes the proof for nonzero vectors $\mathbf{u}$ and $\mathbf{v}$. You should provide the proofs for the cases where one or both vectors are zero. ∎

Comparing Theorem 11(B) with formula (4), we see that the magnitude of the cross product is the same as the area of the parallelogram with adjacent sides $\mathbf{u}$ and $\mathbf{v}$. Note that if either $\mathbf{u}$ or $\mathbf{v}$ is $\mathbf{0}$, or if $\mathbf{u}$ and $\mathbf{v}$ are parallel, then the area of the parallelogram will be 0 and hence, $\mathbf{u} \times \mathbf{v}$ must be $\mathbf{0}$. Conversely, if $\mathbf{u} \times \mathbf{v} = \mathbf{0}$, then either $\mathbf{u}$ or $\mathbf{v}$ must be $\mathbf{0}$ or $\mathbf{u}$ and $\mathbf{v}$ must be parallel.

The geometric properties of the cross product are summarized in the box.

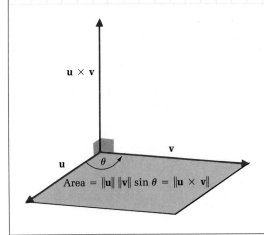

Geometric Properties of the Cross Product

1. $\mathbf{u} \times \mathbf{v}$ is orthogonal to both $\mathbf{u}$ and $\mathbf{v}$.
2. $\mathbf{u}$, $\mathbf{v}$, and $\mathbf{u} \times \mathbf{v}$, in that order, form a right-handed system.
3. The area of the parallelogram determined by $\mathbf{u}$ and $\mathbf{v}$ is

$$\|\mathbf{u} \times \mathbf{v}\| = \|\mathbf{u}\| \, \|\mathbf{v}\| \sin \theta$$

Example 30   Find the area of the parallelogram with vertices at $A = (0, 0, 0)$, $B = (4, 1, 1)$, $C = (1, 3, 2)$, and $D = (5, 4, 3)$.

Solution   First we verify that these points do form a parallelogram. To do this, we must show that $\overrightarrow{AB} = \overrightarrow{CD}$ and $\overrightarrow{AC} = \overrightarrow{BD}$ (see the figure).

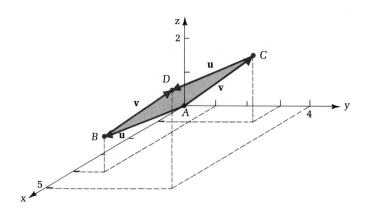

$$\overrightarrow{AB} = (4-0)\mathbf{i} + (1-0)\mathbf{j} + (1-0)\mathbf{k} = 4\mathbf{i} + \mathbf{j} + \mathbf{k}$$
$$\overrightarrow{CD} = (5-1)\mathbf{i} + (4-3)\mathbf{j} + (3-2)\mathbf{k} = 4\mathbf{i} + \mathbf{j} + \mathbf{k}$$

Equal

$$\overrightarrow{AC} = (1-0)\mathbf{i} + (3-0)\mathbf{j} + (2-0)\mathbf{k} = \mathbf{i} + 3\mathbf{j} + 2\mathbf{k}$$
$$\overrightarrow{BD} = (5-4)\mathbf{i} + (4-1)\mathbf{j} + (3-1)\mathbf{k} = \mathbf{i} + 3\mathbf{j} + 2\mathbf{k}$$

Equal

Thus, the points A, B, C, and D form a parallelogram. Letting $\mathbf{u} = \overrightarrow{AB} = \overrightarrow{CD}$ and $\mathbf{v} = \overrightarrow{AC} = \overrightarrow{BD}$, we can find the area of the parallelogram by finding $\|\mathbf{u} \times \mathbf{v}\|$:

$$\mathbf{u} \times \mathbf{v} = \begin{vmatrix} \mathbf{i} & \mathbf{j} & \mathbf{k} \\ 4 & 1 & 1 \\ 1 & 3 & 2 \end{vmatrix}$$

$$= \begin{vmatrix} 1 & 1 \\ 3 & 2 \end{vmatrix} \mathbf{i} - \begin{vmatrix} 4 & 1 \\ 1 & 2 \end{vmatrix} \mathbf{j} + \begin{vmatrix} 4 & 1 \\ 1 & 3 \end{vmatrix} \mathbf{k}$$

$$= -\mathbf{i} - 7\mathbf{j} + 11\mathbf{k}$$

Thus, the area of the parallelogram is

$$\|\mathbf{u} \times \mathbf{v}\| = \sqrt{(-1)^2 + (-7)^2 + 11^2}$$
$$= \sqrt{171}$$
$$= 3\sqrt{19}$$

Problem 30   Find the area of the parallelogram with vertices at $A = (0, 0, 0)$, $B = (2, 4, 3)$, $C = (1, 2, 5)$, and $D = (3, 6, 8)$. ∎

▪ Algebraic Properties

Now we will turn our attention to some of the algebraic properties of the cross product.

Theorem 12

If $\mathbf{u}$, $\mathbf{v}$, and $\mathbf{w}$ are vectors in R^3 and a is a real number, then:

(A) $\mathbf{u} \times \mathbf{v} = -(\mathbf{v} \times \mathbf{u})$
(B) $\mathbf{u} \times \mathbf{u} = \mathbf{0}$
(C) $\mathbf{u} \times (\mathbf{v} + \mathbf{w}) = \mathbf{u} \times \mathbf{v} + \mathbf{u} \times \mathbf{w}$
(D) $(\mathbf{u} + \mathbf{v}) \times \mathbf{w} = \mathbf{u} \times \mathbf{w} + \mathbf{v} \times \mathbf{w}$
(E) $(a\mathbf{u}) \times \mathbf{v} = a(\mathbf{u} \times \mathbf{v}) = \mathbf{u} \times (a\mathbf{v})$
(F) $(\mathbf{u} \times \mathbf{v}) \cdot \mathbf{w} = \mathbf{u} \cdot (\mathbf{v} \times \mathbf{w})$

Proof   All these statements can be proved by using the definition of the cross product and performing some algebraic manipulations. However, some of the proofs are easier if we use the properties of determinants. We will prove parts (A) and (F) to illustrate each method and leave the remainder as exercises.

(A) *Show that* $\mathbf{u} \times \mathbf{v} = -(\mathbf{v} \times \mathbf{u})$:

$$\mathbf{u} \times \mathbf{v} = \begin{vmatrix} \mathbf{i} & \mathbf{j} & \mathbf{k} \\ u_1 & u_2 & u_3 \\ v_1 & v_2 & v_3 \end{vmatrix} \qquad \text{Definition of } \mathbf{u} \times \mathbf{v}$$

$$= - \begin{vmatrix} \mathbf{i} & \mathbf{j} & \mathbf{k} \\ v_1 & v_2 & v_3 \\ u_1 & u_2 & u_3 \end{vmatrix} \qquad \begin{array}{l} \text{Interchanging two rows} \\ \text{changes the sign of a} \\ \text{determinant} \end{array}$$

$$= -(\mathbf{v} \times \mathbf{u})$$

(F) *Show that* $(\mathbf{u} \times \mathbf{v}) \cdot \mathbf{w} = \mathbf{u} \cdot (\mathbf{v} \times \mathbf{w})$:

$$(\mathbf{u} \times \mathbf{v}) \cdot \mathbf{w} = [(u_2v_3 - v_2u_3)\mathbf{i} - (u_1v_3 - v_1u_3)\mathbf{j} + (u_1v_2 - v_1u_2)\mathbf{k}] \cdot (w_1\mathbf{i} + w_2\mathbf{j} + w_3\mathbf{k})$$

$$= (u_2v_3 - v_2u_3)w_1 - (u_1v_3 - v_1u_3)w_2 + (u_1v_2 - v_1u_2)w_3$$

$$= u_2v_3w_1 - v_2u_3w_1 - u_1v_3w_2 + v_1u_3w_2 + u_1v_2w_3 - v_1u_2w_3 \tag{5}$$

$$\mathbf{u} \cdot (\mathbf{v} \times \mathbf{w}) = (u_1\mathbf{i} + u_2\mathbf{j} + u_3\mathbf{k}) \cdot [(v_2w_3 - w_2v_3)\mathbf{i} - (v_1w_3 - w_1v_3)\mathbf{j} + (v_1w_2 - w_1v_2)\mathbf{k}]$$

$$= u_1(v_2w_3 - w_2v_3) - u_2(v_1w_3 - w_1v_3) + u_3(v_1w_2 - w_1v_2)$$

$$= u_1v_2w_3 - u_1w_2v_3 - u_2v_1w_3 + u_2w_1v_3 + u_3v_1w_2 - u_3w_1v_2 \tag{6}$$

Comparing (5) and (6) shows that

$$(\mathbf{u} \times \mathbf{v}) \cdot \mathbf{w} = \mathbf{u} \cdot (\mathbf{v} \times \mathbf{w}) \qquad \blacksquare$$

▪ Scalar Triple Product

In Theorem 12(F) we saw that $\mathbf{u} \cdot (\mathbf{v} \times \mathbf{w})$ and $(\mathbf{u} \times \mathbf{v}) \cdot \mathbf{w}$ represent the same scalar quantity. Each expression is referred to as a *scalar triple product*. We can evaluate either form directly, as in the proof of Theorem 12(F), or by using the third-order determinant in Theorem 13, which is a little more convenient and more easily remembered. We leave the proof of Theorem 13 as an exercise.

Theorem 13	For $\mathbf{u}$, $\mathbf{v}$, and $\mathbf{w}$ any three vectors in R^3, their **scalar triple product** is given by $$\mathbf{u} \cdot (\mathbf{v} \times \mathbf{w}) = (\mathbf{u} \times \mathbf{v}) \cdot \mathbf{w} = \begin{vmatrix} u_1 & u_2 & u_3 \\ v_1 & v_2 & v_3 \\ w_1 & w_2 & w_3 \end{vmatrix}$$

The scalar triple product has an interesting and useful geometric interpretation. If $\mathbf{u}$, $\mathbf{v}$, and $\mathbf{w}$ are nonzero vectors positioned so that their initial points coincide, and if all three vectors are not coplanar (all three do not lie in the same plane), then the three vectors will form adjacent sides of a three-dimensional figure called a *parallelepiped* (opposite faces are parallel parallelograms), as

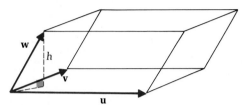

Figure 28 Parallelepiped determined by **u**, **v**, and **w**

shown in Figure 28. If V is the volume of the parallelepiped, then

  V = (Area of base)h

where h is the perpendicular distance from the terminal point of **w** to the base. We saw earlier that the area of the base is given by $\|\mathbf{u} \times \mathbf{v}\|$. To determine h, we need a vector perpendicular to the base, and $\mathbf{u} \times \mathbf{v}$ is just such a vector (see Figure 29). If θ is the angle between $\mathbf{u} \times \mathbf{v}$ and **w**, then

  $h = \|\mathbf{w}\| \, |\cos \theta|$

(In Figure 29, $\cos \theta > 0$, but in some cases it may be negative, so we must use $|\cos \theta|$.)

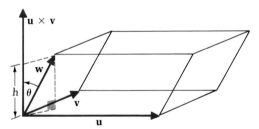

Figure 29

From Theorem 8 in Section 4-3 we know that

$$\cos \theta = \frac{(\mathbf{u} \times \mathbf{v}) \cdot \mathbf{w}}{\|\mathbf{u} \times \mathbf{v}\| \, \|\mathbf{w}\|}$$

Thus,

$$
\begin{aligned}
V &= (\text{Area of base})(\text{Altitude}) \\
&= \|\mathbf{u} \times \mathbf{v}\| h && h = \|\mathbf{w}\| \, |\cos \theta| \\
&= \|\mathbf{u} \times \mathbf{v}\| \, \|\mathbf{w}\| \, |\cos \theta| && \text{Theorem 8} \\
&= \|\mathbf{u} \times \mathbf{v}\| \, \|\mathbf{w}\| \frac{|(\mathbf{u} \times \mathbf{v}) \cdot \mathbf{w}|}{\|\mathbf{u} \times \mathbf{v}\| \, \|\mathbf{w}\|} \\
&= |(\mathbf{u} \times \mathbf{v}) \cdot \mathbf{w}|
\end{aligned}
$$

We combine this result with Theorem 13 for convenient reference.

Geometric Interpretation of $(\mathbf{u} \times \mathbf{v}) \cdot \mathbf{w}$ and $\mathbf{u} \cdot (\mathbf{v} \times \mathbf{w})$

The volume of a parallelepiped determined by $\mathbf{u}$, $\mathbf{v}$, and $\mathbf{w}$ (see Figure 29) is given by the absolute value of the scalar triple product:

$$V = |\mathbf{u} \cdot (\mathbf{v} \times \mathbf{w})| = |(\mathbf{u} \times \mathbf{v}) \cdot \mathbf{w}| = \begin{Vmatrix} u_1 & u_2 & u_3 \\ v_1 & v_2 & v_3 \\ w_1 & w_2 & w_3 \end{Vmatrix}$$

Example 31   Find the volume of the parallelepiped determined by the following vectors:

$$\mathbf{u} = 5\mathbf{i} + \mathbf{j} + \mathbf{k} \qquad \mathbf{v} = \mathbf{i} + 4\mathbf{j} + \mathbf{k} \qquad \mathbf{w} = \mathbf{i} + \mathbf{j} + 3\mathbf{k}$$

Solution $$V = |\mathbf{u} \cdot (\mathbf{v} \times \mathbf{w})| = \begin{Vmatrix} 5 & 1 & 1 \\ 1 & 4 & 1 \\ 1 & 1 & 3 \end{Vmatrix}$$

$$= |50| = 50$$

Problem 31   Find the volume of the parallelepiped determined by the following vectors:

$$\mathbf{u} = \mathbf{i} + \mathbf{j} + 2\mathbf{k} \qquad \mathbf{v} = 2\mathbf{i} + 3\mathbf{j} + \mathbf{k} \qquad \mathbf{w} = \mathbf{i} + 2\mathbf{j} + 4\mathbf{k}$$ ∎

Answers to Matched Problems

27. (A) $16\mathbf{i} + \mathbf{j} - 6\mathbf{k}$ (B) $2\mathbf{i} - 3\mathbf{j} - 7\mathbf{k}$ (C) $-5\mathbf{i} + 13\mathbf{j} - 7\mathbf{k}$

28. $4\mathbf{i} - \mathbf{j} - 3\mathbf{k}$ 29. $-11\mathbf{k}$ 30. $7\sqrt{5}$ 31. 5

∎ **Exercise 4-4**

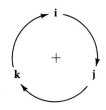

A **1.** Complete the following cross product table for the standard unit vectors:

×	i	j	k
i	0		
j		0	
k			0

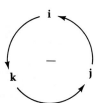

2. The diagram in the margin is a device that can be used to complete the table in Problem 1. The cross product of two consecutive vectors taken in the clockwise direction is the third vector. The cross product of two consecutive vectors taken in the counterclockwise direction is the negative of the third vector. Use this diagram to verify the products computed in Problem 1.

In Problems 3–10, let $\mathbf{u} = \mathbf{i} + 3\mathbf{j} - 2\mathbf{k}$, $\mathbf{v} = 4\mathbf{i} - 2\mathbf{j} + \mathbf{k}$, and $\mathbf{w} = -2\mathbf{i} + \mathbf{j} + 3\mathbf{k}$. Compute the indicated cross products.

3. $\mathbf{u} \times \mathbf{i}$ 4. $\mathbf{j} \times \mathbf{v}$ 5. $\mathbf{w} \times \mathbf{k}$ 6. $\mathbf{u} \times \mathbf{v}$

7. $\mathbf{u} \times \mathbf{w}$ 8. $\mathbf{v} \times \mathbf{w}$ 9. $(\mathbf{u} \times \mathbf{v}) \times \mathbf{w}$ 10. $\mathbf{u} \times (\mathbf{v} \times \mathbf{w})$

B In Problems 11–14, let $\mathbf{u} = \mathbf{i} + \mathbf{j}$, $\mathbf{v} = 2\mathbf{i} - 3\mathbf{j}$, and $\mathbf{w} = 3\mathbf{i} - 2\mathbf{j}$. Compute the indicated cross products.

11. $\mathbf{u} \times \mathbf{v}$ 12. $\mathbf{v} \times \mathbf{w}$ 13. $(\mathbf{u} \times \mathbf{v}) \times \mathbf{w}$ 14. $\mathbf{u} \times (\mathbf{v} \times \mathbf{w})$

In Problems 15–18, find a nonzero vector orthogonal to both $\mathbf{u}$ and $\mathbf{v}$.

15. $\mathbf{u} = (1, 2, -5)$; $\mathbf{v} = (6, -1, 2)$ 16. $\mathbf{u} = (2, -5, 4)$; $\mathbf{v} = (2, 5, -4)$

17. $\mathbf{u} = \mathbf{i} + \mathbf{j}$; $\mathbf{v} = \mathbf{j} + \mathbf{k}$ 18. $\mathbf{u} = \mathbf{j} + 2\mathbf{k}$; $\mathbf{v} = 2\mathbf{j} - \mathbf{k}$

In Problems 19–22, verify that points A, B, C, and D are the vertices of a parallelogram, and find the area of that parallelogram.

19. $A = (0, 0, 0)$; $B = (0, 2, 0)$; $C = (3, 0, 0)$; $D = (3, 2, 0)$
20. $A = (0, 0, 0)$; $B = (1, 3, 0)$; $C = (4, 2, 0)$; $D = (5, 5, 0)$
21. $A = (1, -1, 3)$; $B = (-2, 1, 4)$; $C = (2, 3, -4)$; $D = (-1, 5, -3)$
22. $A = (2, 1, -3)$; $B = (1, 4, 2)$; $C = (-3, -2, 1)$; $D = (-4, 1, 6)$

In Problems 23 and 24, use the formulas

$$\cos \theta = \frac{\mathbf{u} \cdot \mathbf{v}}{\|\mathbf{u}\| \, \|\mathbf{v}\|} \qquad \sin \theta = \frac{\|\mathbf{u} \times \mathbf{v}\|}{\|\mathbf{u}\| \, \|\mathbf{v}\|}$$

to find the cosine and sine of the angle between $\mathbf{u}$ and $\mathbf{v}$. As a check, verify that $\cos^2 \theta + \sin^2 \theta = 1$.

23. $\mathbf{u} = 2\mathbf{i} + 2\mathbf{j} + \mathbf{k}$, $\mathbf{v} = 6\mathbf{i} + 3\mathbf{j} + 2\mathbf{k}$
24. $\mathbf{u} = 7\mathbf{i} + 4\mathbf{j} - 4\mathbf{k}$, $\mathbf{v} = -\mathbf{i} + 8\mathbf{j} + 4\mathbf{k}$

In Problems 25–28, find the volume of the parallelepiped determined by $\mathbf{u}$, $\mathbf{v}$, and $\mathbf{w}$.

25. $\mathbf{u} = 3\mathbf{i}$, $\mathbf{v} = 4\mathbf{j}$, $\mathbf{w} = 5\mathbf{k}$
26. $\mathbf{u} = 2\mathbf{i} + 2\mathbf{j}$, $\mathbf{v} = 2\mathbf{i} - 2\mathbf{j}$, $\mathbf{w} = 4\mathbf{i} + 3\mathbf{k}$
27. $\mathbf{u} = 2\mathbf{i} + 5\mathbf{j} - \mathbf{k}$, $\mathbf{v} = \mathbf{i} + 3\mathbf{j} + 2\mathbf{k}$, $\mathbf{w} = 3\mathbf{i} - 2\mathbf{k} + \mathbf{j}$
28. $\mathbf{u} = \mathbf{i} + 2\mathbf{j} - 3\mathbf{k}$, $\mathbf{v} = -2\mathbf{i} + \mathbf{j} + \mathbf{k}$, $\mathbf{w} = \mathbf{i} + \mathbf{j} - \mathbf{k}$

C 29. Show that the area of the triangle with vertices at points A, B, and C is $\frac{1}{2}\|\overrightarrow{AB} \times \overrightarrow{AC}\|$. [Hint: Consider the parallelogram determined by $\overrightarrow{AB}$ and $\overrightarrow{AC}$.]

In Problems 30–32, use the result in Problem 29 to find the area of the triangle with vertices at A, B, and C.

30. $A = (0, 0, 0)$; $B = (4, 0, 0)$; $C = (0, 7, 0)$
31. $A = (1, 2, -1)$; $B = (4, -3, 2)$; $C = (-2, 1, 5)$
32. $A = (1, 0, -1)$; $B = (2, 3, 2)$; $C = (-1, -2, 1)$

Problems 33–36 refer to statements from Theorem 12 in this section. Use the definition of the cross product and, in some cases, the properties of determinants to prove each statement. Do not refer to Theorem 12 in your proof.

33. *Theorem 12(B).* $\mathbf{u} \times \mathbf{u} = \mathbf{0}$

34. *Theorem 12(C).* $\mathbf{u} \times (\mathbf{v} + \mathbf{w}) = \mathbf{u} \times \mathbf{v} + \mathbf{u} \times \mathbf{w}$

35. *Theorem 12(D).* $(\mathbf{u} + \mathbf{v}) \times \mathbf{w} = \mathbf{u} \times \mathbf{w} + \mathbf{v} \times \mathbf{w}$

36. *Theorem 12(E).* $(a\mathbf{u}) \times \mathbf{v} = a(\mathbf{u} \times \mathbf{v}) = \mathbf{u} \times (a\mathbf{v})$

37. *Theorem 13.* Show that

$$\mathbf{u} \cdot (\mathbf{v} \times \mathbf{w}) = (\mathbf{u} \times \mathbf{v}) \cdot \mathbf{w} = \begin{vmatrix} u_1 & u_2 & u_3 \\ v_1 & v_2 & v_3 \\ w_1 & w_2 & w_3 \end{vmatrix}$$

38. Show that $(\mathbf{w} \times \mathbf{u}) \cdot \mathbf{v} = \mathbf{u} \cdot (\mathbf{v} \times \mathbf{w})$.

39. Show that $\mathbf{u}$ and $\mathbf{v}$ are orthogonal if and only if $\|\mathbf{u} \times \mathbf{v}\| = \|\mathbf{u}\| \, \|\mathbf{v}\|$.

40. Show that $\mathbf{u}$ and $\mathbf{v}$ are parallel if and only if $\mathbf{u} \times \mathbf{v} = \mathbf{0}$.

41. If $\mathbf{u} + \mathbf{v} + \mathbf{w} = \mathbf{0}$, show that $\mathbf{u} \times \mathbf{v} = \mathbf{v} \times \mathbf{w} = \mathbf{w} \times \mathbf{u}$.

42. If $\mathbf{u}$ is a nonzero vector, show that there cannot exist a vector $\mathbf{v}$ with the property that $\mathbf{u} \times \mathbf{v} = \mathbf{u}$.

43. Show that $(\mathbf{u} + \mathbf{v}) \cdot [(\mathbf{v} + \mathbf{w}) \times (\mathbf{w} + \mathbf{u})] = 2[\mathbf{u} \cdot (\mathbf{v} \times \mathbf{w})]$.

▌4-5▐   Lines in Three Dimensions

- Vector Equation for a Line
- Parametric Equations for a Line
- Symmetric Equations for a Line
- Skew Lines
- Distance from a Point to a Line

A basic axiom from geometry states that two distinct points P_1 and P_2 determine a unique line ℓ that passes through these points. If $P_1 = (x_1, y_1)$ and $P_2 = (x_2, y_2)$ are points in the plane, then the familiar slope formula for ℓ and the point–slope form for the equation of ℓ are illustrated in Figure 30. If $x_1 = x_2$, then ℓ is a vertical line, the slope of ℓ is undefined, and the equation of ℓ is $x = x_1$.

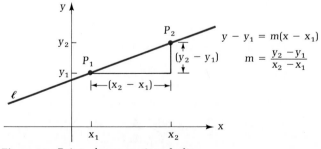

Figure 30   Point–slope equation of a line

Since lines with equal slopes are parallel, the slope determines only the direction of the line. We can find the equation of a line in a plane if we are given its direction (slope) and a point on the line, or if we are given two points on the line. How can we extend these concepts to lines in three dimensions?

■ Vector Equation for a Line

Suppose ℓ is the line determined by the distinct points $P_1 = (x_1, y_1, z_1)$ and $P_2 = (x_2, y_2, z_2)$, and $\mathbf{v}$ is defined by

$$\mathbf{v} = \overrightarrow{P_1P_2} = (x_2 - x_1)\mathbf{i} + (y_2 - y_1)\mathbf{j} + (z_2 - z_1)\mathbf{k}$$

Since $\mathbf{v}$ and ℓ are parallel (in fact, the representative $\overrightarrow{P_1P_2}$ of $\mathbf{v}$ lies on ℓ; see Figure 31), $\mathbf{v}$ determines the direction of ℓ and is referred to as a *direction vector for ℓ*. In general, any nonzero vector that is parallel to a line is called a **direction vector** for that line. Thus, if $\mathbf{v}$ is a direction vector for ℓ, then any nonzero scalar multiple of $\mathbf{v}$ will also be a direction vector for ℓ. Direction vectors are not unique.

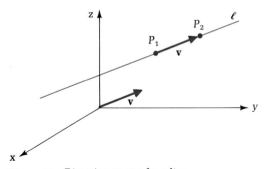

Figure 31   Direction vector for a line

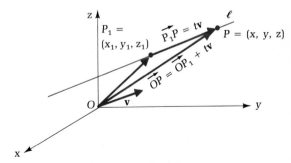

Figure 32   Vector equation for a line

Now, we would like to find a vector equation of a line ℓ that passes through $P_1 = (x_1, y_1, z_1)$ with direction vector $\mathbf{v} = a\mathbf{i} + b\mathbf{j} + c\mathbf{k}$. A point $P = (x, y, z)$ (an arbitrary point in R^3) will be on ℓ if and only if $\overrightarrow{P_1P}$ is parallel to $\mathbf{v}$; that is, if and only if $\overrightarrow{P_1P}$ is a scalar multiple of $\mathbf{v}$ (see Figure 32). Symbolically,

$$\overrightarrow{P_1P} = t\mathbf{v}$$

But

$$\overrightarrow{P_1P} = \overrightarrow{OP} - \overrightarrow{OP_1}$$

Therefore,

$$\overrightarrow{OP} - \overrightarrow{OP_1} = t\mathbf{v}$$
$$\overrightarrow{OP} = \overrightarrow{OP_1} + t\mathbf{v}$$

Thus, we have a **vector equation of the line** ℓ:

$$\overrightarrow{OP} = \overrightarrow{OP_1} + t\mathbf{v} \qquad t \text{ any real number}$$

where $\mathbf{v}$ is a direction vector for ℓ, $P_1 = (x_1, y_1, z_1)$ is a fixed point on ℓ, $P = (x, y, z)$ is an arbitrary point on ℓ, and $O = (0, 0, 0)$ is the origin.

Example 32   Find a vector equation for the line that passes through the point $P_1 = (-4, 6, 9)$ in the direction of $\mathbf{v} = 2\mathbf{i} - \mathbf{j} + 3\mathbf{k}$.

Solution $\overrightarrow{OP} = \underset{\overrightarrow{OP_1}}{\underbrace{}} + t\underset{\mathbf{v}}{\underbrace{}}$

$$= (-4\mathbf{i} + 6\mathbf{j} + 9\mathbf{k}) + t(2\mathbf{i} - \mathbf{j} + 3\mathbf{k})$$

Problem 32   Find a vector equation for the line that passes through the point $P_1 = (2, -4, 7)$ in the direction of $\mathbf{v} = 3\mathbf{i} + 5\mathbf{j} - 6\mathbf{k}$.   ▌▌

Example 33   Find a vector equation for the line that passes through the points $P_1 = (1, 2, -4)$ and $P_2 = (3, -1, 1)$.

Solution   A direction vector for the line is

$$\mathbf{v} = \overrightarrow{P_1P_2} = (3 - 1)\mathbf{i} + (-1 - 2)\mathbf{j} + [1 - (-4)]\mathbf{k}$$
$$= 2\mathbf{i} - 3\mathbf{j} + 5\mathbf{k}$$

Now we can use P_1 or P_2 to write a vector equation for the line (we arbitrarily choose P_1):

$$\overrightarrow{OP} = \overrightarrow{OP_1} + t\mathbf{v}$$
$$= (\mathbf{i} + 2\mathbf{j} - 4\mathbf{k}) + t(2\mathbf{i} - 3\mathbf{j} + 5\mathbf{k})$$

Problem 33   Find the vector equation for the line that passes through the points $P_1 = (-1, 2, 3)$ and $P_2 = (2, -1, 1)$.   ▌▌

▪ Parametric Equations for a Line

If $\mathbf{v} = a\mathbf{i} + b\mathbf{j} + c\mathbf{k}$ is a direction vector for a line ℓ, $P_1 = (x_1, y_1, z_1)$ is a fixed point on ℓ, and $P = (x, y, z)$ is an arbitrary point on ℓ, then the vector equation for ℓ can be written as

$$\overrightarrow{OP} = \overrightarrow{OP_1} + t\mathbf{v}$$
$$x\mathbf{i} + y\mathbf{j} + z\mathbf{k} = (x_1\mathbf{i} + y_1\mathbf{j} + z_1\mathbf{k}) + t(a\mathbf{i} + b\mathbf{j} + c\mathbf{k})$$
$$= (x_1 + at)\mathbf{i} + (y_1 + bt)\mathbf{j} + (z_1 + ct)\mathbf{k}$$

Equating the components of these vectors produces **parametric equations for a line:**

$$x = x_1 + at$$
$$y = y_1 + bt$$
$$z = z_1 + ct \qquad t \text{ any real number}$$

where $\mathbf{v} = a\mathbf{i} + b\mathbf{j} + c\mathbf{k}$ is a direction vector for the line and (x_1, y_1, z_1) is a fixed point on the line.

Example 34   For each of the following, find parametric equations for the line through P_1 in the direction of $\mathbf{v}$.

(A) $P_1 = (-4, 6, 9);$ $\mathbf{v} = 2\mathbf{i} + 4\mathbf{j} + 3\mathbf{k}$
(B) $P_1 = (0, 2, -1);$ $\mathbf{v} = 3\mathbf{i} - 2\mathbf{j}$

Solution   (A) $P_1 = (-4, 6, 9);$ $\mathbf{v} = 2\mathbf{i} + 4\mathbf{j} + 3\mathbf{k}$

$$x = -4 + 2t$$
$$y = 6 + 4t$$
$$z = 9 + 3t$$

For example, if $t = 3$, then $(2, 18, 18)$ is a point on the line; if $t = -1$, then $(-6, 2, 6)$ is a point on the line; and so on.

(B) $P_1 = (0, 2, -1);$ $\mathbf{v} = 3\mathbf{i} - 2\mathbf{j}$

$$x = 3t$$
$$y = 2 - 2t$$
$$z = -1$$

Problem 34   For each of the following, find parametric equations for the line through P_1 in the direction of $\mathbf{v}$.

(A) $P_1 = (2, -1, 3);$ $\mathbf{v} = -\mathbf{i} + 2\mathbf{j} + 5\mathbf{k}$
(B) $P_1 = (4, -1, 0);$ $\mathbf{v} = 2\mathbf{j} + 6\mathbf{k}$ ∎

▪ Symmetric Equations for a Line

If

$$x = x_1 + at$$
$$y = y_1 + bt$$
$$z = z_1 + ct$$

are parametric equations for a line and a, b, and c are all nonzero, then we can solve each equation for t:

$$\frac{x - x_1}{a} = t \qquad \frac{y - y_1}{b} = t \qquad \frac{z - z_1}{c} = t$$

Eliminating t produces **symmetric equations for a line:**

$$\frac{x - x_1}{a} = \frac{y - y_1}{b} = \frac{z - z_1}{c} \qquad a, b, c \text{ nonzero}$$

where $\mathbf{v} = a\mathbf{i} + b\mathbf{j} + c\mathbf{k}$ is a direction vector for the line and (x_1, y_1, z_1) is a fixed point on the line.

If any of the numbers a, b, or c is zero, then the symmetric form must be written as two separate equations. For example, if $a = 0$, the parametric equations are

$$x = x_1$$
$$y = y_1 + bt$$
$$z = z_1 + ct$$

and the symmetric form is

$$x = x_1 \qquad \frac{y - y_1}{b} = \frac{z - z_1}{c} \qquad a = 0, b \neq 0, c \neq 0$$

If both $a = 0$ and $b = 0$, the parametric equations are

$$x = x_1$$
$$y = y_1$$
$$z = z_1 + ct$$

In this case, x and y are both constant and z can take on any value. The symmetric form is

$$x = x_1 \qquad y = y_1 \qquad a = b = 0, c \neq 0$$

There is no need for an equation for z. Similar symmetric forms can be derived for other combinations of zero values of a, b, and c.

Example 35   Find symmetric equations for the line through P_1 in the direction of $\mathbf{v}$.

(A) $P_1 = (-3, 2, 4)$; $\mathbf{v} = 2\mathbf{i} - 5\mathbf{j} + 9\mathbf{k}$
(B) $P_1 = (0, 5, -7)$; $\mathbf{v} = 4\mathbf{i} + 2\mathbf{j}$
(C) $P_1 = (4, 2, 1)$; $\mathbf{v} = \mathbf{j}$

Solution   (A) $P_1 = (-3, 2, 4)$

$$\frac{x + 3}{2} = \frac{y - 2}{-5} = \frac{z - 4}{9}$$

$$\mathbf{v} = 2\mathbf{i} - 5\mathbf{j} + 9\mathbf{k}$$

(B) Since $c = 0$, the symmetric form is

$$\frac{x}{4} = \frac{y - 5}{2} \qquad z = -7$$

(C) Since $a = 0$ and $c = 0$, the symmetric form is

$$x = 4 \qquad z = 1$$

Problem 35   Find symmetric equations for the line through P_1 in the direction of $\mathbf{v}$.

(A) $P = (4, 7, -5);$ $\mathbf{v} = -3\mathbf{i} + 2\mathbf{j} - 6\mathbf{k}$
(B) $P = (4, 0, 2);$ $\mathbf{v} = 2\mathbf{i} - \mathbf{k}$
(C) $P = (6, 10, -2);$ $\mathbf{v} = 4\mathbf{i}$ ▐▐

Example 36   If ℓ is the line through the point $P_1 = (2, 4, 6)$ in the direction of $\mathbf{v} = \mathbf{i} + 2\mathbf{j} + 3\mathbf{k}$, find:

(A) A vector equation for ℓ
(B) Parametric equations for ℓ
(C) Symmetric equations for ℓ

Solution   (A) Let $\overrightarrow{OP_1} = 2\mathbf{i} + 4\mathbf{j} + 6\mathbf{k}$. Then

$$\overrightarrow{OP} = \overrightarrow{OP_1} + t\mathbf{v} = (2\mathbf{i} + 4\mathbf{j} + 6\mathbf{k}) + t(\mathbf{i} + 2\mathbf{j} + 3\mathbf{k})$$

(B) $x = 2 + t$
$y = 4 + 2t$
$z = 6 + 3t$

(C) $x - 2 = \dfrac{y - 4}{2} = \dfrac{z - 6}{3}$

Problem 36   Repeat Example 36 for the line through the point $P_1 = (5, 10, 15)$ in the direction of $\mathbf{v} = 2\mathbf{i} + 3\mathbf{j} + 4\mathbf{k}$.   ▐▐

We summarize the above discussion in the box below.

Equations for a Line

1. If $P_1 = (x_1, y_1, z_1)$ and $P_2 = (x_2, y_2, z_2)$ are distinct points on a line ℓ, then

$$\mathbf{v} = (x_2 - x_1)\mathbf{i} + (y_2 - y_1)\mathbf{j} + (z_2 - z_1)\mathbf{k}$$

is a direction vector for ℓ.

2. Let $\mathbf{v} = a\mathbf{i} + b\mathbf{j} + c\mathbf{k}$ be a direction vector for a line ℓ, $P_1 = (x_1, y_1, z_1)$ be a fixed point on ℓ, and $P = (x, y, z)$ be an arbitrary point on ℓ. Then the three types of equations for ℓ are:

Vector Equation	Parametric Equations	Symmetric Equations
$\overrightarrow{OP} = \overrightarrow{OP_1} + t\mathbf{v}$	$x = x_1 + at$ $y = y_1 + bt$ $z = z_1 + ct$	$\dfrac{x - x_1}{a} = \dfrac{y - y_1}{b} = \dfrac{z - z_1}{c}$ a, b, c nonzero

---

The three forms for the equation of a line in three dimensions are displayed in the box. The following examples illustrate the use of these equations in a variety of situations.

Example 37 Find the symmetric equations of the line ℓ that passes through the point $P_1 = (-2, 1, 4)$ and is parallel to the line ℓ_1 with symmetric equations

$$\frac{x - 3}{2} = \frac{y + 5}{-3} = \frac{z - 2}{7}$$

Solution Comparing the equations for ℓ_1 with the general symmetric form, we see that $\mathbf{v} = 2\mathbf{i} - 3\mathbf{j} + 7\mathbf{k}$ is a direction vector for ℓ_1. Since ℓ is parallel to ℓ_1, $\mathbf{v}$ is also a direction vector for ℓ. Using $\mathbf{v}$ and P_1, the symmetric equations for ℓ are

$$\frac{x + 2}{2} = \frac{y - 1}{-3} = \frac{z - 4}{7}$$

Problem 37 Find parametric equations for the line ℓ that passes through the point $P_1 = (-2, 1, 3)$ and is parallel to the line ℓ_1 with parametric equations

$$x = -7 + 4t$$
$$y = 12 - 6t$$
$$z = 5 + 2t$$ ❚❚

Example 38 The line ℓ has parametric equations

$$x = 2 - 3t$$
$$y = -1 + 4t$$
$$z = 6 - 2t$$

Find the point where ℓ intersects the xy plane.

Solution The z coordinate of the point of intersection of ℓ and the xy plane must be 0. Thus,

$$z = 6 - 2t = 0$$
$$2t = 6$$
$$t = 3$$

Substituting $t = 3$ in the equations for x and y, we have

$$x = 2 - 3(3) = -7 \quad \text{and} \quad y = -1 + 4(3) = 11$$

Thus, ℓ intersects the xy plane at the point $(-7, 11, 0)$.

Problem 38 Find the point of intersection of the line in Example 38 with the xz plane. ❚❚

▪ Skew Lines

If ℓ_1 and ℓ_2 are two lines in the plane, then there are three possibilities for their relative positions: They are parallel and do not intersect, or they intersect in one point, or they coincide and intersect in an infinite number of points. In three dimensions there is a fourth possibility: Two lines are **skew** if they are not parallel and do not intersect. For example, the line ℓ_1 with symmetric equations

$$z = 4 \qquad y = 0$$

is parallel to the xy plane, and the line ℓ_2 with symmetric equations

$$2x = y \qquad z = 0$$

lies in the xy plane. As Figure 33 illustrates, ℓ_1 and ℓ_2 are not parallel and do not intersect. Hence, they are skew lines.

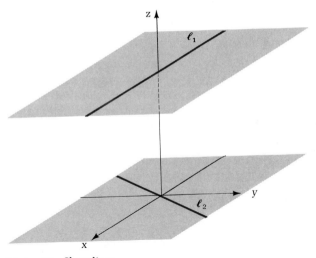

Figure 33 Skew lines

Parallel lines are easy to recognize. Their direction vectors must also be parallel, and hence, one direction vector must be a scalar multiple of the other. If two lines are not parallel, then either they intersect or they are skew. In general, the only way we can determine which is the case is to attempt to find the point of intersection.

Example 39 Determine whether the following pairs of lines are parallel, skew, or intersecting.

(A) ℓ_1: $x = 2 + t$ ℓ_2: $x = 4 + 2s$

 $y = 3 - t$ $y = -2 - 2s$

 $z = 4 + 2t$ $z = 6 + 4s$

(B) ℓ_1: $x = 2 + t$ ℓ_2: $x = \ 7 - 3s$

    $y = 3 - t$ $y = \qquad 2s$

    $z = 4 + 2t$ $z = \ 4 - s$

(C) ℓ_1: $x = 2 + t$ ℓ_2: $x = \ 1 - 2s$

    $y = 3 - t$ $y = -1 + s$

    $z = 4 + 2t$ $z = \ 3 - s$

Solution   (A) $\mathbf{v}_1 = \mathbf{i} - \mathbf{j} + 2\mathbf{k}$ is a direction vector for ℓ_1 and $\mathbf{v}_2 = 2\mathbf{i} - 2\mathbf{j} + 4\mathbf{k}$ is a direction vector for ℓ_2. Since $\mathbf{v}_2 = 2\mathbf{v}_1$, lines ℓ_1 and ℓ_2 are parallel.

(B) The direction vectors $\mathbf{v}_1 = \mathbf{i} - \mathbf{j} + 2\mathbf{k}$ and $\mathbf{v}_2 = -3\mathbf{i} + 2\mathbf{j} - \mathbf{k}$ are not parallel; hence, ℓ_1 and ℓ_2 are not parallel. If ℓ_1 and ℓ_2 do intersect in a point P, then there must be a value of t that will produce this point when substituted in the equations for ℓ_1 and a value of s that will produce the same point when substituted in the equations for ℓ_2. Thus, we must try to find t and s so that

$$2 + t = x = 7 - 3s$$
$$3 - t = y = \qquad 2s$$
$$4 + 2t = z = 4 - s$$

This gives us a system of three equations in two variables:

$$t + 3s = \ 5$$
$$-t - 2s = -3$$
$$2t + s = \ 0$$

The reduced form for this system is (calculations omitted)

$$\begin{bmatrix} 1 & 0 & | & -1 \\ 0 & 1 & | & 2 \\ 0 & 0 & | & 0 \end{bmatrix}$$

Thus, the solution is $t = -1$ and $s = 2$. Substituting $t = -1$ in the equations for ℓ_1 and $s = 2$ in the equations for ℓ_2, we have

ℓ_1: $t = -1$ ℓ_2: $s = 2$

$x = 2 - 1 \qquad = 1$ $x = 7 - 3(2) = 1$

$y = 3 - (-1) = 4$ $y = \qquad 2(2) = 4$

$z = 4 + 2(-1) = 2$ $z = 4 - 2 \quad = 2$

Thus, ℓ_1 and ℓ_2 intersect at the point (1, 4, 2). (Notice that the values of the parameters s and t that produce the intersection point are not equal. This is the reason that we used different parameters in the equations for ℓ_1 and ℓ_2. Using the same parameter in both sets of equations usually will lead to an incorrect solution.)

(C) Once again, examining the direction vectors shows that the lines are not parallel. To determine whether they intersect, we must solve the system

$$
\begin{array}{rl}
2+ \ t=x= & 1-2s \\
3- \ t=y=-1+ \ s \\
4+2t=z= & 3- \ s
\end{array}
\quad \text{or} \quad
\begin{array}{rl}
t+2s= & 1 \\
-t- \ s=-4 \\
2t+ \ s=-1
\end{array}
$$

The reduced form for this system is

$$
\begin{bmatrix}
1 & 0 & \Big| & 9 \\
0 & 1 & \Big| & -5 \\
0 & 0 & \Big| & -14
\end{bmatrix}
$$

which shows that the system is inconsistent. Since ℓ_1 and ℓ_2 do not intersect and are not parallel, they are skew lines.

Problem 39 Determine whether the following pairs of lines are parallel, skew, or intersecting.

(A) ℓ_1: $x=1+3t$ ℓ_2: $x=-4-9s$
        $y=2- \ t$ $y= \ \ 7+3s$
        $z=4+2t$ $z= \ \ 5-6s$

(B) ℓ_1: $x=1+3t$ ℓ_2: $x=-2+ \ s$
        $y=2- \ t$ $y= \ \ 1-2s$
        $z=4+2t$ $z= \ \ 3+3s$

(C) ℓ_1: $x=1+3t$ ℓ_2: $x= \ \ 4+ \ s$
        $y=2- \ t$ $y= \ \ 6-2s$
        $z=4+2t$ $z=-1+3s$

∎

▪ Distance from a Point to a Line

Given a line ℓ and a point Q not on the line, the distance from the point to the line is the length of the perpendicular line segment from Q to ℓ, as shown in Figure 34(A).

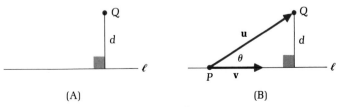

(A)   (B)

Figure 34 Distance from a point to a line

Suppose P is a point on ℓ and $\mathbf{v}$ is a direction vector for ℓ, as shown in Figure 34(B). If we form the vector $\mathbf{u} = \overrightarrow{PQ}$, then

$$d = \|\mathbf{u}\| \sin \theta \tag{1}$$

where θ is the angle between $\mathbf{u}$ and $\mathbf{v}$. From Theorem 11(B) in Section 4-4,

$$\sin \theta = \frac{\|\mathbf{u} \times \mathbf{v}\|}{\|\mathbf{u}\| \, \|\mathbf{v}\|} \tag{2}$$

Substituting (2) into (1), we have

$$d = \|\mathbf{u}\| \frac{\|\mathbf{u} \times \mathbf{v}\|}{\|\mathbf{u}\| \, \|\mathbf{v}\|}$$

$$= \frac{\|\mathbf{u} \times \mathbf{v}\|}{\|\mathbf{v}\|}$$

$$= \frac{\|\overrightarrow{PQ} \times \mathbf{v}\|}{\|\mathbf{v}\|} \qquad \mathbf{u} = \overrightarrow{PQ}$$

These ideas are summarized in the box.

Distance from a Point to a Line

The distance from a point Q to a line ℓ is given by

$$d = \frac{\|\overrightarrow{PQ} \times \mathbf{v}\|}{\|\mathbf{v}\|}$$

where $\mathbf{v}$ is a direction vector for ℓ and P is a point on ℓ.

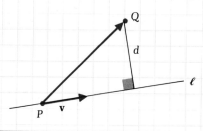

Example 40   Find the distance from the point $Q = (4, 2, 1)$ to the line ℓ with symmetric equations

$$\frac{x - 1}{-1} = \frac{y + 2}{2} = \frac{z - 3}{2}$$

Solution   The point $P = (1, -2, 3)$ is on ℓ, and $\mathbf{v} = -\mathbf{i} + 2\mathbf{j} + 2\mathbf{k}$ is a direction vector for ℓ. Let

$$\overrightarrow{PQ} = 3\mathbf{i} + 4\mathbf{j} - 2\mathbf{k}$$

Then

$$\overrightarrow{PQ} \times \mathbf{v} = \begin{vmatrix} \mathbf{i} & \mathbf{j} & \mathbf{k} \\ 3 & 4 & -2 \\ -1 & 2 & 2 \end{vmatrix}$$

$$= \begin{vmatrix} 4 & -2 \\ 2 & 2 \end{vmatrix} \mathbf{i} - \begin{vmatrix} 3 & -2 \\ -1 & 2 \end{vmatrix} \mathbf{j} + \begin{vmatrix} 3 & 4 \\ -1 & 2 \end{vmatrix} \mathbf{k}$$

$$= 12\mathbf{i} - 4\mathbf{j} + 10\mathbf{k}$$

and

$$d = \frac{\|\overrightarrow{PQ} \times \mathbf{v}\|}{\|\mathbf{v}\|}$$

$$= \frac{\sqrt{12^2 + (-4)^2 + 10^2}}{\sqrt{(-1)^2 + 2^2 + 2^2}}$$

$$= \frac{\sqrt{260}}{\sqrt{9}} = \frac{2}{3}\sqrt{65}$$

Problem 40 Find the distance from the point $Q = (1, 3, -2)$ to the line ℓ with symmetric equations

$$\frac{x-2}{2} = \frac{y+1}{3} = \frac{z-4}{-6}$$ ∎

Answers to Matched Problems

32. $\overrightarrow{OP} = (2\mathbf{i} - 4\mathbf{j} + 7\mathbf{k}) + t(3\mathbf{i} + 5\mathbf{j} - 6\mathbf{k})$

33. $\overrightarrow{OP} = (-\mathbf{i} + 2\mathbf{j} + 3\mathbf{k}) + t(3\mathbf{i} - 3\mathbf{j} - 2\mathbf{k})$

34. (A) $x = 2 - t$ (B) $x = 4$
$y = -1 + 2t$ $y = -1 + 2t$
$z = 3 + 5t$ $z = 6t$

35. (A) $\dfrac{x-4}{-3} = \dfrac{y-7}{2} = \dfrac{z+5}{-6}$ (B) $\dfrac{x-4}{2} = \dfrac{z-2}{-1}, y = 0$

(C) $y = 10, z = -2$

36. (A) $\overrightarrow{OP} = (5\mathbf{i} + 10\mathbf{j} + 15\mathbf{k}) + t(2\mathbf{i} + 3\mathbf{j} + 4\mathbf{k})$

(B) $x = 5 + 2t$ (C) $\dfrac{x-5}{2} = \dfrac{y-10}{3} = \dfrac{z-15}{4}$
$y = 10 + 3t$
$z = 15 + 4t$

37. $x = -2 + 4t$ 38. $(\frac{5}{4}, 0, \frac{11}{2})$
$y = 1 - 6t$
$z = 3 + 2t$

39. (A) Parallel   (B) Skew   (C) Intersect at $(7, 0, 8)$ 40. $\frac{1}{7}\sqrt{481}$

‖ Exercise 4-5

A   In Problems 1–6, find the indicated form of the equation for the line through P_1 with direction vector **v**.

1. $P_1 = (2, -1, 3)$; $\mathbf{v} = 4\mathbf{i} + 6\mathbf{j} - 5\mathbf{k}$;   vector form
2. $P_1 = (-1, 0, 2)$; $\mathbf{v} = -\mathbf{i} + \mathbf{j} - 2\mathbf{k}$;   vector form
3. $P_1 = (0, 0, 0)$; $\mathbf{v} = \mathbf{i} + 2\mathbf{j} + 3\mathbf{k}$;   parametric form
4. $P_1 = (-1, 2, 4)$; $\mathbf{v} = 2\mathbf{i} - 5\mathbf{k}$;   parametric form
5. $P_1 = (1, 2, 3)$; $\mathbf{v} = 4\mathbf{i} + 5\mathbf{j} + 6\mathbf{k}$;   symmetric form
6. $P_1 = (-1, 1, -2)$; $\mathbf{v} = 2\mathbf{j} + 5\mathbf{k}$;   symmetric form

In Problems 7–12, find a point on the line and a direction vector for the line.

7. $\overrightarrow{OP} = (-2\mathbf{i} + 4\mathbf{j} + \mathbf{k}) + t(\mathbf{i} - 3\mathbf{j} + 2\mathbf{k})$
8. $\overrightarrow{OP} = (5\mathbf{i} + \mathbf{j} - 6\mathbf{k}) + t(2\mathbf{i} - 5\mathbf{k})$
9. $x = 3 + 2t$, $y = -5t$, $z = 7$
10. $x = -5 - 3t$, $y = 2 + 4t$, $z = -8 + 2t$
11. $\dfrac{x + 10}{5} = \dfrac{z - 15}{10}$, $y = 5$
12. $\dfrac{x - 2}{6} = \dfrac{y + 2}{-6} = \dfrac{z}{7}$

B   In Problems 13–18, find the indicated form of the equation of the line through P_1 and P_2.

13. $P_1 = (2, 1, 3)$; $P_2 = (5, 7, 8)$;   vector form
14. $P_1 = (-1, 1, -1)$; $P_2 = (2, -2, -2)$;   vector form
15. $P_1 = (-6, 4, 2)$; $P_2 = (-4, 4, -2)$;   parametric form
16. $P_1 = (0, 1, 0)$; $P_2 = (2, 2, 0)$;   parametric form
17. $P_1 = (1, 2, -1)$; $P_2 = (1, 5, 1)$;   symmetric form
18. $P_1 = (5, 10, 15)$; $P_2 = (5, 10, 20)$;   symmetric form

In Problems 19–24, find the indicated form of the equation of the line through P_1 parallel to ℓ.

19. $P_1 = (2, -1, 4)$; ℓ: $\overrightarrow{OP} = (\mathbf{i} + 3\mathbf{j} - \mathbf{k}) + t(4\mathbf{i} - 5\mathbf{j} + 6\mathbf{k})$;   vector form
20. $P_1 = (-1, 0, 0)$; ℓ: $x = 1 + 2t$, $y = 1$, $z = 1 - 2t$;   vector form
21. $P_1 = (2, 1, -3)$; ℓ: $x = 1$, $y = 2 - 4t$, $z = 5 + 3t$;   parametric form
22. $P_1 = (1, 0, 1)$; ℓ: $x = 1 + t$, $y = 2$, $z = -1 + 2t$; symmetric form
23. $P_1 = (2, 3, -5)$; ℓ: $\dfrac{x - 2}{2} = \dfrac{y - 3}{3} = \dfrac{z - 5}{5}$;   symmetric form
24. $P_1 = (2, 0, 0)$; ℓ: $x = 4$, $y = 2$;   vector form

In Problems 25 and 26, find the intersection of ℓ with each of the coordinate planes.

25. ℓ: $x = 2 - t$, $y = 6 + 2t$, $z = 5 - 2t$
26. ℓ: $x = 1 + 2t$, $y = 4 - t$, $z = 1 - 3t$

In Problems 27– 30, find the distance from the point Q to the line ℓ.

27. $Q = (4, -1, -2)$; ℓ: $\overrightarrow{OP} = (5\mathbf{i} + 2\mathbf{j} + 3\mathbf{k}) + t(3\mathbf{i} + \mathbf{j} + 4\mathbf{k})$

28. $Q = (6, 8, 3)$; ℓ: $x = 5 + 6t$, $y = -1 - 3t$, $z = -7 - 4t$

29. $Q = (0, 0, 0)$; ℓ: $\dfrac{x - 1}{2} = \dfrac{y + 1}{-1} = \dfrac{z - 2}{2}$

30. $Q = (2, 1, -1)$; ℓ: $\dfrac{x - 5}{1} = \dfrac{y - 7}{2} = \dfrac{z - 8}{3}$

C   In Problems 31 and 32, ℓ_1 and ℓ_2 are parallel lines. Find the distance between ℓ_1 and ℓ_2.

31. ℓ_1: $\overrightarrow{OP} = (\mathbf{i} - 2\mathbf{j} + \mathbf{k}) + t(2\mathbf{i} - \mathbf{j} + \mathbf{k})$;
  ℓ_2: $\overrightarrow{OP} = (4\mathbf{i} - 5\mathbf{j} + 4\mathbf{k}) + t(2\mathbf{i} - \mathbf{j} + \mathbf{k})$

32. ℓ_1: $\dfrac{x - 6}{1} = \dfrac{y - 5}{1} = \dfrac{z - 5}{-3}$; ℓ_2: $\dfrac{x - 3}{1} = \dfrac{y + 1}{1} = \dfrac{z - 2}{-3}$

In Problems 33–38, determine whether ℓ_1 and ℓ_2 are parallel, skew, or intersecting lines.

33. ℓ_1: $x = 3 + 2t$, $y = 4 - 6t$, $z = -2 + 4t$;
  ℓ_2: $x = -11 - s$, $y = 14 + 3s$, $z = 15 - 2s$

34. ℓ_1: $x = 4 - t$, $y = 1 + t$, $z = -3t$;
  ℓ_2: $x = 2 + 2s$, $y = 6 - s$, $z = -3 + 4s$

35. ℓ_1: $x = 9 + 2t$, $y = 5 - t$, $z = 8 + 3t$;
  ℓ_2: $x = 11 - 2s$, $y = 4 + s$, $z = -7 + 3s$

36. ℓ_1: $\dfrac{x + 9}{3} = \dfrac{y - 4}{-2} = \dfrac{z - 5}{1}$; ℓ_2: $\dfrac{x}{-2} = \dfrac{y - 6}{4} = \dfrac{z - 1}{-3}$

37. ℓ_1: $\dfrac{x - 2}{1} = \dfrac{y - 1}{2}$, $z = 4$; ℓ_2: $\dfrac{x + 3}{2} = \dfrac{z - 1}{-3}$, $y = 3$

38. ℓ_1: $\dfrac{x - 2}{3} = \dfrac{y + 1}{4} = \dfrac{z - 3}{-1}$; ℓ_2: $\dfrac{x + 7}{-6} = \dfrac{y - 9}{-8} = \dfrac{z + 2}{2}$

In Problems 39–42, let ℓ be a line in the xy plane with equation $y = mx + b$.

39. Find parametric equations for ℓ.

40. Find a vector equation for ℓ.

41. Show that the distance from the origin to ℓ is

$$d = \frac{|b|}{\sqrt{1 + m^2}}$$

42. Show that the distance from (x_0, y_0) to ℓ is

$$d = \frac{|y_0 - mx_0 - b|}{\sqrt{1 + m^2}}$$

43. Show that $\overrightarrow{P_1P} \times \mathbf{v} = \mathbf{0}$ is a vector equation of a line that passes through P_1 with direction vector $\mathbf{v}$. The point P is an arbitrary point in R^3.

44. Use the result of Problem 43 to derive the symmetric form for the equation of a line. Assume $a, b, c \neq 0$ for $\mathbf{v} = a\mathbf{i} + b\mathbf{j} + c\mathbf{k}$.

|4-6| Planes

- Equation of a Plane
- Planes and Linear Systems
- Distance from a Point to a Plane
- Applications to Geometry

▪ Equation of a Plane

If $P_0 = (x_0, y_0, z_0)$ is a point in R^3 and ℓ is a line passing through P_0, then the set p of all lines passing through P_0 and perpendicular to ℓ is a **plane** (Figure 35).

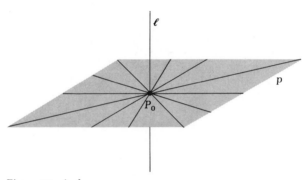

Figure 35   A plane

We want to find an equation that is satisfied by all the points on the plane p and only those points. To begin, any direction vector for the line ℓ, or equivalently, any nonzero vector perpendicular to the plane p, is called a **normal vector** for the plane (Figure 36).

Let $\mathbf{n} = a\mathbf{i} + b\mathbf{j} + c\mathbf{k}$ be a normal vector for the plane and let $P = (x, y, z)$ be any point in R^3. The point P is on the plane if and only if $\overrightarrow{P_0P}$ is orthogonal to $\mathbf{n}$ (see Figure 36); that is, if and only if

$$\mathbf{n} \cdot \overrightarrow{P_0P} = 0 \tag{1}$$

Equation (1) is a **vector equation of a plane** passing through P_0 with normal $\mathbf{n}$. Substituting

$$\mathbf{n} = a\mathbf{i} + b\mathbf{j} + c\mathbf{k}$$

and

$$\overrightarrow{P_0P} = (x - x_0)\mathbf{i} + (y - y_0)\mathbf{j} + (z - z_0)\mathbf{k}$$

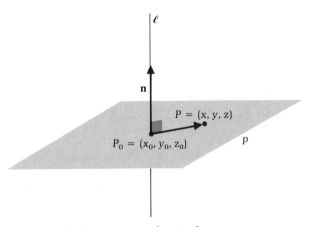

Figure 36 Vector **n** is a normal vector for p

into (1) and computing the dot product, we obtain the **point–normal form of the equation of a plane:**

$$(a\mathbf{i} + b\mathbf{j} + c\mathbf{k}) \cdot [(x - x_0)\mathbf{i} + (y - y_0)\mathbf{j} + (z - z_0)\mathbf{k}] = 0$$

$$a(x - x_0) + b(y - y_0) + c(z - z_0) = 0 \tag{2}$$

By collecting all the constant terms (terms not involving the variables x, y, and z) on the right side, equation (2) is written in the **general form**

$$ax + by + cz = d \tag{3}$$

The general form (3) is a linear equation in three variables like those studied in Chapter 1. It is useful to remember that the coefficients of the variables in (3) are the components of a **normal vector for the plane,**

$$\mathbf{n} = a\mathbf{i} + b\mathbf{j} + c\mathbf{k} \tag{4}$$

Example 41   Find the point–normal and general forms of an equation for the plane passing through the point (1, 2, 5) with normal vector $\mathbf{n} = 3\mathbf{i} + 2\mathbf{j} + \mathbf{k}$.

Solution   Substituting in (2), we have

$$3(x - 1) + 2(y - 2) + (z - 5) = 0 \qquad \text{Point–normal form}$$
$$3x - 3 + 2y - 4 + z - 5 = 0$$
$$3x + 2y + z = 12 \qquad \text{General form}$$

Problem 41   Find the point–normal and general forms of an equation for the plane passing through the point (2, 3, −1) with normal vector $\mathbf{n} = 2\mathbf{i} - \mathbf{j} + 4\mathbf{k}$. ▌▌

Example 42   Find an equation for the plane passing through the point (4, 3, 1) and perpendicular to the line ℓ with parametric equations

$$x = 5 + 2t$$
$$y = 4$$
$$z = 6 + 2t$$

Solution The vector $\mathbf{n} = 2\mathbf{i} + 2\mathbf{k}$ is a direction vector for ℓ. If p is the plane perpendicular to ℓ passing through $P_0 = (4, 3, 1)$, then $\mathbf{n}$ is also a normal vector for p. Using (2), the equation for p is

$$2(x - 4) + 0(y - 3) + 2(z - 1) = 0$$

or

$$2x + 2z = 10$$

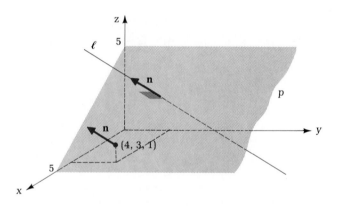

Problem 42 Find an equation for the plane passing through the point $(1, 2, -3)$ and perpendicular to the line with symmetric equations

$$\frac{x - 2}{3} = \frac{y + 1}{-2} = \frac{z - 4}{1}$$ ▌▌

Example 43 Find a vector equation for the line ℓ passing through the point $P_1 = (5, 4, 9)$ and perpendicular to the plane p with equation $2x + y + 4z = 8$.

Solution From (4), a normal vector for p is

$$\mathbf{n} = 2\mathbf{i} + \mathbf{j} + 4\mathbf{k}$$

Since ℓ is perpendicular to p, $\mathbf{n}$ is also a direction vector for ℓ. Thus, a vector equation for ℓ is

$$\overrightarrow{OP} = \overrightarrow{OP_1} + t\mathbf{n}$$
$$= (5\mathbf{i} + 4\mathbf{j} + 9\mathbf{k}) + t(2\mathbf{i} + \mathbf{j} + 4\mathbf{k})$$

Problem 43 Find a vector equation for the line passing through the point $(2, 3, -4)$ and perpendicular to the plane with equation $x - 2y + 5z = 7$. ▌▌

In the beginning of this section we defined a plane as a collection of lines perpendicular to a given line at a given point. Another characterization of a plane states that any three points that are noncollinear (that is, do not lie on the same line) determine a unique plane containing these points. If P_0, P_1, and P_2 are noncollinear points, then the vectors $\mathbf{u} = \overrightarrow{P_0P_1}$ and $\mathbf{v} = \overrightarrow{P_0P_2}$ lie in the plane determined by these points. Since the cross product of two vectors is perpendicular to each vector in the product, it follows that $\mathbf{n} = \mathbf{u} \times \mathbf{v}$ is a normal vector for this plane (Figure 37).

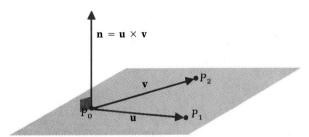

Figure 37 Normal vector for the plane through P_0, P_1, and P_2

Example 44 Find an equation for the plane passing through the points $P_0 = (1, 2, 1)$, $P_1 = (3, 5, 2)$, and $P_2 = (2, 4, 3)$.

Solution Let

$$\mathbf{u} = \overrightarrow{P_0P_1} = 2\mathbf{i} + 3\mathbf{j} + \mathbf{k}$$

and

$$\mathbf{v} = \overrightarrow{P_0P_2} = \mathbf{i} + 2\mathbf{j} + 2\mathbf{k}$$

If P_0, P_1, and P_2 were collinear, then $\mathbf{u}$ would be a scalar multiple of $\mathbf{v}$. Since this is not the case, these points must be noncollinear and determine a plane p. A normal vector for p is

$$\mathbf{n} = \mathbf{u} \times \mathbf{v} = \begin{vmatrix} \mathbf{i} & \mathbf{j} & \mathbf{k} \\ 2 & 3 & 1 \\ 1 & 2 & 2 \end{vmatrix} = 4\mathbf{i} - 3\mathbf{j} + \mathbf{k}$$

Using P_0 and $\mathbf{n}$, an equation for p is

$$4(x - 1) - 3(y - 2) + (z - 1) = 0$$

or

$$4x - 3y + z = -1$$

As a check you should verify that P_0, P_1, and P_2 satisfy this equation.

Problem 44   Find an equation for the plane passing through.the points $P_0 = (-1, 1, 4)$, $P_1 = (2, 3, 6)$, and $P_2 = (1, 1, 2)$.   **▮▮**

The important facts related to equations of a plane are summarized in the box for easy reference.

Equations of a Plane

1. If $P_0 = (x_0, y_0, z_0)$ is a point on a plane and $\mathbf{n} = a\mathbf{i} + b\mathbf{j} + c\mathbf{k}$ is a normal vector for the plane, then the **point–normal form** of the equation for the plane is

$$a(x - x_0) + b(y - y_0) + c(z - z_0) = 0$$

2. The linear equation in three variables

$$ax + by + cz = d$$

is the **general form** of the equation for a plane with normal vector

$$\mathbf{n} = a\mathbf{i} + b\mathbf{j} + c\mathbf{k}$$

3. If P_0, P_1, and P_2 are three noncollinear points on a plane, then a normal vector for the plane is

$$\mathbf{n} = \overrightarrow{P_0P_1} \times \overrightarrow{P_0P_2}$$

▪ Planes and Linear Systems

As we saw in Chapter 1, a system of linear equations in two variables can be represented geometrically as a collection of lines in R^2. Similarly, a system of linear equations in three variables can now be represented as a collection of planes in R^3. We can use the geometry of R^3 to help us understand the nature of solutions of such systems. For example, the geometric representation of a system of two linear equations in three variables is a pair of planes. If the planes are not parallel and do not coincide, then they must intersect in a line. The methods we used to solve linear systems in Chapter 1 will produce the parametric equations of the line of intersection.

Example 45   Find the line of intersection of the planes $x - y + 2z = 4$ and $2x - y + z = 3$.

Solution   The normal vectors for these planes are

$$\mathbf{i} - \mathbf{j} + 2\mathbf{k} \qquad \text{and} \qquad 2\mathbf{i} - \mathbf{j} + \mathbf{k}$$

Since these vectors are not parallel, the two planes are not parallel and must intersect in a line. The line of intersection is the set of points that satisfy both equations. Thus, we must solve the system of equations

$$x - y + 2z = 4$$
$$2x - y + z = 3$$

The augmented coefficient matrix for this system is

$$\begin{bmatrix} 1 & -1 & 2 & \Big| & 4 \\ 2 & -1 & 1 & \Big| & 3 \end{bmatrix}$$

Omitting the calculations, the reduced form of this matrix is

$$\begin{bmatrix} 1 & 0 & -1 & \Big| & -1 \\ 0 & 1 & -3 & \Big| & -5 \end{bmatrix}$$

The solution of this system is

$$x = -1 + t$$
$$y = -5 + 3t$$
$$z = t$$

These are the parametric equations for the line of intersection.

Check:

$$x - y + 2z = 4 \qquad\qquad\qquad 2x - y + z = 3$$
$$(-1 + t) - (-5 + 3t) + 2(t) \overset{?}{=} 4 \qquad 2(-1 + t) - (-5 + 3t) + t \overset{?}{=} 3$$
$$-1 + t + 5 - 3t + 2t \overset{?}{=} 4 \qquad -2 + 2t + 5 - 3t + t \overset{?}{=} 3$$
$$4 \overset{\checkmark}{=} 4 \qquad\qquad\qquad 3 \overset{\checkmark}{=} 3$$

Problem 45 Find the line of intersection of the planes $x + 2y + z = 4$ and $2x + 3y - z = 7$.

∎

Earlier, we proved that any system of linear equations must have no solution, a unique solution, or an infinite number of solutions. No other possibility exists. If a system has three variables, then each equation represents a plane, and we can illustrate the system's solution set geometrically. Figure 38 (page 264) illustrates some of the possibilities.

▪ Distance from a Point to a Plane

If p is a plane and Q is a point not on p, then the **distance from the point to the plane** is the length of the perpendicular line segment from Q to p, as shown in Figure 39(A).

If P is any point on the plane and **n** is a normal vector to the plane, then

$$d = \|\overrightarrow{PQ}\| \, |\cos \theta|$$

where θ is the angle between **n** and $\overrightarrow{PQ}$. [In Figure 39(B), θ is an acute angle and $\cos \theta > 0$. However, if $\theta > \pi/2$, then $\cos \theta < 0$. To ensure that the distance d is

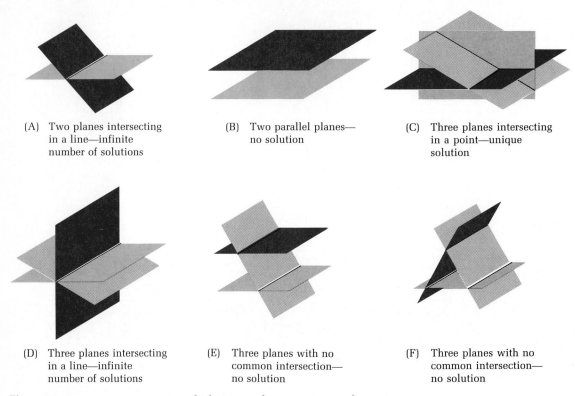

(A)   Two planes intersecting in a line—infinite number of solutions

(B)   Two parallel planes— no solution

(C)   Three planes intersecting in a point—unique solution

(D)   Three planes intersecting in a line—infinite number of solutions

(E)   Three planes with no common intersection— no solution

(F)   Three planes with no common intersection— no solution

Figure 38   Geometric representation of solution sets for some systems of equations

positive, we use the absolute value of cos θ.] From Theorem 8 in Section 4-3,

$$\cos \theta = \frac{\mathbf{n} \cdot \overrightarrow{PQ}}{\|\mathbf{n}\| \, \|\overrightarrow{PQ}\|}$$

Thus,

$$d = \|\overrightarrow{PQ}\| \, |\cos \theta| = \frac{|\mathbf{n} \cdot \overrightarrow{PQ}|}{\|\mathbf{n}\|}$$

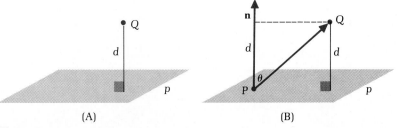

(A)

(B)

Figure 39   Distance from a point to a plane

Distance from a Point to a Plane

The distance from the point Q to the plane p is given by

$$d = \frac{|\mathbf{n} \cdot \overrightarrow{PQ}|}{\|\mathbf{n}\|}$$

where $\mathbf{n}$ is a normal vector for p and P is a point on p.

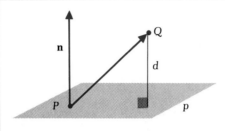

Example 46   Find the distance from the point $Q = (5, 4, 9)$ to the plane $2x + y + 4z = 8$.

Solution   A normal vector for the plane is $\mathbf{n} = 2\mathbf{i} + \mathbf{j} + 4\mathbf{k}$. Substituting $y = 0$ and $z = 0$ in the equation for the plane and solving for x shows that $P = (4, 0, 0)$ is a point on the plane. Thus,

$$\overrightarrow{PQ} = \mathbf{i} + 4\mathbf{j} + 9\mathbf{k}$$

and

$$d = \frac{|\mathbf{n} \cdot \overrightarrow{PQ}|}{\|\mathbf{n}\|}$$

$$= \frac{|(2\mathbf{i} + \mathbf{j} + 4\mathbf{k}) \cdot (\mathbf{i} + 4\mathbf{j} + 9\mathbf{k})|}{\|2\mathbf{i} + \mathbf{j} + 4\mathbf{k}\|}$$

$$= \frac{|2 + 4 + 36|}{\sqrt{21}} = \frac{42}{\sqrt{21}} = 2\sqrt{21}$$

Problem 46   Find the distance from the point $Q = (2, 1, 4)$ to the plane $x + 2y - z = 8$.   ▋▌

▪ Applications to Geometry

The following examples show how the material in this and the preceding section can be used to solve some typical geometry problems.

Example 47   Find an equation for the line ℓ through the point $P_0 = (3, 1, 4)$ and perpendicular to and intersecting the line

$$\ell_1: \quad x = 2 + t$$
$$y = 3 - 2t$$
$$z = 5 + 2t$$

Solution   We begin by sketching the given point and line, as shown in Figure (A). We let P_1 be the point where ℓ (the line we are trying to find) intersects ℓ_1 (the given line), as shown in Figure (B). If we can find the coordinates of P_1, then we will have

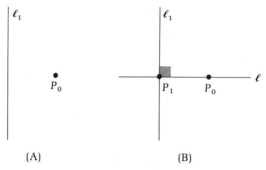

(A)   (B)

two points on ℓ (P_0 and P_1) and we can easily find the equation of ℓ. But how can we find P_1? Recall that we defined a plane as a collection of lines passing through a given point and perpendicular to a given line. Let p be the plane passing through the given point P_0 and perpendicular to the given line ℓ_1, as shown in Figure (C).

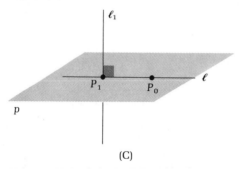

(C)

The line ℓ lies in this plane, and P_1 is the point of intersection of ℓ_1 and p. Thus, this problem can be solved as follows:

Step 1.   Find the equation of the plane p passing through P_0 and perpendicular to ℓ_1.

Step 2.   Find P_1, the point of intersection of ℓ_1 and p.

Step 3.   Find ℓ, the line through P_0 and P_1.

Step 1.   *Find the equation of the plane p passing through P_0 and perpendicular to ℓ_1.* From the parametric equations for ℓ_1, we see that $\mathbf{n}_1 = \mathbf{i} - 2\mathbf{j} + 2\mathbf{k}$ is a direction vector for ℓ_1. Since ℓ_1 is perpendicular to p, $\mathbf{n}_1$ is also a normal vector for p. Since $P_0 = (3, 1, 4)$ is a point on p, the equation of p is

$$(x - 3) - 2(y - 1) + 2(z - 4) = 0$$

or

$$x - 2y + 2z = 9$$

Step 2. *Find P_1, the point of intersection of ℓ_1 and p. To find P_1, we substitute the parametric equations for ℓ_1 into the equation for p and solve for t:*

$$x - 2y \qquad + 2z \qquad = 9$$
$$(2 + t) - 2(3 - 2t) + 2(5 + 2t) = 9$$
$$2 + t - 6 + 4t \quad + 10 + 4t \; = 9$$
$$9t = 3$$
$$t = \tfrac{1}{3}$$

The coordinates of P_1 can be found by substituting $t = \tfrac{1}{3}$ in the parametric equations for ℓ_1:

$$x = 2 + \tfrac{1}{3} = \tfrac{7}{3}$$
$$y = 3 - \tfrac{2}{3} = \tfrac{7}{3}$$
$$z = 5 + \tfrac{2}{3} = \tfrac{17}{3}$$

Thus, $P_1 = (\tfrac{7}{3}, \tfrac{7}{3}, \tfrac{17}{3})$.

Step 3. *Find ℓ, the line through P_0 and P_1. Since ℓ passes through P_0 and P_1, a direction vector for ℓ is*

$$\overrightarrow{P_0P_1} = (\tfrac{7}{3} - 3)\mathbf{i} + (\tfrac{7}{3} - 1)\mathbf{j} + (\tfrac{17}{3} - 4)\mathbf{k}$$
$$= -\tfrac{2}{3}\mathbf{i} + \tfrac{4}{3}\mathbf{j} + \tfrac{5}{3}\mathbf{k}$$

Finally, using the point P_0 and the direction vector $\overrightarrow{P_0P_1}$, the parametric equations for ℓ are

$$x = 3 - \tfrac{2}{3}s \qquad P_0 = (3, 1, 4)$$
$$y = 1 + \tfrac{4}{3}s \qquad \overrightarrow{P_0P_1} = -\tfrac{2}{3}\mathbf{i} + \tfrac{4}{3}\mathbf{j} + \tfrac{5}{3}\mathbf{k}$$
$$z = 4 + \tfrac{5}{3}s$$

Problem 47 Find an equation for the line ℓ through the point $P_0 = (3, 3, 5)$ and perpendicular to and intersecting the line

$$\ell_1: \quad x = \quad 1 + t$$
$$y = \quad 2 - t$$
$$z = -2 + 3t$$

∎

Theorem 14 will be useful in the work that follows. The proof is omitted.

Theorem 14

Relationship Between a Line and a Plane

Let ℓ be a line with direction vector **v** and let p be a plane with normal vector **n**.

(A) ℓ is parallel to (or lies in) p if and only if $\mathbf{n} \cdot \mathbf{v} = 0$.

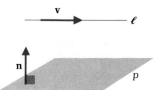

(B) ℓ is perpendicular to p if and only if $\mathbf{n} = k\mathbf{v}$ for some nonzero scalar k.

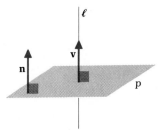

(C) ℓ intersects p at an oblique angle if and only if $\mathbf{n} \cdot \mathbf{v} \neq 0$ and $\mathbf{n} \neq k\mathbf{v}$ for any nonzero scalar k.

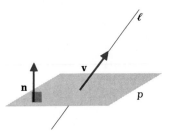

Example 48   If ℓ is the line with symmetric equations

$$\frac{x-1}{1} = \frac{y+2}{-3} = \frac{z-2}{2}$$

and p is the plane with equation

$$2x - y + z = 5$$

find an equation for the plane that contains ℓ and is perpendicular to p.

Solution Examining the equations for ℓ and p, we have

$$\mathbf{v} = \mathbf{i} - 3\mathbf{j} + 2\mathbf{k} \qquad \text{Direction vector for } \ell$$

and

$$\mathbf{n} = 2\mathbf{i} - \mathbf{j} + \mathbf{k} \qquad \text{Normal vector for } p$$

Since $\mathbf{n} \cdot \mathbf{v} = 7 \neq 0$ and $\mathbf{n} \neq k\mathbf{v}$, Theorem 14(C) tells us that ℓ intersects p at an oblique angle. If p_1 is the plane containing ℓ and perpendicular to p, then we must find a normal vector for p_1. Any vector perpendicular to p_1 would have to be perpendicular to $\mathbf{n}$, since p and p_1 are perpendicular, and to $\mathbf{v}$, since ℓ lies in p_1 (see the figure).

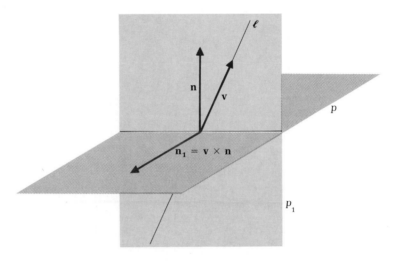

Recall that the cross product of two vectors is a vector perpendicular to both vectors in the product. Thus, a normal vector for p_1 is

$$\mathbf{n}_1 = \mathbf{v} \times \mathbf{n} = \begin{vmatrix} \mathbf{i} & \mathbf{j} & \mathbf{k} \\ 1 & -3 & 2 \\ 2 & -1 & 1 \end{vmatrix}$$

$$= \begin{vmatrix} -3 & 2 \\ -1 & 1 \end{vmatrix} \mathbf{i} - \begin{vmatrix} 1 & 2 \\ 2 & 1 \end{vmatrix} \mathbf{j} + \begin{vmatrix} 1 & -3 \\ 2 & -1 \end{vmatrix} \mathbf{k}$$

$$= -\mathbf{i} + 3\mathbf{j} + 5\mathbf{k}$$

From the given equations for ℓ, the point $P_0 = (1, -2, 2)$ is on ℓ, so it must be a point in p_1. Using $\mathbf{n}_1$ and P_0, the equation for p_1 is

$$-(x - 1) + 3(y + 2) + 5(z - 2) = 0 \qquad P_0 = (1, -2, 2)$$

or

$$-x + 3y + 5z = 3 \qquad \mathbf{n}_1 = -\mathbf{i} + 3\mathbf{j} + 5\mathbf{k}$$

Problem 48   If ℓ is the line with symmetric equations

$$\frac{x+1}{3} = \frac{y-4}{-2} = \frac{z-1}{2}$$

and p is the plane with equation

$$2x + y - 4z = 6$$

find an equation for the plane containing ℓ and perpendicular to p.   ▌▌

Referring to Example 48, what would happen if ℓ were perpendicular to p? In this case, any plane containing ℓ would also be perpendicular to p (see Figure 40). Thus, the problem would have an infinite number of solutions. This is the reason that the first thing we did in the solution of Example 48 was to determine the relationship between p and ℓ.

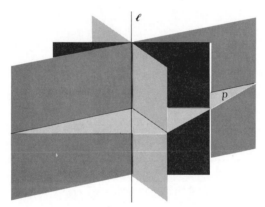

Figure 40   Any plane containing ℓ is perpendicular to p if ℓ is perpendicular to p

Answers to Matched Problems

41. Point–normal form: $2(x-2) - (y-3) + 4(z+1) = 0$; general form: $2x - y + 4z = -3$

42. $3(x-1) - 2(y-2) + (z+3) = 0$ **43.** $\mathbf{x} = (2\mathbf{i} + 3\mathbf{j} - 4\mathbf{k}) + t(\mathbf{i} - 2\mathbf{j} + 5\mathbf{k})$

44. $4x - 10y + 4z = 2$ **45.** $x = 2 + 5t,\ y = 1 - 3t,\ z = t$

46. $8/\sqrt{6}$ **47.** $x = 3,\ y = 3 - 3s,\ z = 5 - s$ **48.** $6x + 16y + 7z = 65$

▌▌ Exercise 4-6

A   *In Problems 1–4, find a normal vector for the given plane.*

1. $2x - 3y + 5z = 7$ **2.** $-x + 7y - z = 15$

3. $2x - 5z = 10$ **4.** $y = 11$

In Problems 5–8, find the point–normal and general forms of an equation for the plane passing through P_0 with normal vector $\mathbf{n}$.

5. $P_0 = (2, 1, 3);$ $\mathbf{n} = \mathbf{i} + 2\mathbf{j} + 4\mathbf{k}$ **6.** $P_0 = (-1, 2, 0);$ $\mathbf{n} = -2\mathbf{i} + 3\mathbf{j} - 5\mathbf{k}$

7. $P_0 = (0, 0, 0);$ $\mathbf{n} = 3\mathbf{i} + 4\mathbf{j} - 7\mathbf{k}$ **8.** $P_0 = (4, -6, 7);$ $\mathbf{n} = \mathbf{i} + \mathbf{j}$

In Problems 9–12, find an equation for the plane passing through P_0 and perpendicular to ℓ.

9. $P_0 = (-2, 1, -3);$ $\ell:$ $\dfrac{x-1}{2} = \dfrac{y+5}{-1} = \dfrac{z-2}{4}$

10. $P_0 = (5, -5, 15);$ $\ell:$ $\dfrac{x-10}{5} = \dfrac{y+20}{2} = \dfrac{z-5}{-1}$

11. $P_0 = (0, 0, 0);$ $\ell:$ $\dfrac{x}{2} = \dfrac{y}{3} = \dfrac{z}{6}$

12. $P_0 = (0, 1, 0);$ $\ell:$ $\dfrac{x+5}{2} = \dfrac{z-10}{-3},$ $y = 11$

B   In Problems 13–16, find an equation for the plane passing through $P_0, P_1,$ and P_2.

13. $P_0 = (2, 4, 4);$ $P_1 = (4, 2, 2);$ $P_2 = (2, 6, 3)$

14. $P_0 = (1, 2, 1);$ $P_1 = (2, 1, 1);$ $P_2 = (3, 2, 0)$

15. $P_0 = (1, -2, 2);$ $P_1 = (-2, -3, 4);$ $P_2 = (4, -2, -4)$

16. $P_0 = (4, 2, 1);$ $P_1 = (4, 4, -2);$ $P_2 = (4, 3, 7)$

In Problems 17 and 18, find a vector equation for the line passing through P_0 and perpendicular to p.

17. $P_0 = (1, -1, 1);$ p: $2x + 3y - 6z = 18$

18. $P_0 = (0, 2, -3);$ p: $x + 3z = 11$

In Problems 19 and 20, find parametric equations for the line of intersection of p_1 and p_2.

19. $p_1:$ $x - 2y + z = 4;$ $p_2:$ $2x - 3y - 2z = 11$

20. $p_1:$ $2x + y - z = 8;$ $p_2:$ $3x + y - 3z = 13$

In Problems 21 and 22, find the points of intersection of the given plane with each of the coordinate axes.

21. $2x - 3y + 4z = 24$ **22.** $x + 2y + 5z = 15$

In Problems 23–26, determine whether ℓ is parallel to p, lies in p, is perpendicular to p, or intersects p at an oblique angle. If ℓ intersects p (perpendicularly or obliquely), find the point of intersection.

23. p: $3x - 4y + 2z = 15;$ $\ell:$ $\dfrac{x-4}{2} = \dfrac{y+3}{1} = \dfrac{z-7}{-1}$

24. p: $x + 2y - z = 11;$ $\ell:$ $\dfrac{x-1}{2} = \dfrac{y+3}{4} = \dfrac{z-2}{-2}$

25. p: $-x + 2y + 3z = 5$; ℓ: $\dfrac{x-1}{1} = \dfrac{y-3}{2} = \dfrac{z-2}{3}$

26. p: $2x - y - 3z = 8$; ℓ: $\dfrac{x-1}{5} = \dfrac{y+3}{1} = \dfrac{z+1}{3}$

In Problems 27 and 28, find the distance from the origin to the given plane.

27. $2(x-1) + 2(y+2) - (z-3) = 0$
28. $11(x-1) - 10(y-5) + 2(z+3) = 0$

In Problems 29 and 30, find the distance from the point Q to the plane p.

29. $Q = (4, 2, 6)$;   p: $5x - 2y + z = 2$
30. $Q = (4, -8, -4)$;   p: $x - y - 4z = 4$

The angle between two planes is defined to be the angle between their normal vectors. In Problems 31–36, find the angle between each pair of planes and determine whether the planes are parallel, perpendicular, identical, or none of these.

31. $5x + 4y + 7z = 10$,
$\quad\;\; 5x - 2y + 4z = \;\; 7$

32. $2x - \;\; y + 4z = \;\;\; 9$,
$\quad\;\; 7x + 2y - 3z = -4$

33. $\quad 3x - \;\; y + 2z = \;\;\; 3$,
$\quad -9x + 3y - 6z = -9$

34. $3x - 8y - 5z = \;\;\; 2$,
$\quad\;\; x + 9y - 4z = -1$

35. $-x + 3y + 5z = \;\;\; 6$,
$\quad\;\; x - 3y + 2z = 11$

36. $\quad 2x - \;\; y + \;\; z = 10$,
$\quad -4x + 2y - 2z = \;\; 5$

C   **37.** Find an equation for the plane passing through the point $(1, -2, 4)$ and parallel to the plane with equation $2x - 4y + 5z = 17$.

38. Find an equation for the plane that contains the line

$$\frac{x+1}{-2} = \frac{y-3}{1} = \frac{z+2}{4}$$

and is perpendicular to the plane

$$2x + y - z = 27$$

39. Find a symmetric form for the equation of the line passing through the point $(4, 5, 8)$ and perpendicular to and intersecting the line

$$\frac{x+7}{3} = \frac{y-5}{-2} = \frac{z+1}{1}$$

40. Find an equation for the plane through the line

$$\frac{x-2}{2} = \frac{y+1}{-1} = \frac{z-4}{3}$$

and parallel to the line

$$\frac{x+1}{-2} = \frac{y-4}{4} = \frac{z-1}{1}$$

41. Find an equation for the plane through the point $(2, 1, -1)$ and containing the line with parametric equations

$$x = \quad 4 + 3t$$
$$y = \quad 2 + \quad t$$
$$z = -1 + \quad t$$

42. Find an equation for the plane containing the parallel lines

$$\ell_1 : \quad \frac{x-2}{1} = \frac{y+3}{-1} = \frac{z-2}{2} \quad \text{and} \quad \ell_2 : \quad \frac{x+1}{1} = \frac{y-4}{-1} = \frac{z+5}{2}$$

43. Given the skew lines

ℓ_1: line through $P_1 = (0, 0, 1)$ and $P_2 = (-1, 2, -1)$
ℓ_2: line through $P_3 = (1, 2, 1)$ and $P_4 = (0, -1, -1)$

find equations of two parallel planes p_1 and p_2 such that p_1 contains ℓ_1 and p_2 contains ℓ_2.

44. Given the skew lines

ℓ_1: $x = 2t, \quad y = 2 - t, \quad z = 5 + 3t$
ℓ_2: $x = 1 + t, \quad y = 1 - t, \quad z = 2t$

find equations of two parallel planes p_1 and p_2 such that p_1 contains ℓ_1 and p_2 contains ℓ_2.

45. Find the distance between the skew lines ℓ_1 and ℓ_2 given by

ℓ_1: line through $P_1 = (0, 0, 1)$ and $P_2 = (-1, 2, -1)$
ℓ_2: line through $P_3 = (1, 2, 1)$ and $P_4 = (0, -1, -1)$

[*Hint:* See Problem 43.]

46. Find the distance between the skew lines ℓ_1 and ℓ_2 given by

ℓ_1: $x = 2t, \quad y = 2 - t, \quad z = 5 + 3t$
ℓ_2: $x = 1 + t, \quad y = 1 - t, \quad z = 2t$

[*Hint:* See Problem 44.]

47. Show that the distance from the point (x_1, y_1, z_1) to the plane with equation $ax + by + cz = D$ is given by

$$d = \frac{|ax_1 + by_1 + cz_1 - D|}{\sqrt{a^2 + b^2 + c^2}}$$

48. Show that the distance between the parallel planes

$$ax + by + cz = D_1 \qquad \text{and} \qquad ax + by + cz = D_2$$

is given by

$$d = \frac{|D_1 - D_2|}{\sqrt{a^2 + b^2 + c^2}}$$

|4-7| Chapter Review

Important Terms and
Symbols

4-1. *Vectors in the plane.* Vector, directed line segment, initial point, terminal point, length, representative, position vector, standard position, algebraic definition of a vector, components, equality of vectors, magnitude, norm, length, zero vector, vector addition and subtraction, parallelogram law, scalar product, unit vector, standard unit vectors, algebraic properties of R^2, $\overrightarrow{AB}$, $|\overrightarrow{AB}|$, $\mathbf{v}$, $\|\mathbf{v}\|$, $\mathbf{i}$, $\mathbf{j}$

4-2. *Vectors in three dimensions.* Coordinate axes, coordinate planes, right-hand and left-hand coordinate systems, distance formula, components, equality of vectors, length, unit vector, vector addition and subtraction, scalar product, standard unit vectors, algebraic properties of R^3, $\mathbf{i}$, $\mathbf{j}$, $\mathbf{k}$

4-3. *The dot product.* Dot product, algebraic properties of the dot product, angle between two vectors, orthogonal (or perpendicular) vectors, parallel vectors, cosine of the angle between two vectors, direction angles, direction cosines, projection of $\mathbf{u}$ onto $\mathbf{v}$, component of $\mathbf{u}$ orthogonal to $\mathbf{v}$, orthogonal decomposition, $\mathbf{u} \cdot \mathbf{v}$

4-4. *The cross product.* Cross product of two vectors, symbolic determinant, area of a parallelogram, geometric properties of the cross product, algebraic properties of the cross product, scalar triple product, volume of a parallelepiped,

$$\mathbf{u} \times \mathbf{v} = \begin{vmatrix} \mathbf{i} & \mathbf{j} & \mathbf{k} \\ u_1 & u_2 & u_3 \\ v_1 & v_2 & v_3 \end{vmatrix}, \quad (\mathbf{u} \times \mathbf{v}) \cdot \mathbf{w} = \begin{vmatrix} u_1 & u_2 & u_3 \\ v_1 & v_2 & v_3 \\ w_1 & w_2 & w_3 \end{vmatrix}$$

4-5. *Lines in three dimensions.* Direction vector, vector equation, parametric equations, symmetric equations, parallel lines, intersecting lines, skew lines, distance from a point to a line

4-6. *Planes.* Plane, normal vector, vector equation of a plane, point–normal equation of a plane, general form, geometric representation of systems of equations in three variables, distance from a point to a plane, relationship between a line and a plane

▌ Exercise 4-7   Chapter Review

Work through all the problems in this chapter review and check your answers in the back of the book. (Answers to most review problems are there.) Where weaknesses show up, review appropriate sections in the text.

A   **1.** If **v** is the vector represented by $\overrightarrow{AB}$ where $A = (3, 6)$ and $B = (10, 8)$:

    (A) Find the length of **v**.
    (B) Find a representative of **v** with its initial point at the origin.
    (C) Graph both representatives of **v** on the same set of axes.

  2. Represent the vector with initial point $A = (-1, 5)$ and terminal point $B = (3, 1)$ as an ordered pair. Then represent this vector in terms of the standard unit vectors **i** and **j**.

  3. Find a representative of the vector $\mathbf{v} = (-4, 2)$ with its initial point at $A = (3, -5)$.

  4. If $\mathbf{u} = (6, 1)$ and $\mathbf{v} = (2, 4)$, find $\mathbf{u} + \mathbf{v}$ and $\mathbf{u} - \mathbf{v}$. Graph **u**, **v**, $\mathbf{u} + \mathbf{v}$, and $\mathbf{u} - \mathbf{v}$ on the same set of axes.

  5. Locate the points $P = (3, 6, 7)$ and $Q = (8, 16, 2)$ in a three-dimensional coordinate system. Find the distance between P and Q.

  6. If **v** is the vector represented by $\overrightarrow{AB}$ where $A = (3, 6, 4)$ and $B = (5, 10, 9)$:

    (A) Find the length of **v**.
    (B) Find a representative of **v** with its initial point at the origin.
    (C) Graph both representatives of **v** on the same set of axes.

  7. Find a representative of the vector $\mathbf{v} = (2, -1, 4)$ with its initial point at $A = (-4, 3, 6)$.

  8. Represent the vector with initial point $A = (5, -2, 3)$ and terminal point $B = (7, 1, -4)$ in terms of the standard unit vectors **i**, **j**, and **k**.

  9. If $\mathbf{u} = 8\mathbf{i} + 2\mathbf{j} + 7\mathbf{k}$ and $\mathbf{v} = 4\mathbf{i} + 10\mathbf{j} + 9\mathbf{k}$, find $\mathbf{u} + \mathbf{v}$ and $\mathbf{u} - \mathbf{v}$.

In Problems 10–13, find the indicated dot product.

  10. $(2, -4) \cdot (2, 1)$ **11.** $(3\mathbf{i} + \mathbf{j}) \cdot (3\mathbf{i} + \mathbf{j})$

  12. $(1, -2, 1) \cdot (-1, 2, -1)$ **13.** $(2\mathbf{i} + \mathbf{j} - \mathbf{k}) \cdot (3\mathbf{i} + 2\mathbf{j} + 7\mathbf{k})$

In Problems 14–17, find the indicated cross product.

  14. $(2, 1, -3) \times (4, -2, 5)$ **15.** $(1, 0, 1) \times (1, 1, 0)$

  16. $(2\mathbf{i} + \mathbf{j} - 3\mathbf{k}) \times (-4\mathbf{i} - 2\mathbf{j} + 6\mathbf{k})$ **17.** $(5\mathbf{i} + 3\mathbf{j}) \times (\mathbf{i} + 2\mathbf{j})$

In Problems 18–20, find a point on the line and a direction vector for the line.

  18. $\overrightarrow{OP} = (2\mathbf{i} - \mathbf{j} + 3\mathbf{k}) + t(3\mathbf{i} - 4\mathbf{j} + 7\mathbf{k})$

  19. $x = -4 + 2t, \quad y = 6 - 5t, \quad z = 2 + 3t$

  20. $\dfrac{x + 3}{4} = \dfrac{y - 2}{5} = \dfrac{z - 1}{-6}$

21. If ℓ is the line through the point $P_1 = (2, 5, 7)$ with direction vector $\mathbf{v} = 3\mathbf{i} + 6\mathbf{j} + 4\mathbf{k}$, find:

    (A) A vector equation for ℓ
    (B) Parametric equations for ℓ
    (C) Symmetric equations for ℓ

22. Find a normal vector for the plane with equation $2x - 3y + 4z = 10$.

23. Find an equation for the plane passing through the point $P_0 = (2, 5, 6)$ with normal vector $\mathbf{n} = 3\mathbf{i} + 4\mathbf{j} + \mathbf{k}$.

B   *In Problems 24 and 25, find a unit vector in the direction of* $\mathbf{v}$.

24. $\mathbf{v} = 2\mathbf{i} - 4\mathbf{j}$ **25.** $\mathbf{v} = (\frac{2}{5}, -\frac{2}{5}, -\frac{1}{5})$

In Problems 26 and 27, find a vector of length 10 that has the opposite direction as $\mathbf{v}$.

26. $\mathbf{v} = (3, -4)$ **27.** $\mathbf{v} = 9\mathbf{i} - 6\mathbf{j} - 2\mathbf{k}$

In Problems 28–31, find the cosine of the angle between $\mathbf{u}$ *and* $\mathbf{v}$. *Then find the angle itself.*

28. $\mathbf{u} = 2\mathbf{i} - \mathbf{j} + 4\mathbf{k}$; $\mathbf{v} = \mathbf{i} - 2\mathbf{j} - \mathbf{k}$
29. $\mathbf{u} = (-5, 3)$; $\mathbf{v} = (4, 1)$
30. $\mathbf{u} = \mathbf{i} - 2\mathbf{j}$; $\mathbf{v} = 2\mathbf{i} - 4\mathbf{j}$
31. $\mathbf{u} = (2, -1, 1)$; $\mathbf{v} = (3, -4, 5)$

In Problems 32 and 33, determine whether $\mathbf{u}$ *and* $\mathbf{v}$ *are orthogonal.*

32. $\mathbf{u} = (7, 1, -2)$; $\mathbf{v} = (1, -1, 3)$
33. $\mathbf{u} = \mathbf{i} + 2\mathbf{j} - 3\mathbf{k}$; $\mathbf{v} = 2\mathbf{i} + \mathbf{j} + \mathbf{k}$
34. Find the direction cosines for $\mathbf{u} = (12, -4, 3)$.

In Problems 35 and 36, find the value(s) of t that will make $\mathbf{u}$ *and* $\mathbf{v}$ *orthogonal.*

35. $\mathbf{u} = 3\mathbf{i} + 2\mathbf{j}$; $\mathbf{v} = 4\mathbf{i} + t\mathbf{j}$
36. $\mathbf{u} = (t, 3t, -1)$; $\mathbf{v} = (t, -1, -2)$
37. Let $\mathbf{u} = (5, 5)$ and $\mathbf{v} = (9, 3)$. Find $\mathbf{p}$, the projection of $\mathbf{u}$ onto $\mathbf{v}$, and $\mathbf{q}$, the component of $\mathbf{u}$ orthogonal to $\mathbf{v}$. Graph $\mathbf{u}, \mathbf{v}, \mathbf{p}$, and $\mathbf{q}$ on the same set of axes.
38. Let $\mathbf{u} = \mathbf{i} - 2\mathbf{j} + 3\mathbf{k}$, $\mathbf{v} = 2\mathbf{i} + \mathbf{j} - \mathbf{k}$, and $\mathbf{w} = -\mathbf{i} + 2\mathbf{j} + 4\mathbf{k}$. Find:

    (A) $\mathbf{u} \times \mathbf{v}$ (B) $\mathbf{v} \times \mathbf{w}$ (C) $(\mathbf{u} \times \mathbf{v}) \times \mathbf{w}$
    (D) $\mathbf{u} \times (\mathbf{v} \times \mathbf{w})$ (E) $(\mathbf{u} \times \mathbf{v}) \cdot \mathbf{w}$ (F) $\mathbf{u} \cdot (\mathbf{v} \times \mathbf{w})$

39. Find a vector that is orthogonal to $\mathbf{u} = (1, 2, -1)$ and $\mathbf{v} = (-1, 0, 1)$.

40. Verify that the points $A = (0, 0, 0)$, $B = (2, 2, 1)$, $C = (6, 3, 2)$, and $D = (8, 5, 3)$ form the vertices of a parallelogram, and find the area of the parallelogram.

41. Find the volume of the parallelepiped determined by $\mathbf{u} = 6\mathbf{i} - 2\mathbf{j} + \mathbf{k}$, $\mathbf{v} = 2\mathbf{i} + 5\mathbf{j} + 2\mathbf{k}$, and $\mathbf{w} = \mathbf{i} + 2\mathbf{j} + 7\mathbf{k}$.

42. Find parametric equations for the line through the points $(-2, 1, 4)$ and $(3, -2, 6)$.

43. Find symmetric equations for the line through the point $(1, -2, 0)$ and parallel to the line

$$\frac{x+4}{7} = \frac{y-5}{-5} = \frac{z-6}{8}$$

44. Find an equation for the plane passing through the points $(1, -2, 0)$, $(2, 3, -2)$, and $(4, 1, -5)$.

45. Find a vector equation for the line passing through the point $(2, -1, 4)$ and perpendicular to the plane $3x + y - z = 10$.

46. Find parametric equations for the line of intersection of the planes $2x + y - z = 4$ and $x + 3y + 2z = 7$.

47. Find the point of intersection of the line

$$\frac{x-1}{2} = \frac{y+1}{-1} = \frac{z}{4}$$

and the plane $2x + 3y + z = 9$.

48. Find the distance from the point $(2, 1, 3)$ to the line $\overrightarrow{OP} = (2\mathbf{i} - 3\mathbf{j} - 5\mathbf{k}) + t(\mathbf{i} + \mathbf{j} + \mathbf{k})$.

49. Find the distance from the origin to the plane $2x + 5y + 14z = 30$.

C **50.** Find two unit vectors orthogonal to $\mathbf{u} = (3, -4)$.

51. Find two unit vectors orthogonal to both $\mathbf{u} = (12, -8, -9)$ and $\mathbf{v} = (-5, 4, 3)$.

52. Show that $(u_1\mathbf{i} + u_2\mathbf{j}) \times (v_1\mathbf{i} + v_2\mathbf{j}) = (u_1v_2 - u_2v_1)\mathbf{k}$.

In Problems 53–56, show that each statement is valid for all vectors $\mathbf{u}$ and $\mathbf{v}$ in R^3.

53. $\|\mathbf{u} + \mathbf{v}\|^2 = \|\mathbf{u}\|^2 + 2\mathbf{u} \cdot \mathbf{v} + \|\mathbf{v}\|^2$

54. $\|\mathbf{u} + \mathbf{v}\|^2 = \|\mathbf{u}\|^2 + \|\mathbf{v}\|^2$ if and only if $\mathbf{u}$ and $\mathbf{v}$ are orthogonal.

55. $(\mathbf{u} + \mathbf{v}) \cdot (\mathbf{u} - \mathbf{v}) = \mathbf{u} \cdot \mathbf{u} - \mathbf{v} \cdot \mathbf{v}$

56. $(\mathbf{u} + \mathbf{v}) \times (\mathbf{u} - \mathbf{v}) = 2(\mathbf{v} \times \mathbf{u})$

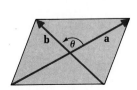

57. If $\mathbf{a}$ and $\mathbf{b}$ are the diagonals of a parallelogram and θ is the angle between them (see the figure), show that the area is given by $A = \frac{1}{2}\|\mathbf{a}\| \|\mathbf{b}\| \sin \theta$.

58. Determine whether ℓ_1 and ℓ_2 are parallel, intersecting, or skew lines.

(A) ℓ_1: $x = 2 + t$, $y = 3 - t$, $z = 4 - 2t$;

   ℓ_2: $x = 1 + s$, $y = 1 - 2s$, $z = 2 + 3s$

(B) ℓ_1: $\dfrac{x-2}{3} = \dfrac{y-4}{-1} = \dfrac{z-5}{1}$; ℓ_2: $\dfrac{x-7}{-6} = \dfrac{y-5}{2} = \dfrac{z-11}{-2}$

(C) ℓ_1: $\overrightarrow{OP} = (3\mathbf{i} + 4\mathbf{j} + 5\mathbf{k}) + t(-\mathbf{i} + \mathbf{j} + 2\mathbf{k})$;

   ℓ_2: $\overrightarrow{OP} = (2\mathbf{i} + \mathbf{j} + 2\mathbf{k}) + s(3\mathbf{i} + \mathbf{j} - \mathbf{k})$

59. Find symmetric equations for the line passing through the point (x_0, y_0, z_0) and perpendicular to the yz plane.

60. Find parametric equations for the line passing through the point (x_0, y_0, z_0) and perpendicular to the xz plane.

61. Find a vector equation for the line passing through the point (x_0, y_0, z_0) and perpendicular to the xy plane.

62. Find an equation for the plane passing through the point (x_0, y_0, z_0) and:

    (A) Parallel to the xy plane

    (B) Parallel to the xz plane

    (C) Parallel to the yz plane

63. Find an equation for the plane containing the line

$$\frac{x-1}{4} = \frac{y+6}{2} = \frac{z-3}{3}$$

and perpendicular to the plane $3x + 5y + z = 23$.

64. Find a symmetric form for the equation of the line through the point $(-4, 5, 3)$ and intersecting and perpendicular to the line

$$\frac{x-10}{2} = \frac{y-8}{3} = \frac{z+2}{-1}$$

65. Introduce parameters for y and z to show that the equation of the plane $ax + by + cz = d$, $a \neq 0$, can be written as

$$(x, y, z) = \left(\frac{d}{a}, 0, 0\right) + t\left(-\frac{b}{a}, 1, 0\right) + s\left(-\frac{c}{a}, 0, 1\right)$$

66. If ℓ is the line passing through the point P_1 with direction vector $\mathbf{v}$ and P is any point in R^3, show that P is on ℓ if and only if $\overrightarrow{P_1P} \times \mathbf{v} = \mathbf{0}$.

|5| Vector Spaces

| 5 | Contents

Historically, vectors were first considered to be quantities possessing a magnitude and a direction, and they were used to describe forces acting on objects in the plane or in space. As was emphasized in Chapter 4, the interplay between the geometric definition of a vector (a directed line segment) and the algebraic definition (an ordered pair or ordered triple of real numbers) played an important role in developing and interpreting properties of vectors. Subsequently, mathematicians and scientists began to realize that two or three coordinates were not enough to describe certain mathematical and physical phenomena. For example, the temperature at a point in a room depends not only on location, but also on time; hence, the need for the "space–time" coordinates (x, y, z, t). Thus, the concept of a vector was generalized.

The generalization is best described as a two-stage process. First, it is possible to extend the algebraic definition of vectors to "points" with four, five, or more coordinates and still retain the algebraic properties of two- and three-dimensional vectors. Thus, vectors can be generalized to include ordered n-tuples of real numbers $(a_1, a_2, \ldots, a_n)$. These n-tuples represent "points" in an n-dimensional space, but it is not possible to visualize vectors with more than three coordinates. Nevertheless, by exploiting relationships between algebraic and geometric vectors in two and three dimensions, some geometric ideas, such as length and angle, can be extended to higher dimensions.

The second stage in this generalization process involves a much higher level of abstraction. It is possible to define operations on many different sets of mathematical objects (including certain types of functions and certain types of matrices) so that these objects exhibit the same algebraic structure as two- and three-dimensional vectors under the operations of addition and scalar multiplication. This observation leads to a formal definition of an "abstract" vector space. Ordered n-tuples then become just one of many examples of a vector space.

In this chapter we will formally introduce the concept of an abstract vector space and study its important properties. This is a powerful and efficient way to proceed, since all the properties of the abstract vector space will automatically apply to *any* particular example of the abstract space without further proof. In other words, once we learn about abstract vector spaces, we will automatically know about any particular example of a vector space (in terms of structure, properties, and operations).

You are going to be surprised at the number of different mathematical objects you have already studied that may be categorized as vector spaces; and you will encounter many more examples as you advance in this and future courses. For many of you this will be the most formal and abstract mathematics you have ever been asked to deal with. By proceeding carefully and slowly, and by including many illustrative examples, we will make these abstract concepts as understandable as the familiar two- and three-dimensional vectors considered in Chapter 4.

▌5-1▐ *n*-Dimensional Euclidean Space

- *n*-Space
- *n*-Dimensional Vectors
- Addition and Scalar Multiplication
- Dot Product and Norm
- Standard Unit Vectors in R^n
- Matrix and Vector Notation

We begin the generalization of two- and three-dimensional vectors by introducing n-dimensional vectors. All the definitions and theorems in this section are motivated by their two- and three-dimensional analogs presented in Chapter 4. Since we cannot draw graphs in four or more dimensions, our main emphasis will be on generalizing algebraic properties. However, many geometric concepts, such as point, origin, distance, line, plane, angle, and so on, will be used in higher-dimensional space even though we cannot draw the graphs. In addition, it is often useful to draw figures in two- and three-dimensional space as aids in "visualizing" relationships under discussion in a higher-dimensional space.

▪ n-Space

A point in three dimensions is specified by listing its coordinates in an ordered triple, such as $(1, -2, 4)$. A point in *n-dimensional space* or, more simply, *n-space* is specified in the same manner.

---

n-Space

If $x_1, x_2, \ldots, x_n$ are real numbers, then

$$P = (x_1, x_2, \ldots, x_n)$$

is called an **ordered *n*-tuple.** The collection of all ordered n-tuples for a fixed value of n is referred to as ***n*-space** and is denoted by

$$R^n = \{(x_1, x_2, \ldots, x_n) | x_1, x_2, \ldots, x_n \text{ real numbers}\}$$

Each ordered n-tuple is called a **point** in n-space, and x_i is called the ***i*th coordinate** of the point.

Notice that R^2 and R^3 are the familiar two- and three-dimensional spaces discussed in Chapter 4. If $n = 1$, then

$$R^1 = \{(x_1)|x_1 \text{ a real number}\}$$

Since this set has all the properties of the set of real numbers, it is customary to omit the superscript 1. Thus, R represents both the set of real numbers and 1-space. The context in which R is used will determine its meaning.

The point with all zero coordinates,

$$\overbrace{O = (0, 0, \ldots , 0)}^{n \text{ coordinates}}$$

is referred to as the **origin** of R^n.

Example 1   (A) $(2, -0.5, \pi, \sqrt{2}, 0)$ is a point in R^5.
(B) The point in R^4 whose ith coordinate is given by $x_i = 1/i$ is $(1, \frac{1}{2}, \frac{1}{3}, \frac{1}{4})$.
(C) $(0, 0, 0, 0, 0, 0)$ is the origin of R^6.

Problem 1   (A) Find the second coordinate of $(-1, 4, 3, 2)$.
(B) Find the point in R^5 whose ith coordinate is given by $x_i = i^2$.
(C) Find the origin of R^7.   ▐▐

▪ n-Dimensional Vectors

Recall that an ordered triple (x_1, x_2, x_3) can be considered to be a *point* in R^3 or a *vector* in standard position with initial point at the origin and terminal point at (x_1, x_2, x_3). Just as three-dimensional vectors are identified with points in R^3, we identify *n-dimensional vectors* with points in R^n.

_n_-Dimensional Vectors

A **vector v** in n-space is an ordered n-tuple. If $\mathbf{v} = (v_1, v_2, \ldots , v_n)$, then $v_1, v_2, \ldots , v_n$ are called the **components** of **v**.

Now we want to place some algebraic structure on R^n. Specifically, we want to extend the operations of addition, scalar multiplication, and dot product from R^2 and R^3 to R^n. We begin with the definition of equality.

Equality of Vectors in R^n

The vectors $\mathbf{u} = (u_1, u_2, \ldots , u_n)$ and $\mathbf{v} = (v_1, v_2, \ldots , v_n)$ are said to be **equal** if corresponding components are equal. That is, if

$$u_1 = v_1, \quad u_2 = v_2, \quad \ldots , \quad u_n = v_n$$

Example 2   Determine the values of x and y that will make **u** and **v** equal if
$\mathbf{u} = (x + 1, 2, y - 4, -5)$ and $\mathbf{v} = (4, 2, 0, -5)$.

Solution   Since the second and fourth components of these vectors are already equal, **u** and **v** will be equal if

$x + 1 = 4$ First components

$y - 4 = 0$ Third components

Thus, $x = 3$ and $y = 4$.

Problem 2   Determine the values of x and y that will make **u** and **v** equal if
$\mathbf{u} = (3 - x, 7, 9, y + 5)$ and $\mathbf{v} = (0, 7, 9, 11)$.   ▐▐

▪ Addition and Scalar Multiplication

Next, we define *addition* and *scalar multiplication* in R^n. Notice that these definitions are simple extensions of the definitions in R^2 and R^3.

Addition in R^n

If $\mathbf{u} = (u_1, u_2, \ldots, u_n)$ and $\mathbf{v} = (v_1, v_2, \ldots, v_n)$, then the **sum** of **u** and **v** is defined by

$$\mathbf{u} + \mathbf{v} = (u_1 + v_1, u_2 + v_2, \ldots, u_n + v_n)$$

The **zero vector** is defined to be

$$\mathbf{0} = (0, 0, \ldots, 0)$$

The **negative** of **u** is defined by

$$-\mathbf{u} = (-u_1, -u_2, \ldots, -u_n)$$

The **difference** $\mathbf{u} - \mathbf{v}$ is defined by

$$\mathbf{u} - \mathbf{v} = \mathbf{u} + (-\mathbf{v}) = (u_1 - v_1, u_2 - v_2, \ldots, u_n - v_n)$$

Example 3   (A) $(2, -1, 3, -4, 5) + (-1, 4, -2, 4, 3) = (1, 3, 1, 0, 8)$
(B) $(3, 0, -4, 1) + (0, 0, 0, 0) = (3, 0, -4, 1)$
(C) $-(\frac{1}{2}, \frac{2}{3}, \frac{3}{4}, \frac{4}{5}) = (-\frac{1}{2}, -\frac{2}{3}, -\frac{3}{4}, -\frac{4}{5})$
(D) $(1, 2, 4, 8) - (1, 2, 3, 4) = (0, 0, 1, 4)$
(E) $(1, -1, 2, -2, 3, -3) + (-1, 1, -2, 2, -3, 3) = (0, 0, 0, 0, 0, 0) = \mathbf{0}$

Problem 3   Perform the indicated operations:

(A) $(-1, 2, -3, 4, -5) + (2, 4, 6, 8, 10)$
(B) $-(2, -\frac{1}{2}, \frac{1}{4}, -\frac{1}{8}, \frac{1}{16})$
(C) $(5, 10, 15, 20) - (20, 15, 10, 5)$
(D) $(1, 2, 3, -3, -2, -1) + (-1, -2, -3, 3, 2, 1)$ ▐▐

Scalar Multiplication in R^n

If $\mathbf{u} = (u_1, u_2, \ldots, u_n)$ and k is a scalar (real number), then the **scalar product** of k and $\mathbf{u}$ is defined by

$$k\mathbf{u} = (ku_1, ku_2, \ldots, ku_n)$$

Example 4   (A) $2(1, 2, 3, 4, 5) = (2, 4, 6, 8, 10)$ (B) $-\frac{1}{2}(1, -2, 4, -8) = (-\frac{1}{2}, 1, -2, 4)$

Problem 4   Perform the indicated scalar multiplications:

(A) $-3(-1, 1, -2, 2)$ (B) $\frac{1}{3}(-9, 2, 0, -3, 5)$ ∎

Theorem 1 lists some important properties of addition and scalar multiplication in R^n. This list includes all the properties of vectors in R^2 and in R^3 listed in Theorem 3 in Section 4-1 and in Theorem 6 in Section 4-2, respectively. Furthermore, these properties will form the basis for the general definition of an abstract vector space in the next section. The proof of each statement follows directly from the definitions of addition and scalar multiplication and the properties of real numbers. We will prove Theorem 1(A), part (2), and leave the remaining parts as exercises.

Theorem 1

Algebraic Properties of R^n

(A) The following **addition properties** are satisfied for all vectors $\mathbf{u}$, $\mathbf{v}$, and $\mathbf{w}$ in R^n:

(1) R^n is closed under addition; that is, $\mathbf{u} + \mathbf{v}$ is in R^n	Closure property
(2) $\mathbf{u} + \mathbf{v} = \mathbf{v} + \mathbf{u}$	Commutative property
(3) $\mathbf{u} + (\mathbf{v} + \mathbf{w}) = (\mathbf{u} + \mathbf{v}) + \mathbf{w}$	Associative property
(4) $\mathbf{u} + \mathbf{0} = \mathbf{0} + \mathbf{u} = \mathbf{u}$	Additive identity
(5) $\mathbf{u} + (-\mathbf{u}) = (-\mathbf{u}) + \mathbf{u} = \mathbf{0}$	Additive inverse

(B) The following **scalar multiplication properties** are satisfied for all vectors $\mathbf{u}$ and $\mathbf{v}$ in R^n and all scalars k and ℓ:

(1) R^n is closed under scalar multiplication; that is, $k\mathbf{u}$ is in R^n	Closure property
(2) $k(\ell\mathbf{u}) = (k\ell)\mathbf{u}$	Associative property
(3) $k(\mathbf{u} + \mathbf{v}) = k\mathbf{u} + k\mathbf{v}$	Distributive property
(4) $(k + \ell)\mathbf{u} = k\mathbf{u} + \ell\mathbf{u}$	Distributive property
(5) $1\mathbf{u} = \mathbf{u}$	Multiplicative identity

Proof   To prove part (2) of Theorem 1(A), let $\mathbf{u} = (u_1, u_2, \ldots, u_n)$ and $\mathbf{v} = (v_1, v_2, \ldots, v_n)$. Then

$$\mathbf{u} + \mathbf{v} = (u_1 + v_1, u_2 + v_2, \ldots, u_n + v_n) \qquad \text{Definition of addition in } R^n$$

$$= (v_1 + u_1, v_2 + u_2, \ldots, v_n + u_n) \qquad \text{Commutative property of real numbers}$$

$$= \mathbf{v} + \mathbf{u} \qquad \text{Definition of addition in } R^n \; \blacksquare$$

▪ Dot Product and Norm

The definition of dot product in R^n is a direct generalization of the dot products in R^2 and R^3.

Dot Product in R^n

If $\mathbf{u} = (u_1, u_2, \ldots, u_n)$ and $\mathbf{v} = (v_1, v_2, \ldots, v_n)$, then the **dot product** of $\mathbf{u}$ and $\mathbf{v}$ is defined to be

$$\mathbf{u} \cdot \mathbf{v} = u_1 v_1 + u_2 v_2 + \cdots + u_n v_n$$

Example 5   (A) $(1, -2, 4, 6) \cdot (2, -1, 3, 2) = 1 \cdot 2 + (-2)(-1) + 4 \cdot 3 + 6 \cdot 2 = 28$

(B) $(1, 0, 2, 0, 3, 0) \cdot (0, 1, 0, 2, 0, 3) = 0$

Problem 5   Evaluate the indicated dot products:

(A) $(2, -1, 3, -2, 4) \cdot (1, 2, -1, 3, 5)$ (B) $(1, 2, 3, 4) \cdot (3, -1, 1, -1)$ $\blacksquare$

The properties of the dot product in R^n are listed in Theorem 2. Once again, these are the same as the properties of the dot product in R^2 and R^3 (Theorem 7 in Section 4-3). We will prove Theorem 2(A) and leave the rest as exercises.

Theorem 2

Properties of the Dot Product in R^n

If $\mathbf{u}$, $\mathbf{v}$, and $\mathbf{w}$ are vectors in R^n and k is a scalar, then

(A) $\mathbf{u} \cdot \mathbf{u} \geq 0$
(B) $\mathbf{u} \cdot \mathbf{u} = 0$ if and only if $\mathbf{u} = \mathbf{0}$
(C) $\mathbf{u} \cdot \mathbf{0} = 0$
(D) $\mathbf{u} \cdot \mathbf{v} = \mathbf{v} \cdot \mathbf{u}$
(E) $\mathbf{u} \cdot (\mathbf{v} + \mathbf{w}) = \mathbf{u} \cdot \mathbf{v} + \mathbf{u} \cdot \mathbf{w}$
(F) $(\mathbf{u} + \mathbf{v}) \cdot \mathbf{w} = \mathbf{u} \cdot \mathbf{w} + \mathbf{v} \cdot \mathbf{w}$
(G) $(k\mathbf{u}) \cdot \mathbf{v} = k(\mathbf{u} \cdot \mathbf{v})$
(H) $\mathbf{u} \cdot (k\mathbf{v}) = k(\mathbf{u} \cdot \mathbf{v})$

Proof   (A) Let $\mathbf{u} = (u_1, u_2, \ldots, u_n)$. Then

$$\mathbf{u} \cdot \mathbf{u} = u_1 u_1 + u_2 u_2 + \cdots + u_n u_n$$
$$= u_1^2 + u_2^2 + \cdots + u_n^2 \qquad x^2 \geq 0 \text{ for any real number x}$$
$$\geq 0 \qquad \blacksquare$$

The fact that $\mathbf{u} \cdot \mathbf{u}$ is always nonnegative allows us to use the dot product to define distance in R^n.

Norm and Distance in R^n

If $\mathbf{u} = (u_1, u_2, \ldots, u_n)$ and $\mathbf{v} = (v_1, v_2, \ldots, v_n)$ are vectors in R^n, then the **norm (length** or **magnitude)** of $\mathbf{u}$ is defined to be

$$\|\mathbf{u}\| = \sqrt{\mathbf{u} \cdot \mathbf{u}} = \sqrt{u_1^2 + u_2^2 + \cdots + u_n^2}$$

and the **distance** between $\mathbf{u}$ and $\mathbf{v}$ is defined to be

$$d(\mathbf{u}, \mathbf{v}) = \|\mathbf{u} - \mathbf{v}\| = \sqrt{(u_1 - v_1)^2 + (u_2 - v_2)^2 + \cdots + (u_n - v_n)^2}$$

If $\|\mathbf{u}\| = 1$, then $\mathbf{u}$ is called a **unit vector.** If $\mathbf{v}$ is any nonzero vector, then $\mathbf{u} = (1/\|\mathbf{v}\|)\mathbf{v}$ is a **unit vector in the direction of v.**

Example 6   Let $\mathbf{u} = (4, 3, -3, 1, 5)$ and $\mathbf{v} = (1, -3, 4, 7, -5)$. Find:

(A) $\|\mathbf{v}\|$ (B) $d(\mathbf{u}, \mathbf{v})$
(C) A unit vector in the direction of $\mathbf{v}$

Solution   (A) $\|\mathbf{v}\| = \sqrt{1^2 + (-3)^2 + 4^2 + 7^2 + (-5)^2} = \sqrt{100} = 10$
(B) $d(\mathbf{u}, \mathbf{v}) = \|\mathbf{u} - \mathbf{v}\|$

$$= \sqrt{(4-1)^2 + [3-(-3)]^2 + (-3-4)^2 + (1-7)^2 + [5-(-5)]^2}$$
$$= \sqrt{9 + 36 + 49 + 36 + 100} = \sqrt{230}$$

(C) $\dfrac{1}{\|\mathbf{v}\|} \mathbf{v} = \dfrac{1}{10}(1, -3, 4, 7, -5) = \left(\dfrac{1}{10}, -\dfrac{3}{10}, \dfrac{2}{5}, \dfrac{7}{10}, -\dfrac{1}{2}\right)$

Problem 6   Repeat Example 6 for $\mathbf{u} = (5, -3, 2, -4)$ and $\mathbf{v} = (2, 4, 1, -2)$.   ∎

The set of vectors R^n, with the operations of addition, scalar multiplication, and dot product that we have defined, is often called **Euclidean n-space,** and the distance formula

$$d(\mathbf{u}, \mathbf{v}) = \|\mathbf{u} - \mathbf{v}\| = \sqrt{(u_1 - v_1)^2 + (u_2 - v_2)^2 + \cdots + (u_n - v_n)^2}$$

is referred to as the **Euclidean distance** between $\mathbf{u}$ and $\mathbf{v}$. It is possible to define products other than the dot product which provide different distance formulas for R^n (see Problem 62 in Exercise 5-1). If R^n is equipped with one of these nonstandard distance formulas, it is then referred to as a **non-Euclidean space.** Whenever we refer to R^n, we will assume (unless otherwise noted) that we are working with Euclidean distance and the standard operations defined in this section.

▪ Standard Unit Vectors in R^n

The standard unit vectors **i**, **j**, and **k** provide a convenient way to represent vectors in R^3. Their n-dimensional analogs are defined in the box.

Standard Unit Vectors

The **standard unit vectors** in R^n are defined by

$$\mathbf{e}_1 = (1, 0, 0, \ldots, 0)$$
$$\mathbf{e}_2 = (0, 1, 0, \ldots, 0)$$

$$\vdots$$

$$\mathbf{e}_n = (0, 0, 0, \ldots, 1)$$

Theorem 3 lists some of the properties of the standard unit vectors in R^n. These are comparable to the familiar properties of **i**, **j**, and **k**. The proof is left as an exercise.

Theorem 3

Properties of the Standard Unit Vectors

If $\mathbf{e}_1, \mathbf{e}_2, \ldots, \mathbf{e}_n$ are the standard unit vectors in R^n, then

(A) $\mathbf{e}_i \cdot \mathbf{e}_i = 1$
(B) $\mathbf{e}_i \cdot \mathbf{e}_j = 0$ if $i \neq j$
(C) If $\mathbf{u} = (u_1, u_2, \ldots, u_n)$ is any vector in R^n, then

$$\mathbf{u} = u_1\mathbf{e}_1 + u_2\mathbf{e}_2 + \cdots + u_n\mathbf{e}_n$$

▪ Matrix and Vector Notation

One of the most important applications of vectors in R^n is the analysis of the relationship between the solution set of a system of equations and the coefficient matrix for the system. In studying this relationship, we will find it convenient to use matrix notation for vectors. The vector $\mathbf{x} = (x_1, x_2, \ldots, x_n)$ will sometimes be written as a **row vector** of the form

$$\mathbf{x} = [x_1 \quad x_2 \quad \cdots \quad x_n] \qquad \text{Row vector}$$

and, at other times, as a **column vector** of the form

$$\mathbf{x} = \begin{bmatrix} x_1 \\ x_2 \\ \cdot \\ \cdot \\ \cdot \\ x_n \end{bmatrix} \qquad \text{Column vector}$$

In particular, the system of m linear equations in n variables,

$$a_{11}x_1 + a_{12}x_2 + \cdots + a_{1n}x_n = b_1$$
$$a_{21}x_1 + a_{22}x_2 + \cdots + a_{2n}x_n = b_2$$
$$\cdot \qquad \cdot \qquad\qquad \cdot \qquad \cdot$$
$$\cdot \qquad \cdot \qquad\qquad \cdot \qquad \cdot$$
$$\cdot \qquad \cdot \qquad\qquad \cdot \qquad \cdot$$
$$a_{m1}x_1 + a_{m2}x_2 + \cdots + a_{mn}x_n = b_m$$

will be written as

$$A\mathbf{x} = \mathbf{b}$$

where

$$A = \begin{bmatrix} a_{11} & a_{12} & \cdots & a_{1n} \\ a_{21} & a_{22} & \cdots & a_{2n} \\ \cdot & \cdot & & \cdot \\ \cdot & \cdot & & \cdot \\ \cdot & \cdot & & \cdot \\ a_{m1} & a_{m2} & \cdots & a_{mn} \end{bmatrix} \qquad \mathbf{x} = \begin{bmatrix} x_1 \\ x_2 \\ \cdot \\ \cdot \\ \cdot \\ x_n \end{bmatrix} \qquad \mathbf{b} = \begin{bmatrix} b_1 \\ b_2 \\ \cdot \\ \cdot \\ \cdot \\ b_m \end{bmatrix}$$

Notice that if the system is not square, then $m \neq n$ and $\mathbf{x}$ and $\mathbf{b}$ belong to different spaces, $\mathbf{x}$ to R^n and $\mathbf{b}$ to R^m.

In some situations, we will want to consider the coefficient matrix to be a collection of m row vectors,

$$A = \begin{bmatrix} \mathbf{r}_1 \\ \mathbf{r}_2 \\ \cdot \\ \cdot \\ \cdot \\ \mathbf{r}_m \end{bmatrix} \qquad \begin{aligned} \mathbf{r}_1 &= [a_{11} \quad a_{12} \quad \cdots \quad a_{1n}] \\ \mathbf{r}_2 &= [a_{21} \quad a_{22} \quad \cdots \quad a_{2n}] \\ &\quad \cdot \\ &\quad \cdot \\ &\quad \cdot \\ \mathbf{r}_m &= [a_{m1} \quad a_{m2} \quad \cdots \quad a_{mn}] \end{aligned} \qquad \text{Row vectors for the matrix } A$$

or a collection of n column vectors,

$$A = [\mathbf{c}_1 \quad \mathbf{c}_2 \quad \cdots \quad \mathbf{c}_n]$$

$$\mathbf{c}_1 = \begin{bmatrix} a_{11} \\ a_{21} \\ \cdot \\ \cdot \\ \cdot \\ a_{m1} \end{bmatrix}, \quad \mathbf{c}_2 = \begin{bmatrix} a_{12} \\ a_{22} \\ \cdot \\ \cdot \\ \cdot \\ a_{m2} \end{bmatrix}, \quad \cdots, \quad \mathbf{c}_n = \begin{bmatrix} a_{1n} \\ a_{2n} \\ \cdot \\ \cdot \\ \cdot \\ a_{mn} \end{bmatrix} \qquad \text{Column vectors for the matrix } A$$

Using matrix notation for vectors allows us to express the dot product as matrix multiplication in two different ways, one using row vectors and the other using column vectors. If $\mathbf{u} = (u_1, u_2, \ldots, u_n)$ and $\mathbf{v} = (v_1, v_2, \ldots, v_n)$, then, using row vectors, we have

$$[u_1 \quad u_2 \quad \cdots \quad u_n][v_1 \quad v_2 \quad \cdots \quad v_n]^{\mathrm{T}} \qquad \text{The transpose of a } 1 \times n \text{ matrix is an } n \times 1 \text{ matrix.}$$

$$= [u_1 \quad u_2 \quad \cdots \quad u_n] \begin{bmatrix} v_1 \\ v_2 \\ \cdot \\ \cdot \\ \cdot \\ v_n \end{bmatrix} \qquad \text{The product of a } 1 \times n \text{ and an } n \times 1 \text{ matrix is a } 1 \times 1 \text{ matrix.}$$

$$= [u_1 v_1 + u_2 v_2 + \cdots + u_n v_n] \qquad 1 \times 1 \text{ matrices are identified with real numbers.}^*$$

$$= u_1 v_1 + u_2 v_2 + \cdots + u_n v_n = \mathbf{u} \cdot \mathbf{v}$$

---

* Technically (as was pointed out in Section 2-2),

$$[u_1 \quad u_2 \quad \cdots \quad u_n] \begin{bmatrix} u_1 \\ u_2 \\ \cdot \\ \cdot \\ \cdot \\ u_n \end{bmatrix} \qquad \text{and} \qquad [u_1 \quad u_2 \quad \cdots \quad u_n] \cdot \begin{bmatrix} u_1 \\ u_2 \\ \cdot \\ \cdot \\ \cdot \\ u_n \end{bmatrix}$$

Matrix product   Dot product

are different—the matrix product is a single-element matrix, while the dot product is a single real number. However, since the basic algebraic structure of single-element matrices of the form $[x]$, where x is a real number, and the set of real numbers $x \in R$ is the same, we often use $[x]$ and x interchangeably, and even write $[x] = x$. This is convenient, and no harm is done as long as we understand the distinction.

Using column vectors, we have

$$
\begin{bmatrix} u_1 \\ u_2 \\ \cdot \\ \cdot \\ \cdot \\ u_n \end{bmatrix}^{\mathrm{T}} \begin{bmatrix} v_1 \\ v_2 \\ \cdot \\ \cdot \\ \cdot \\ v_n \end{bmatrix} = \begin{bmatrix} u_1 & u_2 & \cdots & u_n \end{bmatrix} \begin{bmatrix} v_1 \\ v_2 \\ \cdot \\ \cdot \\ \cdot \\ v_n \end{bmatrix}
$$

$$
= [u_1 v_1 + u_2 v_2 + \cdots + u_n v_n]
$$

$$
= \mathbf{u} \cdot \mathbf{v}
$$

Example 7   Given the system

$$
\begin{aligned}
2x_1 - 3x_2 + x_3 + 5x_4 &= 9 \\
-x_1 + 4x_2 + 3x_3 \quad\quad &= -4 \\
7x_1 - 2x_2 - 6x_3 + x_4 &= 8
\end{aligned}
$$

(A) Write this system in the form $A\mathbf{x} = \mathbf{b}$.
(B) Find the row vectors for A.
(C) Find the column vectors for A.

Solution   (A) $\begin{bmatrix} 2 & -3 & 1 & 5 \\ -1 & 4 & 3 & 0 \\ 7 & -2 & -6 & 1 \end{bmatrix} \begin{bmatrix} x_1 \\ x_2 \\ x_3 \\ x_4 \end{bmatrix} = \begin{bmatrix} 9 \\ -4 \\ 8 \end{bmatrix}$

(B) $\mathbf{r}_1 = [2 \quad -3 \quad 1 \quad 5], \quad \mathbf{r}_2 = [-1 \quad 4 \quad 3 \quad 0], \quad \mathbf{r}_3 = [7 \quad -2 \quad -6 \quad 1]$

(C) $\mathbf{c}_1 = \begin{bmatrix} 2 \\ -1 \\ 7 \end{bmatrix}, \quad \mathbf{c}_2 = \begin{bmatrix} -3 \\ 4 \\ -2 \end{bmatrix}, \quad \mathbf{c}_3 = \begin{bmatrix} 1 \\ 3 \\ -6 \end{bmatrix}, \quad \mathbf{c}_4 = \begin{bmatrix} 5 \\ 0 \\ 1 \end{bmatrix}$

Problem 7   Repeat Example 7 for the system

$$
\begin{aligned}
5x_1 - x_2 + 6x_3 &= 4 \\
2x_2 + x_3 &= -3 \\
-x_1 - 2x_2 + 4x_3 &= 6 \\
2x_1 + 3x_2 - x_3 &= 1
\end{aligned}
$$

Answers to   **1.** (A) 4   (B) (1, 4, 9, 16, 25)   (C) (0, 0, 0, 0, 0, 0, 0)
Matched Problems   **2.** $x = 3, y = 6$

3. (A) (1, 6, 3, 12, 5)   (B) $(-2, \frac{1}{2}, -\frac{1}{4}, \frac{1}{8}, -\frac{1}{16})$ (C) $(-15, -5, 5, 15)$
(D) (0, 0, 0, 0, 0, 0)

4. (A) $(3, -3, 6, -6)$ (B) $(-3, \frac{2}{3}, 0, -1, \frac{5}{3})$

5. (A) 11   (B) 0

6. (A) 5   (B) $3\sqrt{7}$ (C) $(\frac{2}{5}, \frac{4}{5}, \frac{1}{5}, -\frac{2}{5})$

7. (A) $\begin{bmatrix} 5 & -1 & 6 \\ 0 & 2 & 1 \\ -1 & -2 & 4 \\ 2 & 3 & -1 \end{bmatrix} \begin{bmatrix} x_1 \\ x_2 \\ x_3 \end{bmatrix} = \begin{bmatrix} 4 \\ -3 \\ 6 \\ 1 \end{bmatrix}$

(B) $\mathbf{r}_1 = [5 \quad -1 \quad 6], \quad \mathbf{r}_2 = [0 \quad 2 \quad 1],$
$\mathbf{r}_3 = [-1 \quad -2 \quad 4], \quad \mathbf{r}_4 = [2 \quad 3 \quad -1]$

(C) $\mathbf{c}_1 = \begin{bmatrix} 5 \\ 0 \\ -1 \\ 2 \end{bmatrix}, \quad \mathbf{c}_2 = \begin{bmatrix} -1 \\ 2 \\ -2 \\ 3 \end{bmatrix}, \quad \mathbf{c}_3 = \begin{bmatrix} 6 \\ 1 \\ 4 \\ -1 \end{bmatrix}$

▌▌ Exercise 5-1

A   **1.** Find the third coordinate of $(2, 1, -3, 4, 6, 9)$.
2. Find the fifth coordinate of $(\sqrt{2}, -\sqrt{2}, \sqrt{3}, -\sqrt{3}, \sqrt{5}, -\sqrt{5})$.
3. Find the point in R^4 whose ith coordinate is $i + 2$.
4. Find the point in R^6 whose ith coordinate is $i^2 + i$.

In Problems 5 and 6, determine the values of x, y, and z for which the vectors **u**
and **v** *will be equal.*

5. $\mathbf{u} = (z, x, y, 4); \quad \mathbf{v} = (z, 3, -1, y + z)$
6. $\mathbf{u} = (3x, y + 1, z - 2, 1); \quad \mathbf{v} = (x^2 + 2, 4, 3, x)$

In Problems 7–14, perform the indicated operation.

7. $(1, -2, 4, 0) + (2, 5, -1, 6)$
8. $(5, -4, 2, 1, 3) + (2, 1, -6, 3, -2)$
9. $(6, 1, 3, 4, -2, -7) - (3, -2, 1, -4, 0, 2)$
10. $(1, 2, 3, 4, 5) - (5, 4, 3, 2, 1)$
11. $2(8, 4, 2, 1, \frac{1}{2}, \frac{1}{4}, \frac{1}{8})$
12. $\frac{1}{3}(1, 3, 5, 7, 9)$
13. $2(1, 2, -3, -1) + 3(3, -2, 0, 2)$
14. $5(0, 1, -1, 2) - \frac{1}{2}(1, -2, -3, 8)$

B   *In Problems 15–18, find* $\|\mathbf{u}\|$.

15. $\mathbf{u} = (6, -5, -4, 2)$ **16.** $\mathbf{u} = (3, -4, 6, 5, -7, 3)$

17. $\mathbf{u} = (-3, 5, 2, -5, 3)$ **18.** $\mathbf{u} = (5, 6, -7, 1, 7)$

In Problems 19–22, find the distance between **u** *and* **v**.

19. $\mathbf{u} = (1, 0, -1, 0); \quad \mathbf{v} = (0, -1, 0, 1)$

20. $\mathbf{u} = (4, -3, 2, 1, 3);$ $\mathbf{v} = (3, -3, 2, 1, 3)$
21. $\mathbf{u} = (-2, -4, 1, 3, -1, 1);$ $\mathbf{v} = (3, 1, 5, 6, 1, 2)$
22. $\mathbf{u} = (1, -1, 3, 5, 2, 1, 0);$ $\mathbf{v} = (2, 1, 2, 3, 1, 0, 3)$

In Problems 23–26, find a unit vector in the direction of $\mathbf{u}$.

23. $\mathbf{u} = (5, 3, -1, 1)$ **24.** $\mathbf{u} = (-8, 5, 3, -1, 1)$

25. $\mathbf{u} = (1, -1, 1, -1, 1, -1)$ **26.** $\mathbf{u} = (2, 2, -2, -1, 1, -1, -1)$

In Problems 27–30, find $\mathbf{u} \cdot \mathbf{v}$.

27. $\mathbf{u} = (1, 2, -1, 3);$ $\mathbf{v} = (1, -1, -2, 2)$
28. $\mathbf{u} = (1, -1, 2, -2, 3);$ $\mathbf{v} = (-1, 1, -1, 1, 2)$
29. $\mathbf{u} = (4, 2, -1, 3, 5, -2);$ $\mathbf{v} = (2, 4, 2, 3, -5, -1)$
30. $\mathbf{u} = (1, 0, 2, -1, 0, 1, 0);$ $\mathbf{v} = (0, 3, 2, 4, 5, 1, 2)$

In Problems 31–34, find the value(s) of t for which $\mathbf{u} \cdot \mathbf{v} = 0$.

31. $\mathbf{u} = (t, 1, 2, 2);$ $\mathbf{v} = (-1, 2, -5, t)$
32. $\mathbf{u} = (t, -3, t, 4, 2);$ $\mathbf{v} = (1, 2, -2, 2, t)$
33. $\mathbf{u} = (1, 2, 2, t);$ $\mathbf{v} = (-1, -2, -2, t)$
34. $\mathbf{u} = (t, 2, -1, 3, -2);$ $\mathbf{v} = (t, 2, 5, 4, 2)$

In Problems 35 and 36, write the system in matrix form and find the row and column vectors of the coefficient matrix.

35. $2x_1 - x_2 + x_3 = 4$
      $x_1 + x_2 - x_3 = 5$

36. $x_1 + 2x_2 - x_3 = 4$
        $2x_1 + x_2 = -1$
       $- x_1 - 3x_2 + x_3 = 2$
        $x_1 - x_2 + 4x_3 = 0$

In Problems 37–40, compute the indicated matrix products.

37. $\begin{bmatrix} 1 & 2 & -1 & 3 \end{bmatrix}\begin{bmatrix} 2 & 3 & 2 & -4 \end{bmatrix}^{\mathrm{T}}$
38. $\begin{bmatrix} 2 & -1 & 3 & 2 & 1 \end{bmatrix}\begin{bmatrix} 4 & 6 & 2 & -2 & -4 \end{bmatrix}^{\mathrm{T}}$

39. $\begin{bmatrix} 2 \\ -1 \\ 1 \\ 2 \end{bmatrix}^{\mathrm{T}} \begin{bmatrix} 3 \\ 4 \\ 4 \\ -3 \end{bmatrix}$

40. $\begin{bmatrix} 3 \\ -1 \\ 1 \\ 2 \\ 5 \end{bmatrix}^{\mathrm{T}} \begin{bmatrix} 2 \\ 4 \\ -3 \\ 2 \\ 1 \end{bmatrix}$

C   In Problems 41–44, prove each statement for $\mathbf{u}$ and $\mathbf{v}$ in R^4.

41. $\|\mathbf{u} + \mathbf{v}\|^2 + \|\mathbf{u} - \mathbf{v}\|^2 = 2\|\mathbf{u}\|^2 + 2\|\mathbf{v}\|^2$
42. $\mathbf{u} \cdot \mathbf{v} = \frac{1}{4}\|\mathbf{u} + \mathbf{v}\|^2 - \frac{1}{4}\|\mathbf{u} - \mathbf{v}\|^2$
43. $\|\mathbf{u} + \mathbf{v}\| = \|\mathbf{u} - \mathbf{v}\|$ if and only if $\mathbf{u} \cdot \mathbf{v} = 0$ [Hint: Use Problem 42.]
44. $\|\mathbf{u} + \mathbf{v}\|^2 = \|\mathbf{u}\|^2 + \|\mathbf{v}\|^2$ if and only if $\mathbf{u} \cdot \mathbf{v} = 0$ [Hint: Use Problems 41–43.]

Problems 45–56 refer to statements from Theorems 1 and 2 in this section. Use the definitions of the operations and properties of real numbers to prove each statement for **u**, **v**, *and* **w** *any vectors in R^n and k and ℓ any scalars. Do not refer to Theorems 1 and 2 in your proof.*

45. Theorem 1(A), part (3). $\mathbf{u} + (\mathbf{v} + \mathbf{w}) = (\mathbf{u} + \mathbf{v}) + \mathbf{w}$
46. Theorem 1(A), part (4). $\mathbf{u} + \mathbf{0} = \mathbf{0} + \mathbf{u} = \mathbf{u}$
47. Theorem 1(A), part (5). $\mathbf{u} + (-\mathbf{u}) = (-\mathbf{u}) + \mathbf{u} = \mathbf{0}$
48. Theorem 1(B), part (2). $k(\ell\mathbf{u}) = (k\ell)\mathbf{u}$
49. Theorem 1(B), part (3). $k(\mathbf{u} + \mathbf{v}) = k\mathbf{u} + k\mathbf{v}$
50. Theorem 1(B), part (4). $(k + \ell)\mathbf{u} = k\mathbf{u} + \ell\mathbf{u}$
51. Theorem 2(B). $\mathbf{u} \cdot \mathbf{u} = 0$ if and only if $\mathbf{u} = \mathbf{0}$
52. Theorem 2(D). $\mathbf{u} \cdot \mathbf{v} = \mathbf{v} \cdot \mathbf{u}$
53. Theorem 2(E). $\mathbf{u} \cdot (\mathbf{v} + \mathbf{w}) = \mathbf{u} \cdot \mathbf{v} + \mathbf{u} \cdot \mathbf{w}$
54. Theorem 2(F). $(\mathbf{u} + \mathbf{v}) \cdot \mathbf{w} = \mathbf{u} \cdot \mathbf{w} + \mathbf{v} \cdot \mathbf{w}$
55. Theorem 2(G). $(k\mathbf{u}) \cdot \mathbf{v} = k(\mathbf{u} \cdot \mathbf{v})$ **56.** Theorem 2(H). $\mathbf{u} \cdot (k\mathbf{v}) = k(\mathbf{u} \cdot \mathbf{v})$

Problems 57–59 refer to statements from Theorem 3 in this section. Prove each statement for $\mathbf{e}_1 = (1, 0, 0, 0), \mathbf{e}_2 = (0, 1, 0, 0), \mathbf{e}_3 = (0, 0, 1, 0),$ *and* $\mathbf{e}_4 = (0, 0, 0, 1)$ *in* R^4. *Do not refer to Theorem 3 in your proof.*

57. Theorem 3(A). $\mathbf{e}_i \cdot \mathbf{e}_i = 1$
58. Theorem 3(B). $\mathbf{e}_i \cdot \mathbf{e}_j = 0$ if $i \neq j$
59. Theorem 3(C). If $\mathbf{u} = (u_1, u_2, u_3, u_4)$ is any vector in R^4, then $\mathbf{u} = u_1\mathbf{e}_1 + u_2\mathbf{e}_2 + u_3\mathbf{e}_3 + u_4\mathbf{e}_4$.

60. If

$$\mathbf{c}_1 = \begin{bmatrix} a_{11} \\ a_{21} \\ \cdot \\ \cdot \\ \cdot \\ a_{n1} \end{bmatrix}, \quad \mathbf{c}_2 = \begin{bmatrix} a_{12} \\ a_{22} \\ \cdot \\ \cdot \\ \cdot \\ a_{n2} \end{bmatrix}, \quad \ldots, \quad \mathbf{c}_n = \begin{bmatrix} a_{1n} \\ a_{2n} \\ \cdot \\ \cdot \\ \cdot \\ a_{nn} \end{bmatrix}$$

are given vectors in R^n and $k_1, k_2, \ldots, k_n$ are unknown scalars, show that the vector equation

$$k_1\mathbf{c}_1 + k_2\mathbf{c}_2 + \cdots + k_n\mathbf{c}_n = \mathbf{0}$$

is equivalent to the system

$$a_{11}k_1 + a_{12}k_2 + \cdots + a_{1n}k_n = 0$$
$$a_{21}k_1 + a_{22}k_2 + \cdots + a_{2n}k_n = 0$$
$$\cdot \qquad \cdot \qquad \cdot$$
$$\cdot \qquad \cdot \qquad \cdot$$
$$\cdot \qquad \cdot \qquad \cdot$$
$$a_{n1}k_1 + a_{n2}k_2 + \cdots + a_{nn}k_n = 0$$

61. Refer to Problem 60. Show that the vector equation $k_1\mathbf{c}_1 + k_2\mathbf{c}_2 + \cdots + k_n\mathbf{c}_n = \mathbf{0}$ has nontrivial solutions (solutions with not all k_i

zero) if and only if

$$\begin{vmatrix} a_{11} & a_{12} & \cdots & a_{1n} \\ a_{21} & a_{22} & \cdots & a_{2n} \\ \cdot & \cdot & & \cdot \\ \cdot & \cdot & & \cdot \\ \cdot & \cdot & & \cdot \\ a_{n1} & a_{n2} & \cdots & a_{nn} \end{vmatrix} = 0$$

62. If $\mathbf{u} = (u_1, u_2)$ and $\mathbf{v} = (v_1, v_2)$ are vectors in R^2, define the operation $\mathbf{u} * \mathbf{v}$ by $\mathbf{u} * \mathbf{v} = u_1 v_1 + 2u_2 v_2$. Show that this operation satisfies the same properties as those listed for the dot product in Theorem 2. That is, show that $\mathbf{u} * \mathbf{u} \geq 0$, that $\mathbf{u} * \mathbf{u} = 0$ if and only if $\mathbf{u} = \mathbf{0}$, and so on.

▌5-2▐   Definition of a Vector Space

- Polynomials
- Matrices
- Definition of a Vector Space
- Examples of Vector Spaces
- Examples of Nonvector Spaces
- Some Properties of Any Vector Space

In this section we will introduce a formal definition of a *vector space*. Before we do this, we will consider two examples of sets of mathematical objects — polynomials and matrices — which appear to be unrelated. However, it will turn out that both of these examples possess an algebraic structure that is the same as the algebraic structure of R^n discussed in Section 5-1. Generalizing from these examples, we will introduce the general definition of a vector space. This definition will provide a unifying theme, enabling us to prove general properties that are valid not only for all these examples, but for many more. Once these properties are established for vector spaces in general, they will be valid for *any* particular example of a vector space without additional proof. In this and following sections you will begin to see the power and efficiency of abstract mathematics.

▪ Polynomials

Our first example will involve polynomials in one variable. We start by reviewing some basic terminology. A (real) **polynomial function in one variable** is a function p defined by an equation of the form

$$p(x) = a_0 + a_1 x + a_2 x^2 + \cdots + a_n x^n \tag{1}$$

where $a_0, a_1, a_2, \ldots, a_n$ are real numbers. The domain of p is the set of real numbers R. The expression

$$a_0 + a_1x + a_2x^2 + \cdots + a_nx^n$$

is often referred to simply as a **polynomial.***

Each expression of the form a_ix^i is called a **term** of the polynomial; a_i is the **coefficient** of the term; and i, the power of x, is a nonnegative integer called the **degree** of the term. The **degree of a polynomial** is the highest degree of the terms in the polynomial with nonzero coefficients. Any nonzero constant function is defined to be a **polynomial of degree 0.** The **zero function** is also considered to be a polynomial, but is not assigned a degree. Some examples of polynomials are

$p(x) = 2 + x - 3x^2 + 5x^3$ Third-degree polynomial

$q(x) = \frac{1}{2}x^2 - \pi x^4 + \sqrt{2}\,x^9$ Ninth-degree polynomial

$m(x) = x$ First-degree polynomial

$r(x) = 4$ Constant polynomial; degree is 0

$z(x) = 0$ Zero polynomial; degree is undefined

For each nonnegative integer n, we define P_n to be the set of all polynomials of degree less than or equal to n, together with the zero polynomial. Thus,

$$P_0 = \{a_0 | a_0 \in R\}$$
$$P_1 = \{a_0 + a_1x | a_0, a_1 \in R\}$$
$$P_2 = \{a_0 + a_1x + a_2x^2 | a_0, a_1, a_2 \in R\}$$

and, in general,

$$\boldsymbol{P_n = \{a_0 + a_1x + a_2x^2 + \cdots + a_nx^n | a_0, a_1, a_2, \ldots, a_n \in R\}}$$

Notice that

$$P_0 \subset P_1 \subset P_2 \subset \cdots \subset P_n$$

Our goal is to place an algebraic structure on P_n by defining operations of addition and scalar multiplication for elements of P_n. Preliminary to defining these operations, however, we must discuss *equality* for elements of P_n. Recall from earlier studies of single-variable functions that we say **function f is equal to function g,** and write $f = g$, if both f and g have the same domain D and $f(x) = g(x)$ for each x in D. Since this definition applies to single-variable functions in general, it applies to polynomial functions in P_n in particular. Thus, **two polynomial functions p and q are equal,** and we write $p = q$, if $p(x) = q(x)$ for each x in R. Stated in another way, two polynomial functions p and q are equal if

---

* To keep discussions on an informal level, when referring to polynomial functions, we will use statements such as "the polynomial function $p(x)$," "the polynomial function $a_0 + a_1x + \cdots + a_nx^n$," "the polynomial function $p(x) = a_0 + a_1x + \cdots + a_nx^n$," and "the polynomial" in place of the more formal "the polynomial p specified by . . . ."

the equation $p(x) = q(x)$ is an identity (as opposed to a conditional equation such as $1 + x = 3 + 2x$, which is true for some values of x and not for others).

Theorem 4, which we state without proof, supplies us with a convenient way of determining equality among polynomials in P_n that will be useful to us as we progress.

Theorem 4

Equality in P_n

Two polynomials in P_n are **equal** if and only if the coefficients of corresponding terms are equal. That is, if

$$p(x) = a_0 + a_1 x + \cdots + a_n x^n$$

and

$$q(x) = b_0 + b_1 x + \cdots + b_n x^n$$

then $p(x) = q(x)$ if and only if $a_0 = b_0, a_1 = b_1, \ldots, a_n = b_n$.

Now we are ready to define operations on P_n.

Operations on P_n

Let

$$p(x) = a_0 + a_1 x + \cdots + a_n x^n$$

and

$$q(x) = b_0 + b_1 x + \cdots + b_n x^n$$

be elements of P_n, and let k be a scalar (real number). Then it follows from real number properties that:

The **sum** of $p(x)$ and $q(x)$ is given by

$$p(x) + q(x) = (a_0 + b_0) + (a_1 + b_1)x + \cdots + (a_n + b_n)x^n \tag{2}$$

The **negative** of $p(x)$ is given by

$$-p(x) = (-a_0) + (-a_1)x + \cdots + (-a_n)x^n \tag{3}$$

The **scalar product** of $p(x)$ and k is given by

$$kp(x) = ka_0 + ka_1 x + \cdots + ka_n x^n \tag{4}$$

The algebraic properties of P_n are listed in Theorem 5. We will prove Theorem 5(A), part (1), and leave Theorem 5(B), part (1), as an exercise. The remaining properties are special cases of real number properties, since scalars k and ℓ are real numbers and $p(x)$, $q(x)$, and $r(x)$ represent real numbers for all real x.

Theorem 5

Algebraic Properties of P_n

(A) The following **addition properties** are satisfied for all $p(x)$, $q(x)$, and $r(x)$ in P_n:
  (1) P_n is closed under addition; that is, $p(x) + q(x)$ is in P_n
  (2) $p(x) + q(x) = q(x) + p(x)$
  (3) $[p(x) + q(x)] + r(x) = p(x) + [q(x) + r(x)]$
  (4) $p(x) + 0 = 0 + p(x) = p(x)$ (where $0 = z(x)$ is the zero polynomial)
  (5) $p(x) + [-p(x)] = [-p(x)] + p(x) = 0$
(B) The following **scalar multiplication properties** are satisfied for all $p(x)$ and $q(x)$ in P_n and all scalars k and ℓ:
  (1) P_n is closed under scalar multiplication; that is, $kp(x)$ is in P_n
  (2) $(k\ell)p(x) = k[\ell p(x)]$
  (3) $k[p(x) + q(x)] = kp(x) + kq(x)$
  (4) $(k + \ell)p(x) = kp(x) + \ell p(x)$
  (5) $1p(x) = p(x)$

Proof   To prove part (1) of Theorem 5(A), we must show that for any two elements in P_n, their sum is also in P_n. If $p(x) = a_0 + a_1x + \cdots + a_nx^n$ and $q(x) = b_0 + b_1x + \cdots + b_nx^n$ are in P_n, then the sum of $p(x)$ and $q(x)$ is given by

$$p(x) + q(x) = (a_0 + b_0) + (a_1 + b_1)x + \cdots + (a_n + b_n)x^n$$

Since P_n consists of *all* polynomial functions of the form

$$c_0 + c_1x + \cdots + c_nx^n$$

where $c_0, c_1, \ldots, c_n$ are any real numbers, and since $p(x) + q(x)$ has this form, it follows that $p(x) + q(x)$ is in P_n. This property may seem somewhat obvious to you at this point, but it is a very important one. Later, we will encounter sets that are not closed under addition. ▮

The properties of P_n listed in Theorem 5 were not selected at random. If you compare these with the algebraic properties of addition and scalar multiplication in R^n (Theorem 1, Section 5-1), you will see that the two lists are identical. Thus, P_n, a collection of polynomial functions, possesses an algebraic structure relative to addition and scalar multiplication* that is identical to the algebraic structure of R^n, a collection of points. We now turn to a second example (a set of objects entirely different from P_n) that possesses the same algebraic structure as R^n.

---

* It is possible to place additional algebraic structure on P_n by considering other operations, such as polynomial multiplication. This is a topic of interest in an abstract algebra course, but we will not discuss it here.

■ Matrices

For a second example, we will consider a set of matrices. For fixed positive integers n and m, let $M_{m\times n}$ denote the set of all $m \times n$ matrices with real entries. Thus,

$$M_{2\times 2} = \left\{ \begin{bmatrix} a & b \\ c & d \end{bmatrix} \middle| a, b, c, d \in R \right\} \qquad M_{2\times 3} = \left\{ \begin{bmatrix} a & b & c \\ d & e & f \end{bmatrix} \middle| a, b, c, d, e, f \in R \right\}$$

and, in general,

$$M_{m\times n} = \left\{ \begin{bmatrix} a_{11} & a_{12} & \cdots & a_{1n} \\ a_{21} & a_{22} & \cdots & a_{2n} \\ \cdot & \cdot & & \cdot \\ \cdot & \cdot & & \cdot \\ \cdot & \cdot & & \cdot \\ a_{m1} & a_{m2} & \cdots & a_{mn} \end{bmatrix} \middle| a_{ij} \in R, 1 \le i \le m, 1 \le j \le n \right\}$$

The operations of addition and scalar multiplication for matrices were defined in Chapter 2 and their properties were discussed at that time. In Theorem 6 we restate these properties in the same format as we used in Theorem 5. The proof is omitted (see Theorems 1 and 2 in Section 2-1).

Theorem 6

Algebraic Properties of $M_{m\times n}$

(A) The following **addition properties** are satisfied for all A, B, and C in $M_{m\times n}$:
  (1) $M_{m\times n}$ is closed under addition; that is, $A + B$ is in $M_{m\times n}$
  (2) $A + B = B + A$
  (3) $A + (B + C) = (A + B) + C$
  (4) $A + 0 = 0 + A = A$ (0 is the $m \times n$ zero matrix)
  (5) $A + (-A) = (-A) + A = 0$
(B) The following **scalar multiplication properties** are satisfied for all A and B in $M_{m\times n}$ and all scalars k and ℓ:
  (1) $M_{m\times n}$ is closed under scalar multiplication; that is, kA is in $M_{m\times n}$
  (2) $k(\ell A) = (k\ell)A$
  (3) $k(A + B) = kA + kB$
  (4) $(k + \ell)A = kA + \ell A$
  (5) $1A = A$

Once again we see that a totally different set of mathematical objects ($m \times n$ matrices) has operations of addition and scalar multiplication with the same properties as R^n. From these concrete examples and observations, we are now ready to define an abstract vector space.

▪ Definition of a Vector Space

We have seen that R^n, P_n, and $M_{m \times n}$ all possess the same algebraic properties with respect to the operations of addition and scalar multiplication. Mathematicians have discovered many other sets of objects with the same properties. This has led to the general definition of a *vector space*.

Vector Space

Let V be a set of objects on which operations of addition and scalar multiplication have been defined. Then V is called a **vector space** with respect to these operations, and the elements of V are called **vectors** if:

(A) The following **addition properties** are satisfied for all $\mathbf{u}$, $\mathbf{v}$, and $\mathbf{w}$ in V:
  (1) The sum of $\mathbf{u}$ and $\mathbf{v}$, denoted $\mathbf{u} + \mathbf{v}$, is in V.
  (2) $\mathbf{u} + \mathbf{v} = \mathbf{v} + \mathbf{u}$
  (3) $(\mathbf{u} + \mathbf{v}) + \mathbf{w} = \mathbf{u} + (\mathbf{v} + \mathbf{w})$
  (4) There is an object $\mathbf{0}$ in V with the property that $\mathbf{u} + \mathbf{0} = \mathbf{0} + \mathbf{u} = \mathbf{u}$.
  (5) For each $\mathbf{u}$ in V there is an element $-\mathbf{u}$ in V with the property that
    $\mathbf{u} + (-\mathbf{u}) = (-\mathbf{u}) + \mathbf{u} = \mathbf{0}$.
(B) The following **scalar multiplication properties** are satisfied for all $\mathbf{u}$ and $\mathbf{v}$ in V and all scalars (real numbers) k and ℓ:
  (1) The scalar multiple of $\mathbf{u}$ by k, denoted $k\mathbf{u}$, is in V.
  (2) $(k\ell)\mathbf{u} = k(\ell\mathbf{u})$
  (3) $k(\mathbf{u} + \mathbf{v}) = k\mathbf{u} + k\mathbf{v}$
  (4) $(k + \ell)\mathbf{u} = k\mathbf{u} + \ell\mathbf{u}$
  (5) $1\mathbf{u} = \mathbf{u}$

Vector spaces are sometimes classified as *real vector spaces* or *complex vector spaces*, depending on whether the scalars are real numbers or complex numbers. Since we are using only real numbers as scalars, all vector spaces considered in this book will be real vector spaces.

▪ Examples of Vector Spaces

The simplest example of a vector space is $V = \{\mathbf{0}\}$, called the **zero vector space.** Addition is defined by

$$\mathbf{0} + \mathbf{0} = \mathbf{0}$$

and scalar multiplication is defined by

$$k\mathbf{0} = \mathbf{0} \qquad \text{for any scalar } k$$

It is easy to show that V is a vector space.

Example 8   The set

$$R^n = \{(x_1, x_2, \ldots, x_n) | x_i \in R, i = 1, 2, \ldots, n\}$$

with the standard operations of addition and scalar multiplication as defined in Section 5-1, is a vector space. Each of the properties in the definition of a vector space was proved for R^n in Theorem 1, Section 5-1.

Problem 8   Is R, the set of real numbers, a vector space? ▌▌

Example 9   The set

$$P_n = \{a_0 + a_1 x + \cdots + a_n x^n | a_0, a_1, \ldots, a_n \in R\}$$

with the standard operations of addition and scalar multiplication discussed earlier in this section, is a vector space. Each of the properties in the definition of a vector space was verified for P_n in Theorem 5.

Problem 9   Let P be the set of *all* polynomials of *any* degree. Thus, $P_n \subset P$ for all n. Is P a vector space? ▌▌

Example 10   The set

$$M_{m \times n} = \{A | A \text{ is an } m \times n \text{ matrix with real entries}\}$$

with the standard operations of matrix addition and scalar multiplication, is a vector space. Theorem 6 shows that $M_{m \times n}$ satisfies all the properties in the definition of a vector space.

Problem 10   Let M be the set of *all* matrices of *any* size. Thus, $M_{m \times n} \subset M$ for all m and n. Is M a vector space? ▌▌

Example 11   Let

$$F = \{f | f \text{ is a real-valued function with domain } R\}$$

be a set of functions with the usual definitions of addition, scalar multiplication, negative, and zero function as follows:

Addition: The sum of two functions f and g in F is the function $f + g$ defined by

$$\begin{pmatrix} \text{Value of} \\ \text{sum function} \end{pmatrix} = \begin{pmatrix} \text{Sum of values} \\ \text{of } f \text{ and } g \end{pmatrix}$$

$$(f + g)(x) = f(x) + g(x) \qquad x \in R$$

Scalar Multiplication: The scalar multiple of a function f in F by a scalar k is the function kf defined by

$$\begin{pmatrix} \text{Value of scalar} \\ \text{multiple function} \end{pmatrix} = \begin{pmatrix} \text{Product of } k \\ \text{and value of } f \end{pmatrix}$$

$$(kf)(x) = kf(x) \qquad x \in R$$

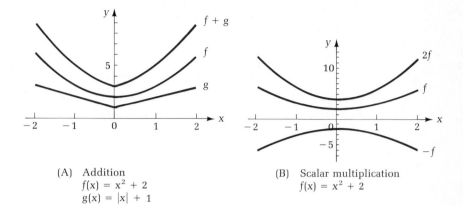

(A)   Addition
$f(x) = x^2 + 2$
$g(x) = |x| + 1$

(B)   Scalar multiplication
$f(x) = x^2 + 2$

Negative: The negative of a function f in F is the function $-f$ defined by

$$(-f)(x) = -f(x) \qquad x \in R$$

Zero Function: The zero function is the function z defined by

$$z(x) = 0 \qquad x \in R$$

It is easy to show, under the above definitions, that the set of functions F forms a vector space. Take the commutative property for addition, for example. To show that addition of functions is a commutative operation, we must show that $f + g$ and $g + f$ are the same function.

$$
\begin{aligned}
(f + g)(x) &= f(x) + g(x) && \text{Definition of addition in } F \\
&= g(x) + f(x) && \text{Commutative property of real numbers} \\
&= (g + f)(x) && \text{Definition of addition in } F
\end{aligned}
$$

Since this holds for all x, $f + g$ and $g + f$ are the same function; that is, $f + g = g + f$. The verification of the remaining properties is left as an exercise.

Problem 11   Let $f(x) = \tfrac{1}{2}x^2 + 2$ and $g(x) = \tfrac{1}{4}x^2 + 1$.

(A) Graph f, g, and $f + g$ on the same set of axes.
(B) Graph f, $2f$, and $-f$ on the same set of axes.   ▌▌

▪ Examples of Nonvector Spaces

Now we want to consider some examples of sets with operations defined on them that are not vector spaces. It is important to realize that a vector space is a set of objects *together with* two operations defined on these objects. If the operations or the set of objects is changed, the result may no longer be a vector space.

Example 12   Let $V = \{(x_1, x_2) | x_1, x_2 \in R\}$ with "addition" defined by

$$(x_1, x_2) \oplus (y_1, y_2) = (x_1 - y_1, x_2 - y_2)$$

and "scalar multiplication" defined by

$$k \otimes (x_1, x_2) = (k^2 x_1, k^2 x_2)$$

(The symbols $\oplus$ and $\otimes$ are used to indicate that these are not the standard operations on this set.) Is V a vector space under these operations?

Solution   We start by checking each of the required properties in the definition of a vector space, and continue until all are verified, or until we find one property that fails.

(A) *Addition properties*
    (1) *Closure property:*

$$(x_1, x_2) \oplus (y_1, y_2) = (x_1 - y_1, x_2 - y_2)$$

    is in V whenever (x_1, x_2) and (y_1, y_2) are in V, since $x_1 - y_1$ and $x_2 - y_2$ are real numbers.

    (2) *Commutative property:* We suspect that this property is not valid, since subtraction of real numbers is not a commutative operation. Thus, we begin by considering some examples:

$$(1, 2) \oplus (3, 4) = (1 - 3, 2 - 4) = (-2, -2)$$
$$(3, 4) \oplus (1, 2) = (3 - 1, 4 - 2) = (2, 2)$$

    Since $(1, 2) \oplus (3, 4) \neq (3, 4) \oplus (1, 2)$, property (A-2) is not satisfied. We can stop now and conclude that V, with the operations $\oplus$ and $\otimes$, is not a vector space.

Problem 12   (A) Show that the addition operation defined in Example 12 does not satisfy property (A-3) in the definition of a vector space.
    (B) Show that the scalar multiplication defined in Example 12 does not satisfy property (B-4) in the definition of a vector space.   ▐

Example 13   Let $V = \{(x_1, x_2) | x_1, x_2 \in R, x_2 \geq 0\}$ with the standard operations of addition and scalar multiplication in R^2. Is V a vector space?

Solution   This time we begin with the scalar multiplication properties. The vector $\mathbf{u} = (1, 2)$ is in V, but the vector $-2\mathbf{u} = (-2, -4)$ is not in V because its second component is negative. Thus, V is not closed under scalar multiplication and is not a vector space.

Problem 13   Find an addition property that is not satisfied by the set in Example 13.   ▐

    Examples 12 and 13 illustrate the two most common reasons that sets with operations defined on them fail to be vector spaces. In Example 12, the set V is the same as R^2, but the operations have been changed and no longer satisfy the necessary requirements. In Example 13, the operations are the standard ones in R^2, but the set V has been changed and is not closed under one of these operations.

▪ Some Properties of Any Vector Space

As we indicated earlier, one of the primary reasons for introducing a general definition of a vector space is to determine which properties are valid for all such spaces and which may hold only for a particular vector space. Much of the work in this chapter will be concerned with discovering properties possessed by all vector spaces. Theorem 7 lists some properties that are true in any vector space. When proving properties of this type, we must be certain not to assume (subconsciously) that we are working with a particular vector space, such as R^n. Note that each statement in the proof of Theorem 7 must be verified by one of the defining properties of a general vector space, a real number property, an already proved part of Theorem 7 or an earlier theorem, or a hypothesis in the theorem.

Theorem 7

If V is a vector space, $\mathbf{u}$ is a vector in V, and k is a scalar, then:

(A) $0\mathbf{u} = \mathbf{0}$
(B) $k\mathbf{0} = \mathbf{0}$
(C) If $k\mathbf{u} = \mathbf{0}$, then $k = 0$ or $\mathbf{u} = \mathbf{0}$ (or both).
(D) $(-1)\mathbf{u} = -\mathbf{u}$

Proof (A) First, notice that

$$0\mathbf{u} = (0 + 0)\mathbf{u} \qquad \text{$0 + 0 = 0$ is a property of real numbers}$$
$$= 0\mathbf{u} + 0\mathbf{u} \qquad \text{Property (B-4)} \qquad\qquad (5)$$

Next, consider

$$\mathbf{0} = 0\mathbf{u} + (-0\mathbf{u}) \qquad \text{Property (A-5)}$$
$$= (0\mathbf{u} + 0\mathbf{u}) + (-0\mathbf{u}) \qquad \text{Substitute, using (5)}$$
$$= 0\mathbf{u} + [0\mathbf{u} + (-0\mathbf{u})] \qquad \text{Property (A-3)}$$
$$= 0\mathbf{u} + \mathbf{0} \qquad \text{Property (A-5)}$$
$$= 0\mathbf{u} \qquad \text{Property (A-4)}$$

Thus, $\mathbf{0} = 0\mathbf{u}$.

(B) $k\mathbf{0} = k(0\mathbf{u}) \qquad \text{Theorem 7(A)}$
$\quad = (k0)\mathbf{u} \qquad \text{Property (B-2)}$
$\quad = 0\mathbf{u} \qquad \text{$k0 = 0$ is a property of real numbers}$
$\quad = \mathbf{0} \qquad \text{Theorem 7(A)}$

(C) Assume that $k\mathbf{u} = \mathbf{0}$. If $k = 0$, then there is nothing to prove. If $k \neq 0$, then

$$\mathbf{0} = \frac{1}{k}\,\mathbf{0}$$ Supply the justification for each step.

$$= \frac{1}{k}\,(k\mathbf{u})$$

$$= \left(\frac{1}{k}k\right)\mathbf{u}$$

$$= 1\mathbf{u}$$

$$= \mathbf{u}$$

(D) $(-1)\mathbf{u} = (-1)\mathbf{u} + \mathbf{0}$ Supply the justification for each step.

$$= (-1)\mathbf{u} + [\mathbf{u} + (-\mathbf{u})]$$

$$= [(-1)\mathbf{u} + \mathbf{u}] + (-\mathbf{u})$$

$$= [(-1)\mathbf{u} + 1\mathbf{u}] + (-\mathbf{u})$$

$$= [(-1) + 1]\mathbf{u} + (-\mathbf{u})$$

$$= 0\mathbf{u} + (-\mathbf{u})$$

$$= \mathbf{0} + (-\mathbf{u})$$

$$= -\mathbf{u}$$ ▌▌

It is important to realize (now that we have finished proving Theorem 7) that all the properties listed in Theorem 7 are valid for any particular vector space, including R^n, P_n, $M_{m \times n}$, and F. It is instructive to interpret each property in Theorem 7 relative to R^n, P_n, $M_{m \times n}$, and F.

Answers to
Matched Problems

8. Yes, R^n is a vector space for any $n > 0$ and $R = R^1$.

9. Yes, it can be shown that each of the properties in the definition of a vector space is satisfied for P.

10. No, matrix addition is not defined for matrices of different sizes.

11. (A)

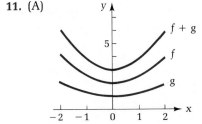

(B)

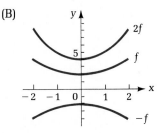

12. (A) $[(1, 2) \oplus (3, 4)] \oplus (5, 6) = (-2, -2) \oplus (5, 6) = (-7, -8)$
$(1, 2) \oplus [(3, 4) \oplus (5, 6)] = (1, 2) \oplus (-2, -2) = (3, 4)$
Since these expressions are not equal, property (A-3) is not satisfied.
(B) $2(1, 1) = (4, 4)$, but $1(1, 1) + 1(1, 1) = (0, 0)$

13. Property (A-5) is not satisfied.

‖ Exercise 5-2

A In Problems 1–8, $p(x) = x^2 - 2x + 3$ and $q(x) = x^2 + 2x + 2$ are polynomials in P_2.

1. Find $2p(x)$. **2.** Find $2q(x)$. **3.** Find $-p(x)$.

4. Find $-q(x)$. **5.** Find $p(x) + q(x)$. **6.** Find $p(x) - q(x)$.

7. Graph $p(x)$, $q(x)$, and $p(x) + q(x)$ on the same set of axes.

8. Graph $p(x)$, $2p(x)$, and $-p(x)$ on the same set of axes.

In Problems 9–18, f and g are two functions in F defined by $f(x) = |x - 1|$ and $g(x) = |x - 2|$.

9. Find $(f + g)(3)$. **10.** Find $(f + g)(-4)$.

11. Find $(2f)(5)$. **12.** Find $(-f)(0)$.

13. Find the function $f + g$.

14. Find the function $f - g$.

15. Find the function $2f$.

16. Find the function $-f$.

17. Graph f, g, and $f + g$ on the same set of axes.

18. Graph f, $2f$, and $-f$ on the same set of axes.

B **19.** Theorem 5(B), part (1). Show that P_n is closed under scalar multiplication.

20. Show that F with the standard operations of function addition and scalar multiplication is a vector space. [Property (A-2) was verified in Example 11.]

In Problems 21 and 22, show that the indicated set (a subset of R^2) is a vector space under the standard operations in R^2.

21. $V = \{(x_1, x_2) | x_2 = 2x_1\}$ **22.** $V = \{(x_1, x_2) | x_1 + x_2 = 0\}$

In Problems 23 and 24, show that the indicated set (a subset of P_2) is a vector space under the standard operations in P_2.

23. $V = \{a_0 + a_2 x^2 | a_0, a_2 \in R\}$ **24.** $V = \{a_0 + a_0 x + a_0 x^2 | a_0 \in R\}$

In Problems 25 and 26, show that the indicated set (a subset of $M_{2 \times 2}$) is a vector space under the standard operations in $M_{2 \times 2}$.

25. $V = \left\{ \begin{bmatrix} a & b \\ 0 & 0 \end{bmatrix} \middle| a, b \in R \right\}$ **26.** $V = \left\{ \begin{bmatrix} a & a+b \\ 0 & b \end{bmatrix} \middle| a, b \in R \right\}$

In Problems 27 and 28, show that the indicated set (a subset of F) is a vector space under the standard operations in F.

27. $V = \{f \in F | f(0) = 0\}$ **28.** $V = \{f \in F | f(0) + f(1) = 0\}$

In Problems 29–36, each set is not a vector space because it fails to satisfy one or both of the closure properties for a vector space [properties (A-1) and (B-1)]. Use the indicated operations for each set to determine which of the closure properties is not satisfied.

29. $V = \{(x_1, x_2)|x_1, x_2 \geq 0\}$ with the standard operations in R^2

30. $V = \{(x_1, x_2)|x_2 = x_1^2\}$ with the standard operations in R^2

31. $V = \{a_0 + a_1x + a_2x^2|a_2 \geq a_0\}$ with the standard operations in P_2

32. $V = \{a_0 + a_1x + x^2|a_0, a_1 \in R\}$ with the standard operations in P_2

33. $V = \left\{ \begin{bmatrix} a & 1 \\ 0 & b \end{bmatrix} \middle| a, b \in R \right\}$ with the standard operations in $M_{2\times2}$

34. $V = \left\{ \begin{bmatrix} a & ab \\ 0 & b \end{bmatrix} \middle| a, b \in R \right\}$ with the standard operations in $M_{2\times2}$

35. $V = \{f \in F|f(0) \geq 0\}$ with the standard operations in F

36. $V = \{f \in F|f(0)f(1) \geq 0\}$ with the standard operations in F

C   *In Problems 37–42, operations of addition and scalar multiplication are defined for each set. Determine whether the set, together with these operations, is a vector space. If the set is not a vector space, indicate which of the properties of a vector space are not satisfied. [Note: Do not assume that the additive identity and the additive inverse are the usual ones in R^2.]*

37. $V = \{(x_1, x_2)|x_1, x_2 \in R\}$ with the operations

$$(x_1, x_2) \oplus (y_1, y_2) = (x_1 + y_1, 0) \quad \text{and} \quad k \otimes (x_1, x_2) = (kx_1, kx_2)$$

38. $V = \{(x_1, x_2)|x_1, x_2 \in R\}$ with the operations

$$(x_1, x_2) \oplus (y_1, y_2) = (x_1 + y_1, x_2 + y_2) \quad \text{and} \quad k \otimes (x_1, x_2) = (0, 0)$$

39. $V = \{(x_1, x_2)|x_1, x_2 \in R\}$ with the operations

$$(x_1, x_2) \oplus (y_1, y_2) = (x_1 + y_1, x_2 + y_2)$$

and

$$k \otimes (x_1, x_2) = (kx_1, x_2)$$

40. $V = \{(x_1, x_2)|x_1 > 0, x_2 > 0\}$ with the operations

$$(x_1, x_2) \oplus (y_1, y_2) = (x_1y_1, x_2y_2)$$

and

$$k \otimes (x_1, x_2) = (x_1^k, x_2^k)$$

41. $V = \{(x_1, x_2)|x_1, x_2 \in R\}$ with the operations

$$(x_1, x_2) \oplus (y_1, y_2) = (x_1 + y_1 + 1, x_2 + y_2 + 1)$$

and

$$k \otimes (x_1, x_2) = (kx_1 + k - 1, kx_2 + k - 1)$$

42. $V = \{(x_1, x_2)|x_1, x_2 \in R\}$ with the operations

$$(x_1, x_2) \oplus (y_1, y_2) = (x_1 + y_1 + 1, x_2 + y_2 + 1)$$

and

$$k \otimes (x_1, x_2) = (kx_1, kx_2)$$

43. *Calculus.* The solution set of the differential equation $y'' - y = 0$ is the set

$$S = \{c_1 e^x + c_2 e^{-x}|c_1, c_2 \in R\}$$

Show that S is a vector space under the standard operations in F.

44. *Calculus.* The solution set of the differential equation $ay'' + by' + cy = 0$, where a, b, and c are constants, can be expressed as

$$S = \{y|ay'' + by' + cy = 0\}$$

Show that S is a vector space under the standard operations in F.

In Problems 45–50, show that each statement is true in an arbitrary vector space V.

45. If $\mathbf{u}$ and $\mathbf{v}$ are in V and k and ℓ are scalars, then $k\mathbf{u} + \ell\mathbf{v}$ is in V.

46. The zero vector is unique. That is, if $\mathbf{z}$ is a vector in V with the property that

$$\mathbf{u} + \mathbf{z} = \mathbf{z} + \mathbf{u} = \mathbf{u}$$

for all $\mathbf{u}$ in V, then $\mathbf{z} = \mathbf{0}$.

47. The negative of a vector is unique. That is, given $\mathbf{u} \in V$, if

$$\mathbf{u} + \mathbf{v} = \mathbf{v} + \mathbf{u} = \mathbf{0}$$

then $\mathbf{v} = -\mathbf{u}$.

48. $-(-\mathbf{u}) = \mathbf{u}$.

49. If $\mathbf{u} + \mathbf{v} = \mathbf{u} + \mathbf{w}$, then $\mathbf{v} = \mathbf{w}$.

50. If $k\mathbf{u} = \ell\mathbf{u}$ and $\mathbf{u} \neq \mathbf{0}$, then $k = \ell$.

|5-3| Subspaces

- Definition of a Subspace
- Characterization of Subspaces
- Examples of Subspaces

In Section 5-2, we saw that P_n (the set consisting of the zero polynomial and all polynomials of degree less than or equal to n) and P (the set of all polynomials) are both vector spaces under the standard operations of addition and scalar multiplication for polynomials. Furthermore, P_n is a subset of P for each n. Thus, each of the sets P_n is a *subset* of the vector space P and is also a *vector space* under the same operations as defined on P. Is this always the case? That is, is every subset of a vector space another vector space? Example 13 in Section 5-2 shows that the answer to this question is no. The set $V = \{(x_1, x_2) | x_2 \geq 0\}$ is a subset of R^2, but V is not a vector space under the standard operations on R^2. In this section we introduce the concept of a *subspace*, a subset of a vector space that is also a vector space.

▪ Definition of a Subspace

> **Subspace**
>
> A nonempty subset W of a vector space V is called a **subspace** of V if W is itself a vector space under the operations of addition and scalar multiplication defined on V.

Example 14   Let

$$W = \{(x, y, z) \in R^3 | ax + by + cz = 0\}$$

Is W a subspace of R^3?

Solution   *Case 1:   a, b, and c are all zero.* If a, b, and c are all zero, then every point in R^3 satisfies the equation

$$0x + 0y = 0z = 0$$

and $W = R^3$. This certainly implies that W is a vector space under the operations defined on R^3. Thus, W is a subspace of R^3.

Case 2:   a, b, and c are not all zero. If any of a, b, and c are nonzero, then W is a plane through the origin in R^3. To show that W is a subspace, we must show that W satisfies all the properties in the definition of a vector space, using the standard operations in R^3.

(A) *Addition properties.* Let $\mathbf{u} = (u_1, u_2, u_3)$ and $\mathbf{v} = (v_1, v_2, v_3)$ be in W.

    (1) Since $\mathbf{u}$ and $\mathbf{v}$ are in W, their components must satisfy the equation that defines W. Thus,

$$au_1 + bu_2 + cu_3 = 0 \qquad \text{and} \qquad av_1 + bv_2 + cv_3 = 0$$

Now, using the operation of addition in R^3,

$$\mathbf{u} + \mathbf{v} = (u_1 + v_1, u_2 + v_2, u_3 + v_3)$$

To show that $\mathbf{u} + \mathbf{v}$ is in W, we must show that its components satisfy the equation that defines W:

$$a(u_1 + v_1) + b(u_2 + v_2) + c(u_3 + v_3)$$
$$= au_1 + av_1 + bu_2 + bv_2 + cu_3 + cv_3$$
$$= (au_1 + bu_2 + cu_3) + (av_1 + bv_2 + cv_3)$$
$$= 0 + 0 = 0$$

Thus, $\mathbf{u} + \mathbf{v}$ is in W, and W is closed under the operation of addition of R^3.

(2) If $\mathbf{u}$ and $\mathbf{v}$ are in W, then they are also in R^3. Since the equation

$$\mathbf{u} + \mathbf{v} = \mathbf{v} + \mathbf{u}$$

is valid for *all* vectors in R^3, it is also valid for all vectors in any *subset* of R^3. Thus, the commutative property holds for W. This situation is often described by saying that W **inherits** the commutative property from R^3.

(3) The associative property is also inherited from R^3.

(4) Since $\mathbf{0} = (0, 0, 0)$ satisfies $a(0) + b(0) + c(0) = 0$, $\mathbf{0}$ is in W.

(5) Since each $\mathbf{u} \in W$ is also in R^3, $\mathbf{u}$ has a negative, $-\mathbf{u} = (-u_1, -u_2, -u_3)$, in R^3 that satisfies

$$\mathbf{u} + (-\mathbf{u}) = (-\mathbf{u}) + \mathbf{u} = \mathbf{0}$$

We must show that $-\mathbf{u}$ is in W:

$$a(-u_1) + b(-u_2) + c(-u_3) = -(au_1 + bu_2 + cu_3)$$
$$= -0 = 0$$

Thus, $-\mathbf{u}$ is in W.

(B) *Multiplication properties*

(1) If $\mathbf{u} = (u_1, u_2, u_3)$ is in W and k is a scalar, then $k\mathbf{u} = (ku_1, ku_2, ku_3)$ satisfies

$$a(ku_1) + b(ku_2) + c(ku_3) = k(au_1 + bu_2 + cu_3)$$
$$= k(0) = 0$$

Thus, $k\mathbf{u}$ is in W, and W is closed under the scalar multiplication operation defined on R^3.

(2)–(5) Since these scalar multiplication properties hold for *all* vectors in R^3, they are inherited by W.

Having verified that W satisfies all the properties in the definition of a vector space under the operations defined on R^3, we can now conclude that W is a subspace of R^3.

Problem 14   Is $W = \{(x, y, z) \in R^3 | ax + by + c = d, d \neq 0\}$ a subspace of R^3?

▪ Characterization of Subspaces

In Example 14, we simplified our work by noting that properties (A-2), (A-3), and (B-2) through (B-5) are inherited by W from R^3. In general, we see that these same properties will be inherited by any nonempty subset of any vector space. Thus, the essential requirements for a subset to be a subspace are that it be closed under addition, property (A-1); contain the zero vector, property (A-4); contain the negative of each of its elements, property (A-5); and be closed under scalar multiplication, property (B-1). As Theorem 8 indicates, it is not even necessary to verify all four of these properties. It is sufficient to show that properties (A-1) and (B-1) are satisfied.

Theorem 8

Characterization of a Subspace

A nonempty subset W of a vector space V is a subspace of V if and only if W is closed under addition and scalar multiplication.

Proof   *Part I.* First, we assume that W is a *subspace* of V. This means that W is a vector space in its own right and must be closed under addition and scalar multiplication.

Part II. Now we assume that W is a *subset* of V that is closed under addition and scalar multiplication. We must show that W is a vector space. As we noted earlier, properties (A-2), (A-3), and (B-2) through (B-5) are automatically inherited by any subset of a vector space. Furthermore, properties (A-1) and (B-1) in the definition of a vector space are assumed to be true as part of our hypothesis. This leaves only properties (A-4) and (A-5) to be verified.

Property (A-4):   *Is the zero vector in W?* Since W is nonempty, we can find a vector $\mathbf{w} \in W$. Since $\mathbf{w}$ is also in V and V is a vector space, Theorem 7(A) in Section 5-2 implies that $0\mathbf{w} = \mathbf{0}$. But W is closed under scalar multiplication. Thus, $\mathbf{0} = 0\mathbf{w} \in W$, and W contains the zero vector.

Property (A-5):   *Is $-\mathbf{w}$ in W whenever $\mathbf{w}$ is in W?* Let $\mathbf{w}$ be any vector in W. Since $\mathbf{w}$ is in the vector space V, $-\mathbf{w}$ is also in V. Theorem 7(D) in Section 5-2 implies that $-\mathbf{w} = (-1)\mathbf{w}$. Since W is closed under scalar multiplication, $-\mathbf{w} = (-1)\mathbf{w}$ is in W.   ▌▌

▪ Examples of Subspaces

Example 15   Let $W = \{(x_1, x_2, 0) | x_1, x_2 \in R\}$. Is W a subspace of R^3?

Solution   If $\mathbf{u} = (u_1, u_2, 0)$ and $\mathbf{v} = (v_1, v_2, 0)$ are in W and k is a scalar, then

$$\mathbf{u} + \mathbf{v} = (u_1, u_2, 0) + (v_1, v_2, 0)$$
$$= (u_1 + v_1, u_2 + v_2, 0)$$

and

$$k\mathbf{u} = k(u_1, u_2, 0)$$
$$= (ku_1, ku_2, 0)$$

Since W is the subset of R^3 consisting of all vectors whose third component is zero, $\mathbf{u} + \mathbf{v}$ and $k\mathbf{u}$ are in W. Thus, Theorem 8 implies that W is a subspace of R^3.

Problem 15   Is $W = \{(x_1, 0, 0)|x_1 \in R\}$ a subspace of R^3?   ▌▌

Example 16   Let $\mathbf{v}$ be a nonzero vector in R^3 and let

$$W = \{\mathbf{x}|\mathbf{x} = t\mathbf{v}, t \in R\}$$

Is W a subspace of R^3?

Solution   You should recognize the equation $\mathbf{x} = t\mathbf{v}$ as the equation of the line in R^3 passing through the origin in the direction of $\mathbf{v}$. If $\mathbf{u}_1$ and $\mathbf{u}_2$ are in W, then there must be scalars t_1 and t_2 such that

$$\mathbf{u}_1 = t_1\mathbf{v} \qquad \text{and} \qquad \mathbf{u}_2 = t_2\mathbf{v}$$

Thus,

$$\mathbf{u}_1 + \mathbf{u}_2 = t_1\mathbf{v} + t_2\mathbf{v} \qquad \text{Use property (B-4), which is valid for all vectors in } R^3.$$
$$= (t_1 + t_2)\mathbf{v}$$

Since W consists of *all* vectors of the form $t\mathbf{v}$ and since $\mathbf{u}_1 + \mathbf{u}_2$ is in this form, it follows that $\mathbf{u}_1 + \mathbf{u}_2$ is in W. Thus, W is closed under addition.
   If $\mathbf{u}$ is in W and k is any scalar, then $\mathbf{u} = t\mathbf{v}$ for some scalar t and

$$k\mathbf{u} = k(t\mathbf{v}) \qquad \text{Use property (B-2), which is valid for all vectors in } R^3.$$
$$= (kt)\mathbf{v}$$

Since kt is a scalar, this shows that $k\mathbf{u}$ is in W. Thus, W is closed under scalar multiplication. Theorem 8 now implies that W is a subspace of R^3.

Problem 16   If $\mathbf{v}$ and $\mathbf{u}$ are nonzero vectors in R^3 and $\mathbf{u}$ is not a scalar multiple of $\mathbf{v}$, is

$$W = \{\mathbf{x}|\mathbf{x} = \mathbf{u} + t\mathbf{v}, t \in R\}$$

a subspace of R^3?   ▌▌

Example 17   Let $M_{2\times2}$ be the vector space of 2×2 matrices with the standard operations of matrix addition and scalar multiplication, and let

$$W = \left\{ \begin{bmatrix} a & b \\ 0 & c \end{bmatrix} \middle| a, b, c \in R \right\}$$

be the subset of upper triangular matrices in $M_{2\times2}$. Is W a subspace of $M_{2\times2}$?

Solution   Let

$$\mathbf{u} = \begin{bmatrix} a & b \\ 0 & c \end{bmatrix} \quad \text{and} \quad \mathbf{v} = \begin{bmatrix} d & e \\ 0 & f \end{bmatrix}$$

be in W and let k be any scalar. Then

$$\mathbf{u} + \mathbf{v} = \begin{bmatrix} a+d & b+e \\ 0 & c+f \end{bmatrix} \quad \text{and} \quad k\mathbf{v} = \begin{bmatrix} ka & kb \\ 0 & kc \end{bmatrix}$$

Since $\mathbf{u} + \mathbf{v}$ and $k\mathbf{v}$ are 2×2 upper triangular matrices, W is closed under addition and scalar multiplication. Thus, W is a subspace of $M_{2\times2}$.

Problem 17   Is

$$W = \left\{ \begin{bmatrix} a & b \\ b & c \end{bmatrix} \middle| a, b, c \in R \right\}$$

a subspace of $M_{2\times2}$?   ▌▌

Example 18   Let W be the subset of P_2 consisting of all quadratic polynomials. That is,

$$W = \{a_0 + a_1x + a_2x^2 | a_0, a_1, a_2 \in R, a_2 \neq 0\}$$

Is W a subspace of P_2?

Solution   Since W consists of all quadratic polynomials, the quadratic polynomials $\mathbf{u} = 1 + x^2$ and $\mathbf{v} = x - x^2$ are in W. But

$$\mathbf{u} + \mathbf{v} = 1 + x^2 + x - x^2 = 1 + x$$

is not a quadratic polynomial, and hence, it is not in W. Since W is not closed under addition, W is not a subspace of P_2.

Problem 18   Let W be the subset of P_2 consisting of all polynomials with integer coefficients. That is,

$$W = \{a_0 + a_1x + a_2x^2 | a_0, a_1, a_2 \text{ integers}\}$$

Is W a subspace of P_2?   ▌▌

Example 19   Is the solution set of the homogeneous system of linear equations given at the top of the next page a subspace of R^n?

$$a_{11}x_1 + a_{12}x_2 + \cdots + a_{1n}x_n = 0$$
$$a_{21}x_1 + a_{22}x_2 + \cdots + a_{2n}x_n = 0$$

$$\cdot \qquad \cdot \qquad \cdot$$
$$\cdot \qquad \cdot \qquad \cdot$$
$$\cdot \qquad \cdot \qquad \cdot$$

$$a_{m1}x_1 + a_{m2}x_2 + \cdots + a_{mn}x_n = 0$$

Solution   If A is the coefficient matrix for this system, then the solution set is

$$W = \{\mathbf{x}|A\mathbf{x} = \mathbf{0}\}$$

where $\mathbf{x} = (x_1, x_2, \ldots, x_n)$ is in R^n and $\mathbf{0}$ is the zero vector in R^m. If $\mathbf{u}$ and $\mathbf{v}$ are in W, then

$$A\mathbf{u} = \mathbf{0} \qquad \text{and} \qquad A\mathbf{v} = \mathbf{0}$$

This implies that

$$A(\mathbf{u} + \mathbf{v}) = A\mathbf{u} + A\mathbf{v} = \mathbf{0} + \mathbf{0} = \mathbf{0}$$

Thus, $\mathbf{u} + \mathbf{v}$ is in W. If k is a scalar, then

$$A(k\mathbf{u}) = k(A\mathbf{u}) = k\mathbf{0} = \mathbf{0}$$

and $k\mathbf{u}$ is in W. Since W is closed under addition and scalar multiplication, W is a subspace of R^n. From now on, we will refer to the solution set of a homogeneous system as the **solution space.**

Problem 19   If A is an $m \times n$ matrix and $\mathbf{b}$ is a nonzero vector in R^m, is the solution set of the nonhomogeneous system $A\mathbf{x} = \mathbf{b}$ a subspace of R^n? ▐▌

Example 20   Let F be the vector space of real-valued functions defined on $(-\infty, \infty)$ and let
Calculus

$$C = \{f \in F | f \text{ is continuous on } (-\infty, \infty)\}$$

Is C a subspace of F?

Solution   In a calculus course, it is shown that if f and g are continuous functions and k is a scalar, then $f + g$ and kf are also continuous functions. Thus, C is a subspace of F.

Problem 20   Is $D = \{f \in F | f \text{ is differentiable on } (-\infty, \infty)\}$ a subspace of F? ▐▌

Answers to   **14.** No; for example, $\mathbf{0} \notin W$.
Matched Problems
15. Yes; $(x_1, 0, 0) + (x_2, 0, 0) = (x_1 + x_2, 0, 0) \in W$ and
$k(x_1, 0, 0) = (kx_1, 0, 0) \in W$.

16. No; for example, $\mathbf{0} \notin W$.

17. Yes; $\begin{bmatrix} a & b \\ b & c \end{bmatrix} + \begin{bmatrix} d & e \\ e & f \end{bmatrix} = \begin{bmatrix} a+d & b+e \\ b+e & c+f \end{bmatrix} \in W$ and

$k\begin{bmatrix} a & b \\ b & c \end{bmatrix} = \begin{bmatrix} ka & kb \\ kb & kc \end{bmatrix} \in W.$

18. No; for example, $\mathbf{u} = 1 + x^2$ is in W, but $\frac{1}{2}\mathbf{u} = \frac{1}{2} + \frac{1}{2}x^2$ is not in W.

19. No; for example, if $A\mathbf{u} = \mathbf{b}$ and $A\mathbf{v} = \mathbf{b}$, then $A(\mathbf{u} + \mathbf{v}) = 2\mathbf{b} \neq \mathbf{b}$.

20. Yes; if f and g are differentiable, then so are $f + g$ and kf.

▌ Exercise 5-3

Most of the problems in this exercise set are concerned with determining whether a subset of a vector space is a subspace of that space. In each such problem, either prove that the subset is a subspace or indicate that the subset is not closed under addition or scalar multiplication (or both).

A   In Problems 1–4, determine whether W is a subspace of R^2.

1. $W = \{(x_1, 0)|x_1 \in R\}$ **2.** $W = \{(x_1, 2)|x_1 \in R\}$

3. $W = \{(x_1, 1 + x_1)|x_1 \in R\}$ **4.** $W = \{(x_1, 2x_1)|x_1 \in R\}$

In Problems 5–8, determine whether W is a subspace of R^3.

5. $W = \{(x_1, x_2, x_3)|x_1x_2x_3 = 0\}$ **6.** $W = \{(x_1, x_2, x_3)|x_1 = x_2 + x_3\}$

7. $W = \{(x_1, x_1, x_1)|x_1 \in R\}$ **8.** $W = \{(x_1, x_1, x_1 + 1)|x_1 \in R\}$

B   In Problems 9–12, determine whether W is a subspace of P_2.

9. $W = \{a_0 - a_0x + 2a_0x^2|a_0 \in R\}$ **10.** $W = \{a_0 + a_0x + x^2|a_0 \in R\}$

11. $W = \{a_0 + a_1x + a_2x^2|a_2 \geq 0\}$ **12.** $W = \{a_1x + a_1x^2|a_1 \in R\}$

In Problems 13–16, determine whether W is a subspace of $M_{2 \times 2}$.

13. $W = \left\{ \begin{bmatrix} a & 0 \\ 0 & b \end{bmatrix} \middle| a, b \in R \right\}$ **14.** $W = \left\{ \begin{bmatrix} a & 0 \\ 0 & -a \end{bmatrix} \middle| a \in R \right\}$

15. $W = \left\{ \begin{bmatrix} a & b \\ c & d \end{bmatrix} \middle| ad - bc = 0 \right\}$ **16.** $W = \left\{ \begin{bmatrix} a & b \\ c & d \end{bmatrix} \middle| ad - bc \neq 0 \right\}$

In Problems 17–20, determine whether W is a subspace of F.

17. $W = \{f \in F|f(0) = 1\}$ **18.** $W = \{f \in F|f(0) = 0\}$

19. $W = \{f \in F|f(0) = f(1)\}$ **20.** $W = \{f \in F|f(0) \geq 0\}$

21. Is $W = \{(x_1, x_2, x_3, x_4)|ax_1 + bx_2 + cx_3 + dx_4 = 0\}$ a subspace of R^4?

22. Is $W = \{(x_1, x_2, x_3, x_4)|ax_1 + bx_2 + cx_3 + dx_4 = 1\}$ a subspace of R^4?

23. Let N_3 be the subset of $M_{3 \times 3}$ consisting of all 3×3 invertible matrices. Is N_3 a subspace of $M_{3 \times 3}$?

24. Let D_3 be the subset of $M_{3\times3}$ consisting of all 3×3 diagonal matrices. Is D_3 a subspace of $M_{3\times3}$?

25. Let U_3 be the subset of $M_{3\times3}$ consisting of all 3×3 upper triangular matrices. Is U_3 a subspace of $M_{3\times3}$? (Recall that a square matrix is upper triangular if all the entries below the diagonal are zero.)

26. Let S_3 be the subset of $M_{3\times3}$ consisting of all 3×3 symmetric matrices. Is S_3 a subspace of $M_{3\times3}$? (Recall that a square matrix A is symmetric if $A = A^T$.)

27. A real-valued function f is called an **even function** if $f(x) = f(-x)$ for all x. Let E be the subset of F consisting of all even functions. Is E a subspace of F?

28. A real-valued function f is called an **odd function** if $f(-x) = -f(x)$ for all x. Let O be the subset of F consisting of all odd functions. Is O a subspace of F?

C 29. A real-valued function f of the form

$$f(x) = a_0 + a_1 \sin x + b_1 \cos x + a_2 \sin 2x + b_2 \cos 2x + \cdots$$
$$+ a_n \sin nx + b_n \cos nx$$

is called a **trigonometric polynomial.** Let T be the subset of F consisting of all trigonometric polynomials. Is T a subspace of F?

30. Is P, the set of all polynomials, a subspace of F?

31. *Calculus.* Let $I = \{f \in F | \int_{-\infty}^{\infty} f(x)\, dx$ exists$\}$.

   (A) Show that I is a subspace of F. [Functions in I are said to be **integrable** on $(-\infty, \infty)$.]

   (B) Is $W = \{f \in I | \int_{-\infty}^{\infty} f(x)\, dx = 0\}$ a subspace of I?

   (C) Is $W = \{f \in I | \int_{-\infty}^{\infty} f(x)\, dx \geq 0\}$ a subspace of I?

32. *Calculus.* Let $D^{(n)} = \{f \in F | f^{(n)}(x)$ exists for $-\infty < x < \infty\}$.

   (A) Show that $D^{(n)}$ is a subspace of F. (Functions in $D^{(n)}$ are called **differentiable functions of order n.**)

   (B) Is $W = \{f \in D^{(n)} | f(0) = f'(0) = \cdots = f^{(n)}(0) = 0\}$ a subspace of $D^{(n)}$? Of F?

   (C) Is P_m a subspace of $D^{(n)}$?

   (D) Is P a subspace of $D^{(n)}$?

33. Let **v** be a vector in a vector space V. Is $W = \{k\mathbf{v} | k \in R\}$ a subspace of V?

34. Let **u** and **v** be vectors in a vector space V. Is $W = \{k\mathbf{u} + \ell\mathbf{v} | k, \ell \in R\}$ a subspace of V?

35. Let V be a vector space.

   (A) Is the zero vector space a subspace of V?

   (B) Is V a subspace of itself?

36. Let R be the vector space of real numbers with the standard operations and let W be a subspace of R. If W is not the zero vector space, is $W = R$?

37. Let W be a subset of the set U and let U be a subset of the vector space V.

   (A) If U is a subspace of V and W is a subspace of U, is W a subspace of V?

   (B) If U is a subspace of V and W is a subspace of V, is W a subspace of U?

38. Let W_1 and W_2 be subspaces of a vector space V.

(A) Is $W_1 \cap W_2$ a subspace of V?

(B) Is $W_1 \cup W_2$ a subspace of V?

39. Let W be a nonempty subset of a vector space V. Show that W is a subspace of V if and only if $k\mathbf{u} + \ell\mathbf{v}$ is in W for all vectors $\mathbf{u}$ and $\mathbf{v}$ in W and all scalars k and ℓ.

40. Let A be an $m \times n$ matrix and let $W = \{\mathbf{x} \in R^n | A\mathbf{x} \neq \mathbf{0}\}$. Is W a subspace of R^n?

41. Let A be an $m \times n$ matrix and let W be the subset of vectors $\mathbf{b} \in R^m$ for which the system $A\mathbf{x} = \mathbf{b}$ has a solution. That is,

$$W = \{\mathbf{b} \in R^m | A\mathbf{x} = \mathbf{b} \text{ for some } \mathbf{x} \in R^n\}$$

Is W a subspace of R^m?

|5-4| Linear Combinations

- Linear Combinations
- The Span of a Set of Vectors
- Remarks

In Sections 5-2 and 5-3, we have seen examples of a variety of different methods for describing vector spaces and their subspaces. The most fundamental method of expressing elements in a vector space is best illustrated using R^3. Each vector $\mathbf{u}$ in R^3 can be expressed in terms of the three standard unit vectors, using the vector space operations of addition and scalar multiplication:

$$\mathbf{u} = (u_1, u_2, u_3) = u_1\mathbf{i} + u_2\mathbf{j} + u_3\mathbf{k}$$

Thus, an alternate description of R^3 is given by

$$R^3 = \{\mathbf{u} | \mathbf{u} = u_1\mathbf{i} + u_2\mathbf{j} + u_3\mathbf{k}; u_1, u_2, u_3 \in R\}$$

In this section, we begin a development of properties of vector spaces which will culminate in showing that many different vector spaces can be described in a manner similar to the above description of R^3.

■ Linear Combinations

We start this discussion with an example.

Example 21   In Example 14 in Section 5-3, we saw that the plane through the origin,

$$W = \{(x, y, z) | ax + by + cz = 0; a, b, \text{ and } c \text{ not all zero}\}$$

is a subspace of R^3. Now we want to consider a different method of describing W. For the purpose of this discussion, we will assume that $a \neq 0$. (The cases $b \neq 0$

and $c \neq 0$ will be covered in Problem 21, following this example.) If we introduce two parameters k and ℓ by letting $y = k$ and $z = \ell$, where k and ℓ are any real numbers, and if we then solve $ax + by + cz = 0$ for x, we obtain the following parametric representation of the plane:

$$x = -\frac{b}{a}k - \frac{c}{a}\ell \qquad a \neq 0$$

$$y = \quad k$$

$$z = \qquad \ell$$

This set of parametric equations can be written as

$$\begin{bmatrix} x \\ y \\ z \end{bmatrix} = \begin{bmatrix} -\dfrac{b}{a}k - \dfrac{c}{a}\ell \\ k + 0 \\ 0 + \ell \end{bmatrix}$$

Column vectors are useful when a vector is to be decomposed into a sum of vectors. The zeros were added to aid in this process.

$$= \begin{bmatrix} -\dfrac{b}{a}k \\ k \\ 0 \end{bmatrix} + \begin{bmatrix} -\dfrac{c}{a}\ell \\ 0 \\ \ell \end{bmatrix}$$

Factor k out of the first vector and ℓ out of the second.

$$= k \begin{bmatrix} -\dfrac{b}{a} \\ 1 \\ 0 \end{bmatrix} + \ell \begin{bmatrix} -\dfrac{c}{a} \\ 0 \\ 1 \end{bmatrix}$$

Switching now to the more customary vector notation, if $\mathbf{x} = (x, y, z)$, $\mathbf{u} = (-b/a, 1, 0)$, and $\mathbf{v} = (-c/a, 0, 1)$, then $\mathbf{x}$ is a point in the plane W if and only if

$$\mathbf{x} = k\mathbf{u} + \ell\mathbf{v}$$

where k and ℓ are arbitrary real numbers. Thus,

$$W = \{k\mathbf{u} + \ell\mathbf{v} \,|\, k, \ell \in R\}$$

Figure 1 illustrates this result geometrically.

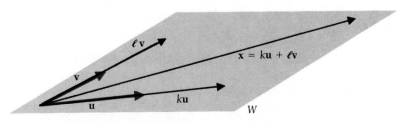

Figure 1 Vector equation of a plane

Problem 21   (A) In Example 21, assume that $b \neq 0$, introduce parameters k and ℓ for x and z, and express each vector in W in the form $k\mathbf{u} + \ell\mathbf{v}$ for appropriate vectors $\mathbf{u}$ and $\mathbf{v}$.

(B) In Example 21, assume that $c \neq 0$, introduce parameters k and ℓ for x and y, and express each vector in W in the form $k\mathbf{u} + \ell\mathbf{v}$ for appropriate vectors $\mathbf{u}$ and $\mathbf{v}$.   ▐▌

Example 21 and Problem 21 show that any plane through the origin in R^3 can be described by an equation of the form

$$\mathbf{x} = k\mathbf{u} + \ell\mathbf{v}$$

The expression on the right is called a *linear combination* of vectors $\mathbf{u}$ and $\mathbf{v}$. Linear combination forms provide an important tool for describing vector spaces and their subspaces.

Linear Combination

If $\mathbf{u}_1, \mathbf{u}_2, \ldots, \mathbf{u}_n$ are vectors in a vector space V and $k_1, k_2, \ldots, k_n$ are scalars, then the vector

$$\mathbf{w} = k_1\mathbf{u}_1 + k_2\mathbf{u}_2 + \cdots + k_n\mathbf{u}_n$$

is called a **linear combination** of $\mathbf{u}_1, \mathbf{u}_2, \ldots, \mathbf{u}_n$.

Since a vector space is closed under addition and scalar multiplication, it follows that any linear combination of vectors from a vector space V is another vector in V.

We have already been using linear combinations in describing certain vector spaces. For example, each vector $\mathbf{u} = (u_1, u_2, u_3)$ in R^3 can be expressed as a linear combination of $\mathbf{i}, \mathbf{j},$ and $\mathbf{k}$ by writing

$$\mathbf{u} = u_1\mathbf{i} + u_2\mathbf{j} + u_3\mathbf{k}$$

Similarly, each vector $\mathbf{u}$ in P_n is defined in terms of a linear combination of the polynomials $1, x, x^2, \ldots, x^n$ by the equation

$$\mathbf{u} = a_0 + a_1x + a_2x^2 + \cdots + a_nx^n$$

Example 22   Show that each vector in R^4 can be expressed as a linear combination of the standard unit vectors $\mathbf{e}_1 = (1, 0, 0, 0)$, $\mathbf{e}_2 = (0, 1, 0, 0)$, $\mathbf{e}_3 = (0, 0, 1, 0)$, and $\mathbf{e}_4 = (0, 0, 0, 1)$.

Solution   Let $\mathbf{u} = (u_1, u_2, u_3, u_4)$ be any vector in R^4. Then

$$\mathbf{u} = \begin{bmatrix} u_1 \\ u_2 \\ u_3 \\ u_4 \end{bmatrix} = \begin{bmatrix} u_1 + 0 + 0 + 0 \\ 0 + u_2 + 0 + 0 \\ 0 + 0 + u_3 + 0 \\ 0 + 0 + 0 + u_4 \end{bmatrix}$$

$$= \begin{bmatrix} u_1 \\ 0 \\ 0 \\ 0 \end{bmatrix} + \begin{bmatrix} 0 \\ u_2 \\ 0 \\ 0 \end{bmatrix} + \begin{bmatrix} 0 \\ 0 \\ u_3 \\ 0 \end{bmatrix} + \begin{bmatrix} 0 \\ 0 \\ 0 \\ u_4 \end{bmatrix}$$

$$= u_1 \begin{bmatrix} 1 \\ 0 \\ 0 \\ 0 \end{bmatrix} + u_2 \begin{bmatrix} 0 \\ 1 \\ 0 \\ 0 \end{bmatrix} + u_3 \begin{bmatrix} 0 \\ 0 \\ 1 \\ 0 \end{bmatrix} + u_4 \begin{bmatrix} 0 \\ 0 \\ 0 \\ 1 \end{bmatrix}$$

$$= u_1\mathbf{e}_1 + u_2\mathbf{e}_2 + u_3\mathbf{e}_3 + u_4\mathbf{e}_4$$

Problem 22   Show that each vector in $M_{2 \times 2}$ can be expressed as a linear combination of the vectors

$$\mathbf{m}_1 = \begin{bmatrix} 1 & 0 \\ 0 & 0 \end{bmatrix} \quad \mathbf{m}_2 = \begin{bmatrix} 0 & 1 \\ 0 & 0 \end{bmatrix} \quad \mathbf{m}_3 = \begin{bmatrix} 0 & 0 \\ 1 & 0 \end{bmatrix} \quad \mathbf{m}_4 = \begin{bmatrix} 0 & 0 \\ 0 & 1 \end{bmatrix} \qquad \blacksquare$$

Example 23   In each of the following, express $\mathbf{w}$ as a linear combination of $\mathbf{u}_1$ and $\mathbf{u}_2$, or $\mathbf{u}_1$, $\mathbf{u}_2$, and $\mathbf{u}_3$, if possible:

(A) $\mathbf{u}_1 = (1, -1, 1)$, $\mathbf{u}_2 = (-2, 1, 3)$, $\mathbf{w} = (4, -3, -1)$

(B) $\mathbf{u}_1 = \begin{bmatrix} 1 & -2 \\ 3 & 1 \end{bmatrix}$, $\mathbf{u}_2 = \begin{bmatrix} 2 & 1 \\ 3 & 1 \end{bmatrix}$, $\mathbf{u}_3 = \begin{bmatrix} -1 & 0 \\ 1 & 2 \end{bmatrix}$, $\mathbf{w} = \begin{bmatrix} 3 & -2 \\ 1 & 4 \end{bmatrix}$

(C) $\mathbf{u}_1 = 1 + 2x$, $\mathbf{u}_2 = -2 + x$, $\mathbf{w} = a_0 + a_1x$

Solution   (A) We must find k_1 and k_2 satisfying

$$k_1\mathbf{u}_1 + k_2\mathbf{u}_2 = \mathbf{w}$$

$$k_1 \begin{bmatrix} 1 \\ -1 \\ 1 \end{bmatrix} + k_2 \begin{bmatrix} -2 \\ 1 \\ 3 \end{bmatrix} = \begin{bmatrix} 4 \\ -3 \\ -1 \end{bmatrix}$$

$$\begin{bmatrix} k_1 \\ -k_1 \\ k_1 \end{bmatrix} + \begin{bmatrix} -2k_2 \\ k_2 \\ 3k_2 \end{bmatrix} = \begin{bmatrix} 4 \\ -3 \\ -1 \end{bmatrix}$$

$$\begin{bmatrix} k_1 - 2k_2 \\ -k_1 + k_2 \\ k_1 + 3k_2 \end{bmatrix} = \begin{bmatrix} 4 \\ -3 \\ -1 \end{bmatrix}$$

Thus, k_1 and k_2 must satisfy the following system of equations:

$$k_1 - 2k_2 = 4$$
$$-k_1 + k_2 = -3$$
$$k_1 + 3k_2 = -1$$

The augmented coefficient matrix for this system is

$$\left[\begin{array}{rr|r} 1 & -2 & 4 \\ -1 & 1 & -3 \\ 1 & 3 & -1 \end{array}\right]$$

The reduced form of this matrix is

$$\left[\begin{array}{rr|r} 1 & 0 & 2 \\ 0 & 1 & -1 \\ 0 & 0 & 0 \end{array}\right]$$

Thus, the solution is $k_1 = 2$ and $k_2 = -1$. That is,

$$\mathbf{w} = 2\mathbf{u}_1 + (-1)\mathbf{u}_2$$

Check: $2(1, -1, 1) + (-1)(-2, 1, 3) \overset{?}{=} (4, -3, -1)$
$$(2, -2, 2) + (2, -1, -3) \overset{?}{=} (4, -3, -1)$$
$$(4, -3, -1) \overset{\checkmark}{=} (4, -3, -1)$$

(B) We must find k_1, k_2, and k_3 satisfying

$$k_1 \begin{bmatrix} 1 & -2 \\ 3 & 1 \end{bmatrix} + k_2 \begin{bmatrix} 2 & 1 \\ 3 & 1 \end{bmatrix} + k_3 \begin{bmatrix} -1 & 0 \\ 1 & 2 \end{bmatrix} = \begin{bmatrix} 3 & -2 \\ 1 & 4 \end{bmatrix}$$

$$\begin{bmatrix} k_1 & -2k_1 \\ 3k_1 & k_1 \end{bmatrix} + \begin{bmatrix} 2k_2 & k_2 \\ 3k_2 & k_2 \end{bmatrix} + \begin{bmatrix} -k_3 & 0 \\ k_3 & 2k_3 \end{bmatrix} = \begin{bmatrix} 3 & -2 \\ 1 & 4 \end{bmatrix}$$

$$\begin{bmatrix} k_1 + 2k_2 - k_3 & -2k_1 + k_2 \\ 3k_1 + 3k_2 + k_3 & k_1 + k_2 + 2k_3 \end{bmatrix} = \begin{bmatrix} 3 & -2 \\ 1 & 4 \end{bmatrix}$$

Thus, k_1, k_2, and k_3 must satisfy

$$k_1 + 2k_2 - k_3 = 3$$
$$-2k_1 + k_2 = -2$$
$$3k_1 + 3k_2 + k_3 = 1$$
$$k_1 + k_2 + 2k_3 = 4$$

The augmented coefficient matrix for this system is

$$\left[\begin{array}{rrr|r} 1 & 2 & -1 & 3 \\ -2 & 1 & 0 & -2 \\ 3 & 3 & 1 & 1 \\ 1 & 1 & 2 & 4 \end{array}\right]$$

which reduces to

$$\begin{bmatrix} 1 & 0 & 0 & | & 1 \\ 0 & 1 & 0 & | & 0 \\ 0 & 0 & 1 & | & -2 \\ 0 & 0 & 0 & | & -7 \end{bmatrix}$$

The last row of the reduced matrix indicates that the system has no solution. Thus, it is not possible to express $\mathbf{w}$ as a linear combination of $\mathbf{u}_1$, $\mathbf{u}_2$, and $\mathbf{u}_3$.

(C) We must find k_1 and k_2 satisfying

$$k_1\mathbf{u}_1 + k_2\mathbf{u}_2 = \mathbf{w}$$

$$k_1(1 + 2x) + k_2(-2 + x) = a_0 + a_1x$$

$$k_1 + 2k_1x - 2k_2 + k_2x = a_0 + a_1x$$

$$(k_1 - 2k_2) + (2k_1 + k_2)x = a_0 + a_1x$$

Equating coefficients, we have

$$k_1 - 2k_2 = a_0$$

$$2k_1 + k_2 = a_1$$

or

$$\begin{bmatrix} 1 & -2 \\ 2 & 1 \end{bmatrix}\begin{bmatrix} k_1 \\ k_2 \end{bmatrix} = \begin{bmatrix} a_0 \\ a_1 \end{bmatrix} \tag{1}$$

Since

$$\begin{vmatrix} 1 & -2 \\ 2 & 1 \end{vmatrix} = 5 \neq 0$$

this system has a unique solution for any a_0 and a_1 (Theorem 10, Section 3-3). Left-multiplying both sides of (1) by the inverse of the coefficient matrix, we have

$$\frac{1}{5}\begin{bmatrix} 1 & 2 \\ -2 & 1 \end{bmatrix}\begin{bmatrix} 1 & -2 \\ 2 & 1 \end{bmatrix}\begin{bmatrix} k_1 \\ k_2 \end{bmatrix} = \frac{1}{5}\begin{bmatrix} 1 & 2 \\ -2 & 1 \end{bmatrix}\begin{bmatrix} a_0 \\ a_1 \end{bmatrix}$$

Verify that

$$\begin{bmatrix} 1 & -2 \\ 2 & 1 \end{bmatrix}^{-1} = \frac{1}{5}\begin{bmatrix} 1 & 2 \\ -2 & 1 \end{bmatrix}$$

$$\begin{bmatrix} k_1 \\ k_2 \end{bmatrix} = \begin{bmatrix} \dfrac{a_0 + 2a_1}{5} \\ \dfrac{-2a_0 + a_1}{5} \end{bmatrix}$$

Thus, $k_1 = (a_0 + 2a_1)/5$ and $k_2 = (-2a_0 + a_1)/5$. Since a_0 and a_1 are arbitrary scalars, this shows that every vector in P_1 can be expressed as a linear combination of $\mathbf{u}_1$ and $\mathbf{u}_2$, as follows:

$$\left(\frac{a_0 + 2a_1}{5}\right)\mathbf{u}_1 + \left(\frac{-2a_0 + a_1}{5}\right)\mathbf{u}_2 = \mathbf{w}$$

Check:

$$\left(\frac{a_0 + 2a_1}{5}\right)(1 + 2x) + \left(\frac{-2a_0 + a_1}{5}\right)(-2 + x) \overset{?}{=} a_0 + a_1 x$$

$$\frac{a_0 + 2a_1}{5} + \frac{2a_0 + 4a_1}{5}x + \frac{4a_0 - 2a_1}{5} + \frac{-2a_0 + a_1}{5}x \overset{?}{=} a_0 + a_1 x$$

$$a_0 + a_1 x \overset{\checkmark}{=} a_0 + a_1 x$$

Problem 23   In each of the following, express $\mathbf{w}$ as a linear combination of $\mathbf{u}_1$ and $\mathbf{u}_2$, or $\mathbf{u}_1$, $\mathbf{u}_2$, and $\mathbf{u}_3$, if possible:

(A) $\mathbf{u}_1 = (-1, 2, 3)$, $\mathbf{u}_2 = (2, 1, -1)$, $\mathbf{w} = (1, 8, 7)$

(B) $\mathbf{u}_1 = \begin{bmatrix} 1 & 0 \\ 0 & 2 \end{bmatrix}$, $\mathbf{u}_2 = \begin{bmatrix} 3 & 0 \\ 0 & 4 \end{bmatrix}$, $\mathbf{w} = \begin{bmatrix} w_1 & 0 \\ 0 & w_2 \end{bmatrix}$

(C) $\mathbf{u}_1 = 1 + 2x + x^2$, $\mathbf{u}_2 = 2 + x + 3x^2$, $\mathbf{u}_3 = 1 - 4x + 3x^2$, $\mathbf{w} = x^2$ ∎

▪ The Span of a Set of Vectors

In Example 23(C), we saw that each vector in P_1 can be expressed as a linear combination of $\mathbf{u}_1 = 1 + 2x$ and $\mathbf{u}_2 = -2 + x$. And since every linear combination of $\mathbf{u}_1 = 1 + 2x$ and $\mathbf{u}_2 = -2 + x$ is a vector in P_1 (Why?), we conclude that P_1 can be described as the set of all possible linear combinations of $\mathbf{u}_1$ and $\mathbf{u}_2$. That is,

$$P_1 = \{k_1(1 + 2x) + k_2(-2 + x) | k_1, k_2 \in R\}$$

In Example 21, we saw that a plane in R^3 can be described as the set of all possible linear combinations of two vectors lying in the plane. Many other vector spaces can be described in this manner. These observations lead to the following definition of the *span* of a finite set of vectors:

The Span of a Set of Vectors

If $S = \{\mathbf{u}_1, \mathbf{u}_2, \ldots, \mathbf{u}_n\}$ is a finite set of vectors in a vector space V, then the **span of S,** denoted span S, is the set of all vectors that can be expressed as linear combinations of vectors in S. That is,

$$\text{span } S = \{\mathbf{w} | \mathbf{w} = k_1\mathbf{u}_1 + k_2\mathbf{u}_2 + \cdots + k_n\mathbf{u}_n\}$$

where $k_1, k_2, \ldots, k_n$ are any scalars. If span $S = V$, then S **is said to span** V.

Using this new terminology, we can write

$$R^3 = \{\mathbf{u} | \mathbf{u} = u_1\mathbf{i} + u_2\mathbf{j} + u_3\mathbf{k}; u_1, u_2, u_3 \in R\}$$
$$= \text{span } \{\mathbf{i}, \mathbf{j}, \mathbf{k}\}$$

Thus, $S = \{\mathbf{i}, \mathbf{j}, \mathbf{k}\}$ spans R^3. In a similar fashion,

$$P_n = \{a_0 + a_1 x + \cdots + a_n x^n | a_0, a_1, \ldots, a_n \in R\}$$
$$= \text{span } \{1, x, \ldots, x^n\}$$

and $S = \{1, x, \ldots, x^n\}$ spans P_n.

Example 24   Let $\mathbf{e}_1 = (1, 0, 0, 0)$, $\mathbf{e}_2 = (0, 1, 0, 0)$, $\mathbf{e}_3 = (0, 0, 1, 0)$, and $\mathbf{e}_4 = (0, 0, 0, 1)$. Show that $S = \{\mathbf{e}_1, \mathbf{e}_2, \mathbf{e}_3, \mathbf{e}_4\}$ spans R^4.

Solution   We must show that

$$R^4 = \text{span } S$$
$$= \{u_1 \mathbf{e}_1 + u_2 \mathbf{e}_2 + u_3 \mathbf{e}_3 + u_4 \mathbf{e}_4 | u_1, u_2, u_3, u_4 \in R\}$$

Since S is a subset of R^4 and since R^4 is closed under addition and scalar multiplication, it follows that any linear combination of vectors in S will be in R^4. Thus, span $S \subset R^4$. In Example 22, we saw that if $\mathbf{u} = (u_1, u_2, u_3, u_4)$ is in R^4, then $\mathbf{u}$ can be expressed as

$$\mathbf{u} = u_1 \mathbf{e}_1 + u_2 \mathbf{e}_2 + u_3 \mathbf{e}_3 + u_4 \mathbf{e}_4$$

which implies that $\mathbf{u} \in$ span S. Thus, $R^4 \subset$ span S. The statements

$$\text{span } S \subset R^4 \qquad \text{and} \qquad R^4 \subset \text{span } S$$

together imply that span $S = R^4$.

Problem 24   Let

$$\mathbf{m}_1 = \begin{bmatrix} 1 & 0 \\ 0 & 0 \end{bmatrix} \qquad \mathbf{m}_2 = \begin{bmatrix} 0 & 1 \\ 0 & 0 \end{bmatrix} \qquad \mathbf{m}_3 = \begin{bmatrix} 0 & 0 \\ 1 & 0 \end{bmatrix} \qquad \mathbf{m}_4 = \begin{bmatrix} 0 & 0 \\ 0 & 1 \end{bmatrix}$$

Show that $S = \{\mathbf{m}_1, \mathbf{m}_2, \mathbf{m}_3, \mathbf{m}_4\}$ spans $M_{2 \times 2}$.   ∎

As was noted in Example 24, the span of a set S in a vector space V is always a *subset* of V. This follows directly from the closure properties of a vector space. Theorem 9 states that span S is also a *subspace* of V.

Theorem 9   If $S = \{\mathbf{u}_1, \mathbf{u}_2, \ldots, \mathbf{u}_n\}$ is a finite set of vectors in a vector space V, then span S is a subspace of V.

Proof   To show that span S is a subspace of V, we must show that span S is closed under addition and scalar multiplication (Theorem 8 in Section 5-3). Let $\mathbf{v}$ and $\mathbf{w}$ be any vectors in span S. Then there must exist scalars $k_1, k_2, \ldots, k_n$ and $\ell_1, \ell_2, \ldots, \ell_n$ such that

$$\mathbf{v} = k_1 \mathbf{u}_1 + k_2 \mathbf{u}_2 + \cdots + k_n \mathbf{u}_n$$

and

$$\mathbf{w} = \ell_1\mathbf{u}_1 + \ell_2\mathbf{u}_2 + \cdots + \ell_n\mathbf{u}_n$$

Then

$$\begin{aligned}\mathbf{v} + \mathbf{w} &= k_1\mathbf{u}_1 + k_2\mathbf{u}_2 + \cdots + k_n\mathbf{u}_n + \ell_1\mathbf{u}_1 + \ell_2\mathbf{u}_2 + \cdots + \ell_n\mathbf{u}_n\\ &= (k_1 + \ell_1)\mathbf{u}_1 + (k_2 + \ell_2)\mathbf{u}_2 + \cdots + (k_n + \ell_n)\mathbf{u}_n\end{aligned}$$

This expresses $\mathbf{v} + \mathbf{w}$ as a linear combination of vectors in S. Since span S is defined to be the set of *all* such linear combinations, it follows that $\mathbf{v} + \mathbf{w} \in$ span S.

Now let t be any scalar. Then

$$\begin{aligned}t\mathbf{v} &= t(k_1\mathbf{u}_1 + k_2\mathbf{u}_2 + \cdots + k_n\mathbf{u}_n)\\ &= tk_1\mathbf{u}_1 + tk_2\mathbf{u}_2 + \cdots + tk_n\mathbf{u}_n\end{aligned}$$

This shows that $t\mathbf{v} \in$ span S. We have shown that span S is closed under addition and scalar multiplication. This implies that span S is a subspace of V. ∎

There are three basic ways we will use the span of a set S, where S is a finite subset of a vector space V:

1. Given S, determine whether span $S = V$.

2. Given S, determine the properties of the subspace spanned by S.

3. Given V, find a set S such that span $S = V$.

The following examples will illustrate each of these uses.

Example 25   In each of the following, determine whether span $S = R^3$:

(A) $S = \{(2, 1, 1), (-1, 3, 1), (1, -2, -3)\}$

(B) $S = \{(2, 1, 1), (-1, 3, 1), (-5, 1, -1)\}$

Solution   (A) Let $\mathbf{w} = (w_1, w_2, w_3)$ be any vector in R^3. In order for span S to be equal to R^3, we must be able to find k_1, k_2, and k_3 so that

$$k_1\begin{bmatrix}2\\1\\1\end{bmatrix} + k_2\begin{bmatrix}-1\\3\\1\end{bmatrix} + k_3\begin{bmatrix}1\\-2\\-3\end{bmatrix} = \begin{bmatrix}w_1\\w_2\\w_3\end{bmatrix}$$

This vector equation is equivalent to the system

$$\begin{aligned}2k_1 - k_2 + k_3 &= w_1\\ k_1 + 3k_2 - 2k_3 &= w_2\\ k_1 + k_2 - 3k_3 &= w_3\end{aligned} \tag{2}$$

with coefficient matrix

$$A = \begin{bmatrix} 2 & -1 & 1 \\ 1 & 3 & -2 \\ 1 & 1 & -3 \end{bmatrix}$$

Since we have not been asked to express **w** in terms of the vectors in S, we do not have to solve this system. It will be sufficient to show that (2) has a solution for any choice of w_1, w_2, and w_3. According to Theorem 10 in Section 3-3, (2) will have a solution for all w_1, w_2, w_3 if and only if $\det(A) \neq 0$. Thus, to determine that (2) always has a solution, we need only evaluate $\det(A)$. Using the cofactor expansion along the first row of A, we have

$$\det(A) = 2 \begin{vmatrix} 3 & -2 \\ 1 & -3 \end{vmatrix} - (-1) \begin{vmatrix} 1 & -2 \\ 1 & -3 \end{vmatrix} + 1 \begin{vmatrix} 1 & 3 \\ 1 & 1 \end{vmatrix}$$

$$= -14 - 1 - 2 = -17$$

Since $\det(A) \neq 0$, (2) has a solution for any choice of w_1, w_2, and w_3. Hence, span $S = R^3$. Of course, if we were interested in actually expressing each $\mathbf{w} \in R^3$ in terms of the vectors in S, we could proceed by using A^{-1} to solve (2), as we did in Example 23(C).

(B) Proceeding as we did in part (A), we must determine the nature of the solution set for the system

$$\begin{aligned} 2k_1 - k_2 - 5k_3 &= w_1 \\ k_1 + 3k_2 + k_3 &= w_2 \\ k_1 + k_2 - k_3 &= w_3 \end{aligned} \tag{3}$$

with coefficient matrix

$$A = \begin{bmatrix} 2 & -1 & -5 \\ 1 & 3 & 1 \\ 1 & 1 & -1 \end{bmatrix}$$

Expanding $\det(A)$ along the first row, we get

$$\det(A) = 2 \begin{vmatrix} 3 & 1 \\ 1 & -1 \end{vmatrix} - (-1) \begin{vmatrix} 1 & 1 \\ 1 & -1 \end{vmatrix} + (-5) \begin{vmatrix} 1 & 3 \\ 1 & 1 \end{vmatrix}$$

$$= -8 - 2 + 10 = 0$$

Since $\det(A) = 0$, (3) will not have a solution for *all* values of w_1, w_2, and w_3. This implies that span $S \neq R^3$. Of course, (3) will have solutions for some values of w_1, w_2, and w_3. For example, if $w_1 = w_2 = w_3 = 0$, then $k_1 = k_2 = k_3 = 0$ is a solution of (3). In fact, the set of vectors for which (3) has a solution is exactly span S and will contain all the vectors that can be expressed as a linear combination of the three vectors in S. Our calculations show only that this does not include all the vectors in R^3.

Problem 25   In each of the following, determine whether span $S = R^3$:

    (A) $S = \{(1, -1, 1), (-2, 3, 2), (2, 4, -5)\}$
    (B) $S = \{(1, -1, 1), (-2, 3, 2), (7, -9, -1)\}$ ❚❚

Example 26   If **u** and **v** are nonzero, nonparallel vectors in R^3, show that span$\{$**u**, **v**$\}$ is a plane through the origin.

Solution   Let $\mathbf{n} = \mathbf{u} \times \mathbf{v} = (a, b, c)$ and let $\mathbf{w} = (x, y, z)$ be any vector in span$\{$**u**, **v**$\}$. Then there exist scalars k and ℓ such that

$$\mathbf{w} = k\mathbf{u} + \ell\mathbf{v}$$

Since **n** is orthogonal to both **u** and **v**,

$$
\begin{aligned}
\mathbf{n} \cdot \mathbf{w} &= \mathbf{n} \cdot (k\mathbf{u} + \ell\mathbf{v}) \\
&= k(\mathbf{n} \cdot \mathbf{u}) + \ell(\mathbf{n} \cdot \mathbf{v}) \\
&= k(0) + \ell(0) \\
&= 0
\end{aligned}
$$

But

$$
\begin{aligned}
\mathbf{n} \cdot \mathbf{w} &= (a, b, c) \cdot (x, y, z) \\
&= ax + by + cz
\end{aligned}
$$

Equating these two expressions for $\mathbf{n} \cdot \mathbf{w}$, we see that the components of **w** satisfy

$$ax + by + cz = 0$$

which is the equation of a plane through the origin in R^3 (see the figure).

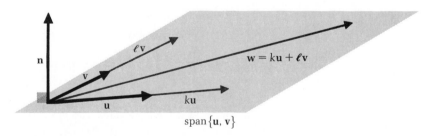

span$\{$**u**, **v**$\}$

Problem 26   If $S = \{\mathbf{u}\}$, where **u** is a nonzero vector in R^3, what is span S?   ❚❚

    If we combine the results in Example 21, Problem 21, and Example 26, we see that W is a plane through the origin in R^3 if and only if W is a subspace of R^3 spanned by two nonzero, nonparallel vectors.

Example 27 Find a set of vectors in R^4 that spans the solution space of the system

$$x_1 + 2x_2 + 3x_3 + x_4 = 0$$
$$x_1 + x_2 + x_3 + 2x_4 = 0$$
$$-x_1 + x_2 + 3x_3 - 4x_4 = 0$$

Solution The augmented coefficient matrix for this system is

$$\begin{bmatrix} 1 & 2 & 3 & 1 & | & 0 \\ 1 & 1 & 1 & 2 & | & 0 \\ -1 & 1 & 3 & -4 & | & 0 \end{bmatrix}$$

The reduced form of this coefficient matrix is

$$\begin{bmatrix} 1 & 0 & -1 & 3 & | & 0 \\ 0 & 1 & 2 & -1 & | & 0 \\ 0 & 0 & 0 & 0 & | & 0 \end{bmatrix}$$

which is the augmented coefficient matrix for the following equivalent system:

$$x_1 - x_3 + 3x_4 = 0$$
$$x_2 + 2x_3 - x_4 = 0$$

Introducing parameters for x_3 and x_4, we have

$$x_1 = t - 3s$$
$$x_2 = -2t + s$$
$$x_3 = t$$
$$x_4 = s$$

Using column vectors, the solution set can be written as

$$\begin{bmatrix} x_1 \\ x_2 \\ x_3 \\ x_4 \end{bmatrix} = \begin{bmatrix} t - 3s \\ -2t + s \\ t + 0 \\ 0 + s \end{bmatrix}$$

$$= \begin{bmatrix} t \\ -2t \\ t \\ 0 \end{bmatrix} + \begin{bmatrix} -3s \\ s \\ 0 \\ s \end{bmatrix}$$

$$= t \begin{bmatrix} 1 \\ -2 \\ 1 \\ 0 \end{bmatrix} + s \begin{bmatrix} -3 \\ 1 \\ 0 \\ 1 \end{bmatrix}$$

Since t and s are arbitrary scalars, this shows that the solution space is spanned by

$$S = \{(1, -2, 1, 0), (-3, 1, 0, 1)\}$$

Problem 27 Find a set of vectors in R^4 that spans the solution space of the system

$$
\begin{array}{rl}
x_1 + x_2 - x_3 + x_4 &= 0 \\
3x_1 + 2x_2 + 2x_3 &= 0 \\
2x_1 + x_2 + 3x_3 - x_4 &= 0
\end{array}
$$

∎

- ## ▪ Remarks

In our definition of the span of a set of vectors, S must be a finite set. In more advanced courses, S is allowed to be an infinite set of vectors. However, linear combinations must always consist of a sum of a finite number of vectors from S. For example, if S is the infinite set consisting of all nonnegative, integral powers of x,

$$S = \{1, x, x^2, \ldots , x^n, x^{n+1}, \ldots\}$$

then any polynomial can be expressed as a linear combination of a finite number of elements of S. In these more advanced treatments, S is said to span P, the vector space of all polynomials.

Answers to
Matched Problems

21. (A) $W = \{k(1, -a/b, 0) + \ell(0, -c/b, 1) | k, \ell \in R\}$
     (B) $W = \{k(1, 0, -a/c) + \ell(0, 1, -b/c) | k, \ell \in R\}$

22. $\begin{bmatrix} a & b \\ c & d \end{bmatrix} = a\mathbf{m}_1 + b\mathbf{m}_2 + c\mathbf{m}_3 + d\mathbf{m}_4$

23. (A) $\mathbf{w} = 3\mathbf{u}_1 + 2\mathbf{u}_2$ (B) $\mathbf{w} = (-2w_1 + \frac{3}{2}w_2)\mathbf{u}_1 + (w_1 - \frac{1}{2}w_2)\mathbf{u}_2$
     (C) Not possible

24. Use Problem 22 to conclude that $M_{2\times2} \subset \text{span}\{\mathbf{m}_1, \mathbf{m}_2, \mathbf{m}_3, \mathbf{m}_4\}$

25. (A) Yes   (B) No

26. $\text{span}\{\mathbf{u}\} = \{t\mathbf{u} | t \in R\}$, a line through the origin in the direction of $\mathbf{u}$

27. $S = \{(-4, 5, 1, 0), (2, -3, 0, 1)\}$

- ## ▌ Exercise 5-4

A   In Problems 1 and 2, express $\mathbf{w}$ as a linear combination of $\mathbf{u}_1 = (1, 1)$ and $\mathbf{u}_2 = (2, -3)$, if possible.

   1. $\mathbf{w} = (9, -1)$ **2.** $\mathbf{w} = (-2, 13)$

In Problems 3–6, express $\mathbf{w}$ as a linear combination of $\mathbf{u}_1 = (1, 0, -1)$ and $\mathbf{u}_2 = (1, 1, 2)$, if possible.

   3. $\mathbf{w} = (7, 4, 5)$ **4.** $\mathbf{w} = (-1, 1, 4)$ **5.** $\mathbf{w} = (1, -3, 1)$ **6.** $\mathbf{w} = (-5, 0, 5)$

In Problems 7–10, express **w** as a linear combination of $\mathbf{u}_1 = 1 + x - x^2$ and $\mathbf{u}_2 = -1 + x + x^2$, if possible.

7. $\mathbf{w} = -1 + 5x + x^2$ **8.** $\mathbf{w} = 1 + x + x^2$

9. $\mathbf{w} = x^2$ **10.** $\mathbf{w} = 1 - 9x - x^2$

In Problems 11–14, express **w** as a linear combination of

$$\mathbf{u}_1 = \begin{bmatrix} 1 & 1 \\ 0 & 1 \end{bmatrix} \quad \text{and} \quad \mathbf{u}_2 = \begin{bmatrix} 1 & -1 \\ 1 & 0 \end{bmatrix}$$

if possible.

11. $\mathbf{w} = \begin{bmatrix} 2 & 0 \\ 1 & 1 \end{bmatrix}$ **12.** $\mathbf{w} = \begin{bmatrix} 0 & 2 \\ -1 & 1 \end{bmatrix}$ **13.** $\mathbf{w} = \begin{bmatrix} 0 & 0 \\ 1 & 1 \end{bmatrix}$

14. $\mathbf{w} = \begin{bmatrix} 1 & 1 \\ 1 & 1 \end{bmatrix}$

B   In Problems 15 and 16, determine whether S spans R^2.

15. $S = \{(3, 4), (4, -3)\}$ **16.** $S = \{(3, 4), (-3, -4)\}$

In Problems 17–20, determine whether S spans R^3.

17. $S = \{(1, 1, 1), (1, -1, -1), (-1, 1, -1)\}$
18. $S = \{(1, 1, 1), (1, -1, -1), (1, 3, 3)\}$
19. $S = \{(1, 0, 1), (1, 1, 0), (2, 1, 1)\}$
20. $S = \{(1, 0, 1), (1, 1, 0), (0, 1, 1)\}$

In Problems 21 and 22, determine whether S spans P_3.

21. $S = \{1, 1 + x, 1 + x + x^2, 1 + x + x^2 + x^3\}$
22. $S = \{1, 1 - x, (1 - x)^2, (1 - x)^3\}$

In Problems 23 and 24, determine whether S spans $M_{2\times2}$.

23. $S = \left\{ \begin{bmatrix} 1 & -1 \\ 1 & 1 \end{bmatrix}, \begin{bmatrix} 1 & 1 \\ -1 & 1 \end{bmatrix}, \begin{bmatrix} 1 & -1 \\ 1 & -1 \end{bmatrix}, \begin{bmatrix} -1 & -1 \\ 1 & 1 \end{bmatrix} \right\}$

24. $S = \left\{ \begin{bmatrix} 1 & 0 \\ 0 & 1 \end{bmatrix}, \begin{bmatrix} 0 & 1 \\ 1 & 0 \end{bmatrix}, \begin{bmatrix} 1 & 0 \\ 0 & -1 \end{bmatrix}, \begin{bmatrix} 0 & 1 \\ -1 & 0 \end{bmatrix} \right\}$

In Problems 25–30, find a set of vectors that spans the solution space of each homogeneous system.

25. $\begin{aligned} x_1 - x_2 + x_3 &= 0 \\ 3x_1 - x_2 - 3x_3 &= 0 \end{aligned}$ **26.** $\begin{aligned} x_1 + 2x_2 - x_3 &= 0 \\ -2x_1 - 4x_2 + 2x_3 &= 0 \end{aligned}$

27. $\begin{aligned} x_1 - x_2 + x_3 &= 0 \\ x_1 + x_2 - x_3 &= 0 \\ x_1 - x_2 - x_3 &= 0 \end{aligned}$ **28.** $\begin{aligned} x_1 + 2x_2 - x_3 &= 0 \\ 3x_1 + 5x_2 - 4x_3 &= 0 \\ 5x_1 + 7x_2 - 8x_3 &= 0 \end{aligned}$

29. $x_1 + 2x_2 + 6x_3 - 7x_4 + 3x_5 = 0$
  $3x_1 + x_2 - 2x_3 + 4x_4 - x_5 = 0$

30. $x_1 + x_2 - 4x_3 - 2x_4 = 0$
  $2x_1 - x_2 + x_3 + 8x_4 = 0$
  $4x_1 + x_2 - 7x_3 + 4x_4 = 0$

In Problems 31 and 32, find two vectors in R^3 that span the plane with the indicated equation.

31. $x + 4y - 5z = 0$

32. $3x + y - 6z = 0$

In Problems 33 and 34, find an equation of the form $ax + by + cz = 0$ for the plane spanned by **u** and **v**.

33. $\mathbf{u} = (2, 1, 0)$; $\mathbf{v} = (0, -1, 3)$

34. $\mathbf{u} = (1, -2, -1)$; $\mathbf{v} = (4, -5, 3)$

C

35. Express $\mathbf{w} = (w_1, w_2)$ as a linear combination of $\mathbf{u}_1 = (4, 3)$ and $\mathbf{u}_2 = (1, 1)$.

36. Express $\mathbf{w} = (w_1, w_2)$ as a linear combination of $\mathbf{u}_1 = (4, 2)$ and $\mathbf{u}_2 = (3, -1)$.

37. Express $\mathbf{w} = a_0 + a_1 x + a_2 x^2$ as a linear combination of $\mathbf{u}_1 = 1 + x$, $\mathbf{u}_2 = x + x^2$, and $\mathbf{u}_3 = 1 + x^2$.

38. Express $\mathbf{w} = a_0 + a_1 x + a_2 x^2$ as a linear combination of $\mathbf{u}_1 = 1 + 2x$, $\mathbf{u}_2 = x + 2x^2$, and $\mathbf{u}_3 = 1 + x - x^2$.

39. Express **w** as a linear combination of $\mathbf{u}_1$, $\mathbf{u}_2$, $\mathbf{u}_3$, and $\mathbf{u}_4$ if

$$\mathbf{w} = \begin{bmatrix} a & b \\ c & d \end{bmatrix} \qquad \mathbf{u}_1 = \begin{bmatrix} 2 & 1 \\ -1 & -1 \end{bmatrix} \qquad \mathbf{u}_2 = \begin{bmatrix} 2 & 2 \\ -2 & -1 \end{bmatrix}$$

$$\mathbf{u}_3 = \begin{bmatrix} -2 & -1 \\ 2 & 1 \end{bmatrix} \qquad \mathbf{u}_4 = \begin{bmatrix} -1 & -1 \\ 1 & 1 \end{bmatrix}$$

40. Express **w** as a linear combination of $\mathbf{u}_1$, $\mathbf{u}_2$, $\mathbf{u}_3$, and $\mathbf{u}_4$ if

$$\mathbf{w} = \begin{bmatrix} a & b \\ c & d \end{bmatrix} \qquad \mathbf{u}_1 = \begin{bmatrix} 2 & 1 \\ 0 & 0 \end{bmatrix} \qquad \mathbf{u}_2 = \begin{bmatrix} 1 & 1 \\ 0 & 0 \end{bmatrix}$$

$$\mathbf{u}_3 = \begin{bmatrix} 0 & 0 \\ 3 & 2 \end{bmatrix} \qquad \mathbf{u}_4 = \begin{bmatrix} 0 & 0 \\ 1 & 1 \end{bmatrix}$$

41. If $\mathbf{u} = (a, b)$ and $\mathbf{v} = (-b, a)$, a and b not both zero, show that $\{\mathbf{u}, \mathbf{v}\}$ spans R^2.

42. If $\mathbf{u} = (a, b, 0)$, $\mathbf{v} = (-b, a, 0)$, and $\mathbf{w} = (0, 0, c)$, where c is not zero and a and b are not both zero, show that $\{\mathbf{u}, \mathbf{v}, \mathbf{w}\}$ spans R^3.

43. Show that

$$\left\{ \begin{bmatrix} 1 & 0 \\ 0 & 0 \end{bmatrix}, \begin{bmatrix} 0 & 0 \\ 0 & 1 \end{bmatrix} \right\} \quad \text{spans} \quad D_2 = \left\{ \begin{bmatrix} a & 0 \\ 0 & b \end{bmatrix} \middle| a, b \in R \right\}$$

44. Show that

$$\left\{ \begin{bmatrix} 1 & 0 \\ 0 & 0 \end{bmatrix}, \begin{bmatrix} 0 & 1 \\ 1 & 0 \end{bmatrix}, \begin{bmatrix} 0 & 0 \\ 0 & 1 \end{bmatrix} \right\} \quad \text{spans} \quad S_2 = \left\{ \begin{bmatrix} a & b \\ b & c \end{bmatrix} \middle| a, b, c \in R \right\}$$

45. Let S be a finite set of vectors in a vector space V. If W is a subspace of V and $S \subset W$, show that span $S \subset W$.

46. Let S and T be finite subsets of a vector space V. If $S \subset T$, show that span $S \subset$ span T.

47. Let $\mathbf{u}_1, \mathbf{u}_2, \mathbf{v}_1$, and $\mathbf{v}_2$ be vectors in a vector space V and assume that $\mathbf{v}_1$ and $\mathbf{v}_2$ are in span$\{\mathbf{u}_1, \mathbf{u}_2\}$. Thus, there exist scalars a, b, c, and d such that

$$\mathbf{v}_1 = a\mathbf{u}_1 + b\mathbf{u}_2$$
$$\mathbf{v}_2 = c\mathbf{u}_1 + d\mathbf{u}_2$$

If the matrix

$$A = \begin{bmatrix} a & b \\ c & d \end{bmatrix}$$

is invertible, show that

$$\text{span}\{\mathbf{v}_1, \mathbf{v}_2\} = \text{span}\{\mathbf{u}_1, \mathbf{u}_2\}$$

48. If $\mathbf{e}_1, \mathbf{e}_2, \ldots, \mathbf{e}_n$ are the standard unit vectors in R^n, show that

$$\text{span}\{\mathbf{e}_1, \mathbf{e}_2, \ldots, \mathbf{e}_n\} = R^n$$

|5-5| Linear Independence

- Linear Dependence and Independence
- Examples
- Basic Properties

In this section, we continue our investigation of the structure of vector spaces and subspaces. We begin by discussing an example from the previous section. Let S_1 and S_2 be subsets of R^3 defined by

$$S_1 = \{(2, 1, 1), (-1, 3, 1), (1, -2, -3)\}$$

and

$$S_2 = \{(2, 1, 1), (-1, 3, 1), (-5, 1, -1)\}$$

In Example 25 of Section 5-4, we saw that

$$\text{span } S_1 = R^3 \quad \text{and} \quad \text{span } S_2 \neq R^3$$

These two sets are very similar. They have the same number of vectors and even have two vectors in common. Yet S_1 spans all of R^3 while S_2 spans a proper subspace of R^3. Why? There must be a difference in the relationship among the vectors in S_1 and those in S_2. Let us take a closer look at S_2. It will be convenient in the discussion that follows to denote the vectors in S_2 by

$$\mathbf{u}_1 = (2, 1, 1) \qquad \mathbf{u}_2 = (-1, 3, 1) \qquad \mathbf{u}_3 = (-5, 1, -1)$$

Suppose that one of the vectors in S_2 can be expressed as a linear combination of the other vectors in S_2, say

$$\mathbf{u}_3 = k\mathbf{u}_1 + \ell\mathbf{u}_2 \tag{1}$$

This implies that $\mathbf{u}_3$ is in span$\{\mathbf{u}_1, \mathbf{u}_2\}$. Furthermore, since span$\{\mathbf{u}_1, \mathbf{u}_2\}$ is a subspace, it follows that span$\{\mathbf{u}_1, \mathbf{u}_2\}$ will contain all linear combinations of the form

$$k_1\mathbf{u}_1 + k_2\mathbf{u}_2 + k_3\mathbf{u}_3 \qquad \mathbf{u}_1 \text{ and } \mathbf{u}_2 \text{ are in span}\{\mathbf{u}_1, \mathbf{u}_2\} \text{ by definition; } \mathbf{u}_3 \text{ is in}$$
$$\text{span}\{\mathbf{u}_1, \mathbf{u}_2\} \text{ by the assumption in (1)}$$

Since span$\{\mathbf{u}_1, \mathbf{u}_2, \mathbf{u}_3\}$ is the set of all linear combinations of the form $k_1\mathbf{u}_1 + k_2\mathbf{u}_2 + k_3\mathbf{u}_3$, we have shown that span$\{\mathbf{u}_1, \mathbf{u}_2, \mathbf{u}_3\}$ is a subset of span$\{\mathbf{u}_1, \mathbf{u}_2\}$. It is always true that span$\{\mathbf{u}_1, \mathbf{u}_2\}$ is a subset of span$\{\mathbf{u}_1, \mathbf{u}_2, \mathbf{u}_3\}$. Thus, we can conclude that these two sets are equal. That is,

$$\mathbf{u}_3 = k\mathbf{u}_1 + \ell\mathbf{u}_2 \qquad \text{implies} \qquad \text{span}\{\mathbf{u}_1, \mathbf{u}_2, \mathbf{u}_3\} = \text{span}\{\mathbf{u}_1, \mathbf{u}_2\}$$

Thus, the subspace of R^3 spanned by $\mathbf{u}_1$, $\mathbf{u}_2$, and $\mathbf{u}_3$ can really be described in terms of $\mathbf{u}_1$ and $\mathbf{u}_2$; we do not need to include $\mathbf{u}_3$. Of course, this discussion is based on the assumption that $\mathbf{u}_3$ is a linear combination of $\mathbf{u}_1$ and $\mathbf{u}_2$. It is easy to verify that

$$\mathbf{u}_3 = -2\mathbf{u}_1 + \mathbf{u}_2$$

which shows that the assumption is valid. It is customary to write this relationship as

$$2\mathbf{u}_1 - \mathbf{u}_2 + \mathbf{u}_3 = \mathbf{0}$$

and to say that $\mathbf{u}_1$, $\mathbf{u}_2$, and $\mathbf{u}_3$ are *linearly dependent* vectors.

▪ Linear Dependence and Independence

We now define the important concepts of *linear dependence* and *independence* for a set of vectors.

Linear Dependence and Independence

Let $S = \{\mathbf{v}_1, \mathbf{v}_2, \ldots, \mathbf{v}_n\}$ be a nonempty, finite set of vectors in a vector space V. The set S is said to be **linearly dependent** if there exist scalars k_1, $k_2, \ldots, k_n$ not all zero, such that

$$k_1\mathbf{v}_1 + k_2\mathbf{v}_2 + \cdots + k_n\mathbf{v}_n = \mathbf{0} \qquad\qquad (2)$$

The set S is said to be **linearly independent** if the only scalars that satisfy (2) are

$$k_1 = k_2 = \cdots = k_n = 0$$

If S is linearly dependent, then any linear combination of the form (2), with $k_1, k_2, \ldots, k_n$ not all zero, is called a **dependency relationship** for S.

We apply this definition to the sets S_1 and S_2 in the next example.

Example 28 Determine whether each of the following sets of vectors in R^3 is linearly dependent or linearly independent. If either set is linearly dependent, find a dependency relationship for that set.

(A) $S_1 = \{(2, 1, 1), (-1, 3, 1), (1, -2, -3)\}$
(B) $S_2 = \{(2, 1, 1), (-1, 3, 1), (-5, 1, -1)\}$

Solution (A) We must determine the nature of the solution set for the equation

$$k_1\mathbf{u}_1 + k_2\mathbf{u}_2 + k_3\mathbf{u}_3 = \mathbf{0} \tag{3}$$

where $\mathbf{u}_1 = (2, 1, 1)$, $\mathbf{u}_2 = (-1, 3, 1)$, and $\mathbf{u}_3 = (1, -2, -3)$. Using column vectors, this can be written as

$$k_1\begin{bmatrix} 2 \\ 1 \\ 1 \end{bmatrix} + k_2\begin{bmatrix} -1 \\ 3 \\ 1 \end{bmatrix} + k_3\begin{bmatrix} 1 \\ -2 \\ -3 \end{bmatrix} = \begin{bmatrix} 0 \\ 0 \\ 0 \end{bmatrix}$$

$$\begin{bmatrix} 2k_1 - k_2 + k_3 \\ k_1 + 3k_2 - 2k_3 \\ k_1 + k_2 - 3k_3 \end{bmatrix} = \begin{bmatrix} 0 \\ 0 \\ 0 \end{bmatrix}$$

Thus, k_1, k_2, and k_3 must satisfy the homogeneous system

$$2k_1 - k_2 + k_3 = 0$$
$$k_1 + 3k_2 - 2k_3 = 0$$
$$k_1 + k_2 - 3k_3 = 0$$

Using matrix notation, this system can be written as

$$A\mathbf{k} = \mathbf{0} \tag{4}$$

where

$$A = \begin{bmatrix} 2 & -1 & 1 \\ 1 & 3 & -2 \\ 1 & 1 & -3 \end{bmatrix} \qquad \mathbf{k} = \begin{bmatrix} k_1 \\ k_2 \\ k_3 \end{bmatrix} \qquad \mathbf{0} = \begin{bmatrix} 0 \\ 0 \\ 0 \end{bmatrix}$$

and

$$\det(A) = 2\begin{vmatrix} 3 & -2 \\ 1 & -3 \end{vmatrix} + \begin{vmatrix} 1 & -2 \\ 1 & -3 \end{vmatrix} + \begin{vmatrix} 1 & 3 \\ 1 & 1 \end{vmatrix}$$

$$= -14 - 1 - 2 = -17$$

Since $\det(A) \neq 0$, Theorem 10 in Section 3-3 implies that the only solution to (4) is the trivial solution, $k_1 = k_2 = k_3 = 0$. This implies that the only solution to the vector equation in (3) is $k_1 = k_2 = k_3 = 0$. Thus, $\mathbf{u}_1$, $\mathbf{u}_2$, and $\mathbf{u}_3$ are linearly independent.

(B) We have already seen (in the discussion at the beginning of this section) that this set is linearly dependent. For completeness, we will show how we arrived at that conclusion. Let $\mathbf{u}_1 = (2, 1, 1)$, $\mathbf{u}_2 = (-1, 3, 1)$, and $\mathbf{u}_3 =$

$(-5, 1, -1)$. Proceeding as we did in part (A), the vector equation

$$k_1\mathbf{u}_1 + k_2\mathbf{u}_2 + k_3\mathbf{u}_3 = \mathbf{0} \tag{5}$$

is equivalent to the matrix equation

$$A\mathbf{k} = \mathbf{0} \tag{6}$$

where

$$A = \begin{bmatrix} 2 & -1 & -5 \\ 1 & 3 & 1 \\ 1 & 1 & -1 \end{bmatrix}$$

and

$$\det(A) = 2 \begin{vmatrix} 3 & 1 \\ 1 & -1 \end{vmatrix} + \begin{vmatrix} 1 & 1 \\ 1 & -1 \end{vmatrix} - 5 \begin{vmatrix} 1 & 3 \\ 1 & 1 \end{vmatrix}$$

$$= -8 - 2 + 10 = 0$$

Since $\det(A) = 0$, Theorem 10 in Section 3-3 implies that equation (6) has nontrivial solutions. Thus, there must exist k_1, k_2, and k_3, not all zero, satisfying (6) and, consequently, (5). This is enough to show that $\mathbf{u}_1$, $\mathbf{u}_2$, and $\mathbf{u}_3$ are linearly dependent, but it does not determine a dependency relationship for $\mathbf{u}_1$, $\mathbf{u}_2$, and $\mathbf{u}_3$. To do this, we must actually find a nontrivial solution of (6). The augmented coefficient matrix for this system is

$$[A|\mathbf{0}] = \begin{bmatrix} 2 & -1 & -5 & \bigg| & 0 \\ 1 & 3 & 1 & \bigg| & 0 \\ 1 & 1 & -1 & \bigg| & 0 \end{bmatrix}$$

which reduces to

$$\begin{bmatrix} 1 & 0 & -2 & \bigg| & 0 \\ 0 & 1 & 1 & \bigg| & 0 \\ 0 & 0 & 0 & \bigg| & 0 \end{bmatrix}$$

Thus, the solution is

$$k_1 = 2t$$
$$k_2 = -t$$
$$k_3 = t$$

where t is a parameter. Any nonzero value of the parameter t will provide a dependency relationship for $\mathbf{u}_1$, $\mathbf{u}_2$, and $\mathbf{u}_3$. For example, $t = 1$ shows that

$$2\mathbf{u}_1 - \mathbf{u}_2 + \mathbf{u}_3 = \mathbf{0}$$

Notice that this dependency relationship implies that each vector in S_2 can be expressed as a linear combination of the remaining vectors in S_2. That is,

$$\mathbf{u}_1 = \tfrac{1}{2}\mathbf{u}_2 - \tfrac{1}{2}\mathbf{u}_3$$
$$\mathbf{u}_2 = 2\mathbf{u}_1 + \mathbf{u}_3$$
$$\mathbf{u}_3 = -2\mathbf{u}_1 + \mathbf{u}_2$$

Problem 28   Determine whether each of the following sets of vectors in R^3 is linearly dependent or linearly independent. If either set is linearly dependent, find a dependency relationship for that set.

(A) $S_1 = \{(1, -1, 1), (-2, 3, 2), (2, 4, -5)\}$
(B) $S_2 = \{(1, -1, 1), (-2, 3, 2), (7, -9, -1)\}$

Example 28 now provides the answer to the question posed at the beginning of this section. There is a fundamental difference in the relationship among the vectors in S_1 and those in S_2. The vectors in S_1 are linearly independent; no one of them can be expressed as a linear combination of the others. The vectors in S_2 are linearly dependent; it is possible to express each vector in terms of the other two. These observations are summarized in Theorem 10. The proof is left as an exercise.

Theorem 10

Let $S = \{\mathbf{v}_1, \mathbf{v}_2, \ldots, \mathbf{v}_n\}$ be a finite set of two or more vectors in a vector space V.

(A) The set S is linearly dependent if and only if it is possible to express at least one vector in S as a linear combination of the other vectors in S.
(B) The set S is linearly independent if and only if it is not possible to express any vector in S as a linear combination of the other vectors in S.

Notice that Theorem 10(A) states that *at least one* vector in S can be expressed as a linear combination of the other vectors in S. It does not state that *each* vector in S can be expressed as a linear combination of the other vectors in S (as was the case in Example 28). For example, if $\mathbf{u}_1 = (1, 0, 0)$, $\mathbf{u}_2 = (0, 1, 0)$, $\mathbf{u}_3 = (0, 0, 1)$, and $\mathbf{u}_4 = (-2, -3, 0)$, then any dependency relationship for these vectors will be of the form

$$2t\mathbf{u}_1 + 3t\mathbf{u}_2 + 0\mathbf{u}_3 + t\mathbf{u}_4 = \mathbf{0}$$

Since the coefficient of $\mathbf{u}_3$ is 0 for any value of the parameter t, it is not possible to use this relationship to express $\mathbf{u}_3$ as a linear combination of $\mathbf{u}_1$, $\mathbf{u}_2$, and $\mathbf{u}_4$.

■ Examples

Example 29   In each of the following, determine whether S is linearly dependent or linearly independent. If S is linearly dependent, find a dependency relationship for the vectors in S.

(A) $S = \{(2, 1, -3), (1, 2, -2), (1, -4, 0), (-1, 7, -1)\} \subset R^3$

(B) $S = \{1 + x + x^2, 1 + 2x + 3x^2, 1 + x - x^2\} \subset P_2$

(C) $S = \left\{ \begin{bmatrix} 1 & 1 \\ 1 & 1 \end{bmatrix}, \begin{bmatrix} -1 & 1 \\ 1 & -1 \end{bmatrix}, \begin{bmatrix} 1 & -1 \\ 1 & -1 \end{bmatrix} \right\} \subset M_{2 \times 2}$

Solution   (A) Using column vectors, we have

$$k_1 \begin{bmatrix} 2 \\ 1 \\ -3 \end{bmatrix} + k_2 \begin{bmatrix} 1 \\ 2 \\ -2 \end{bmatrix} + k_3 \begin{bmatrix} 1 \\ -4 \\ 0 \end{bmatrix} + k_4 \begin{bmatrix} -1 \\ 7 \\ -1 \end{bmatrix} = \begin{bmatrix} 0 \\ 0 \\ 0 \end{bmatrix}$$

which is equivalent to the system

$$\begin{aligned} 2k_1 + k_2 + k_3 - k_4 &= 0 \\ k_1 + 2k_2 - 4k_3 + 7k_4 &= 0 \\ -3k_1 - 2k_2 \qquad\quad - k_4 &= 0 \end{aligned}$$

Since this homogeneous system has more variables than equations, it must have nontrivial solutions (Theorem 4, Section 1-5). Thus, S is linearly dependent. To find a dependency relationship for the vectors in S, we must find one of these nontrivial solutions. The reduced form of the augmented coefficient matrix for this system is

$$\begin{bmatrix} 1 & 0 & 2 & -3 & | & 0 \\ 0 & 1 & -3 & 5 & | & 0 \\ 0 & 0 & 0 & 0 & | & 0 \end{bmatrix}$$

and the solution is

$$\begin{aligned} k_1 &= -2s + 3t \\ k_2 &= 3s - 5t \\ k_3 &= s \\ k_4 &= t \end{aligned}$$

Choosing $s = 1$ and $t = 1$ gives $k_1 = 1$, $k_2 = -2$, $k_3 = 1$, $k_4 = 1$. Thus,

$$(2, 1, -3) - 2(1, 2, -2) + (1, -4, 0) + (-1, 7, -1) = (0, 0, 0)$$

You should verify that this relationship is correct.

(B) $k_1(1 + x + x^2) + k_2(1 + 2x + 3x^2) + k_3(1 + x - x^2)$

$ = (k_1 + k_2 + k_3) + (k_1 + 2k_2 + k_3)x + (k_1 + 3k_2 - k_3)x^2$

$ = 0$

We must determine the nature of the solution set for the system

$$\begin{aligned} k_1 + k_2 + k_3 &= 0 \\ k_1 + 2k_2 + k_3 &= 0 \\ k_1 + 3k_2 - k_3 &= 0 \end{aligned}$$

Since this is a square system, the determinant of the coefficient matrix will tell us whether the system has nontrivial solutions.

$$\begin{vmatrix} 1 & 1 & 1 \\ 1 & 2 & 1 \\ 1 & 3 & -1 \end{vmatrix} = \begin{vmatrix} 2 & 1 \\ 3 & -1 \end{vmatrix} - \begin{vmatrix} 1 & 1 \\ 1 & -1 \end{vmatrix} + \begin{vmatrix} 1 & 2 \\ 1 & 3 \end{vmatrix}$$

$$= -5 + 2 + 1 = -2$$

Since this determinant is nonzero, the system has no nontrivial solutions (Theorem 10, Section 3-3) and S is linearly independent.

(C) The equation

$$k_1 \begin{bmatrix} 1 & 1 \\ 1 & 1 \end{bmatrix} + k_2 \begin{bmatrix} -1 & 1 \\ 1 & -1 \end{bmatrix} + k_3 \begin{bmatrix} 1 & -1 \\ 1 & -1 \end{bmatrix} = \begin{bmatrix} 0 & 0 \\ 0 & 0 \end{bmatrix}$$

is equivalent to the system

$$k_1 - k_2 + k_3 = 0$$
$$k_1 + k_2 - k_3 = 0$$
$$k_1 + k_2 + k_3 = 0$$
$$k_1 - k_2 - k_3 = 0$$

This system has more equations than variables and may or may not have nontrivial solutions. The only way to find out is to solve the system. The reduced form of the augmented coefficient matrix for this system is

$$\begin{bmatrix} 1 & 0 & 0 & | & 0 \\ 0 & 1 & 0 & | & 0 \\ 0 & 0 & 1 & | & 0 \\ 0 & 0 & 0 & | & 0 \end{bmatrix}$$

This shows that the only solution is the trivial one: $k_1 = k_2 = k_3 = 0$. Thus, S is linearly independent.

Problem 29 In each of the following, determine whether S is linearly dependent or linearly independent. If S is linearly dependent, find a dependency relationship for the vectors in S.

(A) $S = \{(1, -1, 1), (2, 1, 1), (-1, 2, -1), (2, 2, 1)\} \subset R^3$

(B) $S = \{1 + x + x^2, 1 + 2x + 3x^2, 1 - x - 2x^2\} \subset P_2$

(C) $S = \left\{ \begin{bmatrix} 1 & 1 \\ 1 & 1 \end{bmatrix}, \begin{bmatrix} -1 & 1 \\ 1 & -1 \end{bmatrix}, \begin{bmatrix} 1 & 2 \\ 2 & 1 \end{bmatrix} \right\} \subset M_{2 \times 2}$ ∎

As Example 29 illustrates, most problems involving linear dependence and independence are related to determining the nature of the solution set for a homogeneous system of linear equations. The important facts concerning the solution sets of such systems are reviewed in the box. (See Theorem 4 in Section 1-5 and Theorem 10 in Section 3-3.)

Solutions of Homogeneous Systems — A Review

Consider the homogeneous system of m linear equations with n variables, $k_1, k_2, \ldots, k_n$:

$$a_{11}k_1 + a_{12}k_2 + \cdots + a_{1n}k_n = 0$$
$$a_{21}k_1 + a_{22}k_2 + \cdots + a_{2n}k_n = 0$$

$$\vdots \qquad \vdots \qquad \qquad \vdots$$

$$a_{m1}k_1 + a_{m2}k_2 + \cdots + a_{mn}k_n = 0$$

(A) $m < n$: If there are more variables than equations, then the system must have nontrivial solutions.

(B) $m = n$: If the system is square, then the system has nontrivial solutions if and only if the determinant of the coefficient matrix is zero.

(C) $m > n$: If there are more equations than variables, then the system may or may not have nontrivial solutions. The only way to find out is to solve the system.

In any of the cases listed in the box, if it is necessary to find a nontrivial solution, then the system must be solved, usually by finding the reduced form of the augmented coefficient matrix.

The next two examples and matched problems show that some of the important sets of vectors discussed previously are linearly independent.

Example 30   Show that $S = \{\mathbf{i}, \mathbf{j}, \mathbf{k}\}$ is a linearly independent subset of R^3.

Solution   Since $\mathbf{i} = (1, 0, 0)$, $\mathbf{j} = (0, 1, 0)$, and $\mathbf{k} = (0, 0, 1)$,

$$k_1\mathbf{i} + k_2\mathbf{j} + k_3\mathbf{k} = (k_1, k_2, k_3) = \mathbf{0}$$

if and only if $k_1 = k_2 = k_3 = 0$.

Problem 30   Show that $S = \{\mathbf{e}_1, \mathbf{e}_2, \mathbf{e}_3, \mathbf{e}_4\}$ is a linearly independent subset of R^4. [Recall that $\mathbf{e}_1 = (1, 0, 0, 0)$, $\mathbf{e}_2 = (0, 1, 0, 0)$, $\mathbf{e}_3 = (0, 0, 1, 0)$, and $\mathbf{e}_4 = (0, 0, 0, 1)$.]   ∎

Example 31   Show that $S = \{1, x, x^2, \ldots, x^n\}$ is a linearly independent subset of P_n.

Solution   Recall that the zero polynomial is the polynomial with all zero coefficients. That is,

$$p(x) = a_0 + a_1x + \cdots + a_nx^n = \mathbf{0}$$

if and only if $a_0 = a_1 = \cdots = a_n = 0$. This implies that the only solution of the equation

$$k_1 \cdot 1 + k_2 x + \cdots + k_n x^n = 0$$

is $k_1 = k_2 = \cdots = k_n = 0$. Thus, S is independent.

Problem 31 Show that

$$S = \left\{ \begin{bmatrix} 1 & 0 \\ 0 & 0 \end{bmatrix}, \begin{bmatrix} 0 & 1 \\ 0 & 0 \end{bmatrix}, \begin{bmatrix} 0 & 0 \\ 1 & 0 \end{bmatrix}, \begin{bmatrix} 0 & 0 \\ 0 & 1 \end{bmatrix} \right\}$$

is a linearly independent subset of $M_{2 \times 2}$.   ▮

▪ Basic Properties

The properties listed in Theorem 11 will be helpful in the work that follows. We will prove part (A) and leave the remaining parts as exercises.

Theorem 11 | Let S be a finite set of vectors in a vector space V.

(A) If $\mathbf{0} \in S$, then S is linearly dependent.
(B) If S has only one element, then S is linearly dependent if and only if that element is the zero vector.
(C) If S has exactly two elements, then S is linearly independent if and only if neither vector is a scalar multiple of the other.

Proof (A) Let $S = \{\mathbf{v}_1, \mathbf{v}_2, \ldots, \mathbf{v}_n\}$ and assume that $\mathbf{v}_1 = \mathbf{0}$. If k_1 is any nonzero scalar, then

$$k_1 \mathbf{v}_1 + 0\mathbf{v}_2 + \cdots + 0\mathbf{v}_n = k_1 \mathbf{0} = \mathbf{0}$$

Thus, S is linearly dependent. (A linear combination of the vectors in S has been found that is zero without all the scalars k_i being zero.)   ▮

Let

$$S_2 = \{(2, 1, 1), (-1, 3, 1), (-5, 1, -1)\}$$

and

$$B = \{(2, 1, 1), (-1, 3, 1)\}$$

At the beginning of this section we saw that S_2 is linearly dependent and that

$$\text{span } S_2 = \text{span } B$$

Furthermore, Theorem 11(C) now implies that B is a linearly independent subset of S_2. Thus, given the linearly dependent set S_2, we were able to find a linearly independent subset B that has the same span as S_2. Theorem 12 states that it is always possible to do this.

Theorem 12

> Let S be a finite set of nonzero vectors in a vector space V. Then there is a linearly independent subset B of S with the property that span $B =$ span S.

Proof   If S is linearly independent, then we can let $B = S$ and conclude that the theorem is true in this case. Now suppose S is linearly dependent. Theorem 10(A) implies that at least one vector in S can be expressed as a linear combination of the other vectors in S. Let

$$S = \{\mathbf{u}_1, \mathbf{u}_2, \ldots, \mathbf{u}_n\}$$

and suppose that

$$\mathbf{u}_1 = k_2\mathbf{u}_2 + \cdots + k_n\mathbf{u}_n \tag{7}$$

Let $S_1 = \{\mathbf{u}_2, \ldots, \mathbf{u}_n\}$. Since $S_1 \subset S$, it follows that span $S_1 \subset$ span S. We want to show that span $S_1 =$ span S. Let $\mathbf{w}$ be any vector in span S. Then there must exist scalars $\ell_1, \ell_2, \ldots, \ell_n$ such that

$$\begin{aligned}
\mathbf{w} &= \ell_1\mathbf{u}_1 + \ell_2\mathbf{u}_2 + \cdots + \ell_n\mathbf{u}_n \qquad \text{Use (7) to eliminate } \mathbf{u}_1. \\
&= \ell_1(k_2\mathbf{u}_2 + k_3\mathbf{u}_3 + \cdots + k_n\mathbf{u}_n) + \ell_2\mathbf{u}_2 + \ell_3\mathbf{u}_3 + \cdots + \ell_n\mathbf{u}_n \\
&= (\ell_1 k_2 + \ell_2)\mathbf{u}_2 + (\ell_1 k_3 + \ell_3)\mathbf{u}_3 + \cdots + (\ell_1 k_n + \ell_n)\mathbf{u}_n
\end{aligned}$$

This expresses $\mathbf{w}$ as a linear combination of vectors in S_1. Thus, $\mathbf{w} \in$ span S_1, and we conclude that span $S \subset$ span S_1. Since span $S_1 \subset$ span S and span $S \subset$ span S_1, it follows that span $S_1 =$ span S. If S_1 is linearly independent, then we have found a linearly independent subset of S that has the same span as S. If S_1 is linearly dependent, then, by applying the same procedure as above, we can find $S_2 \subset S$, with the properties that S_2 has one less element than S_1 and that

$$\text{span } S_2 = \text{span } S_1 = \text{span } S$$

We can continue to apply this procedure until we find the required linearly independent subset. Since, according to Theorem 11(B), a set with one nonzero vector is always linearly independent, this process must eventually terminate. ∎

Example 32   Let $S = \{(-2, 4), (1, -2), (1, 4)\}$. Find a linearly independent subset of S that has the same span as S.

Solution   Let $\mathbf{u}_1 = (-2, 4)$, $\mathbf{u}_2 = (1, -2)$, and $\mathbf{u}_3 = (1, 4)$. Omitting the details, it is easy to verify that $\mathbf{u}_1$, $\mathbf{u}_2$, and $\mathbf{u}_3$ satisfy the dependency relationship

$$\mathbf{u}_1 + 2\mathbf{u}_2 + 0\mathbf{u}_3 = \mathbf{0}$$

Thus, $\mathbf{u}_1$ can be expressed as a linear combination of $\mathbf{u}_2$ and $\mathbf{u}_3$, and

$$\text{span}\{\mathbf{u}_2, \mathbf{u}_3\} = \text{span}\{\mathbf{u}_1, \mathbf{u}_2, \mathbf{u}_3\}$$

Since $\mathbf{u}_2$ and $\mathbf{u}_3$ are not scalar multiples of each other, Theorem 11(C) implies that $\{\mathbf{u}_2, \mathbf{u}_3\}$ is linearly independent. Thus, $B = \{\mathbf{u}_2, \mathbf{u}_3\}$ is a linearly independent subset of S satisfying span $B =$ span S.

Problem 32   Let $S = \{(-6, 3), (4, 2), (4, -2)\}$. Find a linearly independent subset of S that has the same span as S. ▌▌

In Example 32, notice that the dependency relationship can also be used to express $\mathbf{u}_2$ as a linear combination of $\mathbf{u}_1$ and $\mathbf{u}_3$. Thus, $\{\mathbf{u}_1, \mathbf{u}_3\}$ is also a linearly independent subset of S with the same span as S. However, it is not possible to express $\mathbf{u}_3$ as a linear combination of $\mathbf{u}_1$ and $\mathbf{u}_2$ (verify this), so $\{\mathbf{u}_1, \mathbf{u}_2\}$ is not a solution to this problem.

Answers to Matched Problems

28. (A) Linearly independent
    (B) Linearly dependent; $-3(1, -1, 1) + 2(-2, 3, 2) + (7, -9, -1) = \mathbf{0}$
29. (A) Linearly dependent; $(1, -1, 1) + (2, 1, 1) + (-1, 2, -1) - (2, 2, 1) = \mathbf{0}$
    (B) Linearly independent
    (C) Linearly dependent; $-3\begin{bmatrix} 1 & 1 \\ 1 & 1 \end{bmatrix} - \begin{bmatrix} -1 & 1 \\ 1 & -1 \end{bmatrix} + 2\begin{bmatrix} 1 & 2 \\ 2 & 1 \end{bmatrix} = \mathbf{0}$
30. $k_1\mathbf{e}_1 + k_2\mathbf{e}_2 + k_3\mathbf{e}_3 + k_4\mathbf{e}_4 = (k_1, k_2, k_3, k_4) = \mathbf{0}$ implies $k_1 = k_2 = k_3 = k_4 = 0$
31. $k_1\begin{bmatrix} 1 & 0 \\ 0 & 0 \end{bmatrix} + k_2\begin{bmatrix} 0 & 1 \\ 0 & 0 \end{bmatrix} + k_3\begin{bmatrix} 0 & 0 \\ 1 & 0 \end{bmatrix} + k_4\begin{bmatrix} 0 & 0 \\ 0 & 1 \end{bmatrix} = \begin{bmatrix} k_1 & k_2 \\ k_3 & k_4 \end{bmatrix} = \mathbf{0}$
    implies $k_1 = k_2 = k_3 = k_4 = 0$
32. $\{(-6, 3), (4, 2)\}$ or $\{(4, 2), (4, -2)\}$

▌ Exercise 5-5

A   *In Problems 1–20, determine whether the indicated set of vectors is linearly independent or linearly dependent. If a set is linearly dependent, find a dependency relationship for the vectors in the set.*

1. $\{(2, -5)\} \subset R^2$
2. $\{(6, -12), (-4, 8)\} \subset R^2$
3. $\{(0, 0)\} \subset R^2$
4. $\{(2, -5), (3, 7)\} \subset R^2$
5. $\{(1, 0), (0, 1), (-2, 3)\} \subset R^2$
6. $\{(0, 0), (3, 4)\} \subset R^2$

B   7. $\{(2, 0, -1), (3, -1, -4), (-1, 2, 1)\} \subset R^3$
8. $\{(5, 10, -20), (4, 8, -16), (5, 10, 16)\} \subset R^3$
9. $\{(1, 1, 0), (0, 1, -1), (-2, 1, -3), (1, 2, -1)\} \subset R^3$
10. $\{(1, 1, 1), (2, 3, 4), (3, 6, 10)\} \subset R^3$
11. $\{(1, 2, -1, 0), (0, 0, 2, 1), (1, 2, 0, 0), (1, 0, -1, 2)\} \subset R^4$
12. $\{(3, -2, 1, 4), (4, -1, 3, 2), (-5, 2, -3, -4)\} \subset R^4$

13. $\{1 + x + x^2, 1 + x - x^2, x\} \subset P_2$

14. $\{1 + x + x^2, 1 + x - x^2, x^2\} \subset P_2$

15. $\{1 + 2x + 3x^2, 1 - x + x^2, 1 - 4x + x^2\} \subset P_2$

16. $\{1, 1 - x, 1 - x + x^2, 1 + x^2\} \subset P_2$

17. $\left\{ \begin{bmatrix} -4 & 7 \\ -1 & 2 \end{bmatrix}, \begin{bmatrix} 2 & -3 \\ 1 & 0 \end{bmatrix}, \begin{bmatrix} 2 & -1 \\ 3 & 4 \end{bmatrix} \right\} \subset M_{2 \times 2}$

18. $\left\{ \begin{bmatrix} 1 & -1 \\ 2 & -1 \end{bmatrix}, \begin{bmatrix} 1 & -1 \\ 1 & 0 \end{bmatrix}, \begin{bmatrix} 1 & -1 \\ -2 & 3 \end{bmatrix} \right\} \subset M_{2 \times 2}$

19. $\left\{ \begin{bmatrix} 1 & -1 \\ 1 & 0 \end{bmatrix}, \begin{bmatrix} 1 & 2 \\ 1 & -1 \end{bmatrix}, \begin{bmatrix} -1 & 1 \\ 0 & 1 \end{bmatrix}, \begin{bmatrix} 1 & 1 \\ 2 & 1 \end{bmatrix} \right\} \subset M_{2 \times 2}$

20. $\left\{ \begin{bmatrix} 1 & -1 \\ 1 & 0 \end{bmatrix}, \begin{bmatrix} 1 & 2 \\ 1 & -1 \end{bmatrix}, \begin{bmatrix} -1 & 1 \\ 0 & 1 \end{bmatrix}, \begin{bmatrix} 1 & 2 \\ 2 & 0 \end{bmatrix} \right\} \subset M_{2 \times 2}$

In Problems 21–28, find a linearly independent subset of S that has the same span as S.

21. $S = \{(2, -4), (-3, 6)\} \subset R^2$

22. $S = \{(1, 2), (2, -1)\} \subset R^2$

23. $S = \{(1, 2, -1), (-2, -4, 2), (1, -2, 1)\} \subset R^3$

24. $S = \{(2, -4, 6), (-3, 6, -9), (5, -10, 15)\} \subset R^3$

25. $S = \{1, 1 - x, 1 - x + x^2\} \subset P_2$

26. $S = \{1, 1 + x^2, 1 - x^2\} \subset P_2$

27. $S = \left\{ \begin{bmatrix} 1 & 0 \\ 0 & 2 \end{bmatrix}, \begin{bmatrix} -2 & 0 \\ 0 & -4 \end{bmatrix}, \begin{bmatrix} 0 & 3 \\ 6 & 0 \end{bmatrix}, \begin{bmatrix} 0 & -2 \\ -4 & 0 \end{bmatrix} \right\} \subset M_{2 \times 2}$

28. $S = \left\{ \begin{bmatrix} 1 & 0 \\ 0 & 2 \end{bmatrix}, \begin{bmatrix} 2 & 0 \\ 0 & 1 \end{bmatrix}, \begin{bmatrix} 0 & 1 \\ 2 & 0 \end{bmatrix}, \begin{bmatrix} 0 & 2 \\ 1 & 0 \end{bmatrix} \right\} \subset M_{2 \times 2}$

29. Find the value(s) of t for which

$$S = \{(1, t), (t, 1)\}$$

is a linearly dependent subset of R^2.

30. Find the value(s) of t for which

$$S = \{(t, 1, -1), (1, t, -1), (1, -1, t)\}$$

is a linearly dependent subset of R^3.

31. If

$$A = \begin{bmatrix} 1 & 0 & -2 \\ 0 & 1 & 3 \end{bmatrix}$$

then

$$R = \{[1 \quad 0 \quad -2], [0 \quad 1 \quad 3]\}$$

is the set of row vectors of A, and

$$C = \left\{ \begin{bmatrix} 1 \\ 0 \end{bmatrix}, \begin{bmatrix} 0 \\ 1 \end{bmatrix}, \begin{bmatrix} -2 \\ 3 \end{bmatrix} \right\}$$

is the set of column vectors of A. Determine whether R and C are linearly independent or dependent.

32. Repeat Problem 31 for

$$A = \begin{bmatrix} 1 & 0 & 2 & 0 & -3 \\ 0 & 1 & -2 & 0 & 3 \\ 0 & 0 & 0 & 1 & 4 \end{bmatrix}$$

C

33. Show that $\{\sqrt{x^2 - 4x + 4}, |2x - 4|\}$ is a linearly dependent subset of F by finding nonzero scalars k_1 and k_2 so that

$$k_1 \sqrt{x^2 - 4x + 4} + k_2 |2x - 4| = 0$$

for all x. [*Hint:* $\sqrt{x^2} = |x|$ for all x]

34. Show that $\{\cos^2 x, 2 \sin^2 x, 3 \cos 2x\}$ is a linearly dependent subset of F by finding nonzero scalars k_1, k_2, and k_3 so that

$$k_1 \cos^2 x + k_2 (2 \sin^2 x) + k_3 (3 \cos 2x) = 0$$

for all x. [*Hint:* $\cos 2x = \cos^2 x - \sin^2 x$ for all x]

35. Show that a set of three vectors in R^2 must be linearly dependent. [*Hint:* Use Theorem 4 in Section 1-5.]

36. Show that a set of four vectors in R^3 must be linearly dependent. [*Hint:* Use Theorem 4 in Section 1-5.]

Problems 37–40 refer to statements from theorems in this section. Prove each statement without reference to the corresponding statement in the text.

37. *Theorem 10(A).* Let S be a finite set of two or more vectors in a vector space V. The set S is linearly dependent if and only if it is possible to express at least one vector in S as a linear combination of the other vectors in S.

38. *Theorem 10(B).* Let S be a finite set of two or more vectors in a vector space V. The set S is linearly independent if and only if it is not possible to express any vector in S as a linear combination of the other vectors in S.

39. *Theorem 11(B).* If S is a set in a vector space V and S contains only one element, then S is linearly dependent if and only if that element is the zero vector.

40. *Theorem 11(C).* If S is a set in a vector space V and S contains exactly two elements, then S is linearly independent if and only if neither vector is a scalar multiple of the other.

In Problems 41–43, let $S = \{v_1, v_2, v_3, v_4\}$ be a subset of a vector space V. Prove each statement.

41. If S is linearly independent and v is a vector in V that is not in span S, then $S \cup \{v\} = \{v_1, v_2, v_3, v_4, v\}$ is linearly independent.

42. If S is linearly independent, then $\{\mathbf{v}_1, \mathbf{v}_2, \mathbf{v}_3\}$ is linearly independent.

43. If $\{\mathbf{v}_1, \mathbf{v}_2, \mathbf{v}_3\}$ is linearly dependent, then S is linearly dependent.

44. Prove the following generalization of Problem 42: If S is a linearly independent set in a vector space V, then any nonempty subset of S is also linearly independent.

45. Prove the following generalization of Problem 43: If S is a finite set in a vector space V containing a linearly dependent subset, then S is also linearly dependent.

46. *Calculus.* Let D be the vector space of functions that are differentiable on $(-\infty, \infty)$. For f and g in D, the function

$$W(x) = \begin{vmatrix} f(x) & g(x) \\ f'(x) & g'(x) \end{vmatrix}$$

is called the **Wronskian** of f and g. In more advanced courses, it is shown that if $W(x)$ is not the zero function, then $\{f, g\}$ is a linearly independent subset of D. Compute $W(x)$ for each of the following subsets of D and use this result to conclude that these are linearly independent sets:

(A) $\{\cos x, \sin x\}$ (B) $\{e^x, e^{2x}\}$

▌5-6▐   Basis and Dimension

- Definition of a Basis
- Properties of a Basis
- Dimension of a Vector Space

▪ Definition of a Basis

In many of the examples in Section 5-4, a given vector space V was shown to contain a finite subset S with the property that span $S = V$. Theorem 12 in Section 5-5 shows that any finite set of vectors S contains a linearly independent subset B with the property that span $S =$ span B. Now we combine these two concepts to define a set, called a *basis*, which turns out to be the most fundamental way of describing a vector space.

Basis

A set of vectors $B = \{\mathbf{u}_1, \mathbf{u}_2, \ldots, \mathbf{u}_n\}$ in a vector space V is said to be a **basis** for V if:

(A) B is linearly independent
(B) span $B = V$

The vectors in B are called **basis vectors.**

Example 33   Show that $B = \{\mathbf{i}, \mathbf{j}, \mathbf{k}\}$ is a basis for R^3.

Solution   In Example 30 (Section 5-5), we showed that B is linearly independent. In Section 5-4, we noted that B spans R^3. Thus, B satisfies both conditions in the definition of a basis. Set B is called the **standard basis for R^3.**

It is important to observe that *any* set B of linearly independent vectors that spans V is a basis for V. The vector space R^3 has infinitely many bases other than the standard basis (see Example 35 for another basis). However, the basis $B = \{\mathbf{i}, \mathbf{j}, \mathbf{k}\}$ is, in many instances, the simplest and most convenient basis for R^3; hence, the name "standard basis."

Problem 33   Show that $B = \{\mathbf{e}_1, \mathbf{e}_2, \mathbf{e}_3, \mathbf{e}_4\}$ is a basis for R^4. [Recall that $\mathbf{e}_1 = (1, 0, 0, 0)$, $\mathbf{e}_2 = (0, 1, 0, 0)$, $\mathbf{e}_3 = (0, 0, 1, 0)$, and $\mathbf{e}_4 = (0, 0, 0, 1)$.] Set B is called the **standard basis for R^4.** ▐▐

The result in Problem 33 is easily generalized to R^n. The **standard basis for R^n** is $B = \{\mathbf{e}_1, \mathbf{e}_2, \ldots, \mathbf{e}_n\}$ where

$$\mathbf{e}_1 = (1, 0, \ldots, 0)$$
$$\mathbf{e}_2 = (0, 1, \ldots, 0)$$
$$\cdot \qquad \cdot$$
$$\cdot \qquad \cdot$$
$$\cdot \qquad \cdot$$
$$\mathbf{e}_n = (0, 0, \ldots, 1)$$

Example 34   Show that $B = \{1, x, x^2, \ldots, x^n\}$ is a basis for P_n.

Solution   Since a polynomial of degree less than or equal to n is defined to be a linear combination of the vectors in B, it follows that span $B = P_n$. In Example 31 (Section 5-5), we showed that B is linearly independent. Thus, B is a basis for P_n. Set B is called the **standard basis for P_n.**

Problem 34   Show that

$$B = \left\{ \begin{bmatrix} 1 & 0 \\ 0 & 0 \end{bmatrix}, \begin{bmatrix} 0 & 1 \\ 0 & 0 \end{bmatrix}, \begin{bmatrix} 0 & 0 \\ 1 & 0 \end{bmatrix}, \begin{bmatrix} 0 & 0 \\ 0 & 1 \end{bmatrix} \right\}$$

is a basis for $M_{2\times2}$. Set B is called the **standard basis for $M_{2\times2}$.** ▐▐

Example 35   Show that $B = \{(1, 1, 1), (2, -1, 2), (-1, -1, 2)\}$ is a basis for R^3.

Solution   We must verify that both conditions in the definition of a basis are satisfied by set B. Let $\mathbf{x} = (x_1, x_2, x_3)$ be any vector in R^3, and consider the vector equation

$$k_1 \begin{bmatrix} 1 \\ 1 \\ 1 \end{bmatrix} + k_2 \begin{bmatrix} 2 \\ -1 \\ 2 \end{bmatrix} + k_3 \begin{bmatrix} -1 \\ -1 \\ 2 \end{bmatrix} = \begin{bmatrix} x_1 \\ x_2 \\ x_3 \end{bmatrix}$$

This is equivalent to the system

$$k_1 + 2k_2 - k_3 = x_1$$
$$k_1 - k_2 - k_3 = x_2$$
$$k_1 + 2k_2 + 2k_3 = x_3$$

or, using matrix notation,

$$A\mathbf{k} = \mathbf{x}$$

where

$$A = \begin{bmatrix} 1 & 2 & -1 \\ 1 & -1 & -1 \\ 1 & 2 & 2 \end{bmatrix} \qquad \mathbf{k} = \begin{bmatrix} k_1 \\ k_2 \\ k_3 \end{bmatrix} \qquad \mathbf{x} = \begin{bmatrix} x_1 \\ x_2 \\ x_3 \end{bmatrix}$$

To show that B spans R^3, we must show that the system $A\mathbf{k} = \mathbf{x}$ has a solution for any $\mathbf{x} \in R^3$. To show that B is linearly independent, we must show that the only solution of the system $A\mathbf{k} = \mathbf{0}$ is $\mathbf{k} = \mathbf{0}$. But Theorem 10 in Section 3-3 shows that the following statements are equivalent for any square matrix A:

(A) $A\mathbf{k} = \mathbf{x}$ has a solution for any $\mathbf{x}$.
(B) The only solution of $A\mathbf{k} = \mathbf{0}$ is $\mathbf{k} = \mathbf{0}$.
(C) $\det(A) \neq 0$

If we show that $\det(A) \neq 0$, then we can conclude that B spans R^3 and is linearly independent.

$$\det(A) = \begin{vmatrix} 1 & 2 & -1 \\ 1 & -1 & -1 \\ 1 & 2 & 2 \end{vmatrix} = \begin{vmatrix} -1 & -1 \\ 2 & 2 \end{vmatrix} - 2\begin{vmatrix} 1 & -1 \\ 1 & 2 \end{vmatrix} - \begin{vmatrix} 1 & -1 \\ 1 & 2 \end{vmatrix} = -9$$

Thus, B is a basis.

Problem 35   Show that $B = \{(1, 0, 1), (1, 1, 2), (0, 1, 3)\}$ is a basis for R^3.   ▌▌

▪ Properties of a Basis

Since a basis for a vector space spans the space, each vector in the space can be expressed as a linear combination of basis vectors. Theorem 13 shows that this linear combination is unique.

Theorem 13

> If $B = \{\mathbf{u}_1, \mathbf{u}_2, \ldots, \mathbf{u}_n\}$ is a basis for a vector space V and $\mathbf{w}$ is a vector in V, then there is exactly one set of scalars $k_1, k_2, \ldots, k_n$ such that
>
> $$\mathbf{w} = k_1\mathbf{u}_1 + k_2\mathbf{u}_2 + \cdots + k_n\mathbf{u}_n$$

Proof   Suppose there exist two sets of scalars, $k_1, k_2, \ldots, k_n$ and $\ell_1, \ell_2, \ldots, \ell_n$, such that

$$\mathbf{w} = k_1\mathbf{u}_1 + k_2\mathbf{u}_2 + \cdots + k_n\mathbf{u}_n$$

and

$$\mathbf{w} = \ell_1\mathbf{u}_1 + \ell_2\mathbf{u}_2 + \cdots + \ell_n\mathbf{u}_n$$

Then

$$\mathbf{0} = \mathbf{w} - \mathbf{w} = (k_1 - \ell_1)\mathbf{u}_1 + (k_2 - \ell_2)\mathbf{u}_2 + \cdots + (k_n - \ell_n)\mathbf{u}_n$$

The linear independence of B now implies that the scalars in this linear combination are all zero. That is,

$$k_1 - \ell_1 = 0, \quad k_2 - \ell_2 = 0, \quad \ldots, \quad k_n - \ell_n = 0$$

Thus,

$$k_1 = \ell_1, \quad k_2 = \ell_2, \quad \ldots, \quad k_n = \ell_n$$

This shows that there is only one way to express each vector $\mathbf{w}$ in terms of basis vectors.   ∎

Example 36   Refer to Example 35. Express the vector $(-1, 2, 2)$ as a linear combination of the vectors in the following basis for R^3:

$$B = \{(1, 1, 1), (2, -1, 2), (-1, -1, 2)\}$$

Solution   The equation

$$k_1 \begin{bmatrix} 1 \\ 1 \\ 1 \end{bmatrix} + k_2 \begin{bmatrix} 2 \\ -1 \\ 2 \end{bmatrix} + k_3 \begin{bmatrix} -1 \\ -1 \\ 2 \end{bmatrix} = \begin{bmatrix} -1 \\ 2 \\ 2 \end{bmatrix}$$

is equivalent to the system

$$\begin{aligned} k_1 + 2k_2 - k_3 &= -1 \\ k_1 - k_2 - k_3 &= 2 \\ k_1 + 2k_2 + 2k_3 &= 2 \end{aligned}$$

The reduced form of the augmented coefficient matrix for this system is

$$\begin{bmatrix} 1 & 0 & 0 & 2 \\ 0 & 1 & 0 & -1 \\ 0 & 0 & 1 & 1 \end{bmatrix}$$

and the solution is $k_1 = 2$, $k_2 = -1$, $k_3 = 1$. Thus,

$$(-1, 2, 2) = 2(1, 1, 1) - (2, -1, 2) + (-1, -1, 2)$$

and this representation is unique.

Problem 36   Refer to Problem 35. Express the vector (3, 1, 4) as a linear combination of the vectors in the following basis for R^3:

$$B = \{(1, 0, 1), (1, 1, 2), (0, 1, 3)\}$$ ∎

As was pointed out earlier, a vector space V may have many different bases. (See Example 33, Example 35, and Problem 35 for three different bases for R^3.) However, the *number* of vectors in a basis for V provides some very important information about the structure of V. Theorem 14 shows that the number of vectors in a basis for V is the maximum number of linearly independent vectors that can be found in V.

Theorem 14

If $B = \{\mathbf{u}_1, \mathbf{u}_2, \ldots, \mathbf{u}_n\}$ is a basis for a vector space V, $S = \{\mathbf{v}_1, \mathbf{v}_2, \ldots, \mathbf{v}_m\}$ is a set of vectors in V, and $m > n$, then S is linearly dependent.

Proof   Since B is a basis for V, each vector in S can be expressed as a linear combination of basis vectors. Thus, there exist scalars a_{ij}, $1 \le i \le n$, $1 \le j \le m$, such that

$$\mathbf{v}_1 = a_{11}\mathbf{u}_1 + a_{21}\mathbf{u}_2 + \cdots + a_{n1}\mathbf{u}_n$$
$$\mathbf{v}_2 = a_{12}\mathbf{u}_1 + a_{22}\mathbf{u}_2 + \cdots + a_{n2}\mathbf{u}_n$$

The reason for using this form of double subscript notation will become apparent as this proof progresses.

$$\vdots$$

$$\mathbf{v}_m = a_{1m}\mathbf{u}_1 + a_{2m}\mathbf{u}_2 + \cdots + a_{nm}\mathbf{u}_n$$

In order to show that S is linearly dependent, we must consider the vector equation

$$k_1\mathbf{v}_1 + k_2\mathbf{v}_2 + \cdots + k_m\mathbf{v}_m = \mathbf{0} \tag{1}$$

Substituting the expressions given above for $\mathbf{v}_1, \mathbf{v}_2, \ldots, \mathbf{v}_m$ in this vector equation, we have

$$\begin{aligned} & k_1(a_{11}\mathbf{u}_1 + a_{21}\mathbf{u}_2 + \cdots + a_{n1}\mathbf{u}_n) && k_1\mathbf{v}_1 \\ & + k_2(a_{12}\mathbf{u}_1 + a_{22}\mathbf{u}_2 + \cdots + a_{n2}\mathbf{u}_n) + \cdots && + k_2\mathbf{v}_2 + \cdots \\ & + k_m(a_{1m}\mathbf{u}_1 + a_{2m}\mathbf{u}_2 + \cdots + a_{nm}\mathbf{u}_n) && + k_m\mathbf{v}_m \\ & = \mathbf{0} && = \mathbf{0} \end{aligned}$$

Collecting the coefficients of each basis vector produces the following linear combination of basis vectors:

$$\begin{aligned} & (a_{11}k_1 + a_{12}k_2 + \cdots + a_{1m}k_m)\mathbf{u}_1 \\ & + (a_{21}k_1 + a_{22}k_2 + \cdots + a_{2m}k_m)\mathbf{u}_2 + \cdots \\ & + (a_{n1}k_1 + a_{n2}k_2 + \cdots + a_{nm}k_m)\mathbf{u}_n \\ & = \mathbf{0} \end{aligned}$$

Since B is linearly independent, the scalars in this linear combination of basis vectors must all be zero. That is,

$$a_{11}k_1 + a_{12}k_2 + \cdots + a_{1m}k_m = 0$$
$$a_{21}k_1 + a_{22}k_2 + \cdots + a_{2m}k_m = 0$$

Notice that this is the usual notation for a system of n equations in m variables.

.

.

.

$$a_{n1}k_1 + a_{n2}k_2 + \cdots + a_{nm}k_m = 0$$

This is a homogeneous system of n linear equations in the m variables k_1, $k_2, \ldots, k_m$. Since we are given that $m > n$, this system must have nontrivial solutions (Theorem 4, Section 1-5). But the solutions of this system are also solutions of the vector equation (1). This implies that there exist scalars k_1, $k_2, \ldots, k_m$, not all zero, such that

$$k_1\mathbf{v}_1 + k_2\mathbf{v}_2 + \cdots + k_m\mathbf{v}_m = \mathbf{0}$$

Thus, S is linearly dependent.   ∎

   Theorem 14 makes it easy to recognize that certain sets are linearly depen-
dent. For example, since $B = \{\mathbf{i}, \mathbf{j}, \mathbf{k}\}$ is a basis for R^3, any set S in R^3 with more than three elements must be linearly dependent. Be careful that you do not use this theorem incorrectly. If S is a subset of R^3 with three or fewer elements, Theorem 14 does not apply. Such a set may be linearly independent or linearly dependent.
   The different bases we have found for R^3 all have one thing in common: They all have three elements. One of the important consequences of Theorem 14 is that all the bases for a given vector space must have the same number of elements.

Theorem 15

> If $B_1 = \{\mathbf{u}_1, \mathbf{u}_2, \ldots, \mathbf{u}_n\}$ and $B_2 = \{\mathbf{v}_1, \mathbf{v}_2, \ldots, \mathbf{v}_m\}$ are both bases for a vector space V, then B_1 and B_2 have the same number of vectors; that is, $n = m$.

Proof   First, suppose $m > n$. Then B_1 is a basis for V and B_2 is a set in V with more elements than B_1. Theorem 14 implies that B_2 is linearly dependent. But this contradicts the fact that B_2 is a basis. Thus, our assumption must be false, and it must be true that $m \leq n$. By reversing the roles of B_1 and B_2, we can also conclude that $n \leq m$. Thus, we have shown that $m \leq n$ and $n \leq m$. This implies that $m = n$.   ∎

▪ Dimension of a Vector Space

If a vector space V has a finite basis (not all do), then Theorem 15 shows that all bases for V have the same number of elements. This unique number is called the *dimension* of the space.

Dimension of a Vector Space

A vector space V with a finite basis is said to be a **finite-dimensional vector space.** The **dimension** of V, denoted dim(V), is the number of elements in a basis for V. The zero vector space has dimension 0. If a nonzero vector space V does not have a finite basis, then V is said to be **infinite-dimensional.**

The zero vector space is the only finite-dimensional vector space that does not have a basis. This is due to the fact that the zero vector space has no linearly independent subsets. (Remember, the set consisting of just the zero vector is linearly dependent.)

We have already seen that R^n and P_{n-1} have bases with n elements. Thus, they are n-dimensional vector spaces. On the other hand, it can be shown that P, the set of all polynomials, does not have a finite basis. Thus, P is an infinite-dimensional vector space.

Example 37 Show that a plane through the origin in R^3 is a two-dimensional subspace of R^3.

Solution We have already seen that a plane through the origin in R^3 can be described by

$$W = \{k\mathbf{u} + \ell\mathbf{v} \mid k, \ell \in R\}$$

where $\mathbf{u}$ and $\mathbf{v}$ are nonzero, nonparallel vectors in R^3 (Example 21, Section 5-4). We also know that two nonparallel vectors in R^3 are linearly independent (Theorem 11, Section 5-5). Thus, $B = \{\mathbf{u}, \mathbf{v}\}$ is a basis for W and W is a two-dimensional subspace of R^3.

Problem 37 Show that a line through the origin in R^3 is a one-dimensional subspace of R^3.

▌▌

The results in Example 37 and Problem 37 show that our definition of dimension agrees with the familiar use of this term. That is, lines are one-dimensional objects and planes are two-dimensional objects.

Example 38 Find a basis for the solution space of the homogeneous system given at the top of the next page.

$$x_1 + x_2 - x_3 - 2x_4 + x_5 = 0$$
$$2x_1 + 3x_2 - x_3 - 7x_4 + 4x_5 = 0$$
$$2x_1 + 4x_2 \qquad - 10x_4 + 6x_5 = 0$$

What is the dimension of the solution space?

Solution The solution space may have many bases. An effective way to find one basis is as follows: The reduced form of the augmented coefficient matrix for this system is (verify this)

$$\begin{bmatrix} 1 & 0 & -2 & 1 & -1 & | & 0 \\ 0 & 1 & 1 & -3 & 2 & | & 0 \\ 0 & 0 & 0 & 0 & 0 & | & 0 \end{bmatrix}$$

and the solution is given by

$$x_1 = 2r - s + t$$
$$x_2 = -r + 3s - 2t$$
$$x_3 = r$$
$$x_4 = s$$
$$x_5 = t$$

where r, s, and t are parameters. Using column vectors, the solution can be written as

$$\begin{bmatrix} x_1 \\ x_2 \\ x_3 \\ x_4 \\ x_5 \end{bmatrix} = r \begin{bmatrix} 2 \\ -1 \\ 1 \\ 0 \\ 0 \end{bmatrix} + s \begin{bmatrix} -1 \\ 3 \\ 0 \\ 1 \\ 0 \end{bmatrix} + t \begin{bmatrix} 1 \\ -2 \\ 0 \\ 0 \\ 1 \end{bmatrix}$$

Thus, the set

$$B = \{(2, -1, 1, 0, 0), (-1, 3, 0, 1, 0), (1, -2, 0, 0, 1)\}$$

spans the solution space of this system. To determine whether B is a basis, we must consider the vector equation

$$k_1(2, -1, 1, 0, 0) + k_2(-1, 3, 0, 1, 0) + k_3(1, -2, 0, 0, 1) = \mathbf{0}$$

or, equivalently, the system

$$2k_1 - k_2 + k_3 = 0$$
$$-k_1 + 3k_2 - 2k_3 = 0$$
$$k_1 \qquad = 0$$
$$k_2 \qquad = 0$$
$$k_3 = 0$$

The form of this system makes it easy to see that the only solution is $k_1 = k_2 = k_3 = 0$. This shows that B is linearly independent. Thus, B is a basis for the solution space. Since B has three elements in it, the solution space is three-dimensional.

Problem 38   Find a basis for the solution space of the system

$$x_1 - x_2 - 3x_3 - x_4 + 4x_5 = 0$$
$$2x_1 - x_2 - 4x_3 - 3x_4 + 3x_5 = 0$$
$$x_1 + x_2 - x_4 - 4x_5 = 0$$

What is the dimension of the solution space?   ▌▌

The following useful theorem is an immediate consequence of Theorem 14 and the definition of the dimension of a vector space:

Theorem 16   Any set S of m vectors, $m > n$, in an n-dimensional vector space V is linearly dependent.

Proof   From the fact that $\dim(V) = n$, the definition of dimension implies the existence of a basis B for V with exactly n vectors. Since S is a set of m vectors in V with $m > n$, then by Theorem 14, S must be linearly dependent.   ▌▌

Suppose we are given a set B in a vector space V. What do we have to do to show that B is a basis? In general, if we have no information concerning the dimension of V, then we will have to show that B satisfies both conditions in the definition of a basis. That is, we must show that B spans V and that B is linearly independent. This is the approach we used in Example 38. Now suppose that we know that V is n-dimensional and that B has n elements. In this situation, it turns out that it is sufficient to show that B satisfies either one of the conditions for a basis.

Theorem 17   Let $B = \{\mathbf{u}_1, \mathbf{u}_2, \ldots, \mathbf{u}_n\}$ be a set of n vectors in an n-dimensional vector space V.

(A) If B is linearly independent, then B is a basis for V.
(B) If B spans V, then B is a basis for V.

Proof   (A) *Assume B is a linearly independent set of n vectors in an n-dimensional vector space V. To show that B is a basis for V, we must show that B spans V. That is, given any $\mathbf{w} \in V$, we must show that $\mathbf{w}$ can be expressed as a linear combination of elements in B. If $\mathbf{w} \in B$, we are finished, since each element in B can be expressed as a linear combination of the elements in $B(\mathbf{u}_i =$*

$1 \cdot \mathbf{u}_i$). Now assume $\mathbf{w} \notin B$, and let $S = B \cup \{\mathbf{w}\} = \{\mathbf{u}_1, \mathbf{u}_2, \ldots, \mathbf{u}_n, \mathbf{w}\}$. Since S has $n + 1$ elements and $\dim(V) = n$, Theorem 16 implies that S is linearly dependent. Thus, there exist scalars $k_1, k_2, \ldots, k_n, k_{n+1}$, not all zero, such that

$$k_1\mathbf{u}_1 + k_2\mathbf{u}_2 + \cdots + k_n\mathbf{u}_n + k_{n+1}\mathbf{w} = \mathbf{0} \tag{2}$$

Suppose $k_{n+1} = 0$. Then at least one of the other scalars is nonzero. But this implies that we have a nontrivial linear combination of vectors from B equal to the zero vector. This contradicts the fact that B is linearly independent. Thus, $k_{n+1} \neq 0$. Now we can solve (2) for $\mathbf{w}$:

$$\mathbf{w} = -\frac{k_1}{k_{n+1}}\mathbf{u}_1 - \frac{k_2}{k_{n+1}}\mathbf{u}_2 - \cdots - \frac{k_n}{k_{n+1}}\mathbf{u}_n$$

This shows that $\mathbf{w} \in \operatorname{span} B$ and, consequently, that $\operatorname{span} B = V$. Thus, B is a basis for V.

(B) Assume B is a set of n vectors in an n-dimensional vector space V such that $\operatorname{span} B = V$. To show that B is a basis for V, we must show that B is linearly independent. By Theorem 12 in Section 5-5, B contains a linearly independent subset, which we will denote by A, with the property that

$$\operatorname{span} A = \operatorname{span} B = V$$

Since A is a linearly independent subset of V that spans V, A is a basis for V. Since V is n-dimensional, A must have n elements. But A is a subset of B, which also has n elements. The only subset of a finite set with the same number of elements is the set itself. That is, A and B must be the same set. Since A is a basis, this shows that B is a basis. ∎

Do not be misled by Theorem 17. It does not state that we can *always* dispense with one of the conditions for a basis. We can do this only if we know that the number of vectors in the subset B is the same as the dimension of the vector space V. Thus, in order to use Theorem 17, we must know the dimension of V. This means that the first time a vector space is encountered, its dimension is determined by finding a set B that satisfies *both* conditions for a basis. After that, any set that has the required number of elements and satisfies *either* condition in Theorem 17 is also a basis.

Example 39 Show that $B = \{3, 4 - x, 2 + 5x + x^2\}$ is a basis for P_2.

Solution We already know that P_2 is a three-dimensional vector space ($\{1, x, x^2\}$ is the standard basis for P_2). Since B has three elements, Theorem 17 applies, and we can show that B is a basis by showing that B is linearly independent or by showing that $\operatorname{span} B = P_2$. We will show that B is linearly independent. Consider the equation

$$k_1(3) + k_2(4 - x) + k_3(2 + 5x + x^2) = 0$$

or, equivalently, the system

$$3k_1 + 4k_2 + 2k_3 = 0 \qquad \text{Constant terms}$$
$$- \ k_2 + 5k_3 = 0 \qquad \text{Coefficients of x}$$
$$k_3 = 0 \qquad \text{Coefficients of x}^2$$

The only solution to this triangular system is $k_1 = k_2 = k_3 = 0$. Thus, B is linearly independent and forms a basis for P_2.

Problem 39 Show that $B = \{2, 3 + x, 4 - 2x + x^2\}$ is a basis for P_2.   ▊

In Theorem 12 in Section 5-5, we saw that if S spans a vector space V, then S contains a linearly independent subset that also spans V. Thus, any set that spans a vector space contains a basis for that space. Now suppose we have a linearly independent set S that does not span all of V. Theorem 18 shows that we can find a basis for V that contains S. Thus, every linearly independent set in a vector space V is a subset of a basis for V.

Theorem 18

> If $S = \{\mathbf{u}_1, \mathbf{u}_2, \ldots, \mathbf{u}_m\}$ is a linearly independent subset of an n-dimensional vector space V and $m < n$, then there is a basis for V that contains S.

Proof Let $\mathbf{w}_1$ be a nonzero vector in V that is not in span S. Then

$$S_1 = \{\mathbf{u}_1, \mathbf{u}_2, \ldots, \mathbf{u}_m, \mathbf{w}_1\}$$

is linearly independent. (Why?) If span $S_1 = V$, then S_1 is a basis containing S. If span $S_1 \neq V$, then there must be a nonzero vector $\mathbf{w}_2$ in V that is not in span S_1. The set

$$S_2 = \{\mathbf{u}_1, \mathbf{u}_2, \ldots, \mathbf{u}_m, \mathbf{w}_1, \mathbf{w}_2\}$$

is also linearly independent. (Why?) If span $S_2 = V$, then S_2 is a basis containing S. If not, then this process can be repeated until we obtain a set

$$S_{n-m} = \{\mathbf{u}_1, \mathbf{u}_2, \ldots, \mathbf{u}_m, \mathbf{w}_1, \ldots, \mathbf{w}_{n-m}\}$$

of n linearly independent vectors. By Theorem 17, S_{n-m} is a basis for V that contains S.   ▊

Suppose S is a linearly independent m-element subset of an n-dimensional vector space V and $m < n$. Theorem 18 implies that S can always be extended (by adding selected vectors from V) to form a basis for V. But Theorem 18 does not tell us how to carry out this extension. In the next chapter, we will develop a procedure for finding a basis that contains S.

33. Set B spans R^4 (Example 24, Section 5-4) and is linearly independent (Problem 30, Section 5-5).

34. Set B spans $M_{2\times2}$ (Problem 24, Section 5-4) and is linearly independent (Problem 31, Section 5-5).

35. The equivalent system is $A\mathbf{k} = \mathbf{x}$ where $A = \begin{bmatrix} 1 & 1 & 0 \\ 0 & 1 & 1 \\ 1 & 2 & 3 \end{bmatrix}$; $\det(A) = 2 \neq 0$ implies B is a basis.

36. $2(1, 0, 1) + (1, 1, 2) + 0(0, 1, 3)$

37. $B = \{\mathbf{v}\}$, where $\mathbf{v} \neq \mathbf{0}$, is a basis for the line through the origin in the direction of $\mathbf{v}$.

38. $B = \{(4, -3, 2, 1, 0), (3, 1, 2, 0, 1)\}$; 2

39. $k_1(2) + k_2(3 + x) + k_3(4 - 2x + x^2) = 0$ implies $k_1 = k_2 = k_3 = 0$; B is a linearly independent set with three elements; hence, B is a basis.

‖ Exercise 5-6

Many of the problems in this exercise set involve finding a basis for a vector space. Since most vector spaces have many different bases, these problems have many different correct answers. In most cases, if you apply the techniques illustrated in the examples in this section, you should obtain the same basis as we have listed in the answer section. If you find a different basis, be certain to check your work carefully.

A In Problems 1–12, determine whether B is a basis for the indicated vector space.

1. $B = \{(1, -2), (2, 4)\} \subset R^2$

2. $B = \{(1, -2), (-2, 4)\} \subset R^2$

3. $B = \{(0, 0), (1, -2), (2, 4)\} \subset R^2$

4. $B = \{(1, -2)\} \subset R^2$

5. $B = \{(1, 2, -1), (-1, 6, 1)\} \subset R^3$

6. $B = \{(1, 2, -1), (-1, 6, 1), (2, 4, 2)\} \subset R^3$

7. $B = \{(1, 2, -1), (-1, 6, 1), (2, 4, -2)\} \subset R^3$

8. $B = \{(1, 2, -1), (-1, 6, 1), (2, 4, -2), (2, 4, 2)\} \subset R^3$

9. $B = \{1 + 2x - x^2, 1 - x + x^2, 1 + 5x + 2x^2\} \subset P_2$

10. $B = \{1 + 2x - x^2, 1 - x + x^2, 1 + 5x - 3x^2\} \subset P_2$

11. $B = \left\{ \begin{bmatrix} 1 & 1 \\ 1 & 0 \end{bmatrix}, \begin{bmatrix} 0 & 1 \\ 1 & 1 \end{bmatrix}, \begin{bmatrix} 1 & 0 \\ 0 & 1 \end{bmatrix}, \begin{bmatrix} 0 & 1 \\ 1 & 0 \end{bmatrix} \right\} \subset M_{2\times2}$

12. $B = \left\{ \begin{bmatrix} 1 & 0 \\ 0 & 1 \end{bmatrix}, \begin{bmatrix} 1 & 0 \\ 0 & -1 \end{bmatrix}, \begin{bmatrix} 0 & 1 \\ 1 & 0 \end{bmatrix}, \begin{bmatrix} 0 & 1 \\ -1 & 0 \end{bmatrix} \right\} \subset M_{2\times2}$

B In Problems 13 and 14, show that B is a basis for R^2. Express the vector $(1, 2)$ as a linear combination of the vectors in B.

13. $B = \{(2, -1), (4, -1)\}$

14. $B = \{(3, 4), (1, 1)\}$

In Problems 15 and 16, show that B is a basis for P_2. Express the vector $3 + x + 4x^2$ as a linear combination of the vectors in B.

15. $B = \{1 - x, x - x^2, 1 + x^2\}$ **16.** $B = \{1, 1 + x, 1 + x + x^2\}$

In Problems 17–24, find a basis for the indicated subspace of R^3.

17. $\{(x, y, z)|x = 0\}$ **18.** $\{(x, y, z)|x = y\}$

19. $\left\{(x, y, z) \left| \dfrac{x}{2} = \dfrac{y}{-3} = \dfrac{z}{4} \right. \right\}$ **20.** $\left\{(x, y, z) \left| \dfrac{x}{-1} = \dfrac{y}{5} = \dfrac{z}{2} \right. \right\}$

21. $\{(x, y, z)|2x - 4y + 5z = 0\}$ **22.** $\{(x, y, z)|x + 3y - 2z = 0\}$

23. $\text{span}\{(4, -2, 6), (6, -2, 4), (-10, 5, -15)\}$

24. $\text{span}\{(2, -1, 1), (-2, 2, 1), (0, 1, 2)\}$

In Problems 25–28, find a basis for the indicated subspace of P_2.

25. $\{p(x) \in P_2|p(1) = 0\}$ **26.** $\{p(x) \in P_2|p(1) + p(-1) = 0\}$

27. $\text{span}\{2 - x^2, 2 + x^2, x^2\}$ **28.** $\text{span}\{1 + 2x, 2x + 3x^2, 4 + x^2\}$

In Problems 29–34, find a basis for the solution space of the indicated system. Find the dimension of each solution space.

29. $\begin{aligned} x_1 - 3x_2 + 4x_3 &= 0 \\ 2x_1 - 4x_2 + 4x_3 &= 0 \end{aligned}$ **30.** $\begin{aligned} x_1 - 2x_2 + 3x_3 &= 0 \\ 3x_1 - 6x_2 + 10x_3 &= 0 \end{aligned}$

31. $\begin{aligned} x_1 - x_2 + x_3 - 2x_4 &= 0 \\ 2x_1 - 2x_2 + 3x_3 - 4x_4 &= 0 \end{aligned}$ **32.** $\begin{aligned} x_1 - 2x_2 - 3x_3 + 5x_4 &= 0 \\ 2x_1 - 3x_2 - 5x_3 + 6x_4 &= 0 \end{aligned}$

33. $\begin{aligned} x_1 + 2x_2 + 3x_3 \quad\quad - 7x_5 &= 0 \\ x_1 + x_2 - x_3 + x_4 - 5x_5 &= 0 \\ -x_1 + x_2 + 9x_3 - 3x_4 + x_5 &= 0 \end{aligned}$ **34.** $\begin{aligned} x_1 - x_2 + x_3 - x_4 + x_5 &= 0 \\ 3x_1 - 3x_2 + x_3 + 3x_4 + x_5 &= 0 \\ x_1 - x_2 + 2x_3 - 4x_4 + 3x_5 &= 0 \end{aligned}$

35. Find a basis for

$$D_2 = \left\{ \begin{bmatrix} a & 0 \\ 0 & b \end{bmatrix} \middle| a, b \in R \right\}$$

What is the dimension of D_2?

36. Find a basis for

$$S_2 = \left\{ \begin{bmatrix} a & b \\ b & c \end{bmatrix} \middle| a, b, c \in R \right\}$$

What is the dimension of S_2?

37. Find a basis for $M_{2 \times 3}$. What is the dimension of $M_{2 \times 3}$?

38. Find a basis for $M_{3 \times 2}$. What is the dimension of $M_{3 \times 2}$?

C   In Problems 39 and 40, show that B is a basis for R^2. Then express an arbitrary vector $\mathbf{x} = (x_1, x_2)$ as a linear combination of the vectors in B.

39. $B = \{(2, 1), (1, 1)\}$ **40.** $B = \{(2, 5), (3, 8)\}$

In Problems 41 and 42, show that B is a basis for R^3. Then express an arbitrary vector $\mathbf{x} = (x_1, x_2, x_3)$ as a linear combination of the vectors in B.

41. $B = \{(1, 1, 0), (0, 1, 1), (1, 1, 1)\}$
42. $B = \{(1, 0, -1), (1, -1, 0), (0, -2, 1)\}$
43. *Calculus.* Find a basis for $V = \{p(x) \in P_2 | \int_{-1}^{1} p(x)\, dx = 0\}$.
44. *Calculus.* Find a basis for $V = \{p(x) \in P_2 | p'(0) = p''(0)\}$.
45. *Chebyshev polynomials.* Show that

$$T = \{1, x, 2x^2 - 1, 4x^3 - 3x, 8x^4 - 8x^2 + 1\}$$

is a basis for P_4. The functions in T are called *Chebyshev polynomials* and are used in scientific and engineering applications.

46. *Hermite polynomials.* Show that

$$H = \{1, 2x, 4x^2 - 2, 8x^3 - 12x, 16x^4 - 48x^2 + 12\}$$

is a basis for P_4. The functions in H are called *Hermite polynomials* and are used in scientific and engineering applications.

In Problems 47–50, V is a finite-dimensional vector space and W is a subspace of V. Prove each statement.

47. W is finite-dimensional.
48. $\dim(W) \le \dim(V)$
49. If $\dim(W) = \dim(V)$, then $W = V$.
50. If B is a basis for W, then there exists a basis B^* for V with the property that $B \subset B^*$.
51. Show that every subspace of R^3 is either the zero space, a line through the origin, a plane through the origin, or R^3.
52. Show that P, the vector space of all polynomials, is an infinite-dimensional vector space.

|5-7| Chapter Review

Important Terms and Symbols

5-1. *n-Dimensional Euclidean space.* n-Dimensional space, n-space, ordered n-tuple, n-dimensional vectors, equality of vectors, definitions and properties of addition and scalar multiplication, dot product, norm, distance between two vectors, unit vector, Euclidean n-space, standard unit vectors, row and column vectors of a matrix, R^n, $\mathbf{u} \cdot \mathbf{v}$, $\|\mathbf{u}\|$, $d(\mathbf{u}, \mathbf{v})$, $\mathbf{e}_i$

5-2. *Definition of a vector space.* Polynomial operations and properties, matrix operations and properties, vector space, addition and scalar multiplication properties of a vector space, zero vector space, function operations, P_n, $M_{m \times n}$, P, F

5-3. *Subspaces.* Subspace, characterization of a subspace, solution space of a homogeneous system

5-4. *Linear combinations.* Linear combination, span of a set of vectors, span S

5-5. *Linear independence.* Linearly dependent vectors, linearly independent vectors, dependency relationship, solutions of homogeneous systems

5-6. *Basis and dimension.* Basis, basis vectors, standard bases for R^n, P_n, and $M_{2\times2}$, dimension, finite-dimensional vector space, infinite-dimensional vector space, dim(V)

▌▌ Exercise 5-7 Chapter Review

Work through all the problems in this chapter review and check your answers in the back of the book. (Answers to most review problems are there.) Where weaknesses show up, review appropriate sections in the text.

A   **1.** Let $\mathbf{u} = (1, -2, 0, 3)$ and $\mathbf{v} = (2, 1, -4, 2)$.

(A) Find $2\mathbf{u} - 3\mathbf{v}$.   (B) Find $\mathbf{u} \cdot \mathbf{v}$.   (C) Find $\|\mathbf{u}\|$.   (D) Find $\|\mathbf{u} - \mathbf{v}\|$.
(E) Find a unit vector in the direction of $\mathbf{v}$.

2. Let $p(x) = 2 - 2x + x^2$ and $q(x) = 3 + 2x - x^2$.

(A) Find $3p(x) - 5q(x)$.
(B) Graph $p(x)$ and $2p(x)$ on the same set of axes.

3. Let $f(x) = \sqrt{x^2 + 9}$ and $g(x) = |x|$.

(A) Find $(f - g)(0)$.   (B) Find $(2f - 3g)(-4)$.
(C) Graph f, g, and $f + g$ on the same set of axes.

In Problems 4–11, express $\mathbf{w}$ *as a linear combination of the vectors in S, if possible.*

4. $\mathbf{w} = (2, 1)$; $S = \{(4, -5), (-2, 3)\} \subset R^2$
5. $\mathbf{w} = (4, -6)$; $S = \{(4, 5), (-2, 3)\} \subset R^2$
6. $\mathbf{w} = (2, 4, 3)$; $S = \{(1, -1, 0), (2, 0, 1)\} \subset R^3$
7. $\mathbf{w} = (2, 3, -4)$; $S = \{(1, -1, 0), (2, 0, 1)\} \subset R^3$
8. $\mathbf{w} = 3x + 2x^2$; $S = \{2 + 3x, 1 - x^2, 1 + 3x + x^2\} \subset P_2$
9. $\mathbf{w} = -3x + 2x^2$; $S = \{2 + 3x, 1 - x^2, 1 + 3x + x^2\} \subset P_2$
10. $\mathbf{w} = \begin{bmatrix} -1 & 0 \\ 4 & 1 \end{bmatrix}$; $S = \left\{ \begin{bmatrix} 1 & -1 \\ 2 & 1 \end{bmatrix}, \begin{bmatrix} 2 & -1 \\ 1 & 2 \end{bmatrix}, \begin{bmatrix} -1 & 1 \\ 1 & 1 \end{bmatrix} \right\} \subset M_{2\times2}$
11. $\mathbf{w} = \begin{bmatrix} 1 & 1 \\ -4 & 1 \end{bmatrix}$; $S = \left\{ \begin{bmatrix} 1 & -1 \\ 2 & 1 \end{bmatrix}, \begin{bmatrix} 2 & -1 \\ 1 & 2 \end{bmatrix}, \begin{bmatrix} -1 & 1 \\ 1 & 1 \end{bmatrix} \right\} \subset M_{2\times2}$

B   *In Problems 12–19, determine whether W is a subspace of the indicated vector space.*

12. $W = \{(0, x)|x \in R\} \subset R^2$ **13.** $W = \{(3, x)|x \in R\} \subset R^2$

14. $W = \{(x, y, z)|x + y + z = 0\} \subset R^3$ **15.** $W = \{(x, y, z)|xy = 0\} \subset R^3$

16. $W = \{a_0 + a_1x + a_2x^2|a_0 + a_1 = 0\} \subset P_2$

17. $W = \{a_0 + a_1x + a_2x^2|a_0 + a_1 = 1\} \subset P_2$

18. $W = \left\{\begin{bmatrix} 0 & a \\ 1-a & 0 \end{bmatrix}\Big| a \in R\right\} \subset M_{2\times2}$

19. $W = \left\{\begin{bmatrix} 0 & a \\ -a & 0 \end{bmatrix}\Big| a \in R\right\} \subset M_{2\times2}$

In Problems 20–23, determine whether S is a linearly dependent or linearly independent subset of the indicated vector space. If S is linearly dependent, find a dependency relationship for the vectors in S.

20. $S = \{(-2, 0, 3), (1, 1, -1), (2, 4, -1)\} \subset R^3$

21. $S = \{(1, -1, 1), (-3, 2, 4), (2, 1, -1)\} \subset R^3$

22. $S = \{1 + x + x^3, 2 - x + x^2, x + x^2 - x^3\} \subset P_3$

23. $S = \{2 - x + x^3, -1 + 2x - x^2, 1 + 4x - 3x^2 + 2x^3\} \subset P_3$

In Problems 24–31, W is a subspace of the indicated vector space. Find a basis for W. Find the dimension of W. (Remember, there are many different possible bases for most vector spaces.)

24. $W = \{(2x, -x)|x \in R\} \subset R^2$

25. $W = \left\{(x, y, z)|x = \dfrac{y}{2} = \dfrac{z}{-3}\right\} \subset R^3$

26. $W = \{(x, y, z)|x - 2y + 5z = 0\} \subset R^3$

27. $W = \text{span}\{(1, -2, 1), (-3, 6, -3), (2, 0, -1)\} \subset R^3$

28. $W = \{p|p(-x) = -p(x) \text{ for all } x\} \subset P_2$

29. $W = \{p|p(x) = p(-x) \text{ for all } x\} \subset P_2$

30. $W = \left\{\begin{bmatrix} a & -b \\ b & a \end{bmatrix}\Big| a, b \in R\right\} \subset M_{2\times2}$

31. $W = \left\{\begin{bmatrix} a & a \\ b & c \end{bmatrix}\Big| a, b, c \in R\right\} \subset M_{2\times2}$

In Problems 32–35, find a basis for the solution space of each system. Find the dimension of the solution space. (Remember, there are many different possible bases for most vector spaces.)

32. $x_1 + x_2 + x_3 = 0$
$x_1 + 2x_2 + 5x_3 = 0$

33. $x_1 + x_2 + x_3 + x_4 = 0$
$x_1 + 3x_2 + 5x_3 - x_4 = 0$

34. $x_1 + 2x_2 - x_3 - x_4 + 5x_5 = 0$
$2x_1 + 4x_2 + x_3 - 8x_4 + 7x_5 = 0$
$x_1 + 2x_2 + 5x_3 - 13x_4 - x_5 = 0$

35. $x_1 + 2x_2 - x_3 - x_4 + 5x_5 = 0$
$-2x_1 - 3x_2 + 3x_3 + x_4 - 8x_5 = 0$
$-x_1 + x_2 + 5x_3 - 6x_4 - 2x_5 = 0$

36. Show that $B = \{(4, -3), (-3, 2)\}$ is a basis for R^2, and express $\mathbf{w} = (4, -5)$ as a linear combination of the vectors in B.

37. Show that $B = \{1 - 3x + x^2, 2 - 3x + x^2, 3 - 2x + x^2\}$ is a basis for P_2. Express $\mathbf{w} = 1 + x + x^2$ as a linear combination of the vectors in B.

38. Show that

$$B = \left\{\begin{bmatrix} 1 & -1 \\ 1 & 0 \end{bmatrix}, \begin{bmatrix} 0 & 1 \\ 1 & -2 \end{bmatrix}, \begin{bmatrix} 1 & -1 \\ 0 & 1 \end{bmatrix}, \begin{bmatrix} 1 & 0 \\ 1 & 0 \end{bmatrix}\right\}$$

is a basis for $M_{2\times2}$. Express

$$\mathbf{w} = \begin{bmatrix} 4 & 3 \\ 2 & 1 \end{bmatrix}$$

as a linear combination of basis vectors.

C **39.** Show that $S = \{\sin^2 x, \cos^2 x, \tan^2 x, \sec^2 x\}$ is a linearly dependent subset of F. Find a dependency relationship for the vectors in S. [*Hint:* $\sin^2 x + \cos^2 x = 1$ and $1 + \tan^2 x = \sec^2 x$ for all x.]

In Problems 40 and 41, show that the rows of A form a linearly independent set of vectors in R^n, where n is the number of columns in A.

40. $A = \begin{bmatrix} 1 & 0 & 0 & a \\ 0 & 1 & 0 & b \\ 0 & 0 & 1 & c \end{bmatrix}$ **41.** $A = \begin{bmatrix} 1 & a & 0 & 0 & b \\ 0 & 0 & 1 & 0 & c \\ 0 & 0 & 0 & 1 & d \end{bmatrix}$

42. Show that $\{(a, b), (c, d)\}$ is linearly independent if and only if $ad - bc \neq 0$.

43. Let a and b be nonzero real numbers satisfying $a^2 + b^2 = 1$.

(A) Show that $B = \{(a, b), (-b, a)\}$ is a basis for R^2.
(B) If $\mathbf{x} = (x_1, x_2)$ is any vector in R^2, express $\mathbf{x}$ as a linear combination of the basis vectors in B.

44. Let $B = \{\mathbf{v}_1, \mathbf{v}_2, \mathbf{v}_3\}$ be a basis for a vector space V. Determine which of the following sets also form a basis for V:

(A) $\{\mathbf{v}_1, \mathbf{v}_2\}$ (B) $\{\mathbf{0}, \mathbf{v}_1, \mathbf{v}_2, \mathbf{v}_3\}$
(C) $\{\mathbf{v}_1, \mathbf{v}_2, \mathbf{v}_2 + \mathbf{v}_3\}$ (D) $\{\mathbf{v}_1, \mathbf{v}_2, \mathbf{v}_3, \mathbf{v}_2 + \mathbf{v}_3\}$

45. Let S be a finite set of nonzero vectors in a vector space V and let $W = \operatorname{span} S$. Show that S contains a subset B that is a basis for W.

46. *Calculus.* Show that $W = \{p \in P_2 | \int_{-2}^{2} p(x) \, dx = 0\}$ is a subspace of P_2, and find a basis for W.

47. *Calculus.* Let

$$D^{(2)} = \{f \in F | f''(x) \text{ exists for } -\infty < x < \infty\}$$

and

$$W = \{f \in D^{(2)} | f'' - 2f' + 3f = 0\}$$

Show that W is a subspace of $D^{(2)}$.

48. *Laguerre polynomials.* Show that

$$L = \{1, 1 - x, 2 - 4x + x^2, 6 - 18x + 9x^2 - x^3\}$$

is a basis for P_3. The functions in L are called *Laguerre polynomials* and are used in scientific and engineering applications.

Problems 49–51 refer to the set of $n \times n$ skew-symmetric matrices defined by

$$K_n = \{A \in M_{n \times n} | A^T = -A\}$$

49. Show that K_n is a subspace of $M_{n \times n}$. (See Theorem 6 in Section 3-3 for the properties of the transpose operation.)

50. Find a basis for K_2. What is the dimension of this space?

51. Find a basis for K_3. What is the dimension of this space?

52. Let $\mathbf{u}$ be a fixed vector in R^n and let $W = \{\mathbf{w} \in R^n | \mathbf{u} \cdot \mathbf{w} = 0\}$. Show that W is a subspace of R^n.

53. If $\mathbf{u}$ and $\mathbf{v}$ are nonzero vectors in R^n and $\mathbf{u} \cdot \mathbf{v} = 0$, show that $\mathbf{u}$ and $\mathbf{v}$ are linearly independent.

54. Let $W = \text{span}(\mathbf{u}, \mathbf{v})$, where $\mathbf{u}$ and $\mathbf{v}$ are two linearly independent vectors in R^n, and let $\mathbf{w} = \mathbf{u} - (\mathbf{u} \cdot \mathbf{v}/\|\mathbf{v}\|^2)\mathbf{v}$.

(A) Show that $\mathbf{w} \cdot \mathbf{v} = 0$.

(B) Show that $B = \{\mathbf{v}, \mathbf{w}\}$ is a basis for W.

|6| Additional Vector Space Topics

| 6 | Contents

The most fundamental method for describing a finite-dimensional vector space is to exhibit a basis for the space. The basis provides a convenient way to represent each vector in the space and also determines the dimension of the space. In the first two sections of this chapter, we will develop several procedures for finding bases for subspaces in R^n. In Section 6-3, these procedures are generalized to other finite-dimensional vector spaces. In Section 6-4, we will study the relationship between two different bases for the same vector space. Section 6-5 is concerned with generalizing the concept of the dot product to spaces other than R^n. In the last section, we will use this generalized dot product to construct a special type of basis that has many important applications.

| 6-1 | Row and Column Spaces — An Introduction

- Definition of Row and Column Spaces
- Row and Column Spaces of Reduced Matrices
- The Row Space of an Arbitrary Matrix
- Finding a Basis for Span S

■ Definition of Row and Column Spaces

In Chapter 5, we often used the interplay between vector equations and systems of linear equations to answer questions related to linear independence and to spanning sets. We now continue to investigate this relationship by associating two special vector spaces with each matrix A. These vector spaces are called the *row space* of A and the *column space* of A and are defined as follows:

The Row and Column Spaces of a Matrix

Let

$$A = \begin{bmatrix} a_{11} & a_{12} & \cdots & a_{1n} \\ a_{21} & a_{22} & \cdots & a_{2n} \\ \vdots & \vdots & & \vdots \\ a_{m1} & a_{m2} & \cdots & a_{mn} \end{bmatrix}$$

be an $m \times n$ matrix with real elements. Let

$$\mathbf{r}_i = [a_{i1} \quad a_{i2} \quad \cdots \quad a_{in}] \qquad 1 \le i \le m$$

be the row vectors of A, and let

$$\mathbf{c}_j = \begin{bmatrix} a_{1j} \\ a_{2j} \\ \vdots \\ \vdots \\ a_{mj} \end{bmatrix} \qquad 1 \le j \le n$$

be the column vectors of A. The subspace of R^n spanned by $\{\mathbf{r}_1, \mathbf{r}_2, \ldots, \mathbf{r}_m\}$ is called the **row space** of A, and the subspace of R^m spanned by $\{\mathbf{c}_1, \mathbf{c}_2, \ldots, \mathbf{c}_n\}$ is called the **column space** of A.

Since the row and column spaces of a matrix are defined as spans of sets of vectors, it is a simple matter to find these spaces. For example, if

$$A = \begin{matrix} \mathbf{c}_1 & \mathbf{c}_2 & \mathbf{c}_3 \\ \begin{bmatrix} 1 & -2 & 3 \\ 2 & 4 & -1 \end{bmatrix} & & \end{matrix} \begin{matrix} \mathbf{r}_1 \\ \mathbf{r}_2 \end{matrix}$$

then

Row space of $A = \text{span}\{(1, -2, 3), (2, 4, -1)\} \subset R^3$

and

Column space of $A = \text{span} \left\{ \begin{bmatrix} 1 \\ 2 \end{bmatrix}, \begin{bmatrix} -2 \\ 4 \end{bmatrix}, \begin{bmatrix} 3 \\ -1 \end{bmatrix} \right\} \subset R^2$

Unfortunately, these descriptions do not give us much information concerning these two spaces. In particular, we cannot assume that $\{\mathbf{r}_1, \mathbf{r}_2\}$ is a basis for the row space or that $\{\mathbf{c}_1, \mathbf{c}_2, \mathbf{c}_3\}$ is a basis for the column space.

In this and the next section, we will use elementary row operations and properties of reduced matrices to develop systematic procedures for finding bases for the row and column spaces of any matrix. This will provide us with several methods for finding bases of more general vector spaces that work with any finite-dimensional space and that can be implemented on a computer.

We begin by considering some specific examples that will lead to some general observations.

Example 1   Find bases for the row and column spaces of I_4, the 4×4 identity matrix.

Solution   The 4×4 identity matrix is

$$I_4 = \begin{matrix} & \mathbf{c}_1 & \mathbf{c}_2 & \mathbf{c}_3 & \mathbf{c}_4 \\ & \begin{bmatrix} 1 & 0 & 0 & 0 \\ 0 & 1 & 0 & 0 \\ 0 & 0 & 1 & 0 \\ 0 & 0 & 0 & 1 \end{bmatrix} & & & & \begin{matrix} \mathbf{r}_1 \\ \mathbf{r}_2 \\ \mathbf{r}_3 \\ \mathbf{r}_4 \end{matrix} \end{matrix}$$

The row vectors of I_4 are the standard unit vectors for R^4, written as 1×4 row vectors and denoted by

$$\mathbf{r}_1 = [1 \quad 0 \quad 0 \quad 0] = [\mathbf{e}_1] \qquad \mathbf{r}_3 = [0 \quad 0 \quad 1 \quad 0] = [\mathbf{e}_3]$$
$$\mathbf{r}_2 = [0 \quad 1 \quad 0 \quad 0] = [\mathbf{e}_2] \qquad \mathbf{r}_4 = [0 \quad 0 \quad 0 \quad 1] = [\mathbf{e}_4]$$

Thus, $\{\mathbf{r}_1, \mathbf{r}_2, \mathbf{r}_3, \mathbf{r}_4\}$ is a basis for R^4 and

Row space of $I_4 = \text{span}\{\mathbf{r}_1, \mathbf{r}_2, \mathbf{r}_3, \mathbf{r}_4\} = R^4$

The column vectors of I_4 are the same vectors, written as 4×1 column vectors and denoted by

$$\mathbf{c}_1 = \begin{bmatrix} 1 \\ 0 \\ 0 \\ 0 \end{bmatrix} = [\mathbf{e}_1]^T \quad \mathbf{c}_2 = \begin{bmatrix} 0 \\ 1 \\ 0 \\ 0 \end{bmatrix} = [\mathbf{e}_2]^T \quad \mathbf{c}_3 = \begin{bmatrix} 0 \\ 0 \\ 1 \\ 0 \end{bmatrix} = [\mathbf{e}_3]^T \quad \mathbf{c}_4 = \begin{bmatrix} 0 \\ 0 \\ 0 \\ 1 \end{bmatrix} = [\mathbf{e}_4]^T$$

Thus,

Column space of $I_4 = \text{span}\{\mathbf{c}_1, \mathbf{c}_2, \mathbf{c}_3, \mathbf{c}_4\} = R^4$

Problem 1   Find bases for the row and column spaces of

$$A = \begin{bmatrix} 0 & 1 & 0 \\ 1 & 0 & 0 \\ 0 & 0 & 1 \end{bmatrix}$$

■

■ Row and Column Spaces of Reduced Matrices

Now we want to consider the row and column spaces of reduced matrices. The definition of a reduced matrix is restated in the box for convenient reference.

> **Reduced Matrix**
>
> A matrix is in reduced form if:
>
> 1. Each row consisting entirely of 0's is below any row having at least one nonzero element.
> 2. The leftmost nonzero element in each row is 1.
> 3. The column containing the leftmost 1 of a given row has 0's above and below the 1.
> 4. The leftmost 1 in any row is to the right of the leftmost 1 in the preceding row.

Example 2 Find bases for the row and column spaces of

$$
B = \begin{array}{c}
\begin{array}{cccccc} \mathbf{c}_1 & \mathbf{c}_2 & \mathbf{c}_3 & \mathbf{c}_4 & \mathbf{c}_5 & \mathbf{c}_6 \end{array} \\
\left[\begin{array}{cccccc}
1 & -2 & 0 & 0 & 3 & -2 \\
0 & 0 & 1 & 0 & -4 & 1 \\
0 & 0 & 0 & 1 & 5 & -6 \\
0 & 0 & 0 & 0 & 0 & 0
\end{array}\right]
\begin{array}{c} \mathbf{r}_1 \\ \mathbf{r}_2 \\ \mathbf{r}_3 \\ \mathbf{r}_4 \end{array}
\end{array}
$$

Solution   *Column Space.* Matrix B is a 4×6 reduced matrix with three nonzero rows. The set of column vectors of B includes the first three columns of the 4×4 identity matrix:

$$
\mathbf{c}_1 = \begin{bmatrix} 1 \\ 0 \\ 0 \\ 0 \end{bmatrix} \qquad
\mathbf{c}_3 = \begin{bmatrix} 0 \\ 1 \\ 0 \\ 0 \end{bmatrix} \qquad
\mathbf{c}_4 = \begin{bmatrix} 0 \\ 0 \\ 1 \\ 0 \end{bmatrix}
$$

Notice that these are the columns containing the three leftmost 1's in B. All the other columns of B can be expressed as linear combinations of $\mathbf{c}_1$, $\mathbf{c}_3$, and $\mathbf{c}_4$, as follows (verify this):

$$
\begin{aligned}
\mathbf{c}_2 &= -2\mathbf{c}_1 \\
\mathbf{c}_5 &= 3\mathbf{c}_1 - 4\mathbf{c}_3 + 5\mathbf{c}_4 \\
\mathbf{c}_6 &= -2\mathbf{c}_1 + \mathbf{c}_3 - 6\mathbf{c}_4
\end{aligned}
$$

This implies that

$$
\begin{aligned}
\text{Column space of } B &= \text{span}\{\mathbf{c}_1, \mathbf{c}_2, \mathbf{c}_3, \mathbf{c}_4, \mathbf{c}_5, \mathbf{c}_6\} \\
&= \text{span}\{\mathbf{c}_1, \mathbf{c}_3, \mathbf{c}_4\}
\end{aligned}
$$

Furthermore, since $\{\mathbf{c}_1, \mathbf{c}_3, \mathbf{c}_4\}$ is a subset of the standard basis for R^4, it follows that $\{\mathbf{c}_1, \mathbf{c}_3, \mathbf{c}_4\}$ is a linearly independent set. Thus, $\{\mathbf{c}_1, \mathbf{c}_3, \mathbf{c}_4\}$ is a basis for the column space of B.

Row Space. Since r_4 is the zero vector, we have

Row space of $B = \text{span}\{r_1, r_2, r_3\}$

If we can show that $\{r_1, r_2, r_3\}$ is a linearly independent set, then we will have found a basis for the row space of B. Consider the vector equation

$$
\begin{aligned}
0 = {}& k_1 r_1 + k_2 r_2 + k_3 r_3 \\
= {}& \quad k_1(1, \quad -2, \quad 0, \quad 0, \quad 3, -2) \\
& + k_2(0, \qquad 0, \quad 1, \quad 0, -4, \quad 1) \\
& + k_3(0, \qquad 0, \quad 0, \quad 1, \quad 5, -6) \\
= {}& (k_1, -2k_1, k_2, k_3, 3k_1 - 4k_2 + 5k_3, -2k_1 + k_2 - 6k_3)
\end{aligned}
$$

Each nonzero row contains a leftmost 1, and the other rows have 0's in that component.

The only solution to this equation is (verify this)

$k_1 = k_2 = k_3 = 0$

Thus, $\{r_1, r_2, r_3\}$ is linearly independent and forms a basis for the row space of A.

Remark. Notice that the row and column spaces of B are both three-dimensional. Also, 3 is the number of nonzero rows in B and is the number of different columns of the 4×4 identity in B.

Problem 2　Find bases for the row and column spaces of B. What is the dimension of each space?

$$
B = \begin{bmatrix} 1 & 0 & -2 & 0 \\ 0 & 1 & 1 & 3 \\ 0 & 0 & 0 & 0 \end{bmatrix}
$$

∎

Using the definition of reduced form, the conclusions reached in the solution of Example 2 can be extended to any reduced matrix. These results are summarized in Theorem 1. The proof is discussed in the exercises.

Theorem 1

> **Row and Column Spaces of a Reduced Matrix**
>
> Let B be an $m \times n$ reduced matrix with k nonzero rows.
>
> (A) The set of column vectors of B contains the first k columns of the $m \times m$ identity matrix. These columns form a basis for the column space of B.
>
> (B) The nonzero rows of B form a basis for the row space of B.
>
> (C) dim(Row space of B) = k = dim(Column space of B)

Example 3   Find bases for the row and column spaces of B. What is the dimension of these spaces?

$$B = \begin{array}{c} \begin{array}{ccccccc} \mathbf{c}_1 & \mathbf{c}_2 & \mathbf{c}_3 & \mathbf{c}_4 & \mathbf{c}_5 & \mathbf{c}_6 & \mathbf{c}_7 \end{array} \\ \left[\begin{array}{ccccccc} 1 & 1 & 0 & 0 & 0 & 0 & -1 \\ 0 & 0 & 1 & 1 & 0 & 0 & -2 \\ 0 & 0 & 0 & 0 & 1 & 0 & 2 \\ 0 & 0 & 0 & 0 & 0 & 1 & 3 \\ 0 & 0 & 0 & 0 & 0 & 0 & 0 \end{array}\right] \begin{array}{c} \mathbf{r}_1 \\ \mathbf{r}_2 \\ \mathbf{r}_3 \\ \mathbf{r}_4 \\ \mathbf{r}_5 \end{array} \end{array}$$

Solution   Matrix B is a 5×7 reduced matrix with four nonzero rows. The set of column vectors of B includes

$$\mathbf{c}_1 = \begin{bmatrix} 1 \\ 0 \\ 0 \\ 0 \\ 0 \end{bmatrix} \qquad \mathbf{c}_3 = \begin{bmatrix} 0 \\ 1 \\ 0 \\ 0 \\ 0 \end{bmatrix} \qquad \mathbf{c}_5 = \begin{bmatrix} 0 \\ 0 \\ 1 \\ 0 \\ 0 \end{bmatrix} \qquad \mathbf{c}_6 = \begin{bmatrix} 0 \\ 0 \\ 0 \\ 1 \\ 0 \end{bmatrix}$$

which are the first four columns of the 5×5 identity matrix. Thus, $\{\mathbf{c}_1, \mathbf{c}_3, \mathbf{c}_5, \mathbf{c}_6\}$ is a basis for the column space of B. The set $\{\mathbf{r}_1, \mathbf{r}_2, \mathbf{r}_3, \mathbf{r}_4\}$ of nonzero row vectors forms a basis for the row space of B. Both spaces have dimension 4.

Remark. In this example, notice that the first and second columns of the 5×5 identity matrix both appear twice in the collection of columns of B ($\mathbf{c}_1 = \mathbf{c}_2$ and $\mathbf{c}_3 = \mathbf{c}_4$). There is nothing in the definition of reduced form or in the statement of Theorem 1 that precludes the possibility of repeated columns of the identity. However, if a reduced matrix B has k nonzero rows, then Theorem 1 does guarantee that the columns of B will include the first k (distinct) columns of the identity. When selecting columns for a basis, you must be certain to select k distinct columns of the identity matrix.

Problem 3   Find bases for the row and column spaces of B. What is the dimension of these spaces?

$$B = \begin{bmatrix} 1 & 0 & 0 & -1 & 0 & 3 \\ 0 & 1 & 1 & -2 & 0 & 4 \\ 0 & 0 & 0 & 0 & 1 & 2 \\ 0 & 0 & 0 & 0 & 0 & 0 \\ 0 & 0 & 0 & 0 & 0 & 0 \end{bmatrix}$$

■

▪ The Row Space of an Arbitrary Matrix

Now we want to consider the row space of a matrix that is not in reduced form.

Let

$$A = \begin{bmatrix} 1 & 0 & -3 \\ 0 & 2 & 4 \end{bmatrix}$$

This is not a reduced matrix, but it can be transformed to reduced form by performing the elementary row operation $\frac{1}{2}R_2 \rightarrow R_2$:

$$A = \begin{bmatrix} 1 & 0 & -3 \\ 0 & 2 & 4 \end{bmatrix} \underset{\frac{1}{2}R_2 \rightarrow R_2}{\sim} \begin{bmatrix} 1 & 0 & -3 \\ 0 & 1 & 2 \end{bmatrix} = B$$

What is the relationship between the row spaces of A and B? If we let

$$\mathbf{r}_1 = \begin{bmatrix} 1 & 0 & -3 \end{bmatrix} \quad \text{and} \quad \mathbf{r}_2 = \begin{bmatrix} 0 & 2 & 4 \end{bmatrix}$$

be the row vectors of A, then $\mathbf{r}_1$ and $\frac{1}{2}\mathbf{r}_2$ are the row vectors of B. It follows that (verify this)

$$\text{span}\{\mathbf{r}_1, \mathbf{r}_2\} = \text{span}\{\mathbf{r}_1, \tfrac{1}{2}\mathbf{r}_2\}$$

or

Row space of A = Row space of B

Thus, performing the elementary row operation $\frac{1}{2}R_2 \rightarrow R_2$ on A produces a row equivalent matrix B with the same row space. Since B is a reduced matrix, Theorem 1 implies that the nonzero rows of B form a basis for the row space of A.

In general, performing any elementary row operation on a matrix will not change the row space of that matrix. This result is stated in Theorem 2. The proof is discussed in the exercises.

Theorem 2

If A and B are row equivalent matrices, then A and B have the same row space.

If B is the reduced form of A, then Theorem 2 implies that A and B have the same row space. Theorem 1 then implies that the nonzero rows of B must form a basis for the row space of A. Thus, we have just proved the following important result.

Theorem 3

If B is the reduced form of A, then the nonzero rows of B form a basis for the row space of A.

Example 4   Find a basis for the row space of

$$A = \begin{bmatrix} 1 & -1 & 2 & 3 \\ -2 & 3 & -1 & -4 \\ -1 & 2 & 1 & -1 \end{bmatrix}$$

Solution   Omitting the details, the reduced form of A is

$$B = \begin{bmatrix} 1 & 0 & 5 & 5 \\ 0 & 1 & 3 & 2 \\ 0 & 0 & 0 & 0 \end{bmatrix}$$

and

$$\{(1, 0, 5, 5), (0, 1, 3, 2)\}$$

is a basis for the row space of A.

Problem 4   Find a basis for the row space of

$$A = \begin{bmatrix} 1 & 2 & -1 \\ 3 & 7 & 2 \\ 2 & 5 & 3 \\ 1 & 3 & 4 \end{bmatrix}$$ ‖

▪ Finding a Basis for Span S

Theorem 3 can be used to find a basis for the span of any set of vectors in R^n. The procedure for doing this is outlined in the box.

Finding a Basis for Span S—Method 1

Let $S = \{\mathbf{u}_1, \mathbf{u}_2, \ldots, \mathbf{u}_m\}$ be a set of vectors in R^n.

1. Form the $m \times n$ matrix

$$A = \begin{bmatrix} \mathbf{u}_1 \\ \mathbf{u}_2 \\ \cdot \\ \cdot \\ \cdot \\ \mathbf{u}_m \end{bmatrix}$$ The vectors in S are the row vectors of A.

2. Find B, the reduced form of A.

3. The nonzero rows of B form a basis for span S.

Example 5   Use method 1 to find a basis for span S if

$$\overset{\mathbf{u}_1}{} \qquad \overset{\mathbf{u}_2}{} \qquad \overset{\mathbf{u}_3}{} \qquad \overset{\mathbf{u}_4}{}$$
$$S = \{(1, 2, -1, 1), (-2, -3, 6, -1), (1, 1, -5, 0), (0, 1, 5, 2)\}$$

Solution   Let

$$A = \begin{bmatrix} 1 & 2 & -1 & 1 \\ -2 & -3 & 6 & -1 \\ 1 & 1 & -5 & 0 \\ 0 & 1 & 5 & 2 \end{bmatrix} \begin{matrix} \mathbf{u}_1 \\ \mathbf{u}_2 \\ \mathbf{u}_3 \\ \mathbf{u}_4 \end{matrix}$$

The reduced form of A is (verify this)

$$B = \begin{bmatrix} 1 & 0 & 0 & 8 \\ 0 & 1 & 0 & -3 \\ 0 & 0 & 1 & 1 \\ 0 & 0 & 0 & 0 \end{bmatrix}$$

and a basis for span S is

$$\{(1, 0, 0, 8), (0, 1, 0, -3), (0, 0, 1, 1)\}$$

Problem 5   Use method 1 to find a basis for span S if

$$S = \{(1, 2, 1, 3), (1, 3, 4, 12), (1, 2, 2, 8), (2, 4, 5, 21)\} \qquad \blacksquare$$

As Example 5 illustrates, the procedure for finding a basis for span S is very easy to use. However, it does have one drawback. The basis for span S determined by applying method 1 is usually not a subset of S. In some applications, it is necessary to find a basis for span S consisting of vectors in S. In the next section, we will develop a second method for finding bases that will enable us to do this.

In Chapter 1, we noted that a computer is usually used to find the reduced form of large matrices. The program that computes the reduced form of a matrix can now be used to find a basis for the span of a set of vectors in R^n.

Answers to Matched Problems

1. Row space: $\{(0, 1, 0), (1, 0, 0), (0, 0, 1)\}$;

column space: $\left\{ \begin{bmatrix} 0 \\ 1 \\ 0 \end{bmatrix}, \begin{bmatrix} 1 \\ 0 \\ 0 \end{bmatrix}, \begin{bmatrix} 0 \\ 0 \\ 1 \end{bmatrix} \right\}$

2. Row space: $\{(1, 0, -2, 0), (0, 1, 1, 3)\}$;

column space: $\left\{ \begin{bmatrix} 1 \\ 0 \\ 0 \end{bmatrix}, \begin{bmatrix} 0 \\ 1 \\ 0 \end{bmatrix} \right\}$; both spaces are two-dimensional

3. Row space: $\{(1, 0, 0, -1, 0, 3), (0, 1, 1, -2, 0, 4), (0, 0, 0, 0, 1, 2)\}$;

column space: $\left\{ \begin{bmatrix} 1 \\ 0 \\ 0 \\ 0 \\ 0 \end{bmatrix}, \begin{bmatrix} 0 \\ 1 \\ 0 \\ 0 \\ 0 \end{bmatrix}, \begin{bmatrix} 0 \\ 0 \\ 1 \\ 0 \\ 0 \end{bmatrix} \right\}$; both spaces are three-dimensional

4. $\{(1, 0, -11), (0, 1, 5)\}$ **5.** $\{(1, 0, 0, 10), (0, 1, 0, -6), (0, 0, 1, 5)\}$

▮ Exercise 6-1

A   *In Problems 1–8, find bases for the row and column spaces of the indicated reduced matrices.*

1. $\begin{bmatrix} 1 & 0 & -1 \\ 0 & 1 & 2 \\ 0 & 0 & 0 \end{bmatrix}$ **2.** $\begin{bmatrix} 1 & -2 & 0 \\ 0 & 0 & 1 \\ 0 & 0 & 0 \end{bmatrix}$

3. $\begin{bmatrix} 1 & 0 & 0 & -2 \\ 0 & 1 & 0 & 4 \\ 0 & 0 & 1 & -3 \end{bmatrix}$ **4.** $\begin{bmatrix} 1 & 0 & -2 & 0 \\ 0 & 1 & 1 & 0 \\ 0 & 0 & 0 & 0 \end{bmatrix}$

5. $\begin{bmatrix} 1 & 0 & 3 & -2 & 0 & 0 \\ 0 & 1 & 4 & 5 & 0 & 0 \\ 0 & 0 & 0 & 0 & 1 & 0 \\ 0 & 0 & 0 & 0 & 0 & 1 \end{bmatrix}$ **6.** $\begin{bmatrix} 1 & -3 & 0 & 2 & 0 & -1 \\ 0 & 0 & 1 & -4 & 0 & 1 \\ 0 & 0 & 0 & 0 & 1 & 2 \\ 0 & 0 & 0 & 0 & 0 & 0 \end{bmatrix}$

7. $\begin{bmatrix} 1 & 1 & 0 & 0 & 0 & 0 & -1 \\ 0 & 0 & 1 & 0 & 0 & 0 & 1 \\ 0 & 0 & 0 & 1 & 1 & 0 & -2 \\ 0 & 0 & 0 & 0 & 0 & 1 & 3 \\ 0 & 0 & 0 & 0 & 0 & 0 & 0 \end{bmatrix}$ **8.** $\begin{bmatrix} 1 & 0 & 0 & 0 & -1 & 4 & 0 \\ 0 & 1 & 0 & 0 & 1 & -5 & 0 \\ 0 & 0 & 1 & 0 & 2 & 6 & 0 \\ 0 & 0 & 0 & 1 & -3 & -2 & 1 \\ 0 & 0 & 0 & 0 & 0 & 0 & 0 \end{bmatrix}$

B   *In Problems 9–14, find a basis for the row space of each matrix.*

9. $\begin{bmatrix} 0 & 1 & 0 \\ 1 & 0 & 0 \\ 0 & 0 & 1 \end{bmatrix}$ **10.** $\begin{bmatrix} 1 & 0 & 0 \\ -2 & 1 & 0 \\ 0 & 0 & 1 \end{bmatrix}$

11. $\begin{bmatrix} 1 & 2 & 0 & 5 \\ -2 & 1 & 5 & -5 \\ -1 & 3 & 5 & 0 \end{bmatrix}$ **12.** $\begin{bmatrix} 1 & 2 & 2 & -3 \\ -3 & 1 & 1 & 2 \\ 1 & -5 & -5 & 4 \end{bmatrix}$

13.
$$\begin{bmatrix} 0 & 1 & 2 & 3 & 1 \\ 2 & 5 & 8 & 17 & 5 \\ 1 & 3 & 5 & 10 & 3 \\ 1 & 4 & 7 & 13 & 4 \end{bmatrix}$$

14.
$$\begin{bmatrix} 1 & 1 & 0 & 0 & -1 \\ 2 & 1 & -2 & 0 & -2 \\ 3 & 0 & -6 & 3 & -3 \\ 6 & 2 & -8 & 3 & -6 \end{bmatrix}$$

In Problems 15–20, use method 1 to find a basis for span S.

15. $S = \{(1, -1, 1), (-2, 1, 0), (-1, 0, 1)\}$

16. $S = \{(3, -6, -9), (-2, 4, 6), (5, -10, -15)\}$

17. $S = \{(1, -1, 2, 0), (-1, 1, 0, 2), (0, 0, 1, 1), (2, -2, 0, 4)\}$

18. $S = \{(2, -1, 1, 3), (1, 2, -1, 1), (0, 0, 2, 1), (3, 1, 2, 5)\}$

19. $S = \{(1, 1, 1, -1, 1), (1, 0, 1, 0, 1), (2, 0, -1, 1, 0), (0, -1, -3, 1, -2),$
$(1, 2, 4, -3, 3)\}$

20. $S = \{(1, -1, 0, 0, 1), (0, 1, -1, 1, 0), (1, 1, -1, 0, 0), (1, 2, 1, -1, 1), (3, 3, -1, 0, 2)\}$

C **21.** Let $S = \{\mathbf{u}_1, \mathbf{u}_2, \mathbf{u}_3\} \subset R^3$ where $\mathbf{u}_1 = (1, 1, 1)$, $\mathbf{u}_2 = (1, -2, -5)$, and $\mathbf{u}_3 = (2, -1, -4)$.

    (A) Find a basis for span S.

    (B) Express each vector in S as a linear combination of the basis vectors in part (A).

22. Repeat Problem 21 for $\mathbf{u}_1 = (1, -1, 2)$, $\mathbf{u}_2 = (-1, 1, 3)$, and $\mathbf{u}_3 = (2, -2, 5)$.

23. If A is an $m \times n$ matrix, show that the column space of A equals the row space of A^T.

24. If A is an $m \times n$ matrix, show that the row space of A equals the column space of A^T.

Use the result in Problem 23 to find a basis for the column spaces of the matrices in Problems 25 and 26.

25.
$$\begin{bmatrix} 1 & 1 & 2 & 3 \\ 1 & -1 & 1 & 2 \\ 1 & 5 & 4 & 5 \end{bmatrix}$$

26.
$$\begin{bmatrix} 1 & 1 & 3 & 3 \\ 1 & -1 & 1 & -1 \\ 2 & -4 & 0 & -6 \\ -2 & 6 & 2 & 10 \end{bmatrix}$$

Problems 27–32 are representative cases of Theorem 1 in this section. Prove each statement directly, without reference to Theorem 1.

In Problems 27–29, let

$$A = \begin{matrix} & \mathbf{c}_1 & \mathbf{c}_2 & \mathbf{c}_3 & \mathbf{c}_4 & \\ & \begin{bmatrix} 1 & 0 & a_{13} & a_{14} \\ 0 & 1 & a_{23} & a_{24} \\ 0 & 0 & 0 & 0 \end{bmatrix} & \begin{matrix} \mathbf{r}_1 \\ \mathbf{r}_2 \\ \mathbf{r}_3 \end{matrix} \end{matrix}$$

27. *Theorem 1(A).* Prove that $\{\mathbf{c}_1, \mathbf{c}_2\}$ is a basis for the column space of A.

28. *Theorem 1(B).* Prove that $\{\mathbf{r}_1, \mathbf{r}_2\}$ is a basis for the row space of A.

29. *Theorem 1(C).* Prove that dim(Row space of A) = dim(Column space of A).

In Problems 30–32, let

$$
A = \begin{array}{c} \begin{array}{ccccc} \mathbf{c}_1 & \mathbf{c}_2 & \mathbf{c}_3 & \mathbf{c}_4 & \mathbf{c}_5 \end{array} \\ \left[\begin{array}{ccccc} 1 & a_{12} & 0 & 0 & a_{15} \\ 0 & 0 & 1 & 0 & a_{25} \\ 0 & 0 & 0 & 1 & a_{35} \\ 0 & 0 & 0 & 0 & 0 \end{array}\right] \begin{array}{c} \mathbf{r}_1 \\ \mathbf{r}_2 \\ \mathbf{r}_3 \\ \mathbf{r}_4 \end{array} \end{array}
$$

30. *Theorem 1(A).* Prove that $\{\mathbf{c}_1, \mathbf{c}_3, \mathbf{c}_4\}$ is a basis for the column space of A.
31. *Theorem 1(B).* Prove that $\{\mathbf{r}_1, \mathbf{r}_2, \mathbf{r}_3\}$ is a basis for the row space of A.
32. *Theorem 1(C).* Prove that dim(Row space of A) = dim(Column space of A).

Problems 33–38 are related to Theorem 2 in this section. Prove each statement directly, without reference to Theorem 2.

In Problems 33–35, let $\mathbf{u}_1$, $\mathbf{u}_2$, and $\mathbf{u}_3$ be vectors in R^3.

33. Show that span$\{\mathbf{u}_1, \mathbf{u}_2, \mathbf{u}_3\}$ = span$\{k\mathbf{u}_1, \mathbf{u}_2, \mathbf{u}_3\}$ for any nonzero scalar k.
34. Show that span$\{\mathbf{u}_1, \mathbf{u}_2, \mathbf{u}_3\}$ = span$\{\mathbf{u}_1 + k\mathbf{u}_2, \mathbf{u}_2, \mathbf{u}_3\}$ for any scalar k.
35. Show that span$\{\mathbf{u}_1, \mathbf{u}_2, \mathbf{u}_3\}$ = span$\{\mathbf{u}_2, \mathbf{u}_1, \mathbf{u}_3\}$.

Let

$$
A = \begin{bmatrix} a_{11} & a_{12} & a_{13} \\ a_{21} & a_{22} & a_{23} \\ a_{31} & a_{32} & a_{33} \end{bmatrix}
$$

In Problems 36–38, show that B has the same row space as A. [*Hint: Use Problems 33–35.*]

36. $B = \begin{bmatrix} ka_{11} & ka_{12} & ka_{13} \\ a_{21} & a_{22} & a_{23} \\ a_{31} & a_{32} & a_{33} \end{bmatrix}$, k any nonzero scalar

37. $B = \begin{bmatrix} a_{11} + ka_{21} & a_{12} + ka_{22} & a_{13} + ka_{23} \\ a_{21} & a_{22} & a_{23} \\ a_{31} & a_{32} & a_{33} \end{bmatrix}$, k any scalar

38. $B = \begin{bmatrix} a_{21} & a_{22} & a_{23} \\ a_{11} & a_{12} & a_{13} \\ a_{31} & a_{32} & a_{33} \end{bmatrix}$

39. Let A be an $m \times n$ matrix and let E be an $m \times m$ elementary matrix. Show that A and EA have the same row space.
40. Let A be an $m \times n$ matrix and let P be an $m \times m$ invertible matrix. Show that A and PA have the same row space.
41. If A is an $n \times n$ invertible matrix, show that the row space of A is R^n.

I6-2 I Row and Column Spaces and the Rank of a Matrix

- Finding a Basis for the Column Space of a Matrix
- Finding a Basis for Span S
- Extending a Linearly Independent Set to a Basis
- Rank of a Matrix
- Fundamental Theorem—Final Version

■ Finding a Basis for the Column Space of a Matrix

In the preceding section, we saw that performing row operations on a matrix does not change the row space of the matrix. Can we make a similar statement concerning the column space? That is, do row equivalent matrices always have the same column spaces? The following example shows that this is not the case.

Example 6 Let

$$A = \begin{bmatrix} 1 & 1 \\ 0 & 1 \\ 1 & 1 \end{bmatrix} \qquad B = \begin{bmatrix} 1 & 1 \\ 0 & 1 \\ 0 & 0 \end{bmatrix}$$

Since A can be transformed into B by performing the row operation $R_3 + (-1)R_1 \rightarrow R_3$, A and B are row equivalent. Let

$$P = \text{span}\{(1, 0, 1), (1, 1, 1)\} \qquad \text{Column space of } A$$

$$Q = \text{span}\{(1, 0, 0), (1, 1, 0)\} \qquad \text{Column space of } B$$

Since both column vectors of B have zero third components, all vectors in Q will have zero third components. Thus, neither column of A will be in Q. This shows that P and Q are not the same vector space. See Figure 1 for a geometric illustration of these two spaces.

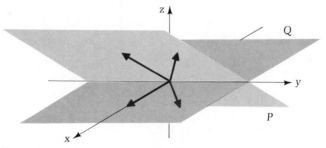

Figure 1

Problem 6 Let

$$A = \begin{bmatrix} 1 & -1 \\ 0 & 1 \end{bmatrix} \qquad B = \begin{bmatrix} 1 & 0 \\ 0 & 1 \end{bmatrix}$$

Do A and B have the same column space?   ▌▐

Even though row equivalent matrices do not necessarily have the same column space, there is an important relationship between the columns of two row equivalent matrices. We will illustrate this relationship for 3×3 matrices and then generalize the results to $m \times n$ matrices.

Let

$$
A = \begin{bmatrix} a_{11} & a_{12} & a_{13} \\ a_{21} & a_{22} & a_{23} \\ a_{31} & a_{32} & a_{33} \end{bmatrix} \qquad B = \begin{bmatrix} b_{11} & b_{12} & b_{13} \\ b_{21} & b_{22} & b_{23} \\ b_{31} & b_{32} & b_{33} \end{bmatrix}
$$

and assume that A and B are row equivalent. The vector equations

$$
k_1 \begin{bmatrix} a_{11} \\ a_{12} \\ a_{13} \end{bmatrix} + k_2 \begin{bmatrix} a_{21} \\ a_{22} \\ a_{23} \end{bmatrix} + k_3 \begin{bmatrix} a_{31} \\ a_{32} \\ a_{33} \end{bmatrix} = \mathbf{0} \tag{1}
$$

$$
k_1 \begin{bmatrix} b_{11} \\ b_{12} \\ b_{13} \end{bmatrix} + k_2 \begin{bmatrix} b_{21} \\ b_{22} \\ b_{23} \end{bmatrix} + k_3 \begin{bmatrix} b_{31} \\ b_{32} \\ b_{33} \end{bmatrix} = \mathbf{0} \tag{2}
$$

are equivalent to the systems

$$
A\mathbf{k} = \mathbf{0} \tag{3}
$$

$$
B\mathbf{k} = \mathbf{0} \tag{4}
$$

where

$$
\mathbf{k} = \begin{bmatrix} k_1 \\ k_2 \\ k_3 \end{bmatrix}
$$

Since A and B are row equivalent matrices, systems (3) and (4) have identical solution sets. Hence, vector equations (1) and (2) also have identical solution sets. We want to investigate the relationship between the independence of the columns of A and the independence of the columns of B.

To begin, suppose all three columns of A are linearly independent. Then the only solution of (1) is $k_1 = k_2 = k_3 = 0$. But this implies that the only solution of (2) is $k_1 = k_2 = k_3 = 0$. Thus, the columns of B are also linearly independent.

Now suppose that just the first two columns of A are linearly independent. Then equation (1) cannot have a solution of the form $(k_1, k_2, 0)$ with both $k_1 \neq 0$ and $k_2 \neq 0$. (Why?) This implies that (2) has no solutions of this form. Hence, the first two columns of B are also linearly independent. A similar argument can be applied to any linearly independent subset of the columns of A. The corresponding columns of B are also linearly independent. This result is generalized to matrices of arbitrary size in Theorem 4. The proof is omitted.

Theorem 4

> If A and B are row equivalent matrices, then a collection of columns of A are linearly independent (dependent) if and only if the corresponding columns of B are linearly independent (dependent).

In Example 6, notice that the columns of A are linearly independent (they span a plane in R^3) and that the columns of B are linearly independent (they span a different plane in R^3).

Theorem 4 has some important applications to finding bases. The first of these is stated in Theorem 5.

Theorem 5

> If B is the reduced form of A, then the columns of A that correspond to the *distinct* columns of the identity in B form a basis for the column space of A.

Proof   Let k be the number of nonzero rows in B. Since B is a reduced matrix, Theorem 1 in Section 6-1 implies that:

1. The columns of B contain the first k columns of the identity.
2. These distinct columns of the identity form a basis for the column space of B.
3. dim(Column space of B) $= k$

Since the k columns of the identity in B form a basis, they are linearly independent. Theorem 4 implies that the corresponding columns of A are linearly independent. Thus,

dim(Column space of A) $\geq k$

Suppose that the dimension of the column space of A is greater than k. Then the columns of A must contain a set of at least $k + 1$ linearly independent vectors. Theorem 4 implies that the corresponding $k + 1$ columns of B are linearly independent. But this contradicts the fact that the column space of B is k-dimensional. Thus,

dim(Column space of A) $= k$

and any linearly independent subset of k columns of A will form a basis for the column space of A. In particular, the columns of A that correspond to the k distinct columns of the identity in B form a basis for the column space of A.   ∎

Example 7   Find a basis for the column space of

$$
\begin{array}{ccccc}
\mathbf{c}_1 & \mathbf{c}_2 & \mathbf{c}_3 & \mathbf{c}_4 & \mathbf{c}_5
\end{array}
$$

$$
A = \begin{bmatrix}
1 & 2 & 3 & 3 & -1 \\
-2 & 1 & 4 & 0 & -2 \\
1 & 0 & -1 & 1 & 1 \\
0 & 3 & 6 & 4 & -2
\end{bmatrix}
$$

Solution   The reduced form of A is

$$
\begin{array}{ccccc}
\mathbf{c}_1 & \mathbf{c}_2 & \mathbf{c}_3 & \mathbf{c}_4 & \mathbf{c}_5
\end{array}
$$

$$
B = \begin{bmatrix}
1 & 0 & -1 & 0 & 0 \\
0 & 1 & 2 & 0 & -2 \\
0 & 0 & 0 & 1 & 1 \\
0 & 0 & 0 & 0 & 0
\end{bmatrix}
$$

Columns $\mathbf{c}_1$, $\mathbf{c}_2$, and $\mathbf{c}_4$ of B are the first three columns of the 4×4 identity matrix. Thus, columns $\mathbf{c}_1$, $\mathbf{c}_2$, and $\mathbf{c}_4$ of A form a basis for the column space of A. That is,

$$
\{(1, -2, 1, 0), (2, 1, 0, 3), (3, 0, 1, 4)\}
$$

is the required basis.

Problem 7   Find a basis for the column space of

$$
A = \begin{bmatrix}
1 & -2 & 1 & 1 & 3 \\
-2 & 4 & 1 & 10 & -3 \\
-1 & 2 & 2 & 11 & 0
\end{bmatrix}
$$

∎

▪ Finding a Basis for Span S

In Section 6-1, we used the reduced form of a matrix to find a basis for the span of a set S of vectors in R^n. However, the basis produced by method 1 will not be a subset of S in most cases. Now we can use Theorem 5 to find a basis for span S that is a subset of S.

Finding a Basis for Span S—Method 2

Let $S = \{\mathbf{u}_1, \mathbf{u}_2, \ldots, \mathbf{u}_n\}$ be a set of vectors in R^m.

1. Form the $m \times n$ matrix

$$
A = [[\mathbf{u}_1]^T \quad [\mathbf{u}_2]^T \quad \cdots \quad [\mathbf{u}_n]^T] \qquad
\begin{array}{l}
\text{The vectors in } S \text{ are the column} \\
\text{vectors of } A.
\end{array}
$$

2. Find B, the reduced form of A.

3. The columns of A corresponding to the distinct columns of the identity in B form a basis for span S. This basis is always a subset of S.

Example 8   Use method 2 to find a basis for span S if

$$
\begin{array}{cccc}
\mathbf{u}_1 & \mathbf{u}_2 & \mathbf{u}_3 & \mathbf{u}_4
\end{array}
$$
$$S = \{(1, 2, -1, 1), (-2, -3, 6, -1), (1, 1, -5, 0), (0, 1, 5, 2)\}$$

Solution   Let

$$
\begin{array}{cccc}
\mathbf{u}_1 & \mathbf{u}_2 & \mathbf{u}_3 & \mathbf{u}_4
\end{array}
$$
$$A = \begin{bmatrix} 1 & -2 & 1 & 0 \\ 2 & -3 & 1 & 1 \\ -1 & 6 & -5 & 5 \\ 1 & -1 & 0 & 2 \end{bmatrix}$$

The reduced form of A is

$$B = \begin{bmatrix} 1 & 0 & -1 & 0 \\ 0 & 1 & -1 & 0 \\ 0 & 0 & 0 & 1 \\ 0 & 0 & 0 & 0 \end{bmatrix}$$

Thus, a basis for span S is composed of the first, second, and fourth column vectors in A:

$$\{(1, 2, -1, 1), (-2, -3, 6, -1), (0, 1, 5, 2)\}$$

Problem 8   Use method 2 to find a basis for span S if

$$S = \{(1, 2, 1, 3), (1, 3, 4, 10), (1, 2, 4, 9), (2, 4, 5, 12)\}$$ ▌▌

In general, the basis provided by method 2 will depend on the order in which the vectors in S are arranged in the matrix A. For example, if we reverse the order of the vectors in Example 8, then

$$A = \begin{bmatrix} 0 & 1 & -2 & 1 \\ 1 & 1 & -3 & 2 \\ 5 & -5 & 6 & -1 \\ 2 & 0 & -1 & 1 \end{bmatrix} \sim \begin{bmatrix} 1 & 0 & 0 & 0 \\ 0 & 1 & 0 & -1 \\ 0 & 0 & 1 & 1 \\ 0 & 0 & 0 & 0 \end{bmatrix} = B$$

and the basis for span S produced by method 2 is

$$\{(0, 1, 5, 2), (1, 1, -5, 0), (-2, -3, 6, -1)\}$$

(In order to have your answers to problems agree with those in the back of the book, you should be careful to keep the vectors in S in the same order as they are given in each problem.)

Now that we have two methods for finding a basis for span S, how do we decide which method to use? Both methods require the same amount of calculation (finding the reduced form of a matrix), so the choice of methods usually depends on the type of basis that is needed in a particular application. Table 1 compares the bases produced by applying both methods to the same set. (See Example 5 in Section 6-1 and Example 8 in this section.)

Table 1
Comparison of Methods 1 and 2

| | Basis | |
Vectors in S	Method 1	Method 2
$(1, 2, -1, 1)$	$(1, 0, 0, 8)$	$(1, 2, -1, 1)$
$(-2, -3, 6, -1)$	$(0, 1, 0, -3)$	$(-2, -3, 6, -1)$
$(1, 1, -5, 0)$	$(0, 0, 1, 1)$	$(0, 1, 5, 2)$
$(0, 1, 5, 2)$		

The basis produced by method 2 is always a subset of S. In some applications, it is necessary to be able to find such a basis. On the other hand, the basis provided by method 1 is simpler; many of the components are 0 or 1. In some applications, this is a desirable feature.

▪ Extending a Linearly Independent Set to a Basis

If S is a linearly independent subset of a finite-dimensional vector space V, then there is a basis for V that contains S (Theorem 18 in Section 5-6). Theorem 5 provides a method for finding this basis.

Consider the set

$$S = \{(1, 1, 1), (2, 1, 1)\} \subset R^3$$

Set S is linearly independent, but does not span R^3. (Why?) We want to find a basis for R^3 that contains S. Consider the following matrix A, formed by using the vectors in S and the 3×3 identity matrix:

$$A = \begin{matrix} \mathbf{c}_1 & \mathbf{c}_2 & \mathbf{c}_3 & \mathbf{c}_4 & \mathbf{c}_5 \\ \begin{bmatrix} 1 & 2 & 1 & 0 & 0 \\ 1 & 1 & 0 & 1 & 0 \\ 1 & 1 & 0 & 0 & 1 \end{bmatrix} \end{matrix}$$

Since the columns of A include the standard basis for R^3, we can be certain that the column space of A is all of R^3. The reduced form of A is

$$B = \begin{bmatrix} 1 & 0 & -1 & 0 & 2 \\ 0 & 1 & 1 & 0 & -1 \\ 0 & 0 & 0 & 1 & -1 \end{bmatrix}$$

Using Theorem 5, we can conclude that columns $\mathbf{c}_1$, $\mathbf{c}_2$, and $\mathbf{c}_4$ of A form a basis for R^3, the column space of A. That is,

$$S$$
$$\overbrace{\phantom{\{(1,\ 1,\ 1),\ (2,\ 1,\ 1),\ (0,\ 1,\ 0)\}}}$$
$$\{(1,\ 1,\ 1),\ (2,\ 1,\ 1),\ (0,\ 1,\ 0)\}$$

is a basis for R^3 containing S.

In general, we say a linearly independent set S in a finite-dimensional vector space V has been **extended** to a basis for V if we have found a basis for V that contains S. The procedure for extending a linearly independent subset of R^n to a basis for R^n is outlined in the box.

Extending a Linearly Independent Set to a Basis

Let $S = \{\mathbf{u}_1, \mathbf{u}_2, \ldots, \mathbf{u}_m\}$ be a linearly independent set of vectors in R^n.

1. Form the $n \times (m + n)$ matrix

$$A = [[\mathbf{u}_1]^T \quad [\mathbf{u}_2]^T \quad \cdots \quad [\mathbf{u}_m]^T \quad I_n]$$

2. Find B, the reduced form of A.

3. The columns of A corresponding to the n distinct columns of the $n \times n$ identity in B form a basis for R^n. This basis will always include all the vectors in S.

If S is not linearly independent, this procedure will still produce a basis for R^n. However, in this case, S cannot be a subset of the resulting basis. (Why?)

Example 9   Extend

$$S = \{\overset{\mathbf{u}_1}{(1, -1, 1, 2)}, \overset{\mathbf{u}_2}{(-1, 2, -2, -4)}\}$$

to a basis for R^4.

Solution   Since neither vector in S is a scalar multiple of the other, S is a linearly independent subset of R^4. Let

$$A = \begin{array}{cc} \begin{array}{cc} \mathbf{u}_1 & \mathbf{u}_2 \end{array} & \quad I_4 \\ \left[\begin{array}{cccccc} 1 & -1 & 1 & 0 & 0 & 0 \\ -1 & 2 & 0 & 1 & 0 & 0 \\ 1 & -2 & 0 & 0 & 1 & 0 \\ 2 & -4 & 0 & 0 & 0 & 1 \end{array}\right] \end{array}$$

The reduced form of A is

$$B = \left[\begin{array}{cccccc} 1 & 0 & 2 & 0 & 0 & -\frac{1}{2} \\ 0 & 1 & 1 & 0 & 0 & -\frac{1}{2} \\ 0 & 0 & 0 & 1 & 0 & \frac{1}{2} \\ 0 & 0 & 0 & 0 & 1 & -\frac{1}{2} \end{array}\right]$$

and

$$\{(1, -1, 1, 2), (-1, 2, -2, -4), (0, 1, 0, 0), (0, 0, 1, 0)\}$$

is a basis for R^4 containing S.

Remark. Notice that the last two vectors in the basis in this example are the second and third columns of the identity from A. These vectors correspond to the third and fourth columns of the identity in B. It is incorrect to use the columns of the identity in B to extend S. You must select a subset of the columns of A.

Problem 9 Extend $S = \{(1, -1, 0), (-1, 0, 1)\}$ to a basis for R^3. ▮

▪ Rank of a Matrix

Theorem 1 in Section 6-1 states that the row and column spaces of a reduced matrix have the same dimension. One of the most fundamental results in linear algebra is that the same statement can be made for any matrix.

Theorem 6

> The row and column spaces of a matrix have the same dimension.

Proof Let A be any matrix, let B be the reduced form of A, and let k be the number of nonzero rows in B. Since the nonzero rows of B form a basis for the row space of A (Theorem 3 in Section 6-1), it follows that

$$\dim(\text{Row space of } A) = k$$

Theorem 1 in Section 6-1 implies that the columns of B include the first k columns of the identity. Theorem 5 then shows that the columns of A corresponding to the k columns of the identity in B form a basis for the column space of A. Thus,

$$\dim(\text{Column space of } A) = k$$

and we can conclude that the row and column spaces of A have the same dimension. ▮

The dimension of the row and column spaces of a matrix A is usually referred to as the *rank* of A.

> **Rank of a Matrix**
>
> The common dimension of the row space and the column space of a matrix A is called the **rank** of A.

Theorem 7 provides a convenient method for finding the rank of a matrix. The proof follows directly from the proof of Theorem 6.

Theorem 7

The rank of A is equal to the number of nonzero rows in the reduced form of A.

Example 10 Find the rank of the following matrices:

(A) $A = \begin{bmatrix} 1 & -2 & 3 & 2 \\ 1 & -3 & 2 & -1 \\ -1 & 1 & -4 & -5 \end{bmatrix}$ (B) $A = \begin{bmatrix} 1 & 4 & -5 \\ -3 & 10 & 9 \\ 2 & 5 & -7 \end{bmatrix}$

Solution (A) The reduced form of A is

$$\begin{bmatrix} 1 & 0 & 5 & 8 \\ 0 & 1 & 1 & 3 \\ 0 & 0 & 0 & 0 \end{bmatrix}$$

Since this reduced matrix has two nonzero rows, the rank of A is 2.

(B) The reduced form of A is

$$\begin{bmatrix} 1 & 0 & 0 \\ 0 & 1 & 0 \\ 0 & 0 & 1 \end{bmatrix}$$

and the rank of A is 3.

Problem 10 Find the rank of the following matrices:

(A) $\begin{bmatrix} 1 & -3 & 4 \\ 2 & -5 & 3 \\ -1 & 1 & 6 \\ 1 & -2 & -1 \end{bmatrix}$ (B) $\begin{bmatrix} 1 & 0 & 1 \\ -1 & 1 & 1 \\ 2 & 1 & 2 \end{bmatrix}$ ∎

▪ Fundamental Theorem — Final Version

The rank of a square matrix A provides some important information about the matrix. An $n \times n$ matrix with rank n will have all the properties listed in Theorem 8, the final version of the fundamental theorem (Theorem 10 in Section 2-4 and Theorem 10 in Section 3-3).

Theorem 8

> **Fundamental Theorem—Final Version**
>
> If A is an $n \times n$ matrix, then the following are equivalent:
>
> (A) A is invertible.
> (B) $A\mathbf{x} = \mathbf{b}$ has a unique solution for any $n \times 1$ matrix $\mathbf{b}$.
> (C) The only solution of $A\mathbf{x} = \mathbf{0}$ is $\mathbf{x} = \mathbf{0}$.
> (D) A is row equivalent to I_n.
> (E) A is a product of elementary matrices.
> (F) $\det(A) \neq 0$
> (G) A has rank n.
> (H) The rows of A are linearly independent.
> (I) The columns of A are linearly independent.

Proof   We have already seen in Sections 2-4 and 3-3 that (A) through (F) are equivalent. We will now show that (D) $\Rightarrow$ (G) $\Rightarrow$ (H) $\Rightarrow$ (I) $\Rightarrow$ (C), which will prove that all nine statements are equivalent.

(D) $\Rightarrow$ (G):   Assume that A is row equivalent to I_n. Then I_n is the reduced form of A. Since I_n has n nonzero rows, Theorem 7 implies that the rank of A is n.

(G) $\Rightarrow$ (H):   Assume that A has rank n. Then the row space of A is n-dimensional. Since A has n rows and these rows span the n-dimensional row space of A, it follows that the rows of A are linearly independent (Theorem 17, Section 5-6).

(H) $\Rightarrow$ (I):   Assume that the rows of A are linearly independent. Then the rows of A form a basis for the row space of A. (They are a linearly independent spanning set.) Thus, the row space of A is n-dimensional. Theorem 6 now implies that the column space of A is also n-dimensional. Since A has n columns that span this n-dimensional column space, the columns of A must be linearly independent (Theorem 17, Section 5-6).

(I) $\Rightarrow$ (C):   Assume that the columns of A are linearly independent. Then the only solution of the vector equation

$$x_1 \begin{bmatrix} a_{11} \\ a_{21} \\ \cdot \\ \cdot \\ \cdot \\ a_{n1} \end{bmatrix} + x_2 \begin{bmatrix} a_{12} \\ a_{22} \\ \cdot \\ \cdot \\ \cdot \\ a_{n2} \end{bmatrix} + \cdots + x_n \begin{bmatrix} a_{1n} \\ a_{2n} \\ \cdot \\ \cdot \\ \cdot \\ a_{nn} \end{bmatrix} = \mathbf{0}$$

is $x_1 = x_2 = \cdots = x_n = 0$. But this vector equation is equivalent to the system $A\mathbf{x} = \mathbf{0}$. Thus, the only solution of the system $A\mathbf{x} = \mathbf{0}$ is $\mathbf{x} = \mathbf{0}$.   ∎

The following example illustrates one application of Theorem 8.

Example 11   Determine whether S is a basis for R^3 if

$$S = \{(1, -1, 3), (2, 1, -4), (6, -4, 3)\}$$

Solution   Let

$$A = \begin{bmatrix} 1 & -1 & 3 \\ 2 & 1 & -4 \\ 6 & -4 & 3 \end{bmatrix}$$ The vectors in S are the row vectors of A.

Since S is a subset of three vectors in a three-dimensional vector space, if S is linearly independent, then it will be a basis for R^3. Thus, we need to show that the rows of A are linearly independent. We can do this by showing that A satisfies any one of the conditions in Theorem 8. The easiest one to use is (F):

$$\det(A) = 1 \begin{vmatrix} 1 & -4 \\ -4 & 3 \end{vmatrix} - (-1) \begin{vmatrix} 2 & -4 \\ 6 & 3 \end{vmatrix} + 3 \begin{vmatrix} 2 & 1 \\ 6 & -4 \end{vmatrix} = -25$$

Since $\det(A) \neq 0$, we have shown that A satisfies condition (F) in Theorem 8. It follows that A satisfies all the conditions in Theorem 8. In particular, the rows of A are linearly independent and S is a basis for R^3.

Problem 11   Determine whether S is a basis for R^3 if

$$S = \{(1, -2, 4), (-2, 3, -3), (-1, 0, 6)\}$$ ∎

Answers to Matched Problems

6. Yes; the column space of both matrices is R^2.
7. $\{(1, -2, -1), (1, 1, 2)\}$ 8. $\{(1, 2, 1, 3), (1, 3, 4, 10), (1, 2, 4, 9)\}$
9. $\{(1, -1, 0), (-1, 0, 1), (1, 0, 0)\}$ 10. (A) 2   (B) 3
11. No; S is linearly dependent.

∎ Exercise 6-2

A   *The matrices in Problems 1–6 are in reduced form. Find the rank of each matrix.*

1. $\begin{bmatrix} 1 & 0 & 0 \\ 0 & 1 & 0 \\ 0 & 0 & 1 \end{bmatrix}$ 2. $\begin{bmatrix} 1 & -1 & 0 \\ 0 & 0 & 1 \\ 0 & 0 & 0 \end{bmatrix}$

3. $\begin{bmatrix} 1 & 0 & -1 & 2 \\ 0 & 1 & 2 & -1 \end{bmatrix}$ 4. $\begin{bmatrix} 1 & -1 \\ 0 & 0 \\ 0 & 0 \end{bmatrix}$

$$5. \begin{bmatrix} 1 & -1 & 0 & -1 \\ 0 & 0 & 1 & 1 \\ 0 & 0 & 0 & 0 \\ 0 & 0 & 0 & 0 \end{bmatrix} \qquad 6. \begin{bmatrix} 1 & -1 & 0 & -2 & 0 \\ 0 & 0 & 1 & 3 & 0 \\ 0 & 0 & 0 & 0 & 1 \\ 0 & 0 & 0 & 0 & 0 \end{bmatrix}$$

In Problems 7–12, B is the reduced form of A. Use B to find a basis for the column space of A and to find the rank of A.

$$7. \ A = \begin{bmatrix} 1 & 1 & 1 \\ 1 & -1 & -3 \\ 3 & 1 & -1 \end{bmatrix}, \quad B = \begin{bmatrix} 1 & 0 & -1 \\ 0 & 1 & 2 \\ 0 & 0 & 0 \end{bmatrix}$$

$$8. \ A = \begin{bmatrix} 1 & -3 & 4 \\ -2 & 6 & -2 \\ 3 & -9 & 1 \end{bmatrix}, \quad B = \begin{bmatrix} 1 & -3 & 0 \\ 0 & 0 & 1 \\ 0 & 0 & 0 \end{bmatrix}$$

$$9. \ A = \begin{bmatrix} 1 & 1 & 0 & 5 & 1 \\ 1 & -1 & -2 & -1 & 2 \\ -2 & 4 & 6 & 8 & 5 \end{bmatrix}, \quad B = \begin{bmatrix} 1 & 0 & -1 & 2 & 0 \\ 0 & 1 & 1 & 3 & 0 \\ 0 & 0 & 0 & 0 & 1 \end{bmatrix}$$

$$10. \ A = \begin{bmatrix} 1 & -1 & 1 & 1 & 2 \\ -2 & 2 & 2 & 3 & 9 \\ -1 & 1 & 2 & 4 & 9 \end{bmatrix}, \quad B = \begin{bmatrix} 1 & -1 & 0 & 0 & -1 \\ 0 & 0 & 1 & 0 & 2 \\ 0 & 0 & 0 & 1 & 1 \end{bmatrix}$$

$$11. \ A = \begin{bmatrix} 1 & 1 & 2 & -2 \\ 1 & -1 & -4 & 6 \\ 1 & 2 & 5 & -6 \\ 3 & 2 & 3 & -2 \end{bmatrix}, \quad B = \begin{bmatrix} 1 & 0 & -1 & 2 \\ 0 & 1 & 3 & -4 \\ 0 & 0 & 0 & 0 \\ 0 & 0 & 0 & 0 \end{bmatrix}$$

$$12. \ A = \begin{bmatrix} 1 & 2 & 1 & -3 \\ -1 & 1 & 2 & 2 \\ 1 & 5 & 4 & -4 \\ 1 & 8 & 7 & -4 \end{bmatrix}, \quad B = \begin{bmatrix} 1 & 0 & -1 & 0 \\ 0 & 1 & 1 & 0 \\ 0 & 0 & 0 & 1 \\ 0 & 0 & 0 & 0 \end{bmatrix}$$

B   In Problems 13–18, find a basis for the column space of A consisting of column vectors from A, and find the rank of A.

$$13. \ A = \begin{bmatrix} 1 & 1 & -1 \\ 1 & -1 & -7 \\ 1 & 3 & 5 \end{bmatrix} \qquad 14. \ A = \begin{bmatrix} 1 & -1 & 1 \\ -2 & 2 & -2 \\ 3 & -3 & 3 \end{bmatrix}$$

$$15. \ A = \begin{bmatrix} 1 & -2 & 1 & 2 \\ 2 & -4 & 1 & 1 \\ 1 & -2 & 0 & -1 \end{bmatrix} \qquad 16. \ A = \begin{bmatrix} 1 & 2 & -3 & -3 \\ 1 & -1 & 4 & 4 \\ 3 & 3 & -1 & -1 \end{bmatrix}$$

$$17. \ A = \begin{bmatrix} 1 & 2 & 1 & 4 & 2 \\ 1 & -2 & -3 & 0 & 1 \\ 2 & 0 & -2 & 4 & 1 \\ 3 & -2 & -5 & 4 & 2 \end{bmatrix} \qquad 18. \ A = \begin{bmatrix} 1 & 1 & 1 & 1 & -1 \\ 2 & 1 & -2 & 5 & 4 \\ 1 & 2 & 5 & -2 & -7 \\ 2 & 1 & -2 & 5 & 4 \end{bmatrix}$$

In Problems 19–24, use method 2 to find a basis for span S consisting of vectors in S.

19. $S = \{(1, 1, 2), (-1, 1, -1), (-1, 9, 3)\} \subset R^3$
20. $S = \{(1, 2, -1), (1, 4, 1), (-1, -5, -2)\} \subset R^3$
21. $S = \{(1, 1, -2, 3), (-2, -2, 4, -6), (1, -1, 1, 5), (1, 3, -5, 1)\} \subset R^4$
22. $S = \{(1, 1, 1, 2), (1, -1, 2, 3), (0, -2, 1, 1), (1, -2, 3, 2)\} \subset R^4$
23. $S = \{(1, 1, 1, 3), (1, -1, 2, 2), (1, -5, 4, 0), (-1, 9, -6, 2), (2, 4, 3, -1)\} \subset R^4$
24. $S = \{(1, 1, 1, 0), (0, 1, 0, 1), (-1, 0, -1, 1), (0, 1, 1, 0), (1, 1, 0, 1)\} \subset R^4$

In Problems 25–28, S is a linearly independent subset of R^3. Extend S to a basis for R^3.

25. $S = \{(1, 1, 1)\}$ **26.** $S = \{(1, 1, 0)\}$

27. $S = \{(1, -2, 1), (-2, 2, -1)\}$ **28.** $S = \{(2, 1, -1), (-1, 2, 1)\}$

In Problems 29–32, S is a linearly independent subset of R^4. Extend S to a basis for R^4.

29. $S = \{(1, -1, 1, 0), (-1, 1, 0, 0)\}$ **30.** $S = \{(1, 2, 0, 3), (0, 2, 0, 3)\}$
31. $S = \{(1, 1, 1, 1), (1, 0, 0, 1), (0, 1, 1, 1)\}$
32. $S = \{(1, -2, 1, 0), (1, 1, -1, -1), (-1, 1, 0, 1)\}$

In Problems 33–36, use Theorem 8 to determine whether S is a basis for R^3.

33. $S = \{(1, -2, 1), (3, -1, 1), (2, 1, 2)\}$
34. $S = \{(-2, 1, 3), (3, -2, -5), (1, -2, -3)\}$
35. $S = \{(4, 2, 1), (3, 4, 1), (6, -2, 1)\}$
36. $S = \{(2, -1, 0), (3, 4, -5), (2, -6, 9)\}$

C   *In Problems 37 and 38, find a basis for the row space of A consisting of row vectors from A. [Hint: Find a basis for the column space of A^T.]*

37. $A = \begin{bmatrix} 1 & 2 & 2 \\ 2 & 1 & 3 \\ 1 & -1 & 1 \\ 3 & -2 & 1 \end{bmatrix}$ **38.** $A = \begin{bmatrix} 1 & -2 & -1 \\ -2 & 4 & 2 \\ 1 & -3 & 0 \\ 1 & 0 & 1 \\ 1 & -1 & 2 \end{bmatrix}$

39. Let A be an $m \times n$ matrix with rank k. Show that $k \le m$ and $k \le n$.
40. Let A be an $m \times n$ matrix.

(A) If $m < n$, show that the columns of A are linearly dependent.
(B) If $n < m$, show that the rows of A are linearly dependent.

41. Let A be an $m \times n$ matrix, and let **b** be an $m \times 1$ column vector. Show that the system $A\mathbf{x} = \mathbf{b}$ has a solution if and only if **b** is in the column space of A.

42. Show that the rank of a matrix A is equal to the rank of A^T.

43. If A is an $m \times n$ matrix and E is an $m \times m$ elementary matrix, show that A and EA have the same rank.

44. If A is an $m \times n$ matrix and P is an invertible matrix, show that A and PA have the same rank.

❙ 6-3 ❙ Coordinate Matrices

- Coordinate Matrices
- Equivalent Representations of Vector Spaces
- Applications

▪ Coordinate Matrices

In this section, we will develop an important relationship between vectors in any n-dimensional vector space and vectors in R^n. This relationship will allow us to apply the procedures developed in the preceding two sections to any finite-dimensional vector space.

Example 12   The ordered pair (3, 4) has been given several different interpretations. If we consider (3, 4) to be a point in 2-space, then 3 and 4 are called the *coordinates* of the point. If we consider $\mathbf{u} = (3, 4)$ to be a vector in R^2, then 3 and 4 are called the *components* of $\mathbf{u}$. If $B = \{(1, 0), (0, 1)\}$ is the standard basis for R^2, then these components can be used to express $\mathbf{u}$ as a linear combination of basis vectors:

$$\mathbf{u} = (3, 4) = 3(1, 0) + 4(0, 1)$$

Thus, the coordinates of the point (3, 4) are the scalars that express the vector $\mathbf{u}$ as a linear combination of standard basis vectors.

Now let $C = \{(3, 5), (1, 2)\}$. This set is also a basis for R^2 (verify this). The vector $\mathbf{u}$ can be expressed as a linear combination of the vectors in C as follows (verify this):

$$\mathbf{u} = (3, 4) = 2(3, 5) + (-3)(1, 2)$$

The scalars 2 and -3 in this linear combination are called the *coordinates of* $\mathbf{u}$ *with respect to the basis* C and are usually denoted as a *coordinate matrix*. Thus, the relationship

$$\mathbf{u} = 2(3, 5) + (-3)(1, 2) \qquad \text{Linear combination of vectors in basis } C$$

is denoted by

$$[\mathbf{u}]_C = \begin{bmatrix} 2 \\ -3 \end{bmatrix} \qquad \text{Coordinate matrix of } \mathbf{u} \text{ with respect to basis } C$$

In a similar fashion, the relationship

$$\mathbf{u} = 3(1, 0) + 4(0, 1) \qquad \text{Linear combination of vectors in basis } B$$

is denoted by

$$[\mathbf{u}]_B = \begin{bmatrix} 3 \\ 4 \end{bmatrix} \qquad \text{Coordinate matrix of } \mathbf{u} \text{ with respect to basis } B$$

Notice that the coordinate matrix of $\mathbf{u}$ with respect to the standard basis B is just the ordered pair (3, 4) written as a column vector.

Problem 12   Refer to Example 12. Find $[\mathbf{u}]_B$ and $[\mathbf{u}]_C$ for $\mathbf{u} = (2, 2)$.   ▮▮

The discussion in Example 12 can be generalized to any finite-dimensional vector space.

Coordinate Matrices

Let $B = \{\mathbf{u}_1, \mathbf{u}_2, \ldots, \mathbf{u}_n\}$ be a basis for a vector space V, let $\mathbf{u}$ be any vector in V, and let $k_1, k_2, \ldots, k_n$ be scalars satisfying

$$\mathbf{u} = k_1\mathbf{u}_1 + k_2\mathbf{u}_2 + \cdots + k_n\mathbf{u}_n$$

The scalars $k_1, k_2, \ldots, k_n$ are called the **coordinates of u with respect to the basis B,** and the matrix

$$[\mathbf{u}]_B = \begin{bmatrix} k_1 \\ k_2 \\ . \\ . \\ . \\ k_n \end{bmatrix}$$

is called the **coordinate matrix of u with respect to the basis B.**

Notice that $[\mathbf{u}]_B$ is an $n \times 1$ column vector in R^n — not a vector in V. We can also use the notation

$$(\mathbf{u})_B = (k_1, k_2, \ldots, k_n) \qquad \text{Ordered } n\text{-tuple}$$

or

$$[\mathbf{u}]_B = [k_1 \ \ k_2 \ \ \ldots \ \ k_n] \qquad 1 \times n \text{ row vector}$$

to represent the coordinates of $\mathbf{u}$ with respect to B. For the types of operations we will be performing, column vectors provide the most convenient method of representing the coordinates of a vector.

Theorem 13 in Section 5-6 states that the scalars used to express a vector as a linear combination of basis vectors are unique. It also follows that the components of any $n \times 1$ column vector in R^n can be used as the scalars in a linear combination of basis vectors. These observations are summarized in Theorem 9. A formal proof is omitted. We will have more to say about the relationship between an n-dimensional vector space V and R^n later in this section.

Theorem 9

> Let V be an n-dimensional vector space. Each vector in V has a unique coordinate matrix in R^n, and each vector in R^n is the coordinate matrix of a vector in V.

Example 13 Let $B = \{(1, -1, 1), (1, 2, -1), (0, -1, 1)\}$ be a basis for R^3.

(A) Find $[\mathbf{u}]_B$ if $\mathbf{u} = (-1, -6, 2)$.   (B) Find $\mathbf{u}$ if $[\mathbf{u}]_B = \begin{bmatrix} 2 \\ -1 \\ 1 \end{bmatrix}$.

Solution (A) The coordinates of $\mathbf{u}$ with respect to B satisfy the equation

$$k_1 \begin{bmatrix} 1 \\ -1 \\ 1 \end{bmatrix} + k_2 \begin{bmatrix} 1 \\ 2 \\ -1 \end{bmatrix} + k_3 \begin{bmatrix} 0 \\ -1 \\ 1 \end{bmatrix} = \begin{bmatrix} -1 \\ -6 \\ 2 \end{bmatrix}$$

or, equivalently, the system

$$\begin{aligned}
k_1 + k_2 \quad\quad &= -1 \\
-k_1 + 2k_2 - k_3 &= -6 \\
k_1 - k_2 + k_3 &= \;\;\;2
\end{aligned}$$

The solution of this system is $k_1 = 3$, $k_2 = -4$, and $k_3 = -5$ (verify this). Thus,

$$[\mathbf{u}]_B = \begin{bmatrix} 3 \\ -4 \\ -5 \end{bmatrix}$$

Check: $3(1, -1, 1) - 4(1, 2, -1) - 5(0, -1, 1) \stackrel{?}{=} \mathbf{u}$

$(-1, -6, 2) \stackrel{?}{=} \mathbf{u}$

(B) If

$$[\mathbf{u}]_B = \begin{bmatrix} 2 \\ -1 \\ 1 \end{bmatrix}$$

then

$$\mathbf{u} = 2(1, -1, 1) + (-1)(1, 2, -1) + 1(0, -1, 1)$$
$$= (1, -5, 4)$$

Problem 13   Refer to the basis B in Example 13.

(A) Find $[\mathbf{u}]_B$ if $\mathbf{u} = (1, 4, -1)$.   (B)   Find $\mathbf{u}$ if $[\mathbf{u}]_B = \begin{bmatrix} 3 \\ -2 \\ 4 \end{bmatrix}$.   ∎

Before considering additional examples, there is a technical point that must be discussed. In Example 13(A), what would happen if we wrote the vectors in B in a different order? Suppose we let

$C = \{(1, 2, -1), (0, -1, 1), (1, -1, 1)\}$

This is the same set of vectors as B, but now if $\mathbf{u} = (-1, -6, 2)$, then

$$[\mathbf{u}]_C = \begin{bmatrix} -4 \\ -5 \\ 3 \end{bmatrix} \neq [\mathbf{u}]_B$$

Thus, changing the order of the vectors in a basis will change the corresponding coordinate matrices. Since coordinate matrices will depend on the order of the vectors in a basis, we will now assume that **all bases are ordered sets of vectors.** Thus, B and C will be considered to be *different* bases, even though they are *equal sets.*

Example 14   Let $B = \{1 + x, x + x^2, 1 + x + x^2\}$ be a basis for P_2.

(A) Find $[\mathbf{u}]_B$ if $\mathbf{u} = 1 + 5x + 2x^2$.

(B) Find $\mathbf{u}$ if $[\mathbf{u}]_B = \begin{bmatrix} 2 \\ -1 \\ 3 \end{bmatrix}$.

Solution   (A) The coordinates of $\mathbf{u}$ with respect to B satisfy the equation

$$k_1 (1 + x) + k_2 (x + x^2) + k_3 (1 + x + x^2) = 1 + 5x + 2x^2$$

or, equivalently, the system

$$\begin{aligned} k_1 \quad\;\; + k_3 &= 1 \\ k_1 + k_2 + k_3 &= 5 \\ k_2 + k_3 &= 2 \end{aligned}$$

The solution of this system is $k_1 = 3$, $k_2 = 4$, and $k_3 = -2$ (verify this). Thus,

$$[\mathbf{u}]_B = \begin{bmatrix} 3 \\ 4 \\ -2 \end{bmatrix}$$

You should check this result.

(B) If

$$[\mathbf{u}]_B = \begin{bmatrix} 2 \\ -1 \\ 3 \end{bmatrix}$$

then

$$\mathbf{u} = 2(1+x) - (x+x^2) + 3(1+x+x^2)$$
$$= 5 + 4x + 2x^2$$

Problem 14   Refer to the basis B in Example 14.

(A) Find $[\mathbf{u}]_B$ if $\mathbf{u} = 1 - x - x^2$.

(B) Find $\mathbf{u}$ if $[\mathbf{u}]_B = \begin{bmatrix} 3 \\ 2 \\ -5 \end{bmatrix}$.

▌▌

Example 15   Let

$$B = \left\{ \begin{bmatrix} 1 & 1 \\ 0 & 0 \end{bmatrix}, \begin{bmatrix} 0 & 1 \\ 0 & 1 \end{bmatrix}, \begin{bmatrix} 0 & 0 \\ 1 & 1 \end{bmatrix}, \begin{bmatrix} 1 & 0 \\ 0 & 1 \end{bmatrix} \right\}$$

be a basis for $M_{2 \times 2}$.

(A) Find $[\mathbf{u}]_B$ if $\mathbf{u} = \begin{bmatrix} 6 & -1 \\ 1 & 2 \end{bmatrix}$.

(B) Find $\mathbf{u}$ if $[\mathbf{u}]_B = \begin{bmatrix} 2 \\ -1 \\ 1 \\ 3 \end{bmatrix}$.

Solution   (A) The coordinates of $\mathbf{u}$ with respect to B satisfy the equation

$$k_1 \begin{bmatrix} 1 & 1 \\ 0 & 0 \end{bmatrix} + k_2 \begin{bmatrix} 0 & 1 \\ 0 & 1 \end{bmatrix} + k_3 \begin{bmatrix} 0 & 0 \\ 1 & 1 \end{bmatrix} + k_4 \begin{bmatrix} 1 & 0 \\ 0 & 1 \end{bmatrix} = \begin{bmatrix} 6 & -1 \\ 1 & 2 \end{bmatrix}$$

or, equivalently, the system

$$
\begin{array}{rcr}
k_1 \quad\quad\; + k_4 &=& 6 \\
k_1 + k_2 \quad\quad\; &=& -1 \\
k_3 \quad\; &=& 1 \\
k_2 + k_3 + k_4 &=& 2
\end{array}
$$

The solution of this system is $k_1 = 2$, $k_2 = -3$, $k_3 = 1$, and $k_4 = 4$ (verify this). Thus,

$$[\mathbf{u}]_B = \begin{bmatrix} 2 \\ -3 \\ 1 \\ 4 \end{bmatrix}$$

You should check this result.

(B) If

$$[\mathbf{u}]_B = \begin{bmatrix} 2 \\ -1 \\ 1 \\ 3 \end{bmatrix}$$

then

$$\mathbf{u} = 2\begin{bmatrix} 1 & 1 \\ 0 & 0 \end{bmatrix} - \begin{bmatrix} 0 & 1 \\ 0 & 1 \end{bmatrix} + \begin{bmatrix} 0 & 0 \\ 1 & 1 \end{bmatrix} + 3\begin{bmatrix} 1 & 0 \\ 0 & 1 \end{bmatrix}$$

$$= \begin{bmatrix} 5 & 1 \\ 1 & 3 \end{bmatrix}$$

Problem 15   Refer to the basis B in Example 15.

(A) Find $[\mathbf{u}]_B$ if $\mathbf{u} = \begin{bmatrix} 4 & -1 \\ -1 & 0 \end{bmatrix}$.

(B) Find $\mathbf{u}$ if $[\mathbf{u}]_B = \begin{bmatrix} 3 \\ -2 \\ 4 \\ 7 \end{bmatrix}$.

▌▌

▪ Equivalent Representations of Vector Spaces

Let

$$U_1 = \{(a, b, c, d) \mid a, b, c, d \in R\}$$
$$U_2 = \{[a \quad b \quad c \quad d] \mid a, b, c, d \in R\}$$
$$U_3 = \left\{ \begin{bmatrix} a \\ b \\ c \\ d \end{bmatrix} \middle| a, b, c, d \in R \right\} \tag{1}$$

It is easy to see that each of these sets is just a different representation of R^4. After all, R^4 is the set of all ordered 4-tuples of real numbers, and it does not matter whether these 4-tuples are written horizontally or vertically, with or without commas.

Now, let $B = \{1, x, x^2, x^3\}$ be the standard basis for P_3. Then P_3 can be represented as

$$P_3 = \{a + bx + cx^2 + dx^3 \mid a, b, c, d \in R\}$$

If we let $[P_3]_B$ denote the set of coordinate matrices with respect to B of all the vectors in P_3, then

$$[P_3]_B = \left\{ \left. \begin{bmatrix} a \\ b \\ c \\ d \end{bmatrix} \right| a, b, c, d \in R \right\} \tag{2}$$

In the same way, if

$$C = \left\{ \begin{bmatrix} 1 & 0 \\ 0 & 0 \end{bmatrix}, \begin{bmatrix} 0 & 1 \\ 0 & 0 \end{bmatrix}, \begin{bmatrix} 0 & 0 \\ 1 & 0 \end{bmatrix}, \begin{bmatrix} 0 & 0 \\ 0 & 1 \end{bmatrix} \right\}$$

is the standard basis for $M_{2 \times 2}$, then

$$M_{2 \times 2} = \left\{ \left. \begin{bmatrix} a & b \\ c & d \end{bmatrix} \right| a, b, c, d \in R \right\}$$

and

$$[M_{2 \times 2}]_C = \left\{ \left. \begin{bmatrix} a \\ b \\ c \\ d \end{bmatrix} \right| a, b, c, d \in R \right\} \tag{3}$$

Comparing (1), (2), and (3), we see that

$$[P_3]_B = [M_{2 \times 2}]_C = U_3 = R^4$$

In a certain sense, we can say that third-degree polynomials and 2×2 matrices are just two more ways of representing ordered 4-tuples of real numbers. In a more advanced course, it is shown that, as vector spaces, P_3, $M_{2 \times 2}$, and R^4 are indistinguishable.* That is, they have identical vector space properties. Of course, P_3 and $M_{2 \times 2}$ have properties that are not related to the vector space operations of addition and scalar multiplication (matrix multiplication in $M_{2 \times 2}$, for example), but we are not concerned with such properties at this point.

In general, if V is any n-dimensional vector space with basis B, then Theorem 9 implies that $[V]_B = R^n$, and it can be shown that, as vector spaces, V and R^n are indistinguishable. A complete discussion of the relationship between V and R^n is beyond the scope of this text. Theorem 10 provides two examples of the correspondence between the vector space properties of V and R^n. We have included this theorem because it allows us to apply the procedures developed for R^n in

---

* The technical term used in this setting is "isomorphic."

Sections 6-1 and 6-2 to any finite-dimensional vector space. The proof is omitted.

Theorem 10

Let V be an n-dimensional vector space with basis B, let S be a set of vectors in V, and let

$$[S]_B = \{[\mathbf{u}]_B | \mathbf{u} \in S\}$$

be the corresponding set of coordinate matrices in R^n.

(A) Set S is a linearly independent (dependent) subset of V if and only if $[S]_B$ is a linearly independent (dependent) subset of R^n.
(B) $\{\mathbf{u}_1, \mathbf{u}_2, \ldots, \mathbf{u}_m\}$ is a basis for span S if and only if $\{[\mathbf{u}_1]_B, [\mathbf{u}_2]_B, \ldots, [\mathbf{u}_m]_B\}$ is a basis for span$[S]_B$.

▪ Applications

Now we can use Theorem 10 to apply the procedures developed in Sections 6-1 and 6-2 to vector spaces other than R^n.

Example 16   Determine whether $S = \{1 + x - x^2, 1 + 2x + x^2, 1 - 3x + x^2\}$ is a linearly independent subset of P_2.

Solution   Let $B = \{1, x, x^2\}$ be the standard basis for P_2, and let

$$\mathbf{u}_1 = 1 + x - x^2 \qquad \mathbf{u}_2 = 1 + 2x + x^2 \qquad \mathbf{u}_3 = 1 - 3x + x^2$$

Then, by inspection, we see that

$$[\mathbf{u}_1]_B = \begin{bmatrix} 1 \\ 1 \\ -1 \end{bmatrix} \qquad [\mathbf{u}_2]_B = \begin{bmatrix} 1 \\ 2 \\ 1 \end{bmatrix} \qquad [\mathbf{u}_3]_B = \begin{bmatrix} 1 \\ -3 \\ 1 \end{bmatrix}$$

Hence,

$$[S]_B = \left\{ \begin{bmatrix} 1 \\ 1 \\ -1 \end{bmatrix}, \begin{bmatrix} 1 \\ 2 \\ 1 \end{bmatrix}, \begin{bmatrix} 1 \\ -3 \\ 1 \end{bmatrix} \right\}$$

To determine whether $[S]_B$ is linearly independent, we form the matrix whose columns are the vectors in $[S]_B$ and find its determinant:

$$\begin{vmatrix} 1 & 1 & 1 \\ 1 & 2 & -3 \\ -1 & 1 & 1 \end{vmatrix} = \begin{vmatrix} 2 & -3 \\ 1 & 1 \end{vmatrix} - \begin{vmatrix} 1 & -3 \\ -1 & 1 \end{vmatrix} + \begin{vmatrix} 1 & 2 \\ -1 & 1 \end{vmatrix} = 10$$

Since this determinant is nonzero, Theorem 8 in Section 6-2 implies that $[S]_B$ is linearly independent. Theorem 10(A) now implies that S is also linearly independent.

Problem 16 Determine whether $S = \{1 + x - x^2, 1 + 2x + x^2, 1 + 3x + 3x^2\}$ is a linearly independent subset of P_2. ▌▌

In Example 16, we could have used any basis for P_2 to form the set $[S]_B$. The standard basis is the preferred choice because coordinate matrices with respect to the standard basis are very easy to find.

Example 17 Let

$$S = \left\{ \begin{bmatrix} 1 & 1 \\ 0 & 0 \end{bmatrix}, \begin{bmatrix} 1 & -1 \\ 1 & 0 \end{bmatrix}, \begin{bmatrix} 0 & 1 \\ 0 & 1 \end{bmatrix}, \begin{bmatrix} 2 & 1 \\ 1 & 1 \end{bmatrix} \right\} \subset M_{2 \times 2}$$

Find a basis for span S consisting of vectors in S.

Solution If

$$B = \left\{ \begin{bmatrix} 1 & 0 \\ 0 & 0 \end{bmatrix}, \begin{bmatrix} 0 & 1 \\ 0 & 0 \end{bmatrix}, \begin{bmatrix} 0 & 0 \\ 1 & 0 \end{bmatrix}, \begin{bmatrix} 0 & 0 \\ 0 & 1 \end{bmatrix} \right\}$$

is the standard basis for $M_{2 \times 2}$, then (verify this)

$$[S]_B = \left\{ \begin{bmatrix} 1 \\ 1 \\ 0 \\ 0 \end{bmatrix}, \begin{bmatrix} 1 \\ -1 \\ 1 \\ 0 \end{bmatrix}, \begin{bmatrix} 0 \\ 1 \\ 0 \\ 1 \end{bmatrix}, \begin{bmatrix} 2 \\ 1 \\ 1 \\ 1 \end{bmatrix} \right\}$$

We now use method 2 in Section 6-2 to find a basis for span$[S]_B$ consisting of vectors in $[S]_B$. We form the matrix whose columns are the vectors in $[S]_B$ and find its reduced form:

$$\begin{bmatrix} 1 & 1 & 0 & 2 \\ 1 & -1 & 1 & 1 \\ 0 & 1 & 0 & 1 \\ 0 & 0 & 1 & 1 \end{bmatrix} \sim \begin{bmatrix} 1 & 0 & 0 & 1 \\ 0 & 1 & 0 & 1 \\ 0 & 0 & 1 & 1 \\ 0 & 0 & 0 & 0 \end{bmatrix}$$

A basis for span$[S]_B$ is

$$\left\{ \begin{bmatrix} 1 \\ 1 \\ 0 \\ 0 \end{bmatrix}, \begin{bmatrix} 1 \\ -1 \\ 1 \\ 0 \end{bmatrix}, \begin{bmatrix} 0 \\ 1 \\ 0 \\ 1 \end{bmatrix} \right\}$$

Theorem 10(B) now implies that a basis for span S is

$$\left\{ \begin{bmatrix} 1 & 1 \\ 0 & 0 \end{bmatrix}, \begin{bmatrix} 1 & -1 \\ 1 & 0 \end{bmatrix}, \begin{bmatrix} 0 & 1 \\ 0 & 1 \end{bmatrix} \right\}$$

Problem 17 Repeat Example 17 for

$$S = \left\{ \begin{bmatrix} -1 & 1 \\ -1 & 0 \end{bmatrix}, \begin{bmatrix} 2 & -1 \\ 1 & 1 \end{bmatrix}, \begin{bmatrix} 1 & 0 \\ 0 & 1 \end{bmatrix}, \begin{bmatrix} 0 & 1 \\ -1 & 1 \end{bmatrix} \right\}$$ ▌▌

Answers to   **12.** $[\mathbf{u}]_B = \begin{bmatrix} 2 \\ 2 \end{bmatrix}$, $[\mathbf{u}]_C = \begin{bmatrix} 2 \\ -4 \end{bmatrix}$ **13.** (A) $\begin{bmatrix} -2 \\ 3 \\ 4 \end{bmatrix}$ (B) $(1, -11, 9)$
Matched Problems

14. (A) $\begin{bmatrix} 0 \\ -2 \\ 1 \end{bmatrix}$ (B) $-2 - 3x^2$ **15.** (A) $\begin{bmatrix} 1 \\ -2 \\ -1 \\ 3 \end{bmatrix}$ (B) $\begin{bmatrix} 10 & 1 \\ 4 & 9 \end{bmatrix}$

16. No; S is linearly dependent.   **17.** $\left\{ \begin{bmatrix} -1 & 1 \\ -1 & 0 \end{bmatrix}, \begin{bmatrix} 2 & -1 \\ 1 & 1 \end{bmatrix} \right\}$

▌ Exercise 6-3

A   Let $B = \{(-1, 2), (2, -3)\}$ be a basis for R^2. In Problems 1–4, find $[\mathbf{u}]_B$.

1. $\mathbf{u} = (1, 0)$ **2.** $\mathbf{u} = (0, 1)$ **3.** $\mathbf{u} = (2, -1)$ **4.** $\mathbf{u} = (-3, 5)$

Let $B = \{(-1, 2), (2, -3)\}$ be a basis for R^2. In Problems 5–8, find $\mathbf{u}$.

5. $[\mathbf{u}]_B = \begin{bmatrix} 1 \\ 0 \end{bmatrix}$ **6.** $[\mathbf{u}]_B = \begin{bmatrix} 0 \\ 1 \end{bmatrix}$ **7.** $[\mathbf{u}]_B = \begin{bmatrix} 2 \\ 3 \end{bmatrix}$ **8.** $[\mathbf{u}]_B = \begin{bmatrix} -2 \\ 5 \end{bmatrix}$

Let $B = \{2 - 3x, -3 + 4x\}$ be a basis for P_1. In Problems 9–12, find $[\mathbf{u}]_B$.

9. $\mathbf{u} = x$ **10.** $\mathbf{u} = 1$

11. $\mathbf{u} = 2 - 5x$ **12.** $\mathbf{u} = -3 + x$

Let $B = \{2 - 3x, -3 + 4x\}$ be a basis for P_1. In Problems 13–16, find $\mathbf{u}$.

13. $[\mathbf{u}]_B = \begin{bmatrix} 2 \\ 0 \end{bmatrix}$ **14.** $[\mathbf{u}]_B = \begin{bmatrix} 0 \\ -1 \end{bmatrix}$

15. $[\mathbf{u}]_B = \begin{bmatrix} 1 \\ 1 \end{bmatrix}$ **16.** $[\mathbf{u}]_B = \begin{bmatrix} 5 \\ 3 \end{bmatrix}$

B   In Problems 17–20, let $B = \{(1, 0, 1), (1, -1, 0), (1, 1, 1)\}$ be a basis for R^3.

17. Find $[\mathbf{u}]_B$ if $\mathbf{u} = (2, 1, -1)$.   **18.** Find $[\mathbf{u}]_B$ if $\mathbf{u} = (3, 2, 4)$.

19. Find $\mathbf{u}$ if $[\mathbf{u}]_B = \begin{bmatrix} 1 \\ 2 \\ 3 \end{bmatrix}$.   **20.** Find $\mathbf{u}$ if $[\mathbf{u}]_B = \begin{bmatrix} -2 \\ 1 \\ 4 \end{bmatrix}$.

In Problems 21–24, let $B = \{1, 1 + x, 1 + x + x^2\}$ be a basis for P_2.

21. Find $[\mathbf{u}]_B$ if $\mathbf{u} = 3 + 2x + x^2$.   **22.** Find $[\mathbf{u}]_B$ if $\mathbf{u} = 1 + 2x + 3x^2$.

23. Find $\mathbf{u}$ if $[\mathbf{u}]_B = \begin{bmatrix} 2 \\ 0 \\ -1 \end{bmatrix}$.   **24.** Find $\mathbf{u}$ if $[\mathbf{u}]_B = \begin{bmatrix} 0 \\ -1 \\ 1 \end{bmatrix}$.

In Problems 25–28, let

$$B = \left\{ \begin{bmatrix} 1 & 1 \\ 0 & 0 \end{bmatrix}, \begin{bmatrix} 1 & 0 \\ 1 & 0 \end{bmatrix}, \begin{bmatrix} 0 & 1 \\ 0 & 1 \end{bmatrix}, \begin{bmatrix} 1 & 1 \\ 0 & 1 \end{bmatrix} \right\}$$

be a basis for $M_{2 \times 2}$.

25. Find $[\mathbf{u}]_B$ if $\mathbf{u} = \begin{bmatrix} 2 & 3 \\ -1 & 4 \end{bmatrix}$.

26. Find $[\mathbf{u}]_B$ if $\mathbf{u} = \begin{bmatrix} -3 & 1 \\ 0 & -2 \end{bmatrix}$.

27. Find $\mathbf{u}$ if $[\mathbf{u}]_B = \begin{bmatrix} -1 \\ 0 \\ -1 \\ 2 \end{bmatrix}$.

28. Find $\mathbf{u}$ if $[\mathbf{u}]_B = \begin{bmatrix} -2 \\ 3 \\ 4 \\ 0 \end{bmatrix}$.

In Problems 29 and 30, use Theorem 10 to determine whether S is a linearly independent or linearly dependent subset of P_2. [Hint: Use coordinate matrices with respect to the standard basis for P_2.]

29. $S = \{1 + 2x - x^2, 3 + 2x - 4x^2, 1 - 2x - 2x^2\}$
30. $S = \{2 + 3x - 5x^2, 3 + 2x^2, 4 - 5x\}$

In Problems 31 and 32, use Theorem 10 to determine whether S is a linearly independent or linearly dependent subset of $M_{2 \times 2}$.

31. $S = \left\{ \begin{bmatrix} 1 & -1 \\ 1 & 2 \end{bmatrix}, \begin{bmatrix} 2 & -1 \\ 1 & 3 \end{bmatrix}, \begin{bmatrix} 0 & 1 \\ -1 & 2 \end{bmatrix}, \begin{bmatrix} -1 & 3 \\ 2 & 4 \end{bmatrix} \right\}$

32. $S = \left\{ \begin{bmatrix} 1 & 2 \\ -1 & 1 \end{bmatrix}, \begin{bmatrix} 2 & -1 \\ 1 & 2 \end{bmatrix}, \begin{bmatrix} 3 & -1 \\ 0 & -2 \end{bmatrix}, \begin{bmatrix} 1 & 4 \\ -1 & 6 \end{bmatrix} \right\}$

In Problems 33–36, use Theorem 10 and method 1 (Section 6-1) to find a basis for span S.

33. $S = \{1 + 2x - x^2 + 3x^3, 2 + 3x + x^2 + 2x^3, 1 + x + 4x^2 - 3x^3,$
$\qquad 1 + 2x - 3x^2 + 5x^3\} \subset P_3$
34. $S = \{1 + 3x - 2x^2 + 3x^3, 2 + x + x^2 + x^3, 1 + 2x - x^2 + 2x^3,$
$\qquad 1 - x + 2x^2 - x^3\} \subset P_3$

35. $S = \left\{ \begin{bmatrix} 1 & 1 \\ 2 & -3 \end{bmatrix}, \begin{bmatrix} 2 & 1 \\ 2 & -6 \end{bmatrix}, \begin{bmatrix} 1 & 2 \\ 4 & -3 \end{bmatrix}, \begin{bmatrix} -1 & 1 \\ 2 & 3 \end{bmatrix} \right\} \subset M_{2 \times 2}$

36. $S = \left\{ \begin{bmatrix} 1 & -1 \\ 2 & 3 \end{bmatrix}, \begin{bmatrix} 1 & -1 \\ -3 & 4 \end{bmatrix}, \begin{bmatrix} -2 & 2 \\ -1 & 1 \end{bmatrix}, \begin{bmatrix} 1 & -1 \\ 4 & -5 \end{bmatrix} \right\} \subset M_{2 \times 2}$

In Problems 37–40, use Theorem 10 and method 2 (Section 6-2) to find a basis for span S consisting of vectors in S.

37. $S = \{1 + x + x^2 + x^3, 2 - x + 2x^2 + 3x^3, 1 - 2x + x^2 + 2x^3,$
$\qquad 1 + 4x + x^2\} \subset P_3$
38. $S = \{1 + x^3, -1 + x + 2x^2, -2 + x + 2x^2 - x^3, 2 - x + x^2 + x^3\} \subset P_3$

39. $S = \left\{ \begin{bmatrix} 1 & 1 \\ -1 & 0 \end{bmatrix}, \begin{bmatrix} 1 & 2 \\ 1 & -3 \end{bmatrix}, \begin{bmatrix} 0 & 1 \\ -1 & 2 \end{bmatrix}, \begin{bmatrix} 2 & 1 \\ -1 & 3 \end{bmatrix} \right\} \subset M_{2\times2}$

40. $S = \left\{ \begin{bmatrix} 1 & -1 \\ 1 & -2 \end{bmatrix}, \begin{bmatrix} -2 & 2 \\ -2 & 4 \end{bmatrix}, \begin{bmatrix} 1 & -2 \\ 0 & -6 \end{bmatrix}, \begin{bmatrix} 0 & 1 \\ 1 & 4 \end{bmatrix} \right\} \subset M_{2\times2}$

In Problems 41–44, S is a linearly independent subset of the indicated vector space. Use Theorem 10 and the procedure discussed in Section 6-2 to extend S to a basis for the entire space.

41. $S = \{1 + 2x - 2x^2,\ 1 + x - x^2\} \subset P_2$

42. $S = \{1 + x + x^2,\ 2 + 3x + 3x^2\} \subset P_2$

43. $S = \left\{ \begin{bmatrix} 1 & 0 \\ 1 & 2 \end{bmatrix}, \begin{bmatrix} 0 & 1 \\ 2 & 4 \end{bmatrix} \right\} \subset M_{2\times2}$

44. $S = \left\{ \begin{bmatrix} 1 & 1 \\ -1 & 2 \end{bmatrix}, \begin{bmatrix} 3 & 2 \\ -2 & 4 \end{bmatrix} \right\} \subset M_{2\times2}$

C   **45.** Let $B = \{(4, -2),\ (3, -1)\}$ be a basis for R^2. Find $[\mathbf{u}]_B$ if:

  (A) $\mathbf{u} = (1, 0)$ (B) $\mathbf{u} = (0, 1)$ (C) $\mathbf{u} = (a, b)$

46. Let $B = \{(3, 2),\ (1, 1)\}$ be a basis for R^2. Find $[\mathbf{u}]_B$ if:

  (A) $\mathbf{u} = (1, 0)$ (B) $\mathbf{u} = (0, 1)$ (C) $\mathbf{u} = (a, b)$

47. Let

$$D_2 = \left\{ \begin{bmatrix} a & 0 \\ 0 & b \end{bmatrix} \middle| a, b \in R \right\}$$

and let

$$B = \left\{ \begin{bmatrix} 2 & 0 \\ 0 & 1 \end{bmatrix}, \begin{bmatrix} 1 & 0 \\ 0 & 1 \end{bmatrix} \right\}$$

be a basis for D_2. Find $[\mathbf{u}]_B$ if:

  (A) $\mathbf{u} = \begin{bmatrix} 1 & 0 \\ 0 & 0 \end{bmatrix}$ (B) $\mathbf{u} = \begin{bmatrix} 0 & 0 \\ 0 & 1 \end{bmatrix}$ (C) $\mathbf{u} = \begin{bmatrix} a & 0 \\ 0 & b \end{bmatrix}$

48. Let

$$S_2 = \left\{ \begin{bmatrix} a & b \\ b & c \end{bmatrix} \middle| a, b, c \in R \right\}$$

and let

$$B = \left\{ \begin{bmatrix} 1 & 1 \\ 1 & 1 \end{bmatrix}, \begin{bmatrix} 1 & 2 \\ 2 & 1 \end{bmatrix}, \begin{bmatrix} 1 & 0 \\ 0 & -1 \end{bmatrix} \right\}$$

be a basis for S_2. Find $[\mathbf{u}]_B$ if:

(A) $\mathbf{u} = \begin{bmatrix} 1 & 0 \\ 0 & 0 \end{bmatrix}$ (B) $\mathbf{u} = \begin{bmatrix} 0 & 1 \\ 1 & 0 \end{bmatrix}$ (C) $\mathbf{u} = \begin{bmatrix} 0 & 0 \\ 0 & 1 \end{bmatrix}$

(D) $\mathbf{u} = \begin{bmatrix} a & b \\ b & c \end{bmatrix}$

49. *Chebyshev polynomials.* Let $T = \{1,\, x,\, 2x^2 - 1,\, 4x^3 - 3x\}$ be the basis of Chebyshev polynomials for P_3. Find $[\mathbf{u}]_T$ if:

(A) $\mathbf{u} = x^2$ (B) $\mathbf{u} = x^3$
(C) $\mathbf{u} = a_0 + a_1x + a_2x^2 + a_3x^3$

50. *Hermite polynomials.* Let $H = \{1,\, 2x,\, 4x^2 - 2,\, 8x^3 - 12x\}$ be the basis of Hermite polynomials for P_3. Find $[\mathbf{u}]_H$ if:

(A) $\mathbf{u} = x^2$ (B) $\mathbf{u} = x^3$
(C) $\mathbf{u} = a_0 + a_1x + a_2x^2 + a_3x^3$

I6-4I   Change of Basis

- Change of Basis and Transition Matrices
- A Property of Transition Matrices
- Rotation of Axes in the Plane

■ Change of Basis and Transition Matrices

Let V be a finite-dimensional vector space with basis B. In some applications, it will be necessary to select a new basis for V with certain specified properties. If C is this new basis, then we must understand the relationship between the old basis B and the new basis C. We begin with an example that will lead to some general observations.

Example 18   Let $B = \{\mathbf{b}_1,\, \mathbf{b}_2\}$ and $C = \{\mathbf{c}_1,\, \mathbf{c}_2\}$ be bases for R^2, where

$$\mathbf{b}_1 = (-1,\, 2) \qquad \mathbf{b}_2 = (-3,\, 5)$$
$$\mathbf{c}_1 = (1,\, -1) \qquad \mathbf{c}_2 = (-2,\, 3)$$

Since C is a basis, each vector in B can be expressed as a linear combination of vectors from C. Omitting the details, we have (verify this)

$$\mathbf{b}_1 = \mathbf{c}_1 + \mathbf{c}_2 \qquad \mathbf{b}_2 = \mathbf{c}_1 + 2\mathbf{c}_2 \qquad\qquad (1)$$

or, using coordinate matrices,

$$[\mathbf{b}_1]_C = \begin{bmatrix} 1 \\ 1 \end{bmatrix} \qquad [\mathbf{b}_2]_C = \begin{bmatrix} 1 \\ 2 \end{bmatrix}$$

Now let **v** be any vector in R^2. Given the coordinate matrix of **v** with respect to B, we want to find the coordinate matrix of **v** with respect to C. That is, we want to *change from a representation with respect to B to a representation with respect to C.* Performing this operation is referred to as **changing bases.** If

$$[\mathbf{v}]_B = \begin{bmatrix} k_1 \\ k_2 \end{bmatrix}$$

then, using (1), we have

$$\begin{aligned}
\mathbf{v} &= k_1\mathbf{b}_1 + k_2\mathbf{b}_2 \\
&= k_1(\mathbf{c}_1 + \mathbf{c}_2) + k_2(\mathbf{c}_1 + 2\mathbf{c}_2) \\
&= (k_1 + k_2)\mathbf{c}_1 + (k_1 + 2k_2)\mathbf{c}_2
\end{aligned}$$

Thus,

$$[\mathbf{v}]_C = \begin{bmatrix} k_1 + k_2 \\ k_1 + 2k_2 \end{bmatrix}$$

The key to establishing a relationship between $[\mathbf{v}]_B$ and $[\mathbf{v}]_C$ is to write $[\mathbf{v}]_C$ as a matrix product:

$$[\mathbf{v}]_C = \begin{bmatrix} k_1 + k_2 \\ k_1 + 2k_2 \end{bmatrix} = \begin{bmatrix} 1 & 1 \\ 1 & 2 \end{bmatrix}\begin{bmatrix} k_1 \\ k_2 \end{bmatrix} = P[\mathbf{v}]_B$$

where

$$P = \begin{bmatrix} 1 & 1 \\ 1 & 2 \end{bmatrix}$$

Thus, we can change from the representation of **v** with respect to B to a representation with respect to C by left-multiplying $[\mathbf{v}]_B$ by P. The matrix P is called the **transition matrix** from B to C. Notice that the columns of P are the coordinate matrices of $\mathbf{b}_1$ and $\mathbf{b}_2$ with respect to C. That is,

$$P = \begin{bmatrix} 1 & 1 \\ 1 & 2 \end{bmatrix} \qquad \begin{bmatrix} 1 \\ 1 \end{bmatrix} = [\mathbf{b}_1]_C, \quad \begin{bmatrix} 1 \\ 2 \end{bmatrix} = [\mathbf{b}_2]_C$$

$$= [[\mathbf{b}_1]_C \quad [\mathbf{b}_2]_C] \qquad \text{Transition matrix } \textit{from } B \textit{ to } C$$

Problem 18   Refer to Example 18. Use the transition matrix P to find $[\mathbf{v}]_C$ if

$$[\mathbf{v}]_B = \begin{bmatrix} 5 \\ -2 \end{bmatrix}$$

‖

The ideas introduced in Example 18 are generalized to an arbitrary n-dimensional vector space in the following definition and theorem:

Transition Matrix

Let

$$B = \{\mathbf{b}_1, \mathbf{b}_2, \ldots, \mathbf{b}_n\}$$

and

$$C = \{\mathbf{c}_1, \mathbf{c}_2, \ldots, \mathbf{c}_n\}$$

be bases for a vector space V. The matrix

$$P = [[\mathbf{b}_1]_C \quad [\mathbf{b}_2]_C \quad \cdots \quad [\mathbf{b}_n]_C]$$

is called the **transition matrix from B to C.**

Theorem 11

Change of Basis Theorem

If $\mathbf{v}$ is any vector in V, then

$$[\mathbf{v}]_C = P[\mathbf{v}]_B$$

Proof   Since C is a basis for V, each $\mathbf{b}_i \in B$ can be expressed as a linear combination of the vectors in C. Thus,

$$\begin{aligned}
\mathbf{b}_1 &= a_{11}\mathbf{c}_1 + a_{21}\mathbf{c}_2 + \cdots + a_{n1}\mathbf{c}_n \\
\mathbf{b}_2 &= a_{12}\mathbf{c}_1 + a_{22}\mathbf{c}_2 + \cdots + a_{n2}\mathbf{c}_n \\
&\;\;\vdots \\
\mathbf{b}_n &= a_{1n}\mathbf{c}_1 + a_{2n}\mathbf{c}_2 + \cdots + a_{nn}\mathbf{c}_n
\end{aligned} \qquad (2)$$

Now let $\mathbf{v}$ be any vector in V and let

$$[\mathbf{v}]_B = \begin{bmatrix} k_1 \\ k_2 \\ \vdots \\ k_n \end{bmatrix}$$

Then

$$\mathbf{v} = k_1\mathbf{b}_1 + k_2\mathbf{b}_2 + \cdots + k_n\mathbf{b}_n \qquad \text{Use (2) to substitute for } \mathbf{b}_1, \mathbf{b}_2, \ldots, \mathbf{b}_n.$$

$$
\begin{aligned}
\mathbf{v} = \quad & k_1(a_{11}\mathbf{c}_1 + a_{21}\mathbf{c}_2 + \cdots + a_{n1}\mathbf{c}_n) \\
+ \; & k_2(a_{12}\mathbf{c}_1 + a_{22}\mathbf{c}_2 + \cdots + a_{n2}\mathbf{c}_n) + \cdots \\
+ \; & k_n(a_{1n}\mathbf{c}_1 + a_{2n}\mathbf{c}_2 + \cdots + a_{nn}\mathbf{c}_n) \\
= \quad & (k_1 a_{11} + k_2 a_{12} + \cdots + k_n a_{1n})\mathbf{c}_1 \\
+ \; & (k_1 a_{21} + k_2 a_{22} + \cdots + k_n a_{2n})\mathbf{c}_2 + \cdots \\
+ \; & (k_1 a_{n1} + k_2 a_{n2} + \cdots + k_n a_{nn})\mathbf{c}_n
\end{aligned}
$$

Multiply and collect the coefficients of $\mathbf{c}_1, \mathbf{c}_2, \ldots, \mathbf{c}_n$.

This expresses $\mathbf{v}$ as a linear combination of the vectors in C.

Thus,

$$
[\mathbf{v}]_C =
\begin{bmatrix}
k_1 a_{11} + k_2 a_{12} + \cdots + k_n a_{1n} \\
k_1 a_{21} + k_2 a_{22} + \cdots + k_n a_{2n} \\
\vdots \qquad \vdots \qquad \qquad \vdots \\
k_1 a_{n1} + k_2 a_{n2} + \cdots + k_n a_{nn}
\end{bmatrix}
$$

Express this column vector as a matrix product.

$$
=
\begin{bmatrix}
a_{11} & a_{12} & \cdots & a_{1n} \\
a_{21} & a_{22} & \cdots & a_{2n} \\
\vdots & \vdots & & \vdots \\
a_{n1} & a_{n2} & \cdots & a_{nn}
\end{bmatrix}
\begin{bmatrix}
k_1 \\ k_2 \\ \vdots \\ k_n
\end{bmatrix}
\qquad
[\mathbf{v}]_B =
\begin{bmatrix}
k_1 \\ k_2 \\ \vdots \\ k_n
\end{bmatrix}
$$

From (2), we have

$$
[\mathbf{b}_1]_C =
\begin{bmatrix}
a_{11} \\ a_{21} \\ \vdots \\ a_{n1}
\end{bmatrix}, \quad
[\mathbf{b}_2]_C =
\begin{bmatrix}
a_{12} \\ a_{22} \\ \vdots \\ a_{n2}
\end{bmatrix}, \quad
\ldots, \quad
[\mathbf{b}_n]_C =
\begin{bmatrix}
a_{1n} \\ a_{2n} \\ \vdots \\ a_{nn}
\end{bmatrix}
$$

Thus,

$$
\begin{aligned}
[\mathbf{v}]_C &= [[\mathbf{b}_1]_C \quad [\mathbf{b}_2]_C \quad \cdots \quad [\mathbf{b}_n]_C][\mathbf{v}]_B \\
&= P[\mathbf{v}]_B
\end{aligned}
$$

∎

The procedure for carrying out a change of basis is summarized in the box.

Changing Basis from B to C

To change from $[\mathbf{v}]_B$ to $[\mathbf{v}]_C$:

Step 1.   Find $[\mathbf{b}_i]_C$ for each $\mathbf{b}_i \in B$.

Step 2.   Form the transition matrix: $P = [[\mathbf{b}_1]_C \quad [\mathbf{b}_2]_C \quad \cdots \quad [\mathbf{b}_n]_C]$

Step 3.   Left-multiply $[\mathbf{v}]_B$ by P: $[\mathbf{v}]_C = P[\mathbf{v}]_B$

Example 19   Let

$$B = \{(1, 1, 0), (1, 0, 0), (1, 3, 1)\}$$

and

$$C = \{(1, 1, 1), (1, 1, -1), (1, -1, 1)\}$$

be bases for R^3. Find P, the transition matrix from B to C, and use P to find $[\mathbf{v}]_C$ if

$$[\mathbf{v}]_B = \begin{bmatrix} 4 \\ -2 \\ 5 \end{bmatrix}$$

Solution   *Step 1.*   *Find* $[\mathbf{b}_i]_C$ *for each* $\mathbf{b}_i \in B$. To find $[\mathbf{b}_1]_C$, we must find k_1, k_2, and k_3 satisfying

$$\overset{\mathbf{c}_1}{k_1(1, 1, 1)} + \overset{\mathbf{c}_2}{k_2(1, 1, -1)} + \overset{\mathbf{c}_3}{k_3(1, -1, 1)} = \overset{\mathbf{b}_1}{(1, 1, 0)}$$

or

$$k_1 + k_2 + k_3 = 1$$
$$k_1 + k_2 - k_3 = 1$$
$$k_1 - k_2 + k_3 = 0$$

The solution to this system is $k_1 = \frac{1}{2}$, $k_2 = \frac{1}{2}$, and $k_3 = 0$ (verify this). Similarly, the solution of

$$\overset{\mathbf{c}_1}{k_1(1, 1, 1)} + \overset{\mathbf{c}_2}{k_2(1, 1, -1)} + \overset{\mathbf{c}_3}{k_3(1, -1, 1)} = \overset{\mathbf{b}_2}{(1, 0, 0)}$$

is $k_1 = 0$, $k_2 = \frac{1}{2}$, and $k_3 = \frac{1}{2}$ (verify this); and the solution of

$$\overset{\mathbf{c}_1}{k_1(1, 1, 1)} + \overset{\mathbf{c}_2}{k_2(1, 1, -1)} + \overset{\mathbf{c}_3}{k_3(1, -1, 1)} = \overset{\mathbf{b}_3}{(1, 3, 1)}$$

is $k_1 = 2$, $k_2 = 0$, and $k_3 = -1$ (verify this). Thus,

$$[\mathbf{b}_1]_C = \begin{bmatrix} \frac{1}{2} \\ \frac{1}{2} \\ 0 \end{bmatrix}$$

$$[\mathbf{b}_2]_C = \begin{bmatrix} 0 \\ \frac{1}{2} \\ \frac{1}{2} \end{bmatrix}$$

$$[\mathbf{b}_3]_C = \begin{bmatrix} 2 \\ 0 \\ -1 \end{bmatrix}$$

Step 2.   *Form the transition matrix.*

$$P = \begin{bmatrix} \frac{1}{2} & 0 & 2 \\ \frac{1}{2} & \frac{1}{2} & 0 \\ 0 & \frac{1}{2} & -1 \end{bmatrix}$$

Step 3.   *Left-multiply* $[\mathbf{v}]_B$ *by P.*

$$[\mathbf{v}]_C = \begin{bmatrix} \frac{1}{2} & 0 & 2 \\ \frac{1}{2} & \frac{1}{2} & 0 \\ 0 & \frac{1}{2} & -1 \end{bmatrix} \begin{bmatrix} 4 \\ -2 \\ 5 \end{bmatrix} = \begin{bmatrix} 12 \\ 1 \\ -6 \end{bmatrix}$$

Check: To check our work, we must show that $[\mathbf{v}]_B$ and $[\mathbf{v}]_C$ represent the same vector in R^3. Using $[\mathbf{v}]_B$, we have

$$\begin{array}{ccc} \mathbf{b}_1 & \mathbf{b}_2 & \mathbf{b}_3 \end{array}$$

$$\mathbf{v} = 4(1, 1, 0) - 2(1, 0, 0) + 5(1, 3, 1) \qquad [\mathbf{v}]_B = \begin{bmatrix} 4 \\ -2 \\ 5 \end{bmatrix}$$
$$= (7, 19, 5)$$

and using $[\mathbf{v}]_C$, we have

$$\begin{array}{ccc} \mathbf{c}_1 & \mathbf{c}_2 & \mathbf{c}_3 \end{array}$$

$$\mathbf{v} = 12(1, 1, 1) + 1(1, 1, -1) - 6(1, -1, 1) \qquad [\mathbf{v}]_C = \begin{bmatrix} 12 \\ 1 \\ -6 \end{bmatrix}$$
$$= (7, 19, 5)$$

Thus, $[\mathbf{v}]_B$ and $[\mathbf{v}]_C$ are both coordinate matrices of $\mathbf{v} = (7, 19, 5)$.

Problem 19   Let

$$B = \{(-1, -1, 1), (5, -1, 1), (1, 1, 5)\}$$

and

$$C = \{(1, -1, 1), (2, 2, -1), (1, 1, 1)\}$$

be bases for R^3. Find P, the transition matrix from B to C, and use P to find $[\mathbf{v}]_C$ if

$$[\mathbf{v}]_B = \begin{bmatrix} 2 \\ -1 \\ 3 \end{bmatrix}$$

Check your results as in Example 19.   ∎

Example 20   Let $B = \{1, x, x^2\}$ and $C = \{1 + x + x^2, 1 - x + 2x^2, 1 + x^2\}$ be bases for P_2.

(A) Find the transition matrix from B to C.
(B) Find the transition matrix from C to B.

Solution   (A) Let

$$\mathbf{b}_1 = 1 \qquad \mathbf{b}_2 = x \qquad \mathbf{b}_3 = x^2$$

and

$$\mathbf{c}_1 = 1 + x + x^2 \qquad \mathbf{c}_2 = 1 - x + 2x^2 \qquad \mathbf{c}_3 = 1 + x^2$$

Then, omitting the details, we have

$$\mathbf{b}_1 = -\mathbf{c}_1 - \mathbf{c}_2 + 3\mathbf{c}_3$$
$$\mathbf{b}_2 = \mathbf{c}_1 \phantom{-\mathbf{c}_2} - \mathbf{c}_3$$
$$\mathbf{b}_3 = \mathbf{c}_1 + \mathbf{c}_2 - 2\mathbf{c}_3$$

Thus,

$$[\mathbf{b}_1]_C = \begin{bmatrix} -1 \\ -1 \\ 3 \end{bmatrix} \qquad [\mathbf{b}_2]_C = \begin{bmatrix} 1 \\ 0 \\ -1 \end{bmatrix} \qquad [\mathbf{b}_3]_C = \begin{bmatrix} 1 \\ 1 \\ -2 \end{bmatrix}$$

and the transition matrix from B to C is

$$P = \begin{bmatrix} -1 & 1 & 1 \\ -1 & 0 & 1 \\ 3 & -1 & -2 \end{bmatrix}$$

(B) The transition matrix from C to B is

$$Q = [[\mathbf{c}_1]_B \quad [\mathbf{c}_2]_B \quad [\mathbf{c}_3]_B]$$

Since B is the standard basis for P_2, the vectors in C are already expressed in terms of the basis vectors in B. Thus,

$$[\mathbf{c}_1]_B = \begin{bmatrix} 1 \\ 1 \\ 1 \end{bmatrix} \qquad [\mathbf{c}_2]_B = \begin{bmatrix} 1 \\ -1 \\ 2 \end{bmatrix} \qquad [\mathbf{c}_3]_B = \begin{bmatrix} 1 \\ 0 \\ 1 \end{bmatrix}$$

and

$$Q = \begin{bmatrix} 1 & 1 & 1 \\ 1 & -1 & 0 \\ 1 & 2 & 1 \end{bmatrix}$$

Problem 20   Let $B = \{1, x\}$ and $C = \{2 + 5x, 3 + 8x\}$ be bases for P_1.

(A) Find the transition matrix from B to C.

(B) Find the transition matrix from C to B. ∎

▪ A Property of Transition Matrices

If P and Q are the transition matrices computed in Example 20, then

$$PQ = \begin{bmatrix} -1 & 1 & 1 \\ -1 & 0 & 1 \\ 3 & -1 & -2 \end{bmatrix} \begin{bmatrix} 1 & 1 & 1 \\ 1 & -1 & 0 \\ 1 & 2 & 1 \end{bmatrix} = \begin{bmatrix} 1 & 0 & 0 \\ 0 & 1 & 0 \\ 0 & 0 & 1 \end{bmatrix} = I$$

This shows that $Q = P^{-1}$. That is, the transition matrix from C to B is the inverse of the transition matrix from B to C. Theorem 12 shows that this is always the case.

Theorem 12   If B and C are bases for a finite-dimensional vector space V and P is the transition matrix from B to C, then P is invertible and P^{-1} is the transition matrix from C to B.

Proof   Let Q be the transition matrix from C to B. Theorem 11 implies that

$$Q[\mathbf{v}]_C = [\mathbf{v}]_B$$

for all $\mathbf{v}$ in V. Left-multiplying this equation by P, we have

$$PQ[\mathbf{v}]_C = P[\mathbf{v}]_B \tag{3}$$

Since P is the transition matrix from B to C, Theorem 11 also implies that $P[\mathbf{v}]_B = [\mathbf{v}]_C$ for all $\mathbf{v}$ in V. Substituting in (3), we have

$$PQ[\mathbf{v}]_C = P[\mathbf{v}]_B$$
$$= [\mathbf{v}]_C$$

for all $\mathbf{v}$ in V. It is left as an exercise to show that this implies that $PQ = I$. Hence, P is invertible, and $P^{-1} = Q$, the transition matrix from C to B. ∎

Example 21   Let

$$B = \left\{ \begin{bmatrix} 1 & 1 \\ 1 & 0 \end{bmatrix}, \begin{bmatrix} 0 & 1 \\ 1 & 1 \end{bmatrix}, \begin{bmatrix} 0 & 0 \\ 1 & 1 \end{bmatrix}, \begin{bmatrix} 1 & 1 \\ 0 & 0 \end{bmatrix} \right\}$$

and

$$C = \left\{ \begin{bmatrix} 1 & 0 \\ 0 & 0 \end{bmatrix}, \begin{bmatrix} 0 & 1 \\ 0 & 0 \end{bmatrix}, \begin{bmatrix} 0 & 0 \\ 1 & 0 \end{bmatrix}, \begin{bmatrix} 0 & 0 \\ 0 & 1 \end{bmatrix} \right\}$$

be bases for $M_{2 \times 2}$.

(A) Find the transition matrix from B to C.
(B) Use matrix inversion to find the transition matrix from C to B.

Solution   (A) Since C is the standard basis for $M_{2 \times 2}$, the coordinate matrices of vectors in B with respect to C can be found by inspection. They are

$$[\mathbf{b}_1]_C = \begin{bmatrix} 1 \\ 1 \\ 1 \\ 0 \end{bmatrix} \qquad [\mathbf{b}_2]_C = \begin{bmatrix} 0 \\ 1 \\ 1 \\ 1 \end{bmatrix} \qquad [\mathbf{b}_3]_C = \begin{bmatrix} 0 \\ 0 \\ 1 \\ 1 \end{bmatrix} \qquad [\mathbf{b}_4]_C = \begin{bmatrix} 1 \\ 1 \\ 0 \\ 0 \end{bmatrix}$$

Thus, the transition matrix from B to C is

$$P = \begin{bmatrix} 1 & 0 & 0 & 1 \\ 1 & 1 & 0 & 1 \\ 1 & 1 & 1 & 0 \\ 0 & 1 & 1 & 0 \end{bmatrix}$$

(B) Coordinate matrices of vectors in C with respect to B cannot be found by inspection. However, Theorem 12 allows us to find the transition matrix from C to B without computing the coordinate matrices. Instead, we will find P^{-1} by using elementary row operations.

$$\begin{bmatrix} 1 & 0 & 0 & 1 & | & 1 & 0 & 0 & 0 \\ 1 & 1 & 0 & 1 & | & 0 & 1 & 0 & 0 \\ 1 & 1 & 1 & 0 & | & 0 & 0 & 1 & 0 \\ 0 & 1 & 1 & 0 & | & 0 & 0 & 0 & 1 \end{bmatrix} \sim \begin{bmatrix} 1 & 0 & 0 & 1 & | & 1 & 0 & 0 & 0 \\ 0 & 1 & 0 & 0 & | & -1 & 1 & 0 & 0 \\ 0 & 1 & 1 & -1 & | & -1 & 0 & 1 & 0 \\ 0 & 1 & 1 & 0 & | & 0 & 0 & 0 & 1 \end{bmatrix}$$

$$\sim \begin{bmatrix} 1 & 0 & 0 & 1 & | & 1 & 0 & 0 & 0 \\ 0 & 1 & 0 & 0 & | & -1 & 1 & 0 & 0 \\ 0 & 0 & 1 & -1 & | & 0 & -1 & 1 & 0 \\ 0 & 0 & 1 & 0 & | & 1 & -1 & 0 & 1 \end{bmatrix} \sim \begin{bmatrix} 1 & 0 & 0 & 1 & | & 1 & 0 & 0 & 0 \\ 0 & 1 & 0 & 0 & | & -1 & 1 & 0 & 0 \\ 0 & 0 & 1 & -1 & | & 0 & -1 & 1 & 0 \\ 0 & 0 & 0 & 1 & | & 1 & 0 & -1 & 1 \end{bmatrix}$$

$$\sim \begin{bmatrix} 1 & 0 & 0 & 0 & | & 0 & 0 & 1 & -1 \\ 0 & 1 & 0 & 0 & | & -1 & 1 & 0 & 0 \\ 0 & 0 & 1 & 0 & | & 1 & -1 & 0 & 1 \\ 0 & 0 & 0 & 1 & | & 1 & 0 & -1 & 1 \end{bmatrix}$$

Thus, the transition matrix from C to B is

$$P^{-1} = \begin{bmatrix} 0 & 0 & 1 & -1 \\ -1 & 1 & 0 & 0 \\ 1 & -1 & 0 & 1 \\ 1 & 0 & -1 & 1 \end{bmatrix}$$

Problem 21   Repeat Example 21 for

$$B = \left\{ \begin{bmatrix} 1 & 2 \\ 0 & 0 \end{bmatrix}, \begin{bmatrix} 0 & 1 \\ 2 & 0 \end{bmatrix}, \begin{bmatrix} 0 & 0 \\ 1 & 2 \end{bmatrix}, \begin{bmatrix} 0 & 0 \\ 1 & 1 \end{bmatrix} \right\}$$

$$C = \left\{ \begin{bmatrix} 1 & 0 \\ 0 & 0 \end{bmatrix}, \begin{bmatrix} 0 & 1 \\ 0 & 0 \end{bmatrix}, \begin{bmatrix} 0 & 0 \\ 1 & 0 \end{bmatrix}, \begin{bmatrix} 0 & 0 \\ 0 & 1 \end{bmatrix} \right\}$$

∎

▪ Rotation of Axes in the Plane

Let $C = \{\mathbf{i}, \mathbf{j}\}$ be the standard basis for R^2, let $A = (x, y)$ be a point in R^2, and let $\mathbf{u} = \overrightarrow{OA}$ be the vector from the origin to A, as shown in Figure 2(A). Since $\mathbf{u} = x\mathbf{i} + y\mathbf{j}$, it follows that

$$[\mathbf{u}]_C = \begin{bmatrix} x \\ y \end{bmatrix}$$

Now suppose a new coordinate system is formed by rotating the x and y axes counterclockwise about the origin through an angle θ, as shown in Figure 2(B). If the new axes are labeled x' and y', then A also has coordinates with respect to this new coordinate system. Our goal is to establish a relationship between the xy coordinates and the x'y' coordinates of A.

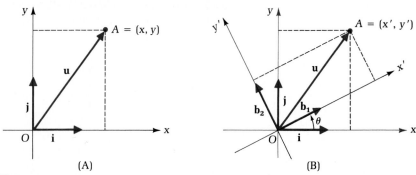

Figure 2

Let $\mathbf{b}_1$ and $\mathbf{b}_2$ be unit vectors along the positive x' and y' axes [see Figure 2(B)]. Then $B = \{\mathbf{b}_1, \mathbf{b}_2\}$ is also a basis for R^2 and

$$[\mathbf{u}]_B = \begin{bmatrix} x' \\ y' \end{bmatrix}$$

If P is the transition matrix from B to C, then the xy and x'y' coordinates of A are related by the equation

$$[\mathbf{u}]_C = P[\mathbf{u}]_B \tag{4}$$

In order to find P, we must find $[\mathbf{b}_1]_C$ and $[\mathbf{b}_2]_C$. From Figure 3, we have

$$\mathbf{b}_1 = \cos\theta\,\mathbf{i} + \sin\theta\,\mathbf{j} \qquad \mathbf{b}_2 = -\sin\theta\,\mathbf{i} + \cos\theta\,\mathbf{j}$$

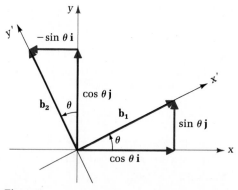

Figure 3

Thus,

$$[\mathbf{b}_1]_C = \begin{bmatrix} \cos\theta \\ \sin\theta \end{bmatrix} \qquad [\mathbf{b}_2]_C = \begin{bmatrix} -\sin\theta \\ \cos\theta \end{bmatrix} \qquad \text{and} \qquad P = \begin{bmatrix} \cos\theta & -\sin\theta \\ \sin\theta & \cos\theta \end{bmatrix}$$

Substituting in (4), we have

$$\begin{bmatrix} x \\ y \end{bmatrix} = \begin{bmatrix} \cos\theta & -\sin\theta \\ \sin\theta & \cos\theta \end{bmatrix} \begin{bmatrix} x' \\ y' \end{bmatrix}$$

$$x = \cos\theta\, x' - \sin\theta\, y'$$
$$y = \sin\theta\, x' + \cos\theta\, y'$$

This expresses the xy coordinates in terms of the $x'y'$ coordinates. To obtain expressions for the $x'y'$ coordinates in terms of the xy coordinates, we use P^{-1}, the transition matrix from C to B:

$$[\mathbf{u}]_B = P^{-1}[\mathbf{u}]_C \qquad P^{-1} = \begin{bmatrix} \cos\theta & \sin\theta \\ -\sin\theta & \cos\theta \end{bmatrix}$$

$$\begin{bmatrix} x' \\ y' \end{bmatrix} = \begin{bmatrix} \cos\theta & \sin\theta \\ -\sin\theta & \cos\theta \end{bmatrix} \begin{bmatrix} x \\ y \end{bmatrix}$$

$$x' = \cos\theta\, x + \sin\theta\, y$$
$$y' = -\sin\theta\, x + \cos\theta\, y$$

These results are summarized in Theorem 13. The proof of Theorem 13 follows directly from the preceding discussion.

Theorem 13

Rotation of Axes in the Plane

If the $x'y'$ coordinate system is formed by rotating the xy coordinate axes counterclockwise through an angle of θ, then the xy and $x'y'$ coordinates are related by:

1. $\begin{bmatrix} x \\ y \end{bmatrix} = \begin{bmatrix} \cos\theta & -\sin\theta \\ \sin\theta & \cos\theta \end{bmatrix} \begin{bmatrix} x' \\ y' \end{bmatrix}$

   or

   $x = x' \cos\theta - y' \sin\theta$
   $y = x' \sin\theta + y' \cos\theta$

   and

   $P = \begin{bmatrix} \cos\theta & -\sin\theta \\ \sin\theta & \cos\theta \end{bmatrix}$

   is the transition matrix from the $x'y'$ system to the xy system

2. $\begin{bmatrix} x' \\ y' \end{bmatrix} = \begin{bmatrix} \cos\theta & \sin\theta \\ -\sin\theta & \cos\theta \end{bmatrix} \begin{bmatrix} x \\ y \end{bmatrix}$

   or

   $x' = x \cos\theta + y \sin\theta$
   $y' = -x \sin\theta + y \cos\theta$

   and

   $P^{-1} = \begin{bmatrix} \cos\theta & \sin\theta \\ -\sin\theta & \cos\theta \end{bmatrix}$

   is the transition matrix from the xy system to the $x'y'$ system

Example 22   The $x'y'$ coordinate system is formed by rotating the xy system counterclockwise through an angle of $\theta = \pi/3$.

(A) Find the transition matrix from the $x'y'$ system to the xy system.
(B) Find the transition matrix from the xy system to the $x'y'$ system.
(C) If $(x', y') = (1, -2)$, find (x, y).
(D) If $(x, y) = (3, 7)$, find (x', y').

Solution   (A) $P = \begin{bmatrix} \cos \dfrac{\pi}{3} & -\sin \dfrac{\pi}{3} \\ \sin \dfrac{\pi}{3} & \cos \dfrac{\pi}{3} \end{bmatrix} = \begin{bmatrix} \dfrac{1}{2} & -\dfrac{\sqrt{3}}{2} \\ \dfrac{\sqrt{3}}{2} & \dfrac{1}{2} \end{bmatrix}$

(B) $P^{-1} = \begin{bmatrix} \dfrac{1}{2} & \dfrac{\sqrt{3}}{2} \\ -\dfrac{\sqrt{3}}{2} & \dfrac{1}{2} \end{bmatrix}$ Check this

(C) $\begin{bmatrix} x \\ y \end{bmatrix} = P \begin{bmatrix} x' \\ y' \end{bmatrix} = \begin{bmatrix} \dfrac{1}{2} & -\dfrac{\sqrt{3}}{2} \\ \dfrac{\sqrt{3}}{2} & \dfrac{1}{2} \end{bmatrix} \begin{bmatrix} 1 \\ -2 \end{bmatrix} = \begin{bmatrix} \dfrac{1}{2} + \sqrt{3} \\ \dfrac{\sqrt{3}}{2} - 1 \end{bmatrix}$

Thus,

$$(x, y) = \left(\frac{1}{2} + \sqrt{3}, \frac{\sqrt{3}}{2} - 1 \right)$$

(D) $\begin{bmatrix} x' \\ y' \end{bmatrix} = P^{-1} \begin{bmatrix} x \\ y \end{bmatrix} = \begin{bmatrix} \dfrac{1}{2} & \dfrac{\sqrt{3}}{2} \\ -\dfrac{\sqrt{3}}{2} & \dfrac{1}{2} \end{bmatrix} \begin{bmatrix} 3 \\ 7 \end{bmatrix} = \begin{bmatrix} \dfrac{3}{2} + \dfrac{7\sqrt{3}}{2} \\ -\dfrac{3\sqrt{3}}{2} + \dfrac{7}{2} \end{bmatrix}$

Thus,

$$(x', y') = \left(\frac{3}{2} + \frac{7\sqrt{3}}{2}, \frac{-3\sqrt{3}}{2} + \frac{7}{2} \right).$$

Problem 22   Repeat Example 22 for $\theta = \pi/4$.   ∎

Answers to Matched Problems   18. $\begin{bmatrix} 3 \\ 1 \end{bmatrix}$ 19. $P = \begin{bmatrix} 0 & 3 & 0 \\ -\frac{2}{3} & \frac{4}{3} & -\frac{4}{3} \\ \frac{1}{3} & -\frac{2}{3} & \frac{11}{3} \end{bmatrix}$, $[\mathbf{v}]_C = \begin{bmatrix} -3 \\ -\frac{20}{3} \\ \frac{37}{3} \end{bmatrix}$

20. (A) $\begin{bmatrix} 8 & -3 \\ -5 & 2 \end{bmatrix}$ (B) $\begin{bmatrix} 2 & 3 \\ 5 & 8 \end{bmatrix}$

21. (A) $\begin{bmatrix} 1 & 0 & 0 & 0 \\ 2 & 1 & 0 & 0 \\ 0 & 2 & 1 & 1 \\ 0 & 0 & 2 & 1 \end{bmatrix}$ (B) $\begin{bmatrix} 1 & 0 & 0 & 0 \\ -2 & 1 & 0 & 0 \\ -4 & 2 & -1 & 1 \\ 8 & -4 & 2 & -1 \end{bmatrix}$

22. (A) $\begin{bmatrix} \sqrt{2}/2 & -\sqrt{2}/2 \\ \sqrt{2}/2 & \sqrt{2}/2 \end{bmatrix}$ (B) $\begin{bmatrix} \sqrt{2}/2 & \sqrt{2}/2 \\ -\sqrt{2}/2 & \sqrt{2}/2 \end{bmatrix}$

(C) $(3\sqrt{2}/2, -\sqrt{2}/2)$ (D) $(5\sqrt{2}, 2\sqrt{2})$

▌▌ Exercise 6-4

A   **1.** Let $B = \{\mathbf{b}_1, \mathbf{b}_2\}$ and $C = \{\mathbf{c}_1, \mathbf{c}_2\}$ be bases for R^2, where

$$\mathbf{b}_1 = (2, -3) \qquad \mathbf{b}_2 = (1, 4) \qquad \mathbf{c}_1 = (1, 0) \qquad \mathbf{c}_2 = (0, 1)$$

(A) Find $[\mathbf{b}_1]_C$.   (B) Find $[\mathbf{b}_2]_C$.
(C) Find P, the transition matrix from B to C.

2. Repeat Problem 1 for

$$\mathbf{b}_1 = (5, -2) \qquad \mathbf{b}_2 = (2, -7) \qquad \mathbf{c}_1 = (1, 0) \qquad \mathbf{c}_2 = (0, 1)$$

3. Let $B = \{\mathbf{b}_1, \mathbf{b}_2\}$ and $C = \{\mathbf{c}_1, \mathbf{c}_2\}$ be bases for R^2, where

$$\mathbf{b}_1 = (3, -1) \qquad \mathbf{b}_2 = (-5, 2) \qquad \mathbf{c}_1 = (2, 1) \qquad \mathbf{c}_2 = (1, 1)$$

(A) Find $[\mathbf{b}_1]_C$.   (B) Find $[\mathbf{b}_2]_C$.
(C) Find P, the transition matrix from B to C.
(D) If

$$[\mathbf{u}]_B = \begin{bmatrix} 5 \\ 3 \end{bmatrix}$$

use P to find $[\mathbf{u}]_C$.
(E) Find $[\mathbf{c}_1]_B$.   (F) Find $[\mathbf{c}_2]_B$.
(G) Use the results in parts (E) and (F) to find Q, the transition matrix from C to B.
(H) If

$$[\mathbf{v}]_C = \begin{bmatrix} -2 \\ 3 \end{bmatrix}$$

use Q to find $[\mathbf{v}]_B$.
(I) Compute PQ.

4. Repeat Problem 3 for

$$\mathbf{b}_1 = (3, -2) \qquad \mathbf{b}_2 = (-1, 1) \qquad \mathbf{c}_1 = (7, -4) \qquad \mathbf{c}_2 = (-5, 3)$$

5. Let $B = \{3 - 2x, -2 + x\}$ and $C = \{-5 + x, 4 - x\}$ be bases for P_1.

(A) Find P, the transition matrix from B to C.
(B) If $\mathbf{u} = 1 - x$, find $[\mathbf{u}]_B$, and then use P to find $[\mathbf{u}]_C$.
(C) Find P^{-1}, the transition matrix from C to B.
(D) If $\mathbf{u} = 1 + x$, find $[\mathbf{u}]_C$, and then use P^{-1} to find $[\mathbf{u}]_B$.

6. Repeat Problem 5 for $B = \{2 - x, -1 + x\}$ and $C = \{4 - 5x, -3 + 4x\}$.

B   **7.** Let $B = \{(2, 1, -1), (1, -1, 3), (2, -1, 3)\}$ and $C = \{(1, 0, 1), (1, 1, 0), (0, 1, 1)\}$ be bases for R^3, and let $\mathbf{u} = (4, -5, 13)$.

(A) Find P, the transition matrix from B to C.
(B) Find $[\mathbf{u}]_B$.   (C) Use P to find $[\mathbf{u}]_C$.

8. Repeat Problem 7 for $B = \{(1, 0, 1), (0, 1, 0), (1, 1, 0)\}$,
$C = \{(1, -1, 1), (0, -1, 1), (2, 1, 0)\}$, and $\mathbf{u} = (1, 2, 2)$.

9. Let $B = \{1 + x + x^2, x + x^2, x^2\}$ and $C = \{1, 1 + x, 1 + x + x^2\}$ be bases for P_2, and let $\mathbf{u} = 1 + 2x + 3x^2$.

(A) Find P, the transition matrix from B to C.
(B) Find $[\mathbf{u}]_B$.   (C) Use P to find $[\mathbf{u}]_C$.

10. Repeat Problem 9 for $B = \{1 + 2x - 3x^2, x + 2x^2, 1 + x^2\}$,
$C = \{1 - x + x^2, 1 - x^2, -1 + x\}$, and $\mathbf{u} = 2 + 3x$.

11. Let $B = \{(1, 2, -1), (0, 1, 3), (1, 2, 2)\}$ and $C = \{(1, 0, 0), (0, 1, 0), (0, 0, 1)\}$ be bases for R^3, and let $\mathbf{u} = (0, 5, 6)$.

(A) Find P, the transition matrix from B to C.
(B) Use matrix inversion to find P^{-1}, the transition matrix from C to B.
(C) Find $[\mathbf{u}]_C$.   (D) Use P^{-1} to find $[\mathbf{u}]_B$.

12. Repeat Problem 11 for $B = \{(1, -1, 3), (-1, 2, -1), (1, -2, 3)\}$,
$C = \{(1, 0, 0), (0, 1, 0), (0, 0, 1)\}$, and $\mathbf{u} = (1, 2, 5)$.

13. Let $B = \{2 + x + x^2, 1 + 2x + x^2, 2 + x^2\}$ and $C = \{1, x, x^2\}$ be bases for P_2, and let $\mathbf{u} = 3 + x + x^2$.

(A) Find P, the transition matrix from B to C.
(B) Use matrix inversion to find P^{-1}, the transition matrix from C to B.
(C) Find $[\mathbf{u}]_C$.   (D) Use P^{-1} to find $[\mathbf{u}]_B$.

14. Repeat Problem 13 for $B = \{x + x^2, 1 - x^2, 1 + x - x^2\}$, $C = \{1, x, x^2\}$, and $\mathbf{u} = 1 + 2x + x^2$.

15. Let

$$B = \left\{ \begin{bmatrix} 1 & 0 \\ 0 & 1 \end{bmatrix}, \begin{bmatrix} 1 & 1 \\ 0 & 0 \end{bmatrix}, \begin{bmatrix} 1 & 0 \\ -1 & 1 \end{bmatrix}, \begin{bmatrix} 1 & 1 \\ -1 & -1 \end{bmatrix} \right\}$$

and

$$C = \left\{ \begin{bmatrix} 1 & 0 \\ 0 & 0 \end{bmatrix}, \begin{bmatrix} 0 & 1 \\ 0 & 0 \end{bmatrix}, \begin{bmatrix} 0 & 0 \\ 1 & 0 \end{bmatrix}, \begin{bmatrix} 0 & 0 \\ 0 & 1 \end{bmatrix} \right\}$$

be bases for $M_{2\times2}$, and let

$$\mathbf{u} = \begin{bmatrix} 1 & 2 \\ 3 & 4 \end{bmatrix}$$

(A) Find P, the transition matrix from B to C.

(B) Use matrix inversion to find P^{-1}, the transition matrix from C to B.

(C) Find $[\mathbf{u}]_C$.   (D) Use P^{-1} to find $[\mathbf{u}]_B$.

16. Repeat Problem 15 for

$$B = \left\{ \begin{bmatrix} 0 & 0 \\ 1 & 1 \end{bmatrix}, \begin{bmatrix} 0 & 1 \\ 0 & 1 \end{bmatrix}, \begin{bmatrix} 1 & 0 \\ 1 & 0 \end{bmatrix}, \begin{bmatrix} 1 & 1 \\ 1 & 0 \end{bmatrix} \right\}$$

$$C = \left\{ \begin{bmatrix} 1 & 0 \\ 0 & 0 \end{bmatrix}, \begin{bmatrix} 0 & 1 \\ 0 & 0 \end{bmatrix}, \begin{bmatrix} 0 & 0 \\ 1 & 0 \end{bmatrix}, \begin{bmatrix} 0 & 0 \\ 0 & 1 \end{bmatrix} \right\}$$

$$\mathbf{u} = \begin{bmatrix} 1 & 2 \\ 3 & 4 \end{bmatrix}$$

C Let $C = \{(1, 0, 0), (0, 1, 0), (0, 0, 1)\}$ be the standard basis for R^3. In Problems 17 and 18, find the transition matrix from C to B, and find $[\mathbf{u}]_B$ for $\mathbf{u} = (u_1, u_2, u_3)$.

17. $B = \{(2, 1, 0), (1, 0, 2), (2, 1, 1)\}$

18. $B = \{(-1, 1, 1), (2, 1, -1), (1, -2, -1)\}$

19. Let

$$B = \left\{ \begin{bmatrix} 2 & 0 \\ 0 & 1 \end{bmatrix}, \begin{bmatrix} 1 & 0 \\ 0 & 1 \end{bmatrix} \right\} \quad \text{and} \quad C = \left\{ \begin{bmatrix} 1 & 0 \\ 0 & 0 \end{bmatrix}, \begin{bmatrix} 0 & 0 \\ 0 & 1 \end{bmatrix} \right\}$$

be bases for

$$D_2 = \left\{ \begin{bmatrix} a & 0 \\ 0 & b \end{bmatrix} \middle| a, b \in R \right\}$$

Find the transition matrix from C to B, and find $[\mathbf{u}]_B$ for

$$\mathbf{u} = \begin{bmatrix} a & 0 \\ 0 & b \end{bmatrix}$$

20. Let

$$B = \left\{ \begin{bmatrix} 1 & 1 \\ 1 & 1 \end{bmatrix}, \begin{bmatrix} 1 & 2 \\ 2 & 1 \end{bmatrix}, \begin{bmatrix} 1 & 0 \\ 0 & -1 \end{bmatrix} \right\}$$

and

$$C = \left\{ \begin{bmatrix} 1 & 0 \\ 0 & 0 \end{bmatrix}, \begin{bmatrix} 0 & 1 \\ 1 & 0 \end{bmatrix}, \begin{bmatrix} 0 & 0 \\ 0 & 1 \end{bmatrix} \right\}$$

be bases for

$$S_2 = \left\{ \begin{bmatrix} a & b \\ b & c \end{bmatrix} \middle| a, b, c \in R \right\}$$

Find the transition matrix from C to B, and find $[\mathbf{u}]_B$ for

$$\mathbf{u} = \begin{bmatrix} a & b \\ b & c \end{bmatrix}$$

21. *Chebyshev polynomials.* Let $T = \{1, x, 2x^2 - 1, 4x^3 - 3x\}$ be the basis of Chebyshev polynomials for P_3, and let $S = \{1, x, x^2, x^3\}$ be the standard basis for P_3. Find the transition matrix from S to T.

22. *Hermite polynomials.* Let $H = \{1, 2x, 4x^2 - 2, 8x^3 - 12x\}$ be the basis of Hermite polynomials for P_3, and let $S = \{1, x, x^2, x^3\}$ be the standard basis for P_3. Find the transition matrix from S to H.

23. An $x'y'$ coordinate system is formed by rotating an xy coordinate system through an angle $\theta = \pi/6$.

   (A) Find the transition matrix from the $x'y'$ system to the xy system.
   (B) Find the transition matrix from the xy system to the $x'y'$ system.
   (C) Find (x, y) if $(x', y') = (-2, 8)$.
   (D) Find (x', y') if $(x, y) = (2, 4)$.

24. Repeat Problem 23 for $\theta = 2\pi/3$.

25. The transition matrix from an $x'y'$ coordinate system to an xy coordinate system is

$$P = \begin{bmatrix} \frac{4}{5} & -\frac{3}{5} \\ \frac{3}{5} & \frac{4}{5} \end{bmatrix}$$

   (A) Find (x, y) if $(x', y') = (5, 0)$.
   (B) Find (x, y) if $(x', y') = (0, 5)$.
   (C) Plot the points found in parts (A) and (B) in the xy coordinate system, and use these points to sketch the $x'y'$ coordinate axes on your graph.

26. Repeat Problem 25 for

$$P = \begin{bmatrix} -\frac{3}{5} & -\frac{4}{5} \\ \frac{4}{5} & -\frac{3}{5} \end{bmatrix}$$

27. *Theorem 12.* Let $B = \{\mathbf{b}_1, \mathbf{b}_2, \ldots, \mathbf{b}_n\}$ be a basis for a vector space V, and let A be an $n \times n$ matrix satisfying

$$A[\mathbf{v}]_B = [\mathbf{v}]_B$$

   for all $\mathbf{v}$ in V. Show that $A = I$, thus completing the proof of Theorem 12. [*Hint:* Let $\mathbf{v} = \mathbf{b}_1, \mathbf{v} = \mathbf{b}_2, \ldots, \mathbf{v} = \mathbf{b}_n$ in $A[\mathbf{v}]_B = [\mathbf{v}]_B$.]

28. If P is the transition matrix from B to C, then Theorem 11 states that $P[\mathbf{v}]_B = [\mathbf{v}]_C$ for all $\mathbf{v}$ in V. Show that P is unique; that is, if T is another matrix satisfying $T[\mathbf{v}]_B = [\mathbf{v}]_C$ for all $\mathbf{v}$ in V, show that $T = P$.

29. If B, C, and D are bases for the same finite-dimensional vector space V, P is the transition matrix from B to C, and Q is the transition matrix from C to D, show that QP is the transition matrix from B to D.

30. The result in Problem 29 can be used to find the transition matrix from B to C by introducing a third basis S (usually one of the standard bases, which make the work easy). If P_B is the transition matrix from B to S, P_C is the transition matrix from C to S, and P is the transition matrix from B to C, use Problem 29 to show that $P = P_C^{-1}P_B$.

In Problems 31–34, introduce the standard basis for R^2 or R^3, find P_B, P_C, and P_C^{-1}, and use the result in Problem 30 to find P, the transition matrix from B to C.

31. $B = \{(3, -1), (-5, 2)\}; \quad C = \{(2, 1), (1, 1)\}$
[Compare with Problem 3.]
32. $B = \{(3, -2), (-1, 1)\}; \quad C = \{(7, -4), (-5, 3)\}$
[Compare with Problem 4.]
33. $B = \{(2, 1, -1), (1, -1, 3), (2, -1, 3)\}; \quad C = \{(1, 0, 1), (1, 1, 0), (0, 1, 1)\}$
[Compare with Problem 7.]
34. $B = \{(1, 0, 1), (0, 1, 0), (1, 1, 0)\}; \quad C = \{(1, -1, 1), (0, -1, 1), (2, 1, 0)\}$
[Compare with Problem 8.]

▌6-5▐ Inner Product Spaces

- ▪ Inner Product
- ▪ Length and Distance
- ▪ The Cauchy–Schwarz Inequality
- ▪ Angle Between Two Vectors
- ▪ Orthogonal Vectors

▪ Inner Product

In Section 4-3, we studied the properties of the dot product in R^2 and R^3. In Section 5-1, the dot product was generalized to R^n and it was shown that the following properties are satisfied (Theorem 2 in Section 5-1):

Properties of the Dot Product in R^n

If $\mathbf{u}$, $\mathbf{v}$, and $\mathbf{w}$ are vectors in R^n and k is a scalar, then

(A) $\mathbf{u} \cdot \mathbf{u} \geq 0$
(B) $\mathbf{u} \cdot \mathbf{u} = 0$ if and only if $\mathbf{u} = \mathbf{0}$
(C) $\mathbf{u} \cdot \mathbf{0} = 0$
(D) $\mathbf{u} \cdot \mathbf{v} = \mathbf{v} \cdot \mathbf{u}$
(E) $\mathbf{u} \cdot (\mathbf{v} + \mathbf{w}) = \mathbf{u} \cdot \mathbf{v} + \mathbf{u} \cdot \mathbf{w}$
(F) $(\mathbf{u} + \mathbf{v}) \cdot \mathbf{w} = \mathbf{u} \cdot \mathbf{w} + \mathbf{v} \cdot \mathbf{w}$
(G) $(k\mathbf{u}) \cdot \mathbf{v} = k(\mathbf{u} \cdot \mathbf{v})$
(H) $\mathbf{u} \cdot (k\mathbf{v}) = k(\mathbf{u} \cdot \mathbf{v})$

Now we want to generalize the concept of dot product to arbitrary vector spaces. The key to understanding this generalization is to realize that the dot product in R^n is a *function* that assigns the real number $\mathbf{u} \cdot \mathbf{v}$ to each pair of vectors $\mathbf{u}$ and $\mathbf{v}$ in R^n. In an arbitrary vector space V, a function that assigns a real number to each pair of vectors in V and has the same properties as the dot product in R^n is called an *inner product* for V. A vector space equipped with an inner product is called an *inner product space*. More precisely, we have the following definition:

Inner Product

An **inner product** on a vector space V is a function that associates a real number, denoted by $\langle \mathbf{u}, \mathbf{v} \rangle$, with each pair of vectors $\mathbf{u}$ and $\mathbf{v}$ in V and that satisfies the following properties for all $\mathbf{u}$, $\mathbf{v}$, and $\mathbf{w}$ in V and all scalars k:

(A) $\langle \mathbf{u}, \mathbf{u} \rangle \geq 0$
(B) $\langle \mathbf{u}, \mathbf{u} \rangle = 0$ if and only if $\mathbf{u} = \mathbf{0}$
(C) $\langle \mathbf{u}, \mathbf{v} \rangle = \langle \mathbf{v}, \mathbf{u} \rangle$
(D) $\langle \mathbf{u}, \mathbf{v} + \mathbf{w} \rangle = \langle \mathbf{u}, \mathbf{v} \rangle + \langle \mathbf{u}, \mathbf{w} \rangle$
(E) $\langle k\mathbf{u}, \mathbf{v} \rangle = k\langle \mathbf{u}, \mathbf{v} \rangle$

A vector space V that has been equipped with an inner product is called an **inner product space.**

If $V = R^n$ and $\langle \mathbf{u}, \mathbf{v} \rangle$ is defined by $\langle \mathbf{u}, \mathbf{v} \rangle = \mathbf{u} \cdot \mathbf{v}$, then all the properties in this definition are included in the properties of the dot product listed in the previous box (Theorem 2, Section 5-1). Thus, the dot product is an inner product on R^n. This inner product is referred to as the **standard inner product** on R^n and also as the **Euclidean inner product** on R^n. When R^n is equipped with this inner product, it is often called **Euclidean n-dimensional space** or **Euclidean n-space.**

Notice that the dot product properties (C), (F), and (H) (Theorem 2, Section 5-1) were not included in the general definition of inner product. Later in this section, we will prove that these additional properties are consequences of the properties in the definition; hence, they do not have to be included in the definition.

As the next example illustrates, the standard inner product is not the only inner product that can be defined on R^n.

Example 23   Let $\mathbf{u} = (u_1, u_2)$ and $\mathbf{v} = (v_1, v_2)$ be in R^2, and let $\langle \mathbf{u}, \mathbf{v} \rangle$ be defined by

$$\langle \mathbf{u}, \mathbf{v} \rangle = 5u_1v_1 + 6u_2v_2 \tag{1}$$

Show that $\langle \mathbf{u}, \mathbf{v} \rangle$ is an inner product on R^2.

Solution   We must show that this function satisfies each of the conditions in the definition of an inner product.

(A) $\langle \mathbf{u}, \mathbf{u} \rangle = 5u_1^2 + 6u_2^2 \geq 0$

(B) $\langle \mathbf{u}, \mathbf{u} \rangle = 5u_1^2 + 6u_2^2 = 0$ if and only if $u_1 = 0$ and $u_2 = 0$; that is, if and only if
$\mathbf{u} = \mathbf{0}$

(C) $\langle \mathbf{u}, \mathbf{v} \rangle = 5u_1v_1 + 6u_2v_2$

$\qquad = 5v_1u_1 + 6v_2u_2$

$\qquad = \langle \mathbf{v}, \mathbf{u} \rangle$

(D) Let $\mathbf{w} = (w_1, w_2)$. Then

$\langle \mathbf{u}, \mathbf{v} + \mathbf{w} \rangle = 5u_1(v_1 + w_1) + 6u_2(v_2 + w_2)$

$\qquad\qquad = 5u_1v_1 + 6u_2v_2 + 5u_1w_1 + 6u_2w_2$

$\qquad\qquad = \langle \mathbf{u}, \mathbf{v} \rangle + \langle \mathbf{u}, \mathbf{w} \rangle$

(E) Let k be a scalar. Then

$\langle k\mathbf{u}, \mathbf{v} \rangle = 5ku_1v_1 + 6ku_2v_2$

$\qquad\quad = k(5u_1v_1 + 6u_2v_2)$

$\qquad\quad = k\langle \mathbf{u}, \mathbf{v} \rangle$

Thus, the function defined in (1) is an inner product on R^2.

Problem 23   Let $\mathbf{u} = (u_1, u_2)$ and $\mathbf{v} = (v_1, v_2)$ be in R^2. Is the function

$\langle \mathbf{u}, \mathbf{v} \rangle = 2u_1v_1 - u_2v_2$

an inner product in R^2?

‖

Example 23 shows that a given vector space may have more than one inner product defined on it. In fact, Theorem 14 shows that each basis for a vector space can be used to define an inner product on the space. The proof is left as an exercise.

Theorem 14

Let V be a finite-dimensional vector space with basis
$B = \{\mathbf{b}_1, \mathbf{b}_2, \ldots, \mathbf{b}_n\}$. If

$$[\mathbf{u}]_B = \begin{bmatrix} k_1 \\ k_2 \\ \cdot \\ \cdot \\ \cdot \\ k_n \end{bmatrix} \qquad [\mathbf{v}]_B = \begin{bmatrix} \ell_1 \\ \ell_2 \\ \cdot \\ \cdot \\ \cdot \\ \ell_n \end{bmatrix}$$

are the coordinate matrices of two vectors $\mathbf{u}$ and $\mathbf{v}$ in V, then the function $\langle \mathbf{u}, \mathbf{v} \rangle$ defined by

$$\langle \mathbf{u}, \mathbf{v} \rangle = [\mathbf{u}]_B^T [\mathbf{v}]_B$$
$$= k_1\ell_1 + k_2\ell_2 + \cdots + k_n\ell_n$$

is an inner product on V.

Theorem 14 shows that there are many different inner products that can be defined on a given vector space V. Thus, it is always necessary to specify clearly the particular inner product to be used on V. The vector spaces R^n, P_n, and $M_{2\times 2}$ all have standard bases. The inner products associated with these standard bases are called the **standard inner products.** These are listed in Table 2.

Table 2

Standard Inner Products

Vector Space	Standard Basis	Standard Inner Product
R^n	$\{\mathbf{e}_1, \mathbf{e}_2, \ldots, \mathbf{e}_n\}$	$\mathbf{u} = (u_1, u_2, \ldots, u_n);\quad \mathbf{v} = (v_1, v_2, \ldots, v_n)$ $\langle \mathbf{u}, \mathbf{v} \rangle = u_1 v_1 + u_2 v_2 + \cdots + u_n v_n$
P_n	$\{1, x, \ldots, x^n\}$	$\mathbf{u} = a_0 + a_1 x + \cdots + a_n x^n;\quad \mathbf{v} = b_0 + b_1 x + \cdots + b_n x^n$ $\langle \mathbf{u}, \mathbf{v} \rangle = a_0 b_0 + a_1 b_1 + \cdots + a_n b_n$
$M_{2\times 2}$	$\left\{ \begin{bmatrix} 1 & 0 \\ 0 & 0 \end{bmatrix}, \begin{bmatrix} 0 & 1 \\ 0 & 0 \end{bmatrix}, \begin{bmatrix} 0 & 0 \\ 1 & 0 \end{bmatrix}, \begin{bmatrix} 0 & 0 \\ 0 & 1 \end{bmatrix} \right\}$	$\mathbf{u} = \begin{bmatrix} a_{11} & a_{12} \\ a_{21} & a_{22} \end{bmatrix};\quad \mathbf{v} = \begin{bmatrix} b_{11} & b_{12} \\ b_{21} & b_{22} \end{bmatrix}$ $\langle \mathbf{u}, \mathbf{v} \rangle = a_{11} b_{11} + a_{12} b_{12} + a_{21} b_{21} + a_{22} b_{22}$

Theorem 15 lists additional properties possessed by all inner product spaces. We will prove part (A) and leave the remaining proofs as exercises.

Theorem 15

Let V be an inner product space; let $\mathbf{u}$, $\mathbf{v}$, and $\mathbf{w}$ be vectors in V; and let k be a scalar. Then:

(A) $\langle \mathbf{u}, k\mathbf{v} \rangle = k\langle \mathbf{u}, \mathbf{v} \rangle$
(B) $\langle \mathbf{u} + \mathbf{v}, \mathbf{w} \rangle = \langle \mathbf{u}, \mathbf{w} \rangle + \langle \mathbf{v}, \mathbf{w} \rangle$
(C) $\langle \mathbf{u}, \mathbf{0} \rangle = 0$

Proof   (A) Using the properties in the definition of an inner product, we have

$$\langle \mathbf{u}, k\mathbf{v} \rangle = \langle k\mathbf{v}, \mathbf{u} \rangle \qquad \text{Property (C)}$$
$$= k\langle \mathbf{v}, \mathbf{u} \rangle \qquad \text{Property (E)}$$
$$= k\langle \mathbf{u}, \mathbf{v} \rangle \qquad \text{Property (C)}$$

∎

Remark (Calculus).   For those who have studied calculus, the definite integral can be used to define a very important type of inner product. If C is the vector space of functions continuous on $(-\infty, \infty)$ and a and b are real numbers, $a < b$, then

$$\langle f, g \rangle = \int_a^b f(x)\, g(x)\, dx$$

is an inner product on C. The proof of this statement is left as an exercise.

▪ Length and Distance

The dot product was used to extend the concepts of length and distance from R^3 to R^n. The inner product can now be used to define length and distance in any inner product space.

Norm and Distance

The **norm** or **length** of a vector **u** in an inner product space V is defined by

$$\|\mathbf{u}\| = \sqrt{\langle \mathbf{u}, \mathbf{u} \rangle}$$

The **distance** between two vectors **u** and **v** in V is defined by

$$d(\mathbf{u}, \mathbf{v}) = \|\mathbf{u} - \mathbf{v}\|$$

If $\|\mathbf{u}\| = 1$, then **u** is called a **unit vector**. If **v** is any nonzero vector, then $\mathbf{u} = (1/\|\mathbf{v}\|)\mathbf{v}$ is a **unit vector in the direction of v**.

Example 24 Let P_2 be equipped with the standard inner product given in Table 2. If $\mathbf{u} = 4 - 2x + 5x^2$ and $\mathbf{v} = 1 + 2x + 2x^2$, find:

(A) $\|\mathbf{v}\|$ (B) $d(\mathbf{u}, \mathbf{v})$
(C) A unit vector in the direction of **v**

Solution   (A) $\|\mathbf{v}\| = \sqrt{\langle \mathbf{v}, \mathbf{v} \rangle}$
$= \sqrt{(1)(1) + (2)(2) + (2)(2)}$
$= \sqrt{9} = 3$

(B) $d(\mathbf{u}, \mathbf{v}) = \|\mathbf{u} - \mathbf{v}\|$ $\mathbf{u} - \mathbf{v} = 3 - 4x + 3x^2$
$= \sqrt{(3)(3) + (-4)(-4) + (3)(3)}$
$= \sqrt{9 + 16 + 9} = \sqrt{34}$

(C) $\dfrac{1}{\|\mathbf{v}\|}\mathbf{v} = \dfrac{1}{3}(1 + 2x + 2x^2) = \dfrac{1}{3} + \dfrac{2}{3}x + \dfrac{2}{3}x^2$

Problem 24 Let $M_{2\times2}$ be equipped with the standard inner product given in Table 2. If

$$\mathbf{u} = \begin{bmatrix} 2 & -3 \\ 4 & 5 \end{bmatrix} \qquad \mathbf{v} = \begin{bmatrix} 1 & -2 \\ 4 & 2 \end{bmatrix}$$

find:

(A) $\|\mathbf{v}\|$ (B) $d(\mathbf{u}, \mathbf{v})$
(C) A unit vector in the direction of **v**

▌▌

■ The Cauchy–Schwarz Inequality

The Cauchy–Schwarz inequality, a fundamental property of inner products, is a basic tool in both applied and theoretical mathematics. Many important inequalities in mathematics are consequences of this result.

Theorem 16

> **Cauchy–Schwarz Inequality**
>
> If $\mathbf{u}$ and $\mathbf{v}$ are vectors in an inner product space V, then
>
> $$|\langle \mathbf{u}, \mathbf{v} \rangle| \leq \|\mathbf{u}\| \, \|\mathbf{v}\|$$
>
> where $|\langle \mathbf{u}, \mathbf{v} \rangle|$ is the absolute value of the real number $\langle \mathbf{u}, \mathbf{v} \rangle$.

Proof   We consider three separate cases.

Case 1. Assume $\mathbf{u}$ and $\mathbf{v}$ are unit vectors. Then

$$\|\mathbf{u} - \mathbf{v}\|^2 = \langle \mathbf{u} - \mathbf{v}, \mathbf{u} - \mathbf{v} \rangle \qquad \text{Expand, using properties in the}$$

definition of inner product and in Theorem 15.

$$= \langle \mathbf{u}, \mathbf{u} \rangle - 2\langle \mathbf{u}, \mathbf{v} \rangle + \langle \mathbf{v}, \mathbf{v} \rangle \qquad \langle \mathbf{u}, \mathbf{u} \rangle = \|\mathbf{u}\|^2,$$
$$\langle \mathbf{v}, \mathbf{v} \rangle = \|\mathbf{v}\|^2$$
$$= \|\mathbf{u}\|^2 - 2\langle \mathbf{u}, \mathbf{v} \rangle + \|\mathbf{v}\|^2 \qquad \|\mathbf{u}\|^2 = \|\mathbf{v}\|^2 = 1, \text{ since we}$$
$$= 2 - 2\langle \mathbf{u}, \mathbf{v} \rangle$$

have assumed that $\mathbf{u}$ and $\mathbf{v}$ are unit vectors

Since $\|\mathbf{u} - \mathbf{v}\|^2 \geq 0$, we can conclude that

$$2 - 2\langle \mathbf{u}, \mathbf{v} \rangle \geq 0$$
$$\langle \mathbf{u}, \mathbf{v} \rangle \leq 1$$

Since $-\mathbf{u}$ is also a unit vector, the same argument shows that

$$\langle -\mathbf{u}, \mathbf{v} \rangle = -\langle \mathbf{u}, \mathbf{v} \rangle \leq 1$$
$$-1 \leq \langle \mathbf{u}, \mathbf{v} \rangle$$

Thus, we have shown that if $\mathbf{u}$ and $\mathbf{v}$ are unit vectors, then

$$-1 \leq \langle \mathbf{u}, \mathbf{v} \rangle \leq 1 \tag{2}$$

Case 2. Assume $\mathbf{u}$ and $\mathbf{v}$ are nonzero vectors. Then $\mathbf{u}/\|\mathbf{u}\|$ and $\mathbf{v}/\|\mathbf{v}\|$ are unit vectors. Applying (2) to these unit vectors, we have

$$-1 \leq \left\langle \frac{\mathbf{u}}{\|\mathbf{u}\|}, \frac{\mathbf{v}}{\|\mathbf{v}\|} \right\rangle \leq 1 \qquad \text{Factor out } 1/\|\mathbf{u}\| \text{ and } 1/\|\mathbf{v}\|.$$

$$-1 \leq \frac{1}{\|\mathbf{u}\|} \frac{1}{\|\mathbf{v}\|} \langle \mathbf{u}, \mathbf{v} \rangle \leq 1 \qquad \text{Multiply by } \|\mathbf{u}\| \, \|\mathbf{v}\|.$$

$$-\|\mathbf{u}\| \, \|\mathbf{v}\| \leq \qquad \langle \mathbf{u}, \mathbf{v} \rangle \qquad \leq \|\mathbf{u}\| \, \|\mathbf{v}\|$$

Thus,

$$|\langle \mathbf{u}, \mathbf{v} \rangle| \le \|\mathbf{u}\| \, \|\mathbf{v}\|$$

and we have established the Cauchy–Schwarz inequality for each pair of nonzero vectors in V.

Case 3.   *Assume that either* $\mathbf{u}$ *or* $\mathbf{v}$ *is the zero vector. Then* $\langle \mathbf{u}, \mathbf{v} \rangle = 0$ and $\|\mathbf{u}\| \, \|\mathbf{v}\| = 0$, and the Cauchy–Schwarz inequality is certainly satisfied $(0 \le 0)$.   ∎

In R^3, the triangle inequality states that

$$\|\mathbf{u} + \mathbf{v}\| \le \|\mathbf{u}\| + \|\mathbf{v}\| \qquad \mathbf{u}, \mathbf{v} \in R^3$$

The Cauchy–Schwarz inequality can be used to extend this important result to any inner product space.

Theorem 17

Triangle Inequality

If $\mathbf{u}$ and $\mathbf{v}$ are vectors in an inner product space V, then

$$\|\mathbf{u} + \mathbf{v}\| \le \|\mathbf{u}\| + \|\mathbf{v}\|$$

Proof

$$\begin{aligned}
\|\mathbf{u} + \mathbf{v}\|^2 &= \langle \mathbf{u} + \mathbf{v}, \mathbf{u} + \mathbf{v} \rangle \\
&= \|\mathbf{u}\|^2 + 2\langle \mathbf{u}, \mathbf{v} \rangle + \|\mathbf{v}\|^2 \qquad \text{The Cauchy–Schwarz inequality} \\
&\qquad\qquad\qquad\qquad\qquad\qquad\quad\; \text{implies that } \langle \mathbf{u}, \mathbf{v} \rangle \le \|\mathbf{u}\| \, \|\mathbf{v}\|. \\
&\le \|\mathbf{u}\|^2 + 2\|\mathbf{u}\| \, \|\mathbf{v}\| + \|\mathbf{v}\|^2 \\
&\le (\|\mathbf{u}\| + \|\mathbf{v}\|)^2
\end{aligned}$$

Thus,

$$\|\mathbf{u} + \mathbf{v}\| \le \|\mathbf{u}\| + \|\mathbf{v}\|$$ ∎

▪ Angle Between Two Vectors

In R^3, the dot product is related to the angle between two nonzero vectors $\mathbf{u}$ and $\mathbf{v}$ by the formula

$$\cos \theta = \frac{\mathbf{u} \cdot \mathbf{v}}{\|\mathbf{u}\| \, \|\mathbf{v}\|}$$

If $\mathbf{u}$ and $\mathbf{v}$ are nonzero vectors in an inner product space, then the Cauchy–Schwarz inequality implies that

$$-1 \le \frac{\langle \mathbf{u}, \mathbf{v} \rangle}{\|\mathbf{u}\| \, \|\mathbf{v}\|} \le 1$$

This allows us to use the quantity $\langle \mathbf{u}, \mathbf{v} \rangle / (\|\mathbf{u}\| \, \|\mathbf{v}\|)$ to define the angle between $\mathbf{u}$ and $\mathbf{v}$.

Angle Between Two Vectors

If **u** and **v** are two nonzero vectors in an inner product space V, then the **angle** between **u** and **v** is the unique angle θ, $0 \le \theta \le \pi$, that satisfies

$$\cos \theta = \frac{\langle \mathbf{u}, \mathbf{v} \rangle}{\|\mathbf{u}\| \, \|\mathbf{v}\|}$$

Example 25   Let $M_{2 \times 2}$ be equipped with the standard inner product given in Table 2. Find the angle between each pair of vectors.

(A) $\mathbf{u} = \begin{bmatrix} 4 & 2 \\ -3 & 5 \end{bmatrix}$, $\mathbf{v} = \begin{bmatrix} -1 & 1 \\ -3 & 4 \end{bmatrix}$ (B) $\mathbf{u} = \begin{bmatrix} 1 & 2 \\ 3 & -1 \end{bmatrix}$, $\mathbf{v} = \begin{bmatrix} -1 & 4 \\ -2 & 1 \end{bmatrix}$

Solution   (A) $\langle \mathbf{u}, \mathbf{v} \rangle = 4(-1) + 2(1) + (-3)(-3) + 5(4) = 27$

$\|\mathbf{u}\| = \sqrt{4^2 + 2^2 + (-3)^2 + 5^2} = \sqrt{54} = 3\sqrt{6}$

$\|\mathbf{v}\| = \sqrt{(-1)^2 + 1^2 + (-3)^2 + 4^2} = \sqrt{27} = 3\sqrt{3}$

$\cos \theta = \dfrac{\langle \mathbf{u}, \mathbf{v} \rangle}{\|\mathbf{u}\| \, \|\mathbf{v}\|} = \dfrac{27}{(3\sqrt{6})(3\sqrt{3})} = \dfrac{1}{\sqrt{2}}$

$\theta = \pi/4$ or $45°$

(B) $\langle \mathbf{u}, \mathbf{v} \rangle = 1(-1) + 2(4) + 3(-2) + (-1)1 = 0$

Thus, $\cos \theta = 0$ and $\theta = \pi/2$ or $90°$.

Problem 25   Let R^4 be equipped with the standard inner product given in Table 2. Find the angle between each pair of vectors.

(A) $\mathbf{u} = (2, -1, 3, 5)$; $\mathbf{v} = (1, -2, 2, -2)$
(B) $\mathbf{u} = (3, 1, 3, -3)$; $\mathbf{v} = (1, -2, 1, -1)$ ∎

▪ Orthogonal Vectors

We are interested primarily in using the angle between two vectors to generalize the concept of orthogonality to any inner product space.

Orthogonal Vectors

Two vectors **u** and **v** in an inner product space V are said to be **orthogonal** if

$\langle \mathbf{u}, \mathbf{v} \rangle = 0$

If W is a subset of V and **u** is orthogonal to each vector **v** in W, then **u** is said to be **orthogonal to the set W.**

Example 26 Let R^4 be equipped with the standard inner product given in Table 2. Determine whether the following pairs of vectors are orthogonal:

(A) $\mathbf{u} = (1, -1, 2, 0); \quad \mathbf{v} = (2, 1, -1, 3)$

(B) $\mathbf{u} = (3, -4, 2, 1); \quad \mathbf{v} = (5, 3, -1, -1)$

Solution (A) $\langle \mathbf{u}, \mathbf{v} \rangle = 1(2) + (-1)1 + 2(-1) + 0(3) = -1$

Thus, $\mathbf{u}$ and $\mathbf{v}$ are not orthogonal.

(B) $\langle \mathbf{u}, \mathbf{v} \rangle = 3(5) + (-4)3 + 2(-1) + 1(-1) = 0$

Thus, $\mathbf{u}$ and $\mathbf{v}$ are orthogonal.

Problem 26 Let $M_{2 \times 2}$ be equipped with the standard inner product given in Table 2. Determine whether the following pairs of vectors are orthogonal:

(A) $\mathbf{u} = \begin{bmatrix} 3 & 1 \\ 4 & 2 \end{bmatrix}, \quad \mathbf{v} = \begin{bmatrix} -3 & 3 \\ 2 & -1 \end{bmatrix}$ (B) $\mathbf{u} = \begin{bmatrix} 2 & -1 \\ 1 & 0 \end{bmatrix}, \quad \mathbf{v} = \begin{bmatrix} 1 & -1 \\ 3 & -2 \end{bmatrix}$ ∎

Example 27 Let P_2 be equipped with the standard inner product given in Table 2, and let

$$W = \text{span}\{1 + x + x^2, 1 - 2x + 3x^2\}$$

Is $\mathbf{u} = 5 - 2x - 3x^2$ orthogonal to W?

Solution Let $\mathbf{v}_1 = 1 + x + x^2$ and $\mathbf{v}_2 = 1 - 2x + 3x^2$. First, we will determine whether $\mathbf{u}$ is orthogonal to each of these vectors:

$$\langle \mathbf{u}, \mathbf{v}_1 \rangle = 5(1) + (-2)1 + (-3)1 = 0$$

$$\langle \mathbf{u}, \mathbf{v}_2 \rangle = 5(1) + (-2)(-2) + (-3)3 = 0$$

Thus, $\mathbf{u}$ is orthogonal to both $\mathbf{v}_1$ and $\mathbf{v}_2$. Now, let $\mathbf{w}$ be any vector in W. Then there exist scalars k_1 and k_2 such that

$$\mathbf{w} = k_1 \mathbf{v}_1 + k_2 \mathbf{v}_2$$

and

$$\begin{aligned} \langle \mathbf{u}, \mathbf{w} \rangle &= \langle \mathbf{u}, k_1 \mathbf{v}_1 + k_2 \mathbf{v}_2 \rangle \\ &= k_1 \langle \mathbf{u}, \mathbf{v}_1 \rangle + k_2 \langle \mathbf{u}, \mathbf{v}_2 \rangle \\ &= k_1(0) + k_2(0) = 0 \end{aligned}$$

Thus, $\mathbf{u}$ is orthogonal to W.

Problem 27 Let P_2 be equipped with the standard inner product given in Table 2, and let

$$W = \text{span}\{1 - x + x^2, 2 + x + x^2\}$$

Is $\mathbf{u} = -2 + x + 3x^2$ orthogonal to W? ∎

❙□❙　**Example 28**　Let C be the vector space of functions continuous on $(-\infty, \infty)$ equipped with the
　　　　Calculus　inner product

$$\langle f, g \rangle = \int_{-1}^{1} f(x)\, g(x)\, dx$$

Determine whether the following pairs of vectors are orthogonal:

(A) $f(x) = 1 + x$,　$g(x) = 1 - x$　　　(B) $f(x) = \sin(\pi x)$,　$g(x) = \cos(\pi x)$

Solution　(A) $\langle f, g \rangle = \displaystyle\int_{-1}^{1} (1 + x)(1 - x)\, dx = \left(x - \frac{1}{3}x^3 \right)\Big|_{-1}^{1} = \frac{4}{3}$

Thus, f and g are not orthogonal.

(B) $\langle f, g \rangle = \displaystyle\int_{-1}^{1} \sin(\pi x)\, \cos(\pi x)\, dx = \frac{1}{2\pi} \sin^2(\pi x)\Big|_{-1}^{1} = 0$

Thus, f and g are orthogonal.

Problem 28　Repeat Example 28 for the following pairs of functions:

(A) $f(x) = 1 + x$,　$g(x) = 1 - 3x$　　　(B) $f(x) = x^2$,　$g(x) = \sqrt{1 + x^3}$　　**❙❙**

Answers to　**23.** No; for example, if $\mathbf{u} = (0, 1)$, then $\langle \mathbf{u}, \mathbf{u} \rangle = -1 \le 0$.
Matched Problems

24. (A) 5　(B) $\sqrt{11}$　(C) $\dfrac{1}{5}\begin{bmatrix} 1 & -2 \\ 4 & 2 \end{bmatrix}$　　**25.** (A) $\pi/2$ or $90°$　(B) $\pi/3$ or $60°$

26. (A) Orthogonal　(B) Not orthogonal

27. Yes　　**28.** (A) Orthogonal　(B) Not orthogonal

❙❙ Exercise 6-5

A　In Problems 1–6, let P_2 be equipped with the standard inner product, and let

$$\mathbf{u} = 2 + x - 2x^2 \qquad \mathbf{v} = -4 + 3x + 5x^2$$

Find each of the following:

1. $\langle \mathbf{u}, \mathbf{v} \rangle$　　　　**2.** $\|\mathbf{u}\|$　　　　**3.** $\|\mathbf{v}\|$　　　　**4.** $d(\mathbf{u}, \mathbf{v})$
5. A unit vector in the direction of $\mathbf{u}$
6. The angle between $\mathbf{u}$ and $\mathbf{v}$

In Problems 7–12, let $M_{2\times 2}$ be equipped with the standard inner product, and let

$$\mathbf{u} = \begin{bmatrix} 1 & -2 \\ 1 & -1 \end{bmatrix} \qquad \mathbf{v} = \begin{bmatrix} 3 & 1 \\ 3 & -3 \end{bmatrix}$$

Find each of the following:

7. $\langle \mathbf{u}, \mathbf{v} \rangle$　　　　**8.** $\|\mathbf{u}\|$　　　　**9.** $\|\mathbf{v}\|$　　　　**10.** $d(\mathbf{u}, \mathbf{v})$

11. A unit vector in the direction of **u**

12. The angle between **u** and **v**

In Problems 13 and 14, let R^4 be equipped with the standard inner product. Determine whether **u** and **v** are orthogonal.

13. $\mathbf{u} = (1, 2, -1, 1); \quad \mathbf{v} = (2, 1, 3, 1)$ **14.** $\mathbf{u} = (3, -1, 1, 2); \quad \mathbf{v} = (1, 2, 3, -2)$

B In Problems 15 and 16, let P_3 be equipped with the standard inner product. Determine the values of k for which **u** and **v** are orthogonal.

15. $\mathbf{u} = 1 + 2x - 5x^2 + kx^3; \quad \mathbf{v} = 4 + x + kx^2 + kx^3$

16. $\mathbf{u} = k + 2x + kx^2 - x^3; \quad \mathbf{v} = -6 + 4x + kx^2 - x^3$

In Problems 17 and 18, let $M_{2\times 2}$ be equipped with the standard inner product. Determine whether **u** is orthogonal to $W = \text{span}\{\mathbf{w}_1, \mathbf{w}_2, \mathbf{w}_3\}$.

17. $\mathbf{u} = \begin{bmatrix} -1 & -4 \\ 4 & 1 \end{bmatrix}; \quad \mathbf{w}_1 = \begin{bmatrix} 1 & 1 \\ 1 & 1 \end{bmatrix}, \quad \mathbf{w}_2 = \begin{bmatrix} 1 & 2 \\ 2 & 1 \end{bmatrix}, \quad \mathbf{w}_3 = \begin{bmatrix} -1 & 2 \\ 1 & 3 \end{bmatrix}$

18. $\mathbf{u} = \begin{bmatrix} -1 & 2 \\ 3 & -1 \end{bmatrix}; \quad \mathbf{w}_1 = \begin{bmatrix} 1 & -1 \\ 2 & 3 \end{bmatrix}, \quad \mathbf{w}_2 = \begin{bmatrix} 3 & -2 \\ 2 & -1 \end{bmatrix}, \quad \mathbf{w}_3 = \begin{bmatrix} 1 & 3 \\ -2 & 1 \end{bmatrix}$

In Problems 19 and 20, let P_2 be equipped with the standard inner product. Find all vectors in P_2 that are orthogonal to both **u** and **v**.

19. $\mathbf{u} = 1 + x; \quad \mathbf{v} = 1 + x^2$ **20.** $\mathbf{u} = 1 + x + x^2; \quad \mathbf{v} = 2 - x - x^2$

In Problems 21 and 22, let R^4 be equipped with the standard inner product. Find all vectors in R^4 that are orthogonal to both **u** and **v**.

21. $\mathbf{u} = (1, -1, -2, 4); \quad \mathbf{v} = (-1, 2, 3, -6)$

22. $\mathbf{u} = (1, -2, 1, 3); \quad \mathbf{v} = (2, -5, 1, 5)$

In Problems 23–26, let R^2 be equipped with the inner product defined by

$$\langle \mathbf{u}, \mathbf{v} \rangle = 4u_1v_1 + 9u_2v_2$$

where $\mathbf{u} = (u_1, u_2)$ and $\mathbf{v} = (v_1, v_2)$.

23. If $\mathbf{u} = (9, 2)$ and $\mathbf{v} = (6, -2)$, find:

    (A) $\langle \mathbf{u}, \mathbf{v} \rangle$ (B) $\|\mathbf{u}\|$ (C) $\|\mathbf{v}\|$ (D) $d(\mathbf{u}, \mathbf{v})$

    (E) A unit vector in the direction of **u**

    (F) The angle between **u** and **v**

24. Repeat Problem 23 for $\mathbf{u} = (15, 2)$ and $\mathbf{v} = (-9, 4)$.

25. Find all vectors in R^2 that are orthogonal to $\mathbf{u} = (1, -1)$.

26. Find all vectors in R^2 that are orthogonal to $\mathbf{u} = (3, 2)$.

In Problems 27–30, let $M_{2\times2}$ be equipped with the inner product defined by

$$\langle \mathbf{u}, \mathbf{v} \rangle = a_{11}b_{11} + 4a_{12}b_{12} + 4a_{21}b_{21} + a_{22}b_{22}$$

where

$$\mathbf{u} = \begin{bmatrix} a_{11} & a_{12} \\ a_{21} & a_{22} \end{bmatrix} \qquad \mathbf{v} = \begin{bmatrix} b_{11} & b_{12} \\ b_{21} & b_{22} \end{bmatrix}$$

27. If

$$\mathbf{u} = \begin{bmatrix} -4 & 5 \\ 0 & -10 \end{bmatrix} \quad \text{and} \quad \mathbf{v} = \begin{bmatrix} 4 & 0 \\ 1 & 2 \end{bmatrix}$$

find:

(A) $\langle \mathbf{u}, \mathbf{v} \rangle$ (B) $\|\mathbf{u}\|$ (C) $\|\mathbf{v}\|$ (D) $d(\mathbf{u}, \mathbf{v})$
(E) A unit vector in the direction of $\mathbf{u}$
(F) The angle between $\mathbf{u}$ and $\mathbf{v}$

28. Repeat Problem 27 for

$$\mathbf{u} = \begin{bmatrix} 4 & 0 \\ 2 & 0 \end{bmatrix} \quad \text{and} \quad \mathbf{v} = \begin{bmatrix} 8 & 2 \\ 5 & 6 \end{bmatrix}$$

29. Find all vectors in $M_{2\times2}$ that are orthogonal to

$$\mathbf{u} = \begin{bmatrix} 1 & -1 \\ -1 & 2 \end{bmatrix}$$

30. Find all vectors in $M_{2\times2}$ that are orthogonal to

$$\mathbf{u} = \begin{bmatrix} 2 & 1 \\ -2 & -4 \end{bmatrix}$$

In Problems 31–38, let C be the vector space of functions continuous on $(-\infty, \infty)$ with the inner product

$$\langle f, g \rangle = \int_{-1}^{1} f(x)\, g(x)\, dx$$

In Problems 31–34, find $\langle f, g \rangle$, $\|f\|$, and $\|g\|$.

31. *Calculus.* $f(x) = 1, \quad g(x) = x^3$
32. *Calculus.* $f(x) = x^2, \quad g(x) = x^4$
33. *Calculus.* $f(x) = x, \quad g(x) = e^x$
34. *Calculus.* $f(x) = e^x + e^{-x}, \quad g(x) = e^x - e^{-x}$

In Problems 35 and 36, find the values of k for which f and g are orthogonal.

35. *Calculus.* $f(x) = x^2, \quad g(x) = 1 + kx^2$
36. *Calculus.* $f(x) = 1 + kx, \quad g(x) = 1 + kx^2$
37. *Calculus.* Show that $f(x) = x^{2n}$ and $g(x) = x^{2m+1}$ are orthogonal for all non-negative integers n and m.

38. *Calculus.* Show that $f(x) = \sin(\pi nx)$ and $g(x) = \cos(\pi mx)$ are orthogonal for all nonzero integers n and m.

C   In Problems 39–42, let $\mathbf{u} = (u_1, u_2)$ and $\mathbf{v} = (v_1, v_2)$ be vectors in R^2. Determine whether the indicated function is an inner product on R^2. If the function is not an inner product, list the properties in the definition that are not satisfied.

39. $\langle \mathbf{u}, \mathbf{v} \rangle = 2u_1v_1 + 3u_2v_2$

40. $\langle \mathbf{u}, \mathbf{v} \rangle = 3u_1v_1 - u_1v_2 - u_2v_1 + u_2v_2$

41. $\langle \mathbf{u}, \mathbf{v} \rangle = u_1v_2 + u_2v_1$ **42.** $\langle \mathbf{u}, \mathbf{v} \rangle = u_1^2v_1^2 + u_2^2v_2^2$

43. Theorem 14. Prove Theorem 14.

44. Let $B = \{(1, 0), (1, 1)\}$ be a basis for R^2.

 (A) Find $[\mathbf{u}]_B$ for $\mathbf{u} = (u_1, u_2)$.

 (B) Theorem 14 implies that $\langle \mathbf{u}, \mathbf{v} \rangle = [\mathbf{u}]_B^T[\mathbf{v}]_B$ is an inner product on R^2. If $\mathbf{u} = (u_1, u_2)$ and $\mathbf{v} = (v_1, v_2)$, express $\langle \mathbf{u}, \mathbf{v} \rangle$ in terms of u_1, u_2, v_1, and v_2.

45. Repeat Problem 44 for $B = \{(2, 1), (1, 1)\}$.

Let $\mathbf{u}, \mathbf{v}$, and $\mathbf{w}$ be vectors in an inner product space V and let k be a scalar. Prove the following statements from Theorem 15:

46. Theorem 15(B). $\langle \mathbf{u} + \mathbf{v}, \mathbf{w} \rangle = \langle \mathbf{u}, \mathbf{w} \rangle + \langle \mathbf{v}, \mathbf{w} \rangle$

47. Theorem 15(C). $\langle \mathbf{u}, \mathbf{0} \rangle = 0$

48. Use the Cauchy–Schwarz inequality to show that

$$|a \cos \theta + b \sin \theta| \le \sqrt{a^2 + b^2}$$

[*Hint:* Let $\mathbf{u} = (a, b)$ and $\mathbf{v} = (\cos \theta, \sin \theta)$.]

In Problems 49–52, let $\mathbf{u}$ and $\mathbf{v}$ be vectors in an inner product space V and let $d(\mathbf{u}, \mathbf{v}) = \|\mathbf{u} - \mathbf{v}\|$ be the distance between $\mathbf{u}$ and $\mathbf{v}$. Prove each statement.

49. $d(\mathbf{u}, \mathbf{v}) \ge 0$

50. $d(\mathbf{u}, \mathbf{v}) = 0$ if and only if $\mathbf{u} = \mathbf{v}$

51. $d(\mathbf{u}, \mathbf{v}) = d(\mathbf{v}, \mathbf{u})$

52. $d(\mathbf{u}, \mathbf{v}) \le d(\mathbf{u}, \mathbf{w}) + d(\mathbf{w}, \mathbf{v})$ for any $\mathbf{w}$ in V

In Problems 53–56, prove each statement for $\mathbf{u}$ and $\mathbf{v}$ in an inner product space V.

53. $\|\mathbf{u} + \mathbf{v}\|^2 + \|\mathbf{u} - \mathbf{v}\|^2 = 2\|\mathbf{u}\|^2 + 2\|\mathbf{v}\|^2$

54. $\langle \mathbf{u}, \mathbf{v} \rangle = \frac{1}{4}\|\mathbf{u} + \mathbf{v}\|^2 - \frac{1}{4}\|\mathbf{u} - \mathbf{v}\|^2$

55. $\|\mathbf{u} + \mathbf{v}\| = \|\mathbf{u} - \mathbf{v}\|$ if and only if $\mathbf{u}$ and $\mathbf{v}$ are orthogonal [*Hint:* Use Problem 54.]

56. $\|\mathbf{u} + \mathbf{v}\|^2 = \|\mathbf{u}\|^2 + \|\mathbf{v}\|^2$ if and only if $\mathbf{u}$ and $\mathbf{v}$ are orthogonal [*Hint:* Use Problems 53–55.]

57. Let $\mathbf{u}$ be a fixed vector in a vector space V and let

$$S = \{\mathbf{v} \in V | \mathbf{v} \text{ is orthogonal to } \mathbf{u}\}$$

Show that S is a subspace of V.

58. Let $\mathbf{u}, \mathbf{v}_1, \mathbf{v}_2, \ldots, \mathbf{v}_n$ be vectors in an inner product space V satisfying $\langle \mathbf{u}, \mathbf{v}_i \rangle = 0$ for $i = 1, 2, \ldots, n$. Show that $\mathbf{u}$ is orthogonal to each vector in $\text{span}\{\mathbf{v}_1, \mathbf{v}_2, \ldots, \mathbf{v}_n\}$.

59. *Calculus.* Let C be the vector space of functions continuous on $(-\infty, \infty)$ and let

$$\langle f, g \rangle = \int_a^b f(x)\, g(x)\, dx$$

for $f, g \in C$ and $a, b \in R$, $a < b$. Show that $\langle f, g \rangle$ is an inner product on C.

▌6-6▐ Orthogonal Bases

- Orthogonal and Orthonormal Sets
- Properties of Orthogonal and Orthonormal Sets
- The Gram–Schmidt Process

▪ Orthogonal and Orthonormal Sets

If R^n is equipped with the standard inner product and if $B = \{\mathbf{e}_1, \mathbf{e}_2, \ldots, \mathbf{e}_n\}$ is its standard basis, then Theorem 3 in Section 5-1 implies that

$$\langle \mathbf{e}_i, \mathbf{e}_j \rangle = \begin{cases} 0 & \text{if } i \neq j \\ 1 & \text{if } i = j \end{cases}$$

In other words, the standard basis for R^n is a collection of mutually orthogonal unit vectors. This is one of the properties of the standard basis that makes it so easy to use in most situations. In this section, we will show that any finite-dimensional inner product space has a basis consisting of mutually orthogonal unit vectors, and we will develop a procedure for finding such a basis. We begin by introducing some terminology that is useful in describing this type of basis.

Orthogonal and Orthonormal Sets

Let W be a subset of an inner product space V.

(A) If every pair of distinct vectors in W is orthogonal, then W is called an **orthogonal set.**

(B) If W is an orthogonal set and if each vector in W is a unit vector, then W is called an **orthonormal set.**

If $W = \{\mathbf{u}_1, \mathbf{u}_2, \ldots, \mathbf{u}_m\}$, then the conditions in (A) and (B) can be stated as follows:

(A) W is an orthogonal set if $\langle \mathbf{u}_i, \mathbf{u}_j \rangle = 0$ for all $i \neq j$.

(B) W is an orthonormal set if $\langle \mathbf{u}_i, \mathbf{u}_j \rangle = \begin{cases} 0 & \text{for all } i \neq j \\ 1 & \text{for } i = j \end{cases}$

Example 29   Let R^3 be equipped with the standard inner product and let $W = \{\mathbf{v}_1, \mathbf{v}_2, \mathbf{v}_3\}$, where

$$\mathbf{v}_1 = (1, -5, 2) \qquad \mathbf{v}_2 = (1, 1, 2) \qquad \mathbf{v}_3 = (2, 0, -1)$$

Is W an orthogonal set?

Solution   To determine whether W is an orthogonal set, we must compute the inner product of each pair of distinct vectors in W:

$$\langle \mathbf{v}_1, \mathbf{v}_2 \rangle = 1(1) + (-5)1 + 2(2) = 0$$
$$\langle \mathbf{v}_1, \mathbf{v}_3 \rangle = 1(2) + (-5)0 + 2(-1) = 0$$
$$\langle \mathbf{v}_2, \mathbf{v}_3 \rangle = 1(2) + 1(0) + 2(-1) = 0$$

Thus, W is an orthogonal set.

Problem 29   Repeat Example 29 for

$$\mathbf{v}_1 = (1, -1, 1) \qquad \mathbf{v}_2 = (1, 1, 0) \qquad \mathbf{v}_3 = (-1, 1, 2)$$ ▐▐

If $\mathbf{u}$ and $\mathbf{v}$ are orthogonal vectors and k and ℓ are scalars, then

$$\langle k\mathbf{u}, \ell\mathbf{v} \rangle = k\ell\langle \mathbf{u}, \mathbf{v} \rangle = 0$$

Thus, $k\mathbf{u}$ and $\ell\mathbf{v}$ are also orthogonal. If W is a set of nonzero orthogonal vectors, then W can be used to form a set of orthonormal vectors by multiplying each vector $\mathbf{w}$ in W by $1/\|\mathbf{w}\|$. This process is referred to as **normalizing** W. (Notice that an orthogonal set can contain the zero vector, but an orthonormal set cannot. Why?)

Example 30   Refer to Example 29. Normalize the orthogonal set

$$W = \{(1, -5, 2), (1, 1, 2), (2, 0, -1)\}$$

Solution $\|\mathbf{v}_1\| = \sqrt{1^2 + (-5)^2 + 2^2} = \sqrt{30} \qquad \mathbf{v}_1 = (1, -5, 2)$

$$\mathbf{u}_1 = \frac{\mathbf{v}_1}{\|\mathbf{v}_1\|} = \left(\frac{1}{\sqrt{30}}, -\frac{5}{\sqrt{30}}, \frac{2}{\sqrt{30}}\right)$$

$\|\mathbf{v}_2\| = \sqrt{1^2 + 1^2 + 2^2} = \sqrt{6} \qquad \mathbf{v}_2 = (1, 1, 2)$

$$\mathbf{u}_2 = \frac{\mathbf{v}_2}{\|\mathbf{v}_2\|} = \left(\frac{1}{\sqrt{6}}, \frac{1}{\sqrt{6}}, \frac{2}{\sqrt{6}}\right)$$

$\|\mathbf{v}_3\| = \sqrt{2^2 + 0^2 + (-1)^2} = \sqrt{5} \qquad \mathbf{v}_3 = (2, 0, -1)$

$$\mathbf{u}_3 = \frac{\mathbf{v}_3}{\|\mathbf{v}_3\|} = \left(\frac{2}{\sqrt{5}}, 0, -\frac{1}{\sqrt{5}}\right)$$

Thus, $B = \{\mathbf{u}_1, \mathbf{u}_2, \mathbf{u}_3\}$ is an orthonormal set.

Problem 30   Refer to Problem 29. Normalize the orthogonal set

$$W = \{(1, -1, 1), (1, 1, 0), (-1, 1, 2)\}$$ ▐▐

▪ Properties of Orthogonal and Orthonormal Sets

Now we want to develop some useful properties of orthogonal and orthonormal sets.

Theorem 18

If $W = \{\mathbf{u}_1, \mathbf{u}_2, \ldots, \mathbf{u}_m\}$ is an orthogonal set of nonzero vectors in an inner product space V, then W is linearly independent.

Proof To show that W is linearly independent, we must show that the only solution to the vector equation

$$k_1\mathbf{u}_1 + k_2\mathbf{u}_2 + \cdots + k_m\mathbf{u}_m = \mathbf{0} \tag{1}$$

is $k_1 = k_2 = \cdots = k_m = 0$. For each i, Theorem 15(C) in Section 6-5 implies that

$$0 = \langle \mathbf{0}, \mathbf{u}_i \rangle \qquad \text{Use (1) to substitute for } \mathbf{0}.$$

$$= \langle \overbrace{k_1\mathbf{u}_1 + \cdots + k_i\mathbf{u}_i + \cdots + k_m\mathbf{u}_m}^{\mathbf{0}}, \mathbf{u}_i \rangle \qquad \begin{array}{l}\text{Expand, using properties of}\\ \text{the inner product.}\end{array}$$

$$= k_1\langle \mathbf{u}_1, \mathbf{u}_i \rangle + \cdots + k_i\langle \mathbf{u}_i, \mathbf{u}_i \rangle + \cdots \qquad \langle \mathbf{u}_j, \mathbf{u}_i \rangle = 0 \text{ for } i \neq j$$
$$\quad + k_m\langle \mathbf{u}_m, \mathbf{u}_i \rangle$$

$$= k_i\langle \mathbf{u}_i, \mathbf{u}_i \rangle$$

Thus, we can conclude that either $k_i = 0$ or $\langle \mathbf{u}_i, \mathbf{u}_i \rangle = 0$. But $\langle \mathbf{u}_i, \mathbf{u}_i \rangle = 0$ would imply that $\mathbf{u}_i = \mathbf{0}$, and we were given that the vectors in W are nonzero. Hence, the only possible conclusion is $k_i = 0$. Since this holds for each i, $1 \leq i \leq m$, we have shown that W is linearly independent. ▮

Theorem 18 and Theorem 17 in Section 5-6 provide a simple characterization of orthogonal bases. The proof is left as an exercise.

Theorem 19

Let V be an n-dimensional inner product space, and let B be an orthogonal set of nonzero vectors in V. Then B is a basis for V if and only if B has exactly n elements.

If $B = \{\mathbf{u}_1, \mathbf{u}_2, \ldots, \mathbf{u}_n\}$ is a basis for a vector space V, then each vector $\mathbf{v}$ in V can be expressed as a linear combination of basis vectors:

$$\mathbf{v} = k_1\mathbf{u}_1 + k_2\mathbf{u}_2 + \cdots + k_n\mathbf{u}_n \tag{2}$$

For most bases, the scalars in (2) are determined by solving a system of linear equations. However, if V is an inner product space and B is an orthonormal basis for V, then Theorem 20 shows that the scalars in (2) can be expressed very simply in terms of inner products.

Theorem 20

If $B = \{\mathbf{u}_1, \mathbf{u}_2, \ldots, \mathbf{u}_n\}$ is an orthonormal basis for an inner product space V and $\mathbf{v}$ is any vector in V, then

$$\mathbf{v} = \langle \mathbf{v}, \mathbf{u}_1 \rangle \mathbf{u}_1 + \langle \mathbf{v}, \mathbf{u}_2 \rangle \mathbf{u}_2 + \cdots + \langle \mathbf{v}, \mathbf{u}_n \rangle \mathbf{u}_n$$

Proof Since B is a basis for V, there exist scalars $k_1, k_2, \ldots, k_n$ satisfying

$$\mathbf{v} = k_1 \mathbf{u}_1 + k_2 \mathbf{u}_2 + \cdots + k_n \mathbf{u}_n \tag{3}$$

Using (3), for each $\mathbf{u}_i$ in B we have

$$
\begin{aligned}
\langle \mathbf{v}, \mathbf{u}_i \rangle &= \langle k_1 \mathbf{u}_1 + \cdots + k_i \mathbf{u}_i + \cdots + k_n \mathbf{u}_n, \mathbf{u}_i \rangle \\
&= k_1 \langle \mathbf{u}_1, \mathbf{u}_i \rangle + \cdots + k_i \langle \mathbf{u}_i, \mathbf{u}_i \rangle + \cdots + k_n \langle \mathbf{u}_n, \mathbf{u}_i \rangle \\
&= k_i
\end{aligned}
$$

$\langle \mathbf{u}_i, \mathbf{u}_i \rangle = 1$ and $\langle \mathbf{u}_j, \mathbf{u}_i \rangle = 0$ for $i \neq j$

Substituting $k_i = \langle \mathbf{v}, \mathbf{u}_i \rangle$, $1 \leq i \leq n$, in (3), we have

$$\mathbf{v} = \langle \mathbf{v}, \mathbf{u}_1 \rangle \mathbf{u}_1 + \langle \mathbf{v}, \mathbf{u}_2 \rangle \mathbf{u}_2 + \cdots + \langle \mathbf{v}, \mathbf{u}_n \rangle \mathbf{u}_n \qquad \blacksquare$$

Example 31 Let R^3 be equipped with the standard inner product. The set $B = \{\mathbf{u}_1, \mathbf{u}_2, \mathbf{u}_3\}$ where

$$\mathbf{u}_1 = \left(\frac{1}{\sqrt{30}}, -\frac{5}{\sqrt{30}}, \frac{2}{\sqrt{30}} \right) \qquad \mathbf{u}_2 = \left(\frac{1}{\sqrt{6}}, \frac{1}{\sqrt{6}}, \frac{2}{\sqrt{6}} \right) \qquad \mathbf{u}_3 = \left(\frac{2}{\sqrt{5}}, 0, -\frac{1}{\sqrt{5}} \right)$$

is an orthonormal subset of R^3 (see Examples 29 and 30). Since B has three elements, Theorem 19 implies that B is a basis for R^3. Now let $\mathbf{v} = (3, -2, 4)$. Then

$$\langle \mathbf{v}, \mathbf{u}_1 \rangle = 3 \left(\frac{1}{\sqrt{30}} \right) + (-2) \left(-\frac{5}{\sqrt{30}} \right) + 4 \left(\frac{2}{\sqrt{30}} \right) = \frac{21}{\sqrt{30}}$$

$$\langle \mathbf{v}, \mathbf{u}_2 \rangle = 3 \left(\frac{1}{\sqrt{6}} \right) + (-2) \left(\frac{1}{\sqrt{6}} \right) + 4 \left(\frac{2}{\sqrt{6}} \right) = \frac{9}{\sqrt{6}}$$

$$\langle \mathbf{v}, \mathbf{u}_3 \rangle = 3 \left(\frac{2}{\sqrt{5}} \right) + (-2)(0) + 4 \left(-\frac{1}{\sqrt{5}} \right) = \frac{2}{\sqrt{5}}$$

Using Theorem 20, we have

$$\overset{\langle \mathbf{v}, \mathbf{u}_1 \rangle \quad\quad \langle \mathbf{v}, \mathbf{u}_2 \rangle \quad\quad \langle \mathbf{v}, \mathbf{u}_3 \rangle}{\mathbf{v} = \frac{21}{\sqrt{30}}\, \mathbf{u}_1 + \frac{9}{\sqrt{6}}\, \mathbf{u}_2 + \frac{2}{\sqrt{5}}\, \mathbf{u}_3}$$

Check: $\dfrac{21}{\sqrt{30}}\, \mathbf{u}_1 + \dfrac{9}{\sqrt{6}}\, \mathbf{u}_2 + \dfrac{2}{\sqrt{5}}\, \mathbf{u}_3$

$$= \left(\frac{21}{30}, -\frac{105}{30}, \frac{42}{30} \right) + \left(\frac{9}{6}, \frac{9}{6}, \frac{18}{6} \right) + \left(\frac{4}{5}, 0, -\frac{2}{5} \right)$$

$$= \left(\frac{90}{30}, -\frac{60}{30}, \frac{120}{30} \right) = (3, -2, 4) = \mathbf{v}$$

Problem 31 Let R^3 be equipped with the standard inner product, and let

$$B = \left\{ \left(\frac{1}{\sqrt{3}}, -\frac{1}{\sqrt{3}}, \frac{1}{\sqrt{3}} \right), \left(\frac{1}{\sqrt{2}}, \frac{1}{\sqrt{2}}, 0 \right), \left(-\frac{1}{\sqrt{6}}, \frac{1}{\sqrt{6}}, \frac{2}{\sqrt{6}} \right) \right\}$$

be an orthonormal basis for R^3 (see Problems 29 and 30). Express $\mathbf{v} = (3, -2, 1)$ as a linear combination of the vectors in B. ∎

Example 32 Let P_2 be equipped with the standard inner product, and let $B = \{\mathbf{u}_1, \mathbf{u}_2, \mathbf{u}_3\}$ where

$$\mathbf{u}_1 = \tfrac{8}{9} + \tfrac{4}{9}x + \tfrac{1}{9}x^2 \qquad \mathbf{u}_2 = \tfrac{1}{9} - \tfrac{4}{9}x + \tfrac{8}{9}x^2 \qquad \mathbf{u}_3 = -\tfrac{4}{9} + \tfrac{7}{9}x + \tfrac{4}{9}x^2$$

Show that B is an orthonormal basis for P_2, and express $\mathbf{v} = 1 + x + x^2$ as a linear combination of the vectors in B.

Solution First, we must show that B is an orthonormal set.

$$\langle \mathbf{u}_1, \mathbf{u}_2 \rangle = \tfrac{8}{81} - \tfrac{16}{81} + \tfrac{8}{81} = 0$$
$$\langle \mathbf{u}_1, \mathbf{u}_3 \rangle = -\tfrac{32}{81} + \tfrac{28}{81} + \tfrac{4}{81} = 0$$
$$\langle \mathbf{u}_2, \mathbf{u}_3 \rangle = -\tfrac{4}{81} - \tfrac{28}{81} + \tfrac{32}{81} = 0$$

These equations show that B is an orthogonal set.

$$\langle \mathbf{u}_1, \mathbf{u}_1 \rangle = \tfrac{64}{81} + \tfrac{16}{81} + \tfrac{1}{81} = 1$$
$$\langle \mathbf{u}_2, \mathbf{u}_2 \rangle = \tfrac{1}{81} + \tfrac{16}{81} + \tfrac{64}{81} = 1$$
$$\langle \mathbf{u}_3, \mathbf{u}_3 \rangle = \tfrac{16}{81} + \tfrac{49}{81} + \tfrac{16}{81} = 1$$

These equations show that the vectors in B are unit vectors. If $\langle \mathbf{u}, \mathbf{u} \rangle = 1$, then $\|\mathbf{u}\| = \sqrt{\langle \mathbf{u}, \mathbf{u} \rangle} = 1$.

Thus, B is an orthonormal set with three elements. Since P_2 is three-dimensional, Theorem 19 implies that B is an orthonormal basis for P_2.

To express $\mathbf{v}$ as a linear combination of basis vectors, we compute $\langle \mathbf{v}, \mathbf{u}_i \rangle$, $i = 1, 2, 3$:

$$\langle \mathbf{v}, \mathbf{u}_1 \rangle = \tfrac{8}{9} + \tfrac{4}{9} + \tfrac{1}{9} = \tfrac{13}{9}$$

$$\langle \mathbf{v}, \mathbf{u}_2 \rangle = \tfrac{1}{9} - \tfrac{4}{9} + \tfrac{8}{9} = \tfrac{5}{9}$$

$$\langle \mathbf{v}, \mathbf{u}_3 \rangle = -\tfrac{4}{9} + \tfrac{7}{9} + \tfrac{4}{9} = \tfrac{7}{9}$$

Thus, using Theorem 20,

$$\mathbf{v} = \tfrac{13}{9}\mathbf{u}_1 + \tfrac{5}{9}\mathbf{u}_2 + \tfrac{7}{9}\mathbf{u}_3$$

You should check this result.

Problem 32 Repeat Example 32 for

$$\mathbf{u}_1 = \tfrac{2}{11} + \tfrac{6}{11}x + \tfrac{9}{11}x^2 \qquad \mathbf{u}_2 = \tfrac{6}{11} + \tfrac{7}{11}x - \tfrac{6}{11}x^2 \qquad \mathbf{u}_3 = \tfrac{9}{11} - \tfrac{6}{11}x + \tfrac{2}{11}x^2 \quad \blacksquare$$

▪ The Gram–Schmidt Process

Now that we have seen that orthonormal sets are convenient to work with, we want to develop a procedure for finding orthonormal sets. First, we note that for the vector spaces R^n, P_n, and $M_{m \times n}$, we would use their standard bases (which are orthonormal) unless there were compelling reasons to do otherwise—and such reasons do occur in certain applications in physics, engineering, and mathematics.

We begin the development of a procedure for finding orthonormal sets by considering an example.

Example 33 Let R^4 be equipped with the standard inner product, and let $V = \text{span}\{\mathbf{u}_1, \mathbf{u}_2, \mathbf{u}_3\}$ where

$$\mathbf{u}_1 = (1, 1, -1, 1) \qquad \mathbf{u}_2 = (2, 1, -1, 0) \qquad \mathbf{u}_3 = (2, -1, 0, 1)$$

Find an orthonormal basis for V.

Solution We first note that $V \neq R^4$. (Why?) Therefore, the standard (orthonormal) basis for R^4 is not a basis for V. The vectors $\mathbf{u}_1$, $\mathbf{u}_2$, and $\mathbf{u}_3$ are linearly independent (verify this); hence, they form a basis for V. However, these vectors are not orthogonal since $\langle \mathbf{u}_1, \mathbf{u}_2 \rangle = 4$, $\langle \mathbf{u}_1, \mathbf{u}_3 \rangle = 2$, and $\langle \mathbf{u}_2, \mathbf{u}_3 \rangle = 3$. We will use these vectors to construct an orthogonal basis for V, and then we will normalize this orthogonal basis to produce an orthonormal basis for V. It turns out that this construction is easier to understand if we require the orthogonal vectors to satisfy some additional conditions. Specifically, we will find orthogonal vectors $\mathbf{w}_1$, $\mathbf{w}_2$, and $\mathbf{w}_3$ satisfying:

1. $\text{span}\{\mathbf{w}_1\} = \text{span}\{\mathbf{u}_1\}$
2. $\text{span}\{\mathbf{w}_1, \mathbf{w}_2\} = \text{span}\{\mathbf{u}_1, \mathbf{u}_2\}$
3. $\text{span}\{\mathbf{w}_1, \mathbf{w}_2, \mathbf{w}_3\} = \text{span}\{\mathbf{u}_1, \mathbf{u}_2, \mathbf{u}_3\} = V$

The step-by-step progression illustrated by these three conditions is the key to understanding this construction.

Step 1. Find $\mathbf{w}_1$ *satisfying span*$\{\mathbf{w}_1\} = span\{\mathbf{u}_1\}$. This condition is easy to satisfy. Simply let $\mathbf{w}_1 = \mathbf{u}_1 = (1, 1, -1, 1)$.

Step 2. Find $\mathbf{w}_2$ *satisfying span*$\{\mathbf{w}_1, \mathbf{w}_2\} = span\{\mathbf{u}_1, \mathbf{u}_2\}$ *and* $\langle \mathbf{w}_1, \mathbf{w}_2 \rangle = 0$. Let

$$\mathbf{w}_2 = \mathbf{u}_2 - k_1\mathbf{w}_1 \tag{4}$$

where k_1 is a scalar. Since $\mathbf{w}_1 = \mathbf{u}_1$, $\mathbf{w}_2$ is in span$\{\mathbf{u}_1, \mathbf{u}_2\}$. (The reason for selecting this particular linear combination form will become apparent as this discussion progresses.) We want to select k_1 so that $\mathbf{w}_1$ and $\mathbf{w}_2$ are orthogonal.

$$\langle \mathbf{w}_2, \mathbf{w}_1 \rangle = 0 \qquad \text{Use (4) to substitute for } \mathbf{w}_2.$$
$$\langle \mathbf{u}_2 - k_1\mathbf{w}_1, \mathbf{w}_1 \rangle = 0$$
$$\langle \mathbf{u}_2, \mathbf{w}_1 \rangle - k_1 \langle \mathbf{w}_1, \mathbf{w}_1 \rangle = 0 \tag{5}$$

Since $\mathbf{u}_2$ is one of the original vectors and $\mathbf{w}_1$ was determined in Step 1, we can compute each of the inner products in (5):

$$\langle \mathbf{u}_2, \mathbf{w}_1 \rangle = 2(1) + 1(1) + (-1)(-1) + 0(1) = 4$$
$$\langle \mathbf{w}_1, \mathbf{w}_1 \rangle = 1(1) + 1(1) + (-1)(-1) + 1(1) = 4$$

Substituting these values in (5), we have

$$4 - 4k_1 = 0$$
$$k_1 = 1$$

Substituting $k_1 = 1$ in (4),

$$\mathbf{w}_2 = \mathbf{u}_2 - \mathbf{u}_1$$
$$= (2, 1, -1, 0) - (1, 1, -1, 1)$$
$$= (1, 0, 0, -1)$$

We compute $\langle \mathbf{w}_1, \mathbf{w}_2 \rangle$ to check our calculations:

$$\langle \mathbf{w}_1, \mathbf{w}_2 \rangle = 1(1) + 1(0) + (-1)0 + 1(-1) = 0$$

Thus, $\{\mathbf{w}_1, \mathbf{w}_2\}$ is a set of two nonzero orthogonal vectors in the two-dimensional space span$\{\mathbf{u}_1, \mathbf{u}_2\}$. Since nonzero orthogonal vectors are linearly independent, $\{\mathbf{w}_1, \mathbf{w}_2\}$ is a basis for span$\{\mathbf{u}_1, \mathbf{u}_2\}$. That is,

$$\text{span}\{\mathbf{w}_1, \mathbf{w}_2\} = \text{span}\{\mathbf{u}_1, \mathbf{u}_2\} \qquad \text{and} \qquad \langle \mathbf{w}_1, \mathbf{w}_2 \rangle = 0$$

Step 3. Find $\mathbf{w}_3$ *satisfying span*$\{\mathbf{w}_1, \mathbf{w}_2, \mathbf{w}_3\} = span\{\mathbf{u}_1, \mathbf{u}_2, \mathbf{u}_3\}$, $\langle \mathbf{w}_3, \mathbf{w}_1 \rangle = 0$, *and* $\langle \mathbf{w}_3, \mathbf{w}_2 \rangle = 0$. Let

$$\mathbf{w}_3 = \mathbf{u}_3 - k_1\mathbf{w}_1 - k_2\mathbf{w}_2 \tag{6}$$

Since span$\{\mathbf{w}_1, \mathbf{w}_2\} = $ span$\{\mathbf{u}_1, \mathbf{u}_2\}$, $-k_1\mathbf{w}_1 - k_2\mathbf{w}_2$ is in span$\{\mathbf{u}_1, \mathbf{u}_2\}$; hence, $\mathbf{w}_3$ is in span$\{\mathbf{u}_1, \mathbf{u}_2, \mathbf{u}_3\}$. First, we use the condition $\langle \mathbf{w}_3, \mathbf{w}_1 \rangle = 0$ to determine k_1:

$$\langle \mathbf{w}_3, \mathbf{w}_1 \rangle = 0 \qquad \text{Use (6) to}$$
$$\langle \mathbf{u}_3 - k_1\mathbf{w}_1 - k_2\mathbf{w}_2, \mathbf{w}_1 \rangle = 0 \qquad \text{substitute for } \mathbf{w}_3.$$
$$\langle \mathbf{u}_3, \mathbf{w}_1 \rangle - k_1\langle \mathbf{w}_1, \mathbf{w}_1 \rangle - k_2\langle \mathbf{w}_2, \mathbf{w}_1 \rangle = 0 \qquad (7)$$

Since $\mathbf{u}_3$ is one of the original vectors and $\mathbf{w}_1$ and $\mathbf{w}_2$ were determined in Steps 1 and 2, we can compute each of the inner products in (7):

$$\langle \mathbf{u}_3, \mathbf{w}_1 \rangle = 2(1) + (-1)1 + 0(-1) + 1(1) = 2$$

$$\langle \mathbf{w}_1, \mathbf{w}_1 \rangle = 4 \qquad\qquad\qquad\qquad \text{Computed in Step 2}$$

$$\langle \mathbf{w}_2, \mathbf{w}_1 \rangle = 0 \qquad\qquad\qquad\qquad \mathbf{w}_1 \text{ and } \mathbf{w}_2 \text{ are orthogonal}$$

Substituting these values in (7), we have

$$2 - 4k_1 - 0k_2 = 0$$
$$k_1 = \tfrac{1}{2}$$

Now, we use the condition $\langle \mathbf{w}_3, \mathbf{w}_2 \rangle = 0$ to determine k_2:

$$\langle \mathbf{w}_3, \mathbf{w}_2 \rangle = 0 \qquad \text{Use (6) to substitute}$$
$$\langle \mathbf{u}_3 - k_1\mathbf{w}_1 - k_2\mathbf{w}_2, \mathbf{w}_2 \rangle = 0 \qquad \text{for } \mathbf{w}_3.$$
$$\langle \mathbf{u}_3, \mathbf{w}_2 \rangle - k_1\langle \mathbf{w}_1, \mathbf{w}_2 \rangle - k_2\langle \mathbf{w}_2, \mathbf{w}_2 \rangle = 0 \qquad (8)$$

Next, we compute the inner products in (8):

$$\langle \mathbf{u}_3, \mathbf{w}_2 \rangle = 2(1) + (-1)0 + 0(0) + 1(-1) = 1$$

$$\langle \mathbf{w}_1, \mathbf{w}_2 \rangle = 0$$

$$\langle \mathbf{w}_2, \mathbf{w}_2 \rangle = 1(1) + 0(0) + 0(0) + (-1)(-1) = 2$$

Substituting in (8), we have

$$1 - 0k_1 - 2k_2 = 0$$
$$k_2 = \tfrac{1}{2}$$

Then, using these values in (6),

$$\mathbf{w}_3 = \mathbf{u}_3 - \tfrac{1}{2}\mathbf{w}_1 - \tfrac{1}{2}\mathbf{w}_2$$
$$= (2, -1, 0, 1) - \tfrac{1}{2}(1, 1, -1, 1) - \tfrac{1}{2}(1, 0, 0, -1)$$
$$= (1, -\tfrac{3}{2}, \tfrac{1}{2}, 1)$$

Check: $\langle \mathbf{w}_3, \mathbf{w}_1 \rangle = 1(1) + (-\tfrac{3}{2})1 + \tfrac{1}{2}(-1) + 1(1) = 0$
$$\langle \mathbf{w}_3, \mathbf{w}_2 \rangle = 1(1) + (-\tfrac{3}{2})0 + \tfrac{1}{2}(0) + 1(-1) = 0$$

Thus, $\{\mathbf{w}_1, \mathbf{w}_2, \mathbf{w}_3\}$ is an orthogonal set of three nonzero vectors in the three-dimensional space V and must be a basis for V. That is,

$$\text{span}\{\mathbf{w}_1, \mathbf{w}_2, \mathbf{w}_3\} = \text{span}\{\mathbf{u}_1, \mathbf{u}_2, \mathbf{u}_3\} = V$$

and

$$\langle \mathbf{w}_1, \mathbf{w}_2 \rangle = 0 \qquad \langle \mathbf{w}_1, \mathbf{w}_3 \rangle = 0 \qquad \langle \mathbf{w}_2, \mathbf{w}_3 \rangle = 0$$

Finally, to find an orthonormal basis for V, we normalize the orthogonal basis $\{\mathbf{w}_1, \mathbf{w}_2, \mathbf{w}_3\}$:

$$\mathbf{v}_1 = \frac{\mathbf{w}_1}{\|\mathbf{w}_1\|} = \frac{1}{2}(1, 1, -1, 1) = \left(\frac{1}{2}, \frac{1}{2}, -\frac{1}{2}, \frac{1}{2}\right)$$

$$\mathbf{v}_2 = \frac{\mathbf{w}_2}{\|\mathbf{w}_2\|} = \frac{1}{\sqrt{2}}(1, 0, 0, -1) = \left(\frac{1}{\sqrt{2}}, 0, 0, -\frac{1}{\sqrt{2}}\right)$$

$$\mathbf{v}_3 = \frac{\mathbf{w}_3}{\|\mathbf{w}_3\|} = \frac{\sqrt{2}}{3}\left(1, -\frac{3}{2}, \frac{1}{2}, 1\right) = \left(\frac{\sqrt{2}}{3}, -\frac{\sqrt{2}}{2}, \frac{\sqrt{2}}{6}, \frac{\sqrt{2}}{3}\right)$$

The set $\{\mathbf{v}_1, \mathbf{v}_2, \mathbf{v}_3\}$ is an orthonormal basis for V.

Problem 33   Let R^3 be equipped with the standard inner product. Find an orthonormal basis for $V = \text{span}\{\mathbf{u}_1, \mathbf{u}_2\}$ where

$$\mathbf{u}_1 = (1, -2, 2) \quad \text{and} \quad \mathbf{u}_2 = (3, 0, 3) \qquad ▮$$

The procedure we followed in Example 33 is easily extended to larger sets of vectors. For example, if a fourth vector $\mathbf{u}_4$ is in the original set of linearly independent vectors, then we let

$$\mathbf{w}_4 = \mathbf{u}_4 - k_1\mathbf{w}_1 - k_2\mathbf{w}_2 - k_3\mathbf{w}_3$$

and use the orthogonality conditions

$$\langle \mathbf{w}_4, \mathbf{w}_1 \rangle = 0 \qquad \langle \mathbf{w}_4, \mathbf{w}_2 \rangle = 0 \qquad \langle \mathbf{w}_4, \mathbf{w}_3 \rangle = 0$$

to determine k_1, k_2, and k_3.

This procedure is called the *Gram–Schmidt orthogonalization process* and is summarized in Theorem 21 on page 439. Some special cases of the proof will be considered in the exercises.

Example 34   Let R^3 be equipped with the standard inner product. Apply the Gram–Schmidt process to the linearly independent set $\{\mathbf{u}_1, \mathbf{u}_2, \mathbf{u}_3\}$ where

$$\mathbf{u}_1 = (1, 1, -1) \qquad \mathbf{u}_2 = (1, 4, -4) \qquad \mathbf{u}_3 = (1, 2, 2)$$

Solution   Step 1.   Let $\mathbf{w}_1 = \mathbf{u}_1 = (1, 1, -1)$.

Step 2.   Let $\mathbf{w}_2 = \mathbf{u}_2 - k_1\mathbf{w}_1$.

$$\begin{aligned} 0 &= \langle \mathbf{w}_2, \mathbf{w}_1 \rangle \\ &= \langle \mathbf{u}_2 - k_1\mathbf{w}_1, \mathbf{w}_1 \rangle \\ &= \langle \mathbf{u}_2, \mathbf{w}_1 \rangle - k_1\langle \mathbf{w}_1, \mathbf{w}_1 \rangle \qquad \langle \mathbf{u}_2, \mathbf{w}_1 \rangle = 9, \langle \mathbf{w}_1, \mathbf{w}_1 \rangle = 3 \\ &= 9 - 3k_1 \end{aligned}$$

$$k_1 = 3$$

$$\mathbf{w}_2 = \mathbf{u}_2 - 3\mathbf{w}_1 = (1, 4, -4) - 3(1, 1, -1) = (-2, 1, -1)$$

Theorem 21

Gram–Schmidt Orthogonalization Process

Let $\{\mathbf{u}_1, \mathbf{u}_2, \ldots, \mathbf{u}_n\}$ be a set of linearly independent vectors in an inner product space V.

Step 1. Let $\mathbf{w}_1 = \mathbf{u}_1$.

Step 2. Let $\mathbf{w}_2 = \mathbf{u}_2 - k_1\mathbf{w}_1$ and use the condition $\langle \mathbf{w}_2, \mathbf{w}_1 \rangle = 0$ to determine k_1.

Step 3. Let $\mathbf{w}_3 = \mathbf{u}_3 - k_1\mathbf{w}_1 - k_2\mathbf{w}_2$ and use the conditions $\langle \mathbf{w}_3, \mathbf{w}_1 \rangle = 0$ and $\langle \mathbf{w}_3, \mathbf{w}_2 \rangle = 0$ to determine k_1 and k_2.

$$\vdots$$

Step n. Let $\mathbf{w}_n = \mathbf{u}_n - k_1\mathbf{w}_1 - k_2\mathbf{w}_2 - \cdots - k_{n-1}\mathbf{w}_{n-1}$ and use the conditions $\langle \mathbf{w}_n, \mathbf{w}_1 \rangle = 0$, $\langle \mathbf{w}_n, \mathbf{w}_2 \rangle = 0$, $\ldots$, $\langle \mathbf{w}_n, \mathbf{w}_{n-1} \rangle = 0$ to determine $k_1, k_2, \ldots, k_{n-1}$.

For each i, $1 \le i \le n$, let $\mathbf{v}_i = \mathbf{w}_i / \|\mathbf{w}_i\|$. Then $\{\mathbf{w}_1, \mathbf{w}_2, \ldots, \mathbf{w}_n\}$ is an orthogonal set of nonzero vectors and $\{\mathbf{v}_1, \mathbf{v}_2, \ldots, \mathbf{v}_n\}$ is an orthonormal set of vectors satisfying

$$\text{span}\{\mathbf{u}_1\} = \text{span}\{\mathbf{w}_1\} = \text{span}\{\mathbf{v}_1\}$$

$$\text{span}\{\mathbf{u}_1, \mathbf{u}_2\} = \text{span}\{\mathbf{w}_1, \mathbf{w}_2\} = \text{span}\{\mathbf{v}_1, \mathbf{v}_2\}$$

$$\vdots$$

$$\text{span}\{\mathbf{u}_1, \mathbf{u}_2, \ldots, \mathbf{u}_n\} = \text{span}\{\mathbf{w}_1, \mathbf{w}_2, \ldots, \mathbf{w}_n\}$$
$$= \text{span}\{\mathbf{v}_1, \mathbf{v}_2, \ldots, \mathbf{v}_n\}$$

Step 3. Let $\mathbf{w}_3 = \mathbf{u}_3 - k_1\mathbf{w}_1 - k_2\mathbf{w}_2$.

$$
\begin{aligned}
0 &= \langle \mathbf{w}_3, \mathbf{w}_1 \rangle \\
&= \langle \mathbf{u}_3 - k_1\mathbf{w}_1 - k_2\mathbf{w}_2, \mathbf{w}_1 \rangle \\
&= \langle \mathbf{u}_3, \mathbf{w}_1 \rangle - k_1\langle \mathbf{w}_1, \mathbf{w}_1 \rangle - k_2\langle \mathbf{w}_2, \mathbf{w}_1 \rangle \\
&= 1 - 3k_1 \\
k_1 &= \tfrac{1}{3} \\
0 &= \langle \mathbf{w}_3, \mathbf{w}_2 \rangle \\
&= \langle \mathbf{u}_3 - k_1\mathbf{w}_1 - k_2\mathbf{w}_2, \mathbf{w}_2 \rangle \\
&= \langle \mathbf{u}_3, \mathbf{w}_2 \rangle - k_1\langle \mathbf{w}_1, \mathbf{w}_2 \rangle - k_2\langle \mathbf{w}_2, \mathbf{w}_2 \rangle \\
&= -2 - 6k_2 \\
k_2 &= -\tfrac{1}{3} \\
\mathbf{w}_3 &= \mathbf{u}_3 - \tfrac{1}{3}\mathbf{w}_1 + \tfrac{1}{3}\mathbf{w}_2 \\
&= (1, 2, 2) - \tfrac{1}{3}(1, 1, -1) + \tfrac{1}{3}(-2, 1, -1) \\
&= (0, 2, 2)
\end{aligned}
$$

Right-side annotations:

$\langle \mathbf{u}_3, \mathbf{w}_1 \rangle = 1$,
$\langle \mathbf{w}_1, \mathbf{w}_1 \rangle = 3$,
$\langle \mathbf{w}_2, \mathbf{w}_1 \rangle = 0$

$\langle \mathbf{u}_3, \mathbf{w}_2 \rangle = -2$,
$\langle \mathbf{w}_1, \mathbf{w}_2 \rangle = 0$,
$\langle \mathbf{w}_2, \mathbf{w}_2 \rangle = 6$

Thus, the orthogonal set produced by the Gram–Schmidt process is

$$\{(1, 1, -1), (-2, 1, -1), (0, 2, 2)\}$$

The corresponding normalized vectors are

$$\mathbf{v}_1 = \frac{\mathbf{w}_1}{\|\mathbf{w}_1\|} = \frac{1}{\sqrt{3}} (1, 1, -1) = \left(\frac{1}{\sqrt{3}}, \frac{1}{\sqrt{3}}, -\frac{1}{\sqrt{3}} \right)$$

$$\mathbf{v}_2 = \frac{\mathbf{w}_2}{\|\mathbf{w}_2\|} = \frac{1}{\sqrt{6}} (-2, 1, -1) = \left(-\frac{2}{\sqrt{6}}, \frac{1}{\sqrt{6}}, -\frac{1}{\sqrt{6}} \right)$$

$$\mathbf{v}_3 = \frac{\mathbf{w}_3}{\|\mathbf{w}_3\|} = \frac{1}{2\sqrt{2}} (0, 2, 2) = \left(0, \frac{1}{\sqrt{2}}, \frac{1}{\sqrt{2}} \right)$$

Problem 34   Repeat Example 34 for

$$\mathbf{u}_1 = (1, 1, 1) \qquad \mathbf{u}_2 = (2, 2, -1) \qquad \mathbf{u}_3 = (0, 2, -1) \qquad \blacksquare$$

We have stated the Gram–Schmidt process as a step-by-step procedure, since that is the most convenient way to use it in hand calculations. However, if the number of vectors is large, then a computer is usually used to apply this process. For computer applications, it is more convenient to state the Gram–Schmidt process as a sequence of formulas, as is done in Theorem 22. Some special cases of the proof of this theorem will be discussed in the exercises.

Theorem 22

Formulas for the Gram–Schmidt Process

Let $\{\mathbf{u}_1, \mathbf{u}_2, \ldots, \mathbf{u}_n\}$ be a set of linearly independent vectors in an inner product space V. The orthogonal vectors generated by the Gram–Schmidt process are also given by the following formulas:

(1) $\mathbf{w}_1 = \mathbf{u}_1$

(2) $\mathbf{w}_2 = \mathbf{u}_2 - \dfrac{\langle \mathbf{u}_2, \mathbf{w}_1 \rangle}{\langle \mathbf{w}_1, \mathbf{w}_1 \rangle} \mathbf{w}_1$

(3) $\mathbf{w}_3 = \mathbf{u}_3 - \dfrac{\langle \mathbf{u}_3, \mathbf{w}_1 \rangle}{\langle \mathbf{w}_1, \mathbf{w}_1 \rangle} \mathbf{w}_1 - \dfrac{\langle \mathbf{u}_3, \mathbf{w}_2 \rangle}{\langle \mathbf{w}_2, \mathbf{w}_2 \rangle} \mathbf{w}_2$

$\cdot$

$\cdot$

$\cdot$

(n) $\mathbf{w}_n = \mathbf{u}_n - \dfrac{\langle \mathbf{u}_n, \mathbf{w}_1 \rangle}{\langle \mathbf{w}_1, \mathbf{w}_1 \rangle} \mathbf{w}_1 - \dfrac{\langle \mathbf{u}_n, \mathbf{w}_2 \rangle}{\langle \mathbf{w}_2, \mathbf{w}_2 \rangle} \mathbf{w}_2 - \cdots - \dfrac{\langle \mathbf{u}_n, \mathbf{w}_{n-1} \rangle}{\langle \mathbf{w}_{n-1}, \mathbf{w}_{n-1} \rangle} \mathbf{w}_{n-1}$

The formulas in Theorem 22 can be memorized and used to apply the Gram–Schmidt process by hand. However, most people find it easier to remember the procedure in Theorem 21.

Finally, if V is an n-dimensional inner product space, then V must have a basis B with n elements. Applying the Gram–Schmidt process to this basis produces an orthonormal set of n vectors in V. Theorem 19 then implies that this orthonormal set is also a basis for V. Thus, we have just proved the following important result:

Theorem 23

Every finite-dimensional inner product space has an orthonormal basis.

Answers to Matched Problems

29. Yes; $\langle \mathbf{v}_1, \mathbf{v}_2 \rangle = \langle \mathbf{v}_1, \mathbf{v}_3 \rangle = \langle \mathbf{v}_2, \mathbf{v}_3 \rangle = 0$

30. $\left\{ \left(\dfrac{1}{\sqrt{3}}, -\dfrac{1}{\sqrt{3}}, \dfrac{1}{\sqrt{3}} \right), \left(\dfrac{1}{\sqrt{2}}, \dfrac{1}{\sqrt{2}}, 0 \right), \left(-\dfrac{1}{\sqrt{6}}, \dfrac{1}{\sqrt{6}}, \dfrac{2}{\sqrt{6}} \right) \right\}$

31. $\dfrac{6}{\sqrt{3}} \left(\dfrac{1}{\sqrt{3}}, -\dfrac{1}{\sqrt{3}}, \dfrac{1}{\sqrt{3}} \right) + \dfrac{1}{\sqrt{2}} \left(\dfrac{1}{\sqrt{2}}, \dfrac{1}{\sqrt{2}}, 0 \right) - \dfrac{3}{\sqrt{6}} \left(-\dfrac{1}{\sqrt{6}}, \dfrac{1}{\sqrt{6}}, \dfrac{2}{\sqrt{6}} \right)$

32. $\frac{17}{11}\mathbf{u}_1 + \frac{7}{11}\mathbf{u}_2 + \frac{5}{11}\mathbf{u}_3$

33. $\left\{ \frac{1}{3}(1, -2, 2), \frac{1}{3}(2, 2, 1) \right\}$

34. $\left\{ \dfrac{1}{\sqrt{3}}(1, 1, 1), \dfrac{1}{\sqrt{6}}(1, 1, -2), \dfrac{1}{\sqrt{2}}(-1, 1, 0) \right\}$

‖ Exercise 6-6

A   In Problems 1–8, let R^2 be equipped with the standard inner product. Classify each set as nonorthogonal, orthogonal, or orthonormal.

1. $\{(0, 0), (1, 0)\}$
2. $\{(0, 0), (1, 0), (0, 1)\}$
3. $\{(1, 0)\}$
4. $\{(1, 0), (0, 1)\}$
5. $\{(5, 6), (6, -5)\}$
6. $\{(5, 6), (-6, -5)\}$

7. $\left\{ \left(\dfrac{1}{\sqrt{2}}, -\dfrac{1}{\sqrt{2}} \right), \left(-\dfrac{1}{\sqrt{2}}, \dfrac{1}{\sqrt{2}} \right) \right\}$

8. $\left\{ \left(\dfrac{1}{\sqrt{3}}, \dfrac{1}{\sqrt{3}} \right), \left(-\dfrac{1}{\sqrt{3}}, \dfrac{1}{\sqrt{3}} \right) \right\}$

In Problems 9 and 10, let R² be equipped with the standard inner product.

9. Let $W = \{(3, 4), (4, -3)\}$.

(A) Show that W is an orthogonal basis for R^2.
(B) Normalize the vectors in W to produce an orthonormal basis for R^2.
(C) Express $\mathbf{u} = (1, 2)$ in terms of these orthonormal basis vectors.

10. Repeat Problem 9 for $W = \{(-5, -12), (12, -5)\}$.

In Problems 11 and 12, let P_1 be equipped with the standard inner product.

11. Let $W = \{1 + x, 1 - x\}$.

(A) Show that W is an orthogonal basis for P_1.
(B) Normalize the vectors in W to produce an orthonormal basis for P_1.
(C) Express $\mathbf{u} = 2 + 3x$ in terms of these orthonormal basis vectors.

12. Repeat Problem 11 for $W = \{1 - 2x, 2 + x\}$.

B *In Problems 13–16, let R^3 be equipped with the standard inner product. Classify each set as nonorthogonal, orthogonal, or orthonormal.*

13. $\left\{ \left(\dfrac{1}{\sqrt{2}}, 0, \dfrac{1}{\sqrt{2}} \right), (0, 1, 0), \left(\dfrac{1}{\sqrt{2}}, 0, -\dfrac{1}{\sqrt{2}} \right) \right\}$

14. $\left\{ (1, 0, 1), \left(-\dfrac{1}{\sqrt{5}}, \dfrac{2}{\sqrt{5}}, \dfrac{1}{\sqrt{5}} \right), \left(\dfrac{1}{\sqrt{3}}, \dfrac{1}{\sqrt{3}}, -\dfrac{1}{\sqrt{3}} \right) \right\}$

15. $\left\{ \left(-\dfrac{7}{11}, -\dfrac{6}{11}, \dfrac{6}{11} \right), \left(-\dfrac{6}{11}, \dfrac{2}{11}, -\dfrac{9}{11} \right), \left(-\dfrac{6}{11}, \dfrac{9}{11}, \dfrac{2}{11} \right) \right\}$

16. $\left\{ \left(\dfrac{7}{11}, \dfrac{6}{11}, -\dfrac{6}{11} \right), \left(\dfrac{6}{11}, \dfrac{2}{11}, \dfrac{9}{11} \right), \left(\dfrac{6}{11}, -\dfrac{9}{11}, -\dfrac{2}{11} \right) \right\}$

In Problems 17–22, let the indicated vector space be equipped with its standard inner product. Show that W is an orthogonal basis for the space, normalize W to obtain an orthonormal basis, and express $\mathbf{u}$ as a linear combination of basis vectors.

17. $W = \{(1, 2, 2), (-2, -1, 2), (2, -2, 1)\} \subset R^3$; $\mathbf{u} = (3, -4, 5)$
18. $W = \{(6, 3, 2), (-3, 2, 6), (2, -6, 3)\} \subset R^3$; $\mathbf{u} = (2, -1, 3)$
19. $W = \{1 + x^2, 1 + x - x^2, 1 - 2x - x^2\} \subset P_2$; $\mathbf{u} = 2 + 3x - 6x^2$
20. $W = \{1 - 4x + x^2, 1 - x^2, 2 + x + 2x^2\} \subset P_2$; $\mathbf{u} = 1 + 2x + 3x^2$

21. $W = \left\{ \begin{bmatrix} 1 & 1 \\ 1 & 1 \end{bmatrix}, \begin{bmatrix} 1 & 1 \\ 3 & -5 \end{bmatrix}, \begin{bmatrix} -5 & 1 \\ 3 & 1 \end{bmatrix}, \begin{bmatrix} 1 & -5 \\ 3 & 1 \end{bmatrix} \right\} \subset M_{2\times2}$; $\mathbf{u} = \begin{bmatrix} 2 & -1 \\ 0 & 1 \end{bmatrix}$

22. $W = \left\{ \begin{bmatrix} 1 & 4 \\ 4 & 4 \end{bmatrix}, \begin{bmatrix} 4 & 2 \\ 2 & -5 \end{bmatrix}, \begin{bmatrix} 4 & -5 \\ 2 & 2 \end{bmatrix}, \begin{bmatrix} 4 & 2 \\ -5 & 2 \end{bmatrix} \right\} \subset M_{2\times2}$; $\mathbf{u} = \begin{bmatrix} -2 & 0 \\ 1 & 3 \end{bmatrix}$

In Problems 23–26, R^n is equipped with its standard inner product and S is a linearly independent subset of R^n. Apply the Gram–Schmidt process to S in order to obtain an orthonormal basis for span S.

23. $S = \{(1, 2, 2), (1, 1, 3)\} \subset R^3$

24. $S = \{(2, 3, 6), (2, 1, 7)\} \subset R^3$

25. $S = \{(1, 1, 0, 1), (1, 0, 1, 2), (1, 1, 3, 1)\} \subset R^4$

26. $S = \{(1, 0, 0, 1), (1, 1, 1, 1), (1, -1, -3, 3)\} \subset R^4$

In Problems 27–34, R^n is equipped with the standard inner product and B is a basis for R^n. Apply the Gram–Schmidt process to B in order to find an orthonormal basis for R^n.

27. $B = \{(1, 2), (1, 3)\} \subset R^2$

28. $B = \{(2, -5), (3, 7)\} \subset R^2$

29. $B = \{(1, -1, 0), (-1, 1, 1), (1, 1, -2)\} \subset R^3$

30. $B = \{(1, 0, -1), (2, -1, 0), (2, 2, 0)\} \subset R^3$

31. $B = \{(0, 1, -2), (1, 0, 1), (1, 1, 1)\} \subset R^3$

32. $B = \{(1, 1, 1), (1, 0, 1), (0, 1, -2)\} \subset R^3$

33. $B = \{(1, -1, 1, -1), (2, 0, 1, -1), (1, 3, -4, 0), (9, 0, 0, 0)\} \subset R^4$

34. $B = \{(1, 1, -1, 1), (2, 1, 0, 1), (1, -1, 3, 1), (2, 1, -3, 2)\} \subset R^4$

In Problems 35 and 36, P_2 is equipped with its standard inner product and B is a basis for P_2. Apply the Gram–Schmidt process to B in order to obtain an orthonormal basis for P_2.

35. $B = \{1 + x + x^2, 1 + 2x, 3\}$ **36.** $B = \{1 + x^2, 1 + 3x^2, x - 2x^2\}$

In Problems 37 and 38, $M_{2\times 2}$ is equipped with its standard inner product and B is a basis for $M_{2\times 2}$. Apply the Gram–Schmidt process to B in order to obtain an orthonormal basis for $M_{2\times 2}$.

37. $B = \left\{ \begin{bmatrix} 1 & 0 \\ 0 & 1 \end{bmatrix}, \begin{bmatrix} 0 & 1 \\ 0 & 1 \end{bmatrix}, \begin{bmatrix} 0 & 0 \\ 1 & 1 \end{bmatrix}, \begin{bmatrix} 0 & 0 \\ 0 & 1 \end{bmatrix} \right\}$

38. $B = \left\{ \begin{bmatrix} 1 & 1 \\ 1 & 1 \end{bmatrix}, \begin{bmatrix} 0 & 1 \\ 0 & 1 \end{bmatrix}, \begin{bmatrix} 0 & 0 \\ 1 & 1 \end{bmatrix}, \begin{bmatrix} 0 & 0 \\ 0 & 1 \end{bmatrix} \right\}$

C   In Problems 39–42, P_n is equipped with the inner product

$$\langle p, q \rangle = \int_{-1}^{1} p(x)\, q(x)\, dx$$

39. *Calculus.* Apply the Gram–Schmidt process to $\{1, x, x^2\}$ in order to obtain an orthonormal basis for P_2. These orthonormal polynomials are called the **Legendre polynomials** and are used in scientific and engineering applications.

40. *Calculus.* Apply the Gram–Schmidt process to $\{1, x, x^2, x^3\}$ in order to obtain an orthonormal basis for P_3. [*Hint:* The first three functions in this orthonormal basis are the same as those computed in Problem 39.]

41. *Calculus.* Express $p(x) = x^2$ as a linear combination of the orthonormal polynomials computed in Problem 39. Use Theorem 20 to compute the coefficients in this linear combination.

42. *Calculus.* Repeat Problem 41 for $p(x) = 1 + x$.

In Problems 43 and 44, let V be an inner product space with orthonormal basis $B = \{\mathbf{v}_1, \mathbf{v}_2, \ldots, \mathbf{v}_n\}$, and let $\mathbf{u}$ and $\mathbf{w}$ be arbitrary vectors in V. Prove each statement.

43. $\|\mathbf{w}\|^2 = \langle \mathbf{w}, \mathbf{v}_1 \rangle^2 + \langle \mathbf{w}, \mathbf{v}_2 \rangle^2 + \cdots + \langle \mathbf{w}, \mathbf{v}_n \rangle^2$

44. $\langle \mathbf{u}, \mathbf{w} \rangle = \langle \mathbf{u}, \mathbf{v}_1 \rangle \langle \mathbf{w}, \mathbf{v}_1 \rangle + \langle \mathbf{u}, \mathbf{v}_2 \rangle \langle \mathbf{w}, \mathbf{v}_2 \rangle + \cdots + \langle \mathbf{u}, \mathbf{v}_n \rangle \langle \mathbf{w}, \mathbf{v}_n \rangle$

In Problems 45 and 46, let $\{\mathbf{u}_1, \mathbf{u}_2, \mathbf{u}_3\}$ be a linearly independent set of vectors in an inner product space V. (These problems are related to the proof of the Gram-Schmidt process in Theorems 21 and 22. These theorems are usually proved simultaneously, using techniques we have not discussed in this book.)

45. Let $\mathbf{w}_1 = \mathbf{u}_1$ and $\mathbf{w}_2 = \mathbf{u}_2 - k_1\mathbf{u}_1$.

    (A) Show that $\mathbf{w}_2 \neq 0$ for any value of k_1.

    (B) Show that $\operatorname{span}\{\mathbf{w}_1, \mathbf{w}_2\} = \operatorname{span}\{\mathbf{u}_1, \mathbf{u}_2\}$.

    (C) Show that if $k_1 = \langle \mathbf{u}_2, \mathbf{w}_1 \rangle / \langle \mathbf{w}_1, \mathbf{w}_1 \rangle$, then $\langle \mathbf{w}_2, \mathbf{w}_1 \rangle = 0$.

46. Let $\mathbf{w}_1 = \mathbf{u}_1$, $\mathbf{w}_2 = \mathbf{u}_2 - (\langle \mathbf{u}_2, \mathbf{w}_1 \rangle / \langle \mathbf{w}_1, \mathbf{w}_1 \rangle)\mathbf{w}_1$, and $\mathbf{w}_3 = \mathbf{u}_3 - k_1\mathbf{u}_1 - k_2\mathbf{u}_2$.

    (A) Show that $\mathbf{w}_3 \neq 0$ for any values of k_1 and k_2.

    (B) Show that $\operatorname{span}\{\mathbf{w}_1, \mathbf{w}_2, \mathbf{w}_3\} = \operatorname{span}\{\mathbf{u}_1, \mathbf{u}_2, \mathbf{u}_3\}$.

    (C) Show that if $k_1 = \langle \mathbf{u}_3, \mathbf{w}_1 \rangle / \langle \mathbf{w}_1, \mathbf{w}_1 \rangle$, then $\langle \mathbf{w}_3, \mathbf{w}_1 \rangle = 0$.

    (D) Show that if $k_2 = \langle \mathbf{u}_3, \mathbf{w}_2 \rangle / \langle \mathbf{w}_2, \mathbf{w}_2 \rangle$, then $\langle \mathbf{w}_3, \mathbf{w}_2 \rangle = 0$.

47. *Theorem 19.* Prove Theorem 19.

48. Let W be a subspace of a finite-dimensional inner product space V. Show that W has an orthonormal basis that is part of an orthonormal basis for V.

❘6-7❘ Chapter Review

Important Terms and Symbols

6-1. *Row and column spaces—an introduction.* Row space, column space, basis for the row space, basis for span S (method 1)

6-2. *Row and column spaces and the rank of a matrix.* Basis for the column space, basis for span S (method 2), extending a linearly independent set to a basis, rank of a matrix, fundamental theorem (final version)

6-3. *Coordinate matrices.* Coordinates and coordinate matrix with respect to a basis, ordered sets of vectors, equivalent representations of vector spaces, $[\mathbf{u}]_B$, $[S]_B$

6-4. *Change of basis.* Changing bases, transition matrix, inverse of a transition matrix, rotation of axes, xy and $x'y'$ coordinate systems

6-5. *Inner product spaces.* Inner product, inner product space, standard inner product, norm, distance between two vectors, unit vector, Cauchy–Schwarz inequality, triangle inequality, angle between two vectors, orthogonal vectors, $\langle \mathbf{u}, \mathbf{v} \rangle$, $\|\mathbf{u}\|$, $d(\mathbf{u}, \mathbf{v})$

6-6. *Orthogonal bases.* Orthogonal set, orthonormal set, normalizing a set of vectors, Gram–Schmidt orthogonalization process

▌ Exercise 6-7   Chapter Review

Work through all the problems in this chapter review and check your answers in the back of the book. (Answers to most review problems are there.) Where weaknesses show up, review appropriate sections in the text.

A   *Problems 1–6 refer to the matrices*

$$A = \begin{bmatrix} 1 & 1 & 2 & -2 \\ 2 & 1 & 1 & 0 \\ 3 & 2 & 3 & -2 \end{bmatrix} \qquad B = \begin{bmatrix} 1 & 0 & -1 & 2 \\ 0 & 1 & 3 & -4 \\ 0 & 0 & 0 & 0 \end{bmatrix}$$

where B is the reduced form of A.

1. Find a basis for the row space of B.
2. Find a basis for the row space of A.
3. Find a basis for the column space of B.
4. Find a basis for the column space of A.
5. Find the rank of B.
6. Find the rank of A.

In Problems 7–12, $B = \{(4, 3), (7, 5)\}$ and $C = \{(1, 0), (0, 1)\}$ are bases for R^2.

7. If $\mathbf{u} = (1, 1)$, find $[\mathbf{u}]_B$.
8. If $[\mathbf{u}]_B = \begin{bmatrix} 6 \\ -2 \end{bmatrix}$, find $\mathbf{u}$.
9. Find P, the transition matrix from B to C.
10. If $[\mathbf{u}]_B = \begin{bmatrix} 4 \\ -1 \end{bmatrix}$, use P to find $[\mathbf{u}]_C$.
11. Find P^{-1}, the transition matrix from C to B.
12. If $[\mathbf{u}]_C = \begin{bmatrix} 3 \\ 2 \end{bmatrix}$, use P^{-1} to find $[\mathbf{u}]_B$.

In Problems 13–18, let R² be equipped with the standard inner product and let

$$\mathbf{u} = (2, -1) \qquad \mathbf{v} = (6, 2)$$

Find each of the following:

13. $\langle \mathbf{u}, \mathbf{v} \rangle$ **14.** $\|\mathbf{u}\|$ **15.** $\|\mathbf{v}\|$ **16.** $d(\mathbf{u}, \mathbf{v})$

17. The angle between $\mathbf{u}$ and $\mathbf{v}$

18. The orthonormal basis obtained by applying the Gram–Schmidt process to $\{\mathbf{u}, \mathbf{v}\}$

B **19.** Let

$$A = \begin{bmatrix} 1 & -1 & 1 & 5 \\ 1 & -1 & -1 & -1 \\ 3 & -3 & -1 & 3 \end{bmatrix}$$

(A) Find a basis for the row space of A.

(B) Find a basis for the column space of A.

(C) Find the rank of A.

20. Let $S = \{\mathbf{u}_1, \mathbf{u}_2, \mathbf{u}_3\}$ where $\mathbf{u}_1 = (1, 2, -1)$, $\mathbf{u}_2 = (-1, 3, 6)$, and $\mathbf{u}_3 = (1, 7, 4)$ are vectors in R^3.

(A) Use method 1 to find a basis for span S.

(B) Express each vector in S as a linear combination of the basis vectors from part (A).

(C) Use method 2 to find a basis for span S consisting of vectors in S.

(D) Express the remaining vectors in S as a linear combination of the basis vectors from part (C).

21. Let $S = \{1 + x + x^2, \ 1 - x - 3x^2, \ 3 + x - x^2\}$ be a subset of P_2.

(A) Use method 1 to find a basis for span S.

(B) Use method 2 to find a basis for span S consisting of vectors in S.

22. Repeat Problem 21 for the following subset of $M_{2\times2}$:

$$S = \left\{ \begin{bmatrix} 1 & 1 \\ 2 & -1 \end{bmatrix}, \begin{bmatrix} -2 & -2 \\ -4 & 2 \end{bmatrix}, \begin{bmatrix} 2 & 3 \\ 7 & 4 \end{bmatrix}, \begin{bmatrix} 1 & 2 \\ 5 & 5 \end{bmatrix} \right\}$$

In Problems 23 and 24, use Theorem 10 to determine whether S is a linearly independent or linearly dependent subset of the indicated vector space.

23. $S = \{2 - x + 3x^2, \ 4 + x + 2x^2, \ -2 - 2x + x^2\} \subset P_2$

24. $S = \left\{ \begin{bmatrix} 1 & -1 \\ 0 & 0 \end{bmatrix}, \begin{bmatrix} 0 & 1 \\ -1 & 0 \end{bmatrix}, \begin{bmatrix} 0 & 0 \\ 1 & -1 \end{bmatrix}, \begin{bmatrix} 0 & 0 \\ 0 & 1 \end{bmatrix} \right\} \subset M_{2\times2}$

In Problems 25–28, S is a linearly independent subset of the indicated vector space. Extend S to a basis for the vector space.

25. $S = \{(1, -1, 1), (1, 1, -1)\} \subset R^3$

26. $S = \{(1, -1, 0, 0), (-1, 2, 1, 0)\} \subset R^4$

27. $S = \{1 - x + x^2 - 2x^3, -3 + 2x - 2x^2 + 4x^3\} \subset P_3$

28. $S = \left\{ \begin{bmatrix} 1 & -1 \\ 0 & 0 \end{bmatrix}, \begin{bmatrix} 1 & 1 \\ 0 & 0 \end{bmatrix}, \begin{bmatrix} 1 & -1 \\ 0 & 1 \end{bmatrix} \right\} \subset M_{2 \times 2}$

29. Let $B = \{(1, 2, 0), (2, 0, -1), (1, -1, -1)\}$ and $C = \{(1, 0, 3), (0, 1, -2), (1, 1, 2)\}$ be bases for R^3.

   (A) If $\mathbf{u} = (1, 2, 1)$, find $[\mathbf{u}]_B$.

   (B) If $[\mathbf{v}]_B = \begin{bmatrix} 5 \\ -3 \\ 2 \end{bmatrix}$, find $\mathbf{v}$.

   (C) Find P, the transition matrix from B to C.

   (D) If $\mathbf{u} = (1, 2, 1)$, use P and $[\mathbf{u}]_B$ from part (A) to find $[\mathbf{u}]_C$.

30. Let $B = \{(1, 2, 0), (2, 0, -1), (1, -1, -1)\}$ and $C = \{(1, 0, 0), (0, 1, 0), (0, 0, 1)\}$ be bases for R^3.

   (A) Find P, the transition matrix from B to C.

   (B) Find P^{-1}, the transition matrix from C to B.

   (C) Use P^{-1} to find $[\mathbf{u}]_B$ for $\mathbf{u} = (1, 2, 1)$.

31. Let $B = \{3 + 2x + x^2, 1 + x + x^2, 2 - x^2\}$ and $C = \{1, x, x^2\}$ be bases for P_2, and let $\mathbf{u} = 1 + x$.

   (A) Find $[\mathbf{u}]_B$.

   (B) Find P, the transition matrix from B to C.

   (C) Find P^{-1}, the transition matrix from C to B.

   (D) Use P^{-1} to find $[\mathbf{u}]_B$.

32. Let

$$B = \left\{ \begin{bmatrix} 1 & 0 \\ 1 & 0 \end{bmatrix}, \begin{bmatrix} 0 & -1 \\ 0 & -1 \end{bmatrix}, \begin{bmatrix} 1 & 0 \\ 0 & 1 \end{bmatrix}, \begin{bmatrix} 1 & 1 \\ 1 & 0 \end{bmatrix} \right\}$$

and

$$C = \left\{ \begin{bmatrix} 1 & 0 \\ 0 & 0 \end{bmatrix}, \begin{bmatrix} 0 & 1 \\ 0 & 0 \end{bmatrix}, \begin{bmatrix} 0 & 0 \\ 1 & 0 \end{bmatrix}, \begin{bmatrix} 0 & 0 \\ 0 & 1 \end{bmatrix} \right\}$$

be bases for $M_{2 \times 2}$, and let

$$\mathbf{u} = \begin{bmatrix} 8 & 2 \\ 5 & 1 \end{bmatrix}$$

   (A) Find $[\mathbf{u}]_B$.

   (B) Find P, the transition matrix from B to C.

   (C) Find P^{-1}, the transition matrix from C to B.

   (D) Use P^{-1} to find $[\mathbf{u}]_B$.

33. Let P_2 be equipped with the standard inner product, and let $\mathbf{u} = 2 - x + x^2$ and $\mathbf{v} = 3 - 4x + 5x^2$. Find each of the following:

(A) $\langle \mathbf{u}, \mathbf{v} \rangle$ (B) $\|\mathbf{u}\|$ (C) $\|\mathbf{v}\|$ (D) $d(\mathbf{u}, \mathbf{v})$
(E) The angle between $\mathbf{u}$ and $\mathbf{v}$

34. Let $M_{2\times 2}$ be equipped with the standard inner product, and let

$$\mathbf{u} = \begin{bmatrix} 1 & 1 \\ -1 & 1 \end{bmatrix} \qquad \mathbf{v} = \begin{bmatrix} 2 & 3 \\ -1 & 4 \end{bmatrix}$$

(A) Find a unit vector in the direction of $\mathbf{u}$.
(B) Find all vectors in $M_{2\times 2}$ that are orthogonal to both $\mathbf{u}$ and $\mathbf{v}$.

35. An $x'y'$ coordinate system is formed by rotating an xy coordinate system through an angle $\theta = 5\pi/6$.

(A) Find the transition matrix from the $x'y'$ system to the xy system.
(B) Find the transition matrix from the xy system to the $x'y'$ system.
(C) Find (x, y) if $(x', y') = (2, -4)$.
(D) Find (x', y') if $(x, y) = (-6, 4)$.

In Problems 36–38, R^n is equipped with the standard inner product and S is a linearly independent subset of R^n. Use the Gram–Schmidt process to find an orthonormal basis for span S.

36. $S = \{(-1, 2, -2), (2, -5, 3)\} \subset R^3$
37. $S = \{(1, 1, -1, -1), (0, 1, 0, 1), (2, -1, 1, 0)\} \subset R^4$
38. $S = \{(1, 0, 0, 0, -1), (0, 1, 0, 0, -1), (0, 0, 1, 0, -1), (0, 0, 0, 1, -1)\} \subset R^5$
39. Let P_2 be equipped with the standard inner product, and let

$$B = \{1 - x^2, x^2, 1 + x + x^2\}$$

be a basis for P_2. Apply the Gram–Schmidt process to B in order to obtain an orthonormal basis for P_2.

40. Let $M_{2\times 2}$ be equipped with the standard inner product, and let

$$B = \left\{ \begin{bmatrix} 1 & 1 \\ 1 & 1 \end{bmatrix}, \begin{bmatrix} 1 & 0 \\ 0 & 1 \end{bmatrix}, \begin{bmatrix} 0 & 1 \\ 0 & 1 \end{bmatrix}, \begin{bmatrix} 0 & 0 \\ 1 & 1 \end{bmatrix} \right\}$$

be a basis for $M_{2\times 2}$. Apply the Gram–Schmidt process to B in order to obtain an orthonormal basis for $M_{2\times 2}$.

C In Problems 41–43, let P_2 be equipped with the inner product defined by

$$\langle p, q \rangle = \int_0^1 x\, p(x)\, q(x)\, dx$$

41. *Calculus.* If $p(x) = x$ and $q(x) = x^2$, find:

(A) $\langle p, q \rangle$ (B) $\|p\|$ (C) $\|q\|$

42. *Calculus.* Determine the value of k for which $p(x) = x$ and $q(x) = 1 + kx$ are orthogonal.

43. *Calculus.* Apply the Gram–Schmidt process to $B = \{1, x, x^2\}$ in order to obtain an orthonormal basis for P_2.

44. If A is an invertible $n \times n$ matrix, show that the rows of A form a basis for R^n.

45. *Laguerre polynomials.* Let P_3 be equipped with the standard inner product, let

$$B = \{1, 1 - x, 2 - 4x + x^2, 6 - 18x + 9x^2 - x^3\}$$

be the basis of Laguerre polynomials for P_3, and let $C = \{1, x, x^2, x^3\}$ be the standard basis for P_3.

(A) Find P, the transition matrix from B to C.

(B) Find P^{-1}, the transition matrix from C to B.

(C) Use P^{-1} to express $p(x) = 1 + x + x^2 + x^3$ as a linear combination of Laguerre polynomials.

Problems 46–49 refer to the function

$$\langle \mathbf{u}, \mathbf{v} \rangle = 2u_1v_1 + u_1v_2 + u_2v_1 + u_2v_2$$

where $\mathbf{u} = (u_1, u_2)$ *and* $\mathbf{v} = (v_1, v_2)$ *are vectors in* R^2.

46. Show that $\langle \mathbf{u}, \mathbf{v} \rangle$ is an inner product on R^2 by showing that $\langle \mathbf{u}, \mathbf{v} \rangle$ satisfies each condition in the definition of an inner product.

47. Let $B = \{(0, 1), (1, -1)\}$ be a basis for R^2.

(A) If $\mathbf{u} = (u_1, u_2)$, find $[\mathbf{u}]_B$.

(B) If $\mathbf{u} = (u_1, u_2)$ and $\mathbf{v} = (v_1, v_2)$, show that

$$\langle \mathbf{u}, \mathbf{v} \rangle = [\mathbf{u}]_B^T[\mathbf{v}]_B$$

[Notice that Theorem 14 and part (B) provide an alternate method for proving that $\langle \mathbf{u}, \mathbf{v} \rangle$ is an inner product.]

48. If $\mathbf{u} = (1, -1)$ and $\mathbf{v} = (3, 1)$, find:

(A) $\langle \mathbf{u}, \mathbf{v} \rangle$ (B) $\|\mathbf{u}\|$

(C) A unit vector in the direction of $\mathbf{v}$

49. Find all vectors in R^2 that are orthogonal to $\mathbf{u} = (1, -2)$.

50. Let A be a 2×2 matrix and let R^2 be equipped with the standard inner product. Show that the rows of A form an orthonormal set in R^2 if and only if $AA^T = I$.

51. Repeat Problem 50 for a 3×3 matrix A.

|7| Eigenvalues and Eigenvectors

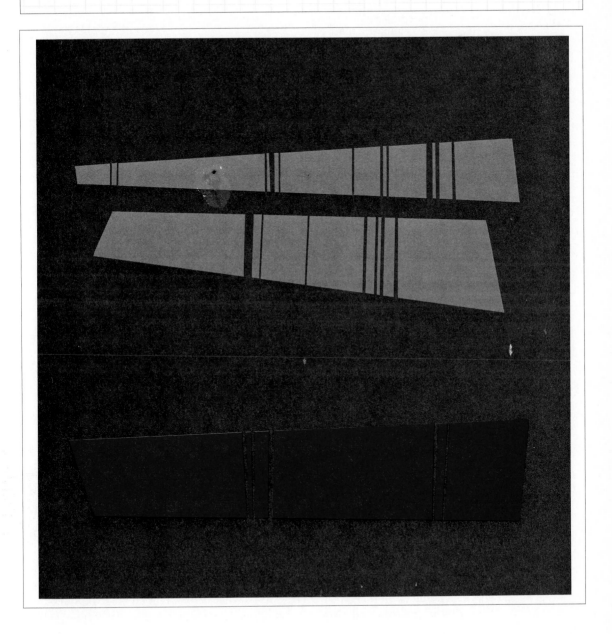

| 7 | Contents

In earlier chapters, we have studied the relationship between a matrix A and its reduced form B, which is obtained by performing elementary row operations on A. Now we want to establish a relationship between certain square matrices and corresponding diagonal matrices. This relationship has far-reaching implications and, in particular, leads to solutions of a wide variety of applied problems. (We will consider some of these applications in Chapter 9.) Elementary row operations will not be used in establishing the relationship between a square matrix A and a diagonal matrix D. Instead, the key idea turns out to be the determination of nonzero vectors $\mathbf{x}$ for which $A\mathbf{x}$ and $\mathbf{x}$ are parallel; that is, $A\mathbf{x}$ is a scalar multiple of $\mathbf{x}$.

| 7-1 | Introduction

- Eigenvalues and Eigenvectors
- Finding Eigenvalues
- Finding Eigenspaces

• Eigenvalues and Eigenvectors

We begin with an example that will lead to some general observations.

Example 1   Find all vectors $\mathbf{x}$ in R^2 for which $A\mathbf{x}$ is a scalar multiple of $\mathbf{x}$ if

$$A = \begin{bmatrix} 1 & -1 \\ 2 & 4 \end{bmatrix}$$

Solution   In order for $A\mathbf{x}$ to be a scalar multiple of $\mathbf{x}$, the vector $\mathbf{x}$ must satisfy the equation

$$A\mathbf{x} = \lambda\mathbf{x} \tag{1}$$

for some scalar λ (the Greek letter "lambda," which is traditionally used in problems of this type). Clearly, $\mathbf{x} = \mathbf{0}$ will satisfy (1) for any matrix A and any scalar λ, so we will restrict our attention to nonzero vectors $\mathbf{x}$. If $\mathbf{x} = (x_1, x_2)$, then (1) can be written as

$$\begin{bmatrix} 1 & -1 \\ 2 & 4 \end{bmatrix}\begin{bmatrix} x_1 \\ x_2 \end{bmatrix} = \lambda\begin{bmatrix} x_1 \\ x_2 \end{bmatrix}$$

or, equivalently, as

$$x_1 - x_2 = \lambda x_1$$
$$2x_1 + 4x_2 = \lambda x_2$$

or (collecting like terms) as

$$(1 - \lambda)x_1 - x_2 = 0$$
$$2x_1 + (4 - \lambda)x_2 = 0$$

(2)

System (2) will have nontrivial solutions if and only if the determinant of the coefficient matrix is zero (Theorem 10, Section 3-3); that is,

$$\begin{vmatrix} 1 - \lambda & -1 \\ 2 & 4 - \lambda \end{vmatrix} = 0$$
$$(1 - \lambda)(4 - \lambda) + 2 = 0$$
$$\lambda^2 - 5\lambda + 6 = 0$$
$$(\lambda - 2)(\lambda - 3) = 0$$
$$\lambda = 2 \quad \text{or} \quad \lambda = 3$$

Thus, the only scalars for which (2), and consequently (1), will have nontrivial solutions are $\lambda = 2$ and $\lambda = 3$. These scalars are called the *eigenvalues* of A. Now we want to find the nontrivial solutions of (2) corresponding to each eigenvalue.

$\lambda = 2$: Substituting $\lambda = 2$ in (2), we have

$$-x_1 - x_2 = 0$$
$$2x_1 + 2x_2 = 0$$

The nontrivial solutions of this system are given by (verify this)

$$\left\{ \begin{bmatrix} -t \\ t \end{bmatrix} \middle| t \text{ any nonzero real number} \right\}$$

The vectors in this set are called the *eigenvectors of A associated with the eigenvalue* $\lambda = 2$.

Check: $A\mathbf{x} = \begin{bmatrix} 1 & -1 \\ 2 & 4 \end{bmatrix} \begin{bmatrix} -t \\ t \end{bmatrix} = \begin{bmatrix} -2t \\ 2t \end{bmatrix} = 2 \begin{bmatrix} -t \\ t \end{bmatrix} = 2\mathbf{x}$

$\lambda = 3$: Substituting $\lambda = 3$ in (2), we have

$$-2x_1 - x_2 = 0$$
$$2x_1 + x_2 = 0$$

The nontrivial solutions of this system are the eigenvectors of A associated with the eigenvalue $\lambda = 3$ and are given by (verify this)

$$\left\{ \begin{bmatrix} -\tfrac{1}{2}t \\ t \end{bmatrix} \middle| t \text{ any nonzero real number} \right\}$$

$$\text{Check:} \quad A\mathbf{x} = \begin{bmatrix} 1 & -1 \\ 2 & 4 \end{bmatrix} \begin{bmatrix} -\frac{1}{2}t \\ t \end{bmatrix} = \begin{bmatrix} -\frac{3}{2}t \\ 3t \end{bmatrix} = 3 \begin{bmatrix} -\frac{1}{2}t \\ t \end{bmatrix} = \lambda\mathbf{x}$$

Problem 1   Repeat Example 1 for $A = \begin{bmatrix} 3 & -1 \\ 1 & 1 \end{bmatrix}$.   ▌▌

We now present a formal definition of the concepts introduced in Example 1.

Eigenvalues and Eigenvectors

Let A be a square matrix of order n. The real number λ is called an **eigenvalue** of A if there exists a nonzero vector $\mathbf{x}$ in R^n satisfying

$$A\mathbf{x} = \lambda\mathbf{x} \tag{3}$$

Any nonzero vector $\mathbf{x}$ that satisfies (3) is called an **eigenvector of A associated with the eigenvalue λ.**

In some texts, eigenvalues are also referred to as *proper values, characteristic values,* or *latent values;* and the corresponding eigenvectors are called *proper vectors, characteristic vectors,* or *latent vectors.* We will not use these terms.

Equation (3), $A\mathbf{x} = \lambda\mathbf{x}$, can be written in a more useful equivalent form by introducing the $n \times n$ identity matrix I. If we substitute $\lambda\mathbf{x} = \lambda I\mathbf{x}$ in (3), we have

$$A\mathbf{x} = \lambda I\mathbf{x}$$
$$(\lambda I - A)\mathbf{x} = \mathbf{0}$$

Thus, λ is a eigenvalue of A if and only if this homogeneous system has nontrivial solutions. We have just proved the following theorem:

Theorem 1   The real number λ is an eigenvalue of the $n \times n$ matrix A if and only if the homogeneous system

$$(\lambda I - A)\mathbf{x} = \mathbf{0} \tag{4}$$

has nontrivial solutions. Furthermore, the nontrivial solutions of (4) are the eigenvectors of A associated with the eigenvalue λ.

As we saw in Example 1, Theorem 10 in Section 3-3 implies that λ is an eigenvalue of A if and only if the determinant of the coefficient matrix in (4) is zero. This gives us Theorem 2.

Theorem 2

> The real number λ is an eigenvalue of A if and only if λ satisfies the equation
>
> $$\det(\lambda I - A) = 0 \qquad\qquad (5)$$

Equation (5) is called the **characteristic equation** for A. If A is a square matrix of order n, then expanding the determinant in (5) will always produce an nth-degree polynomial

$$p(\lambda) = \det(\lambda I - A)$$

which is called the **characteristic polynomial** for A.

Example 2 Find the characteristic polynomial for each of the following matrices:

(A) $A = \begin{bmatrix} 1 & -1 \\ 2 & 4 \end{bmatrix}$ (B) $A = \begin{bmatrix} -1 & 0 & 1 \\ 1 & -1 & -1 \\ -3 & 2 & 3 \end{bmatrix}$

Solution (A) $\lambda I - A = \begin{bmatrix} \lambda & 0 \\ 0 & \lambda \end{bmatrix} - \begin{bmatrix} 1 & -1 \\ 2 & 4 \end{bmatrix} = \begin{bmatrix} \lambda - 1 & 1 \\ -2 & \lambda - 4 \end{bmatrix}$

$p(\lambda) = \det(\lambda I - A) = \begin{vmatrix} \lambda - 1 & 1 \\ -2 & \lambda - 4 \end{vmatrix}$

$\qquad = (\lambda - 1)(\lambda - 4) + 2$

$\qquad = \lambda^2 - 5\lambda + 6 \qquad$ Compare with Example 1.

(B) $\lambda I - A = \begin{bmatrix} \lambda & 0 & 0 \\ 0 & \lambda & 0 \\ 0 & 0 & \lambda \end{bmatrix} - \begin{bmatrix} -1 & 0 & 1 \\ 1 & -1 & -1 \\ -3 & 2 & 3 \end{bmatrix} = \begin{bmatrix} \lambda + 1 & 0 & -1 \\ -1 & \lambda + 1 & 1 \\ 3 & -2 & \lambda - 3 \end{bmatrix}$

$p(\lambda) = \det(\lambda I - A) = \begin{vmatrix} \lambda + 1 & 0 & -1 \\ -1 & \lambda + 1 & 1 \\ 3 & -2 & \lambda - 3 \end{vmatrix}$ Expand along the first row.

$\qquad = (\lambda + 1)\begin{vmatrix} \lambda + 1 & 1 \\ -2 & \lambda - 3 \end{vmatrix} - \begin{vmatrix} -1 & \lambda + 1 \\ 3 & -2 \end{vmatrix}$

$\qquad = (\lambda + 1)[(\lambda + 1)(\lambda - 3) + 2] - [2 - 3(\lambda + 1)]$

$\qquad = \lambda^3 - \lambda^2$

Problem 2 Find the characteristic polynomial for each matrix.

(A) $A = \begin{bmatrix} 3 & -1 \\ 1 & 1 \end{bmatrix}$ (B) $A = \begin{bmatrix} 1 & 2 & 0 \\ 2 & 0 & 2 \\ 0 & 2 & -1 \end{bmatrix}$ ∎

How is the characteristic polynomial for a matrix A related to the eigenvalues of A? Theorem 2 implies that the *real roots r* of the characteristic polynomial are the eigenvalues of A. (In this text, we are interested only in real eigenvalues. Complex eigenvalues will not be discussed.) Also, recall that r is a root of $p(\lambda)$ if and only if $(\lambda - r)$ is a factor of $p(\lambda)$. If $(\lambda - r)^k$ is a factor of $p(\lambda)$ for a positive integer k, then r is called a **root of multiplicity k.** Roots of multiplicity 1 are referred to as *simple roots*, roots of multiplicity 2 are referred to as *double roots*, and so on. We will also apply these terms to eigenvalues. That is, λ is an eigenvalue of multiplicity k if λ is a root of multiplicity k of the characteristic polynomial. Finally, we recall that the fundamental theorem of algebra implies that an nth-degree polynomial can have at most n real roots. Thus, *an $n \times n$ matrix A can have at most n eigenvalues.*

Theorem 3 summarizes the various characterizations of eigenvalues we have discussed. The proof is omitted.

Theorem 3

> Let A be an $n \times n$ matrix. Then the real number λ is an eigenvalue of A if it satisfies any of the following:
>
> (A) $A\mathbf{x} = \lambda\mathbf{x}$ for a nonzero vector $\mathbf{x}$ in R^n.
> (B) The system $(\lambda I - A)\mathbf{x} = \mathbf{0}$ has nontrivial solutions.
> (C) λ is a solution of the characteristic equation $\det(\lambda I - A) = 0$.
> (D) λ is a root of the characteristic polynomial $p(\lambda) = \det(\lambda I - A)$.

▪ Finding Eigenvalues

The characteristic polynomial for an $n \times n$ matrix A has the form

$$p(\lambda) = \lambda^n + a_{n-1}\lambda^{n-1} + \cdots + a_0$$

In order to find the eigenvalues of A, we must find all the real roots of $p(\lambda)$. For large values of n, this can be a difficult task. If $n > 4$, numerical techniques usually are used to find the eigenvalues. (See the computer supplement described in the Preface.) For the matrices we will consider, the roots of the characteristic polynomial will be determined by using factoring techniques and the quadratic formula. The following theorem from algebra, which we state without proof, will be useful in factoring $p(\lambda)$:

Theorem 4

> If the coefficients $a_{n-1}, a_{n-2}, \ldots, a_0$ of the characteristic polynomial
>
> $$p(\lambda) = \lambda^n + a_{n-1}\lambda^{n-1} + a_{n-2}\lambda^{n-2} + \cdots + a_0$$
>
> are all integers and if r is an integer root of $p(\lambda)$, then r must be a factor of the constant term a_0.

Note that Theorem 4 does not imply that a characteristic polynomial with integer coefficients must have any integer roots. It states only that if there are any integer roots, then they must be among the factors of the constant term a_0. We have carefully selected the examples and exercises in this book so that there will always be a sufficient number of integer roots to enable you to factor $p(\lambda)$.

We also state the quadratic formula in Table 1 for reference.

Table 1
The Quadratic Formula

Roots of $a\lambda^2 + b\lambda + c, a \neq 0$	Discriminant	Nature of the Roots
$\lambda = \dfrac{-b \pm \sqrt{b^2 - 4ac}}{2a}$	$b^2 - 4ac > 0$	Two distinct real roots
	$b^2 - 4ac = 0$	One (double) real root
	$b^2 - 4ac < 0$	Two distinct complex roots

Example 3 Find the characteristic polynomial and the eigenvalues for each matrix.

(A) $A = \begin{bmatrix} -1 & 1 \\ -5 & 1 \end{bmatrix}$ (B) $A = \begin{bmatrix} 2 & 1 & -1 \\ 1 & -1 & -1 \\ -1 & -1 & 2 \end{bmatrix}$

Solution (A) $p(\lambda) = \det(\lambda I - A) = \begin{vmatrix} \lambda + 1 & -1 \\ 5 & \lambda - 1 \end{vmatrix} = \lambda^2 + 4$

The characteristic polynomial $p(\lambda)$ is a quadratic polynomial with discriminant

$$b^2 - 4ac = 0 - 4(1)(4) = -16 < 0$$

Thus, $p(\lambda)$ has no real roots and A has no eigenvalues. (It is possible to extend the definition of eigenvalue to include the complex roots of the characteristic polynomial. However, complex eigenvalues always involve complex vector spaces. Since we are not discussing complex vector spaces in this book, we must restrict our attention to the real roots of the characteristic polynomial.)

(B) $p(\lambda) = \det(\lambda I - A) = \begin{vmatrix} \lambda - 2 & -1 & 1 \\ -1 & \lambda + 1 & 1 \\ 1 & 1 & \lambda - 2 \end{vmatrix}$ Expand along the first row.

$$= (\lambda - 2)\begin{vmatrix} \lambda + 1 & 1 \\ 1 & \lambda - 2 \end{vmatrix} + \begin{vmatrix} -1 & 1 \\ 1 & \lambda - 2 \end{vmatrix} + \begin{vmatrix} -1 & \lambda + 1 \\ 1 & 1 \end{vmatrix}$$

$$= \lambda^3 - 3\lambda^2 - 3\lambda + 5$$

According to Theorem 4, the only possible integer roots are the integer factors of the constant term $a_0 = 5$; that is, ± 1 and ± 5. We proceed by evaluating $p(\lambda)$ at $-1, 1, -5$, and 5, stopping if we find an integer root:

$$p(-1) = -1 - 3 + 3 + 5 = 4 \neq 0 \qquad -1 \text{ is not a root}$$

$$p(1) = 1 - 3 - 3 + 5 = 0 \qquad \qquad 1 \text{ is a root}$$

Thus, 1 is a root of $p(\lambda)$ and $(\lambda - 1)$ is a factor of $p(\lambda)$. Polynomial long division can now be used to factor $p(\lambda)$:

$$
\begin{array}{r}
\lambda^2 - 2\lambda - 5 \\
\lambda - 1 \overline{)\lambda^3 - 3\lambda^2 - 3\lambda + 5} \\
\underline{\lambda^3 - \lambda^2} \\
-2\lambda^2 - 3\lambda \\
\underline{-2\lambda^2 + 2\lambda} \\
-5\lambda + 5 \\
\underline{-5\lambda + 5} \\
0
\end{array}
$$

Thus,

$$p(\lambda) = (\lambda - 1)(\lambda^2 - 2\lambda - 5)$$

Using the quadratic formula, the roots of $\lambda^2 - 2\lambda - 5$ are

$$\lambda = \frac{2 \pm \sqrt{24}}{2} = 1 \pm \sqrt{6}$$

and

$$p(\lambda) = (\lambda - 1)(\lambda - 1 - \sqrt{6})(\lambda - 1 + \sqrt{6})$$

This shows that A has three distinct eigenvalues, $\lambda_1 = 1 - \sqrt{6}$, $\lambda_2 = 1$, and $\lambda_3 = 1 + \sqrt{6}$.

Problem 3   Find the characteristic polynomial and the eigenvalues for each matrix.

(A) $\begin{bmatrix} 3 & -1 \\ 5 & -1 \end{bmatrix}$ (B) $\begin{bmatrix} 1 & 1 & 1 \\ 1 & 4 & 1 \\ 1 & 1 & 4 \end{bmatrix}$ ▌▌

▪ Finding Eigenspaces

Now that we have seen how to find the eigenvalues of a matrix, we want to turn our attention to the problem of finding the eigenvectors associated with each eigenvalue.

Eigenspace

If λ is an eigenvalue for the $n \times n$ matrix A, then the **eigenspace** associated with λ, denoted by S_λ, is defined by

$$S_\lambda = \{\mathbf{x} | A\mathbf{x} = \lambda\mathbf{x}\}$$

Since the equations $A\mathbf{x} = \lambda\mathbf{x}$ and $(\lambda I - A)\mathbf{x} = \mathbf{0}$ are equivalent, and since the solution space of a homogeneous $n \times n$ system is a subspace of R^n (Example 19, Section 5-3), it follows that S_λ is a subspace of R^n. Notice that S_λ contains the zero vector, *which is not an eigenvector.* That is,

$S_\lambda = \{$Eigenvectors associated with $\lambda\} \cup \{\mathbf{0}\}$

The zero vector is included in the eigenspace so that it is a subspace of R^n. This will allow us to apply all the theory we have developed for subspaces of a vector space to eigenspaces. In particular, we will be interested in finding a *basis B_λ for each eigenspace S_λ.*

Example 4   Find the eigenvalues of A, and then find a basis for the eigenspace associated with each eigenvalue.

$$A = \begin{bmatrix} -1 & -1 & 1 \\ -1 & -1 & 1 \\ 1 & 1 & -1 \end{bmatrix}$$

Solution $\lambda I - A = \begin{bmatrix} \lambda+1 & 1 & -1 \\ 1 & \lambda+1 & -1 \\ -1 & -1 & \lambda+1 \end{bmatrix}$ \hfill (6)

Omitting the details (which you should supply),

$p(\lambda) = \det(\lambda I - A) = \lambda^2(\lambda + 3)$

Thus, A has two distinct eigenvalues: $\lambda_1 = -3$ is a simple eigenvalue and $\lambda_2 = 0$ is an eigenvalue of multiplicity 2. (Notice that a matrix may have a zero eigenvalue, but it can never have a zero eigenvector.)

$\lambda_1 = -3$:   The eigenspace associated with $\lambda_1 = -3$ is the solution space of the homogeneous system

$$(-3I - A)\mathbf{x} = \mathbf{0} \qquad \text{Substitute } -3 \text{ for } \lambda \text{ in (6).}$$

$$\begin{bmatrix} -2 & 1 & -1 \\ 1 & -2 & -1 \\ -1 & -1 & -2 \end{bmatrix} \begin{bmatrix} x_1 \\ x_2 \\ x_3 \end{bmatrix} = \mathbf{0}$$

This system is equivalent to the following system:

$$\begin{bmatrix} 1 & 0 & 1 \\ 0 & 1 & 1 \\ 0 & 0 & 0 \end{bmatrix} \begin{bmatrix} x_1 \\ x_2 \\ x_3 \end{bmatrix} = \mathbf{0}$$

The general solution is

$$x_1 = -t, \quad x_2 = -t, \quad x_3 = t \qquad t \text{ any real number}$$

Thus, the eigenspace associated with $\lambda_1 = -3$ is

$$S_{-3} = \left\{ t \begin{bmatrix} -1 \\ -1 \\ 1 \end{bmatrix} \middle| \; t \text{ any real number} \right\}$$

and $B_{-3} = \{(-1, -1, 1)\}$ is a basis for S_{-3}.

$\lambda_2 = 0$:   Substituting 0 for λ in (6), we have

$$\begin{bmatrix} 1 & 1 & -1 \\ 1 & 1 & -1 \\ -1 & -1 & 1 \end{bmatrix} \begin{bmatrix} x_1 \\ x_2 \\ x_3 \end{bmatrix} = \mathbf{0}$$

The general solution of this system is (verify this)

$$x_1 = -s + t, \quad x_2 = s, \quad x_3 = t \qquad s, t \text{ any real numbers}$$

or

$$\mathbf{x} = \begin{bmatrix} -s + t \\ s \\ t \end{bmatrix} = s \begin{bmatrix} -1 \\ 1 \\ 0 \end{bmatrix} + t \begin{bmatrix} 1 \\ 0 \\ 1 \end{bmatrix} \qquad s, t \text{ any real numbers}$$

Thus,

$$S_0 = \text{span} \left\{ \begin{bmatrix} -1 \\ 1 \\ 0 \end{bmatrix}, \begin{bmatrix} 1 \\ 0 \\ 1 \end{bmatrix} \right\}$$

Since the vectors $(-1, 1, 0)$ and $(1, 0, 1)$ are linearly independent (Why?), they form a basis for S_0. Thus,

$$B_0 = \left\{ \begin{bmatrix} -1 \\ 1 \\ 0 \end{bmatrix}, \begin{bmatrix} 1 \\ 0 \\ 1 \end{bmatrix} \right\}$$

Check: As a check, we will show that the basis vectors for S_{-3} and S_0 are eigenvectors. For S_{-3}:

$$\begin{bmatrix} -1 & -1 & 1 \\ -1 & -1 & 1 \\ 1 & 1 & -1 \end{bmatrix} \begin{bmatrix} -1 \\ -1 \\ 1 \end{bmatrix} = \begin{bmatrix} 3 \\ 3 \\ -3 \end{bmatrix} = -3 \begin{bmatrix} -1 \\ -1 \\ 1 \end{bmatrix}$$

For S_0:

$$\begin{bmatrix} -1 & -1 & 1 \\ -1 & -1 & 1 \\ 1 & 1 & -1 \end{bmatrix} \begin{bmatrix} -1 \\ 1 \\ 0 \end{bmatrix} = \begin{bmatrix} 0 \\ 0 \\ 0 \end{bmatrix} = 0 \begin{bmatrix} -1 \\ 1 \\ 0 \end{bmatrix}$$

$$\begin{bmatrix} -1 & -1 & 1 \\ -1 & -1 & 1 \\ 1 & 1 & -1 \end{bmatrix} \begin{bmatrix} 1 \\ 0 \\ 1 \end{bmatrix} = \begin{bmatrix} 0 \\ 0 \\ 0 \end{bmatrix} = 0 \begin{bmatrix} -1 \\ 1 \\ 0 \end{bmatrix}$$

Problem 4   Repeat Example 4 for each of the following matrices:

$$\text{(A)} \begin{bmatrix} 2 & 2 & 0 \\ 1 & 1 & 1 \\ -1 & -2 & 1 \end{bmatrix} \qquad \text{(B)} \begin{bmatrix} 4 & 0 & 0 \\ 2 & 0 & -2 \\ -2 & 4 & 6 \end{bmatrix}$$

||

You might have wondered if there is a relationship between the multiplicity of a root of the characteristic polynomial and the dimension of the corresponding eigenspace. There is. It can be shown that if λ is a root of multiplicity k of the characteristic polynomial for an $n \times n$ matrix A and S_λ is the corresponding eigenspace, then

$$\dim(S_\lambda) \leq k$$

We will have more to say about this relationship in the next two sections.

Answers to Matched Problems

1. $\lambda = 2$; $\left\{ \begin{bmatrix} t \\ t \end{bmatrix} \middle| t \text{ any nonzero real number} \right\}$

2. (A) $\lambda^2 - 4\lambda + 4$ (B) $\lambda^3 - 9\lambda$

3. (A) $p(\lambda) = \lambda^2 - 2\lambda + 2$; no eigenvalues
   (B) $p(\lambda) = \lambda^3 - 9\lambda^2 + 21\lambda - 9$; $3 - \sqrt{6}$, 3, $3 + \sqrt{6}$

4. (A) $\lambda_1 = 1$, $\lambda_2 = 2$; $B_1 = \left\{ \begin{bmatrix} -1 \\ \frac{1}{2} \\ 1 \end{bmatrix} \right\}$, $B_2 = \left\{ \begin{bmatrix} -1 \\ 0 \\ 1 \end{bmatrix} \right\}$

   (B) $\lambda_1 = 2$, $\lambda_2 = 4$; $B_2 = \left\{ \begin{bmatrix} 0 \\ -1 \\ 1 \end{bmatrix} \right\}$, $B_4 = \left\{ \begin{bmatrix} 2 \\ 1 \\ 0 \end{bmatrix}, \begin{bmatrix} 1 \\ 0 \\ 1 \end{bmatrix} \right\}$

|| Exercise 7-1

A   In Problems 1–4, verify that $\mathbf{x}_i$ is an eigenvector of A associated with λ_i by showing that $A\mathbf{x}_i = \lambda_i \mathbf{x}_i$.

1. $A = \begin{bmatrix} 4 & -3 \\ 2 & -3 \end{bmatrix}$

   (A) $\lambda_1 = -2$, $\mathbf{x}_1 = \begin{bmatrix} 1 \\ 2 \end{bmatrix}$

   (B) $\lambda_2 = 3$, $\mathbf{x}_2 = \begin{bmatrix} 3 \\ 1 \end{bmatrix}$

2. $A = \begin{bmatrix} 5 & 4 \\ 1 & 2 \end{bmatrix}$

   (A) $\lambda_1 = 1$, $\mathbf{x}_1 = \begin{bmatrix} -1 \\ 1 \end{bmatrix}$

   (B) $\lambda_2 = 6$, $\mathbf{x}_2 = \begin{bmatrix} 4 \\ 1 \end{bmatrix}$

3. $A = \begin{bmatrix} 2 & -1 & -2 \\ -1 & 2 & 2 \\ -2 & 2 & 10 \end{bmatrix}$ **4.** $A = \begin{bmatrix} 1 & 0 & -2 \\ 0 & 1 & -2 \\ -2 & -2 & 8 \end{bmatrix}$

(A) $\lambda_1 = 1$, $\mathbf{x}_1 = \begin{bmatrix} 1 \\ 1 \\ 0 \end{bmatrix}$ (A) $\lambda_1 = 0$, $\mathbf{x}_1 = \begin{bmatrix} 2 \\ 2 \\ 1 \end{bmatrix}$

(B) $\lambda_2 = 2$, $\mathbf{x}_2 = \begin{bmatrix} 2 \\ -2 \\ 1 \end{bmatrix}$ (B) $\lambda_2 = 1$, $\mathbf{x}_2 = \begin{bmatrix} -1 \\ 1 \\ 0 \end{bmatrix}$

(C) $\lambda_3 = 11$, $\mathbf{x}_3 = \begin{bmatrix} -1 \\ 1 \\ 4 \end{bmatrix}$ (C) $\lambda_3 = 9$, $\mathbf{x}_3 = \begin{bmatrix} -1 \\ -1 \\ 4 \end{bmatrix}$

For each matrix in Problems 5–12, find the characteristic polynomial, the eigenvalues, and a basis for each eigenspace.

5. $\begin{bmatrix} 1 & -2 \\ -1 & 2 \end{bmatrix}$ **6.** $\begin{bmatrix} 8 & -9 \\ 1 & 2 \end{bmatrix}$ **7.** $\begin{bmatrix} 1 & 2 \\ 1 & 1 \end{bmatrix}$

8. $\begin{bmatrix} 1 & -2 \\ 4 & 3 \end{bmatrix}$ **9.** $\begin{bmatrix} 3 & -2 \\ 2 & -1 \end{bmatrix}$ **10.** $\begin{bmatrix} 2 & 5 \\ 2 & -1 \end{bmatrix}$

11. $\begin{bmatrix} 3 & -2 \\ 5 & -3 \end{bmatrix}$ **12.** $\begin{bmatrix} 2 & -2 \\ 1 & -2 \end{bmatrix}$

B   *In Problems 13–22, factor each polynomial as far as possible using real numbers and find all real roots.*

13. $\lambda^3 - 2\lambda^2 - 6\lambda + 4$ **14.** $\lambda^3 - \lambda^2 + 3\lambda - 10$

15. $\lambda^3 - 5\lambda^2 + 9\lambda - 9$ **16.** $\lambda^3 - \lambda^2 - 6\lambda - 4$

17. $\lambda^3 - \lambda^2 - 8\lambda + 12$ **18.** $\lambda^3 - 3\lambda^2 - 9\lambda - 5$

19. $\lambda^4 + 4\lambda^3 - 16\lambda - 16$ **20.** $\lambda^4 - 5\lambda^2 + 4$

21. $\lambda^4 - 4\lambda^3 - 2\lambda^2 + 12\lambda + 9$ **22.** $\lambda^4 - 8\lambda^3 + 24\lambda^2 - 32\lambda + 16$

For each matrix in Problems 23–32, find the characteristic polynomial, the eigenvalues, and a basis for each eigenspace.

23. $\begin{bmatrix} 2 & -2 & 4 \\ 2 & -6 & 11 \\ 1 & -4 & 7 \end{bmatrix}$ **24.** $\begin{bmatrix} -1 & 4 & -3 \\ 1 & 5 & -5 \\ 0 & 8 & -7 \end{bmatrix}$ **25.** $\begin{bmatrix} 2 & -1 & -3 \\ 2 & -1 & -2 \\ 0 & 0 & 1 \end{bmatrix}$

26. $\begin{bmatrix} 1 & 2 & -4 \\ 1 & 0 & 1 \\ 1 & -2 & 4 \end{bmatrix}$ **27.** $\begin{bmatrix} 1 & -4 & 6 \\ -3 & 0 & 6 \\ -3 & -4 & 10 \end{bmatrix}$ **28.** $\begin{bmatrix} 1 & -2 & 3 \\ 1 & -2 & 3 \\ -1 & 2 & -3 \end{bmatrix}$

29. $\begin{bmatrix} 1 & 0 & -1 \\ 1 & 3 & 4 \\ -1 & 0 & -1 \end{bmatrix}$ **30.** $\begin{bmatrix} 1 & -2 & 3 \\ 0 & 2 & 0 \\ 1 & 1 & 1 \end{bmatrix}$

31. $\begin{bmatrix} 2 & -1 & 0 & 0 \\ -1 & 2 & 0 & 0 \\ 0 & 0 & 1 & 0 \\ 0 & 0 & 0 & 3 \end{bmatrix}$ **32.** $\begin{bmatrix} 0 & 0 & 2 & 2 \\ 1 & 0 & 0 & 0 \\ 0 & 2 & -1 & -2 \\ 0 & 0 & -2 & -1 \end{bmatrix}$

C   In Problems 33–35, find the eigenvalues of each matrix.

33. $\begin{bmatrix} a_{11} & 0 \\ 0 & a_{22} \end{bmatrix}$ **34.** $\begin{bmatrix} a_{11} & a_{12} & a_{13} \\ 0 & a_{22} & a_{23} \\ 0 & 0 & a_{33} \end{bmatrix}$

35. $\begin{bmatrix} a_{11} & 0 & 0 & 0 \\ a_{21} & a_{22} & 0 & 0 \\ a_{31} & a_{32} & a_{33} & 0 \\ a_{41} & a_{42} & a_{43} & a_{44} \end{bmatrix}$

36. Generalize the results in Problems 33–35 by showing that the eigenvalues of an $n \times n$ triangular matrix $A = [a_{ij}]_{n \times n}$ are $a_{11}, a_{22}, \ldots, a_{nn}$.

37. Let

$$A = \begin{bmatrix} a & b \\ c & d \end{bmatrix}$$

(A) Show that the characteristic equation for A is

$$\lambda^2 - (a + d)\lambda + ad - bc = 0$$

(B) Show that A satisfies its characteristic equation; that is,

$$A^2 - (a + d)A + (ad - bc)I = 0$$

(In more advanced courses, it is shown that every $n \times n$ matrix satisfies its characteristic equation. This result is known as the *Cayley–Hamilton theorem*.)

38. Let

$$A = \begin{bmatrix} a & b \\ c & d \end{bmatrix}$$

Determine conditions on a, b, c, and d so that:

(A) A has two distinct eigenvalues.
(B) A has one (double) eigenvalue.
(C) A has no eigenvalues.

39. Repeat Problem 38 for

$$A = \begin{bmatrix} a & b \\ b & d \end{bmatrix}$$

40. Show that the constant term in the characteristic polynomial for an $n \times n$ matrix A is $(-1)^n \det(A)$. [*Hint:* The constant term in a polynomial $p(\lambda)$ is $p(0)$.]

41. Let A be an $n \times n$ matrix. Show that A and A^T have the same characteristic polynomial.

42. Show that an $n \times n$ matrix A is not invertible if and only if 0 is an eigenvalue of A.

43. If λ is an eigenvalue of an invertible $n \times n$ matrix A, show that $1/\lambda$ is an eigenvalue of A^{-1}.

44. Show that $\lambda = 0$ is the only eigenvalue of the $n \times n$ zero matrix.

45. If λ is an eigenvalue of the $n \times n$ matrix A and k is a positive integer, show that λ^k is an eigenvalue of A^k.

46. An $n \times n$ matrix A is said to be **nilpotent** if $A^k = 0$ for a positive integer k. If A is a nilpotent matrix, show that 0 is the only eigenvalue of A. [*Hint:* Use Problems 44 and 45.]

❙ 7-2 ❙   Diagonalization

- Introduction
- Diagonalizable Matrices
- Procedure for Diagonalizing a Matrix
- Finding Linearly Independent Eigenvectors

■ Introduction

In this section, we will show that certain square matrices can be related to diagonal matrices. We begin with an example that illustrates one application of this very important relationship.

Example 5   Let

$$A = \begin{bmatrix} 1 & 2 \\ -1 & 4 \end{bmatrix} \qquad P = \begin{bmatrix} 2 & 1 \\ 1 & 1 \end{bmatrix} \qquad D = \begin{bmatrix} 2 & 0 \\ 0 & 3 \end{bmatrix}$$

(A) Show that A, P, and D are related by the equation $A = PDP^{-1}$.

(B) Find D^k for any positive integer k.

(C) Find A^k for any positive integer k.

Solution   The associative property for matrix multiplication will be particularly useful in parts (A) and (C). [Recall that $(AB)C = A(BC)$.]

$$\overset{PD}{}\qquad\overset{P^{-1}}{}$$

(A) $PDP^{-1} = \begin{bmatrix} 2 & 1 \\ 1 & 1 \end{bmatrix}\begin{bmatrix} 2 & 0 \\ 0 & 3 \end{bmatrix}\begin{bmatrix} 2 & 1 \\ 1 & 1 \end{bmatrix}^{-1} = \begin{bmatrix} 4 & 3 \\ 2 & 3 \end{bmatrix}\begin{bmatrix} 1 & -1 \\ -1 & 2 \end{bmatrix} = \begin{bmatrix} 1 & 2 \\ -1 & 4 \end{bmatrix} = A$

(B) $D^2 = \begin{bmatrix} 2 & 0 \\ 0 & 3 \end{bmatrix}\begin{bmatrix} 2 & 0 \\ 0 & 3 \end{bmatrix} = \begin{bmatrix} 2^2 & 0 \\ 0 & 3^2 \end{bmatrix}$

$D^3 = DD^2 = \begin{bmatrix} 2 & 0 \\ 0 & 3 \end{bmatrix}\begin{bmatrix} 2^2 & 0 \\ 0 & 3^2 \end{bmatrix} = \begin{bmatrix} 2^3 & 0 \\ 0 & 3^3 \end{bmatrix}$

$$\cdot$$
$$\cdot$$
$$\cdot$$

$D^k = \begin{bmatrix} 2^k & 0 \\ 0 & 3^k \end{bmatrix}$

Note: A formal proof of this result requires mathematical induction, which we omit.

(C) If we proceed by computing successive powers of A, as we did for D in part (B), it is very difficult to determine any pattern in the entries of $A^2, A^3, \ldots$. (Try it!) Instead, we will use the relationship in part (A) to express A^k in terms of P, D^k, and P^{-1}.

$$\overset{A}{}\qquad\overset{A}{}\qquad\qquad\overset{I}{}$$
$$A^2 = (PDP^{-1})(PDP^{-1}) = (PD)(P^{-1}P)(DP^{-1}) = PD^2P^{-1}$$
$$\overset{A}{}\qquad\overset{A^2}{}\qquad\qquad\overset{I}{}$$
$$A^3 = (PDP^{-1})(PD^2P^{-1}) = (PD)(P^{-1}P)(D^2P^{-1}) = PD^3P^{-1}$$

$$\cdot$$
$$\cdot$$
$$\cdot$$

$A^k = PD^kP^{-1}$ Substitute for P, D^k, and P^{-1}.

$= \begin{bmatrix} 2 & 1 \\ 1 & 1 \end{bmatrix}\begin{bmatrix} 2^k & 0 \\ 0 & 3^k \end{bmatrix}\begin{bmatrix} 1 & -1 \\ -1 & 2 \end{bmatrix}$

$= \begin{bmatrix} 2^{k+1} & 3^k \\ 2^k & 3^k \end{bmatrix}\begin{bmatrix} 1 & -1 \\ -1 & 2 \end{bmatrix}$

$= \begin{bmatrix} 2^{k+1} - 3^k & -2^{k+1} + 2 \cdot 3^k \\ 2^k - 3^k & -2^k + 2 \cdot 3^k \end{bmatrix}$

Note: A formal proof of this result requires mathematical induction, which we omit.

Problem 5   Repeat Example 5 for

$$A = \begin{bmatrix} 2 & -1 \\ 2 & 5 \end{bmatrix} \qquad P = \begin{bmatrix} -1 & -1 \\ 2 & 1 \end{bmatrix} \qquad D = \begin{bmatrix} 4 & 0 \\ 0 & 3 \end{bmatrix} \qquad\qquad \blacksquare$$

▪ Diagonalizable Matrices

In Example 5, using the relationship $A = PDP^{-1}$ provided us with a simple method for computing A^k. It turns out that many other problems can be solved by relating a given square matrix to a diagonal matrix. This leads to the following definition:

Diagonalizable Matrices

An $n \times n$ matrix A is **diagonalizable** if there is an invertible matrix P such that

$$D = P^{-1}AP \tag{1}$$

is a diagonal matrix. The matrix P is said to **diagonalize** A.

Left-multiplying both sides of (1) by P, we have

$$PD = AP \tag{2}$$

and right-multiplying both sides of (2) by P^{-1}, we have

$$PDP^{-1} = A \tag{3}$$

which is the relationship we used in Example 5 to compute A^k. Equations (1), (2), and (3) are always equivalent and can be used interchangeably in problems involving diagonalization.

Our goal in this and the next section is to develop procedures that will tell us whether a given square matrix A is diagonalizable (not all are!) and that will enable us to find the diagonal matrix D and the invertible matrix P. Theorem 5 is the key to developing these procedures. Examine the proof of Theorem 5 carefully, since this proof shows how to find D and P.

Theorem 5   An $n \times n$ matrix A is diagonalizable if and only if A has n linearly independent eigenvectors.

Proof   *Part 1.   Assume that A is diagonalizable. Show that A has n linearly independent eigenvectors.* Since A is diagonalizable, using (2), there is an invertible matrix P and a diagonal matrix D such that $AP = PD$. Let

$$D = \begin{bmatrix} \lambda_1 & 0 & \cdots & 0 \\ 0 & \lambda_2 & \cdots & 0 \\ \cdot & \cdot & & \cdot \\ \cdot & \cdot & & \cdot \\ \cdot & \cdot & & \cdot \\ 0 & 0 & \cdots & \lambda_n \end{bmatrix} \qquad P = \begin{bmatrix} \mathbf{p}_1 & \mathbf{p}_2 & \cdots & \mathbf{p}_n \\ p_{11} & p_{12} & \cdots & p_{1n} \\ p_{21} & p_{22} & \cdots & p_{2n} \\ \cdot & \cdot & & \cdot \\ \cdot & \cdot & & \cdot \\ \cdot & \cdot & & \cdot \\ p_{n1} & p_{n2} & \cdots & p_{nn} \end{bmatrix}$$

and let $\mathbf{p}_1, \mathbf{p}_2, \ldots, \mathbf{p}_n$ be the columns of P. (The choice of notation for D is meant to be suggestive. The diagonal elements of D will turn out to be the eigenvalues of A.) Since P is invertible, the columns of P are linearly independent. The key step in this proof is the comparison of the kth column of AP with the kth column of PD. From the definition of matrix multiplication, the kth column of AP is

$$
\begin{bmatrix}
a_{11}p_{1k} + a_{12}p_{2k} + \cdots + a_{1n}p_{nk} \\
a_{21}p_{1k} + a_{22}p_{2k} + \cdots + a_{2n}p_{nk} \\
\cdot \\
\cdot \\
\cdot \\
a_{n1}p_{1k} + a_{n2}p_{2k} + \cdots + a_{nn}p_{nk}
\end{bmatrix}
=
\begin{bmatrix}
a_{11} & a_{12} & \cdots & a_{1n} \\
a_{21} & a_{22} & \cdots & a_{2n} \\
\cdot & \cdot & & \cdot \\
\cdot & \cdot & & \cdot \\
\cdot & \cdot & & \cdot \\
a_{n1} & a_{n2} & \cdots & a_{nn}
\end{bmatrix}
\begin{bmatrix}
p_{1k} \\
p_{2k} \\
\cdot \\
\cdot \\
\cdot \\
p_{nk}
\end{bmatrix}
= A\mathbf{p}_k \qquad (4)
$$

To find the kth column of PD, we note that

$$
PD =
\begin{bmatrix}
p_{11} & p_{12} & \cdots & p_{1n} \\
p_{21} & p_{22} & \cdots & p_{2n} \\
\cdot & \cdot & & \cdot \\
\cdot & \cdot & & \cdot \\
\cdot & \cdot & & \cdot \\
p_{n1} & p_{n2} & \cdots & p_{nn}
\end{bmatrix}
\begin{bmatrix}
\lambda_1 & 0 & \cdots & 0 \\
0 & \lambda_2 & \cdots & 0 \\
\cdot & \cdot & & \cdot \\
\cdot & \cdot & & \cdot \\
\cdot & \cdot & & \cdot \\
0 & 0 & \cdots & \lambda_n
\end{bmatrix}
$$

$$
=
\begin{bmatrix}
p_{11}\lambda_1 & p_{12}\lambda_2 & \cdots & p_{1n}\lambda_n \\
p_{21}\lambda_1 & p_{22}\lambda_2 & \cdots & p_{2n}\lambda_n \\
\cdot & \cdot & & \cdot \\
\cdot & \cdot & & \cdot \\
\cdot & \cdot & & \cdot \\
p_{n1}\lambda_1 & p_{n2}\lambda_2 & \cdots & p_{nn}\lambda_n
\end{bmatrix}
$$

Thus, the kth column of PD is

$$
\begin{bmatrix}
p_{1k}\lambda_k \\
p_{2k}\lambda_k \\
\cdot \\
\cdot \\
\cdot \\
p_{nk}\lambda_k
\end{bmatrix}
= \lambda_k
\begin{bmatrix}
p_{1k} \\
p_{2k} \\
\cdot \\
\cdot \\
\cdot \\
p_{nk}
\end{bmatrix}
= \lambda_k \mathbf{p}_k \qquad (5)
$$

Since $AP = PD$, the kth column of AP must equal the kth column of PD. Using (4) and (5), we have

$$
\begin{pmatrix} \text{kth column} \\ \text{of } AP \end{pmatrix} = \begin{pmatrix} \text{kth column} \\ \text{of } PD \end{pmatrix}
$$

$$
A\mathbf{p}_k = \lambda_k \mathbf{p}_k \qquad (6)
$$

Since $\mathbf{p}_k \neq 0$ (Why?), equation (6) shows that λ_k is an eigenvalue of A and that $\mathbf{p}_k$ is an associated eigenvector. Thus, A has n linearly independent eigenvectors: $\mathbf{p}_1, \mathbf{p}_2, \ldots, \mathbf{p}_n$.

Part 2.   *Assume that A has n linearly independent eigenvectors. Show that A is diagonalizable. Let* $\mathbf{p}_1, \mathbf{p}_2, \ldots, \mathbf{p}_n$ *be n linearly independent eigenvectors of A and let* $\lambda_1, \lambda_2, \ldots, \lambda_n$ *be the corresponding eigenvalues (not necessarily distinct). Form the matrices*

$$P = [\mathbf{p}_1 \quad \mathbf{p}_2 \quad \cdots \quad \mathbf{p}_n] \qquad D = \begin{bmatrix} \lambda_1 & 0 & \cdots & 0 \\ 0 & \lambda_2 & \cdots & 0 \\ \vdots & \vdots & & \vdots \\ 0 & 0 & \cdots & \lambda_n \end{bmatrix}$$

Then, for each k,

$$A\mathbf{p}_k = \lambda_k \mathbf{p}_k$$

which implies that each column of AP is equal to the corresponding column of PD. [Compare (4) and (5) in the first part of this proof.] Thus, $AP = PD$, and A is diagonalizable. ∎

▪ Procedure for Diagonalizing a Matrix

If an $n \times n$ matrix A has n linearly independent eigenvectors, then Theorem 5 not only shows that A is diagonalizable, but also provides a procedure for finding the matrix P that diagonalizes A.

Diagonalizing a Matrix

Let A be an $n \times n$ diagonalizable matrix.

Step 1.   Find n linearly independent eigenvectors $\mathbf{p}_1, \mathbf{p}_2, \ldots, \mathbf{p}_n$.

Step 2.   Form the matrix $P = [\mathbf{p}_1 \quad \mathbf{p}_2 \quad \cdots \quad \mathbf{p}_n]$.

Step 3.   Form the diagonal matrix

$$D = \begin{bmatrix} \lambda_1 & 0 & \cdots & 0 \\ 0 & \lambda_2 & \cdots & 0 \\ \vdots & \vdots & & \vdots \\ 0 & 0 & \cdots & \lambda_n \end{bmatrix}$$

where λ_k is the eigenvalue corresponding to $\mathbf{p}_k$, $k = 1, 2, \ldots, n$.

Then A, P, and D satisfy $P^{-1}AP = D$; that is, P diagonalizes A.

Example 6   Diagonalize the following matrix, if possible:

$$A = \begin{bmatrix} 0 & -1 & 1 \\ -1 & 0 & 1 \\ 1 & 1 & 0 \end{bmatrix}$$

Solution   *Step 1.*   *Find three linearly independent eigenvectors, if possible.* Omitting the details, the characteristic polynomial is

$$p(\lambda) = \det(\lambda I - A) = (\lambda - 1)^2(\lambda + 2)$$

The eigenvalues of A (with due regard to multiplicity) are $\lambda_1 = 1, \lambda_2 = 1,$ and $\lambda_3 = -2$. Omitting the details, the corresponding eigenspaces are

$$S_1 = \text{span} \left\{ \begin{bmatrix} -1 \\ 1 \\ 0 \end{bmatrix}, \begin{bmatrix} 1 \\ 0 \\ 1 \end{bmatrix} \right\} \quad \text{and} \quad S_{-2} = \text{span} \left\{ \begin{bmatrix} -1 \\ -1 \\ 1 \end{bmatrix} \right\}$$

The eigenvectors $\mathbf{p}_1 = (-1, 1, 0)$ and $\mathbf{p}_2 = (1, 0, 1)$ form a basis for S_1 and the eigenvector $\mathbf{p}_3 = (-1, -1, 1)$ forms a basis for S_{-2}. It is left to the reader to show that the set of eigenvectors $\{\mathbf{p}_1, \mathbf{p}_2, \mathbf{p}_3\}$ is linearly independent.

Step 2.   *Form the matrix P:*

$$P = \begin{matrix} \mathbf{p}_1 & \mathbf{p}_2 & \mathbf{p}_3 \\ \begin{bmatrix} -1 & 1 & -1 \\ 1 & 0 & -1 \\ 0 & 1 & 1 \end{bmatrix} \end{matrix}$$

Step 3.   *Form the diagonal matrix D:*

$$D = \begin{matrix} \lambda_1 & \lambda_2 & \lambda_3 \\ \begin{bmatrix} 1 & 0 & 0 \\ 0 & 1 & 0 \\ 0 & 0 & -2 \end{bmatrix} \end{matrix}$$

Then A, P, and D satisfy $P^{-1}AP = D$.

Check: Since the equations $P^{-1}AP = D$ and $AP = PD$ are equivalent, we can check our work by showing that $AP = PD$. (This avoids the calculation of P^{-1}.)

$$\underset{A}{\begin{bmatrix} 0 & -1 & 1 \\ -1 & 0 & 1 \\ 1 & 1 & 0 \end{bmatrix}} \underset{P}{\begin{bmatrix} -1 & 1 & -1 \\ 1 & 0 & -1 \\ 0 & 1 & 1 \end{bmatrix}} = \begin{bmatrix} -1 & 1 & 2 \\ 1 & 0 & 2 \\ 0 & 1 & -2 \end{bmatrix} = \underset{P}{\begin{bmatrix} -1 & 1 & -1 \\ 1 & 0 & -1 \\ 0 & 1 & 1 \end{bmatrix}} \underset{D}{\begin{bmatrix} 1 & 0 & 0 \\ 0 & 1 & 0 \\ 0 & 0 & -2 \end{bmatrix}}$$

Notice that the order in which the eigenvalues are placed on the diagonal in D must agree with the order of the eigenvectors in P. For example, if we reverse the order of the columns of the matrix P in Example 6, then

$$
\begin{array}{ccc} \mathbf{p}_3 & \mathbf{p}_2 & \mathbf{p}_1 \end{array}
$$
$$
P = \begin{bmatrix} -1 & 1 & -1 \\ -1 & 0 & 1 \\ 1 & 1 & 0 \end{bmatrix} \quad \text{and} \quad
\begin{array}{ccc} \lambda_3 & \lambda_2 & \lambda_1 \end{array}
\quad D = \begin{bmatrix} -2 & 0 & 0 \\ 0 & 1 & 0 \\ 0 & 0 & 1 \end{bmatrix}
$$

Problem 6　Diagonalize the following matrix, if possible:

$$
A = \begin{bmatrix} -1 & 0 & -2 \\ 0 & 1 & 0 \\ -2 & 0 & -1 \end{bmatrix}
$$

∎

Example 7　Diagonalize the following matrix, if possible:

$$
A = \begin{bmatrix} 2 & -1 \\ 1 & 4 \end{bmatrix}
$$

Solution　*Step 1.　Find two linearly independent eigenvectors, if possible.*

$$
p(\lambda) = \det(\lambda I - A) = \begin{vmatrix} \lambda - 2 & 1 \\ -1 & \lambda - 4 \end{vmatrix} = (\lambda - 3)^2
$$

The eigenvalues of A (with due regard to multiplicity) are $\lambda_1 = 3$ and $\lambda_2 = 3$. The associated eigenspace is (details omitted)

$$
S_3 = \text{span}\left\{ \begin{bmatrix} -1 \\ 1 \end{bmatrix} \right\}
$$

Since S_3 is a one-dimensional subspace, it is not possible to find two linearly independent eigenvectors in S_3. Since S_3 is the only eigenspace for A, there are no other eigenvectors. Thus, A does not possess two linearly independent eigenvectors and is not diagonalizable.

Problem 7　Diagonalize the following matrix, if possible:

$$
A = \begin{bmatrix} 1 & -2 \\ 2 & -3 \end{bmatrix}
$$

∎

▪ Finding Linearly Independent Eigenvectors

The key step in diagonalizing an $n \times n$ matrix is finding n linearly independent eigenvectors. The following theorem is very helpful in determining whether a matrix has n linearly independent eigenvectors:

Theorem 6

> If $\lambda_1, \lambda_2, \ldots, \lambda_k$ are distinct eigenvalues of an $n \times n$ matrix A and if $\mathbf{x}_i$ is an eigenvector corresponding to λ_i, $i = 1, 2, \ldots, k$, then the set
>
> $$S = \{\mathbf{x}_1, \mathbf{x}_2, \ldots, \mathbf{x}_k\}$$
>
> is linearly independent.

Proof The proof will be by contradiction. That is, we will assume S is linearly dependent and show that this leads to a contradiction. If S is linearly dependent, then there is a linearly independent subset B of S with the property that span $B =$ span S (Theorem 12 in Section 5-5). For convenience, we can assume that

$$B = \{\mathbf{x}_1, \mathbf{x}_2, \ldots, \mathbf{x}_m\} \qquad m < k$$

Since $\mathbf{x}_{m+1}$ is a nonzero vector in span B, there exist scalars $c_1, c_2, \ldots, c_m$, not all zero, such that

$$\mathbf{x}_{m+1} = c_1\mathbf{x}_1 + c_2\mathbf{x}_2 + \cdots + c_m\mathbf{x}_m \tag{7}$$

Left-multiplying both sides of (7) by the matrix A, we have

$$A\mathbf{x}_{m+1} = c_1 A\mathbf{x}_1 + c_2 A\mathbf{x}_2 + \cdots + c_m A\mathbf{x}_m \qquad A\mathbf{x}_i = \lambda_i\mathbf{x}_i, \text{ since } \mathbf{x}_i \text{ is an eigenvector associated with } \lambda_i.$$

$$\lambda_{m+1}\mathbf{x}_{m+1} = c_1\lambda_1\mathbf{x}_1 + c_2\lambda_2\mathbf{x}_2 + \cdots + c_m\lambda_m\mathbf{x}_m \tag{8}$$

Multiplying both sides of (7) by the scalar λ_{m+1}, we also have

$$\lambda_{m+1}\mathbf{x}_{m+1} = c_1\lambda_{m+1}\mathbf{x}_1 + c_2\lambda_{m+1}\mathbf{x}_2 + \cdots + c_m\lambda_{m+1}\mathbf{x}_m \tag{9}$$

Subtracting (9) from (8) yields

$$\mathbf{0} = c_1(\lambda_1 - \lambda_{m+1})\mathbf{x}_1 + c_2(\lambda_2 - \lambda_{m+1})\mathbf{x}_2 + \cdots + c_m(\lambda_m - \lambda_{m+1})\mathbf{x}_m \tag{10}$$

Since the eigenvalues $\lambda_1, \lambda_2, \ldots, \lambda_k$ are distinct by hypothesis, the scalars $\lambda_1 - \lambda_{m+1}, \lambda_2 - \lambda_{m+1}, \ldots, \lambda_m - \lambda_{m+1}$ are all nonzero. We have already noted that the scalars $c_1, c_2, \ldots, c_m$ are not all zero. Thus, we have in (10) a nontrivial linear combination of linearly independent vectors equal to the zero vector. This is a contradiction. Thus, the assumption that S is a linearly dependent set is false and S must be a linearly independent set. ∎

Theorem 7 is an immediate consequence of Theorem 6 and gives us a condition that will ensure that a matrix is diagonalizable. The proof is omitted.

Theorem 7

> If A is an $n \times n$ matrix with n distinct eigenvalues, then A is diagonalizable.

Example 8   Show that the following matrix is diagonalizable:

$$A = \begin{bmatrix} 2 & -1 & 0 \\ -1 & 2 & 0 \\ 0 & 0 & 2 \end{bmatrix}$$

Solution $p(\lambda) = \det(\lambda I - A) = (\lambda - 1)(\lambda - 2)(\lambda - 3)$

Thus, A has three distinct eigenvalues, $\lambda_1 = 1$, $\lambda_2 = 2$, and $\lambda_3 = 3$. Theorem 7 now implies that A is diagonalizable. For example, there must exist an invertible matrix P such that

$$P^{-1}AP = \begin{bmatrix} 1 & 0 & 0 \\ 0 & 2 & 0 \\ 0 & 0 & 3 \end{bmatrix}$$
   The order of the distinct eigenvalues on the diagonal is arbitrary.

If necessary, the matrix P can be determined by finding an eigenvector for each of the eigenvalues. However, in many situations it is not necessary to actually find P. Often, it is sufficient to know that such a matrix exists.

Problem 8   Show that A is diagonalizable, and find a diagonal matrix D satisfying $P^{-1}AP = D$. (Do not find P.)

$$A = \begin{bmatrix} 1 & -2 & 0 \\ -2 & 1 & 0 \\ 0 & 0 & -3 \end{bmatrix}$$ ▌▌

If a matrix does not have n distinct eigenvalues, then it may or may not be diagonalizable. At this point, the only way we can determine whether such a matrix is diagonalizable is to attempt to find n linearly independent eigenvectors. Theorem 8, which is a generalization of Theorem 6, provides a method for doing this. The proof of Theorem 8 is omitted.

Theorem 8

If $\lambda_1, \lambda_2, \ldots, \lambda_k$ are the distinct eigenvalues of an $n \times n$ matrix A, $k \leq n$, and $B_{\lambda_1}, B_{\lambda_2}, \ldots, B_{\lambda_k}$ are bases for the corresponding eigenspaces $S_{\lambda_1}, S_{\lambda_2}, \ldots, S_{\lambda_k}$, then

$$B = B_{\lambda_1} \cup B_{\lambda_2} \cup \cdots \cup B_{\lambda_k}$$

is a linearly independent set of eigenvectors for A. Furthermore, A is diagonalizable if and only if

$$\dim(S_{\lambda_1}) + \dim(S_{\lambda_2}) + \cdots + \dim(S_{\lambda_k}) = n$$

Example 9   Diagonalize the following matrices, if possible:

$$(A) \quad A = \begin{bmatrix} 0 & 2 & 0 & 0 \\ 2 & 0 & 0 & 0 \\ 0 & 0 & -2 & 0 \\ 0 & 0 & 0 & -2 \end{bmatrix} \qquad (B) \quad A = \begin{bmatrix} 0 & 0 & 2 & 0 \\ 2 & 0 & 0 & 0 \\ 0 & 0 & -1 & 0 \\ 0 & 0 & -2 & -1 \end{bmatrix}$$

Solution (A) Omitting the details, the characteristic polynomial is

$$p(\lambda) = (\lambda + 2)^3(\lambda - 2)$$

The distinct eigenvalues are $\lambda_1 = -2$ (multiplicity 3) and $\lambda_2 = 2$. The associated eigenspaces are

$$S_{-2} = \text{span} \left\{ \begin{bmatrix} -1 \\ 1 \\ 0 \\ 0 \end{bmatrix}, \begin{bmatrix} 0 \\ 0 \\ 1 \\ 0 \end{bmatrix}, \begin{bmatrix} 0 \\ 0 \\ 0 \\ 1 \end{bmatrix} \right\}$$

Verify that these vectors are linearly independent.

$$S_2 = \text{span} \left\{ \begin{bmatrix} 1 \\ 1 \\ 0 \\ 0 \end{bmatrix} \right\}$$

Thus,

$$\dim(S_{-2}) + \dim(S_2) = 4$$

Theorem 8 implies that A is diagonalizable and that

$$B = \left\{ \begin{bmatrix} -1 \\ 1 \\ 0 \\ 0 \end{bmatrix}, \begin{bmatrix} 0 \\ 0 \\ 1 \\ 0 \end{bmatrix}, \begin{bmatrix} 0 \\ 0 \\ 0 \\ 1 \end{bmatrix}, \begin{bmatrix} 1 \\ 1 \\ 0 \\ 0 \end{bmatrix} \right\}$$

is a linearly independent set of eigenvectors. If

$$P = \begin{bmatrix} -1 & 0 & 0 & 1 \\ 1 & 0 & 0 & 1 \\ 0 & 1 & 0 & 0 \\ 0 & 0 & 1 & 0 \end{bmatrix}$$

then

$$P^{-1}AP = \begin{bmatrix} -2 & 0 & 0 & 0 \\ 0 & -2 & 0 & 0 \\ 0 & 0 & -2 & 0 \\ 0 & 0 & 0 & 2 \end{bmatrix}$$

(B) $p(\lambda) = \lambda^2(\lambda + 1)^2$ is the characteristic polynomial, $\lambda_1 = -1$ and $\lambda_2 = 0$ are the distinct eigenvalues (both of multiplicity 2), and the eigenspaces are

$$S_{-1} = \operatorname{span}\left\{\begin{bmatrix} 0 \\ 0 \\ 0 \\ 1 \end{bmatrix}\right\} \qquad S_0 = \operatorname{span}\left\{\begin{bmatrix} 0 \\ 1 \\ 0 \\ 0 \end{bmatrix}\right\}$$

Since

$$\dim(S_{-1}) + \dim(S_0) = 2 < 4$$

Theorem 8 implies that A is not diagonalizable.

Problem 9　Diagonalize the following matrices, if possible:

$$(A) \quad A = \begin{bmatrix} 0 & 1 & 0 & 0 \\ 1 & 0 & 0 & 0 \\ 0 & 0 & 1 & 0 \\ 0 & 0 & 0 & -1 \end{bmatrix} \qquad (B) \quad A = \begin{bmatrix} 0 & 0 & 0 & 0 \\ 1 & 0 & 0 & 0 \\ 0 & 0 & -1 & 0 \\ 0 & 0 & 0 & -1 \end{bmatrix}$$ ∎

Be careful to apply Theorem 8 correctly. You must be certain that you have found a basis for each eigenspace before you form the set B. If you select a linearly dependent set of vectors from one of the eigenspaces, then B will not be linearly independent. On the other hand, if you select a set of vectors that does not span the eigenspace, then B cannot contain a sufficient number of vectors to allow you to complete the diagonalization process.

Answers to Matched Problems

5. (B) $D^k = \begin{bmatrix} 4^k & 0 \\ 0 & 3^k \end{bmatrix}$　　(C) $A^k = \begin{bmatrix} -4^k + 2 \cdot 3^k & -4^k + 3^k \\ 2 \cdot 4^k - 2 \cdot 3^k & 2 \cdot 4^k - 3^k \end{bmatrix}$

6. $P = \begin{bmatrix} 0 & -1 & 1 \\ 1 & 0 & 0 \\ 0 & 1 & 1 \end{bmatrix}$, $D = \begin{bmatrix} 1 & 0 & 0 \\ 0 & 1 & 0 \\ 0 & 0 & -3 \end{bmatrix}$ (*Note:* These are not the only possible choices for P and D.)

7. Not possible

8. $D = \begin{bmatrix} -3 & 0 & 0 \\ 0 & -1 & 0 \\ 0 & 0 & 3 \end{bmatrix}$ (*Note:* Order of diagonal elements is arbitrary.)

9. (A) $P = \begin{bmatrix} -1 & 0 & 1 & 0 \\ 1 & 0 & 1 & 0 \\ 0 & 0 & 0 & 1 \\ 0 & 1 & 0 & 0 \end{bmatrix}$, $D = \begin{bmatrix} -1 & 0 & 0 & 0 \\ 0 & -1 & 0 & 0 \\ 0 & 0 & 1 & 0 \\ 0 & 0 & 0 & 1 \end{bmatrix}$

(*Note:* These are not the only possible choices for P and D.)

(B)　Not possible

‖ Exercise 7-2

Many of the problems in this exercise set involve finding matrices P and D satisfying $D = P^{-1}AP$ for a given matrix A. Since P and D are not unique, your answers may not agree with the answers in the back of the book. In this case, you should use the equation $PD = AP$ to check your work.

A For each matrix A in Problems 1–6, find the characteristic polynomial, the eigenvalues, a basis for each eigenspace, and an invertible matrix P and a diagonal matrix D satisfying $P^{-1}AP = D$, if these matrices exist.

1. $A = \begin{bmatrix} -1 & 6 \\ -1 & 4 \end{bmatrix}$
2. $A = \begin{bmatrix} 6 & 6 \\ -1 & 1 \end{bmatrix}$

3. $A = \begin{bmatrix} 5 & 3 \\ -3 & -1 \end{bmatrix}$
4. $A = \begin{bmatrix} 1 & -2 \\ 3 & 5 \end{bmatrix}$

5. $A = \begin{bmatrix} 2 & 1 \\ -1 & 3 \end{bmatrix}$
6. $A = \begin{bmatrix} 1 & -4 \\ 1 & -3 \end{bmatrix}$

In Problems 7 and 8, given P and D, find $A = PDP^{-1}$, and then use this relationship to find A^3.

7. $P = \begin{bmatrix} 4 & -1 \\ -3 & 1 \end{bmatrix}$, $D = \begin{bmatrix} 1 & 0 \\ 0 & 2 \end{bmatrix}$
8. $P = \begin{bmatrix} 3 & 1 \\ 1 & 1 \end{bmatrix}$, $D = \begin{bmatrix} 1 & 0 \\ 0 & 3 \end{bmatrix}$

B In Problems 9 and 10, find an invertible matrix P and a diagonal matrix D satisfying $P^{-1}AP = D$.

9. $A = \begin{bmatrix} -3 & 4 & 0 \\ -2 & 3 & 0 \\ -2 & 2 & 1 \end{bmatrix}$
10. $A = \begin{bmatrix} 1 & 6 & 2 \\ 0 & 3 & 0 \\ -1 & 3 & 4 \end{bmatrix}$

In Problems 11 and 12, use Theorem 7 to show that A is diagonalizable, and find a diagonal matrix D satisfying $P^{-1}AP = D$ for some invertible matrix P. Do not find P.

11. $A = \begin{bmatrix} 3 & 5 & 0 \\ -1 & -3 & 0 \\ 0 & 0 & 1 \end{bmatrix}$
12. $A = \begin{bmatrix} 0 & 0 & 0 \\ 2 & 1 & 0 \\ 4 & 3 & 2 \end{bmatrix}$

In Problems 13–22, determine whether A is diagonalizable, and if it is, find an invertible matrix P and a diagonal matrix D satisfying $P^{-1}AP = D$.

13. $A = \begin{bmatrix} 3 & -2 & 6 \\ -4 & 10 & -15 \\ -3 & 6 & -10 \end{bmatrix}$
14. $A = \begin{bmatrix} 1 & 3 & 6 \\ -1 & 5 & 2 \\ -1 & 1 & 6 \end{bmatrix}$

15. $A = \begin{bmatrix} -1 & 6 & -6 \\ -2 & 6 & -4 \\ -1 & 2 & 0 \end{bmatrix}$
16. $A = \begin{bmatrix} 1 & 0 & 2 \\ 1 & -1 & 1 \\ 3 & -6 & 2 \end{bmatrix}$

17. $A = \begin{bmatrix} -2 & 1 & -1 \\ -2 & -4 & 1 \\ -1 & 0 & -3 \end{bmatrix}$ **18.** $A = \begin{bmatrix} 2 & -7 & 8 \\ 0 & 0 & -2 \\ -1 & 3 & -5 \end{bmatrix}$

19. $A = \begin{bmatrix} 2 & -1 & 0 & 0 \\ -1 & 2 & 0 & 0 \\ 0 & 0 & 1 & 0 \\ 0 & 0 & 0 & 3 \end{bmatrix}$ **20.** $A = \begin{bmatrix} 0 & 0 & 2 & 2 \\ 1 & 0 & 0 & 0 \\ 0 & 2 & -1 & -2 \\ 0 & 0 & -2 & -1 \end{bmatrix}$

21. $A = \begin{bmatrix} 0 & 1 & 0 & 0 \\ 1 & 0 & 0 & 0 \\ 0 & 0 & -1 & -2 \\ 0 & 0 & 2 & 3 \end{bmatrix}$ **22.** $A = \begin{bmatrix} 1 & -1 & 0 & 0 \\ -1 & 1 & 0 & 0 \\ 0 & 0 & 2 & 0 \\ 0 & 0 & 0 & 2 \end{bmatrix}$

C   **23.** A 3×3 matrix A has eigenvalues $\lambda_1 = 1$ and $\lambda_2 = 2$ with associated eigenspaces $S_1 = \text{span}\{(2, 1, 0), (1, 0, 1)\}$ and $S_2 = \text{span}\{(1, 1, 0)\}$. Find A. [*Hint:* Find P and D and use the relationship $A = PDP^{-1}$ to find A.]

24. Repeat Problem 23 if $\lambda_1 = 3$, $\lambda_2 = 4$, $S_3 = \text{span}\{(-1, -1, 1)\}$, and $S_4 = \text{span}\{(0, 1, 0), (-2, 0, 1)\}$.

25. If $A = PDP^{-1}$ where

$$D = \begin{bmatrix} \lambda_1 & 0 & \cdots & 0 \\ 0 & \lambda_2 & \cdots & 0 \\ & \cdot & & \cdot \\ & \cdot & & \cdot \\ & \cdot & & \cdot \\ 0 & 0 & \cdots & \lambda_n \end{bmatrix} \qquad P = [\mathbf{p}_1 \quad \mathbf{p}_2 \quad \cdots \quad \mathbf{p}_n]$$

show that λ_i is an eigenvalue of A with associated eigenvector $\mathbf{p}_i$, $i = 1$, $2, \ldots, n$.

In Problems 26–28, the matrix A is expressed in the form $A = PDP^{-1}$. Use the result in Problem 25 to find the eigenvalues of A and a basis for each eigenspace of A.

$$\qquad\qquad A \qquad\qquad\quad P \qquad\quad D \qquad\quad P^{-1}$$
26. $\begin{bmatrix} 17 & -12 \\ 24 & -17 \end{bmatrix} = \begin{bmatrix} 3 & -2 \\ 4 & -3 \end{bmatrix}\begin{bmatrix} 1 & 0 \\ 0 & -1 \end{bmatrix}\begin{bmatrix} 3 & -2 \\ 4 & -3 \end{bmatrix}$

$$\qquad\qquad A \qquad\qquad\qquad P \qquad\qquad D \qquad\qquad P^{-1}$$
27. $\begin{bmatrix} -1 & 4 & -6 \\ -2 & 5 & -6 \\ -1 & 2 & -2 \end{bmatrix} = \begin{bmatrix} 2 & 2 & -3 \\ 2 & 1 & 0 \\ 1 & 0 & 1 \end{bmatrix}\begin{bmatrix} 0 & 0 & 0 \\ 0 & 1 & 0 \\ 0 & 0 & 1 \end{bmatrix}\begin{bmatrix} 1 & -2 & 3 \\ -2 & 5 & -6 \\ -1 & 2 & -2 \end{bmatrix}$

$$\qquad\qquad A \qquad\qquad\qquad P \qquad\qquad D \qquad\qquad P^{-1}$$
28. $\begin{bmatrix} 7 & -9 & 4 \\ 4 & -6 & 4 \\ -1 & 1 & 2 \end{bmatrix} = \begin{bmatrix} 1 & 1 & -1 \\ 1 & 1 & 0 \\ 1 & 0 & 1 \end{bmatrix}\begin{bmatrix} 2 & 0 & 0 \\ 0 & -2 & 0 \\ 0 & 0 & 3 \end{bmatrix}\begin{bmatrix} 1 & -1 & 1 \\ -1 & 2 & -1 \\ -1 & 1 & 0 \end{bmatrix}$

In Problems 29–32, the matrix A is expressed in the form $A = PDP^{-1}$. If k is a positive integer, find A^k.

29.
$$\overset{A}{\begin{bmatrix} 1 & 1 \\ 0 & 2 \end{bmatrix}} = \overset{P}{\begin{bmatrix} 1 & 1 \\ 0 & 1 \end{bmatrix}} \overset{D}{\begin{bmatrix} 1 & 0 \\ 0 & 2 \end{bmatrix}} \overset{P^{-1}}{\begin{bmatrix} 1 & -1 \\ 0 & 1 \end{bmatrix}}$$

30.
$$\overset{A}{\begin{bmatrix} 2 & 1 \\ 0 & 4 \end{bmatrix}} = \overset{P}{\begin{bmatrix} 1 & 1 \\ 0 & 2 \end{bmatrix}} \overset{D}{\begin{bmatrix} 2 & 0 \\ 0 & 4 \end{bmatrix}} \overset{P^{-1}}{\begin{bmatrix} 1 & -\frac{1}{2} \\ 0 & \frac{1}{2} \end{bmatrix}}$$

31.
$$\overset{A}{\begin{bmatrix} 1 & 1 & -1 \\ 0 & 2 & 0 \\ 0 & 0 & 2 \end{bmatrix}} = \overset{P}{\begin{bmatrix} 1 & 1 & -1 \\ 0 & 1 & 0 \\ 0 & 0 & 1 \end{bmatrix}} \overset{D}{\begin{bmatrix} 1 & 0 & 0 \\ 0 & 2 & 0 \\ 0 & 0 & 2 \end{bmatrix}} \overset{P^{-1}}{\begin{bmatrix} 1 & -1 & 1 \\ 0 & 1 & 0 \\ 0 & 0 & 1 \end{bmatrix}}$$

32.
$$\overset{A}{\begin{bmatrix} 1 & 1 & -1 \\ 0 & 2 & 1 \\ 0 & 0 & 3 \end{bmatrix}} = \overset{P}{\begin{bmatrix} 1 & 1 & 0 \\ 0 & 1 & 1 \\ 0 & 0 & 1 \end{bmatrix}} \overset{D}{\begin{bmatrix} 1 & 0 & 0 \\ 0 & 2 & 0 \\ 0 & 0 & 3 \end{bmatrix}} \overset{P^{-1}}{\begin{bmatrix} 1 & -1 & 1 \\ 0 & 1 & -1 \\ 0 & 0 & 1 \end{bmatrix}}$$

33. If D is an $n \times n$ diagonal matrix with $d_{ii} \geq 0$, $i = 1, 2, \ldots, n$, show that there is a diagonal matrix S with the property that $S^2 = D$. The matrix S is called a **square root** of D.

34. If $A = PDP^{-1}$ and $B = PSP^{-1}$, where D and S are diagonal matrices satisfying $S^2 = D$, show that $B^2 = A$. (That is, B is a square root of A.)

In Problems 35 and 36, find all square roots of D. [Hint: There are four square roots for each matrix.]

35. $D = \begin{bmatrix} 1 & 0 \\ 0 & 4 \end{bmatrix}$

36. $D = \begin{bmatrix} 4 & 0 \\ 0 & 9 \end{bmatrix}$

In Problems 37 and 38, A is expressed in the form $A = PDP^{-1}$. Use the result in Problem 34 and the square roots computed in Problems 35 and 36 to find all square roots of A.

37.
$$\overset{A}{\begin{bmatrix} -2 & 6 \\ -3 & 7 \end{bmatrix}} = \overset{P}{\begin{bmatrix} 2 & 1 \\ 1 & 1 \end{bmatrix}} \overset{D}{\begin{bmatrix} 1 & 0 \\ 0 & 4 \end{bmatrix}} \overset{P^{-1}}{\begin{bmatrix} 1 & -1 \\ -1 & 2 \end{bmatrix}}$$

38.
$$\overset{A}{\begin{bmatrix} -1 & -10 \\ 5 & 14 \end{bmatrix}} = \overset{P}{\begin{bmatrix} 2 & 1 \\ -1 & -1 \end{bmatrix}} \overset{D}{\begin{bmatrix} 4 & 0 \\ 0 & 9 \end{bmatrix}} \overset{P^{-1}}{\begin{bmatrix} 1 & 1 \\ -1 & -2 \end{bmatrix}}$$

39. Let $A = PDP^{-1}$ where D is a diagonal matrix. If A is invertible, then prove each of the following statements:

(A) D is invertible

(B) D^{-1} is a diagonal matrix

(C) $A^{-1} = PD^{-1}P^{-1}$

(D) A^{-1} is diagonalizable

40. If A is a diagonalizable matrix and k is a positive integer, show that A^k is a diagonalizable matrix.

❚7-3❚ Symmetric Matrices

- Symmetric Matrices
- Orthogonal Matrices
- Eigenvectors of Symmetric Matrices
- Diagonalizing a Symmetric Matrix
- Additional Properties of Symmetric Matrices

In the preceding section, we saw that an $n \times n$ matrix A is diagonalizable if it satisfies either of the following conditions:

1. A has n distinct eigenvalues.

2. A has n linearly independent eigenvectors.

To use either of these conditions, it is necessary to compute the characteristic polynomial. This can be a very difficult task for $n > 4$. In this section, we will see that a matrix A that satisfies the simple condition $A = A^T$ is always diagonalizable. This is a very important result, since it allows us to conclude that the matrix A is diagonalizable without having to find the characteristic polynomial. Furthermore, we will see that the invertible matrix P which diagonalizes A has some very nice properties. (If we actually need to find P, it will still be necessary to find n linearly independent eigenvectors, as we did in Section 7-2.)

▪ Symmetric Matrices

We begin with a formal definition of the condition $A = A^T$.

Symmetric Matrices

An $n \times n$ matrix A is called a **symmetric matrix** if $A = A^T$.

If A is a general 3×3 matrix, then A and A^T can be written as

$$A = \begin{bmatrix} a_{11} & a_{12} & a_{13} \\ a_{21} & a_{22} & a_{23} \\ a_{31} & a_{32} & a_{33} \end{bmatrix} \qquad \begin{bmatrix} a_{11} & a_{21} & a_{31} \\ a_{12} & a_{22} & a_{32} \\ a_{13} & a_{23} & a_{33} \end{bmatrix} = A^T$$

In order for A to be symmetric, we must have

$$a_{21} = a_{12} \qquad a_{31} = a_{13} \qquad a_{32} = a_{23}$$

Thus, the diagonal elements are arbitrary, but the elements below the diagonal must be reflections of the elements above the diagonal, as illustrated below:

Similar statements can be made for $n \times n$ matrices.

Example 10 (A) $A = \begin{bmatrix} 1 & 2 & 3 \\ 2 & 4 & 5 \\ 3 & 5 & 6 \end{bmatrix}$ is symmetric

(B) $A = \begin{bmatrix} 1 & 0 & 0 & 0 \\ 1 & 1 & 0 & 0 \\ 0 & 0 & 1 & 0 \\ 0 & 0 & 0 & 1 \end{bmatrix}$ is not symmetric $(a_{21} \neq a_{12})$

Problem 10 Determine whether the following matrices are symmetric:

(A) $\begin{bmatrix} 1 & 0 & 0 \\ 0 & 0 & 0 \\ 0 & 1 & 0 \end{bmatrix}$ (B) $\begin{bmatrix} 1 & -1 & 1 & -1 \\ -1 & 1 & -1 & 1 \\ 1 & -1 & 1 & -1 \\ -1 & 1 & -1 & 1 \end{bmatrix}$ ∎

▪ Orthogonal Matrices

Before we consider the problem of diagonalizing a symmetric matrix, we need to introduce another type of matrix.

Orthogonal Matrices

A matrix P is said to be **orthogonal** if $P^{-1} = P^{T}$.

Example 11 If

$$P = \begin{bmatrix} \frac{3}{5} & \frac{4}{5} \\ -\frac{4}{5} & \frac{3}{5} \end{bmatrix}$$

then

$$PP^{T} = \begin{bmatrix} \frac{3}{5} & \frac{4}{5} \\ -\frac{4}{5} & \frac{3}{5} \end{bmatrix} \begin{bmatrix} \frac{3}{5} & -\frac{4}{5} \\ \frac{4}{5} & \frac{3}{5} \end{bmatrix} = \begin{bmatrix} 1 & 0 \\ 0 & 1 \end{bmatrix}$$

which shows that $P^{-1} = P^{T}$. Thus, P is orthogonal.

Problem 11   Is $P = \begin{bmatrix} 1 & 1 \\ 1 & -1 \end{bmatrix}$ an orthogonal matrix?   ▐▌

In Example 11, if R^2 is equipped with the standard inner product and

$$\mathbf{r}_1 = [\tfrac{3}{5} \quad \tfrac{4}{5}] \qquad \mathbf{r}_2 = [-\tfrac{4}{5} \quad \tfrac{3}{5}]$$

are the row vectors of P, then (verify this)

$$\langle \mathbf{r}_1, \mathbf{r}_1 \rangle = 1 \qquad \langle \mathbf{r}_1, \mathbf{r}_2 \rangle = 0 \qquad \langle \mathbf{r}_2, \mathbf{r}_2 \rangle = 1$$

That is, the rows of P form an orthonormal set of vectors in R^2. In the same way, the column vectors of P,

$$\mathbf{c}_1 = \begin{bmatrix} \tfrac{3}{5} \\ -\tfrac{4}{5} \end{bmatrix} \qquad \mathbf{c}_2 = \begin{bmatrix} \tfrac{4}{5} \\ \tfrac{3}{5} \end{bmatrix}$$

also form an orthonormal set of vectors in R^2 (verify this). Theorem 9 shows that similar statements can be made for any $n \times n$ orthogonal matrix. The proof of Theorem 9 is discussed in the exercises. (Since we are going to be using inner products frequently in what follows, we now assume that R^n is equipped with the standard inner product throughout this section.)

Theorem 9

> Let P be an $n \times n$ matrix. Then the following statements are equivalent:
>
> (A) P is orthogonal.
> (B) The rows of P are an orthonormal subset of R^n.
> (C) The columns of P are an orthonormal subset of R^n.

Example 12   Since the rows (and columns) of

$$P = \begin{bmatrix} \dfrac{1}{\sqrt{2}} & -\dfrac{1}{\sqrt{2}} & 0 \\ 0 & 0 & 1 \\ \dfrac{1}{\sqrt{2}} & \dfrac{1}{\sqrt{2}} & 0 \end{bmatrix}$$

form an orthonormal set in R^3 (verify this), P is an orthogonal matrix and

$$P^{-1} = P^{\mathrm{T}} = \begin{bmatrix} \dfrac{1}{\sqrt{2}} & 0 & \dfrac{1}{\sqrt{2}} \\ -\dfrac{1}{\sqrt{2}} & 0 & \dfrac{1}{\sqrt{2}} \\ 0 & 1 & 0 \end{bmatrix}$$

Problem 12 Show that P is orthogonal and find P^{-1}.

$$P = \begin{bmatrix} \dfrac{1}{\sqrt{3}} & \dfrac{1}{\sqrt{2}} & \dfrac{1}{\sqrt{6}} \\ \dfrac{1}{\sqrt{3}} & 0 & -\dfrac{2}{\sqrt{6}} \\ \dfrac{1}{\sqrt{3}} & -\dfrac{1}{\sqrt{2}} & \dfrac{1}{\sqrt{6}} \end{bmatrix}$$

∎

■ Eigenvectors of Symmetric Matrices

One of the key results in Section 7-2 stated that eigenvectors corresponding to distinct eigenvalues of a matrix A are linearly independent (Theorem 6). If A is a symmetric matrix, then Theorem 10 shows that these eigenvectors satisfy a stronger condition—they are orthogonal.

Theorem 10

> If A is a symmetric matrix, then eigenvectors corresponding to distinct eigenvalues of A are orthogonal.

Proof Let λ_1 and λ_2 be any two distinct eigenvalues of A, and let $\mathbf{x}_1$ and $\mathbf{x}_2$ be eigenvectors associated with λ_1 and λ_2, respectively. To show that $\mathbf{x}_1$ and $\mathbf{x}_2$ are orthogonal, we show that their inner product, $\langle \mathbf{x}_1, \mathbf{x}_2 \rangle$, is zero. We proceed as follows:

$$\langle A\mathbf{x}_1, \mathbf{x}_2 \rangle = \langle \lambda_1 \mathbf{x}_1, \mathbf{x}_2 \rangle \qquad A\mathbf{x}_1 = \lambda_1 \mathbf{x}_1$$
$$= \lambda_1 \langle \mathbf{x}_1, \mathbf{x}_2 \rangle$$

and

$\langle A\mathbf{x}_1, \mathbf{x}_2 \rangle = [A\mathbf{x}_1]^T \mathbf{x}_2$	The standard inner product in R^n can be expressed as a matrix product. See Section 5-1.
$= \mathbf{x}_1^T A^T \mathbf{x}_2$	Theorem 6 in Section 3-3
$= \mathbf{x}_1^T A \mathbf{x}_2$	$A = A^T$ by hypothesis
$= \mathbf{x}_1^T \lambda_2 \mathbf{x}_2$	$A\mathbf{x}_2 = \lambda_2 \mathbf{x}_2$
$= \lambda_2 \mathbf{x}_1^T \mathbf{x}_2$	Theorem 3 in Section 2-2
$= \lambda_2 \langle \mathbf{x}_1, \mathbf{x}_2 \rangle$	Return to inner product notation.

Thus,

$$\lambda_1 \langle \mathbf{x}_1, \mathbf{x}_2 \rangle = \lambda_2 \langle \mathbf{x}_1, \mathbf{x}_2 \rangle$$
$$(\lambda_1 - \lambda_2) \langle \mathbf{x}_1, \mathbf{x}_2 \rangle = 0$$

Since $\lambda_1 - \lambda_2 \neq 0$ by hypothesis, it follows that $\langle \mathbf{x}_1, \mathbf{x}_2 \rangle = 0$. Hence, $\mathbf{x}_1$ and $\mathbf{x}_2$ are orthogonal. ∎

Example 13   Let

$$A = \begin{bmatrix} -1 & 0 & 2 \\ 0 & 1 & 2 \\ 2 & 2 & 0 \end{bmatrix}$$

Notice that A is a symmetric matrix. The characteristic polynomial for A is

$$p(\lambda) = \det(\lambda I - A) = (\lambda + 3)\lambda(\lambda - 3)$$

and the eigenvalues of A are $\lambda_1 = -3$, $\lambda_2 = 0$, $\lambda_3 = 3$. Since A has three distinct eigenvalues, it is diagonalizable (Theorem 7 in Section 7-2). The vectors

$$\mathbf{x}_1 = \begin{bmatrix} -2 \\ -1 \\ 2 \end{bmatrix} \qquad \mathbf{x}_2 = \begin{bmatrix} 2 \\ -2 \\ 1 \end{bmatrix} \qquad \mathbf{x}_3 = \begin{bmatrix} 1 \\ 2 \\ 2 \end{bmatrix}$$

are eigenvectors corresponding to λ_1, λ_2, and λ_3, respectively (verify this). Theorem 10 implies that these vectors are orthogonal. The following calculations verify this fact:

$$\langle \mathbf{x}_1, \mathbf{x}_2 \rangle = (-2)2 + (-1)(-2) + 2(1) = 0$$
$$\langle \mathbf{x}_1, \mathbf{x}_3 \rangle = (-2)1 + (-1)2 + 2(2) = 0$$
$$\langle \mathbf{x}_2, \mathbf{x}_3 \rangle = 2(1) + (-2)2 + 1(2) = 0$$

Since a nonzero scalar multiple of an eigenvector is also an eigenvector, we can normalize $\mathbf{x}_1$, $\mathbf{x}_2$, and $\mathbf{x}_3$ to form an orthonormal set of eigenvectors of A:

$$\mathbf{p}_1 = \begin{bmatrix} -\frac{2}{3} \\ -\frac{1}{3} \\ \frac{2}{3} \end{bmatrix} \qquad \mathbf{p}_2 = \begin{bmatrix} \frac{2}{3} \\ -\frac{2}{3} \\ \frac{1}{3} \end{bmatrix} \qquad \mathbf{p}_3 = \begin{bmatrix} \frac{1}{3} \\ \frac{2}{3} \\ \frac{2}{3} \end{bmatrix}$$

The matrix

$$\begin{array}{ccc} \mathbf{p}_1 & \mathbf{p}_2 & \mathbf{p}_3 \end{array}$$
$$P = \begin{bmatrix} -\frac{2}{3} & \frac{2}{3} & \frac{1}{3} \\ -\frac{1}{3} & -\frac{2}{3} & \frac{2}{3} \\ \frac{2}{3} & \frac{1}{3} & \frac{2}{3} \end{bmatrix}$$

is an orthogonal matrix that diagonalizes A. That is (verify this),

$$P^{-1}AP = P^{\mathsf{T}}AP = \begin{bmatrix} -3 & 0 & 0 \\ 0 & 0 & 0 \\ 0 & 0 & 3 \end{bmatrix}$$

Problem 13   Find three orthogonal eigenvectors for

$$A = \begin{bmatrix} 7 & 0 & 4 \\ 0 & -7 & 4 \\ 4 & 4 & 0 \end{bmatrix}$$

■■

The 3×3 matrix A in Example 13 had three distinct eigenvalues and three orthogonal eigenvectors. What happens if an $n \times n$ symmetric matrix A does not have n distinct eigenvalues? It turns out that A will still have n orthogonal eigenvectors. That is, an $n \times n$ symmetric matrix always has n orthogonal eigenvectors, regardless of the number of distinct eigenvalues. This important result is stated as Theorem 11. The proof is beyond the scope of this book and is omitted.

Theorem 11

> If A is an $n \times n$ symmetric matrix, then A has n orthogonal eigenvectors.

Why are we interested in orthogonal eigenvectors? Recall that nonzero orthogonal vectors are linearly independent (Theorem 18 in Section 6-6). Thus, if an $n \times n$ matrix has n orthogonal eigenvectors, then it has n linearly independent eigenvectors and is diagonalizable. In other words, Theorem 11 implies that every symmetric matrix A is diagonalizable. Furthermore, it can be shown that it is always possible to find an orthogonal matrix P that diagonalizes A. This result is stated as Theorem 12. The details of the proof are omitted.

Theorem 12

> If A is a symmetric matrix, then A is diagonalizable, and there exists an orthogonal matrix P such that
>
> $$P^{-1}AP = P^TAP = D$$
>
> where D is a diagonal matrix.

▪ Diagonalizing a Symmetric Matrix

The procedure for diagonalizing matrices presented in Section 7-2 can be modified to find an orthogonal matrix that diagonalizes a symmetric matrix.

Diagonalizing a Symmetric Matrix

Let A be an $n \times n$ symmetric matrix with distinct eigenvalues λ_1, $\lambda_2, \ldots, \lambda_k$.

Step 1. Find a basis C_{λ_i} for each eigenspace S_{λ_i}.

Step 2. Apply the Gram–Schmidt process to C_{λ_i} in order to find an orthonormal basis B_{λ_i} for each eigenspace S_{λ_i}.

Step 3. Let P be the orthogonal matrix whose columns consist of all the orthonormal vectors found in Step 2.

Then

$$P^{\mathrm{T}}AP = D$$

where D is a diagonal matrix whose diagonal entries are the eigenvalues of A. The number of times each distinct eigenvalue is repeated on the diagonal of D is equal to the dimension of the corresponding eigenspace.

Example 14   Find an orthogonal matrix P that diagonalizes A, and find $P^{\mathrm{T}}AP$ if

$$A = \begin{bmatrix} 1 & 2 & -2 \\ 2 & 1 & -2 \\ -2 & -2 & 1 \end{bmatrix}$$

Solution   *Step 1.*   *Find a basis for each eigenspace.* The characteristic polynomial for A is

$$p(\lambda) = \det(\lambda I - A) = (\lambda + 1)^2(\lambda - 5)$$

and the distinct eigenvalues are $\lambda_1 = -1$ and $\lambda_2 = 5$. Omitting the details, bases for the eigenspaces are

$$C_{-1} = \left\{ \begin{bmatrix} -1 \\ 1 \\ 0 \end{bmatrix}, \begin{bmatrix} 1 \\ 0 \\ 1 \end{bmatrix} \right\} \quad \text{and} \quad C_5 = \left\{ \begin{bmatrix} -1 \\ -1 \\ 1 \end{bmatrix} \right\}$$

Notice that the eigenvectors in C_{-1} are not orthogonal to each other, but each of these vectors is orthogonal to the vector in C_5.

Step 2.   *Find an orthonormal basis for each eigenspace.* Applying the Gram–Schmidt process (details omitted) to C_{-1} and C_5 produces the orthonormal bases

$$B_{-1} = \left\{ \begin{bmatrix} -\dfrac{1}{\sqrt{2}} \\ \dfrac{1}{\sqrt{2}} \\ 0 \end{bmatrix}, \begin{bmatrix} \dfrac{1}{\sqrt{6}} \\ \dfrac{1}{\sqrt{6}} \\ \dfrac{2}{\sqrt{6}} \end{bmatrix} \right\} \quad \text{and} \quad B_5 = \left\{ \begin{bmatrix} -\dfrac{1}{\sqrt{3}} \\ -\dfrac{1}{\sqrt{3}} \\ \dfrac{1}{\sqrt{3}} \end{bmatrix} \right\}$$

You should verify that the vectors in B_{-1} are eigenvectors associated with $\lambda_1 = -1$, are orthogonal to each other, and are orthogonal to the vector in B_5. Notice that since C_5 consisted of a single vector, all we had to do to find B_5 was normalize this vector.

Step 3. Form the orthogonal matrix P.

$$
\overbrace{\phantom{\dfrac{1}{\sqrt{2}} \quad \dfrac{1}{\sqrt{6}}}}^{B_{-1}} \quad \overbrace{\phantom{\dfrac{1}{\sqrt{3}}}}^{B_5}
$$

$$P = \begin{bmatrix} -\dfrac{1}{\sqrt{2}} & \dfrac{1}{\sqrt{6}} & -\dfrac{1}{\sqrt{3}} \\ \dfrac{1}{\sqrt{2}} & \dfrac{1}{\sqrt{6}} & -\dfrac{1}{\sqrt{3}} \\ 0 & \dfrac{2}{\sqrt{6}} & \dfrac{1}{\sqrt{3}} \end{bmatrix}$$

Then

$$
\begin{array}{ccc} \lambda_1 & \lambda_1 & \lambda_2 \end{array}
$$

$$D = P^T A P = \begin{bmatrix} -1 & 0 & 0 \\ 0 & -1 & 0 \\ 0 & 0 & 5 \end{bmatrix}$$

Check this by actually multiplying P^T, A, and P.

Problem 14   Repeat Example 14 for

$$A = \begin{bmatrix} -1 & 2 & 2 \\ 2 & -1 & 2 \\ 2 & 2 & -1 \end{bmatrix}$$

∎

▪ Additional Properties of Symmetric Matrices

If A is a symmetric matrix, then an orthogonal matrix P and a diagonal matrix D satisfying $D = P^T A P$ always exist. In some applications, it is sufficient to find the matrix D, without first finding P. Theorem 13, which we state without proof, provides a method for doing this.

Theorem 13

Let A be an $n \times n$ symmetric matrix with distinct eigenvalues λ_1, $\lambda_2, \ldots, \lambda_k$.

(A) The characteristic polynomial for A can be written in the form

$$p(\lambda) = (\lambda - \lambda_1)^{m_1} (\lambda - \lambda_2)^{m_2} \cdot \cdots \cdot (\lambda - \lambda_k)^{m_k}$$

where $m_1 + m_2 + \cdots + m_k = n$. That is, $p(\lambda)$ has n real roots (not necessarily distinct).

(B) The dimension of each eigenspace is equal to the multiplicity of the corresponding eigenvalue. That is,

$$\dim(S_{\lambda_i}) = m_i$$

(C) If D is a diagonal matrix whose diagonal elements consist of the eigenvalues of A, with each eigenvalue λ_i repeated m_i times, then there exists an orthogonal matrix P such that

$$D = P^{\mathrm{T}} A P$$

Example 15   Let

$$A = \begin{bmatrix} -1 & 0 & 0 & 0 & 0 \\ 0 & 0 & 1 & 0 & 0 \\ 0 & 1 & 0 & 0 & 0 \\ 0 & 0 & 0 & 0 & 1 \\ 0 & 0 & 0 & 1 & 0 \end{bmatrix}$$

The characteristic polynomial for A is

$$p(\lambda) = (\lambda + 1)^3 (\lambda - 1)^2$$

Thus, $\lambda_1 = -1$ is an eigenvalue of multiplicity 3 and $\lambda_2 = 1$ is an eigenvalue of multiplicity 2. Theorem 13 implies that S_{-1} is a three-dimensional space and S_1 is a two-dimensional space. Furthermore, there exists an orthogonal matrix P such that

$$P^{\mathrm{T}}AP = \begin{matrix} \overbrace{}^{m_1 = 3} & \overbrace{}^{m_2 = 2} \\ \begin{bmatrix} -1 & 0 & 0 & 0 & 0 \\ 0 & -1 & 0 & 0 & 0 \\ 0 & 0 & -1 & 0 & 0 \\ 0 & 0 & 0 & 1 & 0 \\ 0 & 0 & 0 & 0 & 1 \end{bmatrix} \end{matrix} = D$$

Problem 15 Let

$$A = \begin{bmatrix} 0 & 1 & -1 & 0 & 0 \\ 1 & 0 & 1 & 0 & 0 \\ -1 & 1 & 0 & 0 & 0 \\ 0 & 0 & 0 & 1 & 0 \\ 0 & 0 & 0 & 0 & 1 \end{bmatrix}$$

(A) Find the characteristic polynomial and the eigenvalues of A.
(B) Find the dimension of each eigenspace.
(C) Find a diagonal matrix D such that $P^T AP = D$. (Do not find P.) **‖**

Answers to
Matched Problems

10. (A) Not symmetric   (B) Symmetric   **11.** No

12. $P^{-1} = P^T = \begin{bmatrix} 1/\sqrt{3} & 1/\sqrt{3} & 1/\sqrt{3} \\ 1/\sqrt{2} & 0 & -1/\sqrt{2} \\ 1/\sqrt{6} & -2/\sqrt{6} & 1/\sqrt{6} \end{bmatrix}$ **13.** $\begin{bmatrix} -\frac{1}{9} \\ -\frac{8}{9} \\ \frac{4}{9} \end{bmatrix}$, $\begin{bmatrix} -\frac{4}{9} \\ \frac{4}{9} \\ \frac{7}{9} \end{bmatrix}$, $\begin{bmatrix} \frac{8}{9} \\ \frac{1}{9} \\ \frac{4}{9} \end{bmatrix}$

14. $P = \begin{bmatrix} -1/\sqrt{2} & 1/\sqrt{6} & 1/\sqrt{3} \\ 1/\sqrt{2} & 1/\sqrt{6} & 1/\sqrt{3} \\ 0 & -2/\sqrt{6} & 1/\sqrt{3} \end{bmatrix}$; $D = \begin{bmatrix} -3 & 0 & 0 \\ 0 & -3 & 0 \\ 0 & 0 & 3 \end{bmatrix}$

15. (A) $p(\lambda) = (\lambda + 2)(\lambda - 1)^4$; $\lambda_1 = -2$ (multiplicity 1), $\lambda_2 = 1$ (multiplicity 4)
    (B) $\dim(S_{-2}) = 1$, $\dim(S_1) = 4$

(C) $D = \begin{bmatrix} -2 & 0 & 0 & 0 & 0 \\ 0 & 1 & 0 & 0 & 0 \\ 0 & 0 & 1 & 0 & 0 \\ 0 & 0 & 0 & 1 & 0 \\ 0 & 0 & 0 & 0 & 1 \end{bmatrix}$

‖ Exercise 7-3

A   *In Problems 1–4, determine whether each matrix is symmetric.*

1. $\begin{bmatrix} 1 & 2 \\ 2 & -1 \end{bmatrix}$ **2.** $\begin{bmatrix} 1 & 2 \\ -2 & 1 \end{bmatrix}$

3. $\begin{bmatrix} 1 & 2 & -1 \\ 2 & 1 & -3 \\ 1 & -3 & 1 \end{bmatrix}$ **4.** $\begin{bmatrix} 1 & -1 & 2 \\ -1 & -1 & 4 \\ 2 & 4 & 1 \end{bmatrix}$

In Problems 5–10, determine whether P is orthogonal. If P is orthogonal, find P^{-1}.

5. $P = \begin{bmatrix} \frac{3}{5} & \frac{4}{5} \\ -\frac{4}{5} & \frac{3}{5} \end{bmatrix}$ **6.** $P = \begin{bmatrix} \frac{5}{13} & \frac{12}{13} \\ -\frac{12}{13} & -\frac{5}{13} \end{bmatrix}$

7. $P = \begin{bmatrix} \dfrac{1}{\sqrt{3}} & -\dfrac{2}{\sqrt{3}} \\ \dfrac{2}{\sqrt{3}} & \dfrac{1}{\sqrt{3}} \end{bmatrix}$

8. $P = \begin{bmatrix} \dfrac{1}{\sqrt{10}} & -\dfrac{3}{\sqrt{10}} \\ -\dfrac{3}{\sqrt{10}} & -\dfrac{1}{\sqrt{10}} \end{bmatrix}$

9. $P = \begin{bmatrix} \frac{2}{7} & \frac{6}{7} & \frac{3}{7} \\ \frac{3}{7} & \frac{2}{7} & -\frac{6}{7} \\ \frac{6}{7} & -\frac{3}{7} & \frac{2}{7} \end{bmatrix}$

10. $P = \begin{bmatrix} \frac{2}{11} & \frac{9}{11} & \frac{6}{11} \\ \frac{6}{11} & -\frac{6}{11} & \frac{7}{11} \\ \frac{9}{11} & \frac{2}{11} & -\frac{6}{11} \end{bmatrix}$

B   In Problems 11–16, find the characteristic polynomial for A. Then use Theorem 13 to find a diagonal matrix D satisfying $D = P^T A P$ for some orthogonal matrix P. Do not find P. (Since D is not unique, your answer may differ from the one in the answer section.)

11. $A = \begin{bmatrix} -3 & 1 & 2 \\ 1 & -3 & 2 \\ 2 & 2 & 0 \end{bmatrix}$

12. $A = \begin{bmatrix} 0 & -4 & 2 \\ -4 & 0 & 2 \\ 2 & 2 & 3 \end{bmatrix}$

13. $A = \begin{bmatrix} 1 & 0 & 0 & 1 \\ 0 & 2 & 0 & 0 \\ 0 & 0 & 2 & 0 \\ 1 & 0 & 0 & 1 \end{bmatrix}$

14. $A = \begin{bmatrix} 1 & 1 & 0 & 0 \\ 1 & 1 & 0 & 0 \\ 0 & 0 & 2 & 0 \\ 0 & 0 & 0 & 0 \end{bmatrix}$

15. $A = \begin{bmatrix} 1 & 0 & 0 & 0 & -1 \\ 0 & 2 & 0 & 0 & 0 \\ 0 & 0 & -1 & 0 & 0 \\ 0 & 0 & 0 & -1 & 0 \\ -1 & 0 & 0 & 0 & 1 \end{bmatrix}$

16. $A = \begin{bmatrix} 1 & 0 & 0 & 0 & 0 \\ 0 & 0 & 0 & 0 & -1 \\ 0 & 0 & 0 & -1 & 0 \\ 0 & 0 & -1 & 0 & 0 \\ 0 & -1 & 0 & 0 & 0 \end{bmatrix}$

In Problems 17–26, find an orthogonal matrix P and a diagonal matrix D satisfying $P^T A P = D$. Compute the product $P^T A P$ to check your answer. (Since P and D are not unique, your answer may differ from the one in the answer section.)

17. $A = \begin{bmatrix} 1 & 2 \\ 2 & 1 \end{bmatrix}$

18. $A = \begin{bmatrix} 1 & 2 \\ 2 & -2 \end{bmatrix}$

19. $A = \begin{bmatrix} -2 & 0 & 4 \\ 0 & 2 & -4 \\ 4 & -4 & 0 \end{bmatrix}$

20. $A = \begin{bmatrix} 7 & 0 & 4 \\ 0 & -7 & 4 \\ 4 & 4 & 0 \end{bmatrix}$

21. $A = \begin{bmatrix} -1 & 1 & 2 \\ 1 & -1 & 2 \\ 2 & 2 & 2 \end{bmatrix}$

22. $A = \begin{bmatrix} 3 & -2 & 2 \\ -2 & 0 & -1 \\ 2 & -1 & 0 \end{bmatrix}$

23. $A = \begin{bmatrix} -1 & 2 & 0 & 0 \\ 2 & 2 & 0 & 0 \\ 0 & 0 & -1 & 0 \\ 0 & 0 & 0 & -1 \end{bmatrix}$

24. $A = \begin{bmatrix} 2 & 1 & 0 & 0 \\ 1 & 2 & 0 & 0 \\ 0 & 0 & 3 & 0 \\ 0 & 0 & 0 & 1 \end{bmatrix}$

25. $A = \begin{bmatrix} 2 & 1 & 0 & 0 \\ 1 & 3 & -1 & -2 \\ 0 & -1 & 2 & 0 \\ 0 & -2 & 0 & 2 \end{bmatrix}$

26. $A = \begin{bmatrix} 1 & 1 & 0 & 0 \\ 1 & 1 & 2 & 2 \\ 0 & 2 & 1 & 0 \\ 0 & 2 & 0 & 1 \end{bmatrix}$

C In Problems 27–30, let

$$P = \begin{matrix} & \mathbf{c}_1 & \mathbf{c}_2 & \mathbf{c}_3 & \\ & \begin{bmatrix} p_{11} & p_{12} & p_{13} \\ p_{21} & p_{22} & p_{23} \\ p_{31} & p_{32} & p_{33} \end{bmatrix} & \begin{matrix} \mathbf{r}_1 \\ \mathbf{r}_2 \\ \mathbf{r}_3 \end{matrix} \end{matrix}$$

Let $\mathbf{r}_1, \mathbf{r}_2$, and $\mathbf{r}_3$ be the row vectors of P and let $\mathbf{c}_1, \mathbf{c}_2$, and $\mathbf{c}_3$ be the column vectors of P. Problems 27–30 provide a proof of Theorem 9 in the case $n = 3$. Prove each statement wihout reference to Theorem 9.

27. Show that

$$PP^T = \begin{bmatrix} \langle \mathbf{r}_1, \mathbf{r}_1 \rangle & \langle \mathbf{r}_1, \mathbf{r}_2 \rangle & \langle \mathbf{r}_1, \mathbf{r}_3 \rangle \\ \langle \mathbf{r}_2, \mathbf{r}_1 \rangle & \langle \mathbf{r}_2, \mathbf{r}_2 \rangle & \langle \mathbf{r}_2, \mathbf{r}_3 \rangle \\ \langle \mathbf{r}_3, \mathbf{r}_1 \rangle & \langle \mathbf{r}_3, \mathbf{r}_2 \rangle & \langle \mathbf{r}_3, \mathbf{r}_3 \rangle \end{bmatrix}$$

28. Show that

$$P^T P = \begin{bmatrix} \langle \mathbf{c}_1, \mathbf{c}_1 \rangle & \langle \mathbf{c}_1, \mathbf{c}_2 \rangle & \langle \mathbf{c}_1, \mathbf{c}_3 \rangle \\ \langle \mathbf{c}_2, \mathbf{c}_1 \rangle & \langle \mathbf{c}_2, \mathbf{c}_2 \rangle & \langle \mathbf{c}_2, \mathbf{c}_3 \rangle \\ \langle \mathbf{c}_3, \mathbf{c}_1 \rangle & \langle \mathbf{c}_3, \mathbf{c}_2 \rangle & \langle \mathbf{c}_3, \mathbf{c}_3 \rangle \end{bmatrix}$$

29. Show that P is orthogonal if and only if the rows of P form an orthonormal set in R^3. [*Hint*: Use the result in Problem 27.]

30. Show that P is orthogonal if and only if the columns of P form an orthonormal set in R^3. [*Hint*: Use the result in Problem 28.]

31. *Theorem 11.* Prove Theorem 11 for $n = 2$. That is, show that every 2×2 symmetric matrix has two orthogonal eigenvectors.

32. If P is an orthogonal $n \times n$ matrix, show that

$$\langle P\mathbf{x}, P\mathbf{y} \rangle = \langle \mathbf{x}, \mathbf{y} \rangle$$

for all $\mathbf{x}$ and $\mathbf{y}$ in R^n. [*Hint*: Express the inner product as matrix multiplication, as in the proof of Theorem 10.]

33. If P and Q are orthogonal $n \times n$ matrices, show that PQ is orthogonal.

34. If P is an orthogonal matrix, show that $\det(P) = \pm 1$.

❙ 7-4❙ Chapter Review

❙❙ Exercise 7-4 Chapter Review

*Work through all the problems in this chapter review and check your answers in
the back of the book. (Answers to most review problems are there.) Where weaknesses show up, review appropriate sections in the text.*

*Problems that involve finding matrices P and D satisfying $P^{-1}AP = D$
do not always have unique answers. You should check carefully any of your
answers that do not agree with those in the back of the book.*

A *Problems 1–4 refer to the matrix A and the vectors $\mathbf{x}_1$, $\mathbf{x}_2$ given below:*

$$A = \begin{bmatrix} 4 & 2 \\ 3 & -1 \end{bmatrix} \qquad \mathbf{x}_1 = \begin{bmatrix} -1 \\ 3 \end{bmatrix} \qquad \mathbf{x}_2 = \begin{bmatrix} 2 \\ 1 \end{bmatrix}$$

1. Find $A\mathbf{x}_1$. Express the answer in the form $\lambda_1 \mathbf{x}_1$.
2. Find $A\mathbf{x}_2$. Express the answer in the form $\lambda_2 \mathbf{x}_2$.
3. Use the results of Problems 1 and 2 to find the eigenvalues of A.
4. Compute $P^{-1}AP$ for

$$P = [\mathbf{x}_1 \quad \mathbf{x}_2] = \begin{bmatrix} -1 & 2 \\ 3 & 1 \end{bmatrix}$$

Problems 5–8 refer to the matrix

$$A = \begin{bmatrix} 3 & -2 \\ -1 & 2 \end{bmatrix}$$

5. Find the characteristic polynomial for A.
6. Find the eigenvalues of A.
7. Find a basis for each eigenspace of A.
8. Find an invertible matrix P and a diagonal matrix D satisfying $P^{-1}AP = D$.

In Problems 9 and 10, determine the values of a, b, and c so that A is a symmetric matrix, if possible.

9. $A = \begin{bmatrix} 1 & a & -4 \\ 3 & -1 & c \\ b & 0 & 2 \end{bmatrix}$

10. $\begin{bmatrix} 3 & -5 & b \\ a & 0 & 6 \\ -2 & -6 & c \end{bmatrix}$

In Problems 11–14, determine whether P is orthogonal. If P is orthogonal, find P^{-1}.

11. $P = \begin{bmatrix} \dfrac{1}{2} & -\dfrac{1}{2} \\ \dfrac{\sqrt{3}}{2} & \dfrac{\sqrt{3}}{2} \end{bmatrix}$

12. $P = \begin{bmatrix} \dfrac{1}{2} & -\dfrac{\sqrt{3}}{2} \\ \dfrac{\sqrt{3}}{2} & \dfrac{1}{2} \end{bmatrix}$

13. $P = \begin{bmatrix} \dfrac{1}{\sqrt{3}} & \dfrac{1}{\sqrt{2}} & \dfrac{1}{\sqrt{6}} \\ \dfrac{1}{\sqrt{3}} & 0 & -\dfrac{2}{\sqrt{6}} \\ \dfrac{1}{\sqrt{3}} & -\dfrac{1}{\sqrt{2}} & \dfrac{1}{\sqrt{6}} \end{bmatrix}$

14. $P = \begin{bmatrix} \dfrac{1}{\sqrt{3}} & \dfrac{1}{\sqrt{2}} & 0 \\ \dfrac{1}{\sqrt{3}} & 0 & 1 \\ \dfrac{1}{\sqrt{3}} & -\dfrac{1}{\sqrt{2}} & 0 \end{bmatrix}$

Problems 15–18 refer to the symmetric matrix

$$A = \begin{bmatrix} -2 & 2 \\ 2 & 1 \end{bmatrix}$$

15. Find the characteristic polynomial for A.
16. Find the eigenvalues of A.
17. Find an orthonormal basis for each eigenspace of A.
18. Find an orthogonal matrix P and a diagonal matrix D satisfying $P^{\mathrm{T}}AP = D$.

B In Problems 19–21, show that A is diagonalizable by showing that one of the following conditions holds: (A) A is symmetric. (B) A has three distinct eigenvalues. (C) A has three linearly independent eigenvectors.

19. $A = \begin{bmatrix} 1 & 0 & 0 \\ -1 & 2 & -2 \\ -1 & 1 & -1 \end{bmatrix}$

20. $A = \begin{bmatrix} 4 & -5 & 6 \\ -5 & 3 & 7 \\ 6 & 7 & 2 \end{bmatrix}$

21. $A = \begin{bmatrix} 0 & 0 & 0 \\ 1 & 0 & 2 \\ 0 & 1 & 2 \end{bmatrix}$

In Problems 22–27, find the characteristic polynomial for A, the eigenvalues of A, and a basis for each eigenspace of A. If A is diagonalizable, find an invertible matrix P and a diagonal matrix D satisfying $P^{-1}AP = D$.

22. $A = \begin{bmatrix} 6 & 6 & -8 \\ 0 & 0 & 0 \\ 4 & 4 & -6 \end{bmatrix}$

23. $A = \begin{bmatrix} 3 & 0 & 0 \\ 4 & 0 & -1 \\ 2 & 1 & 0 \end{bmatrix}$

24. $A = \begin{bmatrix} 0 & 0 & 0 \\ 1 & 1 & 1 \\ -1 & 0 & 1 \end{bmatrix}$ **25.** $A = \begin{bmatrix} 5 & -6 & 0 \\ 3 & -4 & 0 \\ -3 & 6 & 2 \end{bmatrix}$

26. $A = \begin{bmatrix} -1 & 0 & -1 & -1 \\ 0 & -1 & 0 & 0 \\ 0 & -1 & 2 & 1 \\ 0 & 0 & 1 & 2 \end{bmatrix}$ **27.** $A = \begin{bmatrix} 4 & -6 & 0 & 0 \\ 2 & -4 & 0 & 0 \\ 0 & 0 & 2 & 0 \\ 0 & 0 & 0 & -2 \end{bmatrix}$

In Problems 28 and 29, find an orthogonal matrix P and a diagonal matrix D satisfying $P^{T}AP = D$.

28. $A = \begin{bmatrix} 7 & 0 & 6 \\ 0 & -7 & 6 \\ 6 & 6 & 0 \end{bmatrix}$ **29.** $A = \begin{bmatrix} 3 & -1 & 1 \\ -1 & 3 & -1 \\ 1 & -1 & 3 \end{bmatrix}$

30. Find the characteristic polynomial for the matrix A given below, and use Theorem 13 in Section 7-3 to find a diagonal matrix D satisfying $D = P^{T}AP$ for some orthogonal matrix P. Do not find P.

$$A = \begin{bmatrix} 3 & 0 & 0 & 1 \\ 0 & 4 & 0 & 0 \\ 0 & 0 & 4 & 0 \\ 1 & 0 & 0 & 3 \end{bmatrix}$$

C   Problems 31 and 32 refer to the matrix

$$A = \begin{bmatrix} 17 & 16 \\ -8 & -7 \end{bmatrix} = \overset{P}{\begin{bmatrix} -1 & -2 \\ 1 & 1 \end{bmatrix}} \overset{D}{\begin{bmatrix} 1 & 0 \\ 0 & 9 \end{bmatrix}} \overset{P^{-1}}{\begin{bmatrix} 1 & 2 \\ -1 & -1 \end{bmatrix}}$$

31. Find A^{k} for any positive integer k.
32. Find all matrices B satisfying $B^{2} = A$.
33. Find a 3×3 matrix with eigenvalues $\lambda_{1} = 1$, $\lambda_{2} = 2$, and $\lambda_{3} = 3$, and associated eigenvectors

$$\mathbf{x}_{1} = \begin{bmatrix} 1 \\ 1 \\ 0 \end{bmatrix} \qquad \mathbf{x}_{2} = \begin{bmatrix} 0 \\ 1 \\ 0 \end{bmatrix} \qquad \mathbf{x}_{3} = \begin{bmatrix} 3 \\ -1 \\ 2 \end{bmatrix}$$

34. Let $p(\lambda) = \lambda^3 + a_2\lambda^2 + a_1\lambda + a_0$ and let

$$A = \begin{bmatrix} 0 & 0 & -a_0 \\ 1 & 0 & -a_1 \\ 0 & 1 & -a_2 \end{bmatrix}$$

Show that $p(\lambda)$ is the characteristic polynomial for A.

35. Repeat Problem 34 for $p(\lambda) = \lambda^4 + a_3\lambda^3 + a_2\lambda^2 + a_1\lambda + a_0$ and

$$A = \begin{bmatrix} 0 & 0 & 0 & -a_0 \\ 1 & 0 & 0 & -a_1 \\ 0 & 1 & 0 & -a_2 \\ 0 & 0 & 1 & -a_3 \end{bmatrix}$$

36. If A and B are $n \times n$ matrices, show that AB and BA have the same eigenvalues. [*Hint:* Consider two cases, $\lambda = 0$ and $\lambda \neq 0$.]

|8| Linear Transformations

| 8-1 | **Linear Transformations and Linear Operators**

- Functions Defined on Vector Spaces
- Definition of a Linear Transformation
- Additional Examples of Linear Transformations
- Properties of Linear Transformations

▪ Functions Defined on Vector Spaces

The concept of a function plays a fundamental role in the study of mathematics. In this chapter, we will study an important class of functions called *linear transformations*. In addition to being of great theoretical interest to mathematicians, linear transformations are used extensively in applications of linear algebra to areas such as physics, engineering, economics, and biology, as well as other areas in mathematics.

We begin by introducing some terminology related to functions defined on vector spaces. Let V and W be vector spaces. We say that a function T **maps** V into W if T assigns a unique vector $T(\mathbf{v}) \in W$ to each vector $\mathbf{v} \in V$ (see Figure 1). The statement "T maps V into W" is written symbolically as $T: V \to W$. The vector $T(\mathbf{v})$ is called the **image** of $\mathbf{v}$ under T. In familiar function terminology, $\mathbf{v}$ is a domain element of T and $T(\mathbf{v})$ is a range element of T.

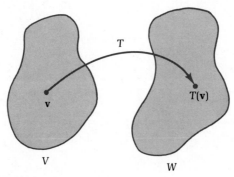

Figure 1 $T: V \to W$

Example 1   Let $\mathbf{v} = (x, y, z) \in R^3$ and let $T: R^3 \to R^2$ be defined by

$$T(\mathbf{v}) = (x + y, y + z)$$

(A) Find the image of $\mathbf{v} = (1, 2, 3)$ under T.

(B) Show that $T(\mathbf{u} + \mathbf{v}) = T(\mathbf{u}) + T(\mathbf{v})$ for all $\mathbf{u}$ and $\mathbf{v}$ in R^3.

Solution   (A) $T(\mathbf{v}) = T(1, 2, 3) = (1 + 2, 2 + 3) = (3, 5)$

The vector $\mathbf{w} = (3, 5)$, an element of R^2, is the image of $\mathbf{v}$, an element of R^3, under T. [Notice that we have omitted one set of parentheses, writing $T(1, 2, 3)$ in place of $T((1, 2, 3))$. This is common practice when writing such expressions.]

(B) If $\mathbf{u} = (x_1, y_1, z_1)$ and $\mathbf{v} = (x_2, y_2, z_2)$ are any two vectors in R^3, then

$$\overset{\text{x}}{} \quad \overset{\text{y}}{} \quad \overset{\text{z}}{}$$

$$T(\mathbf{u} + \mathbf{v}) = T(x_1 + x_2, y_1 + y_2, z_1 + z_2)$$

$$\overset{\text{x + y}}{} \quad \overset{\text{y + z}}{}$$

$$= (x_1 + x_2 + y_1 + y_2, y_1 + y_2 + z_1 + z_2)$$

and

$$T(\mathbf{u}) + T(\mathbf{v}) = (x_1 + y_1, y_1 + z_1) + (x_2 + y_2, y_2 + z_2)$$

$$= (x_1 + y_1 + x_2 + y_2, y_1 + z_1 + y_2 + z_2)$$

$$= (x_1 + x_2 + y_1 + y_2, y_1 + y_2 + z_1 + z_2)$$

Thus, $T(\mathbf{u} + \mathbf{v}) = T(\mathbf{u}) + T(\mathbf{v})$.

Problem 1   Refer to Example 1.

(A) Find the image of $\mathbf{v} = (2, 3, 4)$ under T.

(B) Show that $T(k\mathbf{v}) = kT(\mathbf{v})$ for any $\mathbf{v} \in V$ and any scalar k.   ▮

▪ Definition of a Linear Transformation

Let T be the function defined in Example 1. Example 1(B) shows that T preserves the vector space operation of addition. That is, the image of the sum of two vectors is equal to the sum of the images of the vectors. Problem 1(B) shows that T also preserves scalar multiplication. Functions defined on vector spaces that preserve addition and scalar multiplication are called *linear transformations*.

Linear Transformation

Let V and W be vector spaces, and let T map V into W. The function T is called a **linear transformation from V to W** if:

1. $T(\mathbf{u} + \mathbf{v}) = T(\mathbf{u}) + T(\mathbf{v})$ for all $\mathbf{u}$ and $\mathbf{v}$ in V
2. $T(k\mathbf{v}) = kT(\mathbf{v})$ for all $\mathbf{v}$ in V and all scalars k

If $W = V$, then the linear transformation is called a **linear operator on V**.

Example 1(B) and Problem 1(B) together imply that the function T defined in Example 1 is a linear transformation from $V = R^3$ to $W = R^2$.

Notice that the equations in the definition of a linear transformation involve operations in two different vector spaces:

Addition in V	Addition in W	Scalar multiplication in V	Scalar multiplication in W

$$T(\mathbf{u} + \mathbf{v}) = T(\mathbf{u}) + T(\mathbf{v}) \qquad T(k\mathbf{v}) \quad = \quad kT(\mathbf{v})$$

Example 2   Let $T: P_2 \to R^3$ be defined by

$$T(a_0 + a_1 x + a_2 x^2) = (0, a_1, 2a_2)$$

Is T a linear transformation?

Solution   If $\mathbf{u} = a_0 + a_1 x + a_2 x^2$ and $\mathbf{v} = b_0 + b_1 x + b_2 x^2$ are any two vectors in P_2, then

$$\mathbf{u} + \mathbf{v} = a_0 + b_0 + (a_1 + b_1)x + (a_2 + b_2)x^2$$

and

$$\begin{aligned} T(\mathbf{u} + \mathbf{v}) &= (0, a_1 + b_1, 2(a_2 + b_2)) \\ &= (0, a_1, 2a_2) + (0, b_1, 2b_2) \\ &= T(\mathbf{u}) + T(\mathbf{v}) \end{aligned}$$

If k is a scalar, then

$$k\mathbf{u} + ka_0 + ka_1 x + ka_2 x^2$$

and

$$\begin{aligned} T(k\mathbf{u}) &= (0, ka_1, 2ka_2) \\ &= k(0, a_1, 2a_2) \\ &= kT(\mathbf{u}) \end{aligned}$$

Thus, T is a linear transformation from P_2 to R^3.

Problem 2   Let $T: P_2 \to R^3$ be defined by

$$T(a_0 + a_1 x + a_2 x^2) = (a_0, 2a_1, 3a_2)$$

Is T a linear transformation?

Example 3   Let $T: M_{2\times 2} \to R^5$ be defined by

$$T\left(\begin{bmatrix} a & b \\ c & d \end{bmatrix}\right) = (1, a, b, c, d)$$

Is T a linear transformation?

Solution   Let

$$\mathbf{u} = \begin{bmatrix} a_1 & b_1 \\ c_1 & d_1 \end{bmatrix} \quad \text{and} \quad \mathbf{v} = \begin{bmatrix} a_2 & b_2 \\ c_2 & d_2 \end{bmatrix}$$

be any two vectors in $M_{2\times 2}$. Then

$$\mathbf{u} + \mathbf{v} = \begin{bmatrix} a_1 + a_2 & b_1 + b_2 \\ c_1 + c_2 & d_1 + d_2 \end{bmatrix}$$

and

$$T(\mathbf{u} + \mathbf{v}) = (1, a_1 + a_2, b_1 + b_2, c_1 + c_2, d_1 + d_2)$$

On the other hand,

$$T(\mathbf{u}) + T(\mathbf{v}) = (1, a_1, b_1, c_1, d_1) + (1, a_2, b_2, c_2, d_2)$$
$$= (2, a_1 + a_2, b_1 + b_2, c_1 + c_2, d_1 + d_2)$$

Since the first component of $T(\mathbf{u} + \mathbf{v})$ is 1 and the first component of $T(\mathbf{u}) + T(\mathbf{v})$ is 2, these vectors are not equal. That is,

$$T(\mathbf{u} + \mathbf{v}) \neq T(\mathbf{u}) + T(\mathbf{v})$$

and T is not a linear transformation.

Problem 3   Let $T: M_{2\times 2} \to R^3$ be defined by

$$T\left(\begin{bmatrix} a & b \\ c & d \end{bmatrix} \right) = (a + b, c + d, 2)$$

Is T a linear transformation?   ▌▌

▪ Additional Examples of Linear Transformations

One reason for studying linear transformations is that many of the operations used in linear algebra can be viewed as linear transformations. Developing general properties of linear transformations often provides additional insight into these operations. The following series of examples will illustrate the relationship between linear transformations and some of the operations we have already studied. Additional examples will be considered in the exercises.

Example 4   Let V be any vector space, and let $T: V \to V$ be defined by $T(\mathbf{v}) = a\mathbf{v}$ for a fixed scalar a. [In other words, T is the operation of multiplying a vector by a (fixed) scalar a.] Show that T is a linear transformation from V to V (that is, T is a linear operator on V).

Solution   If $\mathbf{u}$ and $\mathbf{v}$ are in V and k is any scalar, then using the properties of scalar multiplication, we have

$$T(\mathbf{u} + \mathbf{v}) = a(\mathbf{u} + \mathbf{v}) = a\mathbf{u} + a\mathbf{v} = T(\mathbf{u}) + T(\mathbf{v})$$

and

$$T(k\mathbf{u}) = a(k\mathbf{u}) = k(a\mathbf{u}) = \; kT(\mathbf{u})$$

Thus, T is a linear transformation from V to V (that is, a linear operator on V).

In Example 4, if $a > 1$, then T stretches each vector in V and is called a **dilation** [see Figure 2(A) where $V = R^2$]. If $0 < a < 1$, then T shrinks each vector in V and is called a **contraction** [see Figure 2(B)]. If $a = 1$, then $T(\mathbf{v}) = \mathbf{v}$ for all $\mathbf{v}$ in V, and T is called the **identity transformation** on V.

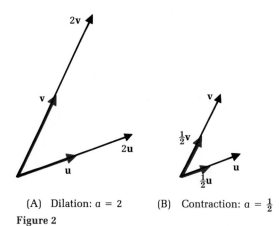

(A) Dilation: $a = 2$ (B) Contraction: $a = \frac{1}{2}$

Figure 2

Problem 4 Let V and W be any vector spaces, and let $T: V \to W$ be defined by $T(\mathbf{v}) = \mathbf{0}$ for all $\mathbf{v}$ in V. Show that T is a linear transformation. ▮▮

The transformation in Problem 4 is called the **zero transformation** from V to W.

Example 5 Let A be an $m \times n$ matrix, and let $T: R^n \to R^m$ be defined by

$$T(\mathbf{x}) = A\mathbf{x} \tag{1}$$

where $\mathbf{x}$ is an $n \times 1$ column vector in R^n. [Notice that $T(\mathbf{x})$ is an $m \times 1$ column vector in R^m.] Show that T is a linear transformation from R^n to R^m.

Solution If $\mathbf{x}$ and $\mathbf{y}$ are any vectors in R^n and k is any scalar, then using the familiar properties of matrix multiplication, we have

$$T(\mathbf{x} + \mathbf{y}) = A(\mathbf{x} + \mathbf{y}) = A\mathbf{x} + A\mathbf{y} = T(\mathbf{x}) + T(\mathbf{y})$$

and

$$T(k\mathbf{x}) = A(k\mathbf{x}) = k(A\mathbf{x}) = kT(\mathbf{x})$$

Thus, T is a linear transformation from R^n to R^m.

Any linear transformation of the form (1) is called a **matrix transformation.** For a specific example of a matrix transformation, let $T: R^2 \rightarrow R^3$ be defined by

$$T(\mathbf{x}) = \begin{bmatrix} 0 & 1 \\ 1 & 0 \\ 1 & 1 \end{bmatrix} \mathbf{x}$$

Then

$$T\left(\begin{bmatrix} 1 \\ 2 \end{bmatrix}\right) = \begin{bmatrix} 0 & 1 \\ 1 & 0 \\ 1 & 1 \end{bmatrix} \begin{bmatrix} 1 \\ 2 \end{bmatrix} = \begin{bmatrix} 2 \\ 1 \\ 3 \end{bmatrix}$$

or, in general,

$$T\left(\begin{bmatrix} x_1 \\ x_2 \end{bmatrix}\right) = \begin{bmatrix} 0 & 1 \\ 1 & 0 \\ 1 & 1 \end{bmatrix} \begin{bmatrix} x_1 \\ x_2 \end{bmatrix} = \begin{bmatrix} x_2 \\ x_1 \\ x_1 + x_2 \end{bmatrix}$$

Problem 5 Let T be the matrix transformation from R^3 to R^2 defined by

$$T(\mathbf{x}) = \begin{bmatrix} 1 & -1 & 2 \\ 3 & -2 & 5 \end{bmatrix} \mathbf{x}$$

Find $T(\mathbf{x})$ if

(A) $\mathbf{x} = \begin{bmatrix} 1 \\ -1 \\ 1 \end{bmatrix}$ (B) $\mathbf{x} = \begin{bmatrix} -1 \\ 1 \\ 1 \end{bmatrix}$ (C) $\mathbf{x} = \begin{bmatrix} x_1 \\ x_2 \\ x_3 \end{bmatrix}$ ∎

Example 6 Let V be a finite-dimensional vector space with basis $B = \{\mathbf{b}_1, \mathbf{b}_2, \ldots, \mathbf{b}_n\}$, let $\mathbf{v}$ be any vector in V, let $[\mathbf{v}]_B$ be the coordinate matrix of $\mathbf{v}$ with respect to B, and let $T: V \rightarrow R^n$ be defined by

$$T(\mathbf{v}) = [\mathbf{v}]_B$$

Show that T is a linear transformation from V to R^n.

Solution If $\mathbf{u}$ and $\mathbf{v}$ are in V, then, relative to the basis B, there exist scalars $k_1, k_2, \ldots, k_n$ and $\ell_1, \ell_2, \ldots, \ell_n$ such that

$$\mathbf{u} = k_1\mathbf{b}_1 + k_2\mathbf{b}_2 + \cdots + k_n\mathbf{b}_n \qquad \text{and} \qquad \mathbf{v} = \ell_1\mathbf{b}_1 + \ell_2\mathbf{b}_2 + \cdots + \ell_n\mathbf{b}_n$$

Thus,

$$T(\mathbf{u}) = [\mathbf{u}]_B = \begin{bmatrix} k_1 \\ k_2 \\ \cdot \\ \cdot \\ \cdot \\ k_n \end{bmatrix} \qquad \text{and} \qquad T(\mathbf{v}) = [\mathbf{v}]_B = \begin{bmatrix} \ell_1 \\ \ell_2 \\ \cdot \\ \cdot \\ \cdot \\ \ell_n \end{bmatrix}$$

Since

$$\mathbf{u} + \mathbf{v} = (k_1 + \ell_1)\mathbf{b}_1 + (k_2 + \ell_2)\mathbf{b}_2 + \cdots + (k_n + \ell_n)\mathbf{b}_n$$

it follows that

$$T(\mathbf{u} + \mathbf{v}) = [\mathbf{u} + \mathbf{v}]_B = \begin{bmatrix} k_1 + \ell_1 \\ k_2 + \ell_2 \\ \cdot \\ \cdot \\ \cdot \\ k_n + \ell_n \end{bmatrix} = \begin{bmatrix} k_1 \\ k_2 \\ \cdot \\ \cdot \\ \cdot \\ k_n \end{bmatrix} + \begin{bmatrix} \ell_1 \\ \ell_2 \\ \cdot \\ \cdot \\ \cdot \\ \ell_n \end{bmatrix} = T(\mathbf{u}) + T(\mathbf{v})$$

In the same way, if t is any scalar, then

$$t\mathbf{u} = tk_1\mathbf{b}_1 + tk_2\mathbf{b}_2 + \cdots + tk_n\mathbf{b}_n$$

implies that

$$T(t\mathbf{u}) = [t\mathbf{u}]_B = \begin{bmatrix} tk_1 \\ tk_2 \\ \cdot \\ \cdot \\ \cdot \\ tk_n \end{bmatrix} = t\begin{bmatrix} k_1 \\ k_2 \\ \cdot \\ \cdot \\ \cdot \\ k_n \end{bmatrix} = tT(\mathbf{u})$$

Thus, T is a linear transformation from V to R^n.

For a specific example of the transformation T in Example 6, let $V = P_2$ and $B = \{1, x, x^2\}$. If $\mathbf{v} = a_0 + a_1x + a_2x^2$, then

$$[\mathbf{v}]_B = \begin{bmatrix} a_0 \\ a_1 \\ a_2 \end{bmatrix}$$

Thus, $T: P_2 \to R^3$ defined by

$$T(a_0 + a_1x + a_2x^2) = \begin{bmatrix} a_0 \\ a_1 \\ a_2 \end{bmatrix}$$

is a linear transformation.

Problem 6   Let $T: M_{2\times 2} \to R^4$ be defined by $T(\mathbf{v}) = [\mathbf{v}]_B$ where

$$B = \left\{ \begin{bmatrix} 1 & 0 \\ 0 & 0 \end{bmatrix}, \begin{bmatrix} 0 & 1 \\ 0 & 0 \end{bmatrix}, \begin{bmatrix} 0 & 0 \\ 1 & 0 \end{bmatrix}, \begin{bmatrix} 0 & 0 \\ 0 & 1 \end{bmatrix} \right\}$$

is the standard basis for $M_{2\times 2}$. Find $T(\mathbf{v})$ if

$$\mathbf{v} = \begin{bmatrix} a & b \\ c & d \end{bmatrix}$$ ▮▮

Example 7   Let R^2 be equipped with the standard inner product, let $\mathbf{w}$ be a fixed unit vector in R^2, and let $T: R^2 \rightarrow R^2$ be defined by

$$T(\mathbf{v}) = \langle \mathbf{v}, \mathbf{w} \rangle \mathbf{w}$$

Show that T is a linear transformation from R^2 to R^2.

Solution   If $\mathbf{u}$ and $\mathbf{v}$ are any vectors in R^2 and k is any scalar, then using the properties of the inner product, we have

$$T(\mathbf{u} + \mathbf{v}) = \langle \mathbf{u} + \mathbf{v}, \mathbf{w} \rangle \mathbf{w} = (\langle \mathbf{u}, \mathbf{w} \rangle + \langle \mathbf{v}, \mathbf{w} \rangle)\mathbf{w}$$
$$= \langle \mathbf{u}, \mathbf{w} \rangle \mathbf{w} + \langle \mathbf{v}, \mathbf{w} \rangle \mathbf{w} = T(\mathbf{u}) + T(\mathbf{v})$$

and

$$T(k\mathbf{v}) = \langle k\mathbf{v}, \mathbf{w} \rangle \mathbf{w} = k\langle \mathbf{v}, \mathbf{w} \rangle \mathbf{w} = kT(\mathbf{v})$$

Thus, T is a linear transformation from R^2 to R^2 (that is, a linear operator on R^2).

Recall from Section 4-3 that the projection of a vector $\mathbf{v}$ onto a unit vector $\mathbf{w}$ is given by

$$\mathbf{p} = \frac{\mathbf{v} \cdot \mathbf{w}}{\|\mathbf{w}\|^2}\mathbf{w} = \langle \mathbf{v}, \mathbf{w} \rangle \mathbf{w} \qquad \begin{array}{l} \|\mathbf{w}\| = 1 \text{ and} \\ \mathbf{v} \cdot \mathbf{w} = \langle \mathbf{v}, \mathbf{w} \rangle \end{array}$$

Thus, the linear operator T in Example 7 projects each vector $\mathbf{v}$ in R^2 onto the fixed unit vector $\mathbf{w}$ (see Figure 3).

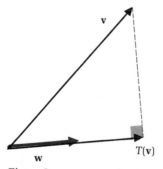

Figure 3   Projection of $\mathbf{v}$ onto a unit vector $\mathbf{w}$

Problem 7   In Example 7, let $\mathbf{w} = (1/\sqrt{2}, 1/\sqrt{2})$.

(A) Find $T(2, 1)$.   (B) Find $T(x, y)$ for any $(x, y) \in R^2$.   ▮▮

■ Properties of Linear Transformations

Theorem 1 lists some general properties of linear transformations. We will prove parts (A) and (C). Proofs of the remaining statements are left as exercises.

Theorem 1

If $T: V \to W$ is a linear transformation, then:

(A) $T(\mathbf{0}) = \mathbf{0}$
(B) $T(-\mathbf{u}) = -T(\mathbf{u})$ for all $\mathbf{u}$ in V
(C) $T(\mathbf{u} - \mathbf{v}) = T(\mathbf{u}) - T(\mathbf{v})$ for all $\mathbf{u}$ and $\mathbf{v}$ in V
(D) $T(k_1\mathbf{u}_1 + k_2\mathbf{u}_2 + \cdots + k_n\mathbf{u}_n) = k_1 T(\mathbf{u}_1) + k_2 T(\mathbf{u}_2) + \cdots + k_n T(\mathbf{u}_n)$
for all $\mathbf{u}_1, \mathbf{u}_2, \ldots, \mathbf{u}_n$ in V and all scalars $k_1, k_2, \ldots, k_n$

Proof (A) Let $\mathbf{u}$ be any vector in V. Then

$$\begin{aligned} T(\mathbf{0}) &= T(0\mathbf{u}) &&\quad 0\mathbf{u} = \mathbf{0} \text{ for any } \mathbf{u} \in V \\ &= 0T(\mathbf{u}) &&\quad T(k\mathbf{v}) = kT(\mathbf{v}) \\ &= \mathbf{0} &&\quad 0\mathbf{w} = \mathbf{0} \text{ for any } \mathbf{w} \in W \end{aligned}$$

(C) $\begin{aligned}[t] T(\mathbf{u} - \mathbf{v}) &= T(\mathbf{u} + (-\mathbf{v})) \\ &= T(\mathbf{u}) + T(-\mathbf{v}) \\ &= T(\mathbf{u}) + [-T(\mathbf{v})] \\ &= T(\mathbf{u}) - T(\mathbf{v}) \end{aligned}$ Supply the justification for each step.

$\blacksquare$

Theorem 1(A) is particularly useful for showing that a function is not a linear transformation. For example, let $T: R^2 \to R^2$ be defined by

$$T(x, y) = (x + 1, y + 2)$$

Since

$$T(\mathbf{0}) = T(0, 0) = (1, 2) \neq \mathbf{0}$$

T does not satisfy Theorem 1(A). Hence, T is not a linear transformation.

Theorem 1(D) also has some important applications, as the next example illustrates.

Example 8 Let $B = \{(1, 0), (0, 1)\}$ be the standard basis for R^2, and let $T: R^2 \to R^3$ be a linear transformation satisfying

$$T(1, 0) = (1, 1, 2) \quad \text{and} \quad T(0, 1) = (2, -1, 3)$$

Find $T(\mathbf{u})$ for any vector $\mathbf{u} = (u_1, u_2) \in R^2$.

Solution $\begin{aligned}[t] \mathbf{u} = (u_1, u_2) &= u_1(1, 0) + u_2(0, 1) \end{aligned}$ Express $\mathbf{u}$ as a linear combination of basis vectors.

$\begin{aligned}[t] T(\mathbf{u}) &= T(u_1(1, 0) + u_2(0, 1)) \end{aligned}$ Apply Theorem 1(D).

$\quad\quad = u_1 T(1, 0) + u_2 T(0, 1)$ Substitute the given images of the basis vectors.

$\quad\quad = u_1(1, 1, 2) + u_2(2, -1, 3)$ Write as a single vector in R^3.

$\quad\quad = (u_1 + 2u_2, u_1 - u_2, 2u_1 + 3u_2)$

Problem 8 If $T: R^2 \to R^2$ is a linear transformation satisfying

$$T(1, 0) = (1, 3) \quad \text{and} \quad T(0, 1) = (2, 1)$$

find $T(\mathbf{u})$ for any vector $\mathbf{u} = (u_1, u_2) \in R^2$. ∎

The procedure used in the solution of Example 8 generalizes to any finite-dimensional vector space V. If B is a basis for V and the image under a linear transformation T of each vector in B is known, then the image of any other vector in V can be determined by using Theorem 1(D). In other words, **a linear transformation is completely determined by its action on a basis.**

Answers to
Matched Problems

1. (A) $(5, 7)$

   (B) $T(k\mathbf{v}) = T(kx, ky, kz) = (kx + ky, ky + kz) = k(x + y, y + z) = kT(\mathbf{v})$

2. Yes   3. No

4. $T(\mathbf{u} + \mathbf{v}) = \mathbf{0} = \mathbf{0} + \mathbf{0} = T(\mathbf{u}) + T(\mathbf{v}); \quad T(k\mathbf{v}) = \mathbf{0} = k\mathbf{0} = kT(\mathbf{v})$

5. (A) $\begin{bmatrix} 4 \\ 10 \end{bmatrix}$ (B) $\begin{bmatrix} 0 \\ 0 \end{bmatrix}$ (C) $\begin{bmatrix} x_1 - x_2 + 2x_3 \\ 3x_1 - 2x_2 + 5x_3 \end{bmatrix}$ 6. $\begin{bmatrix} a \\ b \\ c \\ d \end{bmatrix}$

7. (A) $\left(\dfrac{3}{2}, \dfrac{3}{2}\right)$ (B) $\left(\dfrac{x+y}{2}, \dfrac{x+y}{2}\right)$ 8. $(u_1 + 2u_2, 3u_1 + u_2)$

∎ Exercise 8-1

A In Problems 1 and 2, show that the function $T: R^2 \to R^2$ is a linear transformation.

1. $T(x, y) = (x + y, x - y)$ 2. $T(x, y) = (x + 2y, 2x + y)$

In Problems 3 and 4, show that the function $T: R^2 \to R^2$ is not a linear transformation.

3. $T(x, y) = (x + 1, y + 1)$ 4. $T(x, y) = (x^2, y^2)$

In Problems 5 and 6, show that the function $T: P_2 \to P_1$ is a linear transformation.

5. $T(a_0 + a_1 x + a_2 x^2) = a_0 + a_1 + (a_1 + a_2)x$
6. $T(a_0 + a_1 x + a_2 x^2) = a_1 + (a_2 - a_0)x$

In Problems 7 and 8, show that the function $T: P_2 \to P_1$ is not a linear transformation.

7. $T(a_0 + a_1 x + a_2 x^2) = a_0 + a_1 a_2 x$
8. $T(a_0 + a_1 x + a_2 x^2) = 1 + a_0 + (a_1 - a_2)x$

B In Problems 9–20, determine whether T is a linear transformation.

9. $T: R^3 \to R$ defined by $T(x, y, z) = x + y + z$

10. $T: R^3 \rightarrow R$ defined by $T(x, y, z) = xyz$

11. $T: R^2 \rightarrow R^2$ defined by $T(x, y) = (1, x + y)$

12. $T: R^2 \rightarrow R^2$ defined by $T(x, y) = (x + y, 0)$

13. $T: P_1 \rightarrow R$ defined by $T(a_0 + a_1 x) = 1$

14. $T: P_1 \rightarrow R$ defined by $T(a_0 + a_1 x) = 0$

15. $T: R^3 \rightarrow P_2$ defined by $T(a, b, c) = c + bx + ax^2$

16. $T: R^3 \rightarrow P_2$ defined by $T(a, b, c) = a + bx + cx^2$

17. $T: M_{2 \times 2} \rightarrow R^2$ defined by $T\left(\begin{bmatrix} a & b \\ c & d \end{bmatrix}\right) = (ab, cd)$

18. $T: M_{2 \times 2} \rightarrow R^2$ defined by $T\left(\begin{bmatrix} a & b \\ c & d \end{bmatrix}\right) = (a + b, c + d)$

19. $T: R^4 \rightarrow M_{2 \times 2}$ defined by $T(a, b, c, d) = \begin{bmatrix} a & b \\ c & d \end{bmatrix}$

20. $T: R^3 \rightarrow M_{2 \times 2}$ defined by $T(a, b, c) = \begin{bmatrix} a & b \\ c & 1 \end{bmatrix}$

In Problems 21–24, let $T: R^3 \rightarrow R^2$ be the matrix transformation defined by $T(\mathbf{u}) = A\mathbf{u}$ where

$$A = \begin{bmatrix} 1 & -1 & 2 \\ 1 & 2 & -1 \end{bmatrix}$$

21. Find: $T\left(\begin{bmatrix} 2 \\ 3 \\ 4 \end{bmatrix}\right)$

22. Find: $T\left(\begin{bmatrix} -3 \\ 3 \\ 3 \end{bmatrix}\right)$

23. Find: $T\left(\begin{bmatrix} x \\ y \\ z \end{bmatrix}\right)$

24. Find: $T\left(\begin{bmatrix} -t \\ t \\ t \end{bmatrix}\right)$, t any real number

In Problems 25 and 26, the linear transformation $T: R^2 \rightarrow R^3$ is defined by its action on the standard basis for R^2. Find a formula for T.

25. $T(1, 0) = (1, 1, 0);$ $T(0, 1) = (0, 2, -1)$

26. $T(1, 0) = (2, -1, 3);$ $T(0, 1) = (1, 2, -1)$

In Problems 27 and 28, the linear transformation $T: P_2 \rightarrow P_1$ is defined by its action on the standard basis for P_2. Find a formula for T.

27. $T(1) = 1 + x;$ $T(x) = 1 - x;$ $T(x^2) = x$

28. $T(1) = x;$ $T(x) = 1;$ $T(x^2) = 1 + x$

In Problems 29–32, let $T: R^2 \rightarrow R^2$ be the linear operator that projects each vector in R^2 onto the unit vector $\mathbf{w} = (\frac{3}{5}, \frac{4}{5})$.

29. Find $T(1, 0)$.

30. Find $T(0, 1)$.

31. Find $T(3, -2)$.   **32.** Find $T(x, y)$.

33. Let $D_{2\times2}$ be the vector space of 2×2 diagonal matrices, let

$$B = \left\{ \begin{bmatrix} 1 & 0 \\ 0 & 0 \end{bmatrix}, \begin{bmatrix} 0 & 0 \\ 0 & 1 \end{bmatrix} \right\}$$

be a basis for $D_{2\times2}$, and let $T: D_{2\times2} \to R^2$ be defined by $T(\mathbf{v}) = [\mathbf{v}]_B$. Find a formula for T.

34. Let $S_{2\times2}$ be the vector space of 2×2 symmetric matrices, let

$$B = \left\{ \begin{bmatrix} 1 & 0 \\ 0 & 0 \end{bmatrix}, \begin{bmatrix} 0 & 1 \\ 1 & 0 \end{bmatrix}, \begin{bmatrix} 0 & 0 \\ 0 & 1 \end{bmatrix} \right\}$$

be a basis for $S_{2\times2}$, and let $T: S_{2\times2} \to R^3$ be defined by $T(\mathbf{v}) = [\mathbf{v}]_B$. Find a formula for T.

In Problems 35 and 36, let F be the vector space of functions defined on $(-\infty, \infty)$, let D be the vector space of functions that are differentiable on $(-\infty, \infty)$, and let C be the vector space of functions that are continuous on $(-\infty, \infty)$.

35. *Calculus.* Let $T: D \to F$ be defined by

$$T(f(x)) = f'(x)$$

Show that T is a linear transformation.

36. *Calculus.* Let $T: C \to D$ be defined by

$$T(f(x)) = \int_0^x f(t)\, dt$$

Show that T is a linear transformation.

C   **37.** Let B be a fixed $p \times n$ matrix, and let $T: M_{m\times p} \to M_{m\times n}$ be defined by $T(A) = AB$. Is T a linear transformation? Justify your answer.

38. Let $T: M_{m\times n} \to M_{n\times m}$ be defined by $T(A) = A^T$. Is T a linear transformation? Justify your answer.

39. *Theorem 1(B).* Show that $T(-\mathbf{u}) = -T(\mathbf{u})$ for all $\mathbf{u}$ in V.

40. *Theorem 1(D).* If $\mathbf{u}_1$, $\mathbf{u}_2$, and $\mathbf{u}_3$ are in V and k_1, k_2, and k_3 are scalars, show that

$$T(k_1\mathbf{u}_1 + k_2\mathbf{u}_2 + k_3\mathbf{u}_3) = k_1T(\mathbf{u}_1) + k_2T(\mathbf{u}_2) + k_3T(\mathbf{u}_3)$$

41. Show that $T: V \to W$ is a linear transformation if and only if

$$T(k\mathbf{u} + \ell\mathbf{v}) = kT(\mathbf{u}) + \ell T(\mathbf{v})$$

for all $\mathbf{u}$ and $\mathbf{v}$ in V and all scalars k and ℓ.

42. Let $T_1: V \to W$ and $T_2: V \to W$ be linear transformations, let k_1 and k_2 be scalars, and let $T: V \to W$ be defined by

$$T(\mathbf{v}) = k_1T_1(\mathbf{v}) + k_2T_2(\mathbf{v})$$

Show that T is a linear transformation.

In Problems 43 and 44, let T: V → W be a linear transformation, let $\mathbf{v}_1, \mathbf{v}_2, \ldots,$ $\mathbf{v}_n$ *be vectors in V, and let* $\mathbf{w}_1 = T(\mathbf{v}_1), \mathbf{w}_2 = T(\mathbf{v}_2), \ldots, \mathbf{w}_n = T(\mathbf{v}_n)$ *be the images of these vectors under T.*

43. If $\{\mathbf{w}_1, \mathbf{w}_2, \ldots, \mathbf{w}_n\}$ is a linearly independent subset of W, show that $\{\mathbf{v}_1, \mathbf{v}_2, \ldots, \mathbf{v}_n\}$ is a linearly independent subset of V.

44. If $\{\mathbf{v}_1, \mathbf{v}_2, \ldots, \mathbf{v}_n\}$ is a linearly independent subset of V, show by example that $\{\mathbf{w}_1, \mathbf{w}_2, \ldots, \mathbf{w}_n\}$ need not be a linearly independent subset of W.

▌8-2▌ Kernel and Range

- Kernel of a Linear Transformation
- Range of a Linear Transformation
- Rank and Nullity
- Applications to Matrices and Systems of Linear Equations

▪ Kernel of a Linear Transformation

In this section, we introduce two important subspaces associated with a linear transformation $T: V \rightarrow W$: a subspace of V called the *kernel* of T and a subspace of W called the *range* of T. These subspaces provide fundamental information about the properties of T. In addition, if T is a matrix transformation, then the properties of the kernel and range of T can be used to answer some important questions about systems of linear equations. We begin with an example.

Example 9   Let $T: M_{2\times2} \rightarrow P_2$ be the linear transformation defined by

$$T\left(\begin{bmatrix} a & b \\ c & d \end{bmatrix}\right) = a + (b - c)x + ax^2$$

Find all vectors $\mathbf{u} \in M_{2\times2}$ satisfying $T(\mathbf{u}) = \mathbf{0}$.

Solution   Let

$$\mathbf{u} = \begin{bmatrix} a & b \\ c & d \end{bmatrix}$$

be any vector in $M_{2\times2}$. Then

$$T(\mathbf{u}) = a + (b - c)x + ax^2$$
$$= \mathbf{0}$$

Recall that the zero vector in P_2 is the polynomial with all coefficients equal to zero.

if and only if

$$a = 0 \quad \text{and} \quad b - c = 0$$

Thus, $T(\mathbf{u}) = \mathbf{0}$ if and only if $\mathbf{u}$ has the form

$$\mathbf{u} = \begin{bmatrix} 0 & b \\ b & d \end{bmatrix} = b\begin{bmatrix} 0 & 1 \\ 1 & 0 \end{bmatrix} + d\begin{bmatrix} 0 & 0 \\ 0 & 1 \end{bmatrix}$$

Using set notation, we have

$$\{\mathbf{u} \in M_{2\times2} | T(\mathbf{u}) = \mathbf{0}\} = \text{span}\left\{\begin{bmatrix} 0 & 1 \\ 1 & 0 \end{bmatrix}, \begin{bmatrix} 0 & 0 \\ 0 & 1 \end{bmatrix}\right\}$$

This set is called the *kernel* of T.

Kernel of a Linear Transformation

Let T: $V \to W$ be a linear transformation. The **kernel** of T, denoted by ker(T), is defined to be the subset of all vectors in V that are mapped into the zero vector in W by the transformation T. Using set notation,

$$\text{ker}(T) = \{\mathbf{u} \in V | T(\mathbf{u}) = \mathbf{0}\}$$

Problem 9   Let T: $M_{2\times2} \to P_2$ be the linear transformation defined by

$$T\left(\begin{bmatrix} a & b \\ c & d \end{bmatrix}\right) = (a - d)x + (b - c)x^2$$

Find ker(T).

▌▌

Example 10   Find the kernel of the matrix transformation T: $R^3 \to R^4$ defined by $T(\mathbf{x}) = A\mathbf{x}$ where

$$A = \begin{bmatrix} 1 & 1 & -1 \\ 1 & 2 & -3 \\ 1 & 0 & 1 \\ 1 & 3 & -5 \end{bmatrix}$$

Solution
$$\text{ker}(T) = \{\mathbf{x} \in R^3 | T(\mathbf{x}) = \mathbf{0}\} \qquad T(\mathbf{x}) = A\mathbf{x}$$
$$= \{\mathbf{x} \in R^3 | A\mathbf{x} = \mathbf{0}\}$$

Thus, the kernel of the matrix transformation T is the solution space of the homogeneous system

$$A\mathbf{x} = \begin{bmatrix} 1 & 1 & -1 \\ 1 & 2 & -3 \\ 1 & 0 & 1 \\ 1 & 3 & -5 \end{bmatrix} \begin{bmatrix} x_1 \\ x_2 \\ x_3 \end{bmatrix} = \begin{bmatrix} 0 \\ 0 \\ 0 \\ 0 \end{bmatrix} \qquad (1)$$

The solution of (1) is (verify this)

$$x_1 = -t$$
$$x_2 = 2t$$
$$x_3 = t \qquad t \text{ any real number}$$

Thus, the kernel of T is

$$\ker(T) = \left\{ t \begin{bmatrix} -1 \\ 2 \\ 1 \end{bmatrix} \middle| t \text{ any real number} \right\} = \text{span} \left\{ \begin{bmatrix} -1 \\ 2 \\ 1 \end{bmatrix} \right\}$$

Problem 10 Find the kernel of the matrix transformation $T: R^3 \to R^3$ defined by $T(\mathbf{x}) = A\mathbf{x}$ where

$$A = \begin{bmatrix} 1 & 2 & -1 \\ 1 & 1 & -2 \\ 1 & 3 & 0 \end{bmatrix}$$ ∎

Notice that in Examples 9 and 10, $\ker(T)$ turned out to be a subspace of V. Theorem 2 shows that this is true for any linear transformation T.

Theorem 2

> If $T: V \to W$ is a linear transformation, then $\ker(T)$ is a subspace of V.

Proof To show that $\ker(T)$ is a subspace of V, we must show that $\ker(T)$ is closed under addition and scalar multiplication (Theorem 8, Section 5-3). Let $\mathbf{u}$ and $\mathbf{v}$ be any vectors in $\ker(T)$ and let k be any scalar. Then

$$T(\mathbf{u} + \mathbf{v}) = T(\mathbf{u}) + T(\mathbf{v}) \qquad T(\mathbf{u}) = \mathbf{0} \text{ and } T(\mathbf{v}) = \mathbf{0} \text{ since } \mathbf{u} \text{ and } \mathbf{v} \text{ are in } \ker(T)$$
$$= \mathbf{0} + \mathbf{0} = \mathbf{0}$$

This shows that $\mathbf{u} + \mathbf{v} \in \ker(T)$. Similarly,

$$T(k\mathbf{u}) = kT(\mathbf{u}) = k\mathbf{0} = \mathbf{0}$$

shows that $k\mathbf{u} \in \ker(T)$. Thus, $\ker(T)$ is closed under addition and scalar multiplication and is a subspace of V. ∎

▪ Range of a Linear Transformation

We have just seen that the kernel of a linear transformation $T: V \to W$ is always a subspace of V. Now we want to define a subset of W that is also associated with the linear transformation T. This subset, called the *range* of T, will turn out to be a subspace of W.

> **Range of a Linear Transformation**
>
> Let $T: V \to W$ be a linear transformation. The set of all vectors in W that are images under T of vectors in V is called the **range** of T. Using set notation,
>
> $$\text{range}(T) = \{\mathbf{w} \in W | \mathbf{w} = T(\mathbf{u}) \text{ for some } \mathbf{u} \in V\}$$

Example 11   Find the range of the linear transformation $T: M_{2\times 2} \to P_2$ defined by

$$T\left(\begin{bmatrix} a & b \\ c & d \end{bmatrix}\right) = a + (b - c)x + ax^2$$

Solution   A vector $\mathbf{w} \in P_2$ is in the range of T if and only if there exists a vector

$$\mathbf{u} = \begin{bmatrix} a & b \\ c & d \end{bmatrix} \qquad \mathbf{u} \in M_{2\times 2}$$

such that

$\mathbf{w} = T(\mathbf{u})$
$\quad = a + (b - c)x + ax^2$
$\quad = a(1 + x^2) + (b - c)x$

Since $\mathbf{u} \in M_{2\times 2}$ for any choice of a, b, and c, it follows that

range$(T) = \{a(1 + x^2) + (b - c)x | a, b, c$ any real numbers$\}$
$\qquad\qquad = \text{span}\{1 + x^2, x\}$

Problem 11   Find the range of the linear transformation $T: M_{2\times 2} \to P_2$ defined by

$$T\left(\begin{bmatrix} a & b \\ c & d \end{bmatrix}\right) = (a - d)x + (b - c)x^2 \qquad\qquad \blacksquare$$

Example 12   Find the range of the matrix transformation $T: R^3 \to R^4$ defined by the matrix

$$A = \begin{bmatrix} 1 & 1 & -1 \\ 1 & 2 & -3 \\ 1 & 0 & 1 \\ 1 & 3 & -5 \end{bmatrix}$$

Solution   Using the fact that $T(\mathbf{x}) = A\mathbf{x}$ for a matrix transformation, we can write

range$(T) = \{\mathbf{w} \in R^4 | \mathbf{w} = A\mathbf{x}$ for some $\mathbf{x} \in R^3\}$

Thus, $\mathbf{w}$ is in range(T) if and only if $\mathbf{w}$ is a solution of the system $\mathbf{w} = A\mathbf{x}$ for some $\mathbf{x} \in R^3$.

$$\mathbf{w} = A\mathbf{x} = \begin{array}{ccc} \mathbf{c_1} & \mathbf{c_2} & \mathbf{c_3} \end{array} \qquad\qquad \mathbf{x}$$

$$\mathbf{w} = A\mathbf{x} = \begin{bmatrix} 1 & 1 & -1 \\ 1 & 2 & -3 \\ 1 & 0 & 1 \\ 1 & 3 & -5 \end{bmatrix} \begin{bmatrix} x_1 \\ x_2 \\ x_3 \end{bmatrix}$$

$$= \begin{bmatrix} x_1 + x_2 - x_3 \\ x_1 + 2x_2 - 3x_3 \\ x_1 \qquad + x_3 \\ x_1 + 3x_2 - 5x_3 \end{bmatrix}$$

$$\mathbf{w} = A\mathbf{x} = x_1 \begin{bmatrix} 1 \\ 1 \\ 1 \\ 1 \end{bmatrix} + x_2 \begin{bmatrix} 1 \\ 2 \\ 0 \\ 3 \end{bmatrix} + x_3 \begin{bmatrix} -1 \\ -3 \\ 1 \\ -5 \end{bmatrix}$$

$$= x_1 \mathbf{c}_1 + x_2 \mathbf{c}_2 + x_3 \mathbf{c}_3$$

where $\mathbf{c}_1$, $\mathbf{c}_2$, and $\mathbf{c}_3$ are the column vectors of A. Thus, $\mathbf{w}$ is in range(T) if and only if $\mathbf{w}$ is the column space of A. Since the first two columns of A form a basis for the column space of A (verify this; see Section 6-2), it follows that

$$\text{range}(T) = \text{Column space of } A = \text{span} \left\{ \begin{bmatrix} 1 \\ 1 \\ 1 \\ 1 \end{bmatrix}, \begin{bmatrix} 1 \\ 2 \\ 0 \\ 3 \end{bmatrix} \right\}$$

Problem 12   Find the range of the matrix transformation $T: R^3 \to R^3$ defined by the matrix

$$A = \begin{bmatrix} 1 & 2 & -1 \\ 1 & 1 & -2 \\ 1 & 3 & 0 \end{bmatrix}$$ ∎

We now show that the range of a linear transformation is a subspace.

Theorem 3 | If $T: V \to W$ is a linear transformation, then range(T) is a subspace of W.

Proof   We must show that range(T) is closed under addition and scalar multiplication. Let $\mathbf{w}_1$ and $\mathbf{w}_2$ be in range(T). Then there exist $\mathbf{v}_1$ and $\mathbf{v}_2$ in V such that

$$T(\mathbf{v}_1) = \mathbf{w}_1 \qquad \text{and} \qquad T(\mathbf{v}_2) = \mathbf{w}_2$$

Thus,

$$\mathbf{w}_1 + \mathbf{w}_2 = T(\mathbf{v}_1) + T(\mathbf{v}_2) = T(\mathbf{v}_1 + \mathbf{v}_2)$$

This shows that $\mathbf{w}_1 + \mathbf{w}_2$ is the image under T of $\mathbf{v}_1 + \mathbf{v}_2$; hence, $\mathbf{w}_1 + \mathbf{w}_2$ is in range(T). Now let k be any scalar. Then we can write

$$k\mathbf{w}_1 = kT(\mathbf{v}_1) = T(k\mathbf{v}_1)$$

which shows that $k\mathbf{w}_1$ is the image under T of $k\mathbf{v}_1$. Thus, range(T) is also closed under scalar multiplication and is a subspace of W.   ∎

In Examples 10 and 12, we considered a matrix transformation T defined by a matrix A. The kernel of T turned out to be the solution space of the homogeneous system $A\mathbf{x} = \mathbf{0}$, and the range of T turned out to be the column space of A. These results are generalized in Theorem 4. The proof is left as an exercise.

Theorem 4 | Let A be an $m \times n$ matrix, and let $T: R^n \rightarrow R^m$ be the matrix transformation defined by $T(\mathbf{x}) = A\mathbf{x}$. Then:

(A) $\ker(T) = \{\mathbf{x} \in R^n | A\mathbf{x} = \mathbf{0}\}$
(B) $\text{range}(T) = \text{Column space of } A$

Example 13 Let $T: R^4 \rightarrow R^3$ be the matrix transformation defined by the matrix

$$A = \begin{bmatrix} 1 & 2 & -1 & 1 \\ 2 & 4 & -1 & 4 \\ 1 & 2 & 0 & 3 \end{bmatrix}$$

(A) Find a basis for $\ker(T)$.
(B) Find a basis for $\text{range}(T)$.

Solution The reduced form of A is used in both parts of the solution. Omitting the details, we have

$$A = \begin{bmatrix} 1 & 2 & -1 & 1 \\ 2 & 4 & -1 & 4 \\ 1 & 2 & 0 & 3 \end{bmatrix} \sim \begin{bmatrix} 1 & 2 & 0 & 3 \\ 0 & 0 & 1 & 2 \\ 0 & 0 & 0 & 0 \end{bmatrix} \qquad \text{Reduced form of } A$$

(A) Using the reduced form of A, the solution space of the homogeneous system $A\mathbf{x} = \mathbf{0}$ is

$$\left\{ s\begin{bmatrix} -2 \\ 1 \\ 0 \\ 0 \end{bmatrix} + t\begin{bmatrix} -3 \\ 0 \\ -2 \\ 1 \end{bmatrix} \middle| s, t \text{ any real numbers} \right\}$$

Theorem 4(A) implies that $\ker(T) = \text{span } B$ where

$$B = \left\{ \begin{bmatrix} -2 \\ 1 \\ 0 \\ 0 \end{bmatrix}, \begin{bmatrix} -3 \\ 0 \\ -2 \\ 1 \end{bmatrix} \right\}$$

Since the vectors in B are linearly independent (Why?), B is a basis for $\ker(T)$.

(B) Since the reduced form of A contains two columns of the identity (columns 1 and 3), the corresponding columns of A form a basis for the column space of A (Theorem 5, Section 6-2). Theorem 4(B) now implies that

$$C = \left\{ \begin{bmatrix} 1 \\ 2 \\ 1 \end{bmatrix}, \begin{bmatrix} -1 \\ -1 \\ 0 \end{bmatrix} \right\} \qquad \begin{array}{l} \text{First and third} \\ \text{columns of } A \end{array}$$

is a basis for $\text{range}(T)$.

Problem 13   Repeat Example 13 for the matrix transformation $T: R^3 \to R^2$ defined by the matrix

$$A = \begin{bmatrix} 1 & -2 & 1 \\ -2 & 4 & 2 \end{bmatrix}$$ ∎

In Section 8-3, we will see that every linear transformation from a finite-dimensional vector space V to a vector space W can be expressed as a matrix transformation. Thus, Theorem 4 will provide a method for finding the kernel and range of any linear transformation $T: V \to W$, provided V is finite-dimensional.

▪ Rank and Nullity

In Section 6-2, we defined the rank of a matrix A to be the common dimension of the row and column spaces of A. If T is a matrix transformation defined by the matrix A, then Theorem 4 implies that the rank of A is equal to the dimension of the range of T. The concept of rank is generalized to arbitrary linear transformations in the following definition. In addition to the *rank of* T, we also define the *nullity of* T, an important related concept.

Rank and Nullity

The **rank** and **nullity** of a linear transformation T are defined by

    Rank of $T = \dim(\text{range}(T))$ Nullity of $T = \dim(\ker(T))$

Table 1 lists the rank and nullity of the linear transformations we have studied in this section.

Table 1
Rank and Nullity of $T: V \to W$

Reference	T	V	$\dim(V)$	**Rank of T** $\dim(\text{range}(T))$	**Nullity** $\dim(\ker(T))$
Examples 9, 11	$T: M_{2\times2} \to P_2$	$M_{2\times2}$	4	2	2
Examples 10, 12	$T: R^3 \to R^4$	R^3	3	2	1
Example 13	$T: R^4 \to R^3$	R^4	4	2	2

Examining each line of Table 1, we see that the sum of the rank of T and the nullity of T is always equal to the dimension of V. Theorem 5 shows that this is true for any linear transformation T, provided that V is a finite-dimensional vector space. (The proof of Theorem 5 is long and can be omitted without loss of continuity.)

Theorem 5   If $T: V \rightarrow W$ is a linear transformation and V is a finite-dimensional vector space, then

$$\dim(V) = \dim(\ker(T)) + \dim(\text{range}(T))$$
$$= (\text{Nullity of } T) + (\text{Rank of } T)$$

Proof   If $\dim(V) = n$ and $\dim(\ker(T)) = k$, then since $\ker(T)$ is a subspace of V, it follows that $0 \leq k \leq n$. Since the special cases $k = 0$ and $k = n$ will be considered in the exercises, we will consider only the case $0 < k < n$.

Let $B_k = \{\mathbf{v}_1, \mathbf{v}_2, \ldots, \mathbf{v}_k\}$ be a basis for $\ker(T)$. According to Theorem 18 in Section 5-6, B_k can be extended to a basis for V. That is, there exist vectors $\mathbf{u}_1, \mathbf{u}_2, \ldots, \mathbf{u}_{n-k}$ in V such that

$$B = \{\overbrace{\mathbf{v}_1, \mathbf{v}_2, \ldots, \mathbf{v}_k}^{\text{Basis for } \ker(T)}, \mathbf{u}_1, \mathbf{u}_2, \ldots, \mathbf{u}_{n-k}\}$$

is a basis for V. If we can show that the set of vectors

$$S = \{T(\mathbf{u}_1), T(\mathbf{u}_2), \ldots, T(\mathbf{u}_{n-k})\}$$

is a basis for $\text{range}(T)$, then we can conclude that

$$n \quad = \quad n-k \quad + \quad k$$
$$\dim(V) = \dim(\ker(T)) + \dim(\text{range}(T))$$

In order to show that S is a basis for $\text{range}(T)$, we must show that $\text{range}(T) = \text{span } S$ and that S is a linearly independent subset of W.

Part I: $range(T) = span\ S$. If $\mathbf{w}$ is any vector in $\text{range}(T)$, then there exists a vector $\mathbf{v} \in V$ such that $T(\mathbf{v}) = \mathbf{w}$. Since B is a basis for V, there exist scalars $c_1, c_2, \ldots, c_k, d_1, d_2, \ldots, d_{n-k}$ such that

$$\mathbf{v} = c_1\mathbf{v}_1 + c_2\mathbf{v}_2 + \cdots + c_k\mathbf{v}_k + d_1\mathbf{u}_1 + d_2\mathbf{u}_2 + \cdots + d_{n-k}\mathbf{u}_{n-k}$$

Using Theorem 1(D) in Section 8-1, we can write

$$\mathbf{w} = T(\mathbf{v})$$
$$= c_1 T(\mathbf{v}_1) + c_2 T(\mathbf{v}_2) + \cdots + c_k T(\mathbf{v}_k) \qquad\qquad T(\mathbf{v}_i) = \mathbf{0} \text{ since}$$
$$+ d_1 T(\mathbf{u}_1) + d_2 T(\mathbf{u}_2) + \cdots + d_{n-k} T(\mathbf{u}_{n-k}) \qquad \mathbf{v}_i \in \ker(T)$$
$$= d_1 T(\mathbf{u}_1) + d_2 T(\mathbf{u}_2) + \cdots + d_{n-k} T(\mathbf{u}_{n-k})$$

Thus,

$$\mathbf{w} \in \text{span}\{T(\mathbf{u}_1), T(\mathbf{u}_2), \ldots, T(\mathbf{u}_{n-k})\} = \text{span } S$$

This shows that $\text{range}(T) \subset \text{span } S$. Since $\text{range}(T)$ is a subspace and $S \subset \text{range}(T)$, it follows that $\text{span } S \subset \text{range}(T)$. Thus, we have shown that $\text{span } S = \text{range}(T)$.

Part II: *S is linearly independent.* Let $d_1, d_2, \ldots, d_{n-k}$ be scalars satisfying

$$d_1 T(\mathbf{u}_1) + d_2 T(\mathbf{u}_2) + \cdots + d_{n-k} T(\mathbf{u}_{n-k}) = \mathbf{0} \tag{2}$$

In order to show that the vectors in S are linearly independent, we must show that the scalars $d_1, d_2, \ldots, d_{n-k}$ are all zero. Using Theorem 1(D) in Section 8-1, the vector equation (2) can be written as

$$T(d_1 \mathbf{u}_1 + d_2 \mathbf{u}_2 + \cdots + d_{n-k} \mathbf{u}_{n-k}) = \mathbf{0}$$

This implies that $d_1 \mathbf{u}_1 + d_2 \mathbf{u}_2 + \cdots + d_{n-k} \mathbf{u}_{n-k}$ is in $\ker(T)$. Since B_k is a basis for $\ker(T)$, there exist scalars $c_1, c_2, \ldots, c_k$ satisfying

$$c_1 \mathbf{v}_1 + c_2 \mathbf{v}_2 + \cdots + c_k \mathbf{v}_k = d_1 \mathbf{u}_1 + d_2 \mathbf{u}_2 + \cdots + d_{n-k} \mathbf{u}_{n-k}$$

or, equivalently,

$$c_1 \mathbf{v}_1 + c_2 \mathbf{v}_2 + \cdots + c_k \mathbf{v}_k - d_1 \mathbf{u}_1 - d_2 \mathbf{u}_2 - \cdots - d_{n-k} \mathbf{u}_{n-k} = \mathbf{0} \tag{3}$$

But the left side of equation (3) is a linear combination of vectors in the basis B. Thus, all the scalars in this linear combination must be zero. In particular,

$$d_1 = d_2 = \cdots = d_{n-k} = 0$$

and the vectors $T(\mathbf{u}_1), T(\mathbf{u}_2), \ldots, T(\mathbf{u}_{n-k})$ are linearly independent.

Parts I and II together show that S is a basis for range(T). Consequently, as we noted earlier in this proof, if $\dim(\ker(T)) = k$, then $\dim(\text{range}(T)) = n - k$ and

$$\dim(\ker(T)) + \dim(\text{range}(T)) = k + n - k$$
$$= n = \dim(V) \qquad \blacksquare$$

▪ Applications to Matrices and Systems of Linear Equations

The properties of linear transformations developed in this section have some important applications to matrices and systems of linear equations. To begin, we prove Theorem 6, which was stated without proof as Theorem 2 in Section 1-1.

Theorem 6

Nature of Solutions for a Linear System

Any system of linear equations must have no solution, exactly one solution, or an infinite number of solutions. No other possibility exists.

Proof   Given the linear system

$$A\mathbf{x} = \mathbf{b} \tag{4}$$

where A is an $m \times n$ matrix, $\mathbf{x} \in R^n$, and $\mathbf{b} \in R^m$, let

$$S = \{x|Ax = b\}$$

be the solution set for (4), and let $T: R^n \to R^m$ be the matrix transformation defined by $T(\mathbf{x}) = A\mathbf{x}$.

We know from specific examples considered earlier that each case in Theorem 6 exists. We must show that there are no other possibilities. If S is empty or consists of a single vector, then there is nothing to prove. Thus, it is sufficient to show that if S contains two distinct vectors, then it contains an infinite number of vectors.

Assume that S contains $\mathbf{x}_1$ and $\mathbf{x}_2$, $\mathbf{x}_1 \neq \mathbf{x}_2$. Then

$$T(\mathbf{x}_1) = A\mathbf{x}_1 = \mathbf{b} \qquad T(\mathbf{x}_2) = A\mathbf{x}_2 = \mathbf{b}$$

and

$$T(\mathbf{x}_1 - \mathbf{x}_2) = T(\mathbf{x}_1) - T(\mathbf{x}_2) = \mathbf{b} - \mathbf{b} = \mathbf{0}$$

Thus, $\mathbf{x}_1 - \mathbf{x}_2$ is in ker(T). Since $\mathbf{x}_1 - \mathbf{x}_2 \neq \mathbf{0}$, ker($T$) is a nontrivial subspace of R^n and must contain an infinite number of vectors. In particular, ker(T) contains all vectors of the form $k(\mathbf{x}_1 - \mathbf{x}_2)$, k any real number. Furthermore, if $\mathbf{u} \in$ ker(T), then

$$T(\mathbf{x}_1 + \mathbf{u}) = T(\mathbf{x}_1) + T(\mathbf{u}) = \mathbf{b} + \mathbf{0} = \mathbf{b}$$

Thus, the infinite set of vectors

$$\{\mathbf{x} = \mathbf{x}_1 + \mathbf{u}|\mathbf{u} \in \text{ker}(T)\}$$

is a subset of S. This shows that S contains an infinite number of vectors. ▮

Theorem 7 provides a relationship between the rank of a matrix A and the dimension of the solution space of the homogeneous system $A\mathbf{x} = \mathbf{0}$. The proof follows directly from Theorems 4 and 5 and is left as an exercise.

Theorem 7

If A is an $m \times n$ matrix, then

$$\dim(\{x|Ax = \mathbf{0}\}) + (\text{Rank of } A) = n$$

Answers to Matched Problems

9. $\text{span}\left\{\begin{bmatrix} 1 & 0 \\ 0 & 1 \end{bmatrix}, \begin{bmatrix} 0 & 1 \\ 1 & 0 \end{bmatrix}\right\}$ 10. $\text{span}\left\{\begin{bmatrix} 3 \\ -1 \\ 1 \end{bmatrix}\right\}$ 11. $\text{span}\{x, x^2\}$

12. $\text{span}\left\{\begin{bmatrix} 1 \\ 1 \\ 1 \end{bmatrix}, \begin{bmatrix} 2 \\ 1 \\ 3 \end{bmatrix}\right\}$

13. $\text{ker}(T) = \text{span}\left\{\begin{bmatrix} 2 \\ 1 \\ 0 \end{bmatrix}\right\}$; $\text{range}(T) = \text{span}\left\{\begin{bmatrix} 1 \\ -2 \end{bmatrix}, \begin{bmatrix} 1 \\ 2 \end{bmatrix}\right\}$

▌ Exercise 8-2

A   In Problems 1–8, let T: $R^2 \to R^2$ be the matrix transformation defined by $T(\mathbf{x}) = A\mathbf{x}$ where

$$A = \begin{bmatrix} 1 & -2 \\ -2 & 4 \end{bmatrix}$$

In Problems 1–4, find $T(\mathbf{v})$ and determine whether $\mathbf{v} \in \ker(T)$.

1. $\mathbf{v} = \begin{bmatrix} 0 \\ 0 \end{bmatrix}$ **2.** $\mathbf{v} = \begin{bmatrix} 0 \\ 1 \end{bmatrix}$ **3.** $\mathbf{v} = \begin{bmatrix} 2 \\ -1 \end{bmatrix}$ **4.** $\mathbf{v} = \begin{bmatrix} 2 \\ 1 \end{bmatrix}$

In Problems 5–8, determine whether $\mathbf{w} \in \text{range}(T)$ by solving the equation $T(\mathbf{u}) = \mathbf{w}$ for $\mathbf{u}$, if possible.

5. $\mathbf{w} = \begin{bmatrix} 1 \\ 0 \end{bmatrix}$ **6.** $\mathbf{w} = \begin{bmatrix} 1 \\ -2 \end{bmatrix}$ **7.** $\mathbf{w} = \begin{bmatrix} -2 \\ 4 \end{bmatrix}$ **8.** $\mathbf{w} = \begin{bmatrix} 1 \\ 2 \end{bmatrix}$

B   In Problems 9–20, find the kernel of T, range of T, nullity of T, and rank of T.

9. T: $R^3 \to R^2$ defined by $T(\mathbf{x}) = A\mathbf{x}$,   where $A = \begin{bmatrix} 1 & -1 & 2 \\ -2 & 2 & -4 \end{bmatrix}$

10. T: $R^2 \to R^3$ defined by $T(\mathbf{x}) = A\mathbf{x}$,   where $A = \begin{bmatrix} 1 & -3 \\ -1 & 3 \\ -2 & 6 \end{bmatrix}$

11. T: $R^4 \to R^3$ defined by $T(\mathbf{x}) = A\mathbf{x}$,   where $A = \begin{bmatrix} 1 & -2 & 1 & 2 \\ -2 & 4 & 1 & 5 \\ -1 & 2 & 2 & 7 \end{bmatrix}$

12. T: $R^3 \to R^3$ defined by $T(\mathbf{x}) = A\mathbf{x}$,   where $A = \begin{bmatrix} 1 & -2 & 1 \\ -1 & 3 & -3 \\ 1 & -1 & -1 \end{bmatrix}$

13. T: $P_2 \to P_3$ defined by $T(p(x)) = xp(x)$

14. T: $P_2 \to P_3$ defined by $T(p(x)) = xp(x) - p(1)$

15. T: $M_{2\times2} \to P_3$ defined by

$$T\left(\begin{bmatrix} a & b \\ c & d \end{bmatrix}\right) = a + b + (c - d)x + (c - d)x^2 + (a + b)x^3$$

16. T: $M_{2\times2} \to P_3$ defined by

$$T\left(\begin{bmatrix} a & b \\ c & d \end{bmatrix}\right) = a + (b - a)x + (b - c + d)x^2 + (c - d)x^3$$

17. T: $R^2 \to R^2$ defined by $T(\mathbf{u}) = \langle \mathbf{u}, \mathbf{w} \rangle \mathbf{w}$,   where $\mathbf{w} = (\frac{3}{5}, \frac{4}{5})$
18. T: $R^2 \to R^2$ defined by $T(\mathbf{u}) = \langle \mathbf{u}, \mathbf{w} \rangle \mathbf{w}$,   where $\mathbf{w} = (\frac{5}{13}, -\frac{12}{13})$
19. T: $R^3 \to R^3$ defined by $T(\mathbf{u}) = \langle \mathbf{u}, \mathbf{w} \rangle \mathbf{w}$,   where $\mathbf{w} = (\frac{1}{3}, \frac{2}{3}, \frac{2}{3})$
20. T: $R^3 \to R^3$ defined by $T(\mathbf{u}) = \langle \mathbf{u}, \mathbf{w} \rangle \mathbf{w}$,   where $\mathbf{w} = (\frac{2}{7}, -\frac{3}{7}, -\frac{6}{7})$

*In Problems 21–28, A is a 5 × 5 matrix and T: $R^5 \rightarrow R^5$ is the matrix transformation defined by T(**x**) = A**x**.*

21. If the rank of T is 3, what is the nullity of T?

22. If the nullity of T is 4, what is the rank of T?

23. If dim($\{\mathbf{x}|A\mathbf{x} = \mathbf{0}\}$) = 2, what is the rank of A?

24. If the rank of A is 2, what is dim($\{\mathbf{x}|A\mathbf{x} = \mathbf{0}\}$)?

25. If the nullity of T is 1, what is the dimension of the column space of A?

26. If the range of T is R^5, what is dim($\{\mathbf{x}|A\mathbf{x} = \mathbf{0}\}$)?

27. If the system A**x** = **b** is consistent for all **b** $\in R^5$, what is the nullity of T?

28. If A is invertible, what is the range of T?

29. Let T: $R^3 \rightarrow R^2$ be the linear transformation defined by

$$T(x_1, x_2, x_3) = (x_1 - x_2 + x_3, 2x_1 - x_2 - 3x_2)$$

  (A) Find the kernel of T and the nullity of T.

  (B) Use Theorem 5 to find the rank of T.

  (C) Find the range of T. [*Hint:* Use the result in part (B).]

30. Repeat Problem 29 for T: $R^4 \rightarrow R^2$ defined by

$$T(x_1, x_2, x_3, x_4) = (x_1 + x_2 + x_3 - 3x_4, 2x_1 + x_2 - x_3 - 2x_4)$$

31. *Calculus.* Let T: $P_4 \rightarrow P_3$ be the linear transformation defined by $T(p(x)) = p'(x)$.

  (A) Find the kernel of T and the nullity of T.

  (B) Use Theorem 5 to find the rank of T.

  (C) Find the range of T. [*Hint:* Use the result in part (B).]

32. *Calculus.* Repeat Problem 31 for T: $P_4 \rightarrow P_2$ defined by $T(p(x)) = p''(x)$.

C *In Problems 33–37, A is an m × n matrix and T: $R^n \rightarrow R^m$ is the matrix transformation defined by T(**x**) = A**x**. Prove each statement.*

33. *Theorem 4(A).* $\ker(T) = \{\mathbf{x} \in R^n | A\mathbf{x} = \mathbf{0}\}$

34. *Theorem 4(B).* $\text{range}(T)$ = Column space of A

35. *Theorem 7.* dim($\{\mathbf{x}|A\mathbf{x} = \mathbf{0}\}$) + (Rank of A) = n

36. The system A**x** = **b** is consistent if and only if **b** is in the column space of A.

37. If $n \geq m$, then dim($\{\mathbf{x}|A\mathbf{x} = \mathbf{0}\}$) = $n - m$ if and only if the system A**x** = **b** is consistent for every **b** $\in R^m$.

38. Let T: $V \rightarrow V$ be a linear operator on the finite-dimensional vector space V. Show that range(T) = V if and only if ker(T) = {**0**}.

Let T: $V \rightarrow W$ be a linear transformation defined on the n-dimensional vector space V. Prove the statements in Problems 39 and 40, thus completing the proof of Theorem 5. Do not refer to Theorem 5 in your proof.

39. If dim(ker(T)) = 0, then dim(range(T)) = n.

40. If dim(ker(T)) = n, then dim(range(T)) = 0.

▌8-3▐ Matrix Representation of Linear Transformations

- Standard Matrices of Linear Transformations from R^n to R^m
- Matrices of General Linear Transformations
- Matrices of Linear Operators

▪ Standard Matrices of Linear Transformations from R^n to R^m

We have already seen that every $m \times n$ matrix A determines a matrix transformation $T: R^n \rightarrow R^m$ defined by $T(\mathbf{x}) = A\mathbf{x}$. In this section, we will see that every linear transformation from R^n to R^m is a matrix transformation. That is, given any linear transformation $T: R^n \rightarrow R^m$, there is an $m \times n$ matrix A satisfying $A\mathbf{x} = T(\mathbf{x})$ for all $\mathbf{x}$ in R^n. We begin with an example.

Example 14   Let $T: R^2 \rightarrow R^3$ be the linear transformation defined by

$$T\left(\begin{bmatrix} x_1 \\ x_2 \end{bmatrix}\right) = \begin{bmatrix} x_1 + 2x_2 \\ 3x_1 + 4x_2 \\ 5x_1 + 6x_2 \end{bmatrix}$$

If

$$\mathbf{x} = \begin{bmatrix} x_1 \\ x_2 \end{bmatrix}$$

then we can write

$$T(\mathbf{x}) = \begin{bmatrix} x_1 + 2x_2 \\ 3x_1 + 4x_2 \\ 5x_1 + 6x_2 \end{bmatrix} = \begin{bmatrix} 1 & 2 \\ 3 & 4 \\ 5 & 6 \end{bmatrix} \begin{bmatrix} x_1 \\ x_2 \end{bmatrix} = A\mathbf{x}$$

where

$$A = \begin{bmatrix} 1 & 2 \\ 3 & 4 \\ 5 & 6 \end{bmatrix}$$

Thus, T is the matrix transformation defined by the equation $T(\mathbf{x}) = A\mathbf{x}$. Furthermore, if $B = \{\mathbf{e}_1, \mathbf{e}_2\}$ is the standard basis for R^2, then

$$T(\mathbf{e}_1) = T\left(\begin{bmatrix} 1 \\ 0 \end{bmatrix}\right) = \begin{bmatrix} 1 \\ 3 \\ 5 \end{bmatrix} \qquad \text{First column of } A$$

and

$$T(\mathbf{e}_2) = T\left(\begin{bmatrix} 0 \\ 1 \end{bmatrix}\right) = \begin{bmatrix} 2 \\ 4 \\ 6 \end{bmatrix} \qquad \text{Second column of } A$$

Thus, the columns of A are the images of the standard basis vectors under T. The matrix

$$A = [T(\mathbf{e}_1) \quad T(\mathbf{e}_2)]$$

is called the *standard matrix* for T.

The general definition of the standard matrix for a linear transformation from R^n to R^m is given in the box.

Standard Matrix for $T: R^n \to R^m$

If $T: R^n \to R^m$ is a linear transformation and $B = \{\mathbf{e}_1, \mathbf{e}_2, \ldots, \mathbf{e}_n\}$ is the standard basis for R^n, then the matrix

$$A = [T(\mathbf{e}_1) \quad T(\mathbf{e}_2) \quad \cdots \quad T(\mathbf{e}_n)]$$

is called the **standard matrix** for T.

Problem 14   Find the standard matrix for the linear transformation $T: R^3 \to R^2$ defined by

$$T\left(\begin{bmatrix} x_1 \\ x_2 \\ x_3 \end{bmatrix}\right) = \begin{bmatrix} x_1 - 2x_2 + 3x_3 \\ 6x_1 + 5x_2 - 4x_3 \end{bmatrix}$$ ▮▮

In Example 14, we showed that the linear transformation T was a matrix transformation by showing that $T(\mathbf{x}) = A\mathbf{x}$ for all $\mathbf{x}$ in R^2, where A was the standard matrix for T. This result is generalized in Theorem 8.

Theorem 8   If $T: R^n \to R^m$ is a linear transformation and A is the standard matrix for T, then

$$T(\mathbf{x}) = A\mathbf{x}$$

for all $\mathbf{x}$ in R^n. Thus, every linear transformation from R^n to R^m is a matrix transformation.

Proof   Let $B = \{\mathbf{e}_1, \mathbf{e}_2, \ldots, \mathbf{e}_n\}$ be the standard basis for R^n and let

$$T(\mathbf{e}_1) = \begin{bmatrix} a_{11} \\ a_{21} \\ \vdots \\ \\ a_{m1} \end{bmatrix}, \quad T(\mathbf{e}_2) = \begin{bmatrix} a_{12} \\ a_{22} \\ \vdots \\ \\ a_{m2} \end{bmatrix}, \quad \cdots, \quad T(\mathbf{e}_n) = \begin{bmatrix} a_{1n} \\ a_{2n} \\ \vdots \\ \\ a_{mn} \end{bmatrix}$$

The standard matrix for T is

$$A = [T(\mathbf{e}_1) \quad T(\mathbf{e}_2) \quad \cdots \quad T(\mathbf{e}_n)] = \begin{bmatrix} a_{11} & a_{12} & \cdots & a_{1n} \\ a_{21} & a_{22} & \cdots & a_{2n} \\ \cdot & \cdot & & \cdot \\ \cdot & \cdot & & \cdot \\ \cdot & \cdot & & \cdot \\ a_{m1} & a_{m2} & \cdots & a_{mn} \end{bmatrix}$$

If

$$\mathbf{x} = \begin{bmatrix} k_1 \\ k_2 \\ \cdot \\ \cdot \\ \cdot \\ k_n \end{bmatrix} = k_1\mathbf{e}_1 + k_2\mathbf{e}_2 + \cdots + k_n\mathbf{e}_n$$

is any vector in R^n, then using Theorem 1(D) in Section 8-1, we can write

$$T(\mathbf{x}) = k_1 T(\mathbf{e}_1) + k_2 T(\mathbf{e}_2) + \cdots + k_n T(\mathbf{e}_n)$$

$$= k_1 \begin{bmatrix} a_{11} \\ a_{21} \\ \cdot \\ \cdot \\ \cdot \\ a_{m1} \end{bmatrix} + k_2 \begin{bmatrix} a_{12} \\ a_{22} \\ \cdot \\ \cdot \\ \cdot \\ a_{m2} \end{bmatrix} + \cdots + k_n \begin{bmatrix} a_{1n} \\ a_{2n} \\ \cdot \\ \cdot \\ \cdot \\ a_{mn} \end{bmatrix}$$

$$= \begin{bmatrix} k_1 a_{11} + k_2 a_{12} + \cdots + k_n a_{1n} \\ k_1 a_{21} + k_2 a_{22} + \cdots + k_n a_{2n} \\ \cdot & \cdot & & \cdot \\ \cdot & \cdot & & \cdot \\ \cdot & \cdot & & \cdot \\ k_1 a_{m1} + k_2 a_{m2} + \cdots + k_n a_{mn} \end{bmatrix}$$

$$= \begin{bmatrix} a_{11} & a_{12} & \cdots & a_{1n} \\ a_{21} & a_{22} & \cdots & a_{2n} \\ \cdot & \cdot & & \cdot \\ \cdot & \cdot & & \cdot \\ \cdot & \cdot & & \cdot \\ a_{m1} & a_{m2} & \cdots & a_{mn} \end{bmatrix} \begin{bmatrix} k_1 \\ k_2 \\ \cdot \\ \cdot \\ \cdot \\ k_n \end{bmatrix}$$

$$= A\mathbf{x}$$

Since $T(\mathbf{x}) = A\mathbf{x}$ for all $\mathbf{x}$ in R^n, it follows that T is the matrix transformation defined by A.   ▐

Example 15 Let $\mathbf{w} = (\frac{3}{5}, \frac{4}{5})$ and let $T: R^2 \rightarrow R^2$ be the linear transformation defined by

$$T(\mathbf{u}) = \langle \mathbf{u}, \mathbf{w} \rangle \mathbf{w} \tag{1}$$

(A) Find the standard matrix for T.
(B) Compute $T(3, -2)$ using (1).
(C) Compute $T(3, -2)$ using the standard matrix for T and matrix multiplication.

Solution (A) From (1),

$$\overset{\mathbf{u}}{} \qquad \overset{\mathbf{u}}{} \quad \overset{\mathbf{w}}{} \quad \overset{\mathbf{w}}{}$$
$$T(1, 0) = \langle (1, 0), (\tfrac{3}{5}, \tfrac{4}{5}) \rangle (\tfrac{3}{5}, \tfrac{4}{5})$$
$$= \tfrac{3}{5}(\tfrac{3}{5}, \tfrac{4}{5}) = (\tfrac{9}{25}, \tfrac{12}{25})$$

and

$$\overset{\mathbf{u}}{} \qquad \overset{\mathbf{u}}{} \quad \overset{\mathbf{w}}{} \quad \overset{\mathbf{w}}{}$$
$$T(0, 1) = \langle (0, 1), (\tfrac{3}{5}, \tfrac{4}{5}) \rangle (\tfrac{3}{5}, \tfrac{4}{5})$$
$$= \tfrac{4}{5}(\tfrac{3}{5}, \tfrac{4}{5}) = (\tfrac{12}{25}, \tfrac{16}{25})$$

Writing $T(1, 0)$ and $T(0, 1)$ as column vectors, the standard matrix for T is

$$\begin{array}{cc} T(1, 0) & T(0, 1) \end{array}$$
$$A = \begin{bmatrix} \frac{9}{25} & \frac{12}{25} \\ \frac{12}{25} & \frac{16}{25} \end{bmatrix}$$

If $\mathbf{u} = (u_1, u_2)$ is any vector in R^2, then Theorem 8 implies that

$$T(\mathbf{u}) = A\mathbf{u} = \begin{bmatrix} \frac{9}{25} & \frac{12}{25} \\ \frac{12}{25} & \frac{16}{25} \end{bmatrix} \begin{bmatrix} u_1 \\ u_2 \end{bmatrix} \tag{2}$$

(B) Use (1) to find $T(3, -2)$:

$$T(3, -2) = \langle (3, -2), (\tfrac{3}{5}, \tfrac{4}{5}) \rangle (\tfrac{3}{5}, \tfrac{4}{5})$$
$$= \tfrac{1}{5}(\tfrac{3}{5}, \tfrac{4}{5}) = (\tfrac{3}{25}, \tfrac{4}{25})$$

(C) Use (2) to find $T(3, -2)$:

$$T(3, -2) = \begin{bmatrix} \frac{9}{25} & \frac{12}{25} \\ \frac{12}{25} & \frac{16}{25} \end{bmatrix} \begin{bmatrix} 3 \\ -2 \end{bmatrix} = \begin{bmatrix} \frac{3}{25} \\ \frac{4}{25} \end{bmatrix}$$

Problem 15 Repeat Example 15 for $\mathbf{w} = (\frac{5}{13}, -\frac{12}{13})$. ▮

Example 16 Let $T: R^3 \rightarrow R^2$ be the matrix transformation defined by

$$T(\mathbf{x}) = \begin{bmatrix} 1 & 1 & -1 \\ 2 & 1 & 2 \end{bmatrix} \mathbf{x}$$

Find the standard matrix for T.

Solution $T\left(\begin{bmatrix} 1 \\ 0 \\ 0 \end{bmatrix}\right) = \begin{bmatrix} 1 \\ 2 \end{bmatrix}$ $T\left(\begin{bmatrix} 0 \\ 1 \\ 0 \end{bmatrix}\right) = \begin{bmatrix} 1 \\ 1 \end{bmatrix}$ $T\left(\begin{bmatrix} 0 \\ 0 \\ 1 \end{bmatrix}\right) = \begin{bmatrix} -1 \\ 2 \end{bmatrix}$

Thus, the standard matrix for T is

$$A = \begin{bmatrix} 1 & 1 & -1 \\ 2 & 1 & 2 \end{bmatrix}$$

which is the matrix used to define T.

In general, **the standard matrix for a matrix transformation is the matrix used to define the transformation.** The proof of this statement is left as an exercise.

Problem 16   Find the standard matrix of the matrix transformation $T: R^2 \rightarrow R^2$ defined by

$$T(\mathbf{x}) = \begin{bmatrix} 2 & 1 \\ 1 & 2 \end{bmatrix} \mathbf{x}$$

∎

▪ Matrices of General Linear Transformations

Now that we have shown that every linear transformation from R^n to R^m can be expressed as a matrix transformation, we want to consider linear transformations from an arbitrary finite-dimensional vector space V to another finite-dimensional vector space W. Recall from Section 6-3 that if V is an n-dimensional vector space and B is a basis for V, then

$$[V]_B = \{[\mathbf{v}]_B | \mathbf{v} \in V\} = R^n$$

where $[\mathbf{v}]_B$ is the coordinate matrix of $\mathbf{v}$ with respect to B. Thus, each vector $\mathbf{v} \in V$ corresponds to a unique vector $[\mathbf{v}]_B$ in R^n, and conversely. The relationship between V and R^n can be used to associate a matrix with any linear transformation $T: V \rightarrow W$.

Example 17   Let $T: P_2 \rightarrow P_1$ be defined by

$$T(a_0 + a_1 x + a_2 x^2) = (-a_0 + 2a_1 + a_2) + (3a_0 - a_1 - 3a_2)x$$

and let

$$\begin{matrix} \mathbf{b}_1 & \mathbf{b}_2 & \mathbf{b}_3 & & & \mathbf{c}_1 & \mathbf{c}_2 \\ B = \{1, & x, & x^2\} & & \text{and} & C = \{1, & x\} \end{matrix}$$

be bases for P_2 and P_1, respectively.

(A) Find $[T(\mathbf{b}_i)]_C$ for each basis vector $\mathbf{b}_i \in B$.

(B) Find $[T(p(x))]_C$ for any vector $p(x)$ in P_2, and show that

$$[T(p(x))]_C = A[p(x)]_B$$

where

$$A = [[T(\mathbf{b}_1)]_C \quad [T(\mathbf{b}_2)]_C \quad [T(\mathbf{b}_3)]_C]$$

(C) Use the results in part (B) to find $T(4 + 5x + x^2)$.

(D) Find $T(4 + 5x + x^2)$ directly from the definition of T.

Solution   (A) $T(\mathbf{b}_1) = T(1) = -1 + 3x = -\mathbf{c}_1 + 3\mathbf{c}_2$ $[T(\mathbf{b}_1)]_C = \begin{bmatrix} -1 \\ 3 \end{bmatrix}$

$T(\mathbf{b}_2) = T(x) = 2 - x = 2\mathbf{c}_1 - \mathbf{c}_2$ $[T(\mathbf{b}_2)]_C = \begin{bmatrix} 2 \\ -1 \end{bmatrix}$

$T(\mathbf{b}_3) = T(x^2) = 1 - 3x = \mathbf{c}_1 - 3\mathbf{c}_2$ $[T(\mathbf{b}_3)]_C = \begin{bmatrix} 1 \\ -3 \end{bmatrix}$

(B) Let $p(x)$ be any vector in P_2. Since B is a basis for P_2, there exist scalars k_1, k_2, and k_3 satisfying

$$p(x) = k_1\mathbf{b}_1 + k_2\mathbf{b}_2 + k_3\mathbf{b}_3 \qquad \text{or equivalently} \qquad [p(x)]_B = \begin{bmatrix} k_1 \\ k_2 \\ k_3 \end{bmatrix}$$

Using Theorem 1(D) in Section 8-1, we can write

$$T(p(x)) = k_1 T(\mathbf{b}_1) + k_2 T(\mathbf{b}_2) + k_3 T(\mathbf{b}_3)$$
$$= k_1(-\mathbf{c}_1 + 3\mathbf{c}_2) + k_2(2\mathbf{c}_1 - \mathbf{c}_2) + k_3(\mathbf{c}_1 - 3\mathbf{c}_2)$$
$$= (-k_1 + 2k_2 + k_3)\mathbf{c}_1 + (3k_1 - k_2 - 3k_3)\mathbf{c}_2$$

Thus,

$$[T(p(x))]_C = \begin{bmatrix} -k_1 + 2k_2 + k_3 \\ 3k_1 - k_2 - 3k_3 \end{bmatrix}$$
$$= \begin{bmatrix} -1 & 2 & 1 \\ 3 & -1 & -3 \end{bmatrix} \begin{bmatrix} k_1 \\ k_2 \\ k_3 \end{bmatrix} = A[p(x)]_B$$

where

$$\begin{array}{ccc} [T(\mathbf{b}_1)]_C & [T(\mathbf{b}_2)]_C & [T(\mathbf{b}_3)]_C \end{array}$$
$$A = \begin{bmatrix} -1 & 2 & 1 \\ 3 & -1 & -3 \end{bmatrix}$$

Notice that the columns of A are the coordinate matrices with respect to C of the images under T of the basis vectors in B. The matrix

$$A = [[T(\mathbf{b}_1)]_C \quad [T(\mathbf{b}_2)]_C \quad [T(\mathbf{b}_3)]_C]$$

is called *the matrix of T with respect to the bases B and C.* This matrix can be used to find the image under T of any vector in P_2.

(C) If

$$p(x) = 4 + 5x + x^2$$

then

$$[p(x)]_B = \begin{bmatrix} 4 \\ 5 \\ 1 \end{bmatrix}$$

and

$$[T(p(x))]_C = A[p(x)]_B = \begin{bmatrix} -1 & 2 & 1 \\ 3 & -1 & -3 \end{bmatrix} \begin{bmatrix} 4 \\ 5 \\ 1 \end{bmatrix} = \begin{bmatrix} 7 \\ 4 \end{bmatrix}$$

This implies that $T(p(x)) = 7\mathbf{c}_1 + 4\mathbf{c}_2 = 7 + 4x$.

$$T(a_0 + a_1x + a_2x^2) = (-a_0 + 2a_1 + a_2) + (3a_0 - a_1 - 3a_2)x$$

(D) $T(4 + 5x + x^2) = (-4 + 10 + 1) + (12 - 5 - 3)x$

$$= 7 + 4x$$

Problem 17   Refer to Example 17. Let $p(x) = 4 - 2x + 5x^2$.

(A) Find $[p(x)]_B$.

(B) Use the equation $[T(p(x))]_C = A[p(x)]_B$ to find $[T(p(x))]_C$.

(C) Use $[T(p(x))]_C$ to find $T(p(x))$.

(D) Use the definition of T to find $T(p(x))$ directly.   ▮▮

The concepts introduced in Example 17 are generalized to arbitrary linear transformations in the box.

The Matrix of a Linear Transformation

Let $T: V \rightarrow W$ be a linear transformation, let $B = \{\mathbf{b}_1, \mathbf{b}_2, \ldots, \mathbf{b}_n\}$ be a basis for V, and let $C = \{\mathbf{c}_1, \mathbf{c}_2, \ldots, \mathbf{c}_m\}$ be a basis for W. The $m \times n$ matrix A defined by

$$A = [[T(\mathbf{b}_1)]_C \quad [T(\mathbf{b}_2)]_C \quad \cdots \quad [T(\mathbf{b}_n)]_C]$$

is called the **matrix of T with respect to the bases B and C.**

The relationship between a linear transformation T and its matrix representation A is given in Theorem 9. The proof, which is similar to the proof of Theorem 8, is left as an exercise.

Theorem 9
If A is the matrix of the linear transformation $T: V \to W$ with respect to the bases B and C, then

$$[T(\mathbf{x})]_C = A[\mathbf{x}]_B$$

for all $\mathbf{x}$ in V.

Example 18 Let $T: P_2 \to P_1$ be the linear transformation defined by

$$T(a_0 + a_1 x + a_2 x^2) = (-a_0 + 2a_1 + a_2) + (3a_0 - a_1 - 3a_2)x \qquad (3)$$

and let

$$
\begin{array}{ccc}
\mathbf{b}_1 & \mathbf{b}_2 & \mathbf{b}_3 \\
\end{array}
$$
$$
\begin{array}{ccc}
& & \mathbf{c}_1 \quad \mathbf{c}_2 \\
\end{array}
$$
$$B = \{1 + x,\ 1 + x + x^2,\ 1 + x^2\} \qquad \text{and} \qquad C = \{1 + 2x,\ 2 - x\}$$

be bases for P_2 and P_1, respectively.

(A) Find the matrix of T with respect to B and C.
(B) Use the matrix of T to find $T(p(x))$ if $p(x) = 4 + 5x + x^2$.

Solution (A) First, we use (3) to find the image under T of each vector in B:

$$T(\mathbf{b}_1) = T(1 + x) = 1 + 2x = \mathbf{c}_1$$
$$T(\mathbf{b}_2) = T(1 + x + x^2) = 2 - x = \mathbf{c}_2$$
$$T(\mathbf{b}_3) = T(1 + x^2) = \mathbf{0}$$

Next, we find the coordinate matrices with respect to C of the images of the basis vectors:

$$[T(\mathbf{b}_1)]_C = \begin{bmatrix} 1 \\ 0 \end{bmatrix} \qquad [T(\mathbf{b}_2)]_C = \begin{bmatrix} 0 \\ 1 \end{bmatrix} \qquad [T(\mathbf{b}_3)]_C = \begin{bmatrix} 0 \\ 0 \end{bmatrix}$$

Thus, the matrix of T with respect to B and C is

$$
\begin{array}{ccc}
[T(\mathbf{b}_1)]_C & [T(\mathbf{b}_2)]_C & [T(\mathbf{b}_3)]_C \\
\end{array}
$$
$$A = \begin{bmatrix} 1 & 0 & 0 \\ 0 & 1 & 0 \end{bmatrix}$$

(B) We use a three-step procedure to find $T(p(x))$.

Step 1. Find $[p(x)]_B$. We must find scalars k_1, k_2, and k_3 satisfying

$$4 + 5x + x^2 = k_1(1 + x) + k_2(1 + x + x^2) + k_3(1 + x^2)$$

The solution to this vector equation is $k_1 = 3$, $k_2 = 2$, and $k_3 = -1$ (verify this). Thus,

$$[p(x)]_B = \begin{bmatrix} 3 \\ 2 \\ -1 \end{bmatrix}$$

Step 2. Find $[T(p(x))]_C$. From Theorem 9,

$$[T(p(x))]_C = A[p(x)]_B = \begin{bmatrix} 1 & 0 & 0 \\ 0 & 1 & 0 \end{bmatrix} \begin{bmatrix} 3 \\ 2 \\ -1 \end{bmatrix} = \begin{bmatrix} 3 \\ 2 \end{bmatrix}$$

Step 3. Find $T(p(x))$. If

$$[T(p(x))]_C = \begin{bmatrix} 3 \\ 2 \end{bmatrix}$$

then

$$\begin{aligned} T(p(x)) &= 3\mathbf{c}_1 + 2\mathbf{c}_2 \\ &= 3(1 + 2x) + 2(2 - x) \\ &= 7 + 4x \end{aligned}$$

That is,

$$T(4 + 5x + x^2) = 7 + 4x \qquad T\colon P_2 \to P_1$$

The procedure for using matrix multiplication to compute the image of a vector under a linear transformation is outlined in the box. Notice that in addition to Theorem 9, this procedure uses the relationship between vectors in V and W and the corresponding coordinate matrices in R^n and R^m (see the figure in the box).

Computing $T(\mathbf{x})$ by Matrix Multiplication

Step 1. Find $[\mathbf{x}]_B$.
Step 2. Use the equation $[T(\mathbf{x})]_C = A[\mathbf{x}]_B$ to find $[T(\mathbf{x})]_C$.
Step 3. Use $[T(\mathbf{x})]_C$ to find $T(\mathbf{x})$.

Problem 18   Refer to the linear transformation T in Example 18.

(A) Find the matrix of T with respect to the bases

$$B = \{1, 1 + x^2, x + x^2\} \quad \text{and} \quad C = \{-1 + 3x, 3 - 4x\}$$

(B) Use matrix multiplication to find $T(p(x))$ if $p(x) = 4 - 2x + 5x^2$.   ▐▌

In Examples 17 and 18 and Problem 18, we computed three different matrices for the same linear transformation. Summarizing these results, as shown in Table 2, will illustrate some important ideas concerning matrices of linear transformations.

Table 2
Matrices for $T: P_2 \to P_1$ Defined by
$$T(a_0 + a_1x + a_2x^2) = (-a_0 + 2a_1 + a_2) + (3a_0 - a_1 - 3a_2)x$$

Reference	B	C	A
Example 17	$\{1, x, x^2\}$	$\{1, x\}$	$\begin{bmatrix} -1 & 2 & 1 \\ 3 & -1 & -3 \end{bmatrix}$
Example 18	$\{1 + x, 1 + x + x^2, 1 + x^2\}$	$\{1 + 2x, 2 - x\}$	$\begin{bmatrix} 1 & 0 & 0 \\ 0 & 1 & 0 \end{bmatrix}$
Problem 18	$\{1, 1 + x^2, x + x^2\}$	$\{-1 + 3x, 3 - 4x\}$	$\begin{bmatrix} 1 & 0 & 0 \\ 0 & 0 & 1 \end{bmatrix}$

First, the matrix of a transformation certainly depends on the choice of B and C. Changing either basis will usually change the matrix A. Second, certain choices of B and C will result in a very simple form for A. Later, we will encounter applications involving linear transformations whose solutions depend on selecting B and C so that A has a particular form.

Even if the matrix of a linear transformation has a very simple form, computing $T(\mathbf{x})$ by matrix multiplication seems much more involved than direct evaluation. [Compare the direct evaluation of $T(4 + 5x + x^2)$ in Example 17 with the matrix multiplication computations in both Examples 17 and 18.] In hand calculations, direct computation usually is easier than the matrix multiplication method. However, most real-world applications involving linear transformations require the use of a computer to perform the calculations. Since matrix multiplication is a routine calculation on a computer, it is common practice to use the matrix multiplication method in most problems involving linear transformations.

As the next example illustrates, matrix multiplication methods are convenient to use — even in hand calculations — if the transformation is defined in terms of its action on a basis.

Example 19   Let

$$B = \left\{ \underset{\mathbf{b}_1}{\begin{bmatrix} 1 \\ 1 \end{bmatrix}}, \underset{\mathbf{b}_2}{\begin{bmatrix} 1 \\ -1 \end{bmatrix}} \right\}$$

be a basis for R^2, let

$$
\begin{array}{cccc}
\mathbf{c}_1 & \mathbf{c}_2 & \mathbf{c}_3 & \mathbf{c}_4
\end{array}
$$

$$
C = \left\{ \begin{bmatrix} 1 & 0 \\ 0 & 0 \end{bmatrix}, \begin{bmatrix} 0 & 1 \\ 0 & 0 \end{bmatrix}, \begin{bmatrix} 0 & 0 \\ 1 & 0 \end{bmatrix}, \begin{bmatrix} 0 & 0 \\ 0 & 1 \end{bmatrix} \right\}
$$

be a basis for $M_{2\times2}$, and let $T: R^2 \to M_{2\times2}$ be the linear transformation defined by

$$
T\left(\begin{bmatrix} 1 \\ 1 \end{bmatrix} \right) = \begin{bmatrix} 1 & 0 \\ 0 & 1 \end{bmatrix} \qquad T\left(\begin{bmatrix} 1 \\ -1 \end{bmatrix} \right) = \begin{bmatrix} 0 & 1 \\ -1 & 0 \end{bmatrix}
$$

(A) Find the matrix of T with respect to B and C.
(B) Use matrix multiplication to find $T(\mathbf{x})$ if

$$
\mathbf{x} = \begin{bmatrix} 5 \\ 1 \end{bmatrix}
$$

Solution   (A) $T(\mathbf{b}_1) = \begin{bmatrix} 1 & 0 \\ 0 & 1 \end{bmatrix} = \mathbf{c}_1 + \mathbf{c}_4$ implies $[T(\mathbf{b}_1)]_C = \begin{bmatrix} 1 \\ 0 \\ 0 \\ 1 \end{bmatrix}$

$T(\mathbf{b}_2) = \begin{bmatrix} 0 & 1 \\ -1 & 0 \end{bmatrix} = \mathbf{c}_2 - \mathbf{c}_3$ implies $[T(\mathbf{b}_2)]_C = \begin{bmatrix} 0 \\ 1 \\ -1 \\ 0 \end{bmatrix}$

Thus, the matrix of T with respect to B and C is

$$
A = \begin{bmatrix} 1 & 0 \\ 0 & 1 \\ 0 & -1 \\ 1 & 0 \end{bmatrix}
$$

(B) Step 1.   Find $[\mathbf{x}]_B$. The solution of the vector equation

$$
\mathbf{x} = \begin{bmatrix} 5 \\ 1 \end{bmatrix} = k_1 \begin{bmatrix} 1 \\ 1 \end{bmatrix} + k_2 \begin{bmatrix} 1 \\ -1 \end{bmatrix}
$$

is $k_1 = 3$, $k_2 = 2$ (verify this). Thus,

$$
[\mathbf{x}]_B = \begin{bmatrix} 3 \\ 2 \end{bmatrix}
$$

Step 2.   Find $[T(\mathbf{x})]_C$:

$$
[T(\mathbf{x})]_C = \begin{bmatrix} 1 & 0 \\ 0 & 1 \\ 0 & -1 \\ 1 & 0 \end{bmatrix} \begin{bmatrix} 3 \\ 2 \end{bmatrix} = \begin{bmatrix} 3 \\ 2 \\ -2 \\ 3 \end{bmatrix}
$$

Step 3.   *Find T(x):*

$$T(\mathbf{x}) = 3\mathbf{c}_1 + 2\mathbf{c}_2 - 2\mathbf{c}_3 + 3\mathbf{c}_4 = \begin{bmatrix} 3 & 2 \\ -2 & 3 \end{bmatrix}$$

That is,

$$T\left(\begin{bmatrix} 5 \\ 1 \end{bmatrix}\right) = \begin{bmatrix} 3 & 2 \\ -2 & 3 \end{bmatrix} \qquad T: R^2 \to M_{2\times 2}$$

Problem 19   Repeat Example 19 if

$$T\left(\begin{bmatrix} 1 \\ 1 \end{bmatrix}\right) = \begin{bmatrix} 1 & 0 \\ 2 & 0 \end{bmatrix} \qquad T\left(\begin{bmatrix} 1 \\ -1 \end{bmatrix}\right) = \begin{bmatrix} 0 & 2 \\ 0 & -3 \end{bmatrix} \qquad \blacksquare$$

▪ Matrices of Linear Operators

If T is a linear operator on V (a linear transformation from V to V), it is customary to use a single basis $B = \{\mathbf{b}_1, \mathbf{b}_2, \ldots, \mathbf{b}_n\}$ to determine the matrix of T. In this case, the matrix

$$A = [[T(\mathbf{b}_1)]_B \quad [T(\mathbf{b}_2)]_B \quad \cdots \quad [T(\mathbf{b}_n)]_B]$$

is called the **matrix of T with respect to B.**

Example 20   Let $T: R^2 \to R^2$ be the linear operator defined by

$$T(\mathbf{u}) = \langle \mathbf{u}, \mathbf{w} \rangle \mathbf{w} \qquad \text{where} \qquad \mathbf{w} = (\tfrac{3}{5}, \tfrac{4}{5})$$

Find the matrix of T with respect to the basis

$$\begin{array}{cc} \mathbf{b}_1 & \mathbf{b}_2 \end{array}$$
$$B = \{(3, 4), (4, -3)\}$$

Solution $T(\mathbf{b}_1) = \langle (3, 4), (\tfrac{3}{5}, \tfrac{4}{5}) \rangle (\tfrac{3}{5}, \tfrac{4}{5}) = 5(\tfrac{3}{5}, \tfrac{4}{5}) = (3, 4) = \mathbf{b}_1$

$\qquad\quad$ $T(\mathbf{b}_2) = \langle (4, -3), (\tfrac{3}{5}, \tfrac{4}{5}) \rangle (\tfrac{3}{5}, \tfrac{4}{5}) = 0(\tfrac{3}{5}, \tfrac{4}{5}) = \mathbf{0}$

Thus,

$$[T(\mathbf{b}_1)]_B = \begin{bmatrix} 1 \\ 0 \end{bmatrix} \qquad [T(\mathbf{b}_2)]_B = \begin{bmatrix} 0 \\ 0 \end{bmatrix}$$

and the matrix of T with respect to B is

$$A = \begin{bmatrix} 1 & 0 \\ 0 & 0 \end{bmatrix}$$

Problem 20   Repeat Example 20 if $\mathbf{w} = (\tfrac{5}{13}, -\tfrac{12}{13})$ and $B = \{(12, 5), (5, -12)\}$. $\blacksquare$

Let T be the linear operator considered in Example 20. The standard matrix of T is (see Example 15)

$$\begin{bmatrix} \frac{9}{25} & \frac{12}{25} \\ \frac{12}{25} & \frac{16}{25} \end{bmatrix}$$

which is certainly more complicated than the matrix representation of T obtained in Example 20. In many cases, it is possible to select a basis that will provide a very simple matrix representation of a linear operator. We will have more to say about this in the next section.

Answers to
Matched Problems

14. $\begin{bmatrix} 1 & -2 & 3 \\ 6 & 5 & -4 \end{bmatrix}$ **15.** (A) $\begin{bmatrix} \frac{25}{169} & -\frac{60}{169} \\ -\frac{60}{169} & \frac{144}{169} \end{bmatrix}$ (B)–(C) $\begin{bmatrix} \frac{15}{13} \\ -\frac{36}{13} \end{bmatrix}$

16. $\begin{bmatrix} 2 & 1 \\ 1 & 2 \end{bmatrix}$ **17.** (A) $\begin{bmatrix} 4 \\ -2 \\ 5 \end{bmatrix}$ (B) $\begin{bmatrix} -3 \\ -1 \end{bmatrix}$ (C)–(D) $-3 - x$

18. (A) $\begin{bmatrix} 1 & 0 & 0 \\ 0 & 0 & 1 \end{bmatrix}$ (B) $-3 - x$ **19.** (A) $\begin{bmatrix} 1 & 0 \\ 0 & 2 \\ 2 & 0 \\ 0 & -3 \end{bmatrix}$ (B) $\begin{bmatrix} 3 & 4 \\ 6 & -6 \end{bmatrix}$

20. $\begin{bmatrix} 0 & 0 \\ 0 & 1 \end{bmatrix}$

‖ Exercise 8-3

A   In Problems 1 and 2, find the standard matrix of the linear operator $T: R^2 \rightarrow R^2$.

1. $T\left(\begin{bmatrix} x_1 \\ x_2 \end{bmatrix}\right) = \begin{bmatrix} 2x_1 + 3x_2 \\ 5x_1 + 4x_2 \end{bmatrix}$ **2.** $T\left(\begin{bmatrix} x_1 \\ x_2 \end{bmatrix}\right) = \begin{bmatrix} 3x_1 - 5x_2 \\ -4x_1 + 7x_2 \end{bmatrix}$

In Problems 3 and 4, find the standard matrix of the linear transformation $T: R^2 \rightarrow R^3$.

3. $T\left(\begin{bmatrix} x_1 \\ x_2 \end{bmatrix}\right) = \begin{bmatrix} x_1 + x_2 \\ x_1 \\ x_2 \end{bmatrix}$ **4.** $T\left(\begin{bmatrix} x_1 \\ x_2 \end{bmatrix}\right) = \begin{bmatrix} x_2 \\ 2x_1 + 3x_2 \\ x_1 \end{bmatrix}$

In Problems 5 and 6, find the standard matrix of the linear transformation $T: R^4 \rightarrow R^3$.

5. $T\left(\begin{bmatrix} x_1 \\ x_2 \\ x_3 \\ x_4 \end{bmatrix}\right) = \begin{bmatrix} x_1 - x_2 \\ x_2 - x_3 \\ x_3 - x_4 \end{bmatrix}$ **6.** $T\left(\begin{bmatrix} x_1 \\ x_2 \\ x_3 \\ x_4 \end{bmatrix}\right) = \begin{bmatrix} x_1 + x_4 \\ x_4 + x_3 \\ x_3 + x_2 \end{bmatrix}$

In Problems 7 and 8, find the matrix representation of the linear operator $T: P_1 \rightarrow P_1$ with respect to the basis $B = \{1, x\}$.

7. $T(a_0 + a_1x) = (2a_0 + 5a_1) + (3a_0 + 7a_1)x$

8. $T(a_0 + a_1x) = (-a_0 + 2a_1) + a_0x$

In Problems 9 and 10, find the standard matrix of the linear transformation $T: P_3 \rightarrow P_1$ with respect to the bases $B = \{1, x, x^2, x^3\}$ and $C = \{1, x\}$.

9. $T(a_0 + a_1x + a_2x^2 + a_3x^3) = (a_0 - a_3) + (a_1 + 2a_2)x$

10. $T(a_0 + a_1x + a_2x^2 + a_3x^3) = a_3 + (2a_2 + 3a_1 + 4a_0)x$

B **11.** Let $T: R^2 \rightarrow R^2$ be the linear operator defined by

$$T\left(\begin{bmatrix} x_1 \\ x_2 \end{bmatrix}\right) = \begin{bmatrix} x_1 + 2x_2 \\ -x_1 + 4x_2 \end{bmatrix}$$

let

$$B = \left\{ \begin{bmatrix} 1 \\ 1 \end{bmatrix}, \begin{bmatrix} 2 \\ 1 \end{bmatrix} \right\}$$

be a basis for R^2, and let

$$x = \begin{bmatrix} 4 \\ 1 \end{bmatrix}$$

(A) Find the matrix of T with respect to B.   (B) Find $[\mathbf{x}]_B$.

(C) Use matrix multiplication to compute $[T(\mathbf{x})]_B$.

(D) Use $[T(\mathbf{x})]_B$ to find $T(\mathbf{x})$.

(E) Find $T(\mathbf{x})$ directly from the definition of T.

12. Repeat Problem 11 for

$$T\left(\begin{bmatrix} x_1 \\ x_2 \end{bmatrix}\right) = \begin{bmatrix} 6x_1 + 5x_2 \\ -2x_1 - 5x_2 \end{bmatrix} \qquad B = \left\{ \begin{bmatrix} 1 \\ -2 \end{bmatrix}, \begin{bmatrix} -5 \\ 1 \end{bmatrix} \right\} \qquad x = \begin{bmatrix} -7 \\ 5 \end{bmatrix}$$

13. Let $T: P_1 \rightarrow P_2$ be defined by

$$T(a_0 + a_1x) = a_0 + (a_1 - a_0)x + (a_1 + a_0)x^2$$

let $B = \{1 + x, 2 - x\}$, let $C = \{1, x, x^2\}$, and let $p(x) = 7 + x$.

(A) Find the matrix of T with respect to B and C.   (B) Find $[p(x)]_B$.

(C) Use matrix multiplication to find $[T(p(x))]_C$.

(D) Use $[T(p(x))]_C$ to find $T(p(x))$.

(E) Find $T(p(x))$ directly from the definition of T.

14. Repeat Problem 13 if

$$T(a_0 + a_1x) = a_1 + (a_1 + a_0)x + (a_1 + 2a_0)x^2$$

and $B = \{1 + x, 1 - 2x\}$ [C and $p(x)$ remain unchanged].

15. Let $T: P_2 \rightarrow P_1$ be defined by

$$T(a_0 + a_1x + a_2x^2) = (a_0 + a_1) + (a_1 - a_2)x$$

Find the matrix of T with respect to B and C if:

(A) $B = \{1, x, x^2\}$; $C = \{1, x\}$
(B) $B = \{1 + x, 1 + x^2, 1 - x - x^2\}$; $C = \{2 + x, 1 - x\}$
(C) $B = \{x, 2 - x, 3 + x + x^2\}$; $C = \{1 + x, 1 - x\}$

16. Let $T: P_2 \rightarrow P_1$ be defined by

$$T(a_0 + a_1x + a_2x^2) = (a_0 + 2a_1 - a_2) + (2a_0 + a_1 + a_2)x$$

Find the matrix of T with respect to B and C if:

(A) $B = \{1, x, x^2\}$; $C = \{1, x\}$
(B) $B = \{x + x^2, 2 + x^2, 1 - x - x^2\}$; $C = \{1 + 2x, 1 + 5x\}$
(C) $B = \{1 + x, -2 + 2x, 2x - 2x^2\}$; $C = \{1 + x, 1 - x\}$

17. Let $T: P_2 \rightarrow P_3$ be defined by $T(p(x)) = xp(x)$. Find the matrix of T with respect to B and C if $B = \{1, x, x^2\}$ and $C = \{1, x, x^2, x^3\}$.

18. Repeat Problem 17 if $T(p(x)) = xp(x) - p(1)$.

19. Let $T: R^2 \rightarrow R^2$ be the linear operator defined by

$$T(\mathbf{u}) = \langle \mathbf{u}, \mathbf{w} \rangle \mathbf{w}$$

where $\mathbf{w} = (1/\sqrt{5}, 2/\sqrt{5})$.

(A) Find the standard matrix of T.
(B) Find the matrix of T with respect to the basis $B = \{(1, 2), (2, -1)\}$.
(C) Compute $T(8, -1)$ three ways: directly from the definition of T, using the matrix in part (A), and using the matrix in part (B).

20. Let $T: R^2 \rightarrow R^2$ be the linear operator defined by

$$T(\mathbf{u}) = \langle \mathbf{u}, \mathbf{w} \rangle \mathbf{w}$$

where $\mathbf{w} = (1/\sqrt{2}, -1/\sqrt{2})$

(A) Find the standard matrix of T.
(B) Find the matrix of T with respect to the basis $B = \{(1, 1), (1, -1)\}$.
(C) Compute $T(5, 2)$ three ways: directly from the definition of T, using the matrix in part (A), and using the matrix in part (B).

21. Let $T: R^3 \rightarrow R^3$ be the linear operator defined by

$$T(\mathbf{u}) = \langle \mathbf{u}, \mathbf{v} \rangle \mathbf{v} + \langle \mathbf{u}, \mathbf{w} \rangle \mathbf{w}$$

where $\mathbf{v} = (\frac{1}{3}, \frac{2}{3}, \frac{2}{3})$ and $\mathbf{w} = (\frac{2}{3}, \frac{1}{3}, -\frac{2}{3})$.

(A) Find the standard matrix of T.
(B) Find the matrix of T with respect to the basis

$$B = \{(1, 2, 2), (2, 1, -2), (2, -2, 1)\}$$

(C) Compute $T(5, 1, 1)$ three ways: directly from the definition of T, using the matrix in part (A), and using the matrix in part (B).

22. Let $T: R^3 \to R^3$ be the linear operator defined by

$$T(\mathbf{u}) = \langle \mathbf{u}, \mathbf{v} \rangle \mathbf{v} + \langle \mathbf{u}, \mathbf{w} \rangle \mathbf{w}$$

where $\mathbf{v} = (\frac{2}{7}, -\frac{3}{7}, -\frac{6}{7})$ and $\mathbf{w} = (\frac{3}{7}, \frac{6}{7}, -\frac{2}{7})$.

(A) Find the standard matrix of T.

(B) Find the matrix of T with respect to the basis

$$B = \{(2, -3, -6), (6, -2, 3), (3, 6, -2)\}$$

(C) Compute $T(9, 4, 1)$ three ways: directly from the definition of T, using the matrix in part (A), and using the matrix in part (B).

23. Let

$$B = \left\{ \begin{bmatrix} 1 \\ 2 \end{bmatrix}, \begin{bmatrix} 2 \\ -1 \end{bmatrix} \right\}$$

be a basis for R^2, let

$$C = \left\{ \begin{bmatrix} 1 & 0 \\ 0 & 0 \end{bmatrix}, \begin{bmatrix} 0 & 1 \\ 1 & 0 \end{bmatrix}, \begin{bmatrix} 0 & 0 \\ 0 & 1 \end{bmatrix} \right\}$$

be a basis for $S_{2\times2}$ (the vector space of 2×2 symmetric matrices), and let $T: R^2 \to S_{2\times2}$ be defined by

$$T\left(\begin{bmatrix} 1 \\ 2 \end{bmatrix} \right) = \begin{bmatrix} 1 & 2 \\ 2 & 0 \end{bmatrix} \qquad T\left(\begin{bmatrix} 2 \\ -1 \end{bmatrix} \right) = \begin{bmatrix} 1 & -1 \\ -1 & 2 \end{bmatrix}$$

(A) Find the matrix of T with respect to B and C.

(B) Use matrix multiplication to find $T(\mathbf{x})$ if

$$\mathbf{x} = \begin{bmatrix} 7 \\ 4 \end{bmatrix}$$

24. Repeat Problem 23 if

$$B = \left\{ \begin{bmatrix} 2 \\ 3 \end{bmatrix}, \begin{bmatrix} 3 \\ -2 \end{bmatrix} \right\}$$

and

$$T\left(\begin{bmatrix} 2 \\ 3 \end{bmatrix} \right) = \begin{bmatrix} 3 & -1 \\ -1 & 2 \end{bmatrix} \qquad T\left(\begin{bmatrix} 3 \\ -2 \end{bmatrix} \right) = \begin{bmatrix} 1 & 0 \\ 0 & -1 \end{bmatrix}$$

(Basis C and $\mathbf{x}$ remain unchanged.)

25. Let $B = \{(3, -4), (-2, 3)\}$ and $C = \{1, x, x^2\}$ be bases for R^2 and P_2, respectively, let $T: R^2 \to P_2$ be the linear transformation defined by

$$T(3, -4) = 1 - x^2 \qquad T(-2, 3) = x + 2x^2$$

and let $\mathbf{e}_1 = (1, 0)$, $\mathbf{e}_2 = (0, 1)$.

(A) Find the matrix of T with respect to B and C.

(B) Use matrix multiplication to find $T(\mathbf{e}_1)$ and $T(\mathbf{e}_2)$.

(C) Find $T(a, b)$ for any $(a, b) \in R^2$.

26. Repeat Problem 25 if $B = \{(8, 5), (3, 2)\}$ and T is defined by

$$T(8, 5) = 1 + x \qquad T(3, 2) = 2 + x^2$$

(Basis C remains unchanged.)

27. Let

$$A = \begin{bmatrix} 3 & 0 & 0 \\ 0 & 1 & 0 \\ 0 & 0 & 1 \end{bmatrix}$$

be the matrix of a linear operator $T: R^3 \to R^3$ with respect to the basis $B = \{(-1, 0, 1), (2, 1, 0), (-2, 0, 1)\}$, and let $\mathbf{e}_1 = (1, 0, 0)$, $\mathbf{e}_2 = (0, 1, 0)$, $\mathbf{e}_3 = (0, 0, 1)$.

(A) Find $[\mathbf{e}_i]_B$, $i = 1, 2, 3$.

(B) Find $[T(\mathbf{e}_i)]_B$, $i = 1, 2, 3$.

(C) Find $T(\mathbf{e}_i)$, $i = 1, 2, 3$.

(D) Find the standard matrix of T.

28. Repeat Problem 27 if

$$A = \begin{bmatrix} 2 & 0 & 0 \\ 0 & -1 & 0 \\ 0 & 0 & -1 \end{bmatrix}$$

and $B = \{(0, -2, 1), (1, 1, 0), (1, 0, 1)\}$.

29. *Calculus.* Let $T: P_3 \to P_3$ be the linear operator defined by $T(p(x)) = xp'(x)$. Find the matrix of T with respect to the basis $B = \{1, x, x^2, x^3\}$.

30. *Calculus.* Repeat Problem 29 if $T(p(x)) = p(x) + xp'(x)$.

C **31.** Let V be an n-dimensional vector space, let k be a scalar, and let $T: V \to V$ be defined by $T(\mathbf{v}) = k\mathbf{v}$. Find the matrix of T with respect to any basis of V.

32. Let V be an n-dimensional vector space, let W be an m-dimensional vector space, and let $T: V \to W$ be defined by $T(\mathbf{v}) = \mathbf{0}$. Find the matrix of T with respect to any basis of V and any basis of W.

33. Let A be an $n \times n$ matrix with n linearly independent eigenvectors $\mathbf{x}_1$, $\mathbf{x}_2, \ldots, \mathbf{x}_n$ and associated eigenvalues $\lambda_1, \lambda_2, \ldots, \lambda_n$, and let $T: R^n \to R^n$ be the matrix transformation defined by $T(\mathbf{x}) = A\mathbf{x}$. Find the matrix of T with respect to $B = \{\mathbf{x}_1, \mathbf{x}_2, \ldots, \mathbf{x}_n\}$.

34. Let V be a finite-dimensional vector space with bases B and C, and let $T: V \to V$ be defined by $T(\mathbf{x}) = \mathbf{x}$. Show that the matrix of T with respect to B and C is the transition matrix from B to C. [*Hint:* Compare the definition of the matrix of a linear transformation in this section with the definition of transition matrix in Section 6-4.]

35. Let B be an $m \times n$ matrix, let $T: R^n \to R^m$ be the matrix transformation defined by $T(\mathbf{x}) = B\mathbf{x}$, and let A be the standard matrix of T. Show that $A = B$.

36. *Theorem 9.* Prove Theorem 9.

┃8-4┃ Eigenvalues of Linear Operators

- Eigenvalues and Eigenvectors of Linear Operators
- Diagonalizing Linear Operators
- Similar Matrices

■ Eigenvalues and Eigenvectors of Linear Operators

In this section, we will restrict our attention to *linear operators* $T: V \to V$ where V is a *finite-dimensional* vector space. In Section 8-3, we saw that the matrix representation of T with respect to a basis B is a useful tool for working with linear operators. Now we want to consider the following question: Can we find a basis B for V so that the matrix representation of T with respect to B is a diagonal matrix? It is not surprising that the answer to this question is closely related to the problem of diagonalizing an $n \times n$ matrix, which we studied in Chapter 7. The following example will serve to review some of the basic concepts introduced in Chapter 7 and to motivate the generalization of these concepts to linear operators.

Example 21   Let $B = \{1, x\}$ be a basis for P_1 (we choose the standard basis for ease of computation) and let $T: P_1 \to P_1$ be the linear operator defined by

$$T(a_0 + a_1 x) = (2a_0 + 2a_1) + (-a_0 + 5a_1)x \tag{1}$$

Then

$$T(1) = 2 - x \quad \text{implies} \quad [T(1)]_B = \begin{bmatrix} 2 \\ -1 \end{bmatrix}$$

and

$$T(x) = 2 + 5x \quad \text{implies} \quad [T(x)]_B = \begin{bmatrix} 2 \\ 5 \end{bmatrix}$$

Thus, the matrix of T with respect to B is

$$A = \begin{bmatrix} 2 & 2 \\ -1 & 5 \end{bmatrix}$$

If

$$\mathbf{x}_1 = \begin{bmatrix} 2 \\ 1 \end{bmatrix}$$

then

$$Ax_1 = \begin{bmatrix} 2 & 2 \\ -1 & 5 \end{bmatrix}\begin{bmatrix} 2 \\ 1 \end{bmatrix} = \begin{bmatrix} 6 \\ 3 \end{bmatrix} = 3\begin{bmatrix} 2 \\ 1 \end{bmatrix} = 3x_1$$

Thus, $\lambda_1 = 3$ is an eigenvalue of A and x_1 is an eigenvector of A associated with λ_1. Now let $p_1(x)$ be the vector in P_1 that satisfies

$$[p_1(x)]_B = x_1 = \begin{bmatrix} 2 \\ 1 \end{bmatrix}$$

Then

$$p_1(x) = 2 + x$$

and, using (1),

$$T(p_1(x)) = T(2 + x) = 6 + 3x = 3(2 + x) = 3p_1(x)$$

Thus, it is natural to call the real number $\lambda_1 = 3$ an *eigenvalue* of T and the vector $p_1(x)$ an *eigenvector* of T associated with λ_1.

The results in Example 21 lead to the following definitions:

Eigenvalues and Eigenvectors of a Linear Operator

Let T: $V \rightarrow V$ be a linear operator. The real number λ is called an **eigenvalue** of T if there exists a nonzero vector **v** in V satisfying

$$T(\mathbf{v}) = \lambda\mathbf{v} \tag{2}$$

Any nonzero vector **v** that satisfies (2) is called an **eigenvector** of T associated with the eigenvalue λ. The **eigenspace** of T associated with the eigenvalue λ is the subspace of V defined by

$$S_\lambda = \{\mathbf{v} \in V | T(\mathbf{v}) = \lambda\mathbf{v}\}$$

It is left as an exercise to show that S_λ is, in fact, a subspace of V.

Problem 21   Refer to Example 21.

(A) Show that $\lambda_2 = 4$ is an eigenvalue of A with associated eigenvector

$$x_2 = \begin{bmatrix} 1 \\ 1 \end{bmatrix}$$

(B) Show that $\lambda_2 = 4$ is an eigenvalue of T with associated eigenvector $p_1(x) = 1 + x$. ∎

In Example 21 and Problem 21, the linear operator T and its matrix representation A have the same eigenvalues, $\lambda_1 = 3$ and $\lambda_2 = 4$. Furthermore, the eigenvectors $p_1(x)$ and $p_2(x)$ of T and the eigenvectors x_1 and x_2 of A satisfy

$$[p_1(x)]_B = x_1 \qquad \text{and} \qquad [p_2(x)]_B = x_2$$

Theorem 10 shows that this is always the case. The proof is omitted.

Theorem 10	Let $T: V \to V$ be a linear operator, let B be a basis for V, and let A be the matrix of T with respect to B. Then: (A) T and A have the same eigenvalues. (B) The vector $\mathbf{v} \in V$ is an eigenvector of T associated with λ if and only if the coordinate matrix $[\mathbf{v}]_B$ is an eigenvector of A associated with λ.

Example 22 Find the eigenvalues and bases for the eigenspaces of the linear operator $T: P_2 \to P_2$ defined by

$$T(a_0 + a_1 x + a_2 x^2) = a_0 + (-a_0 - 2a_2)x + (a_0 + a_1 + 3a_2)x^2 \tag{3}$$

Solution **Step 1.** *Find a matrix representation for T.* Since Theorem 10 holds irrespective of the choice of basis for V, we are free to use any basis for P_2 to find a matrix representation of T. To simplify the calculations, we choose the standard basis for P_2, $B = \{1, x, x^2\}$. Then

$$T(1) = 1 - x + x^2 \qquad T(x) = x^2 \qquad T(x^2) = -2x + 3x^2$$

implies that

$$[T(1)]_B = \begin{bmatrix} 1 \\ -1 \\ 1 \end{bmatrix} \qquad [T(x)]_B = \begin{bmatrix} 0 \\ 0 \\ 1 \end{bmatrix} \qquad [T(x^2)]_B = \begin{bmatrix} 0 \\ -2 \\ 3 \end{bmatrix}$$

and the matrix of T with respect to B is

$$A = \begin{bmatrix} 1 & 0 & 0 \\ -1 & 0 & -2 \\ 1 & 1 & 3 \end{bmatrix}$$

Step 2. *Find the eigenvalues of A (see Section 7-1).*

$$p(\lambda) = \det(\lambda I - A) = (\lambda - 1)^2(\lambda - 2)$$

The eigenvalues of A are $\lambda_1 = 1$ and $\lambda_2 = 2$. According to Theorem 10, these are also the eigenvalues of T.

Step 3. *Find bases for the eigenspaces of A.* Omitting the details (which you should supply),

$$B_{\lambda_1} = \left\{ \overset{\mathbf{b}_1}{\begin{bmatrix} -1 \\ 1 \\ 0 \end{bmatrix}}, \overset{\mathbf{b}_2}{\begin{bmatrix} -2 \\ 0 \\ 1 \end{bmatrix}} \right\} \qquad \text{and} \qquad B_{\lambda_2} = \left\{ \overset{\mathbf{b}_3}{\begin{bmatrix} 0 \\ -1 \\ 1 \end{bmatrix}} \right\}$$

are bases for the eigenspaces of A.

Step 4. *Find bases for the eigenspaces of T.* Let $p_1(x)$, $p_2(x)$, and $p_3(x)$ be the vectors in P_2 defined by $[p_i(x)]_B = \mathbf{b}_i$, $i = 1, 2, 3$. That is,

$$p_1(x) = -1 + x \qquad p_2(x) = -2 + x^2 \qquad p_3(x) = -x + x^2$$

Theorem 10 in Section 6-3 implies that

$$\overset{\mathbf{c}_1 \qquad\qquad \mathbf{c}_2}{C_{\lambda_1} = \{-1 + x, -2 + x^2\}} \quad \text{and} \quad \overset{\mathbf{c}_3}{C_{\lambda_2} = \{-x + x^2\}}$$

are bases for the eigenspaces of T.

Steps 1–4 in the solution of Example 22 are summarized in the box for convenient reference.

Finding the Eigenvalues and Eigenspaces of a Linear Operator

Let $T: V \rightarrow V$ be a linear operator.

Step 1. Choose any convenient basis B for V and find A, the matrix of T with respect to B.

Step 2. Find the eigenvalues of A, using $p(\lambda) = \det(\lambda I - A)$. These are also the eigenvalues of T.

Step 3. Find bases for the eigenspaces of A.

Step 4. For each basis

$$B_\lambda = \{\mathbf{b}_1, \mathbf{b}_2, \dots, \mathbf{b}_k\}$$

found in Step 3, let

$$C_\lambda = \{\mathbf{c}_1, \mathbf{c}_2, \dots, \mathbf{c}_k\}$$

where $[\mathbf{c}_i]_B = \mathbf{b}_i$. Then C_λ is a basis for the eigenspace of T associated with λ.

Problem 22   Find the eigenvalues and bases for the eigenspaces of the linear operator $T: P_1 \rightarrow P_1$ defined by

$$T(a_0 + a_1 x) = (4a_0 + 2a_1) + (a_0 + 3a_1)x \qquad\qquad ∥$$

▪ Diagonalizing Linear Operators

A linear operator T can have many different matrix representations. In certain applications, it is important to determine whether T has a *diagonal* matrix representation.

Diagonalizable Linear Operators

A linear operator $T: V \rightarrow V$ is said to be **diagonalizable** if T has a diagonal matrix representation.

Our goal is to develop procedures that will enable us to find a diagonal representation of a linear operator, if such a representation exists. We begin by returning to the linear operator considered in Example 22.

Example 23   Let $T: P_2 \rightarrow P_2$ be the linear operator defined in Example 22 and repeated below:

$$T(a_0 + a_1 x + a_2 x^2) = a_0 + (-a_0 - 2a_2)x + (a_0 + a_1 + 3a_2)x^2 \qquad (3)$$

Let

$$\begin{array}{ccc} \mathbf{c}_1 & \mathbf{c}_2 & \mathbf{c}_3 \end{array}$$
$$C = \{-1 + x, -2 + x^2, -x + x^2\}$$

(A) Show that C is a basis for P_2 consisting of eigenvectors of T.
(B) Find the matrix of T with respect to C.

Solution   (A) From Example 22, bases for the eigenspaces of T are given by

$$\begin{array}{ccc} \mathbf{c}_1 & \mathbf{c}_2 & \mathbf{c}_3 \end{array}$$
$$C_{\lambda_1} = \{-1 + x, -2 + x^2\} \quad \text{and} \quad C_{\lambda_2} = \{-x + x^2\}$$

Thus, the vectors in C are eigenvectors of T. The corresponding bases for the eigenspaces of the matrix representation A are

$$B_{\lambda_1} = \left\{ \begin{bmatrix} -1 \\ 1 \\ 0 \end{bmatrix}, \begin{bmatrix} -2 \\ 0 \\ 1 \end{bmatrix} \right\} \quad \text{and} \quad B_{\lambda_2} = \left\{ \begin{bmatrix} 0 \\ -1 \\ 1 \end{bmatrix} \right\}$$

Theorem 8 in Section 7-2 implies that

$$B = B_{\lambda_1} \cup B_{\lambda_2} = \left\{ \begin{bmatrix} -1 \\ 1 \\ 0 \end{bmatrix}, \begin{bmatrix} -2 \\ 0 \\ 1 \end{bmatrix}, \begin{bmatrix} 0 \\ -1 \\ 1 \end{bmatrix} \right\}$$

is a basis for R^3 consisting of eigenvectors of A. Theorem 10 in Section 6-3 then implies that

$$C = C_{\lambda_1} \cup C_{\lambda_2} = \{-1 + x, -2 + x^2, -x + x^2\}$$

is a basis for P_2.

(B) Using (3),

$$T(\mathbf{c}_1) = T(-1 + x) = -1 + x = \mathbf{c}_1$$
$$T(\mathbf{c}_2) = T(-2 + x^2) = -2 + x^2 = \mathbf{c}_2$$
$$T(\mathbf{c}_3) = T(- x + x^2) = -2x + 2x^2 = 2\mathbf{c}_3$$

Thus,

$$[T(\mathbf{c}_1)]_C = \begin{bmatrix} 1 \\ 0 \\ 0 \end{bmatrix} \qquad [T(\mathbf{c}_2)]_C = \begin{bmatrix} 0 \\ 1 \\ 0 \end{bmatrix} \qquad [T(\mathbf{c}_3)]_C = \begin{bmatrix} 0 \\ 0 \\ 2 \end{bmatrix}$$

and the matrix representation of T with respect to C is

$$D = \begin{bmatrix} 1 & 0 & 0 \\ 0 & 1 & 0 \\ 0 & 0 & 2 \end{bmatrix}$$

Thus, T is a diagonalizable linear operator. Notice that the diagonal entries of D are the eigenvalues of T.

Problem 23   Refer to Problem 22. Let $T: P_1 \rightarrow P_1$ be the linear operator defined by

$$T(a_0 + a_1 x) = (4a_0 + 2a_1) + (a_0 + 3a_1)x$$

and let

$$C = \{1 - x, 2 + x\}$$

(A) Show that C is a basis for P_1 consisting of eigenvectors of T.
(B) Find the matrix of T with respect to C.   ∎

The results illustrated in Example 23 are generalized to arbitrary linear operators in Theorem 11. The proof is omitted.

Theorem 11

Let $T: V \rightarrow V$ be a linear operator, let $n = \dim(V)$, and let A be the matrix of T with respect to any basis S for V.

(A) T is diagonalizable if and only if A is diagonalizable.
(B) If

$$B = \{\mathbf{b}_1, \mathbf{b}_2, \ldots, \mathbf{b}_n\}$$

is a basis for R^n consisting of eigenvectors of A and $\mathbf{c}_1, \mathbf{c}_2, \ldots, \mathbf{c}_n$ are the vectors in V satisfying $[\mathbf{c}_i]_S = \mathbf{b}_i$, $i = 1, 2, \ldots, n$, then

$$C = \{\mathbf{c}_1, \mathbf{c}_2, \ldots, \mathbf{c}_n\}$$

is a basis for V consisting of eigenvectors of T.
(C) If $C = \{\mathbf{c}_1, \mathbf{c}_2, \ldots, \mathbf{c}_n\}$ is a basis for V consisting of eigenvectors of T and $\lambda_1, \lambda_2, \ldots, \lambda_n$ are the corresponding eigenvalues, then the matrix of T with respect to C is

$$D = \begin{bmatrix} \lambda_1 & 0 & \cdots & 0 \\ 0 & \lambda_2 & \cdots & 0 \\ \vdots & \vdots & & \vdots \\ 0 & 0 & \cdots & \lambda_n \end{bmatrix}$$

Theorem 11 reduces the problem of diagonalizing a linear operator to that of diagonalizing a matrix and allows us to apply all techniques developed in Chapter 7 to linear operators. The following examples illustrate some of the applications of Theorem 11.

Example 24 Let $T: R^2 \rightarrow R^2$ be the linear operator defined by

$$T\left(\begin{bmatrix} 1 \\ 1 \end{bmatrix}\right) = \begin{bmatrix} 2 \\ 1 \end{bmatrix} \qquad T\left(\begin{bmatrix} 4 \\ 3 \end{bmatrix}\right) = \begin{bmatrix} 3 \\ 1 \end{bmatrix}$$

Find a diagonal representation of T, if possible.

Solution Since T is defined by its action on the set

$$S = \left\{ \overset{\mathbf{s_1}}{\begin{bmatrix} 1 \\ 1 \end{bmatrix}}, \overset{\mathbf{s_2}}{\begin{bmatrix} 4 \\ 3 \end{bmatrix}} \right\}$$

and since S is a basis for R^2 (Why?), we find the matrix of T with respect to S. To do this, we must express the images of $\mathbf{s_1}$ and $\mathbf{s_2}$ in terms of $\mathbf{s_1}$ and $\mathbf{s_2}$. You should verify that the following equations are correct:

$$T(\mathbf{s_1}) = T\left(\begin{bmatrix} 1 \\ 1 \end{bmatrix}\right) = \begin{bmatrix} 2 \\ 1 \end{bmatrix} = -2\begin{bmatrix} 1 \\ 1 \end{bmatrix} + \begin{bmatrix} 4 \\ 3 \end{bmatrix} = -2\mathbf{s_1} + \mathbf{s_2}$$

$$T(\mathbf{s_2}) = T\left(\begin{bmatrix} 4 \\ 3 \end{bmatrix}\right) = \begin{bmatrix} 3 \\ 1 \end{bmatrix} = -5\begin{bmatrix} 1 \\ 1 \end{bmatrix} + 2\begin{bmatrix} 4 \\ 3 \end{bmatrix} = -5\mathbf{s_1} + 2\mathbf{s_2}$$

Thus,

$$[T(\mathbf{s_1})]_S = \begin{bmatrix} -2 \\ 1 \end{bmatrix} \qquad [T(\mathbf{s_2})]_S = \begin{bmatrix} -5 \\ 2 \end{bmatrix}$$

and the matrix of T with respect to S is

$$A = \begin{bmatrix} -2 & -5 \\ 1 & 2 \end{bmatrix}$$

The characteristic polynomial for A is

$$p(\lambda) = \det(\lambda I - A) = \lambda^2 + 1$$

Since $p(\lambda)$ has no real zeros, A has no eigenvalues and no eigenvectors. Thus, A is not diagonalizable. Theorem 11(A) now implies that T is not diagonalizable. That is, T *does not have any diagonal matrix representations.*

Problem 24 Repeat Example 24 for the linear operator $T: R^2 \rightarrow R^2$ defined by

$$T\left(\begin{bmatrix} 1 \\ 1 \end{bmatrix}\right) = \begin{bmatrix} 2 \\ 1 \end{bmatrix} \qquad T\left(\begin{bmatrix} 5 \\ 4 \end{bmatrix}\right) = \begin{bmatrix} 4 \\ 1 \end{bmatrix}$$

∎

Example 25   Let $T: M_{2\times2} \to M_{2\times2}$ be the linear operator defined by

$$T\left(\begin{bmatrix} a & b \\ c & d \end{bmatrix}\right) = \begin{bmatrix} 2b & 2a \\ -2c & -2d \end{bmatrix}$$

Find a diagonal representation of T, if possible.

Solution   If

$$S = \left\{ \begin{bmatrix} 1 & 0 \\ 0 & 0 \end{bmatrix}, \begin{bmatrix} 0 & 1 \\ 0 & 0 \end{bmatrix}, \begin{bmatrix} 0 & 0 \\ 1 & 0 \end{bmatrix}, \begin{bmatrix} 0 & 0 \\ 0 & 1 \end{bmatrix} \right\}$$

is the standard basis for $M_{2\times2}$, then the matrix of T with respect to S is (verify this)

$$A = \begin{bmatrix} 0 & 2 & 0 & 0 \\ 2 & 0 & 0 & 0 \\ 0 & 0 & -2 & 0 \\ 0 & 0 & 0 & -2 \end{bmatrix}$$

Since A is a symmetric matrix, A is diagonalizable (Theorem 12, Section 7-3). Thus, Theorem 11 implies that T is diagonalizable. The characteristic polynomial for A is

$$p(\lambda) = \det(\lambda I - A) = (\lambda + 2)^3(\lambda - 2)$$

Thus, $\lambda_1 = -2$ is an eigenvalue of multiplicity 3 and $\lambda_2 = 2$ is an eigenvalue of multiplicity 1. Theorem 13 in Section 7-3 implies that the eigenspace of A associated with λ_1 is three-dimensional and the eigenspace of A associated with λ_2 is one-dimensional. Theorem 11 then implies that the corresponding eigenspaces of T have dimensions 3 and 1, respectively. Thus, T has a diagonal representation of the form

$$D = \begin{array}{cccc} \lambda_1 & \lambda_1 & \lambda_1 & \lambda_2 \\ \begin{bmatrix} -2 & 0 & 0 & 0 \\ 0 & -2 & 0 & 0 \\ 0 & 0 & -2 & 0 \\ 0 & 0 & 0 & 2 \end{bmatrix} \end{array}$$

Notice that the order of the eigenvalues on the diagonal of D is not unique, but $\lambda_1 = -2$ must appear three times and $\lambda_2 = 2$ once.

Problem 25   Repeat Example 25 for the linear operator $T: M_{2\times2} \to M_{2\times2}$ defined by

$$T\left(\begin{bmatrix} a & b \\ c & d \end{bmatrix}\right) = \begin{bmatrix} b & a \\ c & -d \end{bmatrix}$$ ▮

The method of solution used in Example 25 produces the diagonal matrix representation of a linear operator, but not the basis of eigenvectors. In some applications, the diagonal matrix representation is sufficient. In others, it will be necessary to find the basis of eigenvectors, as was done in Example 23.

▪ Similar Matrices

The matrix representation of a linear operator with respect to a basis B certainly depends on the choice of the matrix B. In particular, selection of a basis of eigenvectors (if one exists) provides a diagonal representation. Now, there is one final question we want to consider: *Is there any relationship among the various matrix representations of a given linear operator?* Theorem 12 provides the answer to this question (A review of Section 6-4 would be helpful before reading the proof of Theorem 12.)

Theorem 12

> Let $T: V \rightarrow V$ be a linear operator, let A_B be the matrix of T with respect to basis B of V, and let A_C be the matrix of T with respect to basis C of V. Then
>
> $$A_B = P^{-1}A_C P \tag{4}$$
>
> where P is the transition matrix from B to C.

Proof   If $\mathbf{v}$ is any vector in V, then Theorem 9 in Section 8-3 implies that

$$[T(\mathbf{v})]_B = A_B[\mathbf{v}]_B \tag{5}$$

and

$$[T(\mathbf{v})]_C = A_C[\mathbf{v}]_C \tag{6}$$

If P is the transition matrix from B to C, then Theorem 11 in Section 6-4 implies that

$$P[\mathbf{v}]_B = [\mathbf{v}]_C \tag{7}$$

and

$$P[T(\mathbf{v})]_B = [T(\mathbf{v})]_C \tag{8}$$

Since transition matrices are invertible (Theorem 12, Section 6-4), the equation in (8) can be written as

$$[T(\mathbf{v})]_B = P^{-1}[T(\mathbf{v})]_C \tag{9}$$

We now use the relationships in (5), (6), (7), and (9) to show that $P^{-1}A_C P[\mathbf{v}]_B = A_B[\mathbf{v}]_B$:

$$
\begin{aligned}
P^{-1}A_C P[\mathbf{v}]_B &= P^{-1}A_C[\mathbf{v}]_C & &\text{Using (7)} \\
&= P^{-1}[T(\mathbf{v})]_C & &\text{Using (6)} \\
&= [T(\mathbf{v})]_B & &\text{Using (9)} \\
&= A_B[\mathbf{v}]_B & &\text{Using (5)}
\end{aligned}
$$

Thus, we have shown that

$$P^{-1}A_C P[\mathbf{v}]_B = A_B[\mathbf{v}]_B \qquad \text{for all } \mathbf{v} \text{ in } V$$

It is left as an exercise to show that this implies that $P^{-1}A_C P = A_B$.   ▌▌

The relationship between A_B and A_C in (4) is a fundamental one that occurs frequently in linear algebra, including problems that do not involve linear operators or transition matrices. (See Example 5 in Section 7-2 for another example of this relationship.) The general definition is given in the box.

Similar Matrices

An $n \times n$ matrix A is said to be **similar** to an $n \times n$ matrix B if there exists an invertible $n \times n$ matrix P satisfying

$$A = P^{-1}BP$$

If A is similar to B, then it can be shown that B is also similar to A (see Problem 35 in Exercise 8-4). Thus, it is correct to say that A and B are similar if either A is similar to B or B is similar to A. Using this new terminology, Theorem 12 can be restated as: **Any two matrix representations of a linear operator are similar.**

Answers to Matched Problems

21. (A) $\begin{bmatrix} 2 & 2 \\ -1 & 5 \end{bmatrix}\begin{bmatrix} 1 \\ 1 \end{bmatrix} = 4\begin{bmatrix} 1 \\ 1 \end{bmatrix}$ (B) $T(1 + x) = 4(1 + x)$

22. $\lambda_1 = 2, \quad \lambda_2 = 5; \quad C_{\lambda_1} = \{1 - x\}, \quad C_{\lambda_2} = \{2 + x\}$

23. (B) $\begin{bmatrix} 2 & 0 \\ 0 & 5 \end{bmatrix}$ **24.** Not possible   **25.** $D = \begin{bmatrix} 1 & 0 & 0 & 0 \\ 0 & 1 & 0 & 0 \\ 0 & 0 & -1 & 0 \\ 0 & 0 & 0 & -1 \end{bmatrix}$

▌▌ Exercise 8-4

A   In Problems 1 and 2, show that $p(x)$ is an eigenvector of the linear transformation $T: P_1 \rightarrow P_1$, and find the corresponding eigenvalue.

1. $T(a_0 + a_1x) = (4a_0 - 3a_1) + (2a_0 - 3a_1)x$

  (A) $p(x) = 3 + x$ (B) $p(x) = 1 + 2x$

2. $T(a_0 + a_1x) = (5a_0 + 4a_1) + (a_0 + 2a_1)x$

  (A) $p(x) = 4 + x$ (B) $p(x) = -1 + x$

In Problems 3 and 4, show that $\mathbf{v}$ is an eigenvector of the linear transformation $T: M_{2\times2} \rightarrow M_{2\times2}$, and find the corresponding eigenvalue.

3. $T\left(\begin{bmatrix} a & b \\ c & d \end{bmatrix}\right) = \begin{bmatrix} 2b & 2a \\ d & -c \end{bmatrix}$ (A) $\mathbf{v} = \begin{bmatrix} 1 & 1 \\ 0 & 0 \end{bmatrix}$ (B) $\mathbf{v} = \begin{bmatrix} 1 & -1 \\ 0 & 0 \end{bmatrix}$

4. $T\left(\begin{bmatrix} a & b \\ c & d \end{bmatrix}\right) = \begin{bmatrix} b & -a \\ 4d & c \end{bmatrix}$ (A) $\mathbf{v} = \begin{bmatrix} 0 & 0 \\ 2 & 1 \end{bmatrix}$ (B) $\mathbf{v} = \begin{bmatrix} 0 & 0 \\ -2 & 1 \end{bmatrix}$

In Problems 5 and 6, show that C is a basis of eigenvectors for $T: R^3 \to R^3$, and find the matrix of T with respect to C.

5. $T\left(\begin{bmatrix} x_1 \\ x_2 \\ x_3 \end{bmatrix}\right) = \begin{bmatrix} 2x_1 - x_2 - 2x_3 \\ -x_1 + 2x_2 + 2x_3 \\ -2x_1 + 2x_2 + 10x_3 \end{bmatrix}$; $C = \left\{ \begin{bmatrix} 1 \\ 1 \\ 0 \end{bmatrix}, \begin{bmatrix} 2 \\ -2 \\ 1 \end{bmatrix}, \begin{bmatrix} -1 \\ 1 \\ 4 \end{bmatrix} \right\}$

6. $T\left(\begin{bmatrix} x_1 \\ x_2 \\ x_3 \end{bmatrix}\right) = \begin{bmatrix} x_1 - 2x_3 \\ x_2 - 2x_3 \\ -2x_1 - 2x_2 + 8x_3 \end{bmatrix}$; $C = \left\{ \begin{bmatrix} 2 \\ 2 \\ 1 \end{bmatrix}, \begin{bmatrix} -1 \\ 1 \\ 0 \end{bmatrix}, \begin{bmatrix} -1 \\ -1 \\ 4 \end{bmatrix} \right\}$

In Problems 7 and 8, show that C is a basis of eigenvectors for $T: P_2 \to P_2$, and find the matrix of T with respect to C.

7. $T(a_0 + a_1x + a_2x^2) = 4a_0 + (3a_0 + a_1)x + (-3a_0 + 3a_1 + 4a_2)x^2$;
$C = \{-x + x^2, 1 + x, x^2\}$

8. $T(a_0 + a_1x + a_2x^2) = a_0 + (-a_0 + 3a_1 + a_2)x + (a_0 - 2a_1)x^2$;
$C = \{-x + x^2, 2 + x, 1 + x^2\}$

9. Let $T: P_2 \to P_2$ be defined by

$$T(a_0 + a_1x + a_2x^2) = 2a_0 + 2a_1x + (-3a_0 - 3a_1 - a_2)x^2$$

(A) Find A, the matrix of T with respect to the basis $S = \{1, x, x^2\}$.
(B) Find the eigenvalues of A.
(C) Find a basis for each eigenspace of A.
(D) Find a basis for each eigenspace of T.

10. Repeat Problem 9 for $T: P_2 \to P_2$ defined by

$$T(a_0 + a_1x + a_2x^2) = 2a_0 + (a_0 + 2a_1 + a_2)x + (a_0 + 3a_2)x^2$$

B In Problems 11–30, find the eigenvalues of T, find a basis for each eigenspace of T, and find a diagonal representation of T, if one exists.

11. $T: R^2 \to R^2$ defined by $T\left(\begin{bmatrix} x_1 \\ x_2 \end{bmatrix}\right) = \begin{bmatrix} 3x_1 - x_2 \\ x_1 + x_2 \end{bmatrix}$

12. $T: R^2 \to R^2$ defined by $T\left(\begin{bmatrix} x_1 \\ x_2 \end{bmatrix}\right) = \begin{bmatrix} x_1 + 5x_2 \\ -x_1 + 3x_2 \end{bmatrix}$

13. $T: R^2 \to R^2$ defined by $T(\mathbf{x}) = \begin{bmatrix} 1 & 2 \\ -2 & 1 \end{bmatrix}\mathbf{x}$

14. $T: R^2 \to R^2$ defined by $T(\mathbf{x}) = \begin{bmatrix} 1 & 2 \\ -2 & 5 \end{bmatrix}\mathbf{x}$

15. $T: R^2 \to R^2$ defined by $T(\mathbf{u}) = \langle \mathbf{u}, \mathbf{w} \rangle \mathbf{w}$, where $\mathbf{w} = (1/\sqrt{5}, 2/\sqrt{5})$

16. $T: R^2 \to R^2$ defined by $T(\mathbf{u}) = \langle \mathbf{u}, \mathbf{w} \rangle \mathbf{w}$, where $\mathbf{w} = (1/\sqrt{2}, -1/\sqrt{2})$

17. $T: R^3 \rightarrow R^3$ defined by $T(\mathbf{x}) = A\mathbf{x}$, where $A = \begin{bmatrix} -2 & 3 & 0 \\ 0 & -2 & 0 \\ 0 & 1 & -2 \end{bmatrix}$

18. $T: R^3 \rightarrow R^3$ defined by $T(\mathbf{x}) = A\mathbf{x}$, where $A = \begin{bmatrix} 2 & 0 & 0 \\ -3 & 2 & -3 \\ -3 & 0 & -1 \end{bmatrix}$

19. $T: R^3 \rightarrow R^3$ defined by $T\left(\begin{bmatrix} x_1 \\ x_2 \\ x_3 \end{bmatrix}\right) = \begin{bmatrix} 7x_1 - 9x_2 + 6x_3 \\ -2x_2 \\ -3x_1 + 3x_2 - 2x_3 \end{bmatrix}$

20. $T: R^3 \rightarrow R^3$ defined by $T\left(\begin{bmatrix} x_1 \\ x_2 \\ x_3 \end{bmatrix}\right) = \begin{bmatrix} 2x_1 - x_2 + x_3 \\ -2x_1 + x_2 + 5x_3 \\ -x_1 - x_2 + 5x_3 \end{bmatrix}$

21. $T: P_2 \rightarrow P_2$ defined by

$$T(a_0 + a_1 + a_2x^2) = (-2a_0 - a_1 - 2a_2) + (-2a_0 - 2a_1 - 3a_2)x + (a_0 + a_1 + a_2)x^2$$

22. $T: P_2 \rightarrow P_2$ defined by

$$T(a_0 + a_1x + a_2x^2) = (-a_0 - a_1) + (2a_0 + 2a_1)x + (-2a_0 - 2a_1 + 3a_2)x^2$$

23. $T: P_2 \rightarrow P_2$ defined by $T(1) = x^2$, $T(x) = x$, $T(x^2) = 1$
24. $T: P_2 \rightarrow P_2$ defined by $T(1) = x^2$, $T(x) = x$, $T(x^2) = -1$
25. $T: M_{2 \times 2} \rightarrow M_{2 \times 2}$ defined by $T(A) = A^T$
26. $T: M_{2 \times 2} \rightarrow M_{2 \times 2}$ defined by $T(A) = A + A^T$

27. $T: M_{2 \times 2} \rightarrow M_{2 \times 2}$ defined by $T\left(\begin{bmatrix} a & b \\ c & d \end{bmatrix}\right) = \begin{bmatrix} -b & a \\ d & -c \end{bmatrix}$

28. $T: M_{2 \times 2} \rightarrow M_{2 \times 2}$ defined by $T\left(\begin{bmatrix} a & b \\ c & d \end{bmatrix}\right) = \begin{bmatrix} 2c & 2d \\ -a & -b \end{bmatrix}$

29. *Calculus.* $T: P_3 \rightarrow P_3$ defined by $T(p(x)) = xp'(x)$
30. *Calculus.* $T: P_3 \rightarrow P_3$ defined by $T(p(x)) = q(x)$, where

$$q(x) = \frac{1}{x} \int_0^x p(t)\, dt, \quad x \neq 0 \text{ and } q(0) = p(0)$$

C In Problems 31–33, let $T: V \rightarrow V$ be a linear operator defined on the finite-dimensional vector space V.

31. Show that $S_\lambda = \{\mathbf{v} \in V | T(\mathbf{v}) = \lambda\mathbf{v}\}$ is a subspace of V.
32. Show that a nonzero vector $\mathbf{v}$ is in $\ker(T)$ if and only if $\mathbf{v}$ is an eigenvector of T associated with the eigenvalue $\lambda = 0$.
33. If $\lambda = 0$ is an eigenvalue of T, show that $\text{range}(T) \neq V$.
34. *Theorem 12.* Complete the proof of Theorem 12 by showing that if $P^{-1}A_CP[\mathbf{v}]_B = A_B[\mathbf{v}]_B$ for all $\mathbf{v}$ in V, then $P^{-1}A_CP = A_B$.

In Problems 35–40, all matrices are square matrices of order n.

35. If A is similar to B, show that B is similar to A.

36. Show that A is similar to itself.

37. If A is similar to B and B is similar to C, show that A is similar to C.

38. If A is similar to B, show that $\det(A) = \det(B)$.

39. If A is similar to B and A is invertible, show that B is invertible and that B^{-1} is similar to A^{-1}.

40. If A is similar to B, show that A^2 is similar to B^2.

If A and B are $n \times n$ matrices, P is an $n \times n$ orthogonal matrix, and $A = P^T BP$, then A is said to be **orthogonally similar** *to B.*

41. If A is orthogonally similar to B, show that B is orthogonally similar to A.

42. Show that A is orthogonally similar to itself.

43. If A is orthogonally similar to B and B is orthogonally similar to C, show that A is orthogonally similar to C.

44. If A is orthogonally similar to B and B is symmetric, show that A is symmetric.

▌8-5▐   Transformations of the Plane

- Introduction
- Geometric Properties
- Reflections
- Expansions and Contractions
- Shears
- Products of Linear Transformations

▪ Introduction

Up to this point we have been concerned primarily with the algebraic properties of linear transformations. Now we want to consider geometric properties of certain linear transformations. If $T: R^2 \to R^2$ is a linear transformation whose range is all of R^2, then T is said to map R^2 **onto** R^2 and T is called a **linear transformation of the plane.** Theorem 13 provides a simple characterization of linear transformations of the plane. The proof is discussed in the exercises.

Theorem 13

> If $T: R^2 \to R^2$ is a linear transformation and A is the standard matrix for T, then T maps R^2 onto R^2 if and only if A is invertible.

Example 26   Let $T: R^2 \to R^2$ be defined by

$$T\left(\begin{bmatrix} x \\ y \end{bmatrix}\right) = \begin{bmatrix} 2x + 3y \\ x + 2y \end{bmatrix}$$

Then

$$T\left(\begin{bmatrix}1\\0\end{bmatrix}\right)=\begin{bmatrix}2\\1\end{bmatrix}\qquad T\left(\begin{bmatrix}0\\1\end{bmatrix}\right)=\begin{bmatrix}3\\2\end{bmatrix}$$

and the standard matrix for T is

$$A=\begin{bmatrix}2 & 3\\1 & 2\end{bmatrix}$$

Since $\det(A)=(2)(2)-(1)(3)=1\neq 0$, A is invertible. Theorem 13 then implies that T maps R^2 onto R^2. That is,

range$(T)=R^2$

This means that each vector $\mathbf{w}\in R^2$ must be the image under T of another vector $\mathbf{v}\in R^2$. For example, if

$$\mathbf{w}=\begin{bmatrix}3\\5\end{bmatrix}$$

then there must be a vector $\mathbf{v}$ in R^2 such that $T(\mathbf{v})=\mathbf{w}$. Since $T(\mathbf{v})=A\mathbf{v}$ and A is invertible, we can use A^{-1} to find $\mathbf{v}$. If we let

$$\mathbf{v}=A^{-1}\mathbf{w}=\begin{bmatrix}2 & -3\\-1 & 2\end{bmatrix}\begin{bmatrix}3\\5\end{bmatrix}=\begin{bmatrix}-9\\7\end{bmatrix}$$

then

$$T(\mathbf{v})=T\left(\begin{bmatrix}-9\\7\end{bmatrix}\right)=\begin{bmatrix}2(-9)+3(7)\\(-9)+2(7)\end{bmatrix}=\begin{bmatrix}3\\5\end{bmatrix}=\mathbf{w}$$

Problem 26   Refer to Example 26. Find a vector $\mathbf{v}$ satisfying $T(\mathbf{v})=\mathbf{w}$ if:

(A) $\mathbf{w}=\begin{bmatrix}4\\3\end{bmatrix}$ (B) $\mathbf{w}=\begin{bmatrix}0\\0\end{bmatrix}$ (C) $\mathbf{w}=\begin{bmatrix}a\\b\end{bmatrix}$ ∎

▪ Geometric Properties

If T maps R^2 onto R^2, then T also maps every subset of R^2 onto another subset of R^2. The geometric properties of T are best described in terms of the effect T has on certain subsets of the plane. We begin with an example.

Example 27   Let $T\colon R^2\to R^2$ be the linear transformation of the plane defined by

$$T\left(\begin{bmatrix}x\\y\end{bmatrix}\right)=\begin{bmatrix}2x-2y\\x+\ \ y\end{bmatrix}$$

and let ℓ be the line with vector equation

$$\mathbf{u}=\mathbf{u}_0+t\mathbf{u}_1\qquad t\text{ any real number}$$

where $\mathbf{u}_0 = (1, 0)$ and $\mathbf{u}_1 = (1, 2)$. Find the image of ℓ under T.

Solution   The image of ℓ under T can be written as

$$T(\ell) = \{T(\mathbf{u}) | \mathbf{u} = \mathbf{u}_0 + t\mathbf{u}_1, t \text{ any real number}\}$$

Since T is a linear transformation,

$$T(\mathbf{u}) = T(\mathbf{u}_0 + t\mathbf{u}_1) = T(\mathbf{u}_0) + tT(\mathbf{u}_1) \qquad (1)$$

Writing $\mathbf{u}_0$ and $\mathbf{u}_1$ as column vectors,* we have

$$T(\mathbf{u}_0) = T\left(\begin{bmatrix} 1 \\ 0 \end{bmatrix}\right) = \begin{bmatrix} 2 \\ 1 \end{bmatrix}$$

and

$$T(\mathbf{u}_1) = T\left(\begin{bmatrix} 1 \\ 2 \end{bmatrix}\right) = \begin{bmatrix} -2 \\ 3 \end{bmatrix}$$

Reverting back to ordered pair notation and substituting in (1), we have

$$T(\mathbf{u}) = (2, 1) + t(-2, 3)$$

which is a vector equation for the line through $\mathbf{v}_0 = (2, 1)$ in the direction of $\mathbf{v}_1 = (-2, 3)$ (see Figure 4). Thus, the image of ℓ under T is the set of points on this line. That is,

$$T(\ell) = \{(2, 1) + t(-2, 3) | t \text{ any real number}\}$$

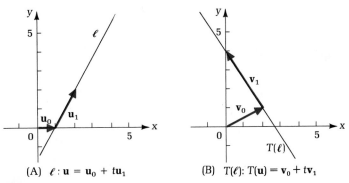

(A) $\ell : \mathbf{u} = \mathbf{u}_0 + t\mathbf{u}_1$ (B) $T(\ell): T(\mathbf{u}) = \mathbf{v}_0 + t\mathbf{v}_1$

Figure 4

Problem 27   Refer to Example 27. Find the image of ℓ under T if a vector equation for ℓ is

$$\mathbf{u} = (0, 1) + t(2, 1)$$ ▮▮

---

* We will continue to express linear transformations in terms of column vectors in order to make use of the relationship between linear transformations and matrix multiplication. However, we will express vectors as ordered pairs when graphing is involved.

Theorem 14 generalizes the ideas illustrated in Example 27. Portions of the proof will be discussed in the exercises.

Theorem 14

If T is a linear transformation of the plane, then:

(A) T maps straight lines onto straight lines.
(B) T maps straight lines through the origin onto straight lines through the origin.
(C) T maps the line segment joining $\mathbf{u}_1$ and $\mathbf{u}_2$ onto the line segment joining $T(\mathbf{u}_1)$ and $T(\mathbf{u}_2)$.
(D) T maps triangles onto triangles.
(E) T maps parallelograms onto parallelograms.

A convenient method for illustrating the geometric effect of a linear transformation is to use Theorem 14 to find the image of a triangle or rectangle under the given transformation. This approach is illustrated in Example 28.

Example 28   Find the image of the rectangle with vertices $(0, 0)$, $(4, 0)$, $(4, 2)$, and $(0, 2)$ under the transformation defined in Example 27. Graph the rectangle and its image under T on separate sets of axes.

Solution
$$T\left(\begin{bmatrix} 0 \\ 0 \end{bmatrix}\right) = \begin{bmatrix} 0 \\ 0 \end{bmatrix} \qquad T\left(\begin{bmatrix} 4 \\ 0 \end{bmatrix}\right) = \begin{bmatrix} 8 \\ 4 \end{bmatrix} \qquad T\left(\begin{bmatrix} 4 \\ 2 \end{bmatrix}\right) = \begin{bmatrix} 4 \\ 6 \end{bmatrix} \qquad T\left(\begin{bmatrix} 0 \\ 2 \end{bmatrix}\right) = \begin{bmatrix} -4 \\ 2 \end{bmatrix}$$

Theorem 14 implies that T maps the rectangle with vertices $(0, 0)$, $(4, 0)$, $(4, 2)$, and $(0, 2)$ onto the parallelogram with vertices $(0, 0)$, $(8, 4)$, $(4, 6)$, and $(-4, 2)$, as shown in Figure 5.

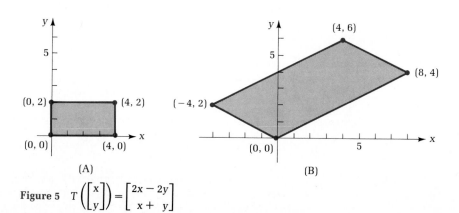

(A)   (B)

Figure 5 $T\left(\begin{bmatrix} x \\ y \end{bmatrix}\right) = \begin{bmatrix} 2x - 2y \\ x + y \end{bmatrix}$

Note that it also follows from Theorem 14 that a linear transformation of the plane maps the interior of a parallelogram onto the interior of the image. This is illustrated by shading the interiors of both parallelograms in Figure 5.

Problem 28   Repeat Example 28 for the square with vertices $(1, 1)$, $(2, 1)$, $(2, 2)$, and $(1, 2)$.

∎

Now, we want to consider some special transformations of the plane. Each of these special transformations is related to a 2×2 elementary matrix. Recall from Section 2-4 that an elementary matrix is a matrix that can be obtained by performing exactly one elementary row operation on an identity matrix. It follows that there are five different 2×2 elementary matrices:

$$
\begin{array}{ccccc}
R_1 \leftrightarrow R_2 & kR_1 \to R_1 & kR_2 \to R_2 & R_1 + kR_2 \to R_1 & R_2 + kR_1 \to R_2 \\[4pt]
\begin{bmatrix} 0 & 1 \\ 1 & 0 \end{bmatrix} &
\begin{bmatrix} k & 0 \\ 0 & 1 \end{bmatrix} &
\begin{bmatrix} 1 & 0 \\ 0 & k \end{bmatrix} &
\begin{bmatrix} 1 & k \\ 0 & 1 \end{bmatrix} &
\begin{bmatrix} 1 & 0 \\ k & 1 \end{bmatrix}
\end{array}
$$

It turns out that studying the geometric properties of the linear transformations defined by these elementary matrices will enable us to give a complete geometric description of any linear transformation of the plane. We now begin a systematic discussion of transformations defined by elementary matrices.

▪ Reflections

Let T_1, T_2, and T_3 be the linear transformations defined by the elementary matrices E_1, E_2, and E_3, respectively, where

$$
\begin{array}{ccc}
R_1 \leftrightarrow R_2 & (-1)R_1 \to R_1 & (-1)R_2 \to R_2 \\[4pt]
E_1 = \begin{bmatrix} 0 & 1 \\ 1 & 0 \end{bmatrix} &
E_2 = \begin{bmatrix} -1 & 0 \\ 0 & 1 \end{bmatrix} &
E_3 = \begin{bmatrix} 1 & 0 \\ 0 & -1 \end{bmatrix}
\end{array}
$$

That is,

$$
T_1\left(\begin{bmatrix} x \\ y \end{bmatrix} \right) = E_1 \begin{bmatrix} x \\ y \end{bmatrix} = \begin{bmatrix} y \\ x \end{bmatrix}
\qquad \text{Reflection about the line } y = x
$$

$$
T_2\left(\begin{bmatrix} x \\ y \end{bmatrix} \right) = E_2 \begin{bmatrix} x \\ y \end{bmatrix} = \begin{bmatrix} -x \\ y \end{bmatrix}
\qquad \text{Reflection about the } y \text{ axis}
$$

$$
T_3\left(\begin{bmatrix} x \\ y \end{bmatrix} \right) = E_3 \begin{bmatrix} x \\ y \end{bmatrix} = \begin{bmatrix} x \\ -y \end{bmatrix}
\qquad \text{Reflection about the } x \text{ axis}
$$

Each of these transformations can be described geometrically as a reflection about a line, as illustrated in Figure 6 at the top of page 554.

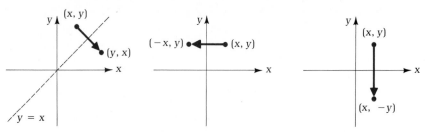

(A) $T_1\left(\begin{bmatrix}x\\y\end{bmatrix}\right) = \begin{bmatrix}y\\x\end{bmatrix}$
Reflection about
the line $y = x$

(B) $T_2\left(\begin{bmatrix}x\\y\end{bmatrix}\right) = \begin{bmatrix}-x\\y\end{bmatrix}$
Reflection about
the y axis

(C) $T_3\left(\begin{bmatrix}x\\y\end{bmatrix}\right) = \begin{bmatrix}x\\-y\end{bmatrix}$
Reflection about
the x axis

Figure 6   Reflections

▪ Expansions and Contractions

Let T_1 and T_2 be the linear transformations defined by the elementary matrices E_1 and E_2 where

$$kR_1 \rightarrow R_1 \qquad\qquad \ell R_2 \rightarrow R_2$$

$$E_1 = \begin{bmatrix}k & 0\\0 & 1\end{bmatrix} \quad k > 0 \quad \text{and} \quad E_2 = \begin{bmatrix}1 & 0\\0 & \ell\end{bmatrix} \quad \ell > 0$$

That is,

$$T_1\left(\begin{bmatrix}x\\y\end{bmatrix}\right) = E_1\begin{bmatrix}x\\y\end{bmatrix} = \begin{bmatrix}kx\\y\end{bmatrix} \qquad \text{Expansion or contraction in the x direction}$$

$$T_2\left(\begin{bmatrix}x\\y\end{bmatrix}\right) = E_2\begin{bmatrix}x\\y\end{bmatrix} = \begin{bmatrix}x\\\ell y\end{bmatrix} \qquad \text{Expansion or contraction in the y direction}$$

The transformation T_1 is called an **expansion in the x direction** with scale factor k if $k > 1$ and a **contraction in the x direction** with scale factor k if $0 < k < 1$. Similarly, T_2 is called an **expansion in the y direction** with scale factor ℓ if $\ell > 1$ and a **contraction in the y direction** with scale factor ℓ if $0 < \ell < 1$.

Example 29   Find the image of the square with vertices $(0, 0)$, $(0, 1)$, $(1, 1)$, and $(1, 0)$ under each transformation:

(A) $T_1\left(\begin{bmatrix}x\\y\end{bmatrix}\right) = \begin{bmatrix}2 & 0\\0 & 1\end{bmatrix}\begin{bmatrix}x\\y\end{bmatrix} = \begin{bmatrix}2x\\y\end{bmatrix}$ (B) $T_2\left(\begin{bmatrix}x\\y\end{bmatrix}\right) = \begin{bmatrix}1 & 0\\0 & \frac{1}{2}\end{bmatrix}\begin{bmatrix}x\\y\end{bmatrix} = \begin{bmatrix}x\\\frac{1}{2}y\end{bmatrix}$

Solution

(A) Expansion in the x
     direction with scale
     factor 2

(B) Contraction in the y
     direction with scale
     factor $\frac{1}{2}$

Problem 29   Repeat Example 29 for

(A) $T_1\left(\begin{bmatrix} x \\ y \end{bmatrix}\right) = \begin{bmatrix} \frac{3}{4}x \\ y \end{bmatrix}$ (B) $T_2\left(\begin{bmatrix} x \\ y \end{bmatrix}\right) = \begin{bmatrix} x \\ \frac{3}{2}y \end{bmatrix}$ ▮▮

▪ Shears

Let T_1 and T_2 be the linear transformations defined by the elementary matrices E_1 and E_2 where

$$R_1 + kR_2 \rightarrow R_1 \qquad\qquad R_2 + \ell R_1 \rightarrow R_2$$

$$E_1 = \begin{bmatrix} 1 & k \\ 0 & 1 \end{bmatrix} \qquad \text{and} \qquad E_2 = \begin{bmatrix} 1 & 0 \\ \ell & 1 \end{bmatrix}$$

That is,

$$T_1\left(\begin{bmatrix} x \\ y \end{bmatrix}\right) = E_1 \begin{bmatrix} x \\ y \end{bmatrix} = \begin{bmatrix} x + ky \\ y \end{bmatrix} \qquad \text{Shear in the x direction with scale factor } k$$

$$T_2\left(\begin{bmatrix} x \\ y \end{bmatrix}\right) = E_2 \begin{bmatrix} x \\ y \end{bmatrix} = \begin{bmatrix} x \\ \ell x + y \end{bmatrix} \qquad \text{Shear in the y direction with scale factor } \ell$$

The transformation T_1 is called a **shear in the x direction** with scale factor k and the transformation T_2 is called a **shear in the y direction** with scale factor ℓ.

Example 30   Find the image of the square with vertices $(0, 0)$, $(0, 1)$, $(1, 1)$, and $(1, 0)$ under each transformation:

(A) $T_1\left(\begin{bmatrix} x \\ y \end{bmatrix}\right) = \begin{bmatrix} 1 & -1 \\ 0 & 1 \end{bmatrix}\begin{bmatrix} x \\ y \end{bmatrix} = \begin{bmatrix} x - y \\ y \end{bmatrix}$

(B) $T_2\left(\begin{bmatrix} x \\ y \end{bmatrix}\right) = \begin{bmatrix} 1 & 0 \\ \frac{1}{2} & 1 \end{bmatrix}\begin{bmatrix} x \\ y \end{bmatrix} = \begin{bmatrix} x \\ \frac{1}{2}x + y \end{bmatrix}$

Solution

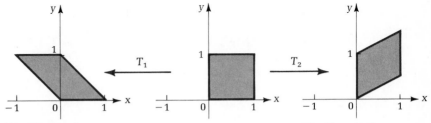

(A) Shear in the x direction with scale factor -1

(B) Shear in the y direction with scale factor $\frac{1}{2}$

Problem 30   Repeat Example 30 for

(A) $T_1\left(\begin{bmatrix} x \\ y \end{bmatrix}\right) = \begin{bmatrix} x + 2y \\ y \end{bmatrix}$ (B) $T_2\left(\begin{bmatrix} x \\ y \end{bmatrix}\right) = \begin{bmatrix} x \\ -\frac{3}{4}x + y \end{bmatrix}$ ▮▮

▪ Products of Linear Transformations

Let T be the linear transformation defined by the elementary matrix E where

$$E = \begin{bmatrix} k & 0 \\ 0 & 1 \end{bmatrix}$$

If $k > 1$, T is an expansion in the x direction; if $0 < k < 1$, T is a contraction in the x direction; and if $k = -1$, T is a reflection about the y axis. (What is T if $k = 1$?) How can we describe the action of T for $-1 < k < 0$ or $k < -1$? We begin by considering a specific example.

Example 31   Let T be the linear transformation defined by the elementary matrix E where

$$E = \begin{bmatrix} -2 & 0 \\ 0 & 1 \end{bmatrix}$$

Then

$$T\left(\begin{bmatrix} x \\ y \end{bmatrix}\right) = E\begin{bmatrix} x \\ y \end{bmatrix} = \begin{bmatrix} -2 & 0 \\ 0 & 1 \end{bmatrix}\begin{bmatrix} x \\ y \end{bmatrix} = \begin{bmatrix} -2x \\ y \end{bmatrix}$$

Now consider the matrix product

$$E = \begin{bmatrix} -2 & 0 \\ 0 & 1 \end{bmatrix} = \overset{E_2}{\begin{bmatrix} -1 & 0 \\ 0 & 1 \end{bmatrix}} \overset{E_1}{\begin{bmatrix} 2 & 0 \\ 0 & 1 \end{bmatrix}} = E_2 E_1$$

Let T_1 and T_2 be the linear transformations defined by E_1 and E_2, respectively. The transformation T_1 is an expansion in the x direction, and the transformation T_2 is a reflection about the y axis. Given any point (x, y) in R^2, we can write

$$\begin{bmatrix} x \\ y \end{bmatrix} \overset{T_1}{\longrightarrow} \begin{bmatrix} 2x \\ y \end{bmatrix} \overset{T_2}{\longrightarrow} \begin{bmatrix} -2x \\ y \end{bmatrix} = T\left(\begin{bmatrix} x \\ y \end{bmatrix}\right)$$

This shows that T has the same effect as T_1 followed by T_2. That is, T can be described as an expansion in the x direction followed by a reflection about the y axis (see Figure 7).

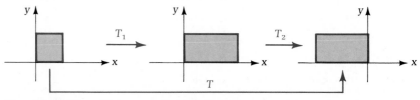

Figure 7   Product of linear transformations: $T = T_2 T_1$

In general, if T_1 and T_2 are linear transformations defined by the matrices A_1 and A_2 and T is the linear transformation defined by the matrix $A = A_2A_1$, then T is called the **product** of T_1 and T_2 and we write $T = T_2T_1$. Geometrically, the effect of $T = T_2T_1$ can be described as the effect of T_1 followed by the effect of T_2.

Notice that the order of the factors in the product T_2T_1 is important. Since matrix multiplication is not a commutative operation, it follows that T_2T_1 is not always equal to T_1T_2 (see Problems 39–42 in Exercise 8-5). We have written the subscripts in the product in decreasing order to emphasize that the geometric effect of the product T_2T_1 is the effect of T_1 *followed* by the effect of T_2. These remarks are generalized to products with an arbitrary number of factors in Theorem 15. The proof is omitted.

Theorem 15

> If $T, T_1, T_2, \ldots, T_k$ are linear transformations defined by the matrices A, $A_1, A_2, \ldots, A_k$ and $A = A_kA_{k-1} \cdot \cdots \cdot A_1$, then the geometric effect of $T = T_kT_{k-1} \cdot \cdots \cdot T_1$ is the same as the effect of T_1, followed by T_2, followed by $T_3, \ldots$, followed by T_k.

Problem 31 Let T be the transformation $T = T_2T_1$ where

$$T_1\left(\begin{bmatrix} x \\ y \end{bmatrix}\right) = \begin{bmatrix} y \\ x \end{bmatrix} \quad \text{and} \quad T_2\left(\begin{bmatrix} x \\ y \end{bmatrix}\right) = \begin{bmatrix} x \\ 2y \end{bmatrix}$$

What is the geometric effect of the transformation T? ▮▮

Theorem 15 enables us to give a geometric description of the transformations defined by the elementary matrices

$$E_1 = \begin{bmatrix} k & 0 \\ 0 & 1 \end{bmatrix} \quad \text{and} \quad E_2 = \begin{bmatrix} 1 & 0 \\ 0 & \ell \end{bmatrix}$$

for all values of k and ℓ. If $k < 0$ and $k \neq -1$, then we can write

$$E_1 = \begin{bmatrix} k & 0 \\ 0 & 1 \end{bmatrix} = \begin{bmatrix} -1 & 0 \\ 0 & 1 \end{bmatrix} \begin{bmatrix} -k & 0 \\ 0 & 1 \end{bmatrix} \qquad \textit{Note: } -k > 0$$

and if $\ell < 0$ and $\ell \neq -1$, then

$$E_2 = \begin{bmatrix} 1 & 0 \\ 0 & \ell \end{bmatrix} = \begin{bmatrix} 1 & 0 \\ 0 & -1 \end{bmatrix} \begin{bmatrix} 1 & 0 \\ 0 & -\ell \end{bmatrix} \qquad \textit{Note: } -\ell > 0$$

In each case, the corresponding transformation can be described as an expansion or contraction followed by a reflection about either the x axis or the y axis. We have now given geometric descriptions of all the transformations defined by 2×2 elementary matrices. These are summarized in Table 3 for convenient reference.

Table 3
Geometric Effect of Transformations Defined by Elementary Matrices

Elementary Matrix	Geometric Effect
$\begin{bmatrix} 0 & 1 \\ 1 & 0 \end{bmatrix}, \begin{bmatrix} -1 & 0 \\ 0 & 1 \end{bmatrix}, \begin{bmatrix} 1 & 0 \\ 0 & -1 \end{bmatrix}$	Reflection
$\begin{bmatrix} k & 0 \\ 0 & 1 \end{bmatrix}, \begin{bmatrix} 1 & 0 \\ 0 & \ell \end{bmatrix}, \ k \neq \pm 1, \ell \neq \pm 1$	Expansion or contraction (followed by a reflection if k or ℓ is negative)
$\begin{bmatrix} 1 & k \\ 0 & 1 \end{bmatrix}, \begin{bmatrix} 1 & 0 \\ \ell & 1 \end{bmatrix}$	Shear

Theorem 10 in Section 2–4 showed that every invertible matrix can be expressed as a product of elementary matrices. Combining this with Theorem 15 provides the geometric description of arbitrary transformations of the plane stated in Theorem 16. A formal proof is omitted.

Theorem 16

If $T: R^2 \rightarrow R^2$ is a linear transformation of the plane, then the geometric effect of T can be expressed as a sequence of reflections, expansions, contractions, and shears.

Example 32   Give a geometric description of the transformation $T: R^2 \rightarrow R^2$ defined by the matrix A where

$$A = \begin{bmatrix} 1 & -1 \\ 3 & -1 \end{bmatrix}$$

Solution   Using the techniques discussed in Section 2-4, matrix A can be expressed as the following product of elementary matrices:

$$A = \begin{bmatrix} 1 & -1 \\ 3 & -1 \end{bmatrix} = \overset{E_3}{\begin{bmatrix} 1 & 0 \\ 3 & 1 \end{bmatrix}} \overset{E_2}{\begin{bmatrix} 1 & 0 \\ 0 & 2 \end{bmatrix}} \overset{E_1}{\begin{bmatrix} 1 & -1 \\ 0 & 1 \end{bmatrix}} = E_3 E_2 E_1$$

Thus, $T = T_3 T_2 T_1$ where

$$T_1 \left(\begin{bmatrix} x \\ y \end{bmatrix} \right) = E_1 \begin{bmatrix} x \\ y \end{bmatrix} = \begin{bmatrix} x - y \\ y \end{bmatrix}$$ A shear in the x direction with scale factor -1

$$T_2 \left(\begin{bmatrix} x \\ y \end{bmatrix} \right) = E_2 \begin{bmatrix} x \\ y \end{bmatrix} = \begin{bmatrix} x \\ 2y \end{bmatrix}$$ An expansion in the y direction with scale factor 2

$$T_3 \left(\begin{bmatrix} x \\ y \end{bmatrix} \right) = E_3 \begin{bmatrix} x \\ y \end{bmatrix} = \begin{bmatrix} x \\ 3x + y \end{bmatrix}$$ A shear in the y direction with scale factor 3

Applying Theorem 15, we conclude that the geometric effect of T is a shear in the x direction with scale factor -1, followed by an expansion in the y direction with scale factor 2, followed by a shear in the y direction with scale factor 3 (see Figure 8).

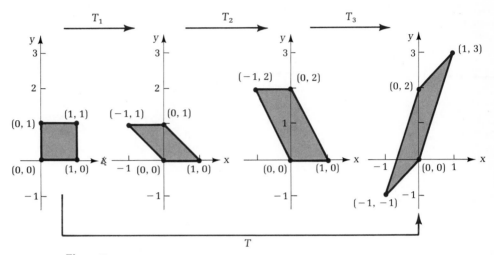

Figure 8

In Example 32, notice that we can trace the effect of the sequence of transformations T_1, T_2, and T_3 on each vertex of the original square in Figure 8 and compare the result with the effect of T. For example,

$$\begin{array}{cccc} & T_1 & T_2 & T_3 \\ (0, 1) \to & (-1, 1) \to & (-1, 2) \to & (-1, -1) \end{array}$$

T

You should perform similar calculations for the vertices $(0, 0)$, $(1, 1)$, and $(1, 0)$ of the original square.

In general, a matrix can be expressed as a product of elementary matrices in many different ways. This implies that the geometric description of a linear transformation is not unique.

Problem 32 Refer to Example 32.

(A) Verify by matrix multiplication that the matrix A can be expressed as the following product of elementary matrices:

$$A = \begin{bmatrix} 1 & -1 \\ 3 & -1 \end{bmatrix} = \begin{bmatrix} 1 & 0 \\ 2 & 1 \end{bmatrix} \begin{bmatrix} 1 & -1 \\ 0 & 1 \end{bmatrix} \begin{bmatrix} 2 & 0 \\ 0 & 1 \end{bmatrix} \begin{bmatrix} 1 & 0 \\ 1 & 1 \end{bmatrix}$$

(B) Use the product of elementary matrices in part (A) to give a geometric description of T.

(C) Construct a figure like Figure 8 to illustrate the geometric effect of the product of the linear transformations defined by the elementary matrices in part (A).   ▮▮

26. (A) $\begin{bmatrix} -1 \\ 2 \end{bmatrix}$ (B) $\begin{bmatrix} 0 \\ 0 \end{bmatrix}$ (C) $\begin{bmatrix} 2a - 3b \\ -a + 2b \end{bmatrix}$

27. $\{(-2, 1) + t(2, 3) \mid t \text{ any real number}\}$

28. The parallelogram with vertices $(0, 2)$, $(2, 3)$, $(0, 4)$, and $(-2, 3)$

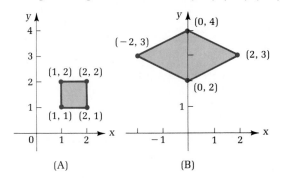

(A)   (B)

29. (A) Contraction in the x direction with scale factor $\frac{3}{4}$

(B) Expansion in the y direction with scale factor $\frac{3}{2}$

30. (A) Shear in the x direction with scale factor 2

(B) Shear in the y direction with scale factor $-\frac{3}{4}$

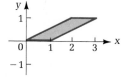

31. A reflection about the line $y = x$ followed by an expansion in the y direction with scale factor 2

32. (B) Shear in the y direction with scale factor 1, followed by an expansion in the x direction with scale factor 2, followed by a shear in the x direction with scale factor -1, followed by a shear in the y direction with scale factor 2

32. (C)

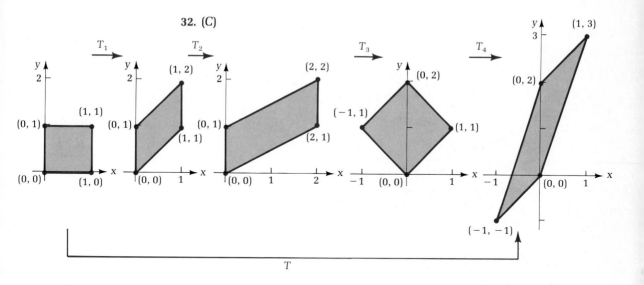

II   Exercise 8-5

A   *In Problems 1–6, let*

$$T\left(\begin{bmatrix} x \\ y \end{bmatrix}\right) = \begin{bmatrix} 2x - 2y \\ 2x + y \end{bmatrix}$$

Graph the image under T of each geometric figure.

1.

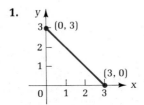

2.

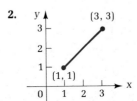

3.

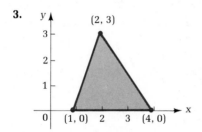

4.

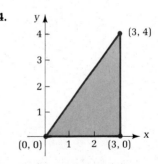

5.

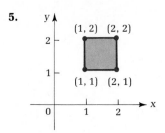

6.

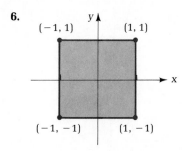

In Problems 7–10, determine whether T is a transformation of the plane.

7. $T\left(\begin{bmatrix} x \\ y \end{bmatrix}\right) = \begin{bmatrix} 3x + 4y \\ 2x - 3y \end{bmatrix}$

8. $T\left(\begin{bmatrix} x \\ y \end{bmatrix}\right) = \begin{bmatrix} 2x - 3y \\ -4x + 6y \end{bmatrix}$

9. $T\left(\begin{bmatrix} x \\ y \end{bmatrix}\right) = \begin{bmatrix} 4x + 3y \\ 12x + 9y \end{bmatrix}$

10. $T\left(\begin{bmatrix} x \\ y \end{bmatrix}\right) = \begin{bmatrix} 3x + 4y \\ 2x - \ y \end{bmatrix}$

B   *In Problems 11 and 12, find the vector equation of the image of the line ℓ under the transformation T.*

11. $T\left(\begin{bmatrix} x \\ y \end{bmatrix}\right) = \begin{bmatrix} x + y \\ x - y \end{bmatrix}$; ℓ: $(2, 1) + t(3, 2)$

12. $T\left(\begin{bmatrix} x \\ y \end{bmatrix}\right) = \begin{bmatrix} x + 2y \\ 3x - \ y \end{bmatrix}$; ℓ: $(-1, 0) + t(-2, 3)$

In Problems 13–18, find the elementary matrix for the linear transformation that performs the indicated action, and graph the image of the rectangle with vertices $(0, 0)$, $(2, 0)$, $(2, 1)$, and $(0, 1)$ under this transformation.

13. Reflection about the line $y = x$
14. Reflection about the y axis
15. Expansion in the y direction with scale factor 3
16. Contraction in the x direction with scale factor $\frac{1}{2}$
17. Shear in the x direction with scale factor -2
18. Shear in the y direction with scale factor $\frac{1}{2}$

In Problems 19–26, find the standard matrix for a transformation that has the indicated geometric effect.

19. A shear in the x direction with scale factor 5
20. A shear in the y direction with scale factor $-\frac{1}{3}$
21. A contraction in the y direction with scale factor $\frac{1}{4}$
22. An expansion in the x direction with scale factor 6
23. A reflection about the x axis, followed by an expansion in the y direction with scale factor 3

24. A contraction in the x direction with scale factor $\frac{1}{5}$, followed by a reflection about the y axis

25. A shear in the x direction with scale factor -3, followed by a shear in the y direction with scale factor 4, followed by a reflection about the line $y = x$

26. A shear in the y direction with scale factor 2, followed by an expansion in the x direction with scale factor 5, followed by a shear in the x direction with scale factor -3

In Problems 27–30, the matrix A of a linear transformation T has been expressed as a product of elementary matrices. Describe the effect of the transformation T in terms of reflections, expansions, contractions, and shears. Illustrate the effect of this sequence of operations on the square with vertices $(0, 0)$, $(1, 0)$, $(1, 1)$, and $(0, 1)$ by constructing a figure like Figure 8 in Example 32.

27. $A = \begin{bmatrix} 1 & 2 \\ -1 & 1 \end{bmatrix} = \begin{bmatrix} 1 & 0 \\ -1 & 1 \end{bmatrix}\begin{bmatrix} 1 & 0 \\ 0 & 3 \end{bmatrix}\begin{bmatrix} 1 & 2 \\ 0 & 1 \end{bmatrix}$

28. $A = \begin{bmatrix} 2 & -1 \\ 1 & -1 \end{bmatrix} = \begin{bmatrix} 0 & 1 \\ 1 & 0 \end{bmatrix}\begin{bmatrix} 1 & 0 \\ 2 & 1 \end{bmatrix}\begin{bmatrix} 1 & -1 \\ 0 & 1 \end{bmatrix}$

29. $A = \begin{bmatrix} 2 & 4 \\ -1 & 2 \end{bmatrix} = \begin{bmatrix} 1 & 0 \\ -\frac{1}{2} & 1 \end{bmatrix}\begin{bmatrix} 1 & 1 \\ 0 & 1 \end{bmatrix}\begin{bmatrix} 2 & 0 \\ 0 & 1 \end{bmatrix}\begin{bmatrix} 1 & 0 \\ 0 & 4 \end{bmatrix}$

30. $A = \begin{bmatrix} -1 & 3 \\ 2 & -2 \end{bmatrix} = \begin{bmatrix} 1 & 0 \\ -2 & 1 \end{bmatrix}\begin{bmatrix} 1 & 0 \\ 0 & 4 \end{bmatrix}\begin{bmatrix} 1 & 3 \\ 0 & 1 \end{bmatrix}\begin{bmatrix} -1 & 0 \\ 0 & 1 \end{bmatrix}$

In Problems 31–34, A is the standard matrix for a linear transformation T. Express A as a product of elementary matrices and then describe the effect of T in terms of reflections, expansions, contractions, and shears. [Note: There are many different correct answers. If your answer differs from the one in the back of the book, use matrix multiplication to check your work.]

31. $A = \begin{bmatrix} 1 & 1 \\ 2 & 3 \end{bmatrix}$ **32.** $A = \begin{bmatrix} 1 & 1 \\ -2 & -1 \end{bmatrix}$

33. $A = \begin{bmatrix} 2 & 1 \\ 4 & 1 \end{bmatrix}$ **34.** $A = \begin{bmatrix} -1 & 2 \\ 2 & -1 \end{bmatrix}$

If θ is a fixed angle, $0 \le \theta < 2\pi$, then it can be shown that the linear transformation defined by the matrix

$$A = \begin{bmatrix} \cos\theta & -\sin\theta \\ \sin\theta & \cos\theta \end{bmatrix}$$

rotates each vector in R^2 through an angle θ in the counterclockwise direction (see Section 6–4). For each angle in Problems 35–38, describe the effect of this transformation in terms of reflections, expansions, contractions, and shears.

35. $\theta = \pi/2$ **36.** $\theta = 3\pi/2$ **37.** $\theta = 5\pi/3$ **38.** $\theta = \pi/4$

In Problems 39–42:
(A) Find the standard matrix for $T_1 T_2$.
(B) Find the standard matrix for $T_2 T_1$.
(C) Determine whether $T_1 T_2 = T_2 T_1$.

39. $T_1\left(\begin{bmatrix} x \\ y \end{bmatrix}\right) = \begin{bmatrix} 2x \\ y \end{bmatrix}$; $T_2\left(\begin{bmatrix} x \\ y \end{bmatrix}\right) = \begin{bmatrix} y \\ x \end{bmatrix}$

40. $T_1\left(\begin{bmatrix} x \\ y \end{bmatrix}\right) = \begin{bmatrix} x \\ 3y \end{bmatrix}$; $T_2\left(\begin{bmatrix} x \\ y \end{bmatrix}\right) = \begin{bmatrix} x \\ -y \end{bmatrix}$

41. $T_1\left(\begin{bmatrix} x \\ y \end{bmatrix}\right) = \begin{bmatrix} x - 2y \\ y \end{bmatrix}$; $T_2\left(\begin{bmatrix} x \\ y \end{bmatrix}\right) = \begin{bmatrix} x + y \\ y \end{bmatrix}$

42. $T_1\left(\begin{bmatrix} x \\ y \end{bmatrix}\right) = \begin{bmatrix} x - 2y \\ y \end{bmatrix}$; $T_2\left(\begin{bmatrix} x \\ y \end{bmatrix}\right) = \begin{bmatrix} x \\ x + y \end{bmatrix}$

C   *In Problems 43–46, let A be the standard matrix for the linear transformation $T: R^2 \rightarrow R^2$. Prove the statements in Problems 43–45 without reference to Theorem 13.*

43. If T maps R^2 onto R^2, then $\ker(T) = \{0\}$.
44. If $\ker(T) = \{0\}$, then A is invertible.
45. If A is invertible, then T maps R^2 onto R^2.
46. *Theorem 13.* Use the statements in Problems 43 and 44 to prove Theorem 13.

In Problems 47 and 48, let T map R^2 onto R^2.

47. *Theorem 14(A).* If ℓ is the line with vector equation

$$\mathbf{u} = \mathbf{u}_0 + t\mathbf{u}_1$$

where $\mathbf{u}_1 \neq \mathbf{0}$ and t is any real number, show that the image of ℓ under T is a line.

48. *Theorem 14(B).* If ℓ is the line with vector equation

$$\mathbf{u} = t\mathbf{u}_1$$

where $\mathbf{u}_1 \neq \mathbf{0}$ and t is any real number, show that the image of ℓ under T is a line through the origin.

| 8-6 | Chapter Review

Important Terms and Symbols

8-1. *Linear transformations and linear operators.* Mapping, image, linear transformation, linear operator, dilation, contraction, identity transformation, zero transformation, matrix transformation, $T: V \rightarrow W$, $T(\mathbf{v})$

8-2. *Kernel and range.* Kernel, range, rank, nullity, ker(T), range(T)

8-3. *Matrix representation of linear transformations.* Standard matrix, matrix of a linear transformation T with respect to the bases B and C, computing $T(\mathbf{x})$

by matrix multiplication, matrix of a linear operator with respect to the basis B, $[[T(\mathbf{b_1})]_C \quad \cdots \quad [T(\mathbf{b_n})]_C]$

8-4. *Eigenvalues of linear operators.* Eigenvalue, eigenvector, eigenspace, diagonalizable linear transformation, similar matrices

8-5. *Transformations of the plane.* Linear transformation of the plane, elementary matrix, reflection, expansion, contraction, shear, product of linear transformations

❚ Exercise 8-6 Chapter Review

Work through all the problems in this chapter review and check your answers in the back of the book. (Answers to most review problems are there.) Where weaknesses show up, review appropriate sections in the text.

A   *In Problems 1–3, let* $T: R^2 \rightarrow R^2$ *be defined by*

$$T(x, y) = (x - 2y, -2x + 4y)$$

1. Show that T is a linear transformation.
2. Determine whether $\mathbf{v}$ is in ker(T).   (A) $\mathbf{v} = (6, 3)$ (B) $\mathbf{v} = (6, -3)$
3. Determine whether $\mathbf{w}$ is in range(T).   (A) $\mathbf{w} = (2, -4)$ (B) $\mathbf{w} = (2, -2)$

In Problems 4–7, let $T: R^2 \rightarrow R^2$ *be the linear operator defined by*

$$T\left(\begin{bmatrix} x_1 \\ x_2 \end{bmatrix}\right) = \begin{bmatrix} 3x_1 + x_2 \\ x_1 + 3x_2 \end{bmatrix}$$

4. Find the standard matrix for T.
5. Show that $\mathbf{u_1}$ and $\mathbf{u_2}$ are eigenvectors of T, and find the corresponding eigenvalues.

   (A) $\mathbf{u_1} = \begin{bmatrix} 1 \\ 1 \end{bmatrix}$ (B) $\mathbf{u_2} = \begin{bmatrix} -1 \\ 1 \end{bmatrix}$

6. Find the matrix of T with respect to the basis $B = \{\mathbf{u_1}, \mathbf{u_2}\}$ where $\mathbf{u_1}$ and $\mathbf{u_2}$ are the eigenvectors defined in Problem 5.
7. Graph the square with vertices $(0, 0)$, $(1, 1)$, $(0, 2)$, and $(-1, 1)$. On the same set of axes, graph the image of this square under T.
8. Let $T: P_2 \rightarrow P_2$ be the linear operator defined by

   $$T(a_0 + a_1 x + a_2 x^2) = (a_0 - a_1) + (a_1 - a_2)x + (a_2 - a_0)x^2$$

   let $B = \{1 + x, 1 + x^2, x + x^2\}$ and $C = \{1, x, x^2\}$ be bases for P_2, and let $p(x) = 3 + 4x + 5x^2$.

   (A) Find the matrix of T with respect to B and C.
   (B) Find $[p(x)]_B$.
   (C) Use matrix multiplication to find $[T(p(x))]_C$.

(D) Use $[T(p(x))]_C$ to find $T(p(x))$.

(E) Find $T(p(x))$ directly from the definition of T.

9. Let $T: P_1 \to P_1$ be the linear operator defined by

$$T(a_0 + a_1 x) = (a_0 + 6a_1) + (-a_0 + 6a_1)x$$

(A) Find A, the matrix of T with respect to the basis $S = \{1, x\}$.

(B) Find the eigenvalues of A and a basis for each eigenspace of A.

(C) Find the eigenvalues of T and a basis for each eigenspace of T.

B   *In Problems 10–15, determine whether T is a linear transformation.*

10. $T: R^2 \to R$ defined by $T(x, y) = 2x + 3y$

11. $T: R^2 \to R^2$ defined by $T(x, y) = 2xy$

12. $T: R^2 \to R^2$ defined by $T(x, y) = (x + 3, y + 4)$

13. $T: R^2 \to R^2$ defined by $T(x, y) = (x - y, 0)$

14. $T: P_2 \to P_3$ defined by $T(a_0 + a_1 x + a_2 x^2) = a_0 x + a_1 x^2 + a_2 x^3$

15. $T: P_2 \to P_3$ defined by $T(a_0 + a_1 x + a_2 x^2) = a_0 + a_1 x + a_2 x^2 + x^3$

In Problems 16–19, find the kernel, range, nullity, and rank of the linear transformation T.

16. $T: R^3 \to R^2$ defined by $T(\mathbf{x}) = A\mathbf{x}$,   where $A = \begin{bmatrix} 1 & -3 & 4 \\ 3 & -7 & 6 \end{bmatrix}$

17. $T: R^3 \to R^2$ defined by $T\left(\begin{bmatrix} x_1 \\ x_2 \\ x_3 \end{bmatrix}\right) = \begin{bmatrix} x_1 - 2x_2 + x_3 \\ -2x_1 + 4x_2 - 2x_3 \end{bmatrix}$

18. $T: P_3 \to P_3$ defined by $T(p(x)) = p(x) + p(-x)$

19. $T: R^3 \to M_{2 \times 2}$ defined by $T(a, b, c) = \begin{bmatrix} a & b \\ c & b \end{bmatrix}$

20. Find the standard matrix for the linear transformation $T: R^3 \to R^2$ defined by

$$T\left(\begin{bmatrix} x_1 \\ x_2 \\ x_3 \end{bmatrix}\right) = \begin{bmatrix} 2x_1 + x_2 \\ x_2 - 3x_3 \end{bmatrix}$$

21. Let $T: R^2 \to R^2$ be the linear operator defined by $T(\mathbf{u}) = \langle \mathbf{u}, \mathbf{w} \rangle \mathbf{w}$ where $\mathbf{w} = (1/\sqrt{10}, 3/\sqrt{10})$.

(A) Find $T(1, 0)$.   (B) Find $T(0, 1)$.   (C) Find $T(a, b)$.

(D) Find ker(T) and the nullity of T.

(E) Find range(T) and the rank of T.

(F) Find the standard matrix for T.

(G) Find the eigenvalues of T and a basis for each eigenspace of T.

22. Let $T: P_2 \to P_1$ be defined by

$$T(a_0 + a_1 x + a_2 x^2) = (2a_0 - a_1) + (3a_1 - 2a_2)x$$

Find the matrix of T with respect to B and C if:

(A) $B = \{1, x, x^2\};$ $C = \{1, x\}$

(B) $B = \{1 + x + x^2, 1 + x^2, 1 + 2x + 3x^2\};$ $C = \{1 + x, 1 - x\}$

23. Let

$$B = \left\{ \begin{bmatrix} 2 \\ 3 \end{bmatrix}, \begin{bmatrix} 3 \\ -2 \end{bmatrix} \right\}$$

be a basis for R^2, let

$$C = \left\{ \begin{bmatrix} 1 & 0 \\ 0 & 0 \end{bmatrix}, \begin{bmatrix} 0 & 1 \\ 1 & 0 \end{bmatrix}, \begin{bmatrix} 0 & 0 \\ 0 & 1 \end{bmatrix} \right\}$$

be a basis for $S_{2\times2}$, the vector space of 2×2 symmetric matrices, and let $T: R^2 \to S_{2\times2}$ be defined by

$$T\left(\begin{bmatrix} 2 \\ 3 \end{bmatrix}\right) = \begin{bmatrix} 2 & 3 \\ 3 & 0 \end{bmatrix} \qquad T\left(\begin{bmatrix} 3 \\ -2 \end{bmatrix}\right) = \begin{bmatrix} 3 & 2 \\ 2 & 4 \end{bmatrix}$$

(A) Find the matrix of T with respect to B and C.

(B) Use matrix multiplication to find $T(\mathbf{x})$ if

$$\mathbf{x} = \begin{bmatrix} 1 \\ -5 \end{bmatrix}$$

24. Let $T: P_2 \to P_2$ be the linear operator defined by

$$T(1) = -1 - 2x - x^2 \qquad T(x) = 3 + 4x + x^2 \qquad T(x^2) = 2x^2$$

(A) Find the matrix of T with respect to the basis $B = \{1, x, x^2\}$.

(B) Find the eigenvalues of T and a basis for each eigenspace of T.

(C) Find a diagonal representation of T, if one exists.

25. Repeat Problem 24 if

$$T(1) = 3 - x \qquad T(x) = 3 - x + x^2 \qquad T(x^2) = 2 - x + x^2$$

Let $B = \{1, x, x^2\}$ and $C = \{1, x\}$ be bases for P_2 and P_1, respectively. In Problems 26 and 27, find the matrix of T with respect to B and C and use matrix multiplication to compute $T(1 + 2x + 3x^2)$. Check your answer by computing $T(1 + 2x + 3x^2)$ directly from the definition of T.

26. $T(a_0 + a_1x + a_2x^2) = (2a_0 - a_1 + a_2) + (a_0 + 3a_1 + a_2)x$

27. $T(a_0 + a_1x + a_2x^2) = (a_0 - a_1 - a_2) + (-a_0 + 2a_1 + a_2)x$

28. Let $T: M_{2\times2} \to M_{2\times2}$ be the linear operator defined by

$$T\left(\begin{bmatrix} a & b \\ c & d \end{bmatrix}\right) = \begin{bmatrix} a & a \\ c & c+d \end{bmatrix}$$

(A) Find the matrix of T with respect to the basis

$$S = \left\{ \begin{bmatrix} 1 & 0 \\ 0 & 0 \end{bmatrix}, \begin{bmatrix} 0 & 1 \\ 0 & 0 \end{bmatrix}, \begin{bmatrix} 0 & 0 \\ 1 & 0 \end{bmatrix}, \begin{bmatrix} 0 & 0 \\ 0 & 1 \end{bmatrix} \right\}$$

(B) Find the eigenvalues of T and a basis for each eigenspace of T.

(C) Find a diagonal representation of T, if one exists.

29. Repeat Problem 28 for

$$T\left(\begin{bmatrix} a & b \\ c & d \end{bmatrix}\right) = \begin{bmatrix} b & 4a \\ 2d & 2c \end{bmatrix}$$

In Problems 30–33, describe the geometric effect of the linear transformation T in terms of reflections, expansions, contractions, and shears. Graph the image under T of the rectangle with vertices (−2, 0), (2, 0), (2, 1), and (−2, 1).

30. $T\left(\begin{bmatrix} x \\ y \end{bmatrix}\right) = \begin{bmatrix} 1 & 0 \\ 0 & 2 \end{bmatrix}\begin{bmatrix} x \\ y \end{bmatrix}$ **31.** $T\left(\begin{bmatrix} x \\ y \end{bmatrix}\right) = \begin{bmatrix} 0 & 1 \\ 1 & 0 \end{bmatrix}\begin{bmatrix} x \\ y \end{bmatrix}$

32. $T\left(\begin{bmatrix} x \\ y \end{bmatrix}\right) = \begin{bmatrix} 1 & 0 \\ \frac{1}{2} & 1 \end{bmatrix}\begin{bmatrix} x \\ y \end{bmatrix}$ **33.** $T\left(\begin{bmatrix} x \\ y \end{bmatrix}\right) = \begin{bmatrix} 1 & -2 \\ 0 & 1 \end{bmatrix}\begin{bmatrix} x \\ y \end{bmatrix}$

34. Let T be the linear transformation defined by the matrix A given below. Describe the geometric effect of T in terms of reflections, expansions, contractions, and shears. Construct a figure that illustrates this sequence of operations on the square with vertices at (0, 0), (1, 0), (1, 1), and (0, 1).

$$A = \begin{bmatrix} 1 & -3 \\ 2 & 1 \end{bmatrix} = \begin{bmatrix} 1 & 0 \\ 2 & 1 \end{bmatrix}\begin{bmatrix} 1 & 0 \\ 0 & 7 \end{bmatrix}\begin{bmatrix} 1 & -3 \\ 0 & 1 \end{bmatrix}$$

35. Let T be the transformation of the plane defined by the matrix

$$A = \begin{bmatrix} 1 & 2 \\ 3 & 5 \end{bmatrix}$$

Express A as a product of elementary matrices and then describe the geometric effect of T in terms of reflections, expansions, contractions, and shears.

In Problems 36 and 37, find the standard matrix for a linear transformation of the plane that has the indicated geometric effect.

36. A shear in the x direction with scale factor 3, followed by an expansion in the y direction with scale factor 2, followed by a reflection about the line $y = x$

37. A contraction in the y direction with scale factor $\frac{1}{2}$, followed by a reflection about the x axis, followed by a shear in the x direction with scale factor -2

38. *Calculus.* Find the kernel of the linear transformation $T: P_2 \to R$ defined by

$$T(p(x)) = \int_{-1}^{1} p(x)\, dx$$

39. *Calculus.* Let $T: P_3 \to P_3$ be defined by

$$T(p(x)) = p''(x) + xp'(x)$$

(A) Find the kernel and the nullity of T.

(B) Find the range and the rank of T.

(C) Find the matrix of T with respect to the basis $B = \{1, x, x^2, x^3\}$.

(D) Find the eigenvalues of T and a basis for each eigenspace of T.

C   In Problems 40–43, A is a 4 × 4 matrix and $T: R^4 \to R^4$ is the matrix transformation defined by $T(\mathbf{x}) = A\mathbf{x}$.

40. If the rank of T is 1, what is the nullity of T?

41. If $\dim\{\mathbf{x}|A\mathbf{x} = \mathbf{0}\} = 2$, what is the nullity of T?

42. If the system $A\mathbf{x} = \mathbf{b}$ is consistent for all $\mathbf{b} \in R^4$, what is the kernel of T?

43. If A is invertible, what is the range of T?

44. Let $\mathbf{w} = (w_1, w_2, w_3)$ be a fixed nonzero vector in R^3 and let $T: R^3 \to R^3$ be defined by

$$T(\mathbf{u}) = \mathbf{u} \times \mathbf{w}$$

(A) Show that T is a linear transformation.

(B) Find the matrix of T with respect to the standard basis for R^3.

(C) Find the eigenvalues of T and a basis for each eigenspace of T.

45. Let $T: R^2 \to R^2$ be defined by the matrix

$$A = \begin{bmatrix} a & -b \\ b & a \end{bmatrix}$$

where $0 < a < 1$ and $a^2 + b^2 = 1$.

(A) Show that the geometric effect of T is a rotation.

(B) Express A as a product of elementary matrices.

(C) Describe the geometric effect of T in terms of reflections, expansions, contractions, and shears.

|9| Additional Applications

Contents

In this chapter, we will consider in detail three of the many applications of linear algebra. (Additional applications are discussed in the computer supplement; see the Preface.) The applications we have selected for this chapter have one thing in common — each involves problems that can be solved by utilizing the matrix diagonalization techniques developed in Chapter 7. In each case, a complicated problem is transformed into a simple one by introducing matrix notation and diagonalizing an appropriate matrix. This is a powerful mathematical technique that is used in many different areas and is one of the primary reasons for studying linear algebra.

Each section in this chapter is independent of the other two and each fully illustrates the application of matrix diagonalization techniques. Thus, it is possible to study these applications in any order and to omit any that are not of interest.

| 9-1 | Quadratic Forms

- Conic Sections
- Translation of Axes
- Rotation of Axes and Orthogonal Matrices
- Quadratic Forms
- Graphing Conic Sections

■ Conic Sections

An equation of the form

$$ax^2 + 2bxy + cy^2 + dx + ey + f = 0 \tag{1}$$

with a, b, and c not all zero, is called a **quadratic equation** in x and y. (The coefficient of xy is written in the form $2b$ for convenience in notation that will be developed later in this section.) The graph of a quadratic equation is a plane curve called a **conic section,**[*] or, more simply, a **conic.** The three basic types of conics are **ellipses** (including **circles** as a special case), **parabolas,** and **hyperbolas.** Certain choices of the coefficients in (1) may produce a **degenerate conic** — that is, a point, a line, or a pair of lines. (See Problems 25 – 30 in Exercise 9-1.)

---

[*] So called because the curve can be obtained by intersecting a cone with a plane.

Finally, it is also possible that there are no points whose coordinates satisfy (1). In this case, the graph is the empty set.

Our goal in this section is to use techniques of linear algebra to help identify and sketch the graphs of conics. We begin by considering the *standard forms* of the three basic types of conics. A (nondegenerate) conic is said to be in **standard position** with respect to the xy coordinate system if its equation can be expressed in one of the **standard forms** given in Figure 1.

(A) Parabola

$y = sx^2, s \neq 0$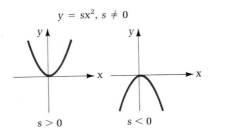

$x = ty^2, t \neq 0$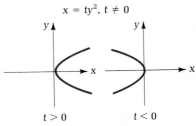

(B) Ellipse

$$\frac{x^2}{s^2} + \frac{y^2}{t^2} = 1, s, t > 0$$

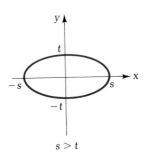 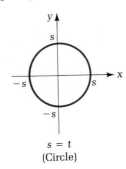 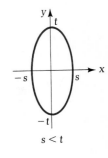

(C) Hyperbola

$$\frac{x^2}{s^2} - \frac{y^2}{t^2} = 1, s, t > 0$$

$$\frac{y^2}{t^2} - \frac{x^2}{s^2} = 1, s, t > 0$$

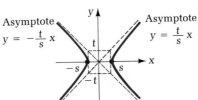 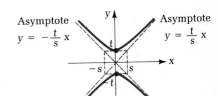

Figure 1 Standard forms of conic sections

Example 1   Identify each conic and sketch its graph.

(A) $\dfrac{x^2}{25} - \dfrac{y^2}{9} = 1$

(B) $x^2 + 2y^2 = 2$

(C) $x^2 + 2y = 0$

Solution   (A) This is the standard form of a hyperbola with $s = 5$ and $t = 3$. Notice that the graph is asymptotic to the lines $y = \pm \frac{3}{5}x$.

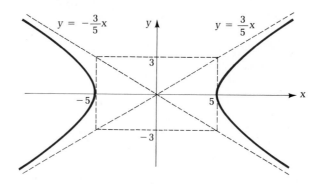

(B) Multiplying both sides of the given equation by $\frac{1}{2}$, we have

$$\frac{x^2}{2} + \frac{y^2}{1} = 1$$

which is the standard form of an ellipse with $s = \sqrt{2}$ and $t = 1$.

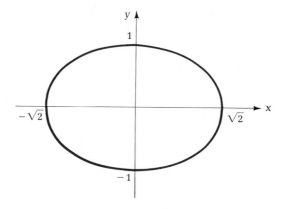

(C) Solving the given equation for y, we have

$$y = -\tfrac{1}{2}x^2$$

which is the standard form of a parabola with $s = -\frac{1}{2}$.

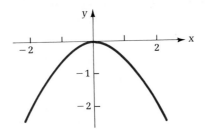

Problem 1 Identify each conic and sketch its graph.

(A) $\dfrac{x^2}{16} + \dfrac{y^2}{25} = 1$ (B) $3y^2 - x = 0$ (C) $9y^2 - 16x^2 = 144$ ▮▮

▪ Translation of Axes

If the equation of a conic cannot be expressed in one of the standard forms, how can we identify and graph the conic? In certain cases, **translating the axes** to a new coordinate system will transform a given quadratic equation into one that is in standard form. This transformation involves an algebraic operation referred to as **completing the square** and is best explained by means of an example.

Example 2 Identify and sketch the graph of

$$9y^2 - 4x^2 - 36y - 24x - 36 = 0$$

Solution We proceed by completing the square relative to x and to y:

$$9y^2 - 36y - 4x^2 - 24x = 36$$
$$9(y^2 - 4y \quad) - 4(x^2 + 6x \quad) = 36$$
$$9(y^2 - 4y + 4) - 4(x^2 + 6x + 9) = 36 + 36 - 36$$
$$9(y - 2)^2 - 4(x + 3)^2 = 36$$
$$\frac{(y - 2)^2}{4} - \frac{(x + 3)^2}{9} = 1$$

If we let $x' = x + 3$ and $y' = y - 2$ in the last equation, we have

$$\frac{y'^2}{4} - \frac{x'^2}{9} = 1$$

which we recognize as the standard form of a hyperbola in the $x'y'$ coordinate system. To graph this hyperbola in the original xy coordinate system, we graph the axes of the $x'y'$ system in the xy system and then graph the hyperbola in the $x'y'$ system, as shown in the figure.

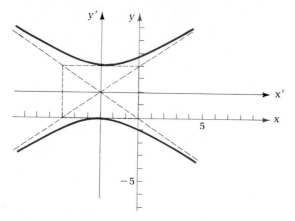

Problem 2   Identify and sketch the graph of

$$x^2 - 2x + 16y^2 + 64y + 49 = 0$$

∎

Translation of axes can be used to identify and graph any conic whose equation does not contain a **mixed term** — a term of the form $2bxy$, $b \neq 0$. If a mixed term is present, then we will see that *rotation of axes* can be used to transform the given equation into one that contains no mixed term. Furthermore, the matrix diagonalization techniques discussed in Chapter 7 can be used to carry out this transformation. We begin by discussing the relationship between rotation of axes and orthogonal matrices.

▪ Rotation of Axes and Orthogonal Matrices

According to Theorem 13 in Section 6-4, if

$$P = \begin{bmatrix} \cos\theta & -\sin\theta \\ \sin\theta & \cos\theta \end{bmatrix} \qquad 0 \leq \theta < 2\pi$$

is the transition matrix from an $x'y'$ coordinate system to an xy coordinate system, then P rotates the xy coordinate system counterclockwise through an angle θ into the $x'y'$ coordinate system. Notice that P is an orthogonal matrix (verify this).

Now, suppose that P is any 2×2 orthogonal matrix and that the xy and $x'y'$ coordinate systems are related by

$$\mathbf{x} = P\mathbf{x}' \qquad \text{where} \quad \mathbf{x} = \begin{bmatrix} x \\ y \end{bmatrix} \quad \text{and} \quad \mathbf{x}' = \begin{bmatrix} x' \\ y' \end{bmatrix}$$

What is the relationship between these coordinate systems? It turns out that the relationship depends on the value of det(P). It can be shown that det(P) = ±1 for any orthogonal matrix P (see Problem 38 in Exercise 3-3). If we assume that det(P) = 1, then Theorem 1 shows that P rotates the xy coordinate system into the x'y' coordinate system.* The proof is omitted.

Theorem 1

If

$$P = \begin{bmatrix} p_{11} & p_{12} \\ p_{21} & p_{22} \end{bmatrix}$$

is an orthogonal matrix with det(P) = 1 and if the xy and x'y' coordinate systems are related by

$$\mathbf{x} = P\mathbf{x}' \qquad \text{where} \quad \mathbf{x} = \begin{bmatrix} x \\ y \end{bmatrix} \quad \text{and} \quad \mathbf{x}' = \begin{bmatrix} x' \\ y' \end{bmatrix}$$

then P rotates the xy coordinate system counterclockwise through an angle θ into the x'y' coordinate system. Furthermore, the columns of P form an orthonormal basis for the x'y' coordinate system (see the figure).

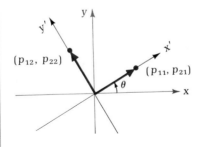

Example 3   If

$$P = \begin{bmatrix} \frac{3}{5} & -\frac{4}{5} \\ \frac{4}{5} & \frac{3}{5} \end{bmatrix}$$

then

$$PP^{\mathrm{T}} = \begin{bmatrix} \frac{3}{5} & -\frac{4}{5} \\ \frac{4}{5} & \frac{3}{5} \end{bmatrix} \begin{bmatrix} \frac{3}{5} & \frac{4}{5} \\ -\frac{4}{5} & \frac{3}{5} \end{bmatrix} = \begin{bmatrix} 1 & 0 \\ 0 & 1 \end{bmatrix}$$

Thus, P is an orthogonal matrix. Since det(P) = 1 (verify this), Theorem 1 implies that P rotates the xy coordinate system into the x'y' coordinate system (see

* For those who have studied Section 8-5, if det(P) = −1, then the geometric effect of P is a reflection about the x axis, followed by a rotation.

the figure). Notice that the columns of P are used to graph the $x'y'$ axes in the xy coordinate system. It is not necessary to find the angle θ.

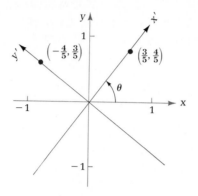

Problem 3 Show that P is an orthogonal matrix with $\det(P) = 1$, and graph the $x'y'$ coordinate system in the xy system.

$$P = \begin{bmatrix} \frac{5}{13} & -\frac{12}{13} \\ \frac{12}{13} & \frac{5}{13} \end{bmatrix}$$

∎

▪ Quadratic Forms

In order to use matrix methods to simplify quadratic equations, we must first introduce some matrix notation. Associated with each quadratic equation

$$ax^2 + 2bxy + cy^2 + dx + ey + f = 0 \tag{2}$$

is the **quadratic form**

$$ax^2 + 2bxy + cy^2 \tag{3}$$

If

$$A = \begin{bmatrix} a & b \\ b & c \end{bmatrix} \quad \text{and} \quad \mathbf{x} = \begin{bmatrix} x \\ y \end{bmatrix}$$

then

$$\begin{aligned}
\mathbf{x}^T A \mathbf{x} &= \begin{bmatrix} x & y \end{bmatrix} \begin{bmatrix} a & b \\ b & c \end{bmatrix} \begin{bmatrix} x \\ y \end{bmatrix} \\
&= \begin{bmatrix} x & y \end{bmatrix} \begin{bmatrix} ax + by \\ bx + cy \end{bmatrix} \\
&= \begin{bmatrix} ax^2 + 2bxy + cy^2 \end{bmatrix}
\end{aligned}$$

Using our convention of identifying 1×1 matrices with real numbers, the quadratic form associated with (2) can be written as $\mathbf{x}^T A \mathbf{x}$ and (2) can be written in matrix form as

$$\mathbf{x}^T A\mathbf{x} + [d \quad e]\mathbf{x} + f = 0$$

The symmetric matrix A is called the **matrix of the quadratic form** (3).

Example 4   The quadratic form associated with the quadratic equation

$$3x^2 + 4xy - 5y^2 + 6x - y + 7 = 0$$

is

$$3x^2 + 4xy - 5y^2$$

The matrix of this quadratic form is

$$A = \begin{bmatrix} 3 & 2 \\ 2 & -5 \end{bmatrix}$$

and the matrix form of the equation is

$$[x \quad y]\begin{bmatrix} 3 & 2 \\ 2 & -5 \end{bmatrix}\begin{bmatrix} x \\ y \end{bmatrix} + [6 \quad -1]\begin{bmatrix} x \\ y \end{bmatrix} + 7 = 0$$

or

$$\mathbf{x}^T A\mathbf{x} + [6 \quad -1]\mathbf{x} + 7 = 0 \qquad \text{where} \quad \mathbf{x} = \begin{bmatrix} x \\ y \end{bmatrix}$$

Problem 4   Find the quadratic form associated with the following equation, find the matrix of the quadratic form, and write the equation in matrix form:

$$7x^2 - 10xy + 9y^2 - 3x + 4y + 9 = 0 \qquad\qquad \blacksquare\blacksquare$$

Since the matrix A of a quadratic form is symmetric, Theorem 12 in Section 7-3 implies that there is an orthogonal matrix P with the property that

$$P^T AP = D = \begin{bmatrix} \lambda_1 & 0 \\ 0 & \lambda_2 \end{bmatrix}$$

where λ_1 and λ_2 are the eigenvalues of A. Now we define an $x'y'$ coordinate system by the equation

$$\mathbf{x} = P\mathbf{x}' \qquad \text{where} \quad \mathbf{x} = \begin{bmatrix} x \\ y \end{bmatrix} \quad \text{and} \quad \mathbf{x}' = \begin{bmatrix} x' \\ y' \end{bmatrix}$$

That is, P is the transition matrix from the $x'y'$ system to the xy system. If we make the substitution $\mathbf{x} = P\mathbf{x}'$ in the quadratic form $\mathbf{x}^T A\mathbf{x}$, we have

$$\begin{aligned}
\mathbf{x}^T A\mathbf{x} &= (P\mathbf{x}')^T A(P\mathbf{x}') \\
&= \mathbf{x}'^T(P^T AP)\mathbf{x}' \\
&= \mathbf{x}'^T D\mathbf{x}' \\
&= [x' \quad y']\begin{bmatrix} \lambda_1 & 0 \\ 0 & \lambda_2 \end{bmatrix}\begin{bmatrix} x' \\ y' \end{bmatrix} \\
&= \lambda_1 x'^2 + \lambda_2 y'^2
\end{aligned}$$

This last expression is a quadratic form in the $x'y'$ coordinate system *that has no mixed term*. We have just proved Theorem 2.

Theorem 2

Principal Axes Theorem for R^2

Let

$$\mathbf{x}^T A \mathbf{x} = ax^2 + 2bxy + cy^2$$

be a quadratic form, and let P be an orthogonal matrix that diagonalizes A. Then the substitution $\mathbf{x} = P\mathbf{x}'$ transforms the quadratic form $\mathbf{x}^T A \mathbf{x}$ into the quadratic form

$$\lambda_1 x'^2 + \lambda_2 y'^2$$

which has no mixed term and where λ_1 and λ_2 are the eigenvalues of A.

▪ Graphing Conic Sections

Theorems 1 and 2 provide a method for identifying and graphing conic sections whose equations contain mixed terms.

Example 5   Identify and graph the conic section with equation

$$5x^2 - 6xy + 5y^2 - 32 = 0 \tag{4}$$

Solution   Since this equation contains a mixed term, it is not one of the standard forms of a conic. We use Theorems 1 and 2 to transform this equation into one that is in standard form. The matrix of the quadratic form

$$5x^2 - 6xy + 5y^2$$

is

$$A = \begin{bmatrix} 5 & -3 \\ -3 & 5 \end{bmatrix}$$

The characteristic polynomial for A is

$$p(\lambda) = \det(\lambda I - A) = (\lambda - 2)(\lambda - 8)$$

The eigenvalues and associated (normalized) eigenvectors for A are

2	8	Eigenvalues
$\begin{bmatrix} \dfrac{1}{\sqrt{2}} \\ \dfrac{1}{\sqrt{2}} \end{bmatrix}$	$\begin{bmatrix} -\dfrac{1}{\sqrt{2}} \\ \dfrac{1}{\sqrt{2}} \end{bmatrix}$	Normalized eigenvectors

(Since the order in which we will use these eigenvalues has not yet been determined, we refrain from identifying them as λ_1 and λ_2.) Recall that the normalized eigenvectors of A are the columns of the orthogonal matrix P that diagonalizes A. The eigenvectors given above provide two possible choices for P:

$$\begin{matrix} 2 & 8 \end{matrix}$$

$$P = \begin{bmatrix} \dfrac{1}{\sqrt{2}} & -\dfrac{1}{\sqrt{2}} \\[2mm] \dfrac{1}{\sqrt{2}} & \dfrac{1}{\sqrt{2}} \end{bmatrix} \quad \begin{matrix} ? \\ \text{or} \end{matrix} \quad P = \begin{bmatrix} -\dfrac{1}{\sqrt{2}} & \dfrac{1}{\sqrt{2}} \\[2mm] \dfrac{1}{\sqrt{2}} & \dfrac{1}{\sqrt{2}} \end{bmatrix}$$

$$\begin{matrix} \quad\quad 8 & 2 \end{matrix}$$

$$\det(P) = 1 \qquad\qquad \det(P) = -1$$

Actually, either form of P could be used to transform equation (4) into one that does not contain a mixed term. However, for graphing purposes, we want P to *rotate* the xy coordinate system into the $x'y'$ system. If we choose P so that $\det(P) = 1$, then Theorem 1 implies that P will be a rotation. Thus, we let $\lambda_1 = 2$, $\lambda_2 = 8$,

$$P = \begin{bmatrix} \dfrac{1}{\sqrt{2}} & -\dfrac{1}{\sqrt{2}} \\[2mm] \dfrac{1}{\sqrt{2}} & \dfrac{1}{\sqrt{2}} \end{bmatrix}$$

and make the substitution $\mathbf{x} = P\mathbf{x}'$ in (4):

$$5x^2 - 6xy + 5y^2 - 32 = 0 \qquad \text{Write the quadratic equation in matrix form.}$$

$$[x \;\; y]\begin{bmatrix} 5 & -3 \\ -3 & 5 \end{bmatrix}\begin{bmatrix} x \\ y \end{bmatrix} - 32 = 0$$

$$\mathbf{x}^\mathsf{T} A\mathbf{x} - 32 = 0 \qquad \text{Substitute } \mathbf{x} = P\mathbf{x}' \text{ and simplify.}$$

$$(P\mathbf{x}')^\mathsf{T} A(P\mathbf{x}') - 32 = 0$$

$$\begin{matrix} & & & & \lambda_1 \;\; \lambda_2 \end{matrix}$$

$$\mathbf{x}'^\mathsf{T}(P^\mathsf{T} A P)\mathbf{x}' - 32 = 0 \qquad P^\mathsf{T} A P = \begin{bmatrix} 2 & 0 \\ 0 & 8 \end{bmatrix}$$

$$[x' \;\; y']\begin{bmatrix} 2 & 0 \\ 0 & 8 \end{bmatrix}\begin{bmatrix} x' \\ y' \end{bmatrix} - 32 = 0$$

$$2x'^2 + 8y'^2 - 32 = 0$$

The last equation can be written as

$$\frac{x'^2}{16} + \frac{y'^2}{4} = 1$$

which is the standard form of an ellipse relative to the $x'y'$ coordinate system.

The graph of (4) in the xy coordinate system is obtained by first graphing the $x'y'$ coordinate axes in the xy coordinate system and then graphing the ellipse relative to the $x'y'$ coordinate system (see the figure at the top of the next page).

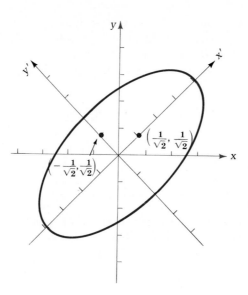

Problem 5   Identify and graph the conic with equation

$$3x^2 + 10xy + 3y^2 + 8 = 0$$ ‖

The matrix P used to simplify the equation of the conic in Example 5 rotated the xy coordinate axes through an angle $\theta = 45°$. The matrix P and the corresponding angle of rotation θ are not unique. In fact, there are three other matrices that could have been used in Example 5 to rotate the xy coordinate system so that the transformed equation contains no mixed term. These are given below and illustrated in Figure 2. The matrices were obtained by using the fact that if $\mathbf{u}$ is a normalized eigenvector of A, then so is $-\mathbf{u}$. You should verify that each matrix diagonalizes the matrix A in Example 5.

(A) $P = \begin{bmatrix} -\dfrac{1}{\sqrt{2}} & -\dfrac{1}{\sqrt{2}} \\[2mm] \dfrac{1}{\sqrt{2}} & -\dfrac{1}{\sqrt{2}} \end{bmatrix}$ (B) $P = \begin{bmatrix} -\dfrac{1}{\sqrt{2}} & \dfrac{1}{\sqrt{2}} \\[2mm] -\dfrac{1}{\sqrt{2}} & -\dfrac{1}{\sqrt{2}} \end{bmatrix}$ (C) $P = \begin{bmatrix} \dfrac{1}{\sqrt{2}} & \dfrac{1}{\sqrt{2}} \\[2mm] -\dfrac{1}{\sqrt{2}} & \dfrac{1}{\sqrt{2}} \end{bmatrix}$

$\dfrac{x'^2}{4} + \dfrac{y'^2}{16} = 1$ $\dfrac{x'^2}{16} + \dfrac{y'^2}{4} = 1$ $\dfrac{x'^2}{4} + \dfrac{y'^2}{16} = 1$

$\theta = 135°$ $\theta = 225°$ $\theta = 315°$

See Figure 2(A).   See Figure 2(B).   See Figure 2(C).

It can be shown that it is always possible to eliminate the mixed term in a quadratic equation by rotating the axes through an angle θ, where $0° < \theta < 90°$. When using matrix methods, this can be accomplished by selecting the two normalized eigenvectors so that one of them lies in the first quadrant of the xy

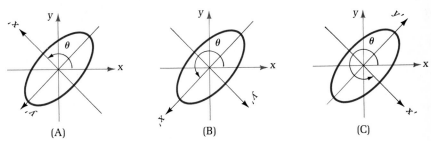

(A) (B) (C)

Figure 2 Additional solutions to Example 5

coordinate system and the other in the second quadrant, as we did in Example 5. We have followed this procedure for selecting eigenvectors in solving the problems in Exercise 9-1 and you should do the same.

In some cases, eliminating the mixed term in a quadratic equation results in an equation of the form

$$\lambda_1 x'^2 + \lambda_2 y'^2 + d'x' + e'y' + f = 0$$

where d' or e' may be nonzero. It may then be necessary to translate the $x'y'$ coordinate axes in order to find the standard form. You will encounter several problems of this type in Exercise 9-1.

Answers to Matched Problems

1. (A) Ellipse

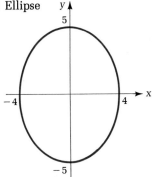

(B) Parabola

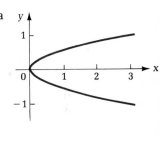

(C) Hyperbola

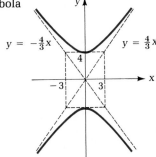

2. $\dfrac{x'^2}{16} + \dfrac{y'^2}{1} = 1$, ellipse

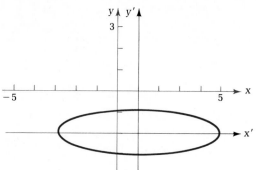

3.

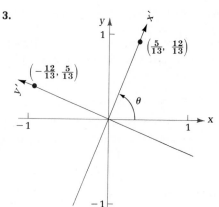

4. $7x^2 - 10xy + 9y^2$; $\begin{bmatrix} 7 & -5 \\ -5 & 9 \end{bmatrix}$;

$[x \quad y]\begin{bmatrix} 7 & -5 \\ -5 & 9 \end{bmatrix}\begin{bmatrix} x \\ y \end{bmatrix} + [-3 \quad 4]\begin{bmatrix} x \\ y \end{bmatrix} + 9 = 0$

5. $\dfrac{y'^2}{4} - \dfrac{x'^2}{1} = 1$; hyperbola

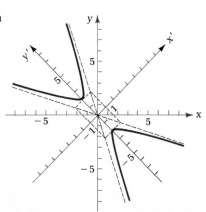

II Exercise 9-1

A In Problems 1–6, identify each conic and sketch its graph.

1. $\dfrac{x^2}{25} + \dfrac{y^2}{4} = 1$

2. $x^2 + \dfrac{y^2}{16} = 1$

3. $x^2 - 4y^2 = 36$

4. $y^2 - 9x^2 = 9$

5. $x^2 - 2y = 0$

6. $x + 4y^2 = 0$

In Problems 7–10, find the quadratic form associated with each equation, find the matrix of this quadratic form, and write the equation in matrix form.

7. $5x^2 - 12xy + 7y^2 - 8x + 9y - 11 = 0$

8. $2x^2 + 6xy - 3y^2 + 4x - y + 14 = 0$

9. $3x^2 + 10xy + 4x - 2y - 7 = 0$

10. $4xy - 3y^2 + 3x - 7y + 9 = 0$

B In Problems 11–16, use a translation of axes to transform each equation into a standard form. Identify and sketch the conic.

11. $x^2 - 2x + y^2 - 2y + 1 = 0$

12. $x^2 - 4x - y^2 - 2y + 2 = 0$

13. $x^2 + 6x - 4y^2 - 8y + 9 = 0$

14. $2x^2 - 4x + y = 0$

15. $2x - y^2 + 4y = 0$

16. $4x^2 - 16x + y^2 - 8y + 16 = 0$

In Problems 17 and 18, P is the transition matrix from the $x'y'$ coordinate system to the xy coordinate system. Show that P is an orthogonal matrix, and graph the x' and y' axes in the xy coordinate system.

17. $P = \begin{bmatrix} \dfrac{3}{\sqrt{10}} & -\dfrac{1}{\sqrt{10}} \\[2ex] \dfrac{1}{\sqrt{10}} & \dfrac{3}{\sqrt{10}} \end{bmatrix}$

18. $P = \begin{bmatrix} \dfrac{1}{2} & -\dfrac{\sqrt{3}}{2} \\[2ex] \dfrac{\sqrt{3}}{2} & \dfrac{1}{2} \end{bmatrix}$

Problems 19–24 involve rotation of axes. In each problem, select normalized eigenvectors lying in the first and second quadrants of the xy coordinate system to ensure that the angle of rotation satisfies $0° < \theta < 90°$ (see page 582).

In Problems 19–22, use a rotation of axes, $0° < \theta < 90°$, to transform each equation into a standard form. Identify and sketch the conic.

19. $8x^2 - 12xy + 17y^2 - 20 = 0$

20. $8x^2 - 4xy + 5y^2 - 36 = 0$

21. $x^2 - 6xy - 7y^2 + 8 = 0$

22. $xy - 2 = 0$

In Problems 23 and 24, use a rotation, $0° < \theta < 90°$, followed by a translation to transform each equation into a standard form. Identify and sketch the conic.

23. $x^2 - 2xy + y^2 + \sqrt{2}\,x - 3\sqrt{2}\,y - 4 = 0$
24. $4x^2 + 4xy + y^2 + 9\sqrt{5}\,x + 2\sqrt{5}\,y + 15 = 0$

The graphs of the equations in Problems 25–30 are degenerate conics (a point, a line, or a pair of lines) or the empty set. Sketch the graph of each equation, if it exists.

25. $x^2 + 4y^2 + 1 = 0$ **26.** $x^2 + 4y^2 = 0$
27. $x^2 - 4y^2 = 0$ **28.** $4x^2 - 4xy + y^2 = 0$
29. $x^2 - 2xy + y^2 - 1 = 0$ **30.** $x^2 - 6x + 4y^2 - 8y + 13 = 0$

C   *In Problems 31–34, let*

$$A = \begin{bmatrix} a & b \\ b & c \end{bmatrix}$$

31. Show that the eigenvalues of A are

$$\lambda_1 = \frac{a + c + \sqrt{(a+c)^2 + 4(b^2 - ac)}}{2} \quad \text{and}$$
$$\lambda_2 = \frac{a + c - \sqrt{(a+c)^2 + 4(b^2 - ac)}}{2}$$

32. If $b^2 - ac < 0$, show that λ_1 and λ_2 have the same sign and neither is zero.
33. If $b^2 - ac = 0$, show that at least one of the eigenvalues of A is zero.
34. If $b^2 - ac > 0$, show that λ_1 and λ_2 have opposite signs and neither is zero.
35. If the general quadratic equation (1) is transformed into the equation

$$\lambda_1 x'^2 + \lambda_2 y'^2 + d'x' + e'y' + f = 0$$

by a rotation of axes, use the results in Problems 32–34 to show that the graph of (1) (excluding degenerate cases) is:

(A) An ellipse if $b^2 - ac < 0$
(B) A parabola if $b^2 - ac = 0$
(C) A hyperbola if $b^2 - ac > 0$

I 9-2I   Recurrence Equations

- First-Order Recurrence Equations
- Systems of First-Order Recurrence Equations
- Second-Order Recurrence Equations
- Remarks

▪ First-Order Recurrence Equations

In this section, we will discuss another type of problem that can be solved by introducing matrix notation and diagonalizing an appropriate matrix. We begin with a simple example that will illustrate some basic concepts.

▌□▌   **Example 6**
Population Growth

The number of bacteria in a culture doubles every hour. If there were 100 bacteria initially, how many will be present after 3 hours? After 10 hours? After n hours?

Solution   Let x_n be the bacteria population after n hours and let x_0 be the initial population. Since the population doubles every hour, we have

$x_0 = 100$	Initial population
$x_1 = 2x_0 = 2 \cdot 100 = 200$	Population after 1 hour
$x_2 = 2x_1 = 2^2 \cdot 100 = 400$	Population after 2 hours
$x_3 = 2x_2 = 2^3 \cdot 100 = 800$	Population after 3 hours

Thus, after 3 hours the culture will contain 800 bacteria.

How can we find the population after 10 hours? We could continue to double the population until we reach x_{10}. Actually, it is easier to develop a general formula for x_n and then evaluate this formula for $n = 10$. To begin this development, notice that the relationship between the bacteria population after $n - 1$ hours (x_{n-1}) and the bacteria population after n hours (x_n) is

$$x_n = 2x_{n-1} \tag{1}$$

Equation (1) is called a **first-order recurrence equation.**[*] The given initial value $x_0 = 100$ is called an **initial condition.** A **solution** to (1) is a formula (function) $x_n = f(n)$ that satisfies (1) for all integers $n > 0$. Examining the sequence of values x_0, x_1, x_2, x_3 computed above suggests that the solution to (1) is

$$x_n = 2^n \cdot 100 \tag{2}$$

A proof that (2) is the only solution to (1) requires techniques that we have not discussed. However, we can verify that this formula is correct by substitution in (1):

$$x_n \overset{?}{=} 2x_{n-1} \qquad \text{From (2), } x_n = 2^n \cdot 100 \text{ and } x_{n-1} = 2^{n-1} \cdot 100.$$
$$2^n \cdot 100 \overset{?}{=} 2(2^{n-1} \cdot 100)$$
$$2^n \cdot 100 \overset{\checkmark}{=} 2^n \cdot 100$$

Thus, the formula in (2) satisfies (1) for all $n > 0$. The bacteria population after n

---

[*] Recurrence equations are special cases of a more general concept called **recursive relations** or simply **recursion.** Recursion is used extensively in certain areas of mathematics and computer science. The special recurrence equations we are considering are also commonly referred to as **linear difference equations.**

hours is therefore $2^n \cdot 100$. In particular, the population after 10 hours is

$$x_{10} = 2^{10} \cdot 100 = 102{,}400$$

The concepts introduced in Example 6 are generalized in Theorem 3. The proof is omitted.

Theorem 3

> The unique solution to the general first-order recurrence equation
>
> $$x_n = ax_{n-1} \qquad n > 0 \tag{3}$$
>
> with initial condition $x_0 = k$ is given by
>
> $$x_n = ka^n$$

Problem 6   Find the solution to the recurrence equation

$$x_n = \tfrac{1}{2}x_{n-1} \qquad n > 0, \quad x_0 = 100 \qquad \blacksquare$$

First-order recurrence equations can be used to describe the behavior of many different quantities, including population growth, growth of money invested at compound interest, depletion of natural resources, the relationship between investment and income, and so on. If an application involves two or more interrelated quantities (such as the population of competing species or the relationship between supply and demand), then it may be necessary to consider a *system of first-order recurrence equations*.

■ Systems of First-Order Recurrence Equations

A system of equations of the form

$$x_n = ax_{n-1} + by_{n-1} \qquad x_0 = k$$

$$y_n = cx_{n-1} + dy_{n-1} \qquad y_0 = \ell \tag{4}$$

is called a **system of first-order recurrence equations.** The **matrix form of system (4)** is

$$\mathbf{u}_n = A\mathbf{u}_{n-1} \qquad \mathbf{u}_0 = \begin{bmatrix} k \\ \ell \end{bmatrix} \tag{5}$$

where

$$A = \begin{bmatrix} a & b \\ c & d \end{bmatrix} \quad \text{and} \quad \mathbf{u}_n = \begin{bmatrix} x_n \\ y_n \end{bmatrix}$$

Notice the similarity between the matrix form (5) for a system of first-order recurrence equations and the general form (3) for a single first-order recurrence equation. Proceeding as we did in the case of a single equation, we have

$$\mathbf{u}_1 = A\mathbf{u}_0$$
$$\mathbf{u}_2 = A\mathbf{u}_1 = A(A\mathbf{u}_0) = A^2\mathbf{u}_0$$
$$\mathbf{u}_3 = A\mathbf{u}_2 = A(A^2\mathbf{u}_0) = A^3\mathbf{u}_0$$

.

.

.

$$\mathbf{u}_n = A\mathbf{u}_{n-1} = A(A^{n-1}\mathbf{u}_0) = A^n\mathbf{u}_0$$

Theorem 4 generalizes Theorem 3 to systems of recurrence equations. The proof of Theorem 4 also involves techniques we have not discussed and is omitted.

Theorem 4   Given the system of first-order recurrence equations

$$x_n = ax_{n-1} + by_{n-1} \qquad x_0 = k$$
$$y_n = cx_{n-1} + dy_{n-1} \qquad y_0 = \ell$$

the **matrix form of the solution** to this system is

$$\mathbf{u}_n = A^n\mathbf{u}_0$$

where

$$A = \begin{bmatrix} a & b \\ c & d \end{bmatrix} \qquad \mathbf{u}_n = \begin{bmatrix} x_n \\ y_n \end{bmatrix} \qquad \mathbf{u}_0 = \begin{bmatrix} k \\ \ell \end{bmatrix}$$

Example 7   Write the system of first-order recurrence equations in matrix form and use Theorem 4 to find the matrix form of the solution:

$$x_n = 4x_{n-1} - y_{n-1} \qquad x_0 = 6$$

(6)

$$y_n = 2x_{n-1} + y_{n-1} \qquad y_0 = 8$$

Solution   The matrix form for this system is

$$\begin{bmatrix} x_n \\ y_n \end{bmatrix} = \begin{bmatrix} 4 & -1 \\ 2 & 1 \end{bmatrix}\begin{bmatrix} x_{n-1} \\ y_{n-1} \end{bmatrix} \qquad \begin{bmatrix} x_0 \\ y_0 \end{bmatrix} = \begin{bmatrix} 6 \\ 8 \end{bmatrix}$$

or

$$\mathbf{u}_n = A\mathbf{u}_{n-1} \qquad \mathbf{u}_0 = \begin{bmatrix} 6 \\ 8 \end{bmatrix}$$

where

$$A = \begin{bmatrix} 4 & -1 \\ 2 & 1 \end{bmatrix} \quad \text{and} \quad \mathbf{u}_n = \begin{bmatrix} x_n \\ y_n \end{bmatrix}$$

Applying Theorem 4, the matrix form for the solution of (6) is

$$\begin{bmatrix} x_n \\ y_n \end{bmatrix} = \mathbf{u}_n = A^n \mathbf{u}_0 = \begin{bmatrix} 4 & -1 \\ 2 & 1 \end{bmatrix}^n \begin{bmatrix} 6 \\ 8 \end{bmatrix}$$

Problem 7   Repeat Example 7 for the system

$$x_n = 6x_{n-1} - y_{n-1} \qquad x_0 = 3$$
$$y_n = 12x_{n-1} - y_{n-1} \qquad y_0 = 10$$

■

The matrix form of the solution to a system of first-order recurrence equations is useful for computing the first few values of x_n and y_n. Referring to the solution to the system in Example 7, we have

$$\begin{array}{ccc} \mathbf{u}_1 & A & \mathbf{u}_0 \end{array}$$
$$\begin{bmatrix} x_1 \\ y_1 \end{bmatrix} = \begin{bmatrix} 4 & -1 \\ 2 & 1 \end{bmatrix} \begin{bmatrix} 6 \\ 8 \end{bmatrix} = \begin{bmatrix} 16 \\ 20 \end{bmatrix}$$

$$\begin{array}{ccc} \mathbf{u}_2 & A & \mathbf{u}_1 \end{array}$$
$$\begin{bmatrix} x_2 \\ y_2 \end{bmatrix} = \begin{bmatrix} 4 & -1 \\ 2 & 1 \end{bmatrix} \begin{bmatrix} 16 \\ 20 \end{bmatrix} = \begin{bmatrix} 44 \\ 52 \end{bmatrix}$$

$$\begin{array}{ccc} \mathbf{u}_3 & A & \mathbf{u}_2 \end{array}$$
$$\begin{bmatrix} x_3 \\ y_3 \end{bmatrix} = \begin{bmatrix} 4 & -1 \\ 2 & 1 \end{bmatrix} \begin{bmatrix} 44 \\ 52 \end{bmatrix} = \begin{bmatrix} 124 \\ 140 \end{bmatrix}$$

and so on. However, the matrix form of the solution does not provide explicit formulas for x_n and y_n. How can we use the matrix form of the solution to find formulas for x_n and y_n?

Let $\mathbf{u}_n = A^n \mathbf{u}_0$ be the matrix form of the solution to an arbitrary system of first-order recurrence equations and suppose that the matrix A is diagonalizable. That is, there exists a diagonal matrix D and an invertible matrix P satisfying

$$A = PDP^{-1}$$

Then the powers of A can be expressed in terms of the powers of D (see Example 5 in Section 7-2):

$$A^2 = (PDP^{-1})(PDP^{-1}) = PD^2P^{-1}$$
$$A^3 = (PDP^{-1})(PD^2P^{-1}) = PD^3P^{-1}$$
$$\cdot$$
$$\cdot$$
$$\cdot$$
$$A^n = (PDP^{-1})(PD^{n-1}P^{-1}) = PD^nP^{-1}$$

Furthermore, since D is a diagonal matrix, the powers of D are easy to compute:

$$D = \begin{bmatrix} \lambda_1 & 0 \\ 0 & \lambda_2 \end{bmatrix}$$

$$D^2 = \begin{bmatrix} \lambda_1^2 & 0 \\ 0 & \lambda_2^2 \end{bmatrix}$$

.
.
.

$$D^n = \begin{bmatrix} \lambda_1^n & 0 \\ 0 & \lambda_2^n \end{bmatrix}$$

Thus, the matrix form of the solution to a system of first-order recurrence equations can be written as

$$\mathbf{u}_n = A^n \mathbf{u}_0 = PD^nP^{-1}\mathbf{u}_0 = P\begin{bmatrix} \lambda_1^n & 0 \\ 0 & \lambda_2^n \end{bmatrix}P^{-1}\mathbf{u}_0 \tag{7}$$

Explicit formulas for x_n and y_n can be obtained by completing the matrix multiplication in (7).

Example 8   Refer to system (6) in Example 7. Find explicit formulas for x_n and y_n.

Solution   The characteristic polynomial for

$$A = \begin{bmatrix} 4 & -1 \\ 2 & 1 \end{bmatrix}$$

is

$$p(\lambda) = \det(\lambda I - A) = \lambda^2 - 5\lambda + 6$$

The eigenvalues and basis for the corresponding eigenspaces are

$$\lambda_1 = 2 \qquad B_{\lambda_1} = \left\{ \begin{bmatrix} 1 \\ 2 \end{bmatrix} \right\} \qquad \text{and} \qquad \lambda_2 = 3 \qquad B_{\lambda_2} = \left\{ \begin{bmatrix} 1 \\ 1 \end{bmatrix} \right\}$$

Thus, we can write

$$A = PDP^{-1} = \begin{bmatrix} 1 & 1 \\ 2 & 1 \end{bmatrix}\begin{bmatrix} 2 & 0 \\ 0 & 3 \end{bmatrix}\begin{bmatrix} -1 & 1 \\ 2 & -1 \end{bmatrix}$$

Using (7), we have

$$\mathbf{u}_n = PD^nP^{-1}\mathbf{u}_0$$

$$= \begin{bmatrix} 1 & 1 \\ 2 & 1 \end{bmatrix}\begin{bmatrix} 2^n & 0 \\ 0 & 3^n \end{bmatrix}\begin{bmatrix} -1 & 1 \\ 2 & -1 \end{bmatrix}\begin{bmatrix} 6 \\ 8 \end{bmatrix} \qquad \mathbf{u}_0 = \begin{bmatrix} 6 \\ 8 \end{bmatrix} \text{ from Example 7}$$

$$= \begin{bmatrix} 2 \cdot 2^n + 4 \cdot 3^n \\ 4 \cdot 2^n + 4 \cdot 3^n \end{bmatrix}$$

Since

$$\mathbf{u}_n = \begin{bmatrix} x_n \\ y_n \end{bmatrix}$$

the explicit solution to system (6) in Example 7 is

$$x_n = 2 \cdot 2^n + 4 \cdot 3^n$$
$$y_n = 4 \cdot 2^n + 4 \cdot 3^n$$

Check: This solution can be checked by direct substitution in (6):

$$x_n \overset{?}{=} 4x_{n-1} - y_{n-1}$$
$$2 \cdot 2^n + 4 \cdot 3^n \overset{?}{=} 4(2 \cdot 2^{n-1} + 4 \cdot 3^{n-1}) - (4 \cdot 2^{n-1} + 4 \cdot 3^{n-1})$$
$$\overset{?}{=} 4 \cdot 2^{n-1} + 12 \cdot 3^{n-1}$$
$$\overset{\checkmark}{=} 2 \cdot 2^n + 4 \cdot 3^n$$

$$y_n \overset{?}{=} 2x_{n-1} + y_{n-1}$$
$$4 \cdot 2^n + 4 \cdot 3^n \overset{?}{=} 2(2 \cdot 2^{n-1} + 4 \cdot 3^{n-1}) + (4 \cdot 2^{n-1} + 4 \cdot 3^{n-1})$$
$$\overset{?}{=} 8 \cdot 2^{n-1} + 12 \cdot 3^{n-1}$$
$$\overset{\checkmark}{=} 4 \cdot 2^n + 4 \cdot 3^n$$

Problem 8   Find the explicit solution to the system in Problem 7.   ∎

Note: When asked to *solve* a system of first-order recurrence equations, you should always find explicit formulas for x_n and y_n, as we did in Example 8. The matrix form of the solution is a preliminary, but essential, step in the solution process.

Example 9
Population
Growth

An ecological system contains two species, rabbits and foxes, that interact with each other in a predator–prey relationship. Suppose that the number of rabbits, x_n (in hundreds), and the number of foxes, y_n (in hundreds), after n years in a given ecological system satisfy the system of recurrence equations

$$x_n = 1.2x_{n-1} - 0.8y_{n-1} \qquad x_0 = 250$$
$$y_n = 0.1x_{n-1} + 0.6y_{n-1} \qquad y_0 = 50$$

(A) Solve the system.
(B) Find the number of rabbits and the number of foxes after 5, 10, 15, and 20 years.

Solution   (A) The matrix form of this system is

$$\mathbf{u}_n = A\mathbf{u}_{n-1} \qquad \mathbf{u}_0 = \begin{bmatrix} 250 \\ 50 \end{bmatrix}$$

where

$$A = \begin{bmatrix} 1.2 & -0.8 \\ 0.1 & 0.6 \end{bmatrix} \quad \text{and} \quad \mathbf{u}_n = \begin{bmatrix} x_n \\ y_n \end{bmatrix}$$

Omitting the details of the diagonalization process (which you should supply), we have

$$
\begin{array}{cccc}
& P & D & P^{-1} \\
A = & \begin{bmatrix} 4 & 2 \\ 1 & 1 \end{bmatrix} & \begin{bmatrix} 1 & 0 \\ 0 & 0.8 \end{bmatrix} & \begin{bmatrix} 0.5 & -1 \\ -0.5 & 2 \end{bmatrix}
\end{array}
$$

and

$$
\begin{array}{ccccc}
& P & D^n & P^{-1} & \mathbf{u}_0 \\
\mathbf{u}_n = A^n \mathbf{u}_0 = & \begin{bmatrix} 4 & 2 \\ 1 & 1 \end{bmatrix} & \begin{bmatrix} 1 & 0 \\ 0 & (0.8)^n \end{bmatrix} & \begin{bmatrix} 0.5 & -1 \\ -0.5 & 2 \end{bmatrix} & \begin{bmatrix} 250 \\ 50 \end{bmatrix}
\end{array}
$$

$$
= \begin{bmatrix} 300 - 50(0.8)^n \\ 75 - 25(0.8)^n \end{bmatrix}
$$

Thus, the solution to the system is

$$x_n = 300 - 50(0.8)^n$$

$$y_n = 75 - 25(0.8)^n$$

(B) The table lists the values of x_n and y_n (rounded to two decimal places) and the corresponding rabbit and fox populations.

n	x_n	y_n	Number of Rabbits	Number of Foxes
0	250	50	25,000	5,000
5	283.62	66.81	28,362	6,681
10	294.63	72.32	29,463	7,232
15	298.24	74.12	29,824	7,412
20	299.42	74.71	29,942	7,471

The values in the table indicate that in 20 years the rabbit population increases from 25,000 to approximately 30,000 and the fox population increases from 5,000 to approximately 7,500. (Those who have studied calculus can verify that $x_n \to 300$ and $y_n \to 75$ as $n \to \infty$.)

Problem 9   Repeat Example 9 if

$$x_n = 1.2x_{n-1} - 0.6y_{n-1} \qquad x_0 = 500$$

$$y_n = 0.2x_{n-1} + 0.4y_{n-1} \qquad y_0 = 20$$

■

▪ Second-Order Recurrence Equations

A recurrence equation of the form

$$x_n = ax_{n-1} + bx_{n-2}$$

is called a **second-order recurrence equation.** One method for solving second-order recurrence equations is to convert the *single* second-order recurrence equation into a *system* of first-order recurrence equations. This method is illustrated in Example 10.

Example 10   Solve the second-order recurrence equation

$$x_n = x_{n-1} + x_{n-2} \qquad n \geq 2 \tag{8}$$

where $x_0 = 1$ and $x_1 = 1$.

Solution   First, notice that a second-order recurrence relation requires two initial conditions, one for x_0 and one for x_1. The first few values of x_n are

1, 1, 2, 3, 5, 8, 13, 21, 34

This is the famous sequence of **Fibonacci numbers** (each number beyond the second is the sum of the preceding two numbers).

In order to use matrix methods to solve (8), we convert this second-order recurrence equation to a system of first-order recurrence equations. If we let $y_n = x_{n-1}$, then $y_{n-1} = x_{n-2}$ and (8) can be written as the system

$$x_n = x_{n-1} + y_{n-1} \qquad y_{n-1} = x_{n-2}$$
$$y_n = x_{n-1}$$

or, equivalently, as the matrix equation

$$\mathbf{u}_n = A\mathbf{u}_{n-1} \tag{9}$$

where

$$A = \begin{bmatrix} 1 & 1 \\ 1 & 0 \end{bmatrix} \quad \text{and} \quad \mathbf{u}_n = \begin{bmatrix} x_n \\ y_n \end{bmatrix}$$

The initial condition for (9) is

$$\mathbf{u}_1 = \begin{bmatrix} x_1 \\ y_1 \end{bmatrix} = \begin{bmatrix} x_1 \\ x_0 \end{bmatrix} = \begin{bmatrix} 1 \\ 1 \end{bmatrix} \qquad y_n = x_{n-1} \text{ implies } y_1 = x_0 = 1$$

(The vector $\mathbf{u}_0$ is undefined since $y_0 = x_{-1}$.) Since the hypothesis of Theorem 4 requires that the initial condition be stated in terms of $\mathbf{u}_0$, we cannot use Theorem 4 directly to solve (9). Instead, we write

$$\mathbf{u}_2 = A\mathbf{u}_1$$
$$\mathbf{u}_3 = A\mathbf{u}_2 = A(A\mathbf{u}_1) = A^2\mathbf{u}_1$$
$$\mathbf{u}_4 = A\mathbf{u}_3 = A(A^2\mathbf{u}_1) = A^3\mathbf{u}_1$$

.

.

.

$$\mathbf{u}_n = A\mathbf{u}_{n-1} = A(A^{n-2}\mathbf{u}_1) = A^{n-1}\mathbf{u}_1$$

and conclude that the matrix form of the solution to (9) is

$$\mathbf{u}_n = A^{n-1}\mathbf{u}_1 \qquad \mathbf{u}_1 = \begin{bmatrix} 1 \\ 1 \end{bmatrix}$$

We can now find the (explicit) solution to (9) by diagonalizing A. Omitting the details (which you should supply), we have

$$
A = \overset{P}{\begin{bmatrix} \dfrac{1+\sqrt{5}}{2} & \dfrac{1-\sqrt{5}}{2} \\ 1 & 1 \end{bmatrix}}
\overset{D}{\begin{bmatrix} \dfrac{1+\sqrt{5}}{2} & 0 \\ 0 & \dfrac{1-\sqrt{5}}{2} \end{bmatrix}}
\overset{P^{-1}}{\begin{bmatrix} \dfrac{1}{\sqrt{5}} & \dfrac{\sqrt{5}-1}{2\sqrt{5}} \\ -\dfrac{1}{\sqrt{5}} & \dfrac{\sqrt{5}+1}{2\sqrt{5}} \end{bmatrix}}
$$

and

$$\mathbf{u}_n = A^{n-1}\mathbf{u}_1 = PD^{n-1}P^{-1}\mathbf{u}_1$$

$$
= \begin{bmatrix} \dfrac{1+\sqrt{5}}{2} & \dfrac{1-\sqrt{5}}{2} \\ 1 & 1 \end{bmatrix}
\begin{bmatrix} \left(\dfrac{1+\sqrt{5}}{2}\right)^{n-1} & 0 \\ 0 & \left(\dfrac{1-\sqrt{5}}{2}\right)^{n-1} \end{bmatrix}
\begin{bmatrix} \dfrac{1}{\sqrt{5}} & \dfrac{\sqrt{5}-1}{2\sqrt{5}} \\ -\dfrac{1}{\sqrt{5}} & \dfrac{\sqrt{5}+1}{2\sqrt{5}} \end{bmatrix}
\begin{bmatrix} 1 \\ 1 \end{bmatrix}
$$

$$
= \begin{bmatrix} \dfrac{1}{\sqrt{5}}\left(\left(\dfrac{1+\sqrt{5}}{2}\right)^{n+1} - \left(\dfrac{1-\sqrt{5}}{2}\right)^{n+1}\right) \\ \dfrac{1}{\sqrt{5}}\left(\left(\dfrac{1+\sqrt{5}}{2}\right)^{n} - \left(\dfrac{1-\sqrt{5}}{2}\right)^{n}\right) \end{bmatrix}
$$

Since

$$\mathbf{u}_n = \begin{bmatrix} x_n \\ y_n \end{bmatrix}$$

it follows that the solution to (8) is given by the first component of $\mathbf{u}_n$. That is, the solution to

$$x_n = x_{n-1} + x_{n-2} \qquad x_0 = 1, \quad x_1 = 1$$

is

$$x_n = \frac{1}{\sqrt{5}}\left(\left(\frac{1+\sqrt{5}}{2}\right)^{n+1} - \left(\frac{1-\sqrt{5}}{2}\right)^{n+1}\right) \tag{10}$$

This formula for x_n is valid for all $n \geq 0$. In particular,

$$x_0 = \frac{1}{\sqrt{5}}\left(\left(\frac{1+\sqrt{5}}{2}\right) - \left(\frac{1-\sqrt{5}}{2}\right)\right) = \frac{1}{\sqrt{5}}(\sqrt{5}) = 1$$

and

$$x_1 = \frac{1}{\sqrt{5}}\left(\left(\frac{1+\sqrt{5}}{2}\right)^2 - \left(\frac{1-\sqrt{5}}{2}\right)^2\right)$$

$$= \frac{1}{\sqrt{5}}\left(\frac{1+2\sqrt{5}+5}{4} - \frac{1-2\sqrt{5}+5}{4}\right) = \frac{1}{\sqrt{5}}(\sqrt{5}) = 1$$

Since x_n is always an integer (Why?), similar simplification must take place in (10) for $n \geq 2$. [Verify this for $n = 2$ by using (10) to compute x_2.] Thus, the formula for the solution to a second-order recurrence equation may involve irrational numbers, even though all the values of x_n are integers.

Problem 10   Rework Example 10 if the initial conditions are changed to $x_0 = 1$ and $x_1 = 3$. [*Hint:* The matrices A, P, and D remain the same. Only the vector $\mathbf{u}_1$ is changed.]

■

▪ Remarks

Two observations must be made about the matrix methods we have presented in this section. First, these methods can be applied only if the matrix A is diagonalizable. If A is not diagonalizable, then other methods of solution must be used. We will leave the discussion of these methods to more advanced texts.

Second, we have restricted our attention to systems with two equations and to second-order equations for simplicity. The methods we have presented are easily generalized to larger systems and higher-order equations. Some 3×3 systems and third-order equations are considered in Exercise 9-2.

Answers to Matched Problems

6. $100(\frac{1}{2})^n$ **7.** $\begin{bmatrix} 6 & -1 \\ 12 & -1 \end{bmatrix}^n \begin{bmatrix} 3 \\ 10 \end{bmatrix}$ **8.** $x_n = 2^n + 2 \cdot 3^n,\ y_n = 4 \cdot 2^n + 6 \cdot 3^n$

9. (A) $x_n = 720 - 220(0.6)^n,\quad y_n = 240 - 220(0.6)^n$

(B) n	x_n	y_n	Number of Rabbits	Number of Foxes
0	500	20	50,000	2,000
5	702.89	222.89	70,289	22,289
10	718.67	238.67	71,867	23,867
15	719.90	239.90	71,990	23,990
20	719.99	239.99	71,999	23,999

10. $x_n = \left(\frac{1+\sqrt{5}}{2}\right)^{n+1} + \left(\frac{1-\sqrt{5}}{2}\right)^{n+1}$

❚❚ Exercise 9-2

A In Problems 1–4, solve the first-order recurrence equation.

1. $x_n = 3x_{n-1}, \quad x_0 = 10$

2. $x_n = \frac{1}{3}x_{n-1}, \quad x_0 = 5$

3. $x_n = -\frac{1}{2}x_{n-1}, \quad x_0 = 3$

4. $x_n = -4x_{n-1}, \quad x_0 = 11$

In Problems 5 and 6, write the system of first-order recurrence equations in matrix form and find (x_1, y_1), (x_2, y_2), and (x_3, y_3).

5. $x_n = 2x_{n-1} + 2y_{n-1}$
$y_n = -x_{n-1} + 5y_{n-1}$
$x_0 = 1, \quad y_0 = 0$

6. $x_n = 11x_{n-1} - 9y_{n-1}$
$y_n = 6x_{n-1} - 4y_{n-1}$
$x_0 = 0, \quad y_0 = 1$

7. Find the matrix form of the solution to the system in Problem 5.

8. Find the matrix form of the solution to the system in Problem 6.

9. Represent the matrix for the system in Problem 5 in the form PDP^{-1} where D is a diagonal matrix.

10. Represent the matrix for the system in Problem 6 in the form PDP^{-1} where D is a diagonal matrix.

11. Find the (explicit) solution to the system in Problem 5.

12. Find the (explicit) solution to the system in Problem 6.

B In Problems 13–18, solve the system of first-order recurrence equations.

13. $x_n = 2x_{n-1} + 2y_{n-1}$
$y_n = x_{n-1} + y_{n-1}$
$x_0 = 2, \quad y_0 = 1$

14. $x_n = x_{n-1} - y_{n-1}$
$y_n = 3x_{n-1} - 3y_{n-1}$
$x_0 = 1, \quad y_0 = 3$

15. $x_n = 5x_{n-1} - 12y_{n-1}$
$y_n = 2x_{n-1} - 5y_{n-1}$
$x_0 = 5, \quad y_0 = 2$

16. $x_n = 3x_{n-1} - 2y_{n-1}$
$y_n = 4x_{n-1} - 3y_{n-1}$
$x_0 = 2, \quad y_0 = 3$

17. $x_n = 2.5x_{n-1} - 2y_{n-1}$
$y_n = x_{n-1} - 0.5y_{n-1}$
$x_0 = 5, \quad y_0 = 3$

18. $x_n = 1.25x_{n-1} - 0.5y_{n-1}$
$y_n = x_{n-1} - 0.25y_{n-1}$
$x_0 = 4, \quad y_0 = 7$

In Problems 19–24, solve the second-order difference equation.

19. $x_n = 3x_{n-1} - 2x_{n-2}, \quad x_0 = 2, \quad x_1 = 3$

20. $x_n = 4x_{n-1} - 3x_{n-2}, \quad x_0 = 3, \quad x_1 = 5$

21. $x_n = 7x_{n-1} - 12x_{n-2}, \quad x_0 = 2, \quad x_1 = 7$

22. $x_n = 5x_{n-1} - 6x_{n-2}, \quad x_0 = 2, \quad x_1 = 5$

23. $x_n = 2x_{n-1} + x_{n-2}, \quad x_0 = 1, \quad x_1 = 1$

24. $x_n = 2x_{n-1} + 2x_{n-2}, \quad x_0 = 1, \quad x_1 = 2$

C If A, P, and D are 3×3 matrices satisfying $A = PDP^{-1}$, then it can be shown that

$$A^n = P \begin{bmatrix} \lambda_1^n & 0 & 0 \\ 0 & \lambda_2^n & 0 \\ 0 & 0 & \lambda_3^n \end{bmatrix} P^{-1} \quad \text{where} \quad D = \begin{bmatrix} \lambda_1 & 0 & 0 \\ 0 & \lambda_2 & 0 \\ 0 & 0 & \lambda_3 \end{bmatrix}$$

Use this fact to solve the systems of first-order recurrence equations in Problems 25–28.

25. $x_n = 2x_{n-1} + 4y_{n-1} - 2z_{n-1}$
$\quad\ y_n = 4x_{n-1} + 8y_{n-1} - 8z_{n-1}$
$\quad\ z_n = 3x_{n-1} + 6y_{n-1} - 5z_{n-1}$
$\quad\ x_0 = 1,\ \ y_0 = 1,\ \ z_0 = 1$

26. $x_n = 3x_{n-1} + 2y_{n-1} - 3z_{n-1}$
$\quad\ y_n = \ \ x_{n-1} + 2y_{n-1} - \ \ z_{n-1}$
$\quad\ z_n = 2x_{n-1} + 2y_{n-1} - 2z_{n-1}$
$\quad\ x_0 = 1,\ \ y_0 = 1,\ \ z_0 = 1$

27. $x_n = \ \ x_{n-1} + 4y_{n-1} - 2z_{n-1}$
$\quad\ y_n = -x_{n-1} + 5y_{n-1} - \ \ z_{n-1}$
$\quad\ z_n = -x_{n-1} + 2y_{n-1} + 2z_{n-1}$
$\quad\ x_0 = 1,\ \ y_0 = 1,\ \ z_0 = 2$

28. $x_n = 3x_{n-1} - y_{n-1} + \ \ z_{n-1}$
$\quad\ y_n = \ \ x_{n-1} + y_{n-1} + \ \ z_{n-1}$
$\quad\ z_n = \ \ x_{n-1} - y_{n-1} + 3z_{n-1}$
$\quad\ x_0 = 1,\ \ y_0 = 2,\ \ z_0 = 3$

29. Show that the third-order recurrence equation

$$x_n = ax_{n-1} + bx_{n-2} + cx_{n-3}$$

with initial conditions $x_0 = k$, $x_1 = \ell$, $x_2 = m$ is equivalent to the system

$$x_n = ax_{n-1} + by_{n-1} + cz_{n-1}$$
$$y_n = x_{n-1}$$
$$z_n = \qquad\quad y_{n-1}$$

with initial conditions $x_2 = m$, $y_2 = \ell$, $z_2 = k$.

30. Show that the matrix form of the solution to the system in Problem 29 is
$\mathbf{u}_n = A^{n-2}\mathbf{u}_2$ where

$$A = \begin{bmatrix} a & b & c \\ 1 & 0 & 0 \\ 0 & 1 & 0 \end{bmatrix} \quad \text{and} \quad \mathbf{u}_2 = \begin{bmatrix} m \\ \ell \\ k \end{bmatrix}$$

31. Show that the characteristic polynomial for the matrix A in Problem 30 is

$$p(\lambda) = \lambda^3 - a\lambda^2 - b\lambda - c$$

32. If λ is an eigenvalue of the matrix A in Problem 30, show that $\mathbf{u}$ is an eigenvector associated with λ, where

$$\mathbf{u} = \begin{bmatrix} \lambda^2 \\ \lambda \\ 1 \end{bmatrix}$$

Use the results in Problems 29–32 to solve the following third-order recurrence equations:

33. $x_n = 6x_{n-1} - 11x_{n-2} + 6x_{n-3},\quad x_0 = 3,\quad x_1 = 6,\quad x_2 = 14$

34. $x_n = 9x_{n-1} - 26x_{n-2} + 24x_{n-3},\quad x_0 = 1,\quad x_1 = 1,\quad x_2 = 1$

▌□▌ Applications

35. *Population growth.* The number of bacteria in a culture increases 50% every hour. If there are 200 bacteria initially, how many will there be after n hours?

36. *Population growth.* Repeat Problem 35 if the number of bacteria increases 10% every hour.

37. *Population growth.* The number of rabbits, x_n (in thousands), and the number of foxes, y_n (in thousands), in an ecological system after n years satisfy the system

$$x_n = 0.8x_{n-1} - 0.3y_{n-1}$$
$$y_n = 0.2x_{n-1} + 0.3y_{n-1}$$

Initially, there are 80,000 rabbits and 60,000 foxes. Solve the system and determine the population of each species after 2, 4, 6, 8, and 10 years. Based on these calculations, what is happening to the population of each species?

38. *Population growth.* Repeat Problem 37 if x_n and y_n satisfy

$$x_n = 0.8x_{n-1} - 0.4y_{n-1}$$
$$y_n = 0.2x_{n-1} + 0.2y_{n-1}$$

and there are 100,000 rabbits and 60,000 foxes initially.

39. *Interrelated markets.* The prices p_n and q_n (in dollars) of two interrelated products after n years satisfy

$$p_n = 0.6p_{n-1} + 0.2q_{n-1}$$
$$q_n = 0.2p_{n-1} + 0.9q_{n-1}$$

The initial prices are $p_0 = 20$ and $q_0 = 90$. Solve the system and determine the prices after 2, 4, 6, 8, and 10 years. Based on these calculations, what is happening to these prices?

40. *Interrelated markets.* Repeat Problem 39 if p_n and q_n satisfy

$$p_n = 0.8p_{n-1} + 0.4q_{n-1}$$
$$q_n = 0.15p_{n-1} + 0.7q_{n-1}$$

and the initial prices are $p_0 = 60$ and $q_0 = 5$.

❙9-3❙ Differential Equations

- First-Order Differential Equations
- Systems of Differential Equations
- Higher-Order Differential Equations

Many important applications of mathematics to physics, engineering, biology, and the social sciences involve equations relating a function and one or more of its *derivatives*. Such equations are called *differential equations*. There are many different types of differential equations and many techniques for solving these equations. In this section, we will consider some differential equations that can be solved by matrix methods. Once again, we will see that introducing matrix notation and diagonalizing an appropriate matrix will change a complicated problem into a simple one. A calculus course including the exponential function is a prerequisite for this section.

▪ First-Order Differential Equations

We begin by reviewing some material from calculus. The equation

$$y' = ky \qquad (1)$$

where k is a (fixed) constant and $y = f(t)$ is an (unknown) function is called a **first-order differential equation.*** The **general solution** of (1) is the function

$$y = Ce^{kt} \qquad (2)$$

where C is an arbitrary constant. A function of the form (2) which also satisfies the **initial condition** $y(t_0) = y_0$ is called a **particular solution** of (1).

Example 11   Find the general solution of the differential equation

$$y' = 2y$$

and find the particular solution that satisfies $y(0) = 5$.

Solution   Using (2) with $k = 2$, the general solution is

$$y = Ce^{2t} \qquad (3)$$

where C is an arbitrary constant. To find the particular solution satisfying $y(0) = 5$, we substitute $y = 5$ and $t = 0$ in (3) and solve for C:

$$y = Ce^{2t}$$
$$5 = Ce^{2(0)} = Ce^0 = C$$

Thus, the particular solution satisfying $y(0) = 5$ is

$$y = 5e^{2t}$$

Problem 11   Find the general solution of the differential equation

$$y' = 3y$$

and find the particular solution that satisfies $y(0) = 8$.   ▮

▪ Systems of Differential Equations

A system of equations of the form

---

* Equation (1) is not the most general form of a first-order differential equation. In a differential equations course, (1) would be called a *homogeneous first-order linear differential equation with constant coefficients*. We will not need this advanced terminology.

$$y_1' = a_{11}y_1 + a_{12}y_2 + \cdots + a_{1n}y_n$$
$$y_2' = a_{21}y_1 + a_{22}y_2 + \cdots + a_{2n}y_n \qquad (4)$$

.

.

.

$$y_n' = a_{n1}y_1 + a_{n2}y_2 + \cdots + a_{nn}y_n$$

where a_{11}, a_{12}, . . . , a_{nn} are (fixed) constants and $y_1 = f_1(t)$, $y_2 = f_2(t)$, . . . , $y_n = f_n(t)$ are (unknown) functions, is called a **system of differential equations** or, more formally, a **linear system of first-order differential equations.** System (4) can be written in **matrix form** as

$$\mathbf{y}' = A\mathbf{y}$$

where

$$A = \begin{bmatrix} a_{11} & a_{12} & \cdots & a_{1n} \\ a_{21} & a_{22} & \cdots & a_{2n} \\ \cdot & \cdot & & \cdot \\ \cdot & \cdot & & \cdot \\ \cdot & \cdot & & \cdot \\ a_{n1} & a_{n2} & \cdots & a_{nn} \end{bmatrix} \quad \text{is the \textbf{matrix of the system,}}$$

$$\mathbf{y} = \begin{bmatrix} y_1 \\ y_2 \\ \cdot \\ \cdot \\ \cdot \\ y_n \end{bmatrix} \quad \text{is an (unknown) \textbf{vector function,} and}$$

$$\mathbf{y}' = \begin{bmatrix} y_1' \\ y_2' \\ \cdot \\ \cdot \\ \cdot \\ y_n' \end{bmatrix} \quad \text{is the \textbf{vector derivative} of } \mathbf{y}.$$

Our goal is to use matrix diagonalization techniques to solve systems of the form in (4). We begin by considering a simple example.

Example 12 (A) Solve the system

$$y_1' = -y_1$$
$$y_2' = y_2 \qquad (5)$$
$$y_3' = 2y_3$$

(B) Write system (5) and its solution in matrix form.

Solution   (A) Each equation in this system is of the form in (1). Thus, using (2), we have

$$y_1 = C_1 e^{-t} \qquad y_2 = C_2 e^t \qquad y_3 = C_3 e^{2t}$$

where C_1, C_2, and C_3 are arbitrary constants.

(B) The matrix form of system (5) is

$$\begin{bmatrix} y_1' \\ y_2' \\ y_3' \end{bmatrix} = \begin{bmatrix} -y_1 \\ y_2 \\ 2y_3 \end{bmatrix} = \begin{bmatrix} -1 & 0 & 0 \\ 0 & 1 & 0 \\ 0 & 0 & 2 \end{bmatrix} \begin{bmatrix} y_1 \\ y_2 \\ y_3 \end{bmatrix}$$

or

$$\mathbf{y}' = A\mathbf{y}$$

where

$$A = \begin{bmatrix} -1 & 0 & 0 \\ 0 & 1 & 0 \\ 0 & 0 & 2 \end{bmatrix} \qquad \mathbf{y} = \begin{bmatrix} y_1 \\ y_2 \\ y_3 \end{bmatrix} \qquad \mathbf{y}' = \begin{bmatrix} y_1' \\ y_2' \\ y_3' \end{bmatrix}$$

Using the results in part (A), the solution to (5) can be written as

$$\mathbf{y} = \begin{bmatrix} C_1 e^{-t} \\ C_2 e^t \\ C_3 e^{2t} \end{bmatrix}$$

Problem 12   Repeat Example 12 for the system

$$y_1' = -y_1$$
$$y_2' = y_2$$
$$y_3' = 3y_3$$

System (5) in Example 12 was easy to solve because the matrix of the system was a diagonal matrix. Theorem 5 (page 603) shows that any system with a diagonal matrix can be solved in the same manner. The proof of Theorem 5 is omitted.

If the matrix of a system of differential equations is not a diagonal matrix, then Theorem 5 cannot be used to solve the system. However, if the matrix of a system is diagonalizable, then it turns out that the system can always be transformed into a new system that can be solved by Theorem 5. The transformation process involves some properties of vector functions which are stated in Theorem 6 without proof.

Theorem 5

The general solution of the system

$$\mathbf{y}' = D\mathbf{y} \qquad \text{where} \quad D = \begin{bmatrix} \lambda_1 & 0 & \cdots & 0 \\ 0 & \lambda_2 & \cdots & 0 \\ \cdot & \cdot & & \cdot \\ \cdot & \cdot & & \cdot \\ \cdot & \cdot & & \cdot \\ 0 & 0 & \cdots & \lambda_n \end{bmatrix}$$

is

$$y_1 = C_1 e^{\lambda_1 t}, \quad y_2 = C_2 e^{\lambda_2 t}, \quad \cdots, \quad y_n = C_n e^{\lambda_n t}$$

or, in matrix form,

$$\mathbf{y} = \begin{bmatrix} C_1 e^{\lambda_1 t} \\ C_2 e^{\lambda_2 t} \\ \cdot \\ \cdot \\ \cdot \\ C_n e^{\lambda_n t} \end{bmatrix}$$

where $C_1, C_2, \ldots, C_n$ are arbitrary constants.

Theorem 6

If the vector functions

$$\mathbf{v} = \begin{bmatrix} v_1 \\ v_2 \\ \cdot \\ \cdot \\ \cdot \\ v_n \end{bmatrix} \qquad \text{and} \qquad \mathbf{y} = \begin{bmatrix} y_1 \\ y_2 \\ \cdot \\ \cdot \\ \cdot \\ y_n \end{bmatrix}$$

where $v_1 = g_1(t), \ldots, v_n = g_n(t)$ and $y_1 = f_1(t), \ldots, y_n = f_n(t)$, are related by either of the equations

$$\mathbf{v} = P^{-1}\mathbf{y} \qquad \text{or} \qquad \mathbf{y} = P\mathbf{v}$$

where P is an $n \times n$ matrix of constants, then the derivatives of $\mathbf{v}$ and $\mathbf{y}$ satisfy

$$\mathbf{v}' = P^{-1}\mathbf{y}' \qquad \text{and} \qquad \mathbf{y}' = P\mathbf{v}'$$

Now, suppose we are given a system of the form

$$\mathbf{y}' = A\mathbf{y} \tag{6}$$

where $A = PDP^{-1}$ and D is a diagonal matrix. We want to transform (6) into a system of the form

$$\mathbf{v}' = D\mathbf{v}$$

The equations in Theorem 6 are the keys to this transformation process:

$$\mathbf{y}' = A\mathbf{y} \qquad \text{Substitute } A = PDP^{-1}.$$
$$\mathbf{y}' = PDP^{-1}\mathbf{y} \qquad \text{Left-multiply by } P^{-1}.$$
$$P^{-1}\mathbf{y}' = DP^{-1}\mathbf{y} \qquad \text{Substitute } \mathbf{v} = P^{-1}\mathbf{y} \text{ and } \mathbf{v}' = P^{-1}\mathbf{y}' \text{ (see Theorem 6).}$$
$$\mathbf{v}' = D\mathbf{v} \tag{7}$$

Since D is a diagonal matrix, system (7) can be solved by applying Theorem 5. The general solution to (7) is

$$\mathbf{v} = \begin{bmatrix} C_1 e^{\lambda_1 t} \\ C_2 e^{\lambda_2 t} \\ \cdot \\ \cdot \\ \cdot \\ C_n e^{\lambda_n t} \end{bmatrix}$$

where

$$D = \begin{bmatrix} \lambda_1 & 0 & \cdots & 0 \\ 0 & \lambda_2 & \cdots & 0 \\ \cdot & \cdot & & \cdot \\ \cdot & \cdot & & \cdot \\ \cdot & \cdot & & \cdot \\ 0 & 0 & \cdots & \lambda_n \end{bmatrix}$$

Since $\mathbf{v} = P^{-1}\mathbf{y}$ implies that $\mathbf{y} = P\mathbf{v}$, it follows that the general solution to (6) is

$$\mathbf{y} = P\mathbf{v} = P\begin{bmatrix} C_1 e^{\lambda_1 t} \\ C_2 e^{\lambda_2 t} \\ \cdot \\ \cdot \\ \cdot \\ C_n e^{\lambda_n t} \end{bmatrix}$$

We have just proved Theorem 7, which is stated at the top of the next page.

Of course, the matrices P and D in Theorem 7 are very familiar. The diagonal entries of D are the eigenvalues of A and the columns of P are the corresponding eigenvectors. Theorem 7 can be used to solve any system of differential equations — provided that the matrix of the system is diagonalizable.

Theorem 7	If $A = PDP^{-1}$, then the general solution of the system

$$\mathbf{y}' = A\mathbf{y}$$

is

$$\mathbf{y} = P \begin{bmatrix} C_1 e^{\lambda_1 t} \\ C_2 e^{\lambda_2 t} \\ \cdot \\ \cdot \\ \cdot \\ C_n e^{\lambda_n t} \end{bmatrix} \qquad \text{where} \quad D = \begin{bmatrix} \lambda_1 & 0 & \cdots & 0 \\ 0 & \lambda_2 & \cdots & 0 \\ \cdot & \cdot & & \cdot \\ \cdot & \cdot & & \cdot \\ \cdot & \cdot & & \cdot \\ 0 & 0 & \cdots & \lambda_n \end{bmatrix}$$

and $C_1, C_2, \ldots, C_n$ are arbitrary constants.

Example 13 Solve the system

$$\begin{aligned} y_1' &= 4y_1 - 6y_2 - 2y_3 \\ y_2' &= 4y_1 - 7y_2 - 4y_3 \\ y_3' &= -3y_1 + 6y_2 + 5y_3 \end{aligned} \qquad (8)$$

Solution The matrix form of system (8) is $\mathbf{y}' = A\mathbf{y}$ where

$$A = \begin{bmatrix} 4 & -6 & -2 \\ 4 & -7 & -4 \\ -3 & 6 & 5 \end{bmatrix} \qquad \mathbf{y} = \begin{bmatrix} y_1 \\ y_2 \\ y_3 \end{bmatrix} \qquad \mathbf{y}' = \begin{bmatrix} y_1' \\ y_2' \\ y_3' \end{bmatrix}$$

The characteristic polynomial for A is

$$p(\lambda) = \det(\lambda I - A) = (\lambda + 1)(\lambda - 1)(\lambda - 2)$$

The eigenvalues and corresponding eigenvectors are

$$\lambda_1 = -1 \qquad\qquad \lambda_2 = 1 \qquad\qquad \lambda_3 = 2$$

$$\mathbf{u}_1 = \begin{bmatrix} -2 \\ -2 \\ 1 \end{bmatrix} \qquad \mathbf{u}_2 = \begin{bmatrix} 2 \\ 1 \\ 0 \end{bmatrix} \qquad \mathbf{u}_3 = \begin{bmatrix} 1 \\ 0 \\ 1 \end{bmatrix}$$

Since A is a 3×3 matrix with three distinct eigenvalues, A is diagonalizable and we can write

$$A = PDP^{-1}$$

where

$$\begin{matrix} \lambda_1 & \lambda_2 & \lambda_3 \end{matrix}$$
$$D = \begin{bmatrix} -1 & 0 & 0 \\ 0 & 1 & 0 \\ 0 & 0 & 2 \end{bmatrix} \qquad \text{and} \qquad \begin{matrix} \mathbf{u}_1 & \mathbf{u}_2 & \mathbf{u}_3 \end{matrix} \\ P = \begin{bmatrix} -2 & 2 & 1 \\ -2 & 1 & 0 \\ 1 & 0 & 1 \end{bmatrix}$$

Applying Theorem 7, the solution to (8) is

$$
\mathbf{y} = P \begin{bmatrix} C_1 e^{\lambda_1 t} \\ C_2 e^{\lambda_2 t} \\ C_3 e^{\lambda_3 t} \end{bmatrix} = \begin{bmatrix} -2 & 2 & 1 \\ -2 & 1 & 0 \\ 1 & 0 & 1 \end{bmatrix} \begin{bmatrix} C_1 e^{-t} \\ C_2 e^{t} \\ C_3 e^{2t} \end{bmatrix}
$$

$$
= \begin{bmatrix} -2C_1 e^{-t} + 2C_2 e^{t} + C_3 e^{2t} \\ -2C_1 e^{-t} + C_2 e^{t} \\ C_1 e^{-t} + C_3 e^{2t} \end{bmatrix}
$$

or

$$
\begin{aligned}
y_1 &= -2C_1 e^{-t} + 2C_2 e^{t} + C_3 e^{2t} \\
y_2 &= -2C_1 e^{-t} + C_2 e^{t} \\
y_3 &= C_1 e^{-t} + C_3 e^{2t}
\end{aligned}
\tag{9}
$$

Check: This solution can be checked by substitution in (8). We will verify that y_1, y_2, and y_3 satisfy the first equation in (8) and leave it for you to verify that they also satisfy the second and third equations in (8).

$$
y_1' \overset{?}{=} 4y_1 - 6y_2 - 2y_3
$$

$$
(-2C_1 e^{-t} + 2C_2 e^{t} + C_3 e^{2t})' \overset{?}{=} 4(-2C_1 e^{-t} + 2C_2 e^{t} + C_3 e^{2t}) - 6(-2C_1 e^{-t} + C_2 e^{t})
$$
$$
- 2(C_1 e^{-t} + C_3 e^{2t})
$$

$$
2C_1 e^{-t} + 2C_2 e^{t} + 2C_3 e^{2t} \overset{\checkmark}{=} 2C_1 e^{-t} + 2C_2 e^{t} + 2C_3 e^{2t}
$$

Problem 13   Solve the system

$$
\begin{aligned}
y_1' &= 5y_1 - 4y_2 + 2y_3 \\
y_2' &= 2y_1 - y_2 + 2y_3 \\
y_3' &= -4y_1 + 4y_2 - y_3
\end{aligned}
$$

∎

According to Theorem 7, the general solution of a system of n differential equations in the n variables $y_1, y_2, \ldots, y_n$ will always involve n arbitrary constants $C_1, C_2, \ldots, C_n$. Equations of the form

$$
y_1(0) = k_1, \quad y_2(0) = k_2, \quad \ldots, \quad y_n(0) = k_n
$$

are called **initial conditions** for the system. Applying the initial conditions to the general solution and solving for $C_1, C_2, \ldots, C_n$ (if possible) produces a **particular solution** of the system.

Example 14   Refer to Example 13. Find the particular solution of system (8) that satisfies the initial conditions

$$
y_1(0) = 1 \qquad y_2(0) = 0 \qquad y_3(0) = 0
$$

Solution   Substituting $t = 0$ in the general solution (9) produces the following system of equations:

$$y_1(0) = -2C_1 + 2C_2 + C_3 = 1$$
$$y_2(0) = -2C_1 + C_2 \qquad = 0$$
$$y_3(0) = \qquad C_1 \qquad + C_3 = 0$$

Solving this ordinary system of linear equations, we have (details omitted)

$$C_1 = 1 \qquad C_2 = 2 \qquad C_3 = -1$$

Thus, the particular solution of (8) satisfying the given initial conditions is

$$y_1 = -2e^{-t} + 4e^t - e^{2t}$$
$$y_2 = -2e^{-t} + 2e^t$$
$$y_3 = e^{-t} - e^{2t}$$

Problem 14   Find the particular solution of the system in Problem 13 satisfying the initial conditions

$$y_1(0) = 1 \qquad y_2(0) = 0 \qquad y_3(0) = 1$$

Example 15
Pollution Control
A company has two 100 gallon holding tanks that are used to control the release of pollutants into a sewage system. Initially, tank I contains 100 gallons of water in which 50 pounds of pollutants have been dissolved and tank II contains 100 gallons of pure water. Pure water is pumped into tank I and mixed with the polluted water. At the same time, polluted water is circulated between the two tanks and also pumped out of tank II. The various rates at which water enters and leaves each tank are given in Figure 3.

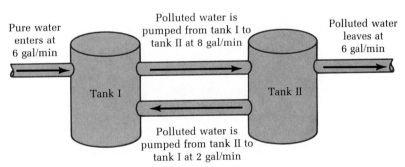

Pure water enters at 6 gal/min

Polluted water is pumped from tank I to tank II at 8 gal/min

Polluted water leaves at 6 gal/min

Tank I

Tank II

Polluted water is pumped from tank II to tank I at 2 gal/min

Figure 3

If $y_1(t)$ and $y_2(t)$ denote the amounts of pollutants in tanks I and II, respectively, t minutes after the pumping begins, it can be shown that y_1 and y_2 satisfy the system

$$y_1' = -0.08y_1 + 0.02y_2$$
$$y_2' = \ \ \ 0.08y_1 - 0.08y_2 \tag{10}$$

with initial conditions

$$y_1(0) = 50 \qquad y_2(0) = 0$$

Solve this system and graph $y_1(t)$ and $y_2(t)$ for $0 \le t \le 60$.

Solution   The matrix form of this system is

$$\mathbf{y}' = A\mathbf{y} \qquad \text{where} \quad A = \begin{bmatrix} -0.08 & 0.02 \\ 0.08 & -0.08 \end{bmatrix}$$

Omitting the details (which you should supply), we have

$$A = PDP^{-1}$$

where

$$P = \begin{bmatrix} 1 & -1 \\ 2 & 2 \end{bmatrix} \quad \text{and} \quad D = \begin{bmatrix} -0.04 & 0 \\ 0 & -0.12 \end{bmatrix}$$

Applying Theorem 7, the general solution of (10) is

$$\mathbf{y} = \begin{bmatrix} 1 & -1 \\ 2 & 2 \end{bmatrix} \begin{bmatrix} C_1 e^{-0.04t} \\ C_2 e^{-0.12t} \end{bmatrix} = \begin{bmatrix} C_1 e^{-0.04t} - C_2 e^{-0.12t} \\ 2C_1 e^{-0.04t} + 2C_2 e^{-0.12t} \end{bmatrix}$$

or

$$y_1 = C_1 e^{-0.04t} - C_2 e^{-0.12t}$$
$$y_2 = 2C_1 e^{-0.04t} + 2C_2 e^{-0.12t}$$

Applying the initial conditions, we have

$$y_1(0) = \ \ C_1 - \ \ C_2 = 50$$
$$y_2(0) = 2C_1 + 2C_2 = \ \ 0$$

Solving for C_1 and C_2, we find $C_1 = 25$ and $C_2 = -25$. Thus, the particular solution of (10) is

$$y_1 = 25e^{-0.04t} + 25e^{-0.12t}$$
$$y_2 = 50e^{-0.04t} - 50e^{-0.12t}$$

The graphs of y_1 and y_2 for $0 \le t \le 60$ are shown in Figure 4. It is interesting to interpret these graphs in terms of the physical process described by system (10). Since pure water is always entering tank I and polluted water is leaving, y_1 (the amount of pollutants in tank I) should be decreasing. The graph of y_1 in Figure 4 shows that this is the case. Since tank II is initially filled with pure water and polluted water is pumped into this tank, y_2 (the amount of pollutants in tank II) should increase initially. However, since the total amount of pollutants in both tanks is decreasing (pure water is always entering and polluted water is leaving), y_2 must eventually begin to decrease. Again, the graph of y_2 in Figure 4 shows that this is the case. In fact, the graphs of y_1 and y_2 intersect at the point where y_2

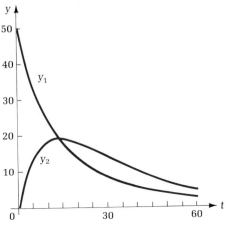

Figure 4

starts decreasing. In words, y_2 increases until both tanks contain the same amount of pollutants and then begins to decrease. This last observation is not intuitively obvious and is typical of the type of information a mathematical model (in this case, a system of differential equations) can provide about a physical process. (The graphs of y_1 and y_2 were obtained by applying standard calculus techniques. You should verify that y_2 assumes its maximum value at the point where $y_1 = y_2$.)

Problem 15 Refer to Example 15. Solve system (10) if tank I initially contains 50 pounds of pollutants and tank II initially contains 100 pounds of pollutants. (This changes the particular solution but not the general solution.) Graph y_1 and y_2 for $0 \le t \le 60$ on the same set of axes and discuss their behavior. ▮

▪ Higher-Order Differential Equations

The **order** of a differential equation is the highest derivative of the (unknown) function that appears in the equation. Certain higher-order equations can be converted to systems of first-order equations and solved by matrix methods. We will illustrate this process for second-order equations. Third-order equations will be considered in Exercise 9-3.

Example 16 Solve the equation

$$y'' - 3y' - 10y = 0 \tag{11}$$

Solution We rewrite equation (11) as

$$y'' = 10y + 3y'$$

and let $y_1 = y$ and $y_2 = y'$. Then

$$y_1' = y' = y_2$$

and

$$y_2' = y'' = 10y + 3y' = 10y_1 + 3y_2$$

If we solve the system

$$\begin{aligned} y_1' &= & y_2 \\ y_2' &= 10y_1 + 3y_2 \end{aligned} \tag{12}$$

then, since $y = y_1$, we will have also solved (11).

The matrix form of (12) is

$$\mathbf{y}' = A\mathbf{y} \quad \text{where} \quad A = \begin{bmatrix} 0 & 1 \\ 10 & 3 \end{bmatrix}$$

Omitting the details, we have

$$A = PDP^{-1}$$

where

$$P = \begin{bmatrix} 1 & 1 \\ -2 & 5 \end{bmatrix} \quad \text{and} \quad D = \begin{bmatrix} -2 & 0 \\ 0 & 5 \end{bmatrix}$$

Applying Theorem (7), the solution to (12) is

$$\mathbf{y} = \begin{bmatrix} 1 & 1 \\ -2 & 5 \end{bmatrix} \begin{bmatrix} C_1 e^{-2t} \\ C_2 e^{5t} \end{bmatrix} = \begin{bmatrix} C_1 e^{-2t} + C_2 e^{5t} \\ -2C_1 e^{-2t} + 5C_2 e^{5t} \end{bmatrix}$$

The first component of this solution is also the solution to (11). That is, the solution to (11) is

$$y = C_1 e^{-2t} + C_2 e^{5t}$$

Check:

$$y'' - 3y' - 10y \overset{?}{=} 0$$

$$(4C_1 e^{-2t} + 25C_2 e^{5t}) - 3(-2C_1 e^{-2t} + 5C_2 e^{5t}) - 10(C_1 e^{-2t} + C_2 e^{5t}) \overset{?}{=} 0$$

$$(4C_1 + 6C_1 - 10C_1)e^{-2t} + (25C_2 - 15C_2 - 10C_2)e^{5t} \overset{?}{=} 0$$

$$0 \overset{\checkmark}{=} 0$$

Problem 16   Solve the equation $y'' + y' - 6y = 0$.   ∎

Answers to Matched Problems

11. $y = Ce^{3t}, \quad y = 8e^{3t}$

12. $y_1 = C_1 e^{-t}, \quad y_2 = C_2 e^t, \quad y_3 = C_3 e^{3t}; \quad \mathbf{y}' = \begin{bmatrix} -1 & 0 & 0 \\ 0 & 1 & 0 \\ 0 & 0 & 3 \end{bmatrix} \mathbf{y}, \quad \mathbf{y} = \begin{bmatrix} C_1 e^{-t} \\ C_2 e^t \\ C_3 e^{3t} \end{bmatrix}$

13. $y_1 = -C_1e^{-t} + C_2e^t - C_3e^{3t}, \quad y_2 = -C_1e^{-t} + C_2e^t, \quad y_3 = C_1e^{-t} + C_3e^{3t}$

14. $y_1 = -2e^{-t} + 2e^t + e^{3t}, \quad y_2 = -2e^{-t} + 2e^t, \quad y_3 = 2e^{-t} - e^{3t}$

15. $y_1 = 50e^{-0.04t}, \quad y_2 = 100e^{-0.04t}$

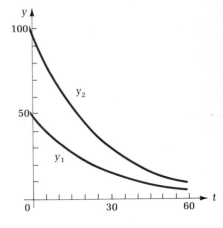

The amount of pollutants in each
tank is decreasing. In addition, the
amount of pollutants in tank II is
always twice the amount in tank I.

16. $y = C_1e^{-3t} + C_2e^{2t}$

‖ Exercise 9-3

A   In Problems 1–4, find the general solution of the indicated first-order equation
and then find the particular solution satisfying the given initial condition.

1. $y' = 0.5y, \quad y(0) = 10$ **2.** $y' = 0.3y, \quad y(0) = 4$

3. $y' = -2y, \quad y(0) = -2$ **4.** $y' = -5y, \quad y(0) = -15$

In Problems 5 and 6, find the general solution of the system. Write the system and
its solution in matrix form.

5. $y_1' = y_1$ **6.** $y_1' = -y_1$
$\quad\; y_2' = 2y_2$ $\quad\; y_2' = 3y_2$

In Problems 7 and 8, find the matrix form of each system.

7. $y_1' = 4y_1 - 3y_2$ **8.** $y_1' = y_1 + 2y_2$
$\quad\; y_2' = 2y_1 - y_2$ $\quad\; y_2' = 2y_1 + y_2$

9. Express the matrix of the system in Problem 7 in the form $A = PDP^{-1}$ where
D is a diagonal matrix.
10. Express the matrix of the system in Problem 8 in the form $A = PDP^{-1}$ where
D is a diagonal matrix.
11. Find the general solution of the system in Problem 7.
12. Find the general solution of the system in Problem 8.
13. Find the particular solution of the system in Problem 7 satisfying the initial
conditions $y_1(0) = 1$ and $y_2(0) = 0$.
14. Find the particular solution of the system in Problem 8 satisfying the initial
conditions $y_1(0) = 3$ and $y_2(0) = 1$.

B   In Problems 15–22, find the general solution of each system of differential equations. Find the particular solution when initial conditions are given.

15. $y_1' = y_1 + y_2$
$\quad y_2' = y_1 + y_2$
$\quad y_1(0) = 1, \quad y_2(0) = 1$

16. $y_1' = y_1 + 2y_2$
$\quad y_2' = y_1 + 2y_2$
$\quad y_1(0) = 2, \quad y_2(0) = -1$

17. $y_1' = \quad y_1 - 0.5y_2$
$\quad y_2' = 2.5y_1 - \quad 2y_2$

18. $y_1' = -y_1 + \quad y_2$
$\quad y_2' = \quad y_1 + 0.5y_2$

19. $y_1' = \qquad y_2 + y_3$
$\quad y_2' = \quad y_1 \qquad - y_3$
$\quad y_3' = -y_1 + y_2 + 2y_3$
$\quad y_1(0) = 3, \quad y_2(0) = 0, \quad y_3(0) = 2$

20. $y_1' = 4y_1 - 6y_2 + 4y_3$
$\quad y_2' = 2y_1 - 3y_2 + 2y_3$
$\quad y_3' = 2y_1 - 4y_2 + 2y_3$
$\quad y_1(0) = 0, \quad y_2(0) = 0, \quad y_3(0) = -1$

21. $y_1' = 3y_1 \qquad - 4y_3$
$\quad y_2' = -y_1 + 4y_2 + 2y_3$
$\quad y_3' = \quad y_1 - \quad y_2 - \quad y_3$

22. $y_1' = \quad 3y_1 - 2y_2 + 2y_3$
$\quad y_2' = \quad 2y_1 - \quad y_2 + 2y_3$
$\quad y_3' = -2y_1 + 2y_2 - \quad y_3$

In Problems 23–26, solve the second-order differential equation.

23. $y'' - 7y' + 12y = 0$

24. $y'' + 5y' + 6y = 0$

25. $y'' - y = 0$

26. $y'' - y' = 0$

C   **27.** Use the substitutions $y_1 = y, y_2 = y', \text{ and } y_3 = y''$ to show that the solution of the third-order equation

$$y''' + ay'' + by' + cy = 0$$

is part of the solution of the system

$$y_1' = \qquad y_2$$
$$y_2' = \qquad\qquad y_3$$
$$y_3' = -cy_1 - by_2 - ay_3$$

28. Show that the matrix of the system in Problem 27 is

$$A = \begin{bmatrix} 0 & 1 & 0 \\ 0 & 0 & 1 \\ -c & -b & -a \end{bmatrix}$$

29. Show that the characteristic polynomial of the matrix A in Problem 28 is

$$p(\lambda) = \lambda^3 + a\lambda^2 + b\lambda + c$$

30. If λ is an eigenvalue of the matrix A in Problem 28, show that $\mathbf{u}$ is an eigenvector associated with λ where

$$\mathbf{u} = \begin{bmatrix} 1 \\ \lambda \\ \lambda^2 \end{bmatrix}$$

Use the results in Problems 27–30 to solve the following third-order differential equations:

31. $y''' - 4y'' - y' + 4y = 0$

32. $y''' + 5y'' - y' - 5y = 0$

❙□❙ Applications

33. *Pollution control.* Refer to Example 15. If pure water is pumped into tank I at 8 gal/min, polluted water is pumped from tank I to tank II at 9 gal/min, polluted water is pumped from tank II back to tank I at 1 gal/min, and polluted water is pumped out of tank II at 8 gal/min, then y_1 (amount of pollutants in tank I) and y_2 (amount of pollutants in tank II) can be shown to satisfy the system

$$y_1' = -0.09y_1 + 0.01y_2$$
$$y_2' = 0.09y_1 - 0.09y_2$$

Initially, the water in tank I contains 50 pounds of pollutants and the water in tank II contains no pollutants. Solve the system and graph y_1 and y_2 on the same set of axes for $0 \le t \le 30$.

34. *Pollution control.* Repeat Problem 33 if both tanks initially contain 30 pounds of pollutants.

35. *Population growth.* The populations y_1 and y_2 (in thousands) of two species that compete for the same food supply satisfy the system

$$y_1' = 0.125y_1 - 0.05y_2 \qquad y_1(0) = 90$$
$$y_2' = -0.05y_1 + 0.05y_2 \qquad y_2(0) = 130$$

Solve the system and graph y_1 and y_2 on the same set of axes for $0 \le t \le 20$.

36. *Population growth.* Repeat Problem 35 if the initial conditions are $y_1(0) = 115$ and $y_2(0) = 255$.

❙9-4❙ Chapter Review

Important Terms and Symbols

9-1. *Quadratic forms.* Quadratic equation, conic section, ellipse, circle, parabola, hyperbola, degenerate conic, standard position, standard form, translation of axes, completing the square, mixed term, rotation of axes, quadratic form, matrix of a quadratic form, principal axes theorem for R^2

9-2. *Recurrence equations.* First-order recurrence equation, initial condition, solution, system of first-order recurrence equations, matrix form of a system, matrix form of the solution of a system, explicit form of the solution of a system, second-order recurrence equation, Fibonacci numbers

9-3. *Differential equations.* First-order differential equation, general solution, initial condition, particular solution, linear system of first-order differential equations, matrix form of a system, matrix of a system, vector function, vector derivative, order of a differential equation, second-order differential equation

▌▌ Exercise 9-4 Chapter Review

Work through all the problems in this chapter review and check your answers in the back of the book. (Answers to all review problems are there.) Where weaknesses show up, review appropriate sections in the text.

A   *In Problems 1–3, identify the conic and sketch its graph.*

1. $\dfrac{x^2}{25} + \dfrac{y^2}{9} = 1$ **2.** $4x^2 - y^2 = 4$ **3.** $4x - y^2 = 0$

4. Solve the first-order recurrence equation $x_n = 5x_{n-1}$, $x_0 = 2$.
5. Solve the first-order differential equation $y' = 5y$, $y(0) = 2$.
6. Given the quadratic equation

$$9x^2 + 12xy + 8y^2 + 10x + 7y + 15 = 0$$

(A) Find the associated quadratic form.
(B) Find the matrix of this quadratic form.
(C) Write the quadratic equation in matrix form.

7. Solve the system of recurrence equations

$$x_n = 2x_{n-1} \qquad x_0 = 5$$
$$y_n = 3y_{n-1} \qquad y_0 = 7$$

8. Solve the system of differential equations

$$y_1' = 2y_1$$
$$y_2' = 3y_2$$
$$y_3' = 4y_3$$

9. Given the system of first-order recurrence equations

$$x_n = 2x_{n-1} - 3y_{n-1} \qquad x_0 = 1$$
$$y_n = 2x_{n-1} + 7y_{n-1} \qquad y_0 = 2$$

(A) Write the system in matrix form.
(B) Find the matrix form of the solution.
(C) Express the matrix of the system in the form PDP^{-1} where D is a diagonal matrix.
(D) Find the (explicit) solution of the system.

10. Given the system of first-order differential equations

$$y_1' = 2y_1 - y_2$$
$$y_2' = 2y_1 + 5y_2$$

(A) Write the system in matrix form.

(B) Express the matrix of the system in the form PDP^{-1} where D is a diagonal matrix.

(C) Find the general solution of the system.

(D) Find the particular solution satisfying the initial conditions $y_1(0) = 1$ and $y_2(0) = 0$.

B **11.** Use a translation of axes to transform the following equation into a standard form. Identify and sketch the conic.

$$x^2 - 4x + y^2 - 2y - 4 = 0$$

12. The matrix P is the transition matrix from the $x'y'$ coordinate system to the xy coordinate system. Show that P is an orthogonal matrix, and graph the x' and y' axes in the xy coordinate system.

$$P = \begin{bmatrix} \dfrac{2}{\sqrt{5}} & -\dfrac{1}{\sqrt{5}} \\ \dfrac{1}{\sqrt{5}} & \dfrac{2}{\sqrt{5}} \end{bmatrix}$$

In Problems 13–16, use a rotation of axes, $0° < \theta < 90°$ (see page 582), followed by a translation, if necessary, to transform each equation into a standard form. Identify and sketch the conic.

13. $5x^2 + 6xy + 5y^2 - 8 = 0$

14. $4x^2 + 6xy - 4y^2 - 5 = 0$

15. $x^2 - 4xy + 4y^2 - 8\sqrt{5}\,x - 4\sqrt{5}\,y = 0$

16. $52x^2 + 72xy + 73y^2 - 280x - 290y + 25 = 0$

In Problems 17 and 18, solve the system of first-order recurrence equations.

17. $x_n = 4x_{n-1} - 3y_{n-1}$
$\quad\ y_n = 2x_{n-1} - y_{n-1}$
$\quad\ x_0 = 4, \quad y_0 = 3$

18. $x_n = 0.5x_{n-1} + y_{n-1}$
$\quad\ y_n = 2x_{n-1} + 1.5y_{n-1}$
$\quad\ x_0 = 6, \quad y_0 = 3$

In Problems 19–22, find the general solution of each system of first-order differential equations. Find the particular solution when initial conditions are given.

19. $y_1' = 4y_1 - 2y_2$
$\quad\ y_2' = 6y_1 - 4y_2$
$\quad\ y_1(0) = -1, \quad y_2(0) = 1$

20. $y_1' = 2.5y_1 + 3y_2$
$\quad\ y_2' = \quad y_1 + 2y_2$
$\quad\ y_1(0) = -1, \quad y_2(0) = 3$

21. $y_1' = \quad 3y_1 + 2y_2 + 2y_3$
$\quad\ y_2' = \quad 2y_1 + 3y_2 + 4y_3$
$\quad\ y_3' = -3y_1 - 3y_2 - 4y_3$

22. $y_1' = \qquad\quad 2y_2 + 2y_3$
$\quad\ y_2' = \quad 2y_1 \qquad - 2y_3$
$\quad\ y_3' = -2y_1 + 2y_2 + 4y_3$

In Problems 23 and 24, solve the second-order recurrence equation.

23. $x_n = 8x_{n-1} - 15x_{n-2}, \quad x_0 = 2, \quad x_1 = 8$

24. $x_n = 2x_{n-1} + 2x_{n-2}, \quad x_0 = 1, \quad x_1 = 1$

In Problems 25 and 26, solve the second-order differential equation.

25. $y'' - 5y' + 4y = 0$ **26.** $y'' - 4y = 0$

C   **27.** Solve the following system of first-order recurrence equations:

$$
\begin{aligned}
x_n &= 4x_{n-1} \qquad\quad + 2z_{n-1} & x_0 &= -1 \\
y_n &= -x_{n-1} + 2y_{n-1} - z_{n-1} & y_0 &= 1 \\
z_n &= x_{n-1} + 2y_{n-1} + 3z_{n-1} & z_0 &= 1
\end{aligned}
$$

28. Solve the third-order recurrence equation $x_n = 12x_{n-1} - 47x_{n-2} + 60x_{n-3}$, $x_0 = 1$, $x_1 = 4$, $x_2 = 18$.

29. Solve the third-order differential equation $y''' - 2y'' - y' + 2y = 0$.

▮□▮   Applications

30. *Population growth.* The number of rabbits, x_n (in thousands), and the number of foxes, y_n (in thousands), in an ecological system after n years satisfy the system of first-order recurrence equations

$$
\begin{aligned}
x_n &= 1.2x_{n-1} - 0.4y_{n-1} \\
y_n &= 0.4x_{n-1} + 0.2y_{n-1}
\end{aligned}
$$

Initially, there are 60,000 rabbits and 75,000 foxes. Solve the system and determine the population of each species after 2, 4, 6, 8, and 10 years. Based on these calculations, what is happening to the population of each species?

31. *Pollution control.* A company has two 100 gallon holding tanks. Tank I contains 100 gallons of water in which 40 pounds of pollutants have been dissolved and tank II contains 100 gallons of pure water. The rates at which water enters and leaves each tank are given in the figure. If y_1 and y_2 denote

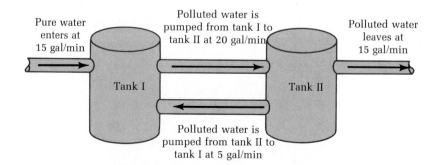

Pure water enters at 15 gal/min

Polluted water is pumped from tank I to tank II at 20 gal/min

Polluted water leaves at 15 gal/min

Tank I   Tank II

Polluted water is pumped from tank II to tank I at 5 gal/min the amounts of pollutants in tank I and tank II, respectively, then it can be shown that y_1 and y_2 satisfy the system of first-order differential equations

$$y_1' = -0.2y_1 + 0.05y_2$$
$$y_2' = 0.2y_1 - 0.2y_2$$

with initial conditions $y_1(0) = 40$, $y_2(0) = 0$. Solve this system and graph y_1 and y_2 on the same set of axes for $0 \leq t \leq 20$.

Answers

Chapter 1 Exercise 1-1

1. Unique solution: $x = 4$, $y = 1$ **3.** No solution

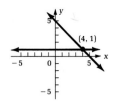

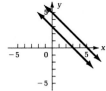

5. Infinite number of solutions: $x = \frac{1}{2} + \frac{1}{2}t$, $y = t$ (t any real number) or $\{(\frac{1}{2} + \frac{1}{2}t, t)|t$ any real number$\}$

7. Unique solution: $x = -2$, $y = -3$

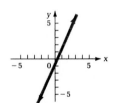

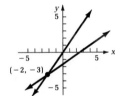

9. $x = 3$, $y = 1$ **11.** No solution   **13.** $x = -2 + 3t$, $y = t$ (t any real number) or $\{(-2 + 3t, t)|t$ any real number$\}$

15. $x = \frac{1}{3}$, $y = \frac{4}{3}$ **17.** $x = 0$, $y = -2$, $z = 1$ **19.** No solution

21. $x = \frac{13}{2} - \frac{3}{2}t$, $y = \frac{5}{2} + \frac{1}{2}t$, $z = t$ (t any real number) or $\{(\frac{13}{2} - \frac{3}{2}t, \frac{5}{2} + \frac{1}{2}t, t)|t$ any real number$\}$

23. $x = \frac{3}{2}$, $y = \frac{16}{3}$, $z = \frac{5}{2}$

25. $x = 25 - 7t$, $y = -5 + 3t$, $z = t$ (t any real number) or $\{(25 - 7t, -5 + 3t, t)|t$ any real number$\}$ **27.** No solution

29. (A)

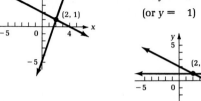

(B) $\begin{aligned} x + 2y &= 4 \\ -7y &= -7 \\ \text{(or } y &= 1) \end{aligned}$

(C) $\begin{aligned} 7x &= 14 \\ \text{(or } x &= 2) \\ 3x - y &= 5 \end{aligned}$

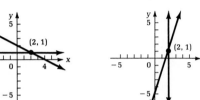

31. (A) $a = 1$, $b \neq 0$ (B) $a = 1$, $b = 0$ (C) $a \neq 1$

35. 10% solution: 24 cubic centimeters; 35% solution: 36 cubic centimeters

37. 50 one-person boats, 200 two-person boats, 100 four-person boats   **39.** Mix A: 80 grams; mix B: 60 grams

Exercise 1-2

1. $\begin{bmatrix} 2 & -3 & 1 & 4 \\ -1 & 4 & -3 & 7 \\ 8 & 2 & 5 & -11 \end{bmatrix}$ **3.** $\begin{bmatrix} 1 & 0 & -1 & 0 \\ 0 & 1 & 1 & 0 \\ -1 & 2 & 0 & 0 \end{bmatrix}$ **5.** $\begin{array}{rcr} 2x_1 - 3x_2 + x_3 &=& 4 \\ -x_1 + 2x_2 - x_3 &=& 5 \\ 4x_1 + 6x_2 + 7x_3 &=& -9 \end{array}$ **7.** $\begin{array}{rcr} 2x_1 \quad + x_3 &=& -1 \\ x_2 - 3x_3 &=& 0 \\ x_1 \quad + 4x_3 &=& 2 \end{array}$

9. $\begin{bmatrix} 2 & 4 & -2 & 6 \\ 1 & -2 & 1 & 3 \\ -3 & 2 & 1 & 5 \end{bmatrix}$ **11.** $\begin{bmatrix} 3 & -6 & 3 & 9 \\ 2 & 4 & -2 & 6 \\ -3 & 2 & 1 & 5 \end{bmatrix}$ **13.** $\begin{bmatrix} 1 & -2 & 1 & 3 \\ 2 & 4 & -2 & 6 \\ 0 & -4 & 4 & 14 \end{bmatrix}$

15. $x_1 = 1, x_2 = 4, x_3 = -2$ or $(1, 4, -2)$

17. $x_1 = 1 + t, x_2 = 1 + 2t, x_3 = t$ (t any real number) or $\{(1 + t, 1 + 2t, t) | t$ any real number$\}$ **19.** Inconsistent

21. $x_1 = 5, x_2 = -5, x_3 = -8$ or $(5, -5, -8)$ **23.** $x_1 = -\frac{1}{2}, x_2 = \frac{3}{2}, x_3 = \frac{1}{2}$ or $(-\frac{1}{2}, \frac{3}{2}, \frac{1}{2})$

25. $x_1 = \frac{1}{2} + t/2, x_2 = -\frac{1}{3} + 2t/3, x_3 = t$ (t any real number) or $\{(\frac{1}{2} + t/2, -\frac{1}{3} + 2t/3, t) | t$ any real number$\}$

27. Inconsistent   **29.** $x_1 = 0, x_2 = 0, x_3 = 0$ or $(0, 0, 0)$ **31.** (A) $R_i \leftrightarrow R_j$ (B) $\left(\dfrac{1}{k}\right)R_i \to R_i$ (C) $R_i + (-k)R_j \to R_i$

37. $x^2 + 1$ **39.** $3x^2 + x - 2$

Exercise 1-3

1. Triangular form   **3.** Not in triangular form, $a_{32} = 1$ **5.** Triangular form

7. Not in triangular form, $a_{21} = 3, a_{32} = 1, a_{43} = 4, a_{54} = 2$ **9.** $\begin{bmatrix} 1 & -1 & 1 & 2 \\ 0 & 3 & 2 & 1 \\ 0 & 0 & 2 & -1 \end{bmatrix}$

11. $\begin{bmatrix} 1 & -1 & 1 & 2 & -1 \\ 0 & 1 & -1 & 1 & 2 \\ 0 & 0 & 3 & -4 & -3 \\ 0 & 0 & 0 & 0 & 0 \end{bmatrix}$ **13.** $\begin{bmatrix} 3 & -1 & 4 & 2 & 5 & -7 \\ 0 & 1 & -1 & 1 & 2 & 2 \\ 0 & 0 & 5 & -4 & -2 & 0 \\ 0 & 0 & 0 & -2 & 5 & 6 \\ 0 & 0 & 0 & 0 & 3 & -5 \end{bmatrix}$ **15.** $x_1 = 6, x_2 = -1, x_3 = 4$ or $(6, -1, 4)$

17. $x_1 = -10, x_2 = 2, x_3 = 7, x_4 = 4$ or $(-10, 2, 7, 4)$ **19.** $x_1 = \frac{3}{2}, x_2 = -\frac{1}{2}, x_3 = -2, x_4 = -\frac{1}{2}$ or $(\frac{3}{2}, -\frac{1}{2}, -2, -\frac{1}{2})$

21. $x_1 = -3, x_2 = 1, x_3 = 2, x_4 = -2, x_5 = 4$ or $(-3, 1, 2, -2, 4)$ **23.** $x_1 = 2, x_2 = -4, x_3 = 3$ or $(2, -4, 3)$

25. $x_1 = \frac{7}{3}, x_2 = \frac{4}{3}, x_3 = \frac{2}{3}$ or $(\frac{7}{3}, \frac{4}{3}, \frac{2}{3})$ **27.** $x_1 = 1, x_2 = 0, x_3 = -1, x_4 = 0$ or $(1, 0, -1, 0)$

29. $x_1 = 1.1, x_2 = -2.2, x_3 = 2.7, x_4 = -0.6$ or $(1.1, -2.2, 2.7, -0.6)$

31. $x_1 = -2, x_2 = 0, x_3 = 1, x_4 = 0, x_5 = 4$ or $(-2, 0, 1, 0, 4)$

33. Upper intersection point: $8°$; lower left-hand intersection point: $16°$; lower right-hand intersection point: $8°$

35. Upper left-hand intersection point: $8°$; upper right-hand intersection point: $24°$; lower left-hand intersection point: $8°$; lower right-hand intersection point: $24°$

Exercise 1-4

1. Yes   **3.** No   **5.** No   **7.** Yes   **9.** $x_1 = -2, x_2 = 3, x_3 = 0$ or $(-2, 3, 0)$

11. $x_1 = 3 + 2t, x_2 = -5 - t, x_3 = t$ (t any real number) or $\{(3 + 2t, -5 - t, t) | t$ any real number$\}$ **13.** Inconsistent

15. $x_1 = -5 + 2s + 3t, x_2 = s, x_3 = 2 - 3t, x_4 = t$ (s and t any real numbers) or $\{(-5 + 2s + 3t, s, 2 - 3t, t) | s$ and t any real numbers$\}$

17. $\begin{bmatrix} 1 & 0 & -7 \\ 0 & 1 & 3 \end{bmatrix}$ **19.** $\begin{bmatrix} 1 & 0 & 0 & -5 \\ 0 & 1 & 0 & 4 \\ 0 & 0 & 1 & -2 \end{bmatrix}$ **21.** $\begin{bmatrix} 1 & 0 & 2 & -\frac{5}{3} \\ 0 & 1 & -2 & \frac{1}{3} \\ 0 & 0 & 0 & 0 \end{bmatrix}$ **23.** $x_1 = -2, x_2 = 3, x_3 = 1$ or $(-2, 3, 1)$

25. $x_1 = 0$, $x_2 = -2$, $x_3 = 2$ or $(0, -2, 2)$

27. $x_1 = 3 + 2t$, $x_2 = -2 + t$, $x_3 = t$ (t any real number) or $\{(3 + 2t, -2 + t, t) | t$ any real number$\}$

29. $x_1 = (-4 - 4t)/7$, $x_2 = (5 + 5t)/7$, $x_3 = t$ (t any real number) or $\{((-4 - 4t)/7, (5 + 5t)/7, t) | t$ any real number$\}$

31. $x_1 = -1$, $x_2 = 2$ or $(-1, 2)$ **33.** Inconsistent   **35.** Inconsistent

37. $x_1 = 17 + 7s - 7t$, $x_2 = -5 - 3s + 3t$, $x_3 = s$, $x_4 = t$ (s and t any real numbers) or
$\{(17 + 7s - 7t, -5 - 3s + 3t, s, t) | s$ and t any real numbers$\}$

39. $x_1 = 3 + 2s - 3t$, $x_2 = 2 + s + 2t$, $x_3 = s$, $x_4 = t$ (s and t any real numbers) or
$\{(3 + 2s - 3t, 2 + s + 2t, s, t) | s$ and t any real numbers$\}$

41. $\begin{bmatrix} 0 & 0 \\ 0 & 0 \end{bmatrix}$, $\begin{bmatrix} 0 & 1 \\ 0 & 0 \end{bmatrix}$, $\begin{bmatrix} 1 & 0 \\ 0 & 1 \end{bmatrix}$, $\begin{bmatrix} 1 & b_{12} \\ 0 & 0 \end{bmatrix}$ **43.** $\begin{bmatrix} 0 & 0 \\ 0 & 0 \\ 0 & 0 \end{bmatrix}$, $\begin{bmatrix} 0 & 1 \\ 0 & 0 \\ 0 & 0 \end{bmatrix}$, $\begin{bmatrix} 1 & b_{12} \\ 0 & 0 \\ 0 & 0 \end{bmatrix}$, $\begin{bmatrix} 1 & 0 \\ 0 & 1 \\ 0 & 0 \end{bmatrix}$

45. $\begin{bmatrix} 1 & 0 & | & (de - bf)/(ad - bc) \\ 0 & 1 & | & (af - ce)/(ad - bc) \end{bmatrix}$ **47.** (A) $a \neq b$ (B) $a = b = c$ (C) $a = b$, $c \neq a$

49. $x_1 =$ Number of containers of 10% solution, $x_2 =$ Number of containers of 30% solution, $x_3 =$ Number of containers of 50% solution; $(x_1, x_2, x_3) = (4, 12, 0)$, $(6, 8, 1)$, $(8, 4, 2)$, or $(10, 0, 3)$

51. 20 one-person boats, 220 two-person boats, 100 four-person boats

53. $(-80 + t)$ one-person boats, $(420 - 2t)$ two-person boats, t four-person boats, $80 \leq t \leq 210$, t an integer

55. No solution; no production schedule will use all the work-hours in all departments

57. 8 ounces food A, 2 ounces food B, 4 ounces food C **59.** No solution

61. 8 ounces of food A, $(10 - 2t)$ ounces of food B, t ounces of food C, $0 \leq t \leq 5$

Exercise 1-5

1. Yes   **3.** No   **5.** Yes   **7.** $x_1 = 6t$, $x_2 = 7t$, $x_3 = t$ (t any real number) or $\{(6t, 7t, t) | t$ any real number$\}$

9. $x_1 = 6t$, $x_2 = 7t$, $x_3 = t$ (t any real number) or $\{(6t, 7t, t) | t$ any real number$\}$ **11.** $x_1 = 0$, $x_2 = 0$ or $(0, 0)$

13. $x_1 = -5s + 7t$, $x_2 = -2s + 3t$, $x_3 = s$, $x_4 = t$ (s and t any real numbers) or
$\{(-5s + 7t, -2s + 3t, s, t) | s$ and t any real numbers$\}$

15. $x_1 = 2r - s + t$, $x_2 = r$, $x_3 = s$, $x_4 = t$ (r, s, and t any real numbers) or $\{(2r - s + t, r, s, t) | r$, s, and t any real numbers$\}$

17. $x_1 = 0$, $x_2 = 0$, $x_3 = 0$, $x_4 = 0$ or $(0, 0, 0, 0)$

19. $x_1 = t$, $x_2 = -t$, $x_3 = t$, $x_4 = -t$, $x_5 = t$ (t any real number) or $\{(t, -t, t, -t, t) | t$ any real number$\}$

21. General solution: $\{(5t/6, t) | t$ any real number$\}$; particular solution: $(\frac{5}{11}, \frac{6}{11})$

23. General solution: $\{(t, 3t/2, t) | t$ any real number$\}$; particular solution: $(\frac{2}{7}, \frac{3}{7}, \frac{2}{7})$ **25.** When $ad - bc = 0$

27. $w_1 = 12t/5$, $w_2 = 2t$, $w_3 = t$ where t is any positive number

29. $w_1 = 10t/9$, $w_2 = 20t/9$, $w_3 = 3t/2$, $w_4 = t$ where t is any positive real number

Exercise 1-6   Chapter Review

1. Unique solution: $x = 1$, $y = -3$ or $(1, -3)$ **2.** No solution

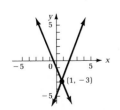

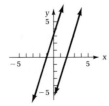

3. Infinite number of solutions: $x = -\frac{1}{2} - t/2$, $y = t$ (t any real number) or $\{(-\frac{1}{2} - t/2,\, t)|t$ any real number$\}$

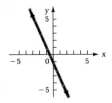

4. $x_1 = -1$, $x_2 = 5$ or $(-1, 5)$ **5.** $x_1 = -\frac{15}{2}$, $x_2 = -\frac{13}{2}$ or $(-\frac{15}{2}, -\frac{13}{2})$ **6.** $x_1 = -1$, $x_2 = 0$, $x_3 = 2$ or $(-1, 0, 2)$
7. $x_1 = 5$, $x_2 = -2$, $x_3 = 1$ or $(5, -2, 1)$ **8.** $x_1 = 2$, $x_2 = 3$ or $(2, 3)$ **9.** Inconsistent
10. $x_1 = 4 - 3t$, $x_2 = 7 + 2t$, $x_3 = t$ (t any real number) or $\{(4 - 3t, 7 + 2t, t)|t$ any real number$\}$

11. $x_1 = 2 + 2t$, $x_2 = t$, $x_3 = 4$ (t any real number) or $\{(2 + 2t, t, 4)|t$ any real number$\}$ **12.** $\begin{bmatrix} 1 & 0 & 6 \\ 0 & 1 & 4 \end{bmatrix}$

13. $\begin{bmatrix} 1 & 0 & 0 & 3 \\ 0 & 1 & 0 & 0 \\ 0 & 0 & 1 & 2 \end{bmatrix}$ **14.** $\begin{bmatrix} 1 & 0 & 0 & 1 & 1 \\ 0 & 1 & 0 & 0 & -2 \\ 0 & 0 & 1 & -1 & 2 \end{bmatrix}$ **15.** $x_1 = -2$, $x_2 = 3$ or $(-2, 3)$ **16.** Inconsistent

17. $x_1 = -6 + 3t$, $x_2 = -2 + t$, $x_3 = t$ (t any real number) or $\{(-6 + 3t, -2 + t, t)|t$ any real number$\}$
18. $x_1 = 11t$, $x_2 = -19t/4$, $x_3 = t$ (t any real number) or $\{(11t, -19t/4, t)|t$ any real number$\}$
19. $x_1 = 4 + 4t$, $x_2 = -\frac{1}{2} - t/2$, $x_3 = t$ (t any real number) or $\{(4 + 4t, -\frac{1}{2} - t/2, t)|t$ any real number$\}$
20. $x_1 = -2t$, $x_2 = t$, $x_3 = 0$ (t any real number) or $\{(-2t, t, 0)|t$ any real number$\}$
21. $x_1 = 0$, $x_2 = 0$, $x_3 = 0$ or $(0, 0, 0)$ **22.** $x_1 = \frac{5}{2}$, $x_2 = \frac{3}{2}$, $x_3 = 2$ or $(\frac{5}{2}, \frac{3}{2}, 2)$
23. $x_1 = 5 + s + 2t$, $x_2 = 6 + 3s + 4t$, $x_3 = s$, $x_4 = t$ (s and t any real numbers) or
$\{(5 + s + 2t, 6 + 3s + 4t, s, t)|s$ and t any real numbers$\}$
24. $x_1 = -3t$, $x_2 = -t$, $x_3 = -t$, $x_4 = t$ (t any real number) or $\{(-3t, -t, -t, t)|t$ any real number$\}$
25. $x_1 = -3s - 3t/5$, $x_2 = s$, $x_3 = 0$, $x_4 = 4t/5$, $x_5 = t$ (s and t any real numbers) or
$\{(-3s - 3t/5, s, 0, 4t/5, t)|s$ and t any real numbers$\}$
26. Inconsistent
27. $x_1 = (b - d)t/(ad - bc)$, $x_2 = (c - a)t/(ad - bc)$, $x_3 = t$ (t any real number) or
$\{((b - d)t/(ad - bc),\ (c - a)t/(ad - bc),\ t)|t$ any real number$\}$
28. (A) $a \neq 0$ (B) $a = 0$, $b \neq 0$ (C) $a = 0$, $b = 0$
31. 60 cubic centimeters of the 20% solution and 40 cubic centimeters of the 45% solution
32. $(-2 + t)$ lecturers, $(8 - 2t)$ instructors, and t assistant professors where $t = 2, 3,$ or 4
33. 2 lecturers and 4 assistant professors
34. Upper left-hand intersection point: 6°; upper right-hand intersection point: 10°; lower left-hand intersection
point: 14°; lower right-hand intersection point: 18°
35. $w_1 = 10t/3$, $w_2 = 4t$, $w_3 = t$ where t is any positive number

Chapter 2   Exercise 2-1

1. 2×2; 1×4 **3.** 2   **5.** $\begin{bmatrix} 0 & 0 \\ 0 & 0 \end{bmatrix}$ **7.** C, D **9.** A, B **11.** $\begin{bmatrix} -1 & 0 \\ 5 & -3 \end{bmatrix}$ **13.** $\begin{bmatrix} -2 \\ 3 \\ 0 \end{bmatrix}$ **15.** $\begin{bmatrix} -1 \\ 6 \\ 5 \end{bmatrix}$

17. $\begin{bmatrix} -15 & 5 \\ 10 & -15 \end{bmatrix}$ **19.** Not defined   **21.** $\begin{bmatrix} 0 & 2 \\ 1 & -1 \end{bmatrix}$ **23.** $\begin{bmatrix} 1 & 3 \\ 1 & 1 \end{bmatrix}$ **25.** $\begin{bmatrix} -4 & 6 \\ 6 & -10 \end{bmatrix}$ **27.** $\begin{bmatrix} 12 & 6 \\ -6 & 18 \end{bmatrix}$

29. $\begin{bmatrix} 0 & 4 \\ 2 & -2 \end{bmatrix}$ **31.** $\begin{bmatrix} 31 & -7 \\ -26 & 60 \end{bmatrix}$ **33.** $\begin{bmatrix} \frac{2}{3} & 1 & 3 \\ -2 & \frac{4}{3} & -4 \end{bmatrix}$ **35.** $\begin{bmatrix} -1 & 0 \\ 3 & 2 \\ 3 & 6 \end{bmatrix}$

47. Guitar Banjo

$$\begin{bmatrix} \$33 & \$26 \\ \$57 & \$77 \end{bmatrix} \begin{matrix} \text{Materials} \\ \text{Labor} \end{matrix}$$

Exercise 2-2

1. 10   **3.** 3   **5.** 11   **7.** $[1 \;\; 5]$ **9.** $\begin{bmatrix} 0 \\ 7 \end{bmatrix}$ **11.** $\begin{bmatrix} 6 & 11 \\ 13 & 18 \end{bmatrix}$ **13.** $\begin{bmatrix} 5 & 10 \\ 13 & 19 \end{bmatrix}$ **15.** $\begin{bmatrix} 8 & 6 \\ 3 & 2 \end{bmatrix}$ **17.** $\begin{bmatrix} 0 & 9 \\ 5 & -4 \end{bmatrix}$

19. $\begin{bmatrix} 5 & 8 & -5 \\ -1 & -3 & 2 \\ -2 & 8 & -6 \end{bmatrix}$ **21.** $\begin{bmatrix} -1 \\ 5 \\ 5 \end{bmatrix}$ **23.** Not defined   **25.** $[32]$ **27.** $\begin{bmatrix} 4 & 8 & 12 \\ 5 & 10 & 15 \\ 6 & 12 & 18 \end{bmatrix}$

33. $B = \begin{bmatrix} a & b \\ b & a \end{bmatrix}$, a and b any real numbers   **47.** (A) \$9   (B) \$24.10   (C)

	Plant I	Plant II	
	\$ 9.00	\$11.00	One-person
	\$14.10	\$17.20	Two-person
	\$19.80	\$24.10	Four-person

Labor costs per boat at each plant

49. (A)

A	B	C	D	E
[16	9	11	11	10]

This is the combined inventory in all three stores.

(B)

W	R
[\$108,300	\$141,340]

This represents the total wholesale and retail values of the total inventory in all three stores.

51. (A)

(B)

		To			
		#1	#2	#3	#4
	#1	0	1	1	0
From	#2	1	0	0	1
	#3	1	0	0	1
	#4	1	1	0	0

(C)

$$S_2 = \begin{bmatrix} 2 & 1 & 1 & 2 \\ 2 & 2 & 1 & 1 \\ 2 & 2 & 1 & 1 \\ 2 & 2 & 1 & 1 \end{bmatrix}$$

All cities can be reached in one or two flights.

53. (A)

Losing player

	A	B	C	D
Winning player A	0	1	0	0
B	0	0	1	0
C	1	0	0	1
D	1	1	0	0

(B)

				Row sum
0	1	1	0	2
1	0	1	1	3
2	2	0	1	5
1	2	1	0	4

Ranking (high to low): Carol, Diane, Barbara, Ann

Exercise 2-3

5. $\begin{bmatrix} 3 & -2 \\ -1 & 1 \end{bmatrix}$ **7.** $\begin{bmatrix} 7 & -3 \\ -2 & 1 \end{bmatrix}$ **9.** (A) $x_1 = -3, x_2 = 2$ (B) $x_1 = -1, x_2 = 2$ (C) $x_1 = -8, x_2 = 3$

11. (A) $x_1 = 17, x_2 = -5$ (B) $x_1 = 7, x_2 = -2$ (C) $x_1 = 24, x_2 = -7$ **13.** $\begin{bmatrix} 7 & 6 & -3 \\ 2 & 2 & -1 \\ -6 & -5 & 3 \end{bmatrix}$

15. $\dfrac{1}{2}\begin{bmatrix} 3 & -1 & -1 \\ -1 & 1 & 1 \\ -3 & 1 & 3 \end{bmatrix}$ **17.** $\begin{bmatrix} 0 & 1 & -1 & 0 \\ 0 & 0 & 1 & -1 \\ 1 & -1 & 0 & 1 \\ -1 & 1 & 0 & 0 \end{bmatrix}$

19. (A) $x_1 = 1, x_2 = 0, x_3 = 0$ (B) $x_1 = -1, x_2 = 0, x_3 = 1$ (C) $x_1 = -1, x_2 = -1, x_3 = 1$
21. (A) $x_1 = 1, x_2 = 1, x_3 = 3$ (B) $x_1 = -1, x_2 = 1, x_3 = -1$ (C) $x_1 = 5, x_2 = -1, x_3 = -5$
23. (A) $x_1 = 0, x_2 = 0, x_3 = 1, x_4 = 0$ (B) $x_1 = 1, x_2 = -1, x_3 = 0, x_4 = 1$ (C) $x_1 = 3, x_2 = 1, x_3 = -3, x_4 = 1$

33. $\begin{bmatrix} 1 & -a & ac-b \\ 0 & 1 & -c \\ 0 & 0 & 1 \end{bmatrix}$ **35.** $A^{-2} = \begin{bmatrix} 2 & -3 \\ -3 & 5 \end{bmatrix}$ **37.** $A^{-1} = \dfrac{1}{ad-bc}\begin{bmatrix} d & -b \\ -c & a \end{bmatrix}$, $ad-bc \neq 0$

45. Allocation 1: 60 model A, 0 model B; allocation 2: 25 model A, 25 model B; allocation 3: 4 model A, 40 model B
47. (A) $I_1 = 4, I_2 = 6, I_3 = 2$ (B) $I_1 = 3, I_2 = 7, I_3 = 4$ (C) $I_1 = 7, I_2 = 8, I_3 = 1$

49. (B) $\dfrac{1}{4}\begin{bmatrix} 2 & 1 & 1 \\ 1 & 2 & 1 \\ 1 & 1 & 2 \end{bmatrix}$ (C) 1: $x_1 = 8°, x_2 = 8°, x_3 = 16°$; 2: $x_1 = 12°, x_2 = 16°, x_3 = 20°$; 3: $x_1 = 16°, x_2 = 18°, x_3 = 22°$

Exercise 2-4

1. Yes   **3.** No   **5.** No   **7.** No   **9.** Yes   **11.** Yes   **13.** $\begin{bmatrix} 0 & 1 \\ 1 & 0 \end{bmatrix} A = \begin{bmatrix} 3 & 6 & 9 \\ 1 & -4 & -1 \end{bmatrix}$

15. $\begin{bmatrix} 1 & 0 \\ -3 & 1 \end{bmatrix} A = \begin{bmatrix} 1 & -4 & -1 \\ 0 & 18 & 12 \end{bmatrix}$ **17.** $B = \begin{bmatrix} 1 & 2 \\ 0 & 0 \end{bmatrix}$, $P = \begin{bmatrix} 1 & 0 \\ -2 & 1 \end{bmatrix}$ **19.** $B = \begin{bmatrix} 1 & 0 & -1 \\ 0 & 1 & 2 \end{bmatrix}$, $P = \begin{bmatrix} -\frac{1}{2} & -2 \\ \frac{1}{2} & 1 \end{bmatrix}$

21. $B = \begin{bmatrix} 1 & 0 & -1 \\ 0 & 1 & 2 \\ 0 & 0 & 0 \end{bmatrix}$, $P = \begin{bmatrix} 1 & 0 & 0 \\ -1 & 1 & 0 \\ -3 & 2 & 1 \end{bmatrix}$ **23.** $\begin{bmatrix} 0 & 1 \\ 1 & 0 \end{bmatrix}$ **25.** $\begin{bmatrix} 1 & 0 \\ -3 & 1 \end{bmatrix}$ **27.** $\begin{bmatrix} 1 & 0 & 0 \\ 0 & 0 & 1 \\ 0 & 1 & 0 \end{bmatrix}$

29. $\begin{bmatrix} 1 & 0 & 0 \\ 5 & 1 & 0 \\ 0 & 0 & 1 \end{bmatrix}$ **31.** $\begin{bmatrix} 1 & 0 \\ -5 & 1 \end{bmatrix}\begin{bmatrix} 1 & -3 \\ 0 & 1 \end{bmatrix} = \begin{bmatrix} 1 & -3 \\ -5 & 16 \end{bmatrix}$ **33.** $\begin{bmatrix} 1 & -2 \\ 0 & 1 \end{bmatrix}\begin{bmatrix} 1 & 0 \\ 0 & \frac{1}{4} \end{bmatrix}\begin{bmatrix} 1 & 0 \\ 1 & 1 \end{bmatrix} = \begin{bmatrix} \frac{1}{2} & -\frac{1}{2} \\ \frac{1}{4} & \frac{1}{4} \end{bmatrix}$

35. $\begin{bmatrix} 1 & 0 & 0 \\ 0 & 1 & 0 \\ 3 & 0 & 1 \end{bmatrix}\begin{bmatrix} 1 & -2 & 0 \\ 0 & 1 & 0 \\ 0 & 0 & 1 \end{bmatrix}\begin{bmatrix} 0 & 1 & 0 \\ 1 & 0 & 0 \\ 0 & 0 & 1 \end{bmatrix} = \begin{bmatrix} -2 & 1 & 0 \\ 1 & 0 & 0 \\ -6 & 3 & 1 \end{bmatrix}$
  37. $\begin{bmatrix} 1 & 0 \\ 3 & 1 \end{bmatrix}\begin{bmatrix} 1 & 2 \\ 0 & 1 \end{bmatrix}$

39. $\begin{bmatrix} 1 & 0 & 0 \\ 0 & 1 & 0 \\ 2 & 0 & 1 \end{bmatrix}\begin{bmatrix} 1 & 0 & 0 \\ 0 & 0 & 1 \\ 0 & 1 & 0 \end{bmatrix}\begin{bmatrix} 1 & 0 & 0 \\ 0 & 1 & 0 \\ 0 & 0 & 3 \end{bmatrix}\begin{bmatrix} 1 & 0 & -2 \\ 0 & 1 & 0 \\ 0 & 0 & 1 \end{bmatrix}$
  41. $\begin{bmatrix} 0 & 1 \\ 1 & 0 \end{bmatrix}, \begin{bmatrix} k & 0 \\ 0 & 1 \end{bmatrix}, \begin{bmatrix} 1 & 0 \\ 0 & k \end{bmatrix}, \begin{bmatrix} 1 & k \\ 0 & 1 \end{bmatrix}, \begin{bmatrix} 1 & 0 \\ k & 1 \end{bmatrix}$

45. $R_i + 2kR_j \rightarrow R_i$

Exercise 2-5 Chapter Review

1. 8 **2.** $[3 \quad 2 \quad -5]$ **3.** $\begin{bmatrix} -1 \\ 1 \\ -2 \end{bmatrix}$ **4.** $\begin{bmatrix} 0 & 2 & 1 \\ 1 & 4 & 0 \\ -1 & -2 & 1 \end{bmatrix}$ **5.** Not defined **6.** $\begin{bmatrix} 15 & -7 \\ 39 & -18 \end{bmatrix}$ **7.** Not defined

8. $\begin{bmatrix} 7 & -1 \\ 16 & 9 \end{bmatrix}$ **9.** Not defined **10.** $\begin{bmatrix} -2 & -2 & 3 \\ 1 & 4 & 0 \end{bmatrix}$ **11.** $\begin{bmatrix} 1 & -3 \\ 20 & 3 \end{bmatrix}$ **12.** Not defined **13.** $\begin{bmatrix} 0 & -3 \\ 2 & -3 \end{bmatrix}$

14. $\begin{bmatrix} -4 & 7 \\ 0 & 7 \end{bmatrix}$ **15.** $\begin{bmatrix} 3 & -2 \\ -4 & 3 \end{bmatrix}$ **16.** (A) $x_1 = -1, x_2 = 3$ (B) $x_1 = 1, x_2 = 2$ (C) $x_1 = 8, x_2 = -10$

17. $\begin{bmatrix} \frac{1}{2} & 0 \\ 0 & 1 \end{bmatrix}\begin{bmatrix} 2 & -4 & 6 \\ 0 & 2 & 5 \end{bmatrix} = \begin{bmatrix} 1 & -2 & 3 \\ 0 & 2 & 5 \end{bmatrix}$ **18.** $\begin{bmatrix} 0 & 1 \\ 1 & 0 \end{bmatrix}\begin{bmatrix} 2 & -4 & 6 \\ 0 & 2 & 5 \end{bmatrix} = \begin{bmatrix} 0 & 2 & 5 \\ 2 & -4 & 6 \end{bmatrix}$

19. $\begin{bmatrix} 1 & 2 \\ 0 & 1 \end{bmatrix}\begin{bmatrix} 2 & -4 & 6 \\ 0 & 2 & 5 \end{bmatrix} = \begin{bmatrix} 2 & 0 & 16 \\ 0 & 2 & 5 \end{bmatrix}$ **20.** B **21.** $A^{-1}BA$ **22.** $I + A^{-1}B$ **23.** $B^{-1} + B^{-1}A^{-1}B$

24. $B^{-1} + A^{-1}$ **25.** $AB^{-1}A^{-1} + A^{-1}$ **26.** $\begin{bmatrix} -\frac{5}{2} & 2 & -\frac{1}{2} \\ 1 & -1 & 1 \\ \frac{1}{2} & 0 & -\frac{1}{2} \end{bmatrix}$ **27.** $\frac{1}{12}\begin{bmatrix} -11 & -1 & 60 \\ 10 & 2 & -48 \\ 1 & -1 & 0 \end{bmatrix}$

28. (A) $x_1 = 2, x_2 = 1, x_3 = -1$ (B) $x_1 = 1, x_2 = -2, x_3 = 1$ (C) $x_1 = -1, x_2 = 2, x_3 = -2$
29. (A) $x_1 = -\frac{8}{3}, x_2 = \frac{7}{3}, x_3 = \frac{1}{3}$ (B) $x_1 = -\frac{253}{12}, x_2 = \frac{103}{6}, x_3 = -\frac{1}{12}$ (C) $x_1 = -\frac{61}{12}, x_2 = \frac{25}{6}, x_3 = -\frac{1}{12}$

30. $B = \begin{bmatrix} 1 & 0 & -5 \\ 0 & 1 & 4 \end{bmatrix}, P = \begin{bmatrix} -1 & -2 \\ 1 & 1 \end{bmatrix}$ **31.** $B = \begin{bmatrix} 1 & 0 & 0 \\ 0 & 1 & 0 \\ 0 & 0 & 1 \end{bmatrix}, P = \begin{bmatrix} \frac{1}{2} & 0 & 0 \\ 1 & 1 & 0 \\ 0 & -2 & 1 \end{bmatrix}$ **32.** $\begin{bmatrix} 1 & -k & k^2-k \\ 0 & 1 & -k \\ 0 & 0 & 1 \end{bmatrix}$

33. $\begin{bmatrix} 1 & k & k^2 & k^3 \\ 0 & 1 & k & k^2 \\ 0 & 0 & 1 & k \\ 0 & 0 & 0 & 1 \end{bmatrix}$ **41.** 250 tons of ore A, 100 tons of ore B

42. (A) 250 tons of ore A, 100 tons of ore B (B) 150 tons of ore A, 40 tons of ore B

43.

		To								
		Akron	Bay City	Canton	Dayton					
	Akron	0	0	0	1		1	1	1	2
From	Bay City	1	0	0	0	$= A, \ S_3 =$	1	1	1	1
	Canton	0	0	0	1		1	1	1	2
	Dayton	0	1	1	0		1	2	2	2

Each city can be reached in three or fewer flights.

44. (B) $\begin{bmatrix} \frac{1}{8} & \frac{3}{8} \\ \frac{3}{8} & \frac{1}{8} \end{bmatrix}$ **(C)** 1: $x_1 = 6°$, $x_2 = 2°$; 2: $x_1 = 12°$, $x_2 = 4°$; 3: $x_1 = 10°$, $x_2 = 6°$

Chapter 3   Exercise 3-1

1. 1   **3.** -2 **5.** $\begin{bmatrix} 1 & -2 \\ -2 & 5 \end{bmatrix}$ **7.** $\begin{bmatrix} \frac{1}{2} & 0 \\ -2 & 1 \end{bmatrix}$ **9.** $\begin{bmatrix} 0.1 & 0.2 \\ -0.3 & 0.4 \end{bmatrix}$ **11.** $x_1 = -2$, $x_2 = 3$ or $(-2, 3)$

13. $x_1 = 4$, $x_2 = 0$ or $(4, 0)$ **15.** $\begin{bmatrix} 5 & 6 \\ 8 & 9 \end{bmatrix}$, $\begin{bmatrix} 4 & 6 \\ 7 & 9 \end{bmatrix}$, $\begin{bmatrix} 4 & 5 \\ 7 & 8 \end{bmatrix}$ **17.** $1(-3) + 2(6) + 3(-3) = 0$ **19.** $-12, 6$

21. $\begin{bmatrix} 1 & 4 & 7 \\ 0 & -2 & 5 \\ -6 & -5 & -7 \end{bmatrix}$, $\begin{bmatrix} 2 & 4 & 7 \\ -1 & -2 & 5 \\ 8 & -5 & -7 \end{bmatrix}$, $\begin{bmatrix} 2 & 1 & 7 \\ -1 & 0 & 5 \\ 8 & -6 & -7 \end{bmatrix}$, $\begin{bmatrix} 2 & 1 & 4 \\ -1 & 0 & -2 \\ 8 & -6 & -5 \end{bmatrix}$

23. $2(-1)^2 \begin{vmatrix} -2 & 5 \\ -5 & -7 \end{vmatrix} + 4(-1)^3 \begin{vmatrix} -1 & 5 \\ 8 & -7 \end{vmatrix} + 7(-1)^4 \begin{vmatrix} -1 & -2 \\ 8 & -5 \end{vmatrix} = 357$

25. $(-1)(-1)^3 \begin{vmatrix} 1 & 4 \\ -6 & -5 \end{vmatrix} + 0(-1)^4 \begin{vmatrix} 2 & 4 \\ 8 & -5 \end{vmatrix} + (-2)(-1)^5 \begin{vmatrix} 2 & 1 \\ 8 & -6 \end{vmatrix} = -21$ **27.** 2   **29.** 0   **31.** 16   **33.** 9

35. 600   **37.** $\lambda = 4$, $\{(2t, t)|t$ any nonzero real number$\}$

39. $\lambda = 4$, $\{(-t, t)|t$ any nonzero real number$\}$; $\lambda = 8$, $\{(3t, t)|t$ any nonzero real number$\}$ **41.** 0, 2   **43.** $-6, 0, 4$

49. abc **51.** 0   **57.** $\lambda = -k$, $\{(t, t)|t$ any nonzero real number$\}$; $\lambda = -3k$, $\{(-t, t)|t$ any nonzero real number$\}$

Exercise 3-2

1. Upper triangular; -40 **3.** Lower triangular; 210   **5.** Diagonal; 1   **7.** None of these; 0   **9.** -10

11. 10   **13.** 0   **15.** 5, -1, 1, -40; det(A) = 200   **17.** 10   **19.** 21   **21.** -12 **23.** -10 **25.** 288

27. 120   **39.** 0, 1   **45.** $12x + 3y - 27 = 0$ or $4x + y - 9 = 0$

51. $6y - 18 - 24x + 6x^2 = 0$ or $y = -x^2 + 4x + 3$ **53.** $12y + 48x - 12x^3 = 0$ or $y = x^3 - 4x$

Exercise 3-3

1. $\begin{bmatrix} 1 & 4 \\ 2 & 5 \\ 3 & 6 \end{bmatrix}$ **3.** $\begin{bmatrix} 1 \\ -1 \\ 0 \\ -4 \end{bmatrix}$ **5.** $\begin{bmatrix} 1 & 2 & 3 \\ 2 & 0 & 4 \\ 3 & 4 & -1 \end{bmatrix}$ **7.** $\begin{bmatrix} 1 & -6 & -5 \\ 1 & 2 & 3 \\ -5 & -2 & 1 \end{bmatrix}$ **9.** 8

11. $\begin{bmatrix} 1 & 1 & -5 \\ -6 & 2 & -2 \\ -5 & 3 & 1 \end{bmatrix}$ **13.** $\begin{bmatrix} 8 & 0 & 0 \\ 0 & 8 & 0 \\ 0 & 0 & 8 \end{bmatrix}$ **15.** $\begin{bmatrix} -1 & 0 & 1 \\ 2 & 1 & -2 \\ -2 & 0 & 1 \end{bmatrix}$ **17.** $\frac{1}{2}\begin{bmatrix} -2 & 0 & 2 \\ -8 & -2 & 8 \\ -16 & -5 & 17 \end{bmatrix} = \begin{bmatrix} -1 & 0 & 1 \\ -4 & -1 & 4 \\ -8 & -2.5 & 8.5 \end{bmatrix}$

19. $\begin{bmatrix} 1 & 0 & 0 & 0 \\ 2 & -1 & 0 & 0 \\ 4 & -2 & 1 & 0 \\ 8 & -4 & 2 & -1 \end{bmatrix}$ **21.** $\frac{1}{4}\begin{bmatrix} 2 & 0 & 0 & 0 \\ 0 & 4 & 0 & -4 \\ -2 & 0 & 2 & 0 \\ 0 & 0 & -2 & 4 \end{bmatrix} = \begin{bmatrix} 0.5 & 0 & 0 & 0 \\ 0 & 1 & 0 & -1 \\ -0.5 & 0 & 0.5 & 0 \\ 0 & 0 & -0.5 & 1 \end{bmatrix}$ **23.** $[x_1^2 + 4x_1x_2 + 3x_2^2]$

25. $\begin{bmatrix} \frac{1}{a_{11}} & 0 & 0 \\ 0 & \frac{1}{a_{22}} & 0 \\ 0 & 0 & \frac{1}{a_{33}} \end{bmatrix}$, $a_{11}a_{22}a_{33} \neq 0$ **27.** $\frac{1}{a_{11}a_{22}a_{33}}\begin{bmatrix} a_{22}a_{33} & -a_{12}a_{33} & a_{12}a_{23} - a_{13}a_{22} \\ 0 & a_{11}a_{33} & -a_{11}a_{23} \\ 0 & 0 & a_{11}a_{22} \end{bmatrix}$, $a_{11}a_{22}a_{33} \neq 0$

29. $\dfrac{1}{1-a^2}\begin{bmatrix} 1 & 0 & -a \\ 0 & 1-a^2 & 0 \\ -a & 0 & 1 \end{bmatrix}$, $a \neq 1, -1$

39. $\begin{bmatrix} 1 & 0 \\ 0 & 1 \end{bmatrix}$, $\begin{bmatrix} -1 & 0 \\ 0 & -1 \end{bmatrix}$, and all matrices of the form $\begin{bmatrix} a & b \\ c & -a \end{bmatrix}$ where $a^2 + bc = 1$

43. (A) $\begin{bmatrix} 1 & -1 & -1 \\ R_1 & R_2 & 0 \\ 0 & -R_2 & R_3 \end{bmatrix}\begin{bmatrix} I_1 \\ I_2 \\ I_3 \end{bmatrix} = \begin{bmatrix} 0 \\ V_1 \\ V_2 \end{bmatrix}$ **(B)** $\dfrac{1}{D}\begin{bmatrix} R_2R_3 & R_2+R_3 & R_2 \\ -R_1R_3 & R_3 & -R_1 \\ -R_1R_2 & R_2 & R_1+R_2 \end{bmatrix}$, $D = R_1R_2 + R_1R_3 + R_2R_3$

    (C) $I_1 = [(R_2 + R_3)V_1 + R_2V_2]/D$, $I_2 = (R_3V_1 - R_1V_2)/D$, $I_3 = [R_2V_1 + (R_1 + R_2)V_2]/D$

45. (A) $\begin{bmatrix} 0 & a & p \\ a & 0 & q \\ p & q & 0 \end{bmatrix}\begin{bmatrix} x \\ y \\ \lambda \end{bmatrix} = \begin{bmatrix} -b \\ -c \\ d \end{bmatrix}$ **(B)** $\dfrac{1}{2apq}\begin{bmatrix} -q^2 & pq & aq \\ pq & -p^2 & ap \\ aq & ap & -a^2 \end{bmatrix}$

    (C) $x = (bq^2 - cpq + adq)/2apq$, $y = (-bpq + cp^2 + adp)/2apq$, $\lambda = (-abq - acp - a^2d)/2apq$

Exercise 3-4

1. $x_1 = 6$, $x_2 = -2$ or $(6, -2)$ **3.** Cramer's Rule does not apply.   **5.** $x_1 = 4$, $x_2 = 3$, $x_3 = -2$ or $(4, 3, -2)$
7. Cramer's Rule does not apply.   **9.** $x_1 = 0$, $x_2 = 2$, $x_3 = 3$ or $(0, 2, 3)$ **11.** Cramer's Rule does not apply.
13. $x_1 = 30$, $x_2 = 20$, $x_3 = -10$ or $(30, 20, -10)$ **15.** Cramer's Rule does not apply.
17. $x_1 = \frac{5}{2}$, $x_2 = \frac{7}{2}$, $x_3 = \frac{13}{2}$ or $(\frac{5}{2}, \frac{7}{2}, \frac{13}{2})$ **19.** $x_1 = -\frac{46}{5} = -9.2$ **21.** $x_1 = 1$ **23.** $x_1 = -\frac{100}{121}$
25. $s \neq 0, -1$; $x_1 = -1/s$, $x_2 = 1/(s+1)$, $x_3 = 1/[s(s+1)]$ **27.** $z = (\cos\theta)u - [(1/r)\sin\theta]v$, $r^2\sin\theta \neq 0$

Exercise 3-5   Chapter Review

1. 2   **2.** $\dfrac{1}{2}\begin{bmatrix} 2 & 3 \\ 2 & 4 \end{bmatrix} = \begin{bmatrix} 1 & 1.5 \\ 1 & 2 \end{bmatrix}$ **3.** $\begin{bmatrix} 4 & -2 \\ -3 & 2 \end{bmatrix}$ **4.** $x_1 = -2$, $x_2 = 3$ or $(-2, 3)$ **5.** $\begin{bmatrix} 1 & 3 & 2 \\ 2 & 0 & 2 \\ -3 & 5 & -2 \end{bmatrix}$

6. $\begin{bmatrix} 0 & 5 \\ 2 & -2 \end{bmatrix}$, $\begin{bmatrix} 3 & 5 \\ 2 & -2 \end{bmatrix}$, $\begin{bmatrix} 3 & 0 \\ 2 & 2 \end{bmatrix}$ **7.** $-10, 16, 6$ **8.** $1(-10) + 2(16) - 3(6) = 4$ **9.** 4

10. $\begin{bmatrix} -10 & 16 & 6 \\ -2 & 4 & 2 \\ 10 & -14 & -6 \end{bmatrix}$ **11.** $\begin{bmatrix} -10 & -2 & 10 \\ 16 & 4 & -14 \\ 6 & 2 & -6 \end{bmatrix}$ **12.** $\dfrac{1}{4}\begin{bmatrix} -10 & -2 & 10 \\ 16 & 4 & -14 \\ 6 & 2 & -6 \end{bmatrix} = \begin{bmatrix} -2.5 & -0.5 & 2.5 \\ 4 & 1 & -3.5 \\ 1.5 & 0.5 & -1.5 \end{bmatrix}$

13. (A) 15   **(B)** 5   **(C)** 5   **14.** 24   **15.** 1   **16.** 3   **17.** -1 **18.** -6 **19.** 40   **20.** 0
21. 90   **22.** 90   **23.** 0   **24.** -126 **25.** -3 **26.** 0   **27.** 120   **28.** -4 **29.** $x_2 = 7.7$ **30.** $x_2 = \frac{1}{2}$

31. $\begin{bmatrix} 17 & 1 & -23 \\ -22 & -1 & 30 \\ -19 & -1 & 26 \end{bmatrix}$ **32.** $\dfrac{1}{2}\begin{bmatrix} 1 & -1 & 4 \\ 3 & -7 & 22 \\ 2 & -4 & 12 \end{bmatrix} = \begin{bmatrix} 0.5 & -0.5 & 2 \\ 1.5 & -3.5 & 11 \\ 1 & -2 & 6 \end{bmatrix}$ **33.** A^{-1} does not exist

34. $\dfrac{1}{10}\begin{bmatrix} 2 & 3 & 1 \\ -8 & 3 & 1 \\ 4 & -4 & 2 \end{bmatrix} = \begin{bmatrix} 0.2 & 0.3 & 0.1 \\ -0.8 & 0.3 & 0.1 \\ 0.4 & -0.4 & 0.2 \end{bmatrix}$ **39.** $\dfrac{1}{ad-bc}\begin{bmatrix} d & 0 & -b \\ 0 & ad-bc & 0 \\ -c & 0 & a \end{bmatrix}$, $ad - bc \neq 0$

40. $\begin{bmatrix} 1 & 0 & 0 \\ -a & 1 & 0 \\ ab & -b & 1 \end{bmatrix}$ **41.** $\lambda = 1$, $\{(3t, t)|t$ any nonzero real number$\}$; $\lambda = -1$, $\{(t, t)|t$ any nonzero real number$\}$

42. 0, 2, 6   **43.** $s \neq 0$, $x_1 = 1/(4 + s^2)$, $x_2 = 2/s$, $x_3 = -s/(s^2 + 4)$

44. $u = (\cos \theta)x - [(1/r) \sin \theta]y$, $v = (\sin \theta)x + [(1/r) \cos \theta]y$, $r \neq 0$

45. (A) 10   (B) 4   (C) 2   (D) 25   (E) $\frac{1}{2}$ (F) $\frac{1}{10}$ **49.** $-8x + 3y + 17 = 0$

50. $6y - 24 + 24x - 6x^2 = 0$ or $y = x^2 - 4x + 4$ **51.** (A) $x^2 + y^2 - 25 = 0$ (B) $x^2 - 2x + y^2 + 4y - 20 = 0$

52. $\lambda = -3k$, $\{(-2t, t)|t$ any nonzero number$\}$; $\lambda = -\frac{1}{2}k$, $\{(t/2, t)|t$ any nonzero number$\}$

53. (A) $\dfrac{1}{D} \begin{bmatrix} R_2R_3 & R_3 & R_2 \\ -R_1R_3 & R_1 + R_3 & -R_1 \\ -R_1R_2 & -R_1 & R_1 + R_2 \end{bmatrix}$, $D = R_1R_2 + R_2R_3 + R_1R_3$

    (B) $I_1 = (R_2 + R_3)V/D$, $I_2 = R_3V/D$, $I_3 = R_2V/D$

Chapter 4   Exercise 4-1

1. $\|\mathbf{v}\| = 13$, $\mathbf{v} = \overrightarrow{OP}$ where $P = (12, 5)$ **3.** $\|\mathbf{v}\| = 3\sqrt{5}$, $\mathbf{v} = \overrightarrow{OP}$ where $P = (-3, 6)$

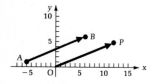

5. $(5, -7)$ **7.** $(3, 0)$

9. $\mathbf{u} + \mathbf{v} = (6, 6)$, $\mathbf{u} - \mathbf{v} = (4, -2)$ **11.** $\mathbf{u} + \mathbf{v} = (3, -1)$, $\mathbf{u} - \mathbf{v} = (1, 7)$

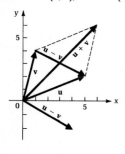

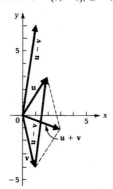

13. $\overrightarrow{AB}$ where $B = (1, 3)$ **15.** $\overrightarrow{AB}$ where $B = (0, 0)$ **17.** $\overrightarrow{AB}$ where $A = (7, 1)$ **19.** $\overrightarrow{AB}$ where $A = (-4, -6)$

21. $\|\mathbf{v}\| = 10$

    (A) $2\mathbf{v} = (12, -16)$; $\|2\mathbf{v}\| = 20$

    (B) $-\frac{1}{2}\mathbf{v} = (-3, 4)$; $\|-\frac{1}{2}\mathbf{v}\| = 5$

23. $4\mathbf{i} - 5\mathbf{j}$ **25.** $5\mathbf{i} - 5\mathbf{j}$

27. $(-\frac{3}{5}, \frac{4}{5})$ **29.** $\dfrac{1}{\sqrt{5}}\mathbf{i} - \dfrac{2}{\sqrt{5}}\mathbf{j}$

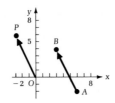

31. $12\mathbf{i} + 7\mathbf{j}$ **33.** $-46\mathbf{j}$

35. $(2\sqrt{10}, -6\sqrt{10})$ **43.** $\{t(-1, 1)|t$ any scalar$\}$ **45.** $\{(5, 0) + t(-2, 1)|t$ any scalar$\}$ **49.** $\mathbf{x} = 6\mathbf{i} - 16\mathbf{j}$

Exercise 4-2

1. $d = 9$

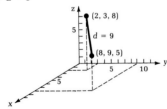

3. $d = \sqrt{110}$

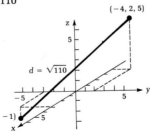
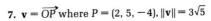
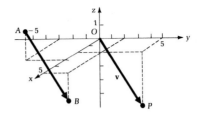

5. $\mathbf{v} = \overrightarrow{OP}$ where $P = (2, 3, 6)$, $\|\mathbf{v}\| = 7$

7. $\mathbf{v} = \overrightarrow{OP}$ where $P = (2, 5, -4)$, $\|\mathbf{v}\| = 3\sqrt{5}$

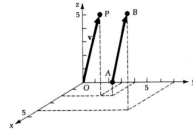

9. $\mathbf{u} + \mathbf{v} = (9, 4, 1)$, $\mathbf{u} - \mathbf{v} = (1, -8, 5)$　**11.** $\mathbf{u} + \mathbf{v} = (5, 7, 2)$, $\mathbf{u} - \mathbf{v} = (5, -5, -2)$
13. $\overrightarrow{AB}$ where $B = (3, -1, -1)$　**15.** $\overrightarrow{AB}$ where $B = (0, 0, 0)$　**17.** $\overrightarrow{AB}$ where $A = (4, 2, 5)$
19. $\overrightarrow{AB}$ where $A = (2, -1, -1)$
21. $\|\mathbf{v}\| = 12$　(A) $2\mathbf{v} = (16, -16, 8)$; $\|2\mathbf{v}\| = 24$　(B) $-\frac{1}{4}\mathbf{v} = (-2, 2, -1)$; $\|-\frac{1}{4}\mathbf{v}\| = 3$
23. $-\mathbf{i} + 2\mathbf{j} - 4\mathbf{k}$　**25.** $-4\mathbf{j}$　**27.** $(\frac{2}{3}, \frac{2}{3}, -\frac{1}{3})$　**29.** $(\sqrt{3}/9)\mathbf{i} - (\sqrt{3}/9)\mathbf{j} - (5\sqrt{3}/9)\mathbf{k}$　**31.** $(\frac{11}{3}, \frac{10}{3}, \frac{2}{3})$
33. $5\mathbf{i} - 3\mathbf{j} + 5\mathbf{k}$　**35.** $\mathbf{0}$　**37.** $5\mathbf{i} + 7\mathbf{j} + \mathbf{k}$　**39.** $\{t(2, -1, 1)|t \text{ any scalar}\}$
41. $\{s(2, 1, 0) + t(-3, 0, 1)|s \text{ and } t \text{ any scalars}\}$　**43.** $\{(2, 4, 0) + t(3, -2, 1)|t \text{ any scalar}\}$

Exercise 4-3

1. (A) 1　(B) 0　(C) 1　**3.** 11　**5.** 0　**7.** 2　**9.** 3　**11.** $\cos \theta = \sqrt{2}/2$; $\theta = \pi/4$ or $45°$
13. $\cos \theta = 0$; $\theta = \pi/2$ or $90°$　**15.** $\cos \theta = \sqrt{3}/2$; $\theta = \pi/6$ or $30°$　**17.** $\cos \theta = \sqrt{10}/5$; $\theta \approx 0.88608$ or $50.77°$
19. Orthogonal　**21.** Not orthogonal　**23.** $(\frac{7}{9}, -\frac{4}{9}, \frac{4}{9})$; $\cos \alpha = \frac{7}{9}$, $\cos \beta = -\frac{4}{9}$, $\cos \gamma = \frac{4}{9}$
25. $(\sqrt{2}/2)\mathbf{i} + (\sqrt{2}/2)\mathbf{k}$; $\cos \alpha = \sqrt{2}/2$, $\cos \beta = 0$, $\cos \gamma = \sqrt{2}/2$　**27.** $\frac{2}{3}$　**29.** t any scalar　**31.** $2, -2$
33. $\{t(5, 1)|t \text{ any scalar}\}$
35. $\mathbf{p} = (5, 10)$, $\mathbf{q} = (4, -2)$　**37.** $\mathbf{p} = \frac{1}{2}\mathbf{i} - \frac{1}{2}\mathbf{j}$, $\mathbf{q} = \frac{1}{2}\mathbf{i} + \frac{1}{2}\mathbf{j} + \mathbf{k}$

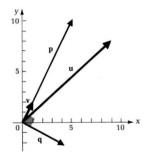
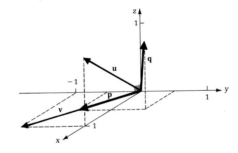

39. Mutually orthogonal   **41.** (A) $x_1 + x_2 - 2x_3 = 0$ (B) $\{t(-3, 5, 1)|t \text{ any scalar}\}$

$$2x_1 + x_2 + x_3 = 0$$

Exercise 4-4

1.

×	i	j	k
i	0	k	−j
j	−k	0	i
k	j	−i	0

3. $-2\mathbf{j} - 3\mathbf{k}$ **5.** $\mathbf{i} + 2\mathbf{j}$ **7.** $11\mathbf{i} + \mathbf{j} + 7\mathbf{k}$ **9.** $-13\mathbf{i} + 31\mathbf{j} - 19\mathbf{k}$ **11.** $-5\mathbf{k}$
13. $-10\mathbf{i} - 15\mathbf{j}$ **15.** $\mathbf{w} = (-1, -32, -13)$ (or any nonzero scalar multiple of $\mathbf{w}$)
17. $\mathbf{w} = \mathbf{i} - \mathbf{j} + \mathbf{k}$ (or any nonzero scalar multiple of $\mathbf{w}$)
19. $\overrightarrow{AB} = 2\mathbf{j} = \overrightarrow{CD}$, $\overrightarrow{AC} = 3\mathbf{i} = \overrightarrow{BD}$; 6

21. $\overrightarrow{AB} = -3\mathbf{i} + 2\mathbf{j} + \mathbf{k} = \overrightarrow{CD}$, $\overrightarrow{AC} = \mathbf{i} + 4\mathbf{j} - 7\mathbf{k} = \overrightarrow{BD}$; $2\sqrt{230}$ **23.** $\cos\theta = \frac{20}{21}$, $\sin\theta = \sqrt{41}/21$ **25.** 60   **27.** 50
31. $9\sqrt{22}/2$

Exercise 4-5

1. $\overrightarrow{OP} = (2\mathbf{i} - \mathbf{j} + 3\mathbf{k}) + t(4\mathbf{i} + 6\mathbf{j} - 5\mathbf{k})$ **3.** $x = t$, $y = 2t$, $z = 3t$ **5.** $(x - 1)/4 = (y - 2)/5 = (z - 3)/6$
7. $P_1 = (-2, 4, 1)$, $\mathbf{v} = \mathbf{i} - 3\mathbf{j} + 2\mathbf{k}$ **9.** $P_1 = (3, 0, 7)$, $\mathbf{v} = 2\mathbf{i} - 5\mathbf{j}$ **11.** $P_1 = (-10, 5, 15)$, $\mathbf{v} = 5\mathbf{i} + 10\mathbf{k}$
13. $\overrightarrow{OP} = (2\mathbf{i} + \mathbf{j} + 3\mathbf{k}) + t(3\mathbf{i} + 6\mathbf{j} + 5\mathbf{k})$ **15.** $x = -6 + 2t$, $y = 4$, $z = 2 - 4t$ **17.** $x = 1$, $(y - 2)/3 = (z + 1)/2$
19. $\overrightarrow{OP} = (2\mathbf{i} - \mathbf{j} + 4\mathbf{k}) + t(4\mathbf{i} - 5\mathbf{j} + 6\mathbf{k})$ **21.** $x = 2$, $y = 1 - 4t$, $z = -3 + 3t$
23. $(x - 2)/2 = (y - 3)/3 = (z + 5)/5$ **25.** $(0, 10, 1)$, $(5, 0, 11)$, $(-\frac{1}{2}, 11, 0)$ **27.** 3   **29.** $\sqrt{5}/3$ **31.** $\sqrt{3}$
33. Parallel   **35.** Intersect at $(5, 7, 2)$ **37.** Skew   **39.** $x = t$, $y = mt + b$ is one set of parametric equations for ℓ.

Exercise 4-6

1. $\mathbf{n} = 2\mathbf{i} - 3\mathbf{j} + 5\mathbf{k}$ **3.** $\mathbf{n} = 2\mathbf{i} - 5\mathbf{k}$ **5.** $(x - 2) + 2(y - 1) + 4(z - 3) = 0$, $x + 2y + 4z = 16$
7. $3x + 4y - 7z = 0$ **9.** $2x - y + 4z = -17$ **11.** $2x + 3y + 6z = 0$ **13.** $3x + y + 2z = 18$
15. $2x - 4y + z = 12$ **17.** $\overrightarrow{OP} = (\mathbf{i} - \mathbf{j} + \mathbf{k}) + t(2\mathbf{i} + 3\mathbf{j} - 6\mathbf{k})$ **19.** $x = 10 + 7t$, $y = 3 + 4t$, $z = t$
21. $(12, 0, 0)$, $(0, -8, 0)$, $(0, 0, 6)$ **23.** Parallel   **25.** Oblique, $(\frac{1}{2}, 2, \frac{1}{2})$ **27.** $\frac{5}{3}$ **29.** $2\sqrt{30}/3$
31. $\pi/4$ or $45°$, none of these   **33.** 0, identical   **35.** $\pi/2$ or $90°$, perpendicular   **37.** $2x - 4y + 5z = 30$
39. $(x - 4)/1 = (y - 5)/3 = (z - 8)/3$ **41.** $-x + 2y + z = -1$ **43.** p_1: $2x - z = -1$; p_2: $2x - z = 1$
45. $d = \frac{2}{5}\sqrt{5}$

Exercise 4-7   Chapter Review

1. (A) $\sqrt{53}$ (B) $\mathbf{v} = \overrightarrow{OP}$ where $P = (7, 2)$ (C)

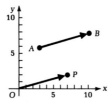

2. $(4, -4)$; $4\mathbf{i} - 4\mathbf{j}$ **3.** $\mathbf{v} = \overrightarrow{AB}$ where $B = (-1, -3)$
4. $\mathbf{u} + \mathbf{v} = (8, 5)$; $\mathbf{u} - \mathbf{v} = (4, -3)$ **5.** $d = 5\sqrt{6}$

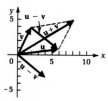

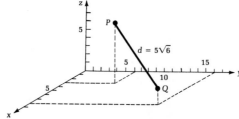

6. (A) $3\sqrt{5}$ (B) $\mathbf{v} = \overrightarrow{OP}$ where $P = (2, 4, 5)$ (C)

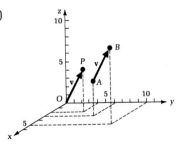

7. $\mathbf{v} = \overrightarrow{AB}$ where $B = (-2, 2, 10)$ **8.** $2\mathbf{i} + 3\mathbf{j} - 7\mathbf{k}$

9. $\mathbf{u} + \mathbf{v} = 12\mathbf{i} + 12\mathbf{j} + 16\mathbf{k}$; $\mathbf{u} - \mathbf{v} = 4\mathbf{i} - 8\mathbf{j} - 2\mathbf{k}$ **10.** 0   **11.** 10   **12.** -6 **13.** 1   **14.** $-\mathbf{i} - 22\mathbf{j} - 8\mathbf{k}$

15. $-\mathbf{i} + \mathbf{j} + \mathbf{k}$ **16.** 0   **17.** $7\mathbf{k}$ **18.** $P_1 = (2, -1, 3)$; $\mathbf{v} = 3\mathbf{i} - 4\mathbf{j} + 7\mathbf{k}$ **19.** $P_1 = (-4, 6, 2)$; $\mathbf{v} = 2\mathbf{i} - 5\mathbf{j} + 3\mathbf{k}$

20. $P_1 = (-3, 2, 1)$; $\mathbf{v} = 4\mathbf{i} + 5\mathbf{j} - 6\mathbf{k}$

21. (A) $\overrightarrow{OP} = (2\mathbf{i} + 5\mathbf{j} + 7\mathbf{k}) + t(3\mathbf{i} + 6\mathbf{j} + 4\mathbf{k})$ (B) $x = 2 + 3t$, $y = 5 + 6t$, $z = 7 + 4t$

     (C) $(x - 2)/3 = (y - 5)/6 = (z - 7)/4$

22. $\mathbf{n} = 2\mathbf{i} - 3\mathbf{j} + 4\mathbf{k}$ **23.** $3x + 4y + z = 32$ **24.** $(\sqrt{5}/5)\mathbf{i} - (2\sqrt{5}/5)\mathbf{j}$ **25.** $(\frac{2}{3}, -\frac{2}{3}, -\frac{1}{3})$ **26.** $(-6, 8)$

27. $-\frac{90}{11}\mathbf{i} + \frac{60}{11}\mathbf{j} + \frac{20}{11}\mathbf{k}$ **28.** $\cos\theta = 0$, $\theta = \pi/2$ or $90°$ **29.** $\cos\theta = -\sqrt{2}/2$, $\theta = 3\pi/4$ or $135°$ **30.** $\cos\theta = 1$, $\theta = 0$

31. $\cos\theta = \sqrt{3}/2$, $\theta = \pi/6$ or $30°$ **32.** Orthogonal   **33.** Not orthogonal   **34.** $\cos\alpha = \frac{12}{13}$, $\cos\beta = -\frac{4}{13}$, $\cos\gamma = \frac{3}{13}$

35. -6 **36.** 1, 2   **37.** $\mathbf{p} = (6, 2)$, $\mathbf{q} = (-1, 3)$

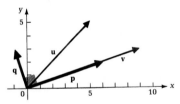

38. (A) $-\mathbf{i} + 7\mathbf{j} + 5\mathbf{k}$ (B) $6\mathbf{i} - 7\mathbf{j} + 5\mathbf{k}$ (C) $18\mathbf{i} - \mathbf{j} + 5\mathbf{k}$ (D) $11\mathbf{i} + 13\mathbf{j} + 5\mathbf{k}$ (E) 35   (F) 35

39. $2\mathbf{i} + 2\mathbf{k}$ **40.** $\overrightarrow{AB} = 2\mathbf{i} + 2\mathbf{j} + \mathbf{k} = \overrightarrow{CD}$, $\overrightarrow{AC} = 6\mathbf{i} + 3\mathbf{j} + 2\mathbf{k} = \overrightarrow{BD}$; $\sqrt{41}$ **41.** 209

42. $x = -2 + 5t$, $y = 1 - 3t$, $z = 4 + 2t$ **43.** $\dfrac{x - 1}{7} = \dfrac{y + 2}{-5} = \dfrac{z}{8}$ **44.** $19x + y + 12z = 17$

45. $\overrightarrow{OP} = (2\mathbf{i} - \mathbf{j} + 4\mathbf{k}) + t(3\mathbf{i} + \mathbf{j} - \mathbf{k})$ **46.** $x = 1 + t$, $y = 2 - t$, $z = t$ **47.** $(5, -3, 8)$

48. $4\sqrt{2}$ **49.** 2   **50.** $(\frac{4}{5}, \frac{3}{5})$, $(-\frac{4}{5}, -\frac{3}{5})$ **51.** $(\frac{12}{17}, \frac{9}{17}, \frac{8}{17})$, $(-\frac{12}{17}, -\frac{9}{17}, -\frac{8}{17})$

58. (A) Skew   (B) Parallel   (C) Intersect at $(5, 2, 1)$ **59.** $y = y_0$, $z = z_0$ **60.** $x = x_0$, $y = y_0 + t$, $z = z_0$

61. $\overrightarrow{OP} = x_0\mathbf{i} + y_0\mathbf{j} + z_0\mathbf{k} + t\mathbf{k}$ **62.** (A) $z = z_0$ (B) $y = y_0$ (C) $x = x_0$ **63.** $13x - 5y - 14z = 1$

64. $\dfrac{x + 4}{8} = \dfrac{y - 5}{-6} = \dfrac{z - 3}{-2}$

Chapter 5   Exercise 5-1

1. -3 **3.** $(3, 4, 5, 6)$ **5.** $x = 3$, $y = -1$, $z = 5$ **7.** $(3, 3, 3, 6)$ **9.** $(3, 3, 2, 8, -2, -9)$

11. $(16, 8, 4, 2, 1, \frac{1}{2}, \frac{1}{4})$ **13.** $(11, -2, -6, 4)$ **15.** 9   **17.** $6\sqrt{2}$ **19.** 2   **21.** $4\sqrt{5}$ **23.** $(\frac{5}{6}, \frac{1}{2}, -\frac{1}{6}, \frac{1}{6})$

25. $(1/\sqrt{6}, -1/\sqrt{6}, 1/\sqrt{6}, -1/\sqrt{6}, 1/\sqrt{6}, -1/\sqrt{6})$ **27.** 7   **29.** 0   **31.** 8   **33.** $3, -3$

35. $\begin{bmatrix} 2 & -1 & 1 \\ 1 & 1 & -1 \end{bmatrix} \begin{bmatrix} x_1 \\ x_2 \\ x_3 \end{bmatrix} = \begin{bmatrix} 4 \\ 5 \end{bmatrix}$; row vectors: $[2 \ \ -1 \ \ 1]$, $[1 \ \ 1 \ \ -1]$; column vectors: $\begin{bmatrix} 2 \\ 1 \end{bmatrix}, \begin{bmatrix} -1 \\ 1 \end{bmatrix}, \begin{bmatrix} 1 \\ -1 \end{bmatrix}$

37. $[-6]$ or -6 **39.** $[0]$ or 0

Exercise 5-2

1. $2x^2 - 4x + 6$ **7.** 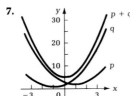   **9.** 3   **17.**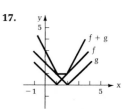
3. $-x^2 + 2x - 3$ **11.** 8
5. $2x^2 + 5$ **13.** $(f + g)(x) = f(x) + g(x) =$
                                              $|x - 1| + |x - 2|$
                                    15. $(2f)(x) = 2f(x) = 2|x - 1|$

29. Closure for scalar multiplication   **31.** Closure for scalar multiplication
33. Closure for both addition and scalar multiplication   **35.** Closure for scalar multiplication
37. V is not a vector space; properties (A-4) and (B-4) are not satisfied.
39. V is not a vector space; property (B-4) is not satisfied.   **41.** V is a vector space.

Exercise 5-3

1. Subspace   **3.** Not closed under either operation   **5.** Not closed under addition   **7.** Subspace
9. Subspace   **11.** Not closed under scalar multiplication   **13.** Subspace   **15.** Not closed under addition
17. Not closed under either operation   **19.** Subspace   **21.** Yes
23. No; N_3 is not closed under either operation.   **25.** Yes   **27.** Yes   **29.** Yes
31. (B) Yes   (C) No; W is not closed under scalar multiplication.   **33.** Yes   **35.** (A) Yes   (B) Yes
37. (A) Yes   (B) No   **41.** Yes

Exercise 5-4

1. $5\mathbf{u}_1 + 2\mathbf{u}_2$ **3.** $3\mathbf{u}_1 + 4\mathbf{u}_2$ **5.** Not possible   **7.** $2\mathbf{u}_1 + 3\mathbf{u}_2$ **9.** Not possible   **11.** $\mathbf{u}_1 + \mathbf{u}_2$
13. Not possible   **15.** Yes   **17.** Yes   **19.** No   **21.** Yes   **23.** No   **25.** $\{(2, 3, 1)\}$ **27.** $\{(0, 0, 0)\}$
29. $\{(2, -4, 1, 0, 0), (-3, 5, 0, 1, 0), (1, -2, 0, 0, 1)\}$ **31.** $(-4, 1, 0)$ and $(5, 0, 1)$ are two of many possible choices.
33. $3x - 6y - 2z = 0$ **35.** $(w_1 - w_2)\mathbf{u}_1 + (-3w_1 + 4w_2)\mathbf{u}_2$
37. $\frac{1}{2}(a_0 + a_1 - a_2)\mathbf{u}_1 + \frac{1}{2}(-a_0 + a_1 + a_2)\mathbf{u}_2 + \frac{1}{2}(a_0 - a_1 + a_2)\mathbf{u}_3$ **39.** $(a + c)\mathbf{u}_1 + (b + d)\mathbf{u}_2 + (b + c)\mathbf{u}_3 + (a + 2d)\mathbf{u}_4$

Exercise 5-5

1. Linearly independent   **3.** Linearly dependent; $k(0, 0) = \mathbf{0}$ for any nonzero scalar k
5. Linearly dependent; $2(1, 0) - 3(0, 1) + (-2, 3) = \mathbf{0}$ **7.** Linearly independent
9. Linearly dependent; $(1, 1, 0) - 4(0, 1, -1) + (-2, 1, -3) + (1, 2, -1) = \mathbf{0}$ **11.** Linearly independent
13. Linearly independent   **15.** Linearly independent

17. Linearly dependent; $2\begin{bmatrix} -4 & 7 \\ -1 & 2 \end{bmatrix} + 5\begin{bmatrix} 2 & -3 \\ 1 & 0 \end{bmatrix} - \begin{bmatrix} 2 & -1 \\ 3 & 4 \end{bmatrix} = \mathbf{0}$

19. Linearly independent   **21.** $\{(2, -4)\}$ or $\{(-3, 6)\}$ **23.** $\{(1, 2, -1), (1, -2, 1)\}$ or $\{(-2, -4, 2), (1, -2, 1)\}$
25. $\{1, 1 - x, 1 - x + x^2\}$

27. $\left\{ \begin{bmatrix} 1 & 0 \\ 0 & 2 \end{bmatrix}, \begin{bmatrix} 0 & 3 \\ 6 & 0 \end{bmatrix} \right\}, \left\{ \begin{bmatrix} 1 & 0 \\ 0 & 2 \end{bmatrix}, \begin{bmatrix} 0 & -2 \\ -4 & 0 \end{bmatrix} \right\}, \left\{ \begin{bmatrix} -2 & 0 \\ 0 & -4 \end{bmatrix}, \begin{bmatrix} 0 & 3 \\ 6 & 0 \end{bmatrix} \right\},$ or $\left\{ \begin{bmatrix} -2 & 0 \\ 0 & -4 \end{bmatrix}, \begin{bmatrix} 0 & -2 \\ -4 & 0 \end{bmatrix} \right\}$

29. $1, -1$ **31.** R is linearly independent; C is linearly dependent   **33.** $k_1 = 2, k_2 = -1$

Exercise 5-6

1. Basis   **3.** Not a basis   **5.** Not a basis   **7.** Not a basis   **9.** Basis   **11.** Not a basis
13. $(1, 2) = -\frac{9}{2}(2, -1) + \frac{5}{2}(4, -1)$ **15.** $3 + x + 4x^2 = -(1 - x) + 0(x - x^2) + 4(1 + x^2)$ **17.** $\{(0, 1, 0), (0, 0, 1)\}$

19. $\{(2, -3, 4)\}$ **21.** $\{(2, 1, 0), (-\frac{5}{2}, 0, 1)\}$ **23.** $\{(4, -2, 6), (6, -2, 4)\}$ or $\{(6, -2, 4), (-10, 5, -15)\}$

25. $\{1 - x^2, \ x - x^2\}, \{1 - x, \ x - x^2\},$ or $\{1 - x, \ 1 - x^2\}$

27. $\{2 - x^2, \ 2 + x^2\}, \{2 - x^2, \ x^2\},$ or $\{2 + x^2, \ x^2\}$

29. $\{(2, 2, 1)\}; 1$ **31.** $\{(1, 1, 0, 0), (2, 0, 0, 1)\}; 2$ **33.** $\{(5, -4, 1, 0, 0), (-2, 1, 0, 1, 0), (3, 2, 0, 0, 1)\}; 3$

35. $\left\{\begin{bmatrix} 1 & 0 \\ 0 & 0 \end{bmatrix}, \begin{bmatrix} 0 & 0 \\ 0 & 1 \end{bmatrix}\right\}; 2$ **37.** $\left\{\begin{bmatrix} 1 & 0 & 0 \\ 0 & 0 & 0 \end{bmatrix}, \begin{bmatrix} 0 & 1 & 0 \\ 0 & 0 & 0 \end{bmatrix}, \begin{bmatrix} 0 & 0 & 1 \\ 0 & 0 & 0 \end{bmatrix}, \begin{bmatrix} 0 & 0 & 0 \\ 1 & 0 & 0 \end{bmatrix}, \begin{bmatrix} 0 & 0 & 0 \\ 0 & 1 & 0 \end{bmatrix}, \begin{bmatrix} 0 & 0 & 0 \\ 0 & 0 & 1 \end{bmatrix}\right\}; 6$

39. $(x_1 - x_2)(2, 1) + (2x_2 - x_1)(1, 1)$ **41.** $(x_2 - x_3)(1, 1, 0) + (x_2 - x_1)(0, 1, 1) + (x_1 - x_2 + x_3)(1, 1, 1)$ **43.** $\{1 - 3x^2, x\}$

Exercise 5-7   Chapter Review

1. (A) $(-4, -7, 12, 0)$ (B) 6   (C) $\sqrt{14}$ (D) $3\sqrt{3}$ (E) $(\frac{2}{5}, \frac{1}{5}, -\frac{4}{5}, \frac{2}{5})$

2. (A) $-9 - 16x + 8x^2$ (B)   **3.** (A) 3   (B) -2 (C)

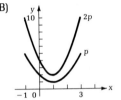   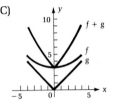

4. $4(4, -5) + 7(-2, 3)$ **5.** $-2(-2, 3)$ **6.** $-4(1, -1, 0) + 3(2, 0, 1)$ **7.** Not possible

8. $3(2 + 3x) - 4(1 - x^2) - 2(1 + 3x + x^2)$ **9.** Not possible   **10.** $2\begin{bmatrix} 1 & -1 \\ 2 & 1 \end{bmatrix} - \begin{bmatrix} 2 & -1 \\ 1 & 2 \end{bmatrix} + \begin{bmatrix} -1 & 1 \\ 1 & 1 \end{bmatrix}$

11. $-3\begin{bmatrix} 1 & -1 \\ 2 & 1 \end{bmatrix} + 2\begin{bmatrix} 2 & -1 \\ 1 & 2 \end{bmatrix}$ **12.** Subspace   **13.** Not a subspace   **14.** Subspace

15. Not a subspace   **16.** Subspace   **17.** Not a subspace   **18.** Not a subspace   **19.** Subspace

20. Linearly dependent; $-(-2, 0, 3) - 4(1, 1, -1) + (2, 4, -1) = \mathbf{0}$ **21.** Linearly independent

22. Linearly independent   **23.** Linearly dependent; $2(2 - x + x^3) + 3(-1 + 2x - x^2) - (1 + 4x - 3x^2 + 2x^3) = \mathbf{0}$

24. $\{(2, -1)\}; 1$ **25.** $\{(1, 2, -3)\}; 1$ **26.** $\{(2, 1, 0), (-5, 0, 1)\}; 2$

27. $\{(1, -2, 1), (2, 0, -1)\}$ or $\{(-3, 6, -3), (2, 0, -1)\}; 2$ **28.** $\{x\}; 1$ **29.** $\{1, x^2\}; 2$ **30.** $\left\{\begin{bmatrix} 1 & 0 \\ 0 & 1 \end{bmatrix}, \begin{bmatrix} 0 & -1 \\ 1 & 0 \end{bmatrix}\right\}; 2$

31. $\left\{\begin{bmatrix} 1 & 1 \\ 0 & 0 \end{bmatrix}, \begin{bmatrix} 0 & 0 \\ 1 & 0 \end{bmatrix}, \begin{bmatrix} 0 & 0 \\ 0 & 1 \end{bmatrix}\right\}; 3$ **32.** $\{(3, -4, 1)\}; 1$ **33.** $\{(1, -2, 1, 0), (-2, 1, 0, 1)\}; 2$

34. $\{(-2, 1, 0, 0, 0), (3, 0, 2, 1, 0), (-4, 0, 1, 0, 1)\}; 3$ **35.** $\{(11, -3, 4, 1, 0), (8, -5, 3, 0, 1)\}; 2$

36. $\mathbf{w} = 7(4, -3) + 8(-3, 2)$ **37.** $\mathbf{w} = 5(1 - 3x + x^2) - 8(2 - 3x + x^2) + 4(3 - 2x + x^2)$

38. $\mathbf{w} = -5\begin{bmatrix} 1 & -1 \\ 1 & 0 \end{bmatrix} + \begin{bmatrix} 0 & 1 \\ 1 & -2 \end{bmatrix} + 3\begin{bmatrix} 1 & -1 \\ 0 & 1 \end{bmatrix} + 6\begin{bmatrix} 1 & 0 \\ 1 & 0 \end{bmatrix}$ **39.** $\sin^2 x + \cos^2 x + \tan^2 x - \sec^2 x = 0$

43. (B) $(ax_1 + bx_2)(a, b) + (ax_2 - bx_1)(-b, a)$ **44.** Only (C) is a basis.   **46.** $\{1 - \frac{3}{4}x^2, x\}$ **50.** $\left\{\begin{bmatrix} 0 & 1 \\ -1 & 0 \end{bmatrix}\right\}; 1$

51. $\left\{\begin{bmatrix} 0 & 1 & 0 \\ -1 & 0 & 0 \\ 0 & 0 & 0 \end{bmatrix}, \begin{bmatrix} 0 & 0 & 1 \\ 0 & 0 & 0 \\ -1 & 0 & 0 \end{bmatrix}, \begin{bmatrix} 0 & 0 & 0 \\ 0 & 0 & 1 \\ 0 & -1 & 0 \end{bmatrix}\right\}; 3$

Chapter 6   Exercise 6-1

1. Row space: $\{(1, 0, -1), (0, 1, 2)\}$;   column space: $\{(1, 0, 0), (0, 1, 0)\}$
3. Row space: $\{(1, 0, 0, -2), (0, 1, 0, 4), (0, 0, 1, -3)\}$;   column space: $\{(1, 0, 0), (0, 1, 0), (0, 0, 1)\}$
5. Row space: $\{(1, 0, 3, -2, 0, 0), (0, 1, 4, 5, 0, 0), (0, 0, 0, 0, 1, 0), (0, 0, 0, 0, 0, 1)\}$;
   column space: $\{(1, 0, 0, 0), (0, 1, 0, 0), (0, 0, 1, 0), (0, 0, 0, 1)\}$
7. Row space: $\{(1, 1, 0, 0, 0, 0, -1), (0, 0, 1, 0, 0, 0, 1), (0, 0, 0, 1, 1, 0, -2), (0, 0, 0, 0, 0, 1, 3)\}$; column
   space: $\{(1, 0, 0, 0, 0), (0, 1, 0, 0, 0), (0, 0, 1, 0, 0), (0, 0, 0, 1, 0)\}$
9. $\{(1, 0, 0), (0, 1, 0), (0, 0, 1)\}$ **11.** $\{(1, 0, -2, 3), (0, 1, 1, 1)\}$ **13.** $\{(1, 0, -1, 1, 0), (0, 1, 2, 3, 1)\}$
15. $\{(1, 0, -1), (0, 1, -2)\}$ **17.** $\{(1, -1, 0, 0), (0, 0, 1, 0), (0, 0, 0, 1)\}$
19. $\{(1, 0, 0, 0, \frac{1}{3}), (0, 1, 0, 0, 0), (0, 0, 1, 0, \frac{2}{3}), (0, 0, 0, 1, 0)\}$

21. (A) $\mathbf{v}_1 = (1, 0, -1), \mathbf{v}_2 = (0, 1, 2)$ (B) $\mathbf{u}_1 = \mathbf{v}_1 + \mathbf{v}_2, \mathbf{u}_2 = \mathbf{v}_1 - 2\mathbf{v}_2, \mathbf{u}_3 = 2\mathbf{v}_1 - \mathbf{v}_2$ **25.** $\left\{ \begin{bmatrix} 1 \\ 0 \\ 3 \end{bmatrix}, \begin{bmatrix} 0 \\ 1 \\ -2 \end{bmatrix} \right\}$

Exercise 6-2

1. 3   **3.** 2   **5.** 2   **7.** $\{(1, 1, 3), (1, -1, 1)\}$; 2   **9.** $\{(1, 1, -2), (1, -1, 4), (1, 2, 5)\}$; 3
11. $\{(1, 1, 1, 3), (1, -1, 2, 2)\}$; 2   **13.** $\{(1, 1, 1), (1, -1, 3)\}$; 2   **15.** $\{(1, 2, 1), (1, 1, 0)\}$; 2
17. $\{(1, 1, 2, 3), (2, -2, 0, -2), (2, 1, 1, 2)\}$; 3   **19.** $\{(1, 1, 2), (-1, 1, -1)\}$ **21.** $\{(1, 1, -2, 3), (1, -1, 1, 5)\}$
23. $\{(1, 1, 1, 3), (1, -1, 2, 2), (2, 4, 3, -1)\}$ **25.** $\{(1, 1, 1), (1, 0, 0), (0, 1, 0)\}$ **27.** $\{(1, -2, 1), (-2, 2, -1), (0, 1, 0)\}$
29. $\{(1, -1, 1, 0), (-1, 1, 0, 0), (1, 0, 0, 0), (0, 0, 0, 1)\}$ **31.** $\{(1, 1, 1, 1), (1, 0, 0, 1), (0, 1, 1, 1), (0, 1, 0, 0)\}$
33. S is a basis   **35.** S is not a basis   **37.** $\{(1, 2, 2), (2, 1, 3), (3, -2, 1)\}$

Exercise 6-3

1. $\begin{bmatrix} 3 \\ 2 \end{bmatrix}$ **3.** $\begin{bmatrix} 4 \\ 3 \end{bmatrix}$ **5.** $(-1, 2)$ **7.** $(4, -5)$ **9.** $\begin{bmatrix} -3 \\ -2 \end{bmatrix}$ **11.** $\begin{bmatrix} 7 \\ 4 \end{bmatrix}$ **13.** $4 - 6x$ **15.** $-1 + x$ **17.** $\begin{bmatrix} -5 \\ 3 \\ 4 \end{bmatrix}$

19. $(6, 1, 4)$ **21.** $\begin{bmatrix} 1 \\ 1 \\ 1 \end{bmatrix}$ **23.** $1 - x - x^2$ **25.** $\begin{bmatrix} -1 \\ -1 \\ 0 \\ 4 \end{bmatrix}$ **27.** $\begin{bmatrix} 1 & 0 \\ 0 & 1 \end{bmatrix}$ **29.** Linearly dependent

31. Linearly independent   **33.** $\{1, x + x^3, x^2 - x^3\}$ **35.** $\left\{ \begin{bmatrix} 1 & 0 \\ 0 & -3 \end{bmatrix}, \begin{bmatrix} 0 & 1 \\ 2 & 0 \end{bmatrix} \right\}$

37. $\{1 + x + x^2 + x^3, 2 - x + 2x^2 + 3x^3\}$ **39.** $\left\{ \begin{bmatrix} 1 & 1 \\ -1 & 0 \end{bmatrix}, \begin{bmatrix} 1 & 2 \\ 1 & -3 \end{bmatrix}, \begin{bmatrix} 0 & 1 \\ -1 & 2 \end{bmatrix}, \begin{bmatrix} 2 & 1 \\ -1 & 3 \end{bmatrix} \right\}$

41. $\{1 + 2x - 2x^2, 1 + x - x^2, x\}$ **43.** $\left\{ \begin{bmatrix} 1 & 0 \\ 1 & 2 \end{bmatrix}, \begin{bmatrix} 0 & 1 \\ 2 & 4 \end{bmatrix}, \begin{bmatrix} 1 & 0 \\ 0 & 0 \end{bmatrix}, \begin{bmatrix} 0 & 0 \\ 1 & 0 \end{bmatrix} \right\}$

45. (A) $\begin{bmatrix} -\frac{1}{2} \\ 1 \end{bmatrix}$ (B) $\begin{bmatrix} -\frac{3}{2} \\ 2 \end{bmatrix}$ (C) $\begin{bmatrix} (-a/2) - (3b/2) \\ a + 2b \end{bmatrix}$ **47.** (A) $\begin{bmatrix} 1 \\ -1 \end{bmatrix}$ (B) $\begin{bmatrix} -1 \\ 2 \end{bmatrix}$ (C) $\begin{bmatrix} a - b \\ -a + 2b \end{bmatrix}$

49. (A) $\begin{bmatrix} \frac{1}{2} \\ 0 \\ \frac{1}{2} \\ 0 \end{bmatrix}$ (B) $\begin{bmatrix} 0 \\ \frac{3}{4} \\ 0 \\ \frac{1}{4} \end{bmatrix}$ (C) $\begin{bmatrix} a_0 + (a_2/2) \\ a_1 + (3a_3/4) \\ a_2/2 \\ a_3/4 \end{bmatrix}$

Exercise 6-4

1. (A) $\begin{bmatrix} 2 \\ -3 \end{bmatrix}$ (B) $\begin{bmatrix} 1 \\ 4 \end{bmatrix}$ (C) $\begin{bmatrix} 2 & 1 \\ -3 & 4 \end{bmatrix}$

3. (A) $\begin{bmatrix} 4 \\ -5 \end{bmatrix}$ (B) $\begin{bmatrix} -7 \\ 9 \end{bmatrix}$ (C) $\begin{bmatrix} 4 & -7 \\ -5 & 9 \end{bmatrix}$ (D) $\begin{bmatrix} -1 \\ 2 \end{bmatrix}$ (E) $\begin{bmatrix} 9 \\ 5 \end{bmatrix}$ (F) $\begin{bmatrix} 7 \\ 4 \end{bmatrix}$ (G) $\begin{bmatrix} 9 & 7 \\ 5 & 4 \end{bmatrix}$

(H) $\begin{bmatrix} 3 \\ 2 \end{bmatrix}$ (I) $\begin{bmatrix} 1 & 0 \\ 0 & 1 \end{bmatrix}$

5. (A) $\begin{bmatrix} 5 & -2 \\ 7 & -3 \end{bmatrix}$ (B) $\begin{bmatrix} 1 \\ 1 \end{bmatrix}; \begin{bmatrix} 3 \\ 4 \end{bmatrix}$ (C) $\begin{bmatrix} 3 & -2 \\ 7 & -5 \end{bmatrix}$ (D) $\begin{bmatrix} -5 \\ -6 \end{bmatrix}; \begin{bmatrix} -3 \\ -5 \end{bmatrix}$

7. (A) $\begin{bmatrix} 0 & \frac{5}{2} & 3 \\ 2 & -\frac{3}{2} & -1 \\ -1 & \frac{1}{2} & 0 \end{bmatrix}$ (B) $\begin{bmatrix} -1 \\ 2 \\ 2 \end{bmatrix}$ (C) $\begin{bmatrix} 11 \\ -7 \\ 2 \end{bmatrix}$ **9.** (A) $\begin{bmatrix} 0 & -1 & 0 \\ 0 & 0 & -1 \\ 1 & 1 & 1 \end{bmatrix}$ (B) $\begin{bmatrix} 1 \\ 1 \\ 1 \end{bmatrix}$ (C) $\begin{bmatrix} -1 \\ -1 \\ 3 \end{bmatrix}$

11. (A) $\begin{bmatrix} 1 & 0 & 1 \\ 2 & 1 & 2 \\ -1 & 3 & 2 \end{bmatrix}$ (B) $\begin{bmatrix} -\frac{4}{3} & 1 & -\frac{1}{3} \\ -2 & 1 & 0 \\ \frac{7}{3} & -1 & \frac{1}{3} \end{bmatrix}$ (C) $\begin{bmatrix} 0 \\ 5 \\ 6 \end{bmatrix}$ (D) $\begin{bmatrix} 3 \\ 5 \\ -3 \end{bmatrix}$

13. (A) $\begin{bmatrix} 2 & 1 & 2 \\ 1 & 2 & 0 \\ 1 & 1 & 1 \end{bmatrix}$ (B) $\begin{bmatrix} 2 & 1 & -4 \\ -1 & 0 & 2 \\ -1 & -1 & 3 \end{bmatrix}$ (C) $\begin{bmatrix} 3 \\ 1 \\ 1 \end{bmatrix}$ (D) $\begin{bmatrix} 3 \\ -1 \\ -1 \end{bmatrix}$

15. (A) $\begin{bmatrix} 1 & 1 & 1 & 1 \\ 0 & 1 & 0 & 1 \\ 0 & 0 & -1 & -1 \\ 1 & 0 & 1 & -1 \end{bmatrix}$ (B) $\begin{bmatrix} 2 & -2 & 1 & -1 \\ -1 & 2 & 0 & 1 \\ -1 & 1 & -1 & 1 \\ 1 & -1 & 0 & -1 \end{bmatrix}$ (C) $\begin{bmatrix} 1 \\ 2 \\ 3 \\ 4 \end{bmatrix}$ (D) $\begin{bmatrix} -3 \\ 7 \\ 2 \\ -5 \end{bmatrix}$

17. $\begin{bmatrix} 2 & -3 & -1 \\ 1 & -2 & 0 \\ -2 & 4 & 1 \end{bmatrix}; \begin{bmatrix} 2u_1 - 3u_2 - u_3 \\ u_1 - 2u_2 \\ -2u_1 + 4u_2 + u_3 \end{bmatrix}$ **19.** $\begin{bmatrix} 1 & -1 \\ -1 & 2 \end{bmatrix}; \begin{bmatrix} a-b \\ -a+2b \end{bmatrix}$ **21.** $\begin{bmatrix} 1 & 0 & \frac{1}{2} & 0 \\ 0 & 1 & 0 & \frac{3}{4} \\ 0 & 0 & \frac{1}{2} & 0 \\ 0 & 0 & 0 & \frac{1}{4} \end{bmatrix}$

23. (A) $\begin{bmatrix} \sqrt{3}/2 & -1/2 \\ 1/2 & \sqrt{3}/2 \end{bmatrix}$ (B) $\begin{bmatrix} \sqrt{3}/2 & 1/2 \\ -1/2 & \sqrt{3}/2 \end{bmatrix}$ (C) $(-4 - \sqrt{3}, -1 + 4\sqrt{3})$ (D) $(2 + \sqrt{3}, -1 + 2\sqrt{3})$

25. (A) $(4, 3)$ (B) $(-3, 4)$ (C)

31. $P_B = \begin{bmatrix} 3 & -5 \\ -1 & 2 \end{bmatrix}, P_C = \begin{bmatrix} 2 & 1 \\ 1 & 1 \end{bmatrix}, P_C^{-1} = \begin{bmatrix} 1 & -1 \\ -1 & 2 \end{bmatrix}, P = \begin{bmatrix} 4 & -7 \\ -5 & 9 \end{bmatrix}$

33. $P_B = \begin{bmatrix} 2 & 1 & 2 \\ 1 & -1 & -1 \\ -1 & 3 & 3 \end{bmatrix}, P_C = \begin{bmatrix} 1 & 1 & 0 \\ 0 & 1 & 1 \\ 1 & 0 & 1 \end{bmatrix}, P_C^{-1} = \begin{bmatrix} \frac{1}{2} & -\frac{1}{2} & \frac{1}{2} \\ \frac{1}{2} & \frac{1}{2} & -\frac{1}{2} \\ -\frac{1}{2} & \frac{1}{2} & \frac{1}{2} \end{bmatrix}, P = \begin{bmatrix} 0 & \frac{5}{2} & 3 \\ 2 & -\frac{3}{2} & -1 \\ -1 & \frac{1}{2} & 0 \end{bmatrix}$

Exercise 6-5

1. -15 **3.** $5\sqrt{2}$ **5.** $\frac{2}{3} + \frac{1}{3}x - \frac{2}{3}x^2$ **7.** 7   **9.** $2\sqrt{7}$ **11.** $\begin{bmatrix} 1/\sqrt{7} & -2/\sqrt{7} \\ 1/\sqrt{7} & -1/\sqrt{7} \end{bmatrix}$ **13.** Not orthogonal   **15.** 2, 3

17. Yes; $\langle \mathbf{u}, \mathbf{w}_i \rangle = 0$, $i = 1, 2, 3$ **19.** $\{t(1 - x - x^2) | t$ any real number$\}$

21. $\{s(1, -1, 1, 0) + t(-2, 2, 0, 1) | s, t$ any real numbers$\}$

23. (A) 180   (B) $6\sqrt{10}$ (C) $6\sqrt{5}$ (D) $6\sqrt{5}$ (E) $(1/\sqrt{10})(\frac{3}{2}, \frac{1}{3})$ (F) $\pi/4$ or $45°$

25. $\{t(9, 4) | t$ any real number$\}$

27. (A) -36 (B) $6\sqrt{6}$ (C) $2\sqrt{6}$ (D) $2\sqrt{78}$ (E) $\frac{\sqrt{6}}{36} \begin{bmatrix} -4 & 5 \\ 0 & -10 \end{bmatrix}$ (F) $2\pi/3$ or $120°$

29. $\left\{ r \begin{bmatrix} 4 & 1 \\ 0 & 0 \end{bmatrix} + s \begin{bmatrix} 4 & 0 \\ 1 & 0 \end{bmatrix} + t \begin{bmatrix} -2 & 0 \\ 0 & 1 \end{bmatrix} \middle| r, s, t \text{ any real numbers} \right\}$

31. $\langle f, g \rangle = 0$, $\|f\| = \sqrt{2}$, $\|g\| = \sqrt{14}/7$ **33.** $\langle f, g \rangle = 2/e$, $\|f\| = \sqrt{6}/3$, $\|g\| = \sqrt{(e^2 - e^{-2})/2}$ **35.** $-\frac{5}{3}$

39. Inner product   **41.** Not an inner product; $\langle \mathbf{u}, \mathbf{u} \rangle$ is not always nonnegative and can be zero for nonzero $\mathbf{u}$.

45. (A) $\begin{bmatrix} u_1 - u_2 \\ -u_1 + 2u_2 \end{bmatrix}$ (B) $2u_1v_1 - 3u_1v_2 - 3u_2v_1 + 5u_2v_2$

Exercise 6-6

1. Orthogonal   **3.** Orthonormal   **5.** Orthogonal   **7.** Nonorthogonal

9. (B) $\mathbf{v}_1 = (\frac{3}{5}, \frac{4}{5})$, $\mathbf{v}_2 = (\frac{4}{5}, -\frac{3}{5})$ (C) $\frac{11}{5}\mathbf{v}_1 - \frac{2}{5}\mathbf{v}_2$

11. (B) $\mathbf{v}_1 = (1/\sqrt{2})(1 + x)$, $\mathbf{v}_2 = (1/\sqrt{2})(1 - x)$ (C) $(5/\sqrt{2})\mathbf{v}_1 - (1/\sqrt{2})\mathbf{v}_2$ **13.** Orthonormal   **15.** Nonorthogonal

17. $\mathbf{v}_1 = (\frac{1}{3}, \frac{2}{3}, \frac{2}{3})$, $\mathbf{v}_2 = (-\frac{2}{3}, -\frac{1}{3}, \frac{2}{3})$, $\mathbf{v}_3 = (\frac{2}{3}, -\frac{2}{3}, \frac{1}{3})$; $\mathbf{u} = \frac{5}{3}\mathbf{v}_1 + \frac{8}{3}\mathbf{v}_2 + \frac{19}{3}\mathbf{v}_3$

19. $\mathbf{v}_1 = (1/\sqrt{2})(1 + x^2)$, $\mathbf{v}_2 = (1/\sqrt{3})(1 + x - x^2)$, $\mathbf{v}_3 = (1/\sqrt{6})(1 - 2x - x^2)$; $\mathbf{u} = -2\sqrt{2}\mathbf{v}_1 + (11/\sqrt{3})\mathbf{v}_2 + (2/\sqrt{6})\mathbf{v}_3$

21. $\mathbf{v}_1 = \frac{1}{2}\begin{bmatrix} 1 & 1 \\ 1 & 1 \end{bmatrix}$, $\mathbf{v}_2 = \frac{1}{6}\begin{bmatrix} 1 & 1 \\ 3 & -5 \end{bmatrix}$, $\mathbf{v}_3 = \frac{1}{6}\begin{bmatrix} -5 & 1 \\ 3 & 1 \end{bmatrix}$, $\mathbf{v}_4 = \frac{1}{6}\begin{bmatrix} 1 & -5 \\ 3 & 1 \end{bmatrix}$; $\mathbf{u} = \mathbf{v}_1 - \frac{2}{3}\mathbf{v}_2 - \frac{5}{3}\mathbf{v}_3 + \frac{4}{3}\mathbf{v}_4$

23. $\left\{ \frac{1}{3}(1, 2, 2), \frac{1}{\sqrt{2}}(0, -1, 1) \right\}$ **25.** $\left\{ \frac{1}{\sqrt{3}}(1, 1, 0, 1), \frac{1}{\sqrt{3}}(0, -1, 1, 1), \frac{1}{\sqrt{6}}(0, 1, 2, -1) \right\}$ **27.** $\left\{ \frac{1}{\sqrt{5}}(1, 2), \frac{1}{\sqrt{5}}(-2, 1) \right\}$

29. $\left\{ \frac{1}{\sqrt{2}}(1, -1, 0), (0, 0, 1), \frac{1}{\sqrt{2}}(1, 1, 0) \right\}$ **31.** $\left\{ \frac{1}{\sqrt{5}}(0, 1, -2), \sqrt{\frac{5}{6}}\left(1, \frac{2}{5}, \frac{1}{5}\right), \sqrt{\frac{3}{2}}\left(-\frac{1}{3}, \frac{2}{3}, \frac{1}{3}\right) \right\}$

33. $\left\{ \frac{1}{2}(1, -1, 1, -1), \frac{1}{\sqrt{2}}(1, 1, 0, 0), \frac{1}{3}\left(\frac{1}{2}, -\frac{1}{2}, -\frac{5}{2}, -\frac{3}{2}\right), \frac{1}{3\sqrt{2}}(2, -2, -1, 3) \right\}$

35. $\left\{ \frac{1}{\sqrt{3}}(1 + x + x^2), \frac{1}{\sqrt{2}}(x - x^2), \frac{1}{\sqrt{6}}(2 - x - x^2) \right\}$ **37.** $\left\{ \frac{1}{\sqrt{2}}\begin{bmatrix} 1 & 0 \\ 0 & 1 \end{bmatrix}, \sqrt{\frac{2}{3}}\begin{bmatrix} -\frac{1}{2} & 1 \\ 0 & \frac{1}{2} \end{bmatrix}, \frac{\sqrt{3}}{2}\begin{bmatrix} -\frac{1}{3} & -\frac{1}{3} \\ 1 & \frac{1}{3} \end{bmatrix}, 2\begin{bmatrix} -\frac{1}{4} & -\frac{1}{4} \\ -\frac{1}{4} & \frac{1}{4} \end{bmatrix} \right\}$

39. $\left\{ \frac{1}{\sqrt{2}}, \sqrt{\frac{3}{2}} x, \frac{\sqrt{5}}{2\sqrt{2}}(3x^2 - 1) \right\}$ **41.** $\frac{\sqrt{2}}{3}\left(\frac{1}{\sqrt{2}}\right) + \frac{4\sqrt{5}}{15\sqrt{2}}\left[\frac{\sqrt{5}}{2\sqrt{2}}(3x^2 - 1)\right]$

Exercise 6-7   Chapter Review

1. $\{(1, 0, -1, 2), (0, 1, 3, -4)\}$ **2.** $\{(1, 0, -1, 2), (0, 1, 3, -4)\}$ **3.** $\{(1, 0, 0), (0, 1, 0)\}$ **4.** $\{(1, 2, 3), (1, 1, 2)\}$

5. 2   **6.** 2   **7.** $\begin{bmatrix} 2 \\ -1 \end{bmatrix}$ **8.** (10, 8)   **9.** $\begin{bmatrix} 4 & 7 \\ 3 & 5 \end{bmatrix}$ **10.** $\begin{bmatrix} 9 \\ 7 \end{bmatrix}$ **11.** $\begin{bmatrix} -5 & 7 \\ 3 & -4 \end{bmatrix}$ **12.** $\begin{bmatrix} -1 \\ 1 \end{bmatrix}$ **13.** 10   **14.** $\sqrt{5}$

15. $2\sqrt{10}$ **16.** 5   **17.** $\theta = \pi/4$ or $45°$ **18.** $\left\{ \frac{1}{\sqrt{5}}(2, -1), \frac{1}{2\sqrt{5}}(2, 4) \right\}$

19. (A) $\{(1, -1, 0, 2), (0, 0, 1, 3)\}$ (B) $\{(1, 1, 3), (1, -1, -1)\}$ (C) 2

20. (A) $\{\mathbf{v}_1, \mathbf{v}_2\}$ where $\mathbf{v}_1 = (1, 0, -3)$, $\mathbf{v}_2 = (0, 1, 1)$ (B) $\mathbf{u}_1 = \mathbf{v}_1 + 2\mathbf{v}_2$, $\mathbf{u}_2 = -\mathbf{v}_1 + 3\mathbf{v}_2$, $\mathbf{u}_3 = \mathbf{v}_1 + 7\mathbf{v}_2$
(C) $\{\mathbf{u}_1, \mathbf{u}_2\}$ (D) $\mathbf{u}_3 = 2\mathbf{u}_1 + \mathbf{u}_2$

21. (A) $\{1 - x^2, x + 2x^2\}$ (B) $\{1 + x + x^2, 1 - x - 3x^2\}$ **22.** (A) $\left\{\begin{bmatrix} 1 & 0 \\ -1 & -7 \end{bmatrix}, \begin{bmatrix} 0 & 1 \\ 3 & 6 \end{bmatrix}\right\}$ (B) $\left\{\begin{bmatrix} 1 & 1 \\ 2 & -1 \end{bmatrix}, \begin{bmatrix} 2 & 3 \\ 7 & 4 \end{bmatrix}\right\}$

23. Linearly dependent   **24.** Linearly independent   **25.** $\{(1, -1, 1), (1, 1, -1), (0, 1, 0)\}$

26. $\{(1, -1, 0, 0), (-1, 2, 1, 0), (1, 0, 0, 0), (0, 0, 0, 1)\}$ **27.** $\{1 - x + x^2 - 2x^3, -3 + 2x - 2x^2 + 4x^3, x, x^2\}$

28. $\left\{\begin{bmatrix} 1 & -1 \\ 0 & 0 \end{bmatrix}, \begin{bmatrix} 1 & 1 \\ 0 & 0 \end{bmatrix}, \begin{bmatrix} 1 & -1 \\ 0 & 1 \end{bmatrix}, \begin{bmatrix} 0 & 0 \\ 1 & 0 \end{bmatrix}\right\}$

29. (A) $\begin{bmatrix} -1 \\ 3 \\ -4 \end{bmatrix}$ (B) $(1, 8, 1)$ (C) $\begin{bmatrix} 0 & 9 & 7 \\ 1 & 7 & 5 \\ 1 & -7 & -6 \end{bmatrix}$ (D) $\begin{bmatrix} -1 \\ 0 \\ 2 \end{bmatrix}$

30. (A) $\begin{bmatrix} 1 & 2 & 1 \\ 2 & 0 & -1 \\ 0 & -1 & -1 \end{bmatrix}$ (B) $\begin{bmatrix} -1 & 1 & -2 \\ 2 & -1 & 3 \\ -2 & 1 & -4 \end{bmatrix}$ (C) $\begin{bmatrix} -1 \\ 3 \\ -4 \end{bmatrix}$

31. (A) $\begin{bmatrix} 2 \\ -3 \\ -1 \end{bmatrix}$ (B) $\begin{bmatrix} 3 & 1 & 2 \\ 2 & 1 & 0 \\ 1 & 1 & -1 \end{bmatrix}$ (C) $\begin{bmatrix} -1 & 3 & -2 \\ 2 & -5 & 4 \\ 1 & -2 & 1 \end{bmatrix}$ (D) $\begin{bmatrix} 2 \\ -3 \\ -1 \end{bmatrix}$

32. (A) $\begin{bmatrix} 1 \\ 2 \\ 3 \\ 4 \end{bmatrix}$ (B) $\begin{bmatrix} 1 & 0 & 1 & 1 \\ 0 & -1 & 0 & 1 \\ 1 & 0 & 0 & 1 \\ 0 & -1 & 1 & 0 \end{bmatrix}$ (C) $\begin{bmatrix} -1 & -1 & 2 & 1 \\ 1 & 0 & -1 & -1 \\ 1 & 0 & -1 & 0 \\ 1 & 1 & -1 & -1 \end{bmatrix}$ (D) $\begin{bmatrix} 1 \\ 2 \\ 3 \\ 4 \end{bmatrix}$

33. (A) 15   (B) $\sqrt{6}$ (C) $5\sqrt{2}$ (D) $\sqrt{26}$ (E) $\pi/6$ or $30°$

34. (A) $\begin{bmatrix} \frac{1}{2} & \frac{1}{2} \\ -\frac{1}{2} & \frac{1}{2} \end{bmatrix}$ (B) $\left\{s\begin{bmatrix} 2 & -1 \\ 1 & 0 \end{bmatrix} + t\begin{bmatrix} 1 & -2 \\ 0 & 1 \end{bmatrix}\middle| s, t \text{ any real numbers}\right\}$

35. (A) $\begin{bmatrix} -\sqrt{3}/2 & -\frac{1}{2} \\ \frac{1}{2} & -\sqrt{3}/2 \end{bmatrix}$ (B) $\begin{bmatrix} -\sqrt{3}/2 & \frac{1}{2} \\ -\frac{1}{2} & -\sqrt{3}/2 \end{bmatrix}$ (C) $(2 - \sqrt{3}, 1 + 2\sqrt{3})$ (D) $(2 + 3\sqrt{3}, 3 - 2\sqrt{3})$

36. $\left\{\frac{1}{3}(-1, 2, -2), \frac{1}{\sqrt{2}}(0, -1, -1)\right\}$ **37.** $\left\{\frac{1}{2}(1, 1, -1, -1), \frac{1}{\sqrt{2}}(0, 1, 0, 1), \frac{2}{\sqrt{22}}\left(2, -\frac{1}{2}, 1, \frac{1}{2}\right)\right\}$

38. $\left\{\frac{1}{\sqrt{2}}(1, 0, 0, 0, -1), \sqrt{\frac{2}{3}}\left(-\frac{1}{2}, 1, 0, 0, -\frac{1}{2}\right), \frac{\sqrt{3}}{2}\left(-\frac{1}{3}, -\frac{1}{3}, 1, 0, -\frac{1}{3}\right), \frac{2}{\sqrt{5}}\left(-\frac{1}{4}, -\frac{1}{4}, -\frac{1}{4}, 1, -\frac{1}{4}\right)\right\}$

39. $\left\{\frac{1}{\sqrt{2}}(1 - x^2), \sqrt{2}\left(\frac{1}{2} + \frac{1}{2}x^2\right), x\right\}$ **40.** $\left\{\frac{1}{2}\begin{bmatrix} 1 & 1 \\ 1 & 1 \end{bmatrix}, \begin{bmatrix} \frac{1}{2} & -\frac{1}{2} \\ -\frac{1}{2} & \frac{1}{2} \end{bmatrix}, \begin{bmatrix} -\frac{1}{2} & \frac{1}{2} \\ -\frac{1}{2} & \frac{1}{2} \end{bmatrix}, \begin{bmatrix} -\frac{1}{2} & -\frac{1}{2} \\ \frac{1}{2} & \frac{1}{2} \end{bmatrix}\right\}$

41. (A) $\frac{1}{5}$ (B) $\frac{1}{2}$ (C) $1/\sqrt{6}$ **42.** $-\frac{4}{3}$ **43.** $\{\sqrt{2}, 6(-\frac{2}{3} + x), 10\sqrt{6}(\frac{3}{10} - \frac{6}{5}x + x^2)\}$

45. (A) $\begin{bmatrix} 1 & 1 & 2 & 6 \\ 0 & -1 & -4 & -18 \\ 0 & 0 & 1 & 9 \\ 0 & 0 & 0 & -1 \end{bmatrix}$ (B) $\begin{bmatrix} 1 & 1 & 2 & 6 \\ 0 & -1 & -4 & -18 \\ 0 & 0 & 1 & 9 \\ 0 & 0 & 0 & -1 \end{bmatrix}$

     (C) $10(1) - 23(1 - x) + 10(2 - 4x + x^2) - (6 - 18x + 9x^2 - x^3)$

47. (A) $\begin{bmatrix} u_1 + u_2 \\ u_1 \end{bmatrix}$ **48.** (A) 3   (B) 1   (C) $(\frac{3}{5}, \frac{1}{5})$ **49.** $\{(t, 0)| t \text{ any real number}\}$

Chapter 7   Exercise 7-1

5. $\lambda^2 - 3\lambda$; $\lambda_1 = 0$, $\lambda_2 = 3$; $B_0 = \left\{ \begin{bmatrix} 2 \\ 1 \end{bmatrix} \right\}$, $B_3 = \left\{ \begin{bmatrix} -1 \\ 1 \end{bmatrix} \right\}$

7. $\lambda^2 - 2\lambda - 1$; $\lambda_1 = 1 - \sqrt{2}$, $\lambda_2 = 1 + \sqrt{2}$; $B_{1-\sqrt{2}} = \left\{ \begin{bmatrix} -\sqrt{2} \\ 1 \end{bmatrix} \right\}$, $B_{1+\sqrt{2}} = \left\{ \begin{bmatrix} \sqrt{2} \\ 1 \end{bmatrix} \right\}$

9. $\lambda^2 - 2\lambda + 1$; $\lambda_1 = 1$ (double root); $B_1 = \left\{ \begin{bmatrix} 1 \\ 1 \end{bmatrix} \right\}$ **11.** $\lambda^2 + 1$; no real eigenvalues

13. $(\lambda - 2 - \sqrt{2})(\lambda - 2 + \sqrt{2})(\lambda + 2)$; -2, $2 - \sqrt{2}$, $2 + \sqrt{2}$ **15.** $(\lambda - 3)(\lambda^2 - 2\lambda + 3)$; 3
17. $(\lambda - 2)^2(\lambda + 3)$; -3, 2 (double root)   **19.** $(\lambda + 2)^3(\lambda - 2)$; -2 (triple root), 2

21. $(\lambda + 1)^2(\lambda - 3)^2$; -1 (double root), 3 (double root)   **23.** $(\lambda - 1)(\lambda^2 - 2\lambda + 2)$; $\lambda_1 = 1$; $B_1 = \left\{ \begin{bmatrix} -2 \\ 1 \\ 1 \end{bmatrix} \right\}$

25. $\lambda(\lambda - 1)^2$; $\lambda_1 = 0$, $\lambda_2 = 1$ (multiplicity 2); $B_0 = \left\{ \begin{bmatrix} \frac{1}{2} \\ 1 \\ 0 \end{bmatrix} \right\}$, $B_1 = \left\{ \begin{bmatrix} 1 \\ 1 \\ 0 \end{bmatrix} \right\}$

27. $(\lambda - 3)(\lambda - 4)^2$; $\lambda_1 = 3$, $\lambda_2 = 4$ (multiplicity 2); $B_3 = \left\{ \begin{bmatrix} 1 \\ 1 \\ 1 \end{bmatrix} \right\}$, $B_4 = \left\{ \begin{bmatrix} -\frac{4}{3} \\ 1 \\ 0 \end{bmatrix}, \begin{bmatrix} 2 \\ 0 \\ 1 \end{bmatrix} \right\}$

29. $(\lambda + \sqrt{2})(\lambda - \sqrt{2})(\lambda - 3)$; $\lambda_1 = -\sqrt{2}$, $\lambda_2 = \sqrt{2}$, $\lambda_3 = 3$; $B_{-\sqrt{2}} = \left\{ \begin{bmatrix} \sqrt{2} - 1 \\ -1 \\ 1 \end{bmatrix} \right\}$,

$B_{\sqrt{2}} = \left\{ \begin{bmatrix} -\sqrt{2} - 1 \\ -1 \\ 1 \end{bmatrix} \right\}$, $B_3 = \left\{ \begin{bmatrix} 0 \\ 1 \\ 0 \end{bmatrix} \right\}$

31. $(\lambda - 1)^2(\lambda - 3)^2$; $\lambda_1 = 1$ (multiplicity 2), $\lambda_2 = 3$ (multiplicity 2); $B_1 = \left\{ \begin{bmatrix} 1 \\ 1 \\ 0 \\ 0 \end{bmatrix}, \begin{bmatrix} 0 \\ 0 \\ 1 \\ 0 \end{bmatrix} \right\}$, $B_3 = \left\{ \begin{bmatrix} -1 \\ 1 \\ 0 \\ 0 \end{bmatrix}, \begin{bmatrix} 0 \\ 0 \\ 0 \\ 1 \end{bmatrix} \right\}$

33. a_{11}, a_{22} **35.** $a_{11}, a_{22}, a_{33}, a_{44}$ **39.** (A) $(a - d)^2 + 4b^2 > 0$ (B) $a = d$, $b = 0$ (C) Not possible

Exercise 7-2

1. $p(\lambda) = (\lambda - 1)(\lambda - 2)$; $\lambda_1 = 1$, $\lambda_2 = 2$; $B_1 = \left\{ \begin{bmatrix} 3 \\ 1 \end{bmatrix} \right\}$, $B_2 = \left\{ \begin{bmatrix} 2 \\ 1 \end{bmatrix} \right\}$; $P = \begin{bmatrix} 3 & 2 \\ 1 & 1 \end{bmatrix}$, $D = \begin{bmatrix} 1 & 0 \\ 0 & 2 \end{bmatrix}$

3. $p(\lambda) = (\lambda - 2)^2$; $\lambda_1 = 2$ (double root); $B_2 = \left\{ \begin{bmatrix} -1 \\ 1 \end{bmatrix} \right\}$; P and D do not exist

5. $p(\lambda) = \lambda^2 - 5\lambda + 7$; no real eigenvalues; P and D do not exist   **7.** $A = \begin{bmatrix} -2 & -4 \\ 3 & 5 \end{bmatrix}$; $A^3 = \begin{bmatrix} -20 & -28 \\ 21 & 29 \end{bmatrix}$

9. $P = \begin{bmatrix} 2 & 1 & 0 \\ 1 & 1 & 0 \\ 1 & 0 & 1 \end{bmatrix}$; $D = \begin{bmatrix} -1 & 0 & 0 \\ 0 & 1 & 0 \\ 0 & 0 & 1 \end{bmatrix}$

11. A has three distinct eigenvalues: $\lambda_1 = -2, \lambda_2 = 1, \lambda_3 = 2$; $D = \begin{bmatrix} -2 & 0 & 0 \\ 0 & 1 & 0 \\ 0 & 0 & 2 \end{bmatrix}$

13. Not diagonalizable   **15.** $P = \begin{bmatrix} 3 & 2 & -2 \\ 2 & 1 & 0 \\ 1 & 0 & 1 \end{bmatrix}$; $D = \begin{bmatrix} 1 & 0 & 0 \\ 0 & 2 & 0 \\ 0 & 0 & 2 \end{bmatrix}$ **17.** Not diagonalizable

19. $P = \begin{bmatrix} 1 & 0 & 1 & 0 \\ 1 & 0 & -1 & 0 \\ 0 & 1 & 0 & 0 \\ 0 & 0 & 0 & 1 \end{bmatrix}$; $D = \begin{bmatrix} 1 & 0 & 0 & 0 \\ 0 & 1 & 0 & 0 \\ 0 & 0 & 3 & 0 \\ 0 & 0 & 0 & 3 \end{bmatrix}$ **21.** Not diagonalizable   **23.** $A = \begin{bmatrix} 0 & 2 & 1 \\ -1 & 3 & 1 \\ 0 & 0 & 1 \end{bmatrix}$

27. $\lambda_1 = 0, \lambda_2 = 1$; $B_0 = \{(2, 2, 1)\}$, $B_1 = \{(2, 1, 0), (-3, 0, 1)\}$ **29.** $A^k = \begin{bmatrix} 1 & 2^k - 1 \\ 0 & 2^k \end{bmatrix}$ **31.** $A^k = \begin{bmatrix} 1 & 2^k - 1 & 1 - 2^k \\ 0 & 2^k & 0 \\ 0 & 0 & 2^k \end{bmatrix}$

35. $\begin{bmatrix} 1 & 0 \\ 0 & 2 \end{bmatrix}, \begin{bmatrix} 1 & 0 \\ 0 & -2 \end{bmatrix}, \begin{bmatrix} -1 & 0 \\ 0 & 2 \end{bmatrix}, \begin{bmatrix} -1 & 0 \\ 0 & -2 \end{bmatrix}$ **37.** $\begin{bmatrix} 0 & 2 \\ -1 & 3 \end{bmatrix}, \begin{bmatrix} 0 & -2 \\ 1 & -3 \end{bmatrix}, \begin{bmatrix} 4 & -6 \\ 3 & -5 \end{bmatrix}, \begin{bmatrix} -4 & 6 \\ -3 & 5 \end{bmatrix}$

Exercise 7-3

1. Symmetric   **3.** Not symmetric   **5.** Orthogonal; $P^{-1} = P^T = \begin{bmatrix} \frac{3}{5} & -\frac{4}{5} \\ \frac{4}{5} & \frac{3}{5} \end{bmatrix}$ **7.** Not orthogonal

9. Orthogonal; $P^{-1} = P^T = \begin{bmatrix} \frac{2}{7} & \frac{3}{7} & \frac{6}{7} \\ \frac{6}{7} & \frac{2}{7} & -\frac{3}{7} \\ \frac{3}{7} & -\frac{6}{7} & \frac{2}{7} \end{bmatrix}$ **11.** $p(\lambda) = (\lambda + 4)^2(\lambda - 2)$; $D = \begin{bmatrix} -4 & 0 & 0 \\ 0 & -4 & 0 \\ 0 & 0 & 2 \end{bmatrix}$

13. $p(\lambda) = \lambda(\lambda - 2)^3$; $D = \begin{bmatrix} 0 & 0 & 0 & 0 \\ 0 & 2 & 0 & 0 \\ 0 & 0 & 2 & 0 \\ 0 & 0 & 0 & 2 \end{bmatrix}$ **15.** $p(\lambda) = (\lambda + 1)^2\lambda(\lambda - 2)^2$; $D = \begin{bmatrix} -1 & 0 & 0 & 0 & 0 \\ 0 & -1 & 0 & 0 & 0 \\ 0 & 0 & 0 & 0 & 0 \\ 0 & 0 & 0 & 2 & 0 \\ 0 & 0 & 0 & 0 & 2 \end{bmatrix}$

17. $P = \begin{bmatrix} -1/\sqrt{2} & 1/\sqrt{2} \\ 1/\sqrt{2} & 1/\sqrt{2} \end{bmatrix}$; $D = \begin{bmatrix} -1 & 0 \\ 0 & 3 \end{bmatrix}$ **19.** $P = \begin{bmatrix} -\frac{2}{3} & \frac{2}{3} & \frac{1}{3} \\ \frac{1}{3} & \frac{2}{3} & -\frac{2}{3} \\ \frac{2}{3} & \frac{1}{3} & \frac{2}{3} \end{bmatrix}$; $D = \begin{bmatrix} -6 & 0 & 0 \\ 0 & 0 & 0 \\ 0 & 0 & 6 \end{bmatrix}$

21. $P = \begin{bmatrix} -1/\sqrt{2} & 1/\sqrt{3} & 1/\sqrt{6} \\ 1/\sqrt{2} & 1/\sqrt{3} & 1/\sqrt{6} \\ 0 & -1/\sqrt{3} & 2/\sqrt{6} \end{bmatrix}$; $D = \begin{bmatrix} -2 & 0 & 0 \\ 0 & -2 & 0 \\ 0 & 0 & 4 \end{bmatrix}$

23. $P = \begin{bmatrix} -2/\sqrt{5} & 0 & 0 & 1/\sqrt{5} \\ 1/\sqrt{5} & 0 & 0 & 2/\sqrt{5} \\ 0 & 1 & 0 & 0 \\ 0 & 0 & 1 & 0 \end{bmatrix}$; $D = \begin{bmatrix} -2 & 0 & 0 & 0 \\ 0 & -1 & 0 & 0 \\ 0 & 0 & -1 & 0 \\ 0 & 0 & 0 & 3 \end{bmatrix}$

25. $P = \begin{bmatrix} -1/\sqrt{10} & 1/\sqrt{2} & 1/\sqrt{3} & -1/\sqrt{15} \\ 2/\sqrt{10} & 0 & 0 & -3/\sqrt{15} \\ 1/\sqrt{10} & 1/\sqrt{2} & -1/\sqrt{3} & 1/\sqrt{15} \\ 2/\sqrt{10} & 0 & 1/\sqrt{3} & 2/\sqrt{15} \end{bmatrix}$; $D = \begin{bmatrix} 0 & 0 & 0 & 0 \\ 0 & 2 & 0 & 0 \\ 0 & 0 & 2 & 0 \\ 0 & 0 & 0 & 5 \end{bmatrix}$

Exercise 7-4　Chapter Review

1. $-2\begin{bmatrix} -1 \\ 3 \end{bmatrix}$　**2.** $5\begin{bmatrix} 2 \\ 1 \end{bmatrix}$　**3.** $\lambda_1 = -2, \lambda_2 = 5$　**4.** $\begin{bmatrix} -2 & 0 \\ 0 & 5 \end{bmatrix}$　**5.** $(\lambda - 1)(\lambda - 4)$　**6.** $\lambda_1 = 1, \lambda_2 = 4$

7. $B_1 = \left\{ \begin{bmatrix} 1 \\ 1 \end{bmatrix} \right\}, B_4 = \left\{ \begin{bmatrix} -2 \\ 1 \end{bmatrix} \right\}$　**8.** $P = \begin{bmatrix} 1 & -2 \\ 1 & 1 \end{bmatrix}; D = \begin{bmatrix} 1 & 0 \\ 0 & 4 \end{bmatrix}$　**9.** $a = 3, b = -4, c = 0$　**10.** Not possible

11. Not orthogonal　**12.** Orthogonal; $P^{-1} = \begin{bmatrix} 1/2 & \sqrt{3}/2 \\ -\sqrt{3}/2 & 1/2 \end{bmatrix}$　**13.** Orthogonal; $P^{-1} = \begin{bmatrix} 1/\sqrt{3} & 1/\sqrt{3} & 1/\sqrt{3} \\ 1/\sqrt{2} & 0 & -1/\sqrt{2} \\ 1/\sqrt{6} & -2/\sqrt{6} & 1/\sqrt{6} \end{bmatrix}$

14. Not orthogonal　**15.** $(\lambda + 3)(\lambda - 2)$　**16.** $\lambda_1 = -3, \lambda_2 = 2$　**17.** $B_{-3} = \left\{ \begin{bmatrix} -2/\sqrt{5} \\ 1/\sqrt{5} \end{bmatrix} \right\}, B_2 = \left\{ \begin{bmatrix} 1/\sqrt{5} \\ 2/\sqrt{5} \end{bmatrix} \right\}$

18. $P = \begin{bmatrix} -2/\sqrt{5} & 1/\sqrt{5} \\ 1/\sqrt{5} & 2/\sqrt{5} \end{bmatrix}; D = \begin{bmatrix} -3 & 0 \\ 0 & 2 \end{bmatrix}$

19. A has three linearly independent eigenvectors: $\mathbf{x}_1 = \begin{bmatrix} 0 \\ 1 \\ 1 \end{bmatrix}$ $(\lambda_1 = 0)$, $\mathbf{x}_2 = \begin{bmatrix} 1 \\ 1 \\ 0 \end{bmatrix}$ $(\lambda_2 = 1)$, and $\mathbf{x}_3 = \begin{bmatrix} -2 \\ 0 \\ 1 \end{bmatrix}$ $(\lambda_3 = 1)$

20. A is symmetric　**21.** A has three distinct eigenvalues: $\lambda_1 = 1 - \sqrt{3}, \lambda_2 = 0, \lambda_3 = 1 + \sqrt{3}$

22. $p(\lambda) = (\lambda + 2)\lambda(\lambda - 2); \lambda_1 = -2, \lambda_2 = 0, \lambda_3 = 2$;

$B_{-2} = \left\{ \begin{bmatrix} 1 \\ 0 \\ 1 \end{bmatrix} \right\}, B_0 = \left\{ \begin{bmatrix} -1 \\ 1 \\ 0 \end{bmatrix} \right\}, B_2 = \left\{ \begin{bmatrix} 2 \\ 0 \\ 1 \end{bmatrix} \right\}; P = \begin{bmatrix} 1 & -1 & 2 \\ 0 & 1 & 0 \\ 1 & 0 & 1 \end{bmatrix}; D = \begin{bmatrix} -2 & 0 & 0 \\ 0 & 0 & 0 \\ 0 & 0 & 2 \end{bmatrix}$

23. $p(\lambda) = (\lambda - 3)(\lambda^2 + 1); \lambda_1 = 3; B_3 = \left\{ \begin{bmatrix} 1 \\ 1 \\ 1 \end{bmatrix} \right\}$; A is not diagonalizable

24. $p(\lambda) = \lambda(\lambda - 1)^2; \lambda_1 = 0, \lambda_2 = 1; B_0 = \left\{ \begin{bmatrix} 1 \\ -2 \\ 1 \end{bmatrix} \right\}, B_1 = \left\{ \begin{bmatrix} 0 \\ 1 \\ 0 \end{bmatrix} \right\}$; A is not diagonalizable

25. $p(\lambda) = (\lambda + 1)(\lambda - 2)^2; \lambda_1 = -1, \lambda_2 = 2$;

$B_{-1} = \left\{ \begin{bmatrix} -1 \\ -1 \\ 1 \end{bmatrix} \right\}, B_2 = \left\{ \begin{bmatrix} 2 \\ 1 \\ 0 \end{bmatrix}, \begin{bmatrix} 0 \\ 0 \\ 1 \end{bmatrix} \right\}; P = \begin{bmatrix} -1 & 2 & 0 \\ -1 & 1 & 0 \\ 1 & 0 & 1 \end{bmatrix}; D = \begin{bmatrix} -1 & 0 & 0 \\ 0 & 2 & 0 \\ 0 & 0 & 2 \end{bmatrix}$

26. $p(\lambda) = (\lambda + 1)^2(\lambda - 1)(\lambda - 3); \lambda_1 = -1, \lambda_2 = 1, \lambda_3 = 3$;

$B_{-1} = \left\{ \begin{bmatrix} 1 \\ 0 \\ 0 \\ 0 \end{bmatrix} \right\}, B_1 = \left\{ \begin{bmatrix} 0 \\ 0 \\ -1 \\ 1 \end{bmatrix} \right\}, B_3 = \left\{ \begin{bmatrix} -\frac{1}{2} \\ 0 \\ 1 \\ 1 \end{bmatrix} \right\}$; A is not diagonalizable

27. $p(\lambda) = (\lambda + 2)^2(\lambda - 2)^2; \lambda_1 = -2, \lambda_2 = 2$;

$B_{-2} = \left\{ \begin{bmatrix} 1 \\ 1 \\ 0 \\ 0 \end{bmatrix}, \begin{bmatrix} 0 \\ 0 \\ 0 \\ 1 \end{bmatrix} \right\}, B_2 = \left\{ \begin{bmatrix} 3 \\ 1 \\ 0 \\ 0 \end{bmatrix}, \begin{bmatrix} 0 \\ 0 \\ 1 \\ 0 \end{bmatrix} \right\}; P = \begin{bmatrix} 1 & 0 & 3 & 0 \\ 1 & 0 & 1 & 0 \\ 0 & 0 & 0 & 1 \\ 0 & 1 & 0 & 0 \end{bmatrix}; D = \begin{bmatrix} -2 & 0 & 0 & 0 \\ 0 & -2 & 0 & 0 \\ 0 & 0 & 2 & 0 \\ 0 & 0 & 0 & 2 \end{bmatrix}$

28. $P = \begin{bmatrix} -\frac{2}{11} & -\frac{6}{11} & \frac{9}{11} \\ -\frac{9}{11} & \frac{6}{11} & \frac{2}{11} \\ \frac{6}{11} & \frac{7}{11} & \frac{6}{11} \end{bmatrix}; D = \begin{bmatrix} -11 & 0 & 0 \\ 0 & 0 & 0 \\ 0 & 0 & 11 \end{bmatrix}$ **29.** $P = \begin{bmatrix} 1/\sqrt{2} & 1/\sqrt{6} & 1/\sqrt{3} \\ 1/\sqrt{2} & -1/\sqrt{6} & -1/\sqrt{3} \\ 0 & -2/\sqrt{6} & 1/\sqrt{3} \end{bmatrix}; D = \begin{bmatrix} 2 & 0 & 0 \\ 0 & 2 & 0 \\ 0 & 0 & 5 \end{bmatrix}$

30. $p(\lambda) = (\lambda - 2)(\lambda - 4)^3$; $D = \begin{bmatrix} 2 & 0 & 0 & 0 \\ 0 & 4 & 0 & 0 \\ 0 & 0 & 4 & 0 \\ 0 & 0 & 0 & 4 \end{bmatrix}$ **31.** $\begin{bmatrix} -1 + 2 \cdot 9^k & -2 + 2 \cdot 9^k \\ 1 - 9^k & 2 - 9^k \end{bmatrix}$

32. $\begin{bmatrix} 5 & 4 \\ -2 & -1 \end{bmatrix}, \begin{bmatrix} -5 & -4 \\ 2 & 1 \end{bmatrix}, \begin{bmatrix} 7 & 8 \\ -4 & -5 \end{bmatrix}, \begin{bmatrix} -7 & -8 \\ 4 & 5 \end{bmatrix}$ **33.** $\begin{bmatrix} 1 & 0 & 3 \\ -1 & 2 & 1 \\ 0 & 0 & 3 \end{bmatrix}$

Chapter 8   Exercise 8-1

9. Yes   **11.** No   **13.** No   **15.** Yes   **17.** No   **19.** Yes   **21.** $\begin{bmatrix} 7 \\ 4 \end{bmatrix}$ **23.** $\begin{bmatrix} x - y + 2z \\ x + 2y - z \end{bmatrix}$

25. $T(x, y) = (x, x + 2y, -y)$ **27.** $T(a_0 + a_1 x + a_2 x^2) = a_0 + a_1 + (a_0 - a_1 + a_2)x$ **29.** $\left(\frac{9}{25}, \frac{12}{25}\right)$ **31.** $\left(\frac{3}{25}, \frac{4}{25}\right)$

33. $T\left(\begin{bmatrix} a & 0 \\ 0 & b \end{bmatrix}\right) = \begin{bmatrix} a \\ b \end{bmatrix}$ **37.** Yes

Exercise 8-2

1. $T(\mathbf{v}) = \begin{bmatrix} 0 \\ 0 \end{bmatrix}$; $\mathbf{v} \in \ker(T)$ **3.** $T(\mathbf{v}) = \begin{bmatrix} 4 \\ -8 \end{bmatrix}$; $\mathbf{v} \notin \ker(T)$ **5.** No solution; $\mathbf{w} \notin \text{range}(T)$

7. For any real number t, $T\left(\begin{bmatrix} -2 + 2t \\ t \end{bmatrix}\right) = \begin{bmatrix} -2 \\ 4 \end{bmatrix}$; $\mathbf{w} \in \text{range}(T)$

9. $\text{span}\left\{\begin{bmatrix} 1 \\ 1 \\ 0 \end{bmatrix}, \begin{bmatrix} -2 \\ 0 \\ 1 \end{bmatrix}\right\}$; $\text{span}\left\{\begin{bmatrix} 1 \\ -2 \end{bmatrix}\right\}$; 2; 1

11. $\text{span}\left\{\begin{bmatrix} 2 \\ 1 \\ 0 \\ 0 \end{bmatrix}, \begin{bmatrix} 1 \\ 0 \\ -3 \\ 1 \end{bmatrix}\right\}$; $\text{span}\left\{\begin{bmatrix} 1 \\ -2 \\ -1 \end{bmatrix}, \begin{bmatrix} 1 \\ 1 \\ 2 \end{bmatrix}\right\}$; 2; 2

13. $\{\mathbf{0}\}$; $\text{span}\{x, x^2, x^3\}$; 0; 3

15. $\text{span}\left\{\begin{bmatrix} 1 & -1 \\ 0 & 0 \end{bmatrix}, \begin{bmatrix} 0 & 0 \\ 1 & 1 \end{bmatrix}\right\}$; $\text{span}\{1 + x^3, x + x^2\}$; 2; 2   **17.** $\text{span}\{(-4, 3)\}$; $\text{span}\{(3, 4)\}$; 1; 1

19. $\text{span}\{(-2, 1, 0), (-2, 0, 1)\}$; $\text{span}\{(1, 2, 2)\}$; 2; 1   **21.** 2   **23.** 3   **25.** 4   **27.** 0

29. (A) $\text{span}\{(4, 5, 1)\}$; 1   (B) 2   (C) R^2 **31.** (A) $\text{span}\{1\}$; 1   (B) 4   (C) P_3

Exercise 8-3

1. $\begin{bmatrix} 2 & 3 \\ 5 & 4 \end{bmatrix}$ **3.** $\begin{bmatrix} 1 & 1 \\ 1 & 0 \\ 0 & 1 \end{bmatrix}$ **5.** $\begin{bmatrix} 1 & -1 & 0 & 0 \\ 0 & 1 & -1 & 0 \\ 0 & 0 & 1 & -1 \end{bmatrix}$ **7.** $\begin{bmatrix} 2 & 5 \\ 3 & 7 \end{bmatrix}$ **9.** $\begin{bmatrix} 1 & 0 & 0 & -1 \\ 0 & 1 & 2 & 0 \end{bmatrix}$

11. (A) $\begin{bmatrix} 3 & 0 \\ 0 & 2 \end{bmatrix}$ (B) $\begin{bmatrix} -2 \\ 3 \end{bmatrix}$ (C) $\begin{bmatrix} -6 \\ 6 \end{bmatrix}$ (D) $\begin{bmatrix} 6 \\ 0 \end{bmatrix}$ (E) $\begin{bmatrix} 6 \\ 0 \end{bmatrix}$

13. (A) $\begin{bmatrix} 1 & 2 \\ 0 & -3 \\ 2 & 1 \end{bmatrix}$ (B) $\begin{bmatrix} 3 \\ 2 \end{bmatrix}$ (C) $\begin{bmatrix} 7 \\ -6 \\ 8 \end{bmatrix}$ (D) $7 - 6x + 8x^2$ (E) $7 - 6x + 8x^2$

15. (A) $\begin{bmatrix} 1 & 1 & 0 \\ 0 & 1 & -1 \end{bmatrix}$ (B) $\begin{bmatrix} 1 & 0 & 0 \\ 0 & 1 & 0 \end{bmatrix}$ (C) $\begin{bmatrix} 1 & 0 & 2 \\ 0 & 1 & 2 \end{bmatrix}$ **17.** $\begin{bmatrix} 0 & 0 & 0 \\ 1 & 0 & 0 \\ 0 & 1 & 0 \\ 0 & 0 & 1 \end{bmatrix}$

19. (A) $\dfrac{1}{5}\begin{bmatrix} 1 & 2 \\ 2 & 4 \end{bmatrix}$ (B) $\begin{bmatrix} 1 & 0 \\ 0 & 0 \end{bmatrix}$ (C) $(\frac{6}{5}, \frac{12}{5})$ **21.** (A) $\dfrac{1}{9}\begin{bmatrix} 5 & 4 & -2 \\ 4 & 5 & 2 \\ -2 & 2 & 8 \end{bmatrix}$ (B) $\begin{bmatrix} 1 & 0 & 0 \\ 0 & 1 & 0 \\ 0 & 0 & 0 \end{bmatrix}$ (C) $(3, 3, 0)$

23. (A) $\begin{bmatrix} 1 & 1 \\ 2 & -1 \\ 0 & 2 \end{bmatrix}$ (B) $\begin{bmatrix} 5 & 4 \\ 4 & 4 \end{bmatrix}$

25. (A) $\begin{bmatrix} 1 & 0 \\ 0 & 1 \\ -1 & 2 \end{bmatrix}$ (B) $T(e_1) = 3 + 4x + 5x^2$; $T(e_2) = 2 + 3x + 4x^2$ (C) $(3a + 2b) + (4a + 3b)x + (5a + 4b)x^2$

27. (A) $[e_1]_B = \begin{bmatrix} 1 \\ 0 \\ -1 \end{bmatrix}$; $[e_2]_B = \begin{bmatrix} -2 \\ 1 \\ 2 \end{bmatrix}$; $[e_3]_B = \begin{bmatrix} 2 \\ 0 \\ -1 \end{bmatrix}$ (B) $[T(e_1)]_B = \begin{bmatrix} 3 \\ 0 \\ -1 \end{bmatrix}$; $[T(e_2)]_B = \begin{bmatrix} -6 \\ 1 \\ 2 \end{bmatrix}$; $[T(e_3)]_B = \begin{bmatrix} 6 \\ 0 \\ -1 \end{bmatrix}$

(C) $T(e_1) = (-1, 0, 2)$; $T(e_2) = (4, 1, -4)$; $T(e_3) = (-4, 0, 5)$ (D) $\begin{bmatrix} -1 & 4 & -4 \\ 0 & 1 & 0 \\ 2 & -4 & 5 \end{bmatrix}$

29. $\begin{bmatrix} 0 & 0 & 0 & 0 \\ 0 & 1 & 0 & 0 \\ 0 & 0 & 2 & 0 \\ 0 & 0 & 0 & 3 \end{bmatrix}$ **31.** kI_n where I_n is the $n \times n$ identity matrix

33. $\begin{bmatrix} \lambda_1 & 0 & \cdots & 0 \\ 0 & \lambda_2 & \cdots & 0 \\ \cdot & \cdot & & \cdot \\ \cdot & \cdot & & \cdot \\ \cdot & \cdot & & \cdot \\ 0 & 0 & \cdots & \lambda_n \end{bmatrix}$

Exercise 8-4

1. (A) 3   (B) -2 **3.** (A) 2   (B) -2 **5.** $\begin{bmatrix} 1 & 0 & 0 \\ 0 & 2 & 0 \\ 0 & 0 & 11 \end{bmatrix}$ **7.** $\begin{bmatrix} 1 & 0 & 0 \\ 0 & 4 & 0 \\ 0 & 0 & 4 \end{bmatrix}$

9. (A) $\begin{bmatrix} 2 & 0 & 0 \\ 0 & 2 & 0 \\ -3 & -3 & -1 \end{bmatrix}$ (B) $\lambda_1 = -1, \lambda_2 = 2$

9. (C) $B_{\lambda_1} = \left\{ \begin{bmatrix} 0 \\ 0 \\ 1 \end{bmatrix} \right\}$, $B_{\lambda_2} = \left\{ \begin{bmatrix} -1 \\ 1 \\ 0 \end{bmatrix}, \begin{bmatrix} -1 \\ 0 \\ 1 \end{bmatrix} \right\}$ (D) $C_{\lambda_1} = \{x^2\}$, $C_{\lambda_2} = \{-1 + x, -1 + x^2\}$

11. $\lambda_1 = 2$ (multiplicity 2); $C_{\lambda_1} = \left\{ \begin{bmatrix} 1 \\ 1 \end{bmatrix} \right\}$; not diagonalizable   **13.** No real eigenvalues; not diagonalizable

15. $\lambda_1 = 0$, $\lambda_2 = 1$; $C_{\lambda_1} = \{(-2, 1)\}$, $C_{\lambda_2} = \{(1, 2)\}$; $\begin{bmatrix} 0 & 0 \\ 0 & 1 \end{bmatrix}$

17. $\lambda_1 = -2$; $C_{\lambda_1} = \left\{ \begin{bmatrix} 1 \\ 0 \\ 0 \end{bmatrix}, \begin{bmatrix} 0 \\ 0 \\ 1 \end{bmatrix} \right\}$; not diagonalizable

19. $\lambda_1 = -2$, $\lambda_2 = 1$, $\lambda_3 = 4$; $C_{\lambda_1} = \left\{ \begin{bmatrix} 1 \\ 1 \\ 0 \end{bmatrix} \right\}$, $C_{\lambda_2} = \left\{ \begin{bmatrix} -1 \\ 0 \\ 1 \end{bmatrix} \right\}$, $C_{\lambda_3} = \left\{ \begin{bmatrix} -2 \\ 0 \\ 1 \end{bmatrix} \right\}$; $\begin{bmatrix} -2 & 0 & 0 \\ 0 & 1 & 0 \\ 0 & 0 & 4 \end{bmatrix}$

21. $\lambda_1 = -1$; $C_{\lambda_1} = \{-1 - x + x^2\}$; not diagonalizable

23. $\lambda_1 = -1$, $\lambda_2 = 1$; $C_{\lambda_1} = \{-1 + x^2\}$, $C_{\lambda_2} = \{1 + x^2, x\}$; $\begin{bmatrix} -1 & 0 & 0 \\ 0 & 1 & 0 \\ 0 & 0 & 1 \end{bmatrix}$

25. $\lambda_1 = -1$, $\lambda_2 = 1$; $C_{\lambda_1} = \left\{ \begin{bmatrix} 0 & -1 \\ 1 & 0 \end{bmatrix} \right\}$, $C_{\lambda_2} = \left\{ \begin{bmatrix} 1 & 0 \\ 0 & 0 \end{bmatrix}, \begin{bmatrix} 0 & 1 \\ 1 & 0 \end{bmatrix}, \begin{bmatrix} 0 & 0 \\ 0 & 1 \end{bmatrix} \right\}$; $\begin{bmatrix} -1 & 0 & 0 & 0 \\ 0 & 1 & 0 & 0 \\ 0 & 0 & 1 & 0 \\ 0 & 0 & 0 & 1 \end{bmatrix}$

27. No real eigenvalues; not diagonalizable

29. $\lambda_1 = 0$, $\lambda_2 = 1$, $\lambda_3 = 2$, $\lambda_4 = 3$; $C_{\lambda_1} = \{1\}$, $C_{\lambda_2} = \{x\}$, $C_{\lambda_3} = \{x^2\}$, $C_{\lambda_4} = \{x^3\}$; $\begin{bmatrix} 0 & 0 & 0 & 0 \\ 0 & 1 & 0 & 0 \\ 0 & 0 & 2 & 0 \\ 0 & 0 & 0 & 3 \end{bmatrix}$

Exercise 8-5

1.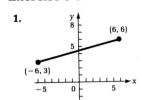

3.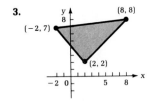

5.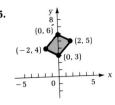

7. Yes   **9.** No   **11.** $(3, 1) + t(5, 1)$, t any real number

13. $\begin{bmatrix} 0 & 1 \\ 1 & 0 \end{bmatrix}$

15. $\begin{bmatrix} 1 & 0 \\ 0 & 3 \end{bmatrix}$

17. $\begin{bmatrix} 1 & -2 \\ 0 & 1 \end{bmatrix}$

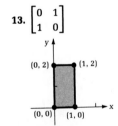

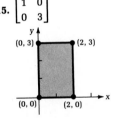

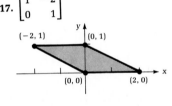

19. $\begin{bmatrix} 1 & 5 \\ 0 & 1 \end{bmatrix}$ **21.** $\begin{bmatrix} 1 & 0 \\ 0 & \frac{1}{4} \end{bmatrix}$ **23.** $\begin{bmatrix} 1 & 0 \\ 0 & -3 \end{bmatrix}$ **25.** $\begin{bmatrix} 4 & -11 \\ 1 & -3 \end{bmatrix}$

27. A shear in the x direction with scale factor 2, followed by an expansion in the y direction with scale factor 3, followed by a shear in the y direction with scale factor −1

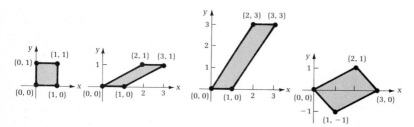

29. An expansion in the y direction with scale factor 4, followed by an expansion in the x direction with scale factor 2, followed by a shear in the x direction with scale factor 1, followed by a shear in the y direction with scale factor $-\frac{1}{2}$

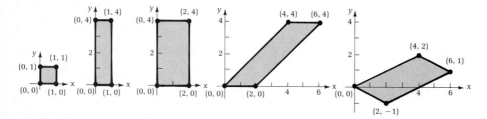

31. $\begin{bmatrix} 1 & 0 \\ 2 & 1 \end{bmatrix}\begin{bmatrix} 1 & 1 \\ 0 & 1 \end{bmatrix}$; a shear in the x direction with scale factor 1, followed by a shear in the y direction with scale factor 2

33. $\begin{bmatrix} 2 & 0 \\ 0 & 1 \end{bmatrix}\begin{bmatrix} 1 & 0 \\ 4 & 1 \end{bmatrix}\begin{bmatrix} 1 & 0 \\ 0 & -1 \end{bmatrix}\begin{bmatrix} 1 & \frac{1}{2} \\ 0 & 1 \end{bmatrix}$; a shear in the x direction with scale factor $\frac{1}{2}$, followed by a reflection about the x axis, followed by a shear in the y direction with scale factor 4, followed by an expansion in the x direction with scale factor 2

35. A reflection about the x axis, followed by a reflection about the line $y = x$

37. A shear in the x direction with scale factor $\sqrt{3}$, followed by an expansion in the y direction with scale factor 2, followed by a shear in the y direction with scale factor $-\sqrt{3}/2$, followed by a contraction in the x direction with scale factor $\frac{1}{2}$

39. (A) $\begin{bmatrix} 0 & 2 \\ 1 & 0 \end{bmatrix}$ (B) $\begin{bmatrix} 0 & 1 \\ 2 & 0 \end{bmatrix}$ (C) No   **41.** (A) $\begin{bmatrix} 1 & -1 \\ 0 & 1 \end{bmatrix}$ (B) $\begin{bmatrix} 1 & -1 \\ 0 & 1 \end{bmatrix}$ (C) Yes

Exercise 8-6   Chapter Review

2. (A) Yes; $T(6, 3) = \mathbf{0}$ (B) No; $T(6, -3) = (12, -24) \neq \mathbf{0}$

3. (A) Yes; for example, $T(2, 0) = (2, -4)$ (B) No; the equation $T(x, y) = (2, -2)$ has no solution

4. $\begin{bmatrix} 3 & 1 \\ 1 & 3 \end{bmatrix}$ **5.** (A) $\lambda_1 = 4$ (B) $\lambda_2 = 2$ **6.** $\begin{bmatrix} 4 & 0 \\ 0 & 2 \end{bmatrix}$

7.

8. (A) $\begin{bmatrix} 0 & 1 & -1 \\ 1 & -1 & 0 \\ -1 & 0 & 1 \end{bmatrix}$ (B) $\begin{bmatrix} 1 \\ 2 \\ 3 \end{bmatrix}$

(C) $\begin{bmatrix} -1 \\ -1 \\ 2 \end{bmatrix}$ (D) $-1 - x + 2x^2$
(E) $-1 - x + 2x^2$

9. (A) $\begin{bmatrix} 1 & 6 \\ -1 & 6 \end{bmatrix}$ (B) $\lambda_1 = 3, B_{\lambda_1} = \left\{ \begin{bmatrix} 3 \\ 1 \end{bmatrix} \right\}; \lambda_2 = 4, B_{\lambda_2} = \left\{ \begin{bmatrix} 2 \\ 1 \end{bmatrix} \right\}$ (C) $\lambda_1 = 3, C_{\lambda_1} = \{3 + x\}; \lambda_2 = 4, C_{\lambda_2} = \{2 + x\}$

10. Linear **11.** Nonlinear **12.** Nonlinear **13.** Linear **14.** Linear **15.** Nonlinear

16. span $\left\{ \begin{bmatrix} 5 \\ 3 \\ 1 \end{bmatrix} \right\}$; span $\left\{ \begin{bmatrix} 1 \\ 3 \end{bmatrix}, \begin{bmatrix} -3 \\ -7 \end{bmatrix} \right\}$; 1; 2 **17.** span $\left\{ \begin{bmatrix} 2 \\ 1 \\ 0 \end{bmatrix}, \begin{bmatrix} -1 \\ 0 \\ 1 \end{bmatrix} \right\}$; span $\left\{ \begin{bmatrix} 1 \\ -2 \end{bmatrix} \right\}$; 2; 1

18. span$\{x, x^3\}$; span$\{1, x^2\}$; 2; 2 **19.** $\{\mathbf{0}\}$; span $\left\{ \begin{bmatrix} 1 & 0 \\ 0 & 0 \end{bmatrix}, \begin{bmatrix} 0 & 1 \\ 0 & 1 \end{bmatrix}, \begin{bmatrix} 0 & 0 \\ 1 & 0 \end{bmatrix} \right\}$; 0; 3 **20.** $\begin{bmatrix} 2 & 1 & 0 \\ 0 & 1 & -3 \end{bmatrix}$

21. (A) $(\frac{1}{10}, \frac{3}{10})$ (B) $(\frac{3}{10}, \frac{9}{10})$ (C) $\frac{1}{10}(a + 3b, 3a + 9b)$ (D) span$\{(-3, 1)\}$; 1 (E) span$\{(1, 3)\}$; 1
(F) $\frac{1}{10}\begin{bmatrix} 1 & 3 \\ 3 & 9 \end{bmatrix}$ (G) $\lambda_1 = 0, B_{\lambda_1} = \{(-3, 1)\}; \lambda_2 = 1, B_{\lambda_2} = \{(1, 3)\}$

22. (A) $\begin{bmatrix} 2 & -1 & 0 \\ 0 & 3 & -2 \end{bmatrix}$ (B) $\begin{bmatrix} 1 & 0 & 0 \\ 0 & 2 & 0 \end{bmatrix}$ **23.** (A) $\begin{bmatrix} 2 & 3 \\ 3 & 2 \\ 0 & 4 \end{bmatrix}$ (B) $\begin{bmatrix} 1 & -1 \\ -1 & 4 \end{bmatrix}$

24. (A) $\begin{bmatrix} -1 & 3 & 0 \\ -2 & 4 & 0 \\ -1 & 1 & 2 \end{bmatrix}$ (B) $\lambda_1 = 1, C_{\lambda_1} = \{3 + 2x + x^2\}; \lambda_2 = 2, C_{\lambda_2} = \{1 + x, x^2\}$ (C) $\begin{bmatrix} 1 & 0 & 0 \\ 0 & 2 & 0 \\ 0 & 0 & 2 \end{bmatrix}$

25. (A) $\begin{bmatrix} 3 & 3 & 2 \\ -1 & -1 & -1 \\ 0 & 1 & 1 \end{bmatrix}$ (B) $\lambda_1 = 1, C_{\lambda_1} = \{-1 + x^2\}$ (C) T is not diagonalizable

26. $\begin{bmatrix} 2 & -1 & 1 \\ 1 & 3 & 1 \end{bmatrix}$; $3 + 10x$ **27.** $\begin{bmatrix} 1 & -1 & -1 \\ -1 & 2 & 1 \end{bmatrix}$; $-4 + 6x$

28. (A) $\begin{bmatrix} 1 & 0 & 0 & 0 \\ 1 & 0 & 0 & 0 \\ 0 & 0 & 1 & 0 \\ 0 & 0 & 1 & 1 \end{bmatrix}$ (B) $\lambda_1 = 0, C_{\lambda_1} = \left\{ \begin{bmatrix} 0 & 1 \\ 0 & 0 \end{bmatrix} \right\}; \lambda_2 = 1, C_{\lambda_2} = \left\{ \begin{bmatrix} 1 & 1 \\ 0 & 0 \end{bmatrix}, \begin{bmatrix} 0 & 0 \\ 0 & 1 \end{bmatrix} \right\}$

(C) T is not diagonalizable

29. (A) $\begin{bmatrix} 0 & 1 & 0 & 0 \\ 4 & 0 & 0 & 0 \\ 0 & 0 & 0 & 2 \\ 0 & 0 & 2 & 0 \end{bmatrix}$ (B) $\lambda_1 = -2, C_{\lambda_1} = \left\{ \begin{bmatrix} 1 & -2 \\ 0 & 0 \end{bmatrix}, \begin{bmatrix} 0 & 0 \\ 1 & -1 \end{bmatrix} \right\}; \quad \lambda_2 = 2, C_{\lambda_2} = \left\{ \begin{bmatrix} 1 & 2 \\ 0 & 0 \end{bmatrix}, \begin{bmatrix} 0 & 0 \\ 1 & 1 \end{bmatrix} \right\}$

(C) $\begin{bmatrix} -2 & 0 & 0 & 0 \\ 0 & -2 & 0 & 0 \\ 0 & 0 & 2 & 0 \\ 0 & 0 & 0 & 2 \end{bmatrix}$

30. An expansion in the y direction with scale factor 2

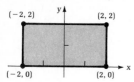

31. A reflection about the line $y = x$

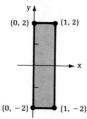

32. A shear in the y direction with scale factor $\frac{1}{2}$

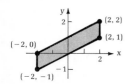

33. A shear in the x direction with scale factor -2

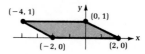

34. A shear in the x direction with scale factor -3, followed by an expansion in the y direction with scale factor 7, followed by a shear in the y direction with scale factor 2

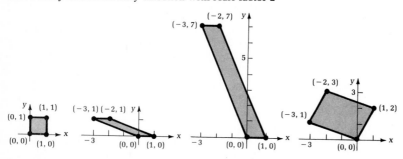

35. $\begin{bmatrix} 1 & 0 \\ 3 & 1 \end{bmatrix}\begin{bmatrix} 1 & 0 \\ 0 & -1 \end{bmatrix}\begin{bmatrix} 1 & 2 \\ 0 & 1 \end{bmatrix}$; a shear in the x direction with scale factor 2, followed by a reflection about the x axis, followed by a shear in the y direction with scale factor 3

36. $\begin{bmatrix} 0 & 2 \\ 1 & 3 \end{bmatrix}$ **37.** $\begin{bmatrix} 1 & 1 \\ 0 & -\frac{1}{2} \end{bmatrix}$ **38.** span$\{1 - 3x^2, x\}$

39. (A) span$\{1\}$; 1   (B) span$\{1 + x^2, 2x + x^3, x\}$; 3

(C) $\begin{bmatrix} 0 & 0 & 2 & 0 \\ 0 & 1 & 0 & 6 \\ 0 & 0 & 2 & 0 \\ 0 & 0 & 0 & 3 \end{bmatrix}$ (D) $\lambda_1 = 0$, $C_{\lambda_1} = \{1\}$; $\lambda_2 = 1$, $C_{\lambda_2} = \{x\}$; $\lambda_3 = 2$, $C_{\lambda_3} = \{1 + x^2\}$; $\lambda_4 = 3$, $C_{\lambda_4} = \{3x + x^3\}$

40. 3   **41.** 2   **42.** $\{0\}$ **43.** $\mathbf{R}^4$ **44.** (B) $\begin{bmatrix} 0 & w_3 & -w_2 \\ -w_3 & 0 & w_1 \\ w_2 & -w_1 & 0 \end{bmatrix}$ (C) $\lambda_1 = 0$, $B_{\lambda_1} = \{w\}$

45. (B) $\begin{bmatrix} a & 0 \\ 0 & 1 \end{bmatrix}\begin{bmatrix} 1 & 0 \\ b & 1 \end{bmatrix}\begin{bmatrix} 1 & -b \\ 0 & 1 \end{bmatrix}\begin{bmatrix} 1 & 0 \\ 0 & 1/a \end{bmatrix}$

(C) An expansion in the y direction with scale factor $1/a$, followed by a shear in the x direction with scale factor $-b$, followed by a shear in the y direction with scale factor b, followed by a contraction in the x direction with scale factor a

Chapter 9   Exercise 9-1

1. Ellipse

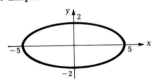

3. Hyperbola

5. Parabola

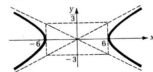

7. $5x^2 - 12xy + 7y^2$; $\begin{bmatrix} 5 & -6 \\ -6 & 7 \end{bmatrix}$; $[x \ \ y]\begin{bmatrix} 5 & -6 \\ -6 & 7 \end{bmatrix}\begin{bmatrix} x \\ y \end{bmatrix} + [-8 \ \ 9]\begin{bmatrix} x \\ y \end{bmatrix} - 11 = 0$

9. $3x^2 + 10xy$; $\begin{bmatrix} 3 & 5 \\ 5 & 0 \end{bmatrix}$; $[x \ \ y]\begin{bmatrix} 3 & 5 \\ 5 & 0 \end{bmatrix}\begin{bmatrix} x \\ y \end{bmatrix} + [4 \ \ -2]\begin{bmatrix} x \\ y \end{bmatrix} - 7 = 0$

11. $x'^2 + y'^2 = 1$; circle   **13.** $y'^2 - \dfrac{x'^2}{4} = 1$; hyperbola   **15.** $x' = \dfrac{y'^2}{2}$; parabola

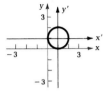

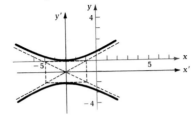

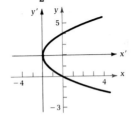

17.

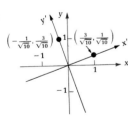

19. $\dfrac{x'^2}{4} + y'^2 = 1$; ellipse

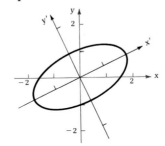

21. $x'^2 - \dfrac{y'^2}{4} = 1$; hyperbola

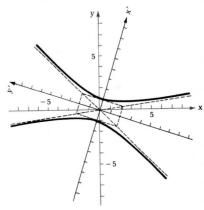

23. $y''^2 = x''$; parabola

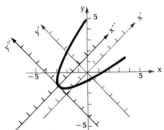

25. No graph exists

27. Two intersecting lines, $x = 2y$ and $x = -2y$

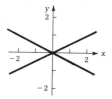

29. Two parallel lines, $x = y + 1$ and $x = y - 1$

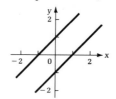

Exercise 9-2

1. $x_n = 10 \cdot 3^n$ **3.** $x_n = 3(-\tfrac{1}{2})^n$ **5.** $\begin{bmatrix} x_n \\ y_n \end{bmatrix} = \begin{bmatrix} 2 & 2 \\ -1 & 5 \end{bmatrix}\begin{bmatrix} x_{n-1} \\ y_{n-1} \end{bmatrix}$; $(2, -1), (2, -7), (-10, -37)$

7. $\begin{bmatrix} x_n \\ y_n \end{bmatrix} = \begin{bmatrix} 2 & 2 \\ -1 & 5 \end{bmatrix}^n \begin{bmatrix} 1 \\ 0 \end{bmatrix}$ **9.** $\begin{bmatrix} 2 & 2 \\ -1 & 5 \end{bmatrix} = \begin{bmatrix} 2 & 1 \\ 1 & 1 \end{bmatrix}\begin{bmatrix} 3 & 0 \\ 0 & 4 \end{bmatrix}\begin{bmatrix} 1 & -1 \\ -1 & 2 \end{bmatrix}$

11. $x_n = 2 \cdot 3^n - 4^n$, $y_n = 3^n - 4^n$ **13.** $x_n = 2 \cdot 3^n$, $y_n = 3^n$ **15.** $x_n = 3 + 2(-1)^n$, $y_n = 1 + (-1)^n$

17. $x_n = (0.5)^n + 4(1.5)^n$, $y_n = (0.5)^n + 2(1.5)^n$ **19.** $x_n = 1 + 2^n$ **21.** $x_n = 3^n + 4^n$

23. $x_n = \tfrac{1}{2}(1 + \sqrt{2})^n + \tfrac{1}{2}(1 - \sqrt{2})^n$ **25.** $x_n = 4^n$, $y_n = 4^n$, $z_n = 4^n$ **27.** $x_n = 2^{n+1} - 3^n$, $y_n = 2^n$, $z_n = 2^n + 3^n$

33. $x_n = 1 + 2^n + 3^n$ **35.** $200(1.5)^n$

37. $x_n = 20(0.5)^n + 60(0.6)^n$, $y_n = 20(0.5)^n + 40(0.6)^n$

39. $p_n = 40 - 20(0.5)^n$, $q_n = 80 + 10(0.5)^n$

n	Rabbits	Foxes
0	80,000	60,000
2	26,600	19,400
4	9,026	6,434
6	3,112	2,179
8	1,086	750
10	382	261

The population of each species decreases to 0.

n	p_n	q_n
0	$20	$90
2	35	82.50
4	38.75	80.63
6	39.69	80.16
8	39.92	80.04
10	39.98	80.01

The first price increases to \$40; the second decreases to \$80.

Exercise 9-3

1. $y = Ce^{0.5t}$; $y = 10e^{0.5t}$ **3.** $y = Ce^{-2t}$; $y = -2e^{-2t}$ **5.** $y_1 = C_1 e^t$, $y_2 = C_2 e^{2t}$; $\mathbf{y}' = \begin{bmatrix} 1 & 0 \\ 0 & 2 \end{bmatrix}\mathbf{y}$, $\mathbf{y} = \begin{bmatrix} C_1 e^t \\ C_2 e^{2t} \end{bmatrix}$

7. $\mathbf{y}' = \begin{bmatrix} 4 & -3 \\ 2 & -1 \end{bmatrix}\mathbf{y}$ **9.** $\begin{bmatrix} 4 & -3 \\ 2 & -1 \end{bmatrix} = \begin{bmatrix} 1 & 3 \\ 1 & 2 \end{bmatrix}\begin{bmatrix} 1 & 0 \\ 0 & 2 \end{bmatrix}\begin{bmatrix} -2 & 3 \\ 1 & -1 \end{bmatrix}$

11. $y_1 = C_1 e^t + 3C_2 e^{2t}$, $y_2 = C_1 e^t + 2C_2 e^{2t}$ **13.** $y_1 = -2e^t + 3e^{2t}$, $y_2 = -2e^t + 2e^{2t}$

15. $y_1 = C_1 + C_2 e^{2t}$, $y_2 = -C_1 + C_2 e^{2t}$; $y_1 = e^{2t}$, $y_2 = e^{2t}$ **17.** $y_1 = C_1 e^{-1.5t} + C_2 e^{0.5t}$, $y_2 = 5C_1 e^{-1.5t} + C_2 e^{0.5t}$

19. $y_1 = C_1 + (C_2 + C_3)e^t$, $y_2 = -C_1 + C_3 e^t$, $y_3 = C_1 + C_2 e^t$; $y_1 = 1 + 2e^t$, $y_2 = -1 + e^t$, $y_3 = 1 + e^t$

21. $y_1 = 2C_1 e^t + 4C_2 e^{2t} + C_3 e^{3t}$, $y_2 = C_2 e^{2t} + C_3 e^{3t}$, $y_3 = C_1 e^t + C_2 e^{2t}$ **23.** $y = C_1 e^{3t} + C_2 e^{4t}$ **25.** $y = C_1 e^{-t} + C_2 e^t$

31. $y = C_1 e^{-t} + C_2 e^t + C_3 e^{4t}$

33. $y_1 = 25e^{-0.12t} + 25e^{-0.06t}$, $y_2 = -75e^{-0.12t} + 75e^{-0.06t}$ **35.** $y_1 = 70e^{0.025t} + 20e^{0.15t}$, $y_2 = 140e^{0.025t} - 10e^{0.15t}$

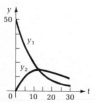

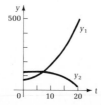

Exercise 9-4 Chapter Review

1. Ellipse

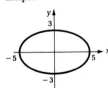

2. Hyperbola

3. Parabola

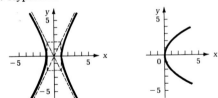

4. $x_n = 2(5)^n$ **5.** $y = 2e^{5t}$

6. (A) $9x^2 + 12xy + 8y^2$ (B) $\begin{bmatrix} 9 & 6 \\ 6 & 8 \end{bmatrix}$ (C) $[x \;\; y]\begin{bmatrix} 9 & 6 \\ 6 & 8 \end{bmatrix}\begin{bmatrix} x \\ y \end{bmatrix} + [10 \;\; 7]\begin{bmatrix} x \\ y \end{bmatrix} + 15 = 0$

7. $x_n = 5 \cdot 2^n$, $y_n = 7 \cdot 3^n$ **8.** $y_1 = C_1 e^{2t}$, $y_2 = C_2 e^{3t}$, $y_3 = C_3 e^{4t}$

9. (A) $\begin{bmatrix} x_n \\ y_n \end{bmatrix} = \begin{bmatrix} 2 & -3 \\ 2 & 7 \end{bmatrix}\begin{bmatrix} x_{n-1} \\ y_{n-1} \end{bmatrix}$ (B) $\begin{bmatrix} x_n \\ y_n \end{bmatrix} = \begin{bmatrix} 2 & -3 \\ 2 & 7 \end{bmatrix}^n \begin{bmatrix} 1 \\ 2 \end{bmatrix}$

(C) $\begin{bmatrix} 2 & -3 \\ 2 & 7 \end{bmatrix} = \begin{bmatrix} 3 & 1 \\ -2 & -1 \end{bmatrix}\begin{bmatrix} 4 & 0 \\ 0 & 5 \end{bmatrix}\begin{bmatrix} 1 & 1 \\ -2 & -3 \end{bmatrix}$ (D) $x_n = 9 \cdot 4^n - 8 \cdot 5^n$, $y_n = -6 \cdot 4^n + 8 \cdot 5^n$

10. (A) $\mathbf{y}' = \begin{bmatrix} 2 & -1 \\ 2 & 5 \end{bmatrix}\mathbf{y}$ (B) $\begin{bmatrix} 2 & -1 \\ 2 & 5 \end{bmatrix} = \begin{bmatrix} 1 & 1 \\ -1 & -2 \end{bmatrix}\begin{bmatrix} 3 & 0 \\ 0 & 4 \end{bmatrix}\begin{bmatrix} 2 & 1 \\ -1 & -1 \end{bmatrix}$

(C) $y_1 = C_1 e^{3t} + C_2 e^{4t}$, $y_2 = -C_1 e^{3t} - 2C_2 e^{4t}$ (D) $y_1 = 2e^{3t} - e^{4t}$, $y_2 = -2e^{3t} + 2e^{4t}$

11. $x'^2 + y'^2 = 9$; circle   **12.**   **13.** $x'^2 + \dfrac{y'^2}{4} = 1$; ellipse

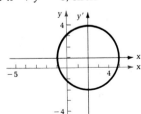 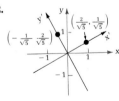 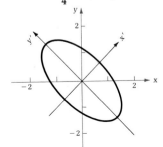

14. $x'^2 - y'^2 = 1$; hyperbola   **15.** $x' = \frac{1}{4}y'^2$; parabola   **16.** $\dfrac{x''^2}{4} + \dfrac{y''^2}{16} = 1$; ellipse

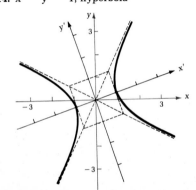 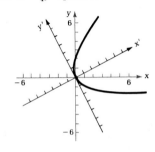 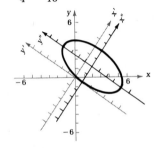

17. $x_n = 1 + 3 \cdot 2^n$, $y_n = 1 + 2 \cdot 2^n$ **18.** $x_n = 3(-0.5)^n + 3(2.5)^n$, $y_n = -3(-0.5)^n + 6(2.5)^n$

19. $y_1 = C_1 e^{-2t} + C_2 e^{2t}$, $y_2 = 3C_1 e^{-2t} + C_2 e^{2t}$; $y_1 = e^{-2t} - 2e^{2t}$, $y_2 = 3e^{-2t} - 2e^{2t}$

20. $y_1 = 3C_1 e^{0.5t} + 2C_2 e^{4t}$, $y_2 = -2C_1 e^{0.5t} + C_2 e^{4t}$; $y_1 = -3e^{0.5t} + 2e^{4t}$, $y_2 = 2e^{0.5t} + e^{4t}$

21. $y_1 = -C_2 e^t - 2C_3 e^{2t}$, $y_2 = -C_1 e^{-t} + C_2 e^t$, $y_3 = C_1 e^{-t} + C_3 e^{2t}$

22. $y_1 = C_1 + C_2 e^{2t} + C_3 e^{2t}$, $y_2 = -C_1 + C_2 e^{2t}$, $y_3 = C_1 + C_3 e^{2t}$

23. $x_n = 3^n + 5^n$ **24.** $x_n = \frac{1}{2}(1 - \sqrt{3})^n + \frac{1}{2}(1 + \sqrt{3})^n$ **25.** $y = C_1 e^t + C_2 e^{4t}$ **26.** $y = C_1 e^{-2t} + C_2 e^{2t}$

27. $x_n = 2^n - 4 \cdot 3^n + 2 \cdot 4^n$, $y_n = 2 \cdot 3^n - 4^n$, $z_n = -2^n + 2 \cdot 3^n$ **28.** $x_n = 3^n - 4^n + 5^n$ **29.** $y = C_1 e^{-t} + C_2 e^t + C_3 e^{2t}$

30. $x_n = 30 + 30(0.4)^n$, $y_n = 15 + 60(0.4)^n$ **31.** $y_1 = 20e^{-0.3t} + 20e^{-0.1t}$, $y_2 = -40e^{-0.3t} + 40e^{-0.1t}$

n	Rabbits	Foxes
0	60,000	75,000
2	34,800	24,600
4	30,768	16,536
6	30,123	15,246
8	30,020	15,039
10	30,003	15,006

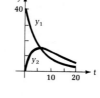

The rabbit population decreases to approximately 30,000; the fox population decreases to approximately 15,000.

Index

Designer: Janet Bollow
Cover designer: John Williams
Cover photographer: John Jensen
Technical artist: Carl Brown
Production coordinator: Phyllis Niklas
Typesetter: Progressive Typographers
Printer and binder: R. R. Donnelley & Sons